Companion Encyclopedia
of the
History and Philosophy
of the
Mathematical Sciences

Volume 2

Edited by
I. GRATTAN-GUINNESS

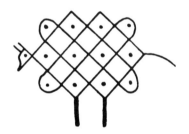
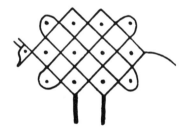

LONDON AND NEW YORK

First published in 1994
by Routledge
11 New Fetter Lane, London EC4P 4EE

Simultaneously published in the USA and Canada
by Routledge Inc.
29 West 35th Street, New York, NY 10001

Printed in Great Britain by
Clay Ltd, St Ives plc
Typeset in 10/12½pt Times Compugraphic by
Mathematical Composition Setters Ltd, Salisbury, UK

Printed on acid-free paper

British Library Cataloguing in Publication Data

A catalogue record for this book is available from the British Library

Library of Congress Cataloging-in-Publication Data

Companion encyclopedia of the history and philosophy of the mathematical
sciences / edited by Ivor Grattan-Guinness.
    p.   cm.
Includes bibliographical references and index.
ISBN 0–415–03785–9
    1. Mathematics—History. 2. Mathematics—Philosophy. I. Grattan-
Guinness, I.
QA21.E57   1992
510′.9—dc20   92–13707
                                                                      CIP

ISBN 0–415–03785–9 (set)
         0–415–09238–8 (Vol. 1)
         0–415–09239–6 (Vol. 2)

Illustration on title page:
Antelopes drawn by the (Ts)Chokwe people of Angola in their tradition
of monolinear art; the main block of the pattern is composed of one
continuous line, and is so designed that a regular lattice of dots can be
inserted in the spaces (see P. Gerdes, *Lusona: Geometrical Recreations of
Africa*, 1991, Maputo, Mozambique: Eduardo Mondlare University Press, 15).
On this tradition, see §1.8 by C. Zaslavsky, Section 3.

Contents

Part 7
Geometries and Topology

7.0

Introduction

As in the previous Part, the plural 'geometries' in the title is deliberately intended to refer to a variety of approaches and theories. §7.1–7.2 work strictly within the Euclidean realm, but the spread begins to emerge in §7.3 when non-Euclidean geometries make their bow. The next three articles (§7.4–7.6) are basically Euclidean in context but handle more general properties; §3.4 on differential geometry should also be noted. Corresponding in role to §6.9, the general philosophical situation as of the end of the nineteenth century is appraised in §7.7.

The next four articles are taken up with mathematical developments since around 1870. The most substantial addition is topology, of which the principal early features are described in §7.10–7.11. Finally, with finite vector spaces and graph theory, two aspects (rather than branches) of geometry are treated in §7.12–7.13; the overlaps with algebra are quite substantial.

One pleasing result of the development of geometry has been the construction of models of curves, surfaces and solids. Various books give instructions on their manufacture, and some departments of mathematics have collections. The history of this development is not treated here, but the pertinent theories are touched upon, and also in certain articles in Parts 1 and 12.

BIBLIOGRAPHY

Coolidge, J. L. *1940, A History of Geometrical Methods*, Oxford: Oxford University Press. [Repr. 1963, New York: Dover.]
Simon, M. *1906, Über die Entwicklung der Elementar-Geometrie im XIX. Jahrhundert*, Leipzig: Teubner.

7.1

Algebraic and analytic geometry

J. J. GRAY

1 COORDINATES IN THE PLANE

The present topic is also called 'Cartesian geometry', and few names are more appropriate in the history of mathematics, for the contribution of René Descartes was decisive in bringing together the methods of algebra and the study of geometry. Indeed, his *La Géométrie* (*1637*) can be read as the first modern mathematics book.

As the name 'algebraic geometry' suggests, the subject studied is geometry and the methods employed are algebraic. They are brought together by the idea of coordinates. Let us take the case of plane geometry. An arbitrary point is chosen, called the 'origin', and labelled O. Through it are drawn two lines at right angles called the *x*- and *y*-axes, respectively. Distances are measured along each axis according to a choice of scale. To give coordinates to a point P in the plane, one draws from P lines parallel to the *x*-axis and the *y*-axis, meeting the *x*-axis at A, say, and the *y*-axis at B. The first, or *x*-coordinate of the point P is the length OA; the second, or *y*-coordinate of the point P is the length OB.

Coordinates may be negative in the modern convention, for the scales along each axis measure positive to the right and up, negative to the left and down, as is usual in the representation of negative quantities. Amusingly enough, the sign convention, which still causes trouble to beginners, was also obscure to Descartes. The curve called the folium of Descartes, which has equation $x^3 + y^3 = 3xy$, actually looks as in Figure 1(a). By drawing it correctly in the first quadrant but incorrectly generalizing to the other quadrants, Descartes thought the curve looked as in Figure 1(b), whence the name 'folium', meaning a leaf-shaped curve.

Once points are given coordinates, it is possible to specify families of points whose coordinates satisfy certain equations, and to identify these families geometrically, thus turning geometry into algebra. For example,

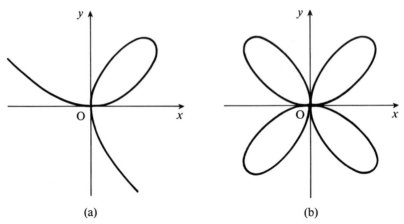

Figure 1 The folium of Descartes, (a) correctly drawn and (b) as Descartes mistakenly drew it

the points P whose coordinates (x, y) are related by the equation $x = 2y$ form a line through the origin; those satisfying the equation $x = y^2$ lie on a parabola (Figure 2). Conversely, given a curve in the plane and a choice of coordinate axes, the coordinates of points will be related by an equation: the sine curve, for example, has equation $y = \sin x$. It is usually to one's advantage to find coordinate axes with respect to which a given curve has a particularly simple equation.

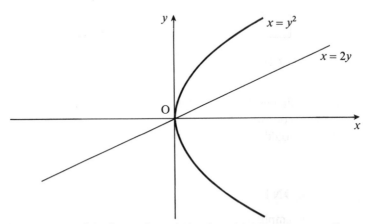

Figure 2 A straight line and parabola plotted in Cartesian coordinates

Coordinate methods can be used to find where two curves meet. To find where the line with equation $y = 2x - 1$ meets the circle with equation

848

$x^2 + y^2 = 1$, one proceeds algebraically. Eliminating y between the two equations, one obtains $5x^2 - 4x = 0$, whose solutions are $x = 0$ and $x = \frac{4}{5}$. The corresponding y-values are $y = -1$ and $y = \frac{3}{5}$, respectively. So the points of intersection are $(0, -1)$ and $(\frac{4}{5}, \frac{3}{5})$. Pursuing this argument as it applies to lines with equations $by = ax - b$, which all pass through the point $(0, -1)$ on the circle, one finds that the other point of intersection is

$$\left(\frac{2ab}{a^2 + b^2}, \frac{a^2 - b^2}{a^2 + b^2} \right).$$

This is an algebraic parametric representation of the circle; parametric representations are often a powerful method of studying curves.

Coordinate methods are also used to find the equation of curves satisfying given conditions. Let us find the equation of the circle with centre $(2, 3)$ and radius 5. We shall need to know the distance of a point $A(a, b)$ from another $C(c, d)$ (Figure 3); here $a = 2$ and $b = 3$, so we want the points (x, y) such that $(x - 2)^2 + (y - 3)^2 = 5^2$. This simplifies to $x^2 - 4x + y^2 - 6y - 12 = 0$, which is the equation we seek.

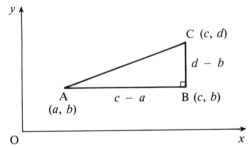

Figure 3 Use of coordinates to establish an equation

Cartesian coordinates are by no means the only possible ones. Indeed, there are so many coordinate systems that it is not possible to describe them here; the reader should consult Coolidge's wide-ranging historical account *1940*.

2 THE CONTRIBUTION OF RENÉ DESCARTES

The motivating example for Descartes was a problem that had come down from the Hellenistic writer Pappus (§1.3). He had described a problem in which four lines and four angles are given, and points P are then sought which satisfy the following rather complicated condition. Lines are drawn from P to each given line l_i, meeting the given lines at the given angles α_i;

four distances d_i are thus obtained (Figure 4). The problem is to find the locus of points for which the product of the first two distances is proportional to the product of the remaining two. Pappus was able to show that the curve so defined was generally a conic section. However, the analogous problem can be formulated for larger numbers of lines and then there was, he said, no known solution. Descartes was able to show that the locus to four lines could be found rather easily by his methods, and that the locus to any number of lines was not in principle any harder to find. His success in this matter, of which he was very proud, was one of the reasons for his confidence in the general efficacy of the new approach.

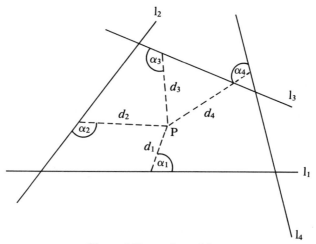

Figure 4 Pappus's problem

However, all was not so simple. In Descartes's time, a solution to a problem in geometry had naturally to be given in geometric terms. As H. J. M. Bos shows in two important papers (*1981, 1984*), this historical circumstance had a determining effect on how Descartes presented his ideas. For Pappus's problem, it was possible to describe the answer in geometrical terms: the solution curve is a conic section. But what could be said about a curve given only by an equation of higher degree than 2? One could hope to plot a finite number of points on it, but this fell well short of contemporary standards. To Descartes, the advocate of basing all reasoning on ideas clear and immediate to the mind, this was a problem. He invested considerable effort in pursuing the idea that for every equation there was an idealized machine that would draw its curve, just as a compass draws a circle. His hope was that simple curves would somehow generate more complicated ones, which in turn would generate still more complicated ones, and so on, complexity being measured by the degree of the equation. He never succeeded, but subsequent generations ceased to worry and accepted that a curve can be presented to the mind via an equation. Indeed, without admitting this it is hard to study curves at all: the ancient Greeks had bequeathed to posterity a mere dozen or so that could be analysed in any detail.

Descartes's aim was to take a geometrical problem, transcribe it into the language of algebra, solve it there, and then translate the solution back into geometrical terms. Even when the solution is the coordinates of a point, difficulties can arise, for it is by no means clear that *any* system of algebraic equations can be solved. This was always to remain a problem, and in practical situations where a numerical solution was required, approximate solutions were (and often still are) all that can be found. But there was a deeper problem, for by 'algebraic methods' Descartes meant those of finite polynomial algebra. When the solution sought is a curve, it is not true that the curve necessarily has an equation of this type. The sine curve does not; nor, as his critics pointed out, does the cycloid, the curve traced by a point on the rim of a wheel rolling along a straight line (Figure 5). Points on it have coordinates $(t - \sin t, 1 - \cos t)$, and there is no merely algebraic connection between the x- and y-coordinates. This mattered in its day, because Descartes regarded finding tangents to curves as the most important problem in mathematics. His critic Gilles Personne de Roberval had a simple argument by motion to find the tangent to a cycloid at a given point; but, as he pointed out, the cycloid necessarily eluded Descartes's method for finding tangents.

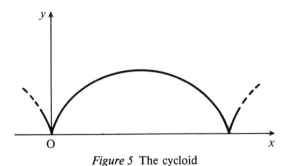

Figure 5 The cycloid

Another critic of Descartes was Isaac Newton, who increasingly found fault with all the Frenchman's ideas, be they in mathematics or physics. In particular, he disputed Descartes's idea that a curve was simple if its equation was simple, and gave the example of the cycloid. Newton's ideas struck home, but the resolution of the matter would have satisfied neither man. Later generations took their cue from the masterly Leonhard Euler, the second volume of whose *Introductio in analysin infinitorum* (*1748*) made algebraic geometry seem so easy and natural that the need for a criterion for the simplicity of curves was no longer felt, and algebra reigned supreme.

In summary, Descartes associated curves and equations via the idea of a coordinate system, but he maintained the contemporary preference for geometrical answers to geometrical questions. In that way his algebraic geometry was more cumbersome than ours, but in another way it was more flexible. Our coordinate axes are usually at right angles to one another; his were not so restricted. This helps to simplify the equations that arise, but complicates their interpretation. The modern perpendicular system became standard with, and because of, Euler.

Was Descartes really the first to do this? Coordinate systems have been claimed for the Greeks (by Coolidge *1940*), and even for the Egyptians. There are fragmented pictures that suggest curves were discussed in a coordinate-based way, much as one might describe a sloping roof: 'this far along and this far up'. But such insights did not translate into an algebraic method, for the simple reason that algebra − reasoning with letters − did not exist in Egyptian mathematical culture (§1.2). Coolidge's case for the Greeks, which rests on Apollonius's profound study of conic sections, is much stronger. What is to be found in Apollonius's *Conics* is the idea that a conic section has axes, with respect to which the proportion (not, strictly, the equation) defining its shape can best be studied. This is a wonderful idea, but it falls well short of Descartes's realization that axes can be chosen in the plane with respect to which any curve can be profitably studied. And again, one could argue that the Greeks did not have algebra in anything like the modern sense.

A high point of algebraic development came with François Viète. He sought to purify algebra, extend it to solve all problems (many posed geometrically) and establish that this was how the Greeks had really proceeded (§6.9). One cannot credit Viète with the invention of coordinate geometry, for Viète's problems involve only one unknown, and there are no curves described by equations. Nor, indeed, does there seem to have been any influence of Viète on Descartes. However, Descartes's contemporary Pierre de Fermat did read Viète and did develop the coordinate idea in the language of Viète's algebra for the study of conic sections. He did this before Descartes, but he did not publish it, and he did not present the coordinate method in clear generality. The honour does go to Descartes.

Before turning to modern developments, the name 'analytic geometry' should also be explained. To analyse something in the mathematical terminology of the Greeks and the seventeenth century was to take something apart, much as one speaks of a chemical analysis today. In the Cartesian approach, figures are analysed in this sense: points are given coordinates, curves are successively given equations; above all, the unknown point or points are given coordinates and these are treated on a par with the known quantities until they can be found from certain equations that this process of analysis yields. In short, the Cartesian method is a discovery method. One starts knowing very little about the unknowns, and the method leads one to discover their values or some equations they satisfy. From its inception, this was thought to be a great advance over the synthetic methods of the Greeks, which sought to explain why an answer is correct. Compare chemistry: to synthesize a chemical is to check that one's prior analysis of its composition was correct, but synthesis is no way to discover the structure of a given substance.

3 THE DEVELOPMENT OF COORDINATE GEOMETRY

Coordinate geometry can be used to turn geometrical problems into algebraic ones and, conversely, to provide a geometrical interpretation of algebraic problems. The former case was extended by many writers to include most problems with a mechanical origin; the latter is helpful as an aid to thought and when approximate solutions are sought. Almost every aspect of eighteenth-century mathematics, both pure and applied, bears witness to the fertility of the union. But what really transformed mathematics and augmented its role in the growth of science was the yoking together of coordinate geometry and the calculus (§3.2 and §4.3). Invented in the 1660s and 1670s by Newton and Gottfried Wilhelm Leibniz independently, the calculus existed for over a hundred years in two superficially distinct formulations, each remarkably successful at reducing problems about tangents to curves and areas under curves to systematic calculation. Each was also well adapted to dealing with curves defined by equations, and so the calculus promoted just the vision of geometry that had most bothered Descartes and his contemporaries. In the hands of Euler, who wrote three great expository treatises on the subjects of curves, the differential and the integral calculus (*1748, 1755, 1768–70*), the result was a body of techniques seemingly easy to apply and almost certain of success.

The era of rational mechanics, the analysis of nature along the lines of Newton's *Principia* (*1687*), but conducted in the language of Cartesian geometry and the Leibnizian calculus, occupied the second half of the eighteenth century. Such are its breadth and its unity, encompassing everything from the geometry of curves to hydrodynamics, that it is difficult to select topics which belong to analytic geometry in the strict sense, but a few do stand out. In plane geometry, Newton was the first to take up the study of curves defined by equations of degree 3. Beginning in the 1660s, when he set himself this task in order to master the Cartesian methods, and culminating in his presentation as an appendix to his *Opticks* (*1704*), Newton completely classified curves of this type. His classification was cumbersome, for the problem is full of technicalities, and it was reworked by Euler in the second volume of his *Introductio* (*1748*). Euler, Gabriel Cramer *1750* and others also began the study of the singular points of curves. These are typically points where a curve either crosses itself or else has a cusp (Figure 6), although there are more complicated possibilities.

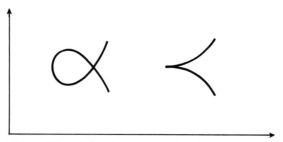

Figure 6 Singular points: a crossing-point and a cusp

The natural extension of Cartesian methods to three dimensions, although hinted at by Descartes, had to wait until the eighteenth century, when it was set forth by the 18-year-old Alexis Clairaut *1731*. A single equation in three variables generally represents a surface; Clairaut considered the case of cubic surfaces (surfaces defined by an equation of degree 3) and their intersections by planes. The simpler case of equations of degree 2 yields the so-called 'quadric surfaces'. These include the ellipsoid, the hyperboloids of one and two sheets, and the cone.

Curves in space were (and are) much harder to represent. The most usual way was to realize a curve as the common points of two surfaces, and so as given by two equations. The alternative was to represent it parametrically, but this could be made to work in only a limited number of cases.

It is the fate of some vigorous branches of any subject to split into topics that between them exhaust the parent discipline. Such was the case with coordinate geometry. From it grew such subjects as differential geometry and algebraic projective geometry, but there is no such subject as 'advanced coordinate geometry'. Differential geometry has many roots (§3.4): one lies in the study of surfaces, specifically in the study of plane sections through each point; another lies in the study of curves in space, where it proved useful to look at the way a moving coordinate system varied as it travelled along a curve (Figure 7).

Algebraic projective geometry (§7.6) resumed at the start of the nineteenth century, when a reaction set in to the preference – then well established – for algebra and calculus over all other modes of mathematics. In France this move was led by Gaspard Monge, a highly influential founder-member of the Ecole Polytechnique (§11.1). In Germany, algebraic geometry was reinvigorated by August Ferdinand Möbius and, most notably, by Julius Plücker. Their work forms the natural generalization of the studies of Newton, Cramer and others, but the complexities of the subject inevitably pushed Cartesian geometry into projective geometry.

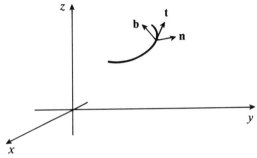

Figure 7 A space curve

Typical of the richness of analytic geometry in the nineteenth century is the study of three surfaces: the cubic surface, Fresnel's wave surface and Kummer's quartic surface (for a good modern reference, full of photographs, see Fischer *1986*). The cubic surface was studied by many authors, but the English mathematician Arthur Cayley and his Irish colleague George Salmon made the most remarkable discovery: the general cubic surface contains 27 straight lines, all of which can be real. (Several of them are visible in Figure 8, which shows a late-nineteenth-century model.) It is surprising that a curved surface should carry straight lines, and their interrelation proved to be a fascinating object of study.

Augustin Jean Fresnel was one of the architects of the classical wave theory of light (§9.1). One problem he set himself was to determine the image of a point seen from a fixed position through a biaxial crystal. As Buchwald *1989* has carefully described, this question led Fresnel to describe his wave surface, which is the locus, up to unit time, of all the plane waves initially emitted from the given point. The determination of the equation of

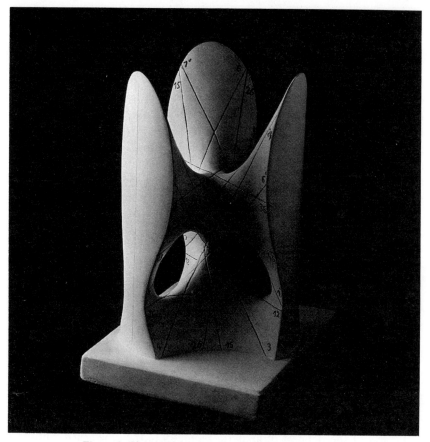

Figure 8 Clebsch's model of a general cubic surface

this surface on optical grounds was not easy; it turned out to be

$$\frac{x^2}{r^2 - a^2} + \frac{y^2}{r^2 - b^2} + \frac{z^2}{r^2 - c^2} = 1, \tag{1}$$

where a, b and c are constants determined by the crystal, and $r^2 = x^2 + y^2 + z^2$. It is therefore a quartic surface (i.e. one of degree 4: see Figure 9). The differential geometry of curves on this surface interested mathematicians, but an even more remarkable fact soon caught their attention. Ernst Kummer, a leading light in the revival of mathematics in Berlin, investigated in the 1850s those quartic surfaces having the maximum possible number of singular points (16 double points, most of them visible in Figure 10). This purely geometrical line of enquiry turned out to yield a

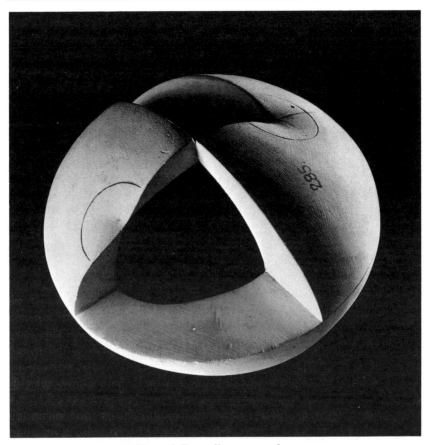

Figure 9 Fresnel's wave surface

family of surfaces containing the wave surface as a special case, a surprising coincidence that stimulated much further work.

4 COORDINATE TRANSFORMATIONS

Although there is no such subject as advanced coordinate geometry, there is a mainstream development of Descartes's ideas that fills university courses and is a staple technique of both pure and applied mathematics: coordinate transformations. Coordinate geometry, whether in the plane, in space or in higher dimensions, lends itself naturally to a description by vectors (§7.12). Changing the origin of the coordinate system or moving the coordinate axes changes the equation of the curve, opening the prospect of finding a simplest possible equation for a curve. For this reason, in the

857

Figure 10 Kummer's quartic surface with 16 double points

nineteenth century there was developed a powerful theory for describing coordinate transformations in the language of vector spaces, as Hawkins (*1975*, *1977*) has described. This theory, which has an independent interest, is discussed in the articles on matrices and invariants (§6.7–6.8).

BIBLIOGRAPHY

Apollonius *1891–3*, *Apollonii Pergaei quae Graece existant cum commentariis antiquis*, 2 vols (ed. J. L. Heiberg), Leipzig: Teubner.

——*1896*, *Treatise on Conic Sections* (ed. T. L. Heath), Cambridge: Cambridge University Press.

Bos, H. J. M. *1981*, 'On the representation of curves in Descartes's *Géométrie*', *Archive for History of Exact Sciences*, **24**, 295–338.

—— *1984*, 'Arguments on motivation in the rise and decline of a mathematical theory; the "construction of equations", 1637–*ca.* 1750', *Archive for History of Exact Sciences*, **30**, 331–80.

Boyer, C. B. *1956*, *History of Analytic Geometry*, New York: Scripta mathematica.

Buchwald, J. Z. 1989, *The Rise of the Wave Theory of Light*, Chicago, IL: University of Chicago Press.

Clairaut, A. C. *1731*, 'Sur les courbes que l'on forme en coupant une surface courbe quelconque par un plan donné de position', *Histoire de l'Académie Royale des Sciences*, 183–93.

Coolidge, J. L. *1940*, *A History of Geometrical Methods*, Oxford: Oxford University Press. [Repr. 1963, New York: Dover.]

Cramer, G. *1750*, *Introduction à l'analyse des lignes courbes algébriques*, Geneva: Cramer.

Descartes, R. *1637*, *La Géométrie* [one of the appendices to *Discours de la méthode*], Leiden: Maire. [Many subsequent editions. Repr. in D. E. Smith and M. L. Latham (transls), 1954, *The Geometry of René Descartes*, New York: Dover.]

Euler, L. *1748*, *Introductio in analysin infinitorum*, 2 vols, Lausanne: Bousquet. [Also *Opera omnia*, Series 1, Vols 8 and 9. English transl. by J. D. Blanton, 1988–90, *Introduction to the Analysis of the Infinite*, New York: Springer.]

—— *1755*, *Institutiones calculi differentialis*, St Petersburg: Academy. [Also *Opera omnia*, Series 1, Vol. 10.]

—— *1768–70*, *Institutiones calculi integralis*, 3 vols, St Petersburg: Academy. [Also *Opera omnia*, Series 1, Vols 11–13.]

Fischer, G. *1986*, *Mathematische Modelle/Mathematical Models*, 2 vols [Vol. 1 pictures, Vol. 2 commentary], Braunschweig: Vieweg.

Hawkins, T. *1975*, 'The theory of matrices in the 19th century', in *Proceedings of the International Congress of Mathematicians*, Vancouver, 1974, [no place]: Canadian Mathematical Congress, 561–70.

—— *1977*, 'Another look at Cayley and the theory of matrices', *Archives Internationales d'Histoire des Sciences*, **26**, 82–112.

Newton, I. *1687*, *Philosophiae naturalis principia mathematica*, 1st edn, London: Streater. [English transl. by A. Motte, 1729; rev. F. Cajori, 1934, as *The Mathematical Principles of Natural Philosophy* [. . .], 2 vols, Berkeley, CA: University of California Press.]

—— *1704*, *Opticks*, London: Smith and Walford. [See also 'Enumeration of the lines of the third order' (*circa* 1695), in *The Mathematical Papers of Isaac Newton* (ed. and transl. D. T. Whiteside), Cambridge: Cambridge University Press, Vol. 7, 1976, 589–645.]

7.2

Curves

J. J. GRAY

1 CURVES IN GREEK MATHEMATICS

A good, naive, definition of a curve would be the path traced out by a moving point. While examples are legion, a curve can enter mathematics only when a mathematician has something specific to say about it, and the general concept arises only when there is something worthwhile and general to say. Yet it still comes as a surprise that the Greeks had no general concept, and few specific examples to study. Euclid's *Elements* discusses properties of the straight line and circle; following him Archimedes and Apollonius studied the conic sections (the ellipse, parabola and hyperbola), but only a few other curves were known, such as the spiral of Archimedes (§1.3). This paucity can be ascribed partly to the means available for defining and treating curves, partly to the range of problems discussed in Classical times. So, for example, the conic sections were defined as particular kinds of section of a cone by a plane. Their study was extremely difficult with the means available; that an extensive theory of the conics was obtained is eloquent testimony to the brilliance of Archimedes and Apollonius. No less a modern mathematician than B. L. van der Waerden has described Apollonius's *Conics* as a masterpiece whose author was 'a virtuoso in hiding his original line of thought [which] is what makes his work hard to understand' (*1961*: 248).

The conic sections and most of the other curves known to the Greeks were used by them to tackle significant problems, notably those known today under the collective title of 'the three classical problems'. These were: to trisect an angle (i.e. given any angle, to find an angle of one-third the size); to duplicate the cube (i.e. given a cube, to find one of twice the volume); and to square the circle (i.e. given a circle, to find a square of equal area). It seems that the Greeks despaired of finding solutions to any of these problems by straightedge and compass alone, and indeed it was shown rigorously in the nineteenth century that under those restrictions it cannot be done. So they set about solving these problems in other ways.

Trisecting an angle may be difficult, but trisecting a length is trivial, and the solution to the angle problem offered by Hippias of Elia in the fifth century reduced it to the trisection of an interval. He described a curve in this way (Figure 1). In a square ABCD of side a, one rod AR of length a is attached at A, where it is free to pivot, while a second, horizontal rod $C'D'$ of the same length is free to move up and down. Initially the first rod is vertical and the second rod lies along CD. They are set in motion together, so as to move uniformly and at such speeds that when AR has rotated through a right angle, $C'D'$ has reached AB. Hippias's curve is traced by X, the intersection of the two rods; and AS = 2SB.

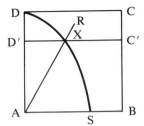

Figure 1 The trisectrix

To use the curve to trisect a given angle of q_0, one has merely to allow the rod AR to rotate through that amount, to AR_0 say (Figure 2), and locate the corresponding position $C_0'D_0'$ of the horizontal rod. One then trisects the segment DD_0', say at D_1', and finds the corresponding position of the rod AR, say AR_1. The angle DAR_1 is one-third of the angle DAR_0. For this reason the curve is sometimes called the 'trisectrix'.

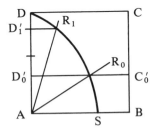

Figure 2 The trisection of angle DAR_0. $DD_1' = DD_0'/3$

The trisectrix has another use: it helps to solve the problem of squaring the circle. It meets the side AB at the point S in Figure 2, for which

$AS = 2a/\pi$. A circle of radius r has an area of πr^2; for a circle of unit radius this reduces to π. With $a = 1/2$, the trisectrix gives a length of $1/\pi$, which can easily be made to yield a length of π, thus squaring the circle. For this reason the trisectrix is also called the 'quadratrix' (it has brought about the quadrature of the circle). At a time when it had to be shown rigorously that the concept of area, first introduced for rectilinear figures, applies to the circle, this was a considerable achievement.

Duplicating the cube recalls the problem of duplicating the square, but whereas the latter problem only requires $\sqrt{2}$ to be found, and was solved in Plato's dialogue the *Meno*, duplicating the cube requires $\sqrt[3]{2}$ to be found, which is much harder. Two solutions associated with the name of Menaechmus involve conic sections. Another, due to Nicomedes, introduced a curve called the conchoid. (It can also be used to trisect the angle.) He considered a line l, and a point O not on it, and looked at all the lines through O. Let the points P_1 and P_2 lie on such a line that meets l at Q. The conchoid is composed of the points P_1 and P_2 for which $P_1Q = QP_2 = a$ a fixed distance a (Figure 3). Yet another solution was proposed by Diocles. This curve, called the cissoid, was obtained as follows (Figure 4). Let AB and CD be perpendicular diameters of a circle, and let EB and BF be equal arcs. Draw FH perpendicular to CD and let it meet EC at P. As the point E moves round the circle, the point P traces out the cissoid. For a full account of these curves and their uses, see Heath *1921*: Vol. 1.

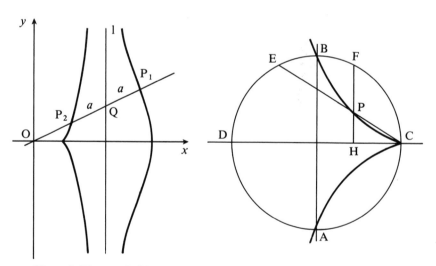

Figure 3 The conchoid of
Nicomedes

Figure 4 The cissoid of Diocles

2 CURVES IN CARTESIAN GEOMETRY

Such elaborate constructions make it plain that very few curves can be introduced in this way. Much less did they elicit general concepts; the idea of a tangent to a curve does not seem to have been defined in general, although tangents to the circle, the conic sections, and the spiral were discussed. What is lacking is the concept of one quantity varying with another, so that the two between them define a moving point tracing out a curve. This idea was introduced into mathematics by René Descartes (§7.1). If, for each value of a variable x, one has one or more values of a variable y, then the points (x, y) define a curve which can be represented in the Cartesian plane. At a stroke one has the capacity to define infinitely many curves, including all those for which the relationship between x and y is given by an algebraic equation, and all those for which y is a known function of x. So in the seventeenth century it became possible to ask for the first time for a general method for finding the tangent to a given curve at a given point. Of course, such a method was soon provided by the rival versions of the calculus put forward by Isaac Newton and Gottfried Wilhelm Leibniz (§3.2).

The period from 1650 to 1850 was the heyday of the study of curves by Cartesian methods. It is usually easy to find the Cartesian equation of a curve defined in some other way. For example, the conchoid of Nicomedes has the equation $(x^2 + y^2)(x - b)^2 = a^2 x^2$, and the cissoid of Diocles has the equation $y^2(a + x) = (a - x)^3$. When this cannot be done the curve can still usually be represented parametrically, that is, in the form $(x(t), y(t))$ as t varies. For example, the cycloid has the parametric representation $(t - \sin t, 1 - \cos t)$. Two major books defined the theme: Euler *1748* and Cramer *1750* (see §7.1 for details).

3 UNEXPECTED PROPERTIES OF CURVES

With the rigorization of the calculus in the nineteenth century (§3.3), the question arose of what sort of behaviour a curve could have. For example, is the plausible claim that every curve has a tangent at every point always true, except where it is palpably false (where the curve crosses itself or suddenly reverses its direction at a cusp)? To everyone's initial surprise, it proved possible to construct continuous functions that never have a tangent at any point. After the first probable cases of such a phenomenon were brought to light by Bernhard Riemann in the 1850s, rigorous examples were found by Karl Weierstrass in the 1860s. It was then possible to define continuous curves that never have a tangent. At first these curves caused distress in some circles. Some regarded them as pathological, and wished to reject them; others, such as Henri Poincaré, eagerly accepted them. Rather surprisingly, it emerged from the work of Albert Einstein and others that the path of a particle undergoing Brownian motion was usually continuous but nowhere differentiable.

A further challenge to naive intuition about curves came when Georg Cantor showed in the 1880s that there is a 1–1 correspondence between the points of a curve and the points of a plane (§3.6). This seems to counter the notion of a curve as one-dimensional and a plane as two-dimensional; order was restored only when it was shown that there is no such 1–1 correspondence which is continuous in each direction. Nevertheless, there are continuous curves which pass through every point of a square or, for that matter, the whole plane. The first of these was defined in formulas derived by the Italian mathematician and logician Giuseppe Peano, and later illustrated by the American mathematician E. H. Moore and the German David Hilbert (§3.8). When the concept of dimension was later stretched by Felix Hausdorff to allow for spaces of non-integer dimension, it was soon found that these unexpected types of curve can have any dimension.

4 COMPLEX CURVES

At the same time, the concept of curve was being extended in another way. An algebraic relationship between two variables x and y can naturally be thought of as defining a curve. For example, the relationship $x^2 + y^2 = 1$ defines a circle. It was the profound idea of Riemann to extend this to the case where the variables x and y are complex (§7.9). Since each complex variable can be thought of as a pair of real ones, the relationship now defines a surface in a space of four real dimensions. This is not intuitively easy to grasp, but it proved possible to proceed as if an equation between two complex variables defined a curve (something of real dimension 2, but *complex* dimension 1), today called an algebraic curve or a Riemann surface. Riemann's approach allows the full resources of algebra to be exploited in studying curves defined by algebraic equations. Since then it has often proved worthwhile to think of an algebraic equation as defining a curve whatever field one is working in, a point of view that has abundantly proved its worth in number theory.

BIBLIOGRAPHY

Apollonius *1891–3*, *Apollonii Pergaei quae Graece existant cum commentariis antiquis*, 2 vols (ed. J. L. Heiberg), Leipzig: Teubner.
—— *1896*, *Treatise on Conic Sections* (ed. T. L. Heath), Cambridge: Cambridge University Press.

Brieskorn, E. and Knörrer, H. *1986*, *Plane Algebraic Curves*, Basel: Birkhäuser. [Part 1 is historical.]

Cramer, G. *1750*, *Introduction à l'analyse des lignes courbes algébriques*, Geneva: Cramer.

Descartes, R. *1637*, *La Géométrie* [one of the appendixes to *Discours de la méthode*], Leiden: Maire. [Repr. in D. E. Smith and M. L. Latham (transls), 1954, *The Geometry of René Descartes*, New York: Dover.]

Euler, L. *1748*, *Introductio in analysin infinitorum*, 2 vols, Lausanne: Bousquet. [Also *Opera omnia*, Series 1, Vols 8 and 9. English transl. by J. D. Blanton, 1988–90, *Introduction to the Analysis of the Infinite*, New York: Springer.]

Heath, T. L. *1921*, *A History of Greek Mathematics*, 2 vols, Oxford: Clarendon Press. [2nd edn 1981, New York: Dover.]

van der Waerden, B. L. *1961*, *Science Awakening* (transl. A. Dresden), New York: Oxford University Press.

7.3

Regular polyhedra

BRANKO GRÜNBAUM

The idea of singling out certain polyhedra with unusual symmetry properties first occurred during the heyday of Greek geometry. Exact information
about the origin of the choice of five specific polyhedra (shown in Figure 1)
which we call 'regular' is not available, and neither is there any certainty
about the first discovery of individual regular polyhedra. The regular polyhedra were discussed in Plato's Academy, where they were attributed
to Theaetetus. The fanciful importance Plato bestowed upon them, as
building blocks of the universe, led to the name 'Platonic solids'. The
regular polyhedra are the crowning achievement of Euclid's *Elements*,
Book 13, where it is proved that there are exactly five different regular polyhedra. In a certain sense, the specific determination of the five polyhedra
may be considered less important than the generation of the concept of
regularity; apparently this should be credited to Theaetetus (see Waterhouse
1972 for a stimulating discussion of this aspect). Illustrations and accounts
of the properties of these polyhedra (and of most of the polyhedra mentioned in this article) can be found, for example, in Coxeter *1948* or Cundy
and Rollett *1951*; a nice introduction to various aspects of polyhedra, their
history and literature is given in the collection Senechal and Fleck *1988*.

1 SHORTCOMINGS OF THE EUCLIDEAN TREATMENT

From the point of view of logic, Euclid's account of the regular polyhedra
presents a striking contrast between good intentions and actual execution.
Although axiomatic geometry enjoyed through the ages the reputation of
presenting logical thought at its best, this was undeserved, even in the case
of the frequently discussed topic of regular polyhedra. In fact, the validity
of the stated characterizations of regular polyhedra requires additional,
unstated assumptions. The basic logical difficulty is that Euclid never
defined (or informally explained) what is a 'solid' or 'figure' – although
he was aiming to show that 'apart from the said five figures, there cannot
be constructed any other figure which is contained by equilateral and

Figure 1 Drawings of the five Platonic solids, attributed to Leonardo da Vinci (from Luca Pacioli, *De divina proportione*, 1498)

equiangular figures equal to one another'. Sympathetic commentators have explained that Euclid must have had only convex polyhedra in mind, but even then there are at least two deficiencies. First, convexity itself is not unambiguously defined: only in recent decades has there been a tendency to a very restrictive interpretation of this word – in fact, the regular polyhedra found by Johannes Kepler and (later) by Louis Poinsot, which are discussed next, were considered throughout the nineteenth century to be 'convex'. Second, even with interpretation of 'figure' as a convex polyhedron in the most restrictive sense, Euclid's statement is plainly wrong: there are five additional 'deltahedra', non-regular convex polyhedra bounded by congruent equilateral triangles.

To salvage Euclid's enumeration it is therefore necessary to impose additional restrictions, besides convexity in the narrow sense. Over the centuries, various conditions have been proposed, such as: all solid angles should be regular; all solid angles should be congruent; all dihedral angles should be equal; equal numbers of faces should meet at each vertex; there exist three concentric spheres, one through all vertices, one touching all edges, and one touching all faces. Although logically distinct, these and other conditions are mutually equivalent in the context under discussion, and indeed yield only the five Platonic solids among strictly convex polyhedra. More recently, it has become customary to define regularity in terms of equivalence under symmetries. The most frequent formulation, which in essence goes back to a paper by Augustin Louis Cauchy (published in 1810) and was apparently first formulated explicitly by Ernst Steinitz in his encyclopedic survey of polyhedra (1916), defines a polyhedron to be regular if its group of symmetries acts transitively on its 'flags', a 'flag' being a triplet consisting of a vertex, an edge and a face, all mutually incident.

2 POLYHEDRA OF KEPLER AND POINSOT

The lack of precision in Euclid's formulation has had the positive effect of admitting other interpretations that led to new investigations and to the designation of additional polyhedra as 'regular'; this helped to break the stranglehold of ossified traditions. The first such step was taken by Kepler in 1619. He started by observing that, if a polygon is interpreted as a circuit of edges, meeting in pairs at common end-points, then the pentagram (and other star polygons) are regular in the sense of having all sides equal as well as all angles equal. The fact that the sides cross each other has no relevance to the question of regularity. Since a similar reasoning applies to polyhedra, and since it is possible to combine pentagrams as faces to obtain polyhedra which satisfy the Euclidean conditions of regularity, Kepler claimed the status of regular polyhedra for two of his inventions (see Figure 2, taken

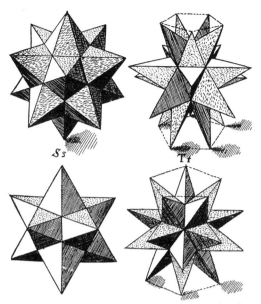

Figure 2 Two views of each of the two regular polyhedra found by Kepler (from his *Harmonice mundi*, 1619)

from Kepler's work). In fact, they are regular even under the flag-transitivity criterion. It may be mentioned that star-shaped solids resembling one of Kepler's are found in the work of some earlier artists; but, as with Theaetetus, Kepler's principal merit was the insight that these polyhedra deserve to be considered regular, rather then just having an attractive shape.

Kepler's discovery lay largely forgotten for two centuries, and the next development is due to Poinsot in *1809*. Apparently unaware of Kepler's polyhedra, Poinsot followed a similar argument to describe them, as well as two additional regular polyhedra in which either triangles or pentagons (without self-intersections) are arranged around each vertex in a star-shaped way. These four polyhedra – usually known as the 'Kepler–Poinsot polyhedra' – are shown in Figure 3. Poinsot's discovery was followed in 1810 by Cauchy's proof of the completeness of that enumeration. However, Poinsot's work suffers from an internal inconsistency. In following Poinsot uncritically, as in other instances where he was insensitive to geometric subtleties, Cauchy made a mistake that has been noticed only recently; as a consequence of this error, additional restrictions are needed to make his result valid. (The discovery of the shortcoming was made during the preparation of this article and was first presented in a course that I was giving in spring 1990 at the University of Washington.)

Figure 3 The four Kepler–Poinsot polyhedra (from Wiener *1865*: Plate 3; this was the first published illustration of these four polyhedra)

3 ERRORS IN POINSOT'S WORK

The problem arises from Poinsot's disregard of his own definition of regular polygons, which – freely but faithfully translated – is as follows. To obtain a polygon, let m points a, b, c, ... be arbitrarily given in the plane, joined by m segments ab, bc, cd, ..., so that the resulting figure is closed. A polygon is regular if its edges are equal and its angles are equal. (Poinsot gives a completely satisfactory definition of equality of angles in a polygon.) He goes on to say (correctly) that if h is an integer relatively

870

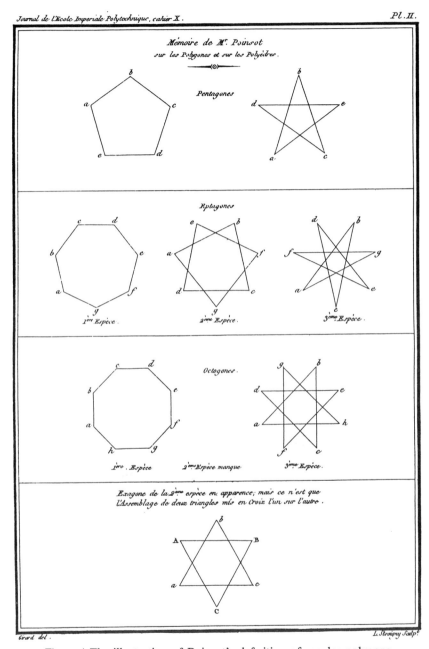

Journal de L'Ecole Imperiale Polytechnique, cahier X.   Pl. II.

Mémoire de M. Poinsot
sur les Polygones et sur les Polyèdres.

Pentagones

Eptagones

1ère Espèce.   2ème Espèce.   3ème Espèce.

Octogones.

1ère. Espèce   2ème Espèce manque   3ème Espèce.

Exagone de la 2ème espèce en apparence; mais ce n'est que
L'Assemblage de deux triangles mis en Croix l'un sur l'autre.

Figure 4 The illustration of Poinsot's definition of regular polygons
(from his 1809)

prime to m, and if m points, equidistributed on a circle, are connected by segments each of which spans h of the arcs determined by the points, then a regular polygon is obtained. In modern notation, such a polygon would be designated $\{m/h\}$; convex polygons correspond to $h = 1$, and the pentagram $\{5/2\}$ to $m = 5$ and $h = 2$ (see Figure 4, taken from Poinsot's paper). The last row in Figure 4 illustrates (for $m = 6$ and $h = 2$) Poinsot's (correct) statement that this construction does not produce a regular polygon if m and h are not relatively prime; in such a case the resulting figure is composed of several regular polygons, and hence is not itself a regular polygon.

The logical error committed by Poinsot and uncritically accepted by all later writers is the assumption that this reasoning proves the non-existence of regular polygons (as defined) other than the ones corresponding to relatively prime m and h. In fact, it is fully in accordance with Poinsot's definition (as well as with more modern ones) to consider as regular polygons obtained by starting from one point on a circle, connecting it to a second point by a segment spanning an arc which is h/m times the length of the circle's perimeter, and continuing by such steps until, after m steps, the original point is reached again. If m and h are relatively prime, the mth step is the first that leads to the starting-point; otherwise, there will have been

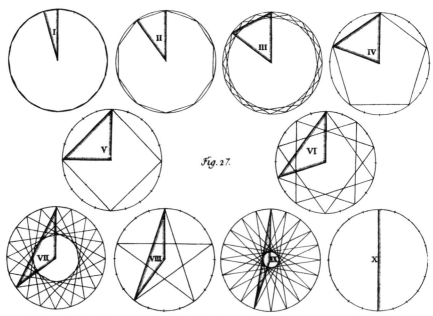

Figure 5 An illustration of the definition of regular polygons, showing all the different polygons with 20 vertices (from Meister *1770*: Plate 6)

previous visits — but this is irrelevant to regularity. It is most remarkable and ironic that precisely this interpretation was originally advanced by A. F. L. Meister in *1770*; Figure 5 shows his illustration for $m = 20$. However, although Meister's article is often mentioned in accounts of the history of polygons, it is apparently rarely, if ever, read. The only discussion of some details of Meister's work, by the historian of mathematics Siegmund Günther (in 1876), misquotes Meister at the critical place, giving the false impression that Meister made the same error as Poinsot.

Clearly, with consistent application of the definitions some points may correspond to several vertices of a regular polygon — a possibility which has not been excluded in any way (except possibly by arbitrary and unstated fiat). A variety of additional polyhedra can now be considered regular, with the requirement of flag transitivity under symmetries fulfilled. A simple example is shown in Figure 6(a), where each vertex of the cube represents two vertices of a regular polyhedron which has 16 vertices and 6 octagons

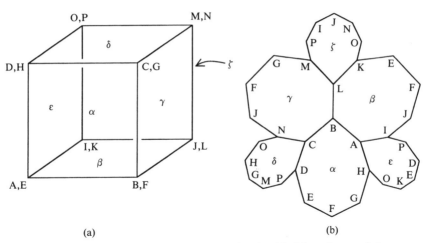

(a)   (b)

Figure 6 (a) A regular polyhedron {8/2, 3} with 16 vertices and 6 octagonal faces:

$$\alpha = [A, B, C, D, E, F, G, H],$$
$$\beta = [A, I, J, F, E, K, L, B],$$
$$\gamma = [B, L, M, G, F, J, N, C],$$
$$\delta = [C, N, O, H, G, M, P, D],$$
$$\varepsilon = [A, H, O, K, E, D, P, I],$$
$$\zeta = [I, P, M, L, K, O, N, J]$$

(b) A topologically regular map of genus 2, which is isomorphic to the regular polyhedron {8/2, 3}; this map is denoted by {4 + 4, 3} in Coxeter and Moser (*1957*: Section 8.8), where references to various papers discussing it may be found

873

of type {8/2} as faces. This polyhedron can be interpreted as a polyhedral realization of the well-known, topologically regular map of genus 2, which is shown in Figure 6(b).

The complete enumeration of regular polyhedra is still outstanding.

In the 1880s the concept of regular polyhedra was extended to higher-dimensional spaces. Related concepts were studied in hyperbolic spaces, and in other settings as well. Space here does not permit these and other developments to be presented; accounts of them can be found in Coxeter *1948*, *1974*, and Fejes Tóth *1964*.

4 OTHER SPECIAL POLYHEDRA

Since Antiquity, various classes of polyhedra that are slightly less special than the regular polyhedra have also been considered. In several cases logical errors as blatant as the one discussed above have been widely accepted; these errors are briefly mentioned in discussing some of these classes of polyhedra.

Archimedean polyhedra are usually defined as strictly convex polyhedra, all faces of which are regular polygons (not necessarily of the same kind), and all vertices of which are congruent (this means that the figure formed by the faces containing one vertex is congruent to the figure analogously formed at any other vertex). A related concept is that of a uniform polyhedron defined to be a regular-faced isogonal polyhedron; some writers confuse 'Archimedean' with 'uniform'. ('Isogonal' means that the vertices are all equivalent under symmetries of the polyhedron.) The enumeration of Archimedean polyhedra is believed to have been carried out by Archimedes, but, if so, no manuscript survived. In any case, from the Renaissance onwards there have been frequent claims that, besides the regular polyhedra, the prisms and the antiprisms, there are precisely 13 other Archimedean polyhedra. While this is correct if applied to uniform polyhedra, there is one additional Archimedean polyhedron which is not uniform. That exceptional polyhedron is the pseudorhombicuboctahedron (Figure 7), first constructed by J. C. P. Miller in the 1920s and described by H. S. M. Coxeter in 1930. It was previously missed by geometers because they assumed – with no logical basis – that if the faces of an Archimedean polyhedron are arranged around a vertex in a certain way, no other Archimedean polyhedron can be formed with this arrangement of faces.

During the last century, isogonal polyhedra have been studied without restriction to those with regular faces. The convex ones among them have the interesting property that each is isomorphic to one of the uniform polyhedra; this holds even for the more general class of those not necessarily convex isogonal polyhedra which are topologically equivalent to a solid

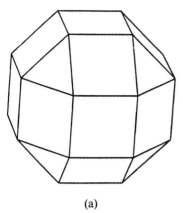 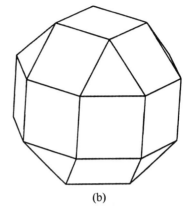

(a) (b)

Figure 7 Two distinct Archimedean polyhedra with the same arrangement of faces around each vertex: (a) the rhombicuboctahedron and (b) the pseudorhombicuboctahedron; (a) is uniform, (b) is not

sphere. However, in most writings it is (explicitly or implicitly) assumed that all isogonal polyhedra (without self-intersections) are 'spherical' in that sense. Only recently, Grünbaum and Shephard *1984* observed that this assumption is not justified: there exist toroidal isogonal polyhedra, as well as isogonal polyhedra of genus 3, 5, 7, 11 or 19. These non-spherical isogonal polyhedra also show that the frequently made claims about 'natural duality' of isogonal polyhedra with isohedral polyhedra (i.e. those in which all faces are equivalent under symmetries) are mistaken: there exist no isohedral polyhedra (without self-intersection) which are dual to the toroidal isogonal polyhedra, or to those of higher genus. The logical error here is due to a widespread confusion between the projective duality which relates points and planes, and the topological duality which relates cells and vertices.

These (and other) facts about regular polyhedra and their generalizations show that, in the distant past as well as in modern times, geometers have been less adept at investigative and deductive geometry and critical thinking than at accepting traditional teachings, even if these are not supported by logic.

BIBLIOGRAPHY

Coxeter, H. S. M. *1948*, *Regular Polytopes*, London: Methuen. [3rd edn repr. 1973, New York: Dover.]
—— *1974*, *Regular Complex Polytopes*, Cambridge: Cambridge University Press. [2nd edn 1991.]

Coxeter, H. S. M. and Moser, W. O. J. *1957, Generators and Relations for Discrete Groups*, Berlin: Springer. [4th edn 1980.]

Cundy, H. M. and Rollett, A. P. *1951, Mathematical Models*, Oxford: Clarendon Press. [2nd edn 1961.]

Fejes Tóth, L. *1964, Regular Figures*, London: Pergamon.

Grünbaum, B. and Shephard, G. C. *1984*, 'Polyhedra with transitivity properties', *Société Royale de Canada. Académie des Sciences. Comptes Rendus Mathématiques*, **6**, 61–6.

Meister, A. F. L. *1770*, 'Generalia de genesi figuram planarum [. . .]', *Novi commentarii Societatis Regiae Scientarum Goettingensis*, **1**, 144–80.

Poinsot, L. *1809*, 'Mémoire sur les polygones et les polyèdres', *Journal de l'Ecole Polytechnique*, **4**, cahier 10, 16–49.

Senechal, M. and Fleck, G. (eds) *1988, Shaping Space: A Polyhedral Approach*, Boston, MA: Birkhäuser.

Waterhouse, W. C. *1972*, 'The discovery of the regular solids', *Archive for History of Exact Sciences*, **9**, 212–21.

Wiener, C. *1865, Über Vielecke und Vielfläche*, Leipzig: Teubner.

7.4

Euclidean and non-Euclidean geometry

J. J. GRAY

1 EUCLIDEAN GEOMETRY

Euclidean geometry is named after the Greek mathematician Euclid (*circa* 300 BC), who wrote what became the definitive account of the elementary part of the subject in his *Elements* (§1.3). This work, in 13 books, begins by describing the properties of plane figures composed of straight lines and circles, starting from appropriate basic definitions, common notions and postulates. Common notions are undemonstrated truths of general applicability, such as 'the whole is greater than the part', while the postulates are specific mathematical statements their author saw fit to assume. Most of these are uncontroversial, for example 'all right angles are equal to one another', but one of them came to be increasingly discussed – the parallel postulate. This is the claim 'that, if a straight line falling on two straight lines make the interior angles on the same side less than two right angles, the two straight lines, if produced indefinitely, meet on that side on which the angles are less than two right angles'. While the other assumptions of Book I alone enabled Euclid to prove some results, such as the isosceles-triangle theorem and the congruence theorems of plane geometry, the parallel postulate was needed to prove such results as Pythagoras's theorem, and that the angles of a triangle add up to two right angles. It follows that most of the rest of the *Elements* also required the postulate.

On the other hand, the postulate seemed open to objection on two opposing grounds: it was not as obviously true as the other assumptions made by Euclid, and it seemed more like a result that could be proved on the basis of the other assumptions alone. Indeed, a number of later Greek writers attempted proofs along these lines, for example the astronomer Ptolemy and the fifth-century commentator Proclus. The common result of these attempts and many that came to be made down the ages was either to lapse into a fallacy, or else to establish, explicitly or implicitly, the equivalence of the parallel postulate with some other assumption. These discoveries are interesting in themselves, and shed light on what Euclidean geometry was taken to be. For more detailed accounts, see Bonola *1912* and Gray *1989*.

2 ATTEMPTS ON THE PARALLEL POSTULATE

Most commonly, it was a property of distance that came to be invoked. Proclus's attempt rests on the claim, not discussed by him, that two parallel lines remain everywhere a bounded distance apart. Others, such as Ibn al-Haytham (*circa* AD 1000), endeavoured to prove that the locus of points equidistant from a straight line is itself straight, and on that basis to prove the postulate (§1.6). Weaker spirits merely assumed that Ibn al-Haytham's claim was true.

One of the most scrupulous attempts was by John Wallis in 1663. He showed that the parallel postulate was equivalent to the assumption that similar figures exist which are not congruent: that is, that a copy of a given figure can be made which has the same shape but not the same size. Wallis did not claim to have established the postulate as a theorem, merely to have reduced the task to that of defending the possibility of making scale copies of figures, for which he gave a different, philosophical argument. The assertions of Ibn al-Haytham and Wallis are, of course, theorems in the *Elements*; they can be proved once the parallel postulate is assumed. What is at stake is whether they are more elementary than the postulate, and so either a more natural initial assumption to make or easier to prove as theorems themselves.

Ibn al-Haytham's case is typical of many in resting on claims about the straight line, defined rather unhelpfully in the *Elements* as 'the curve that lies evenly upon itself'. It is probably no longer possible to determine what Euclid meant by this phrase, but it is likely that he was doing no more than hinting at a concept that was supposed to be grasped intuitively. Aristotle, writing rather earlier, made several comments about lines, including the remarkable one that 'if a line is what we recognize it to be from our physical intuition, then the angle sum of a triangle is two right angles'. It can be argued that to the Greeks, geometry was an abstraction from familiar, physical reality, and that its theorems were in some direct way true of the world. Or it can be argued that the statements on the *Elements* were intended to be taken more formally, the better to display the logical coherence of the work, and that the question of interpretation was to be kept at one remove. Certainly every attempt was made to indicate explicitly when extra assumptions had to be made, and no reliance in the work is placed on what we know from experience other than in the framing of the definitions.

But if this leaves open the question of the extent to which the *Elements* was taken to be true of the world by Greek writers, it is hard to deny that it was generally so interpreted by writers after AD 1500.

The most full-blooded attempt to prove the parallel postulate was made by Girolamo Saccheri and published in 1733, the year of his death. He distinguished between three kinds of geometry, in which the angle sum of triangles was greater than, equal to or less than two right angles, and he aimed to show that of these only the second kind was internally consistent. He succeeded in showing that the first kind 'destroyed itself', as he put it, but his account of the third kind was flawed. Significantly, it rested on establishing a result that for him was irreconcilable with the nature of the straight line. He was followed by Johann Heinrich Lambert, who in the 1760s also explored Saccheri's trichotomy. He did not discover any results that struck him as self-contradictory, finding only those which were contrary to his intuition, so he may be regarded as the first to describe theorems in a geometry other than Euclid's. But, perhaps because his investigations were inconclusive, he did not publish them, and they were first published posthumously by John III Bernoulli in 1786. They were reprinted in Engel and Stäckel *1895*.

Lambert accepted that two straight lines cannot enclose an area, which he regarded as an assumption also made by Euclid (although it is probably an Arab interpolation). This principle was taken by Franz Taurinus, writing in 1825 (also reprinted in Engel and Stäckel *1895*), to rule out of consideration the geometry on a sphere in which great circles are taken as straight lines, and in which indeed the angle sum of every triangle exceeds two right angles. The naturalness of this geometry strongly suggests that in investigating the parallel postulate mathematicians were trying not to devise formal systems differing in some way from Euclid's, but to describe the real world in mathematical terms. Spherical geometry differs from Euclidean geometry, but it is irrelevant if the aim is to describe the universe, because that cannot be like a sphere once it is agreed that two lines cannot enclose an area.

Lambert was a contemporary of Immanuel Kant, with whom he corresponded on the nature of geometry. Kant's view that statements of Euclidean geometry were synthetic *a priori* truths (synthetic because they could in principle be false, *a priori* because independent of any particular experience) seems not to have convinced him, perhaps because of his deep study of the parallel postulate. However, Kant's view that human intuition presented space to us as Euclidean did gain wide acceptance. The result was that, by the start of the nineteenth century, the widespread view was that Euclidean geometry correctly describes the world, and that it is built up logically from axioms which themselves are incontrovertible truths. All this was to change, with consequences not just for the *Elements* but for all of geometry and our very understanding of the nature of mathematics. For a discussion of the philosophical implications, see Torretti *1978*.

3 THE DISCOVERY OF NON-EUCLIDEAN GEOMETRY

The first to doubt the truth of Euclidean geometry was Carl Friedrich Gauss, who investigated the question from a variety of standpoints throughout his life but published almost nothing. Whether he lacked the final insight to clarify the question to his own satisfaction or merely wished to avoid public controversy is almost impossible to decide; both interpretations may be true. But his failure to resolve the matter means that the honour of being the first to describe a geometry other than Euclid's goes to the Russian Nikolai Lobachevsky and the Hungarian János Bolyai, writing independently in the 1830s. The degree of overlap in their descriptions is considerable. Both described a three-dimensional geometry in which coplanar lines λ and ν meeting a line μ at angles of α, which is less than a right angle, and γ, a right angle, may be asymptotic and therefore considered parallel (Figure 1). On the basis of this assumption, and the independent result that the theorems of spherical trigonometry do not depend on the parallel postulate, both men obtained trigonometric formulas relating the sides of a triangle. On the strength of these formulas, both proclaimed that their new geometries made sense and, being different from Euclid's, called into question the truthfulness of Euclidean geometry.

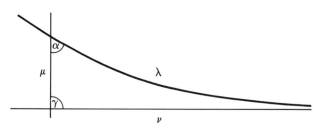

Figure 1 Parallels in the non-Euclidean geometry
of Bolyai and Lobachevsky

Reactions to the new geometry of Bolyai and Lobachevsky, called 'non-Euclidean geometry' because its assumptions differ from those of Euclidean geometry (although only in respect of the parallel postulate), were indifferent or even hostile in their lifetimes. Gauss accepted their findings, but did little in public to defend their cause. Lobachevsky's work was derided by some of his fellow Russians, and Bolyai's was left to languish in the obscurity of its original publication – an appendix to his father's treatise on geometry. Moreover, by the strictest canons, the work of Bolyai and Lobachevsky was flawed, for no proof of the consistency of their original assumption about parallel lines was given. The possibility remained that their theory was nothing more than some formulas in analysis, spuriously dressed up as geometry.

Matters changed decisively with the publication of Eugenio Beltrami's description of non-Euclidean geometry in 1868. He presented a description of Lobachevsky's non-Euclidean geometry in terms of the points inside a disc. By defining the distance between two points appropriately, Beltrami produced a map of non-Euclidean two-dimensional space in which straight lines in the space appeared as straight lines in the disc. This is a map in exactly the sense in which pages of an atlas are maps of the curved surface of the Earth, and so it resolved conclusively and affirmatively the question of whether Lobachevsky's definition of parallel lines was a tenable one. This flaw in the original presentation was undoubtedly one reason for the delay in accepting it, just as the posthumous publication of Gauss's views after 1855 helped to gain it acceptance. But the main reasons for the delay were the enormity of the consequences that accepting it entailed, and the difficulty in mastering the new approach to geometry that it required. The very existence of a new geometry seemed to challenge all one's intuitions about mathematics and space.

Given this challenge to the role of intuition in mathematical thought, it is interesting that both the discovery and the reception of non-Euclidean geometry illustrate a curious feature of mathematical discourse: its symbolism. By casting geometry into trigonometrical formulas, Bolyai and Lobachevsky both produced a language for analysing the behaviour of lines that evades the weight of tradition. Had non-Euclidean geometry turned out to be impossible after all, their formulas would still have been valid exercises in the hyperbolic trigonometric functions that they introduced (functions introduced originally by Lambert, curiously enough: see §4.2). Their formulas were ontologically obscure, if not actually ambiguous. By distancing mathematicians from spatial intuition, they made it possible for their successors to study the fundamental concepts of geometry in an original way, culminating in the reformulations by Beltrami and Bernhard Riemann. For exactly the opposite reason – a refusal to indulge in uninterpreted symbols – the new geometry was often hotly denounced by philosophers. The most remarkable of these denouncements was by Gottlob Frege, whose attitude to definitions and to the concept of number in particular was usually sharp (§5.2), but who could never accept non-Euclidean

geometry because he felt it was precluded by the existence of a unique physical world.

4 THE MEANING OF NON-EUCLIDEAN GEOMETRY

The formulas of Bolyai and Lobachevsky did not remain obscure for long. Beltrami's interpretation fitted very well with the views of Riemann on the way geometry should be reformulated, namely as the study of sets of points forming a surface or higher-dimensional object upon which the notion of length along a curve makes sense. From this point of view there are three exceptionally interesting two-dimensional geometries: those on the plane, on the sphere, and on the surface of constant negative curvature that had been known since the time of Gauss but was best described by Beltrami. They are distinguished mathematically by being the geometries on surfaces of constant curvature. Curvature is a technical concept, but its intuitive meaning is straightforward. Applied to surfaces in space, the curvature of a plane is zero, that of a sphere of radius R is $1/R^2$ and that of a saddle-shaped surface is negative.

From a Riemannian point of view, one starts with a surface and defines a sense of distance upon it; this yields a value for the curvature at each point. The surface with constant, zero curvature is a plane with the familiar Euclidean geometry. The geometry which has constant positive curvature is also familiar: it is that of the sphere, where a 'line' is defined as a great circle. Spherical geometry differs from Euclidean geometry in two fundamental ways: the parallel postulate is false because there are no parallel lines, and lines cannot be extended indefinitely. For this reason its existence does not invalidate Saccheri's and Lambert's proofs that a geometry in which the angle sum of a triangle exceeds two right angles cannot exist, although the angle sum of a spherical triangle always exceeds two right angles. Authors differ over whether spherical geometry should be called a non-Euclidean geometry or simply a geometry different from Euclid's. The simplest solution is to call only the geometry discovered by Bolyai and Lobachevsky non-Euclidean, and that has been adopted here. What Beltrami showed was that non-Euclidean geometry (in which there are many parallels to a given line through a given point not on the line, and in which the angle sum of a triangle is always less than two right angles) is identical to the geometry on a surface of constant negative curvature.

The three geometries on surfaces of constant curvature stand out because figures may be moved around freely on the surfaces without distortion, as we believe to be the case in (two-dimensional) physical space. This gives them all a degree of physical plausibility. Moreover, the crucial property of curvature is that it is intrinsic: it can be defined and measured entirely without reference to any three-dimensional space in which the surface may be placed. The idea of curvature can be generalized to higher dimensions, when there are many more interesting special cases. Again, the new geometries are intrinsically defined, so there is no need for an *a priori* Euclidean space for them to exist in. This makes it clear that, from the Riemannian point of view, Euclidean geometry is just one example of many. In this picture, which was widely adopted during the second half of the nineteenth century, Euclidean and non-Euclidean geometry become part of differential geometry (§3.4).

The implications of non-Euclidean geometry were soon taken up. Since it, like Euclid's, is a mathematically valid, physically plausible geometry (i.e. one which is consistent and in which basic terms like distance and angle make sense), people asked which, if either, was true. To what extent had geometry become an empirical science? Gauss, Friedrich Wilhelm Bessel, Lobachevsky and Bolyai were clear that measuring the angle sums of triangles would in principle serve to distinguish between Euclidean and non-Euclidean geometry, and that to this extent geometry was henceforth empirical. But once that is granted, what is one to make of two thousand years' worth of belief in the intuitive naturalness of the Euclidean terms? What, indeed, is a straight line? In the formulation proposed by Riemann that question had an answer for the first time: a straight line can be defined as a geodesic (curve of shortest length between its points) in a surface of zero curvature. What Beltrami showed, and Riemann himself hinted at, is that non-Euclidean geometry is the intrinsic geometry on a surface of constant negative curvature.

The empirical question was never to be resolved, and is today subsumed in theories of general relativity that interpret gravity as a variable curvature in a four-dimensional spacetime. But for mathematicians the problems posed by the discovery of non-Euclidean geometry were by no means over. By the 1880s projective geometry had come to assume a higher status than that of Euclidean geometry: its assumptions were fewer, and it contained Euclidean geometry as a special case. So although projective geometry was not physically plausible (it contains no concept of length or angle), it is logically prior to Euclid's. Moreover, there are no parallel lines in projective geometry. What, then, of the nature of a line? The intuitive concept of the Euclidean straight line has been refined (at the very least) or replaced (at worst) by that of the projective line. From a metrical point of view, should not the primitive intuition have been that of the non-Euclidean line? Mathematicians' intuition was no longer unproblematic; it had led them into an unwarranted confidence in Euclid. To anyone who based mathematics on some kind of intuitive apprehension of reality, this raised the question of what guarantee there could be that even mathematical truth could be discovered. The first to tackle this question was Moritz Pasch, who discovered a number of tacit assumptions made by Euclid that no one had remarked upon before because they rest on seemingly self-evident propositions (Nagel *1939*).

5 AXIOMATIC GEOMETRY

Pasch may well have hoped to refine our intuition until it became reliable. The far more radical step of denying mathematical terms any meaning and relying completely on formal rules of inference was taken by Guiseppe Peano and, independently, by David Hilbert. (Hilbert's route to this discovery has recently been described in Toepell *1986*.) Hilbert proposed that the geometric terms 'point', 'line', 'plane', and so forth be controlled by a system of axioms that determine what one may say about them, but which makes no attempt to say what they are. A self-consistent set of axioms would be taken to guarantee the existence of the objects it provided for, but all intepretations of them would be equally valid. Self-consistency was assured by any model exemplifying them. So, plane Cartesian geometry is an algebraic model assuring the consistency of the Euclidean axioms, and Beltrami's disc can be seen as a model of the consistency of the non-Euclidean axioms. It also establishes that Euclidean and non-Euclidean geometry are relatively consistent, since it can be interpreted as a picture in Euclidean geometry. This means that attempts to refute non-Euclidean geometry would, if successful, have shown Euclidean geometry to be impossible also − a result no one can have wanted.

Hilbert's axioms for geometry began by proposing three kinds of object (points, lines and planes) with properties to be described by words such as 'lie', 'between' and 'congruent'. These were controlled by five families of axioms: those of incidence, order, congruence, parallelism and continuity. For example, the first axiom is 'For every two points A and B, there exists a line *a* that contains each of the points A and B.' Pasch's axiom appeared among the axioms of order, and Euclid's parallel postulate was given in the equivalent form: 'Let *a* be any line, and A any point not on it. Then there is at most one line in the plane, determined by *a* and A, that passes through A and does not intersect *a*.' Hilbert then showed how the familiar theorems of Euclidean geometry could be derived within his formal system by chains of reasoning that followed the axioms he had laid down.

Hilbert's proposal neatly side-stepped the question of how intuition had led people astray by avoiding all questions of meaning. His intentions, however, were not to render mathematics meaningless, but to make it clear; he continued to assert the importance of intuition in guiding and developing one's thought. Indeed, he might well not have put his abstract axiomatics forward had he not discovered they were of use in generating new theorems. Examples were already known of geometries where the coordinates are taken from finite fields; Hilbert's work led to the discovery of geometries that cannot be described in terms of coordinates at all. His approach to geometry therefore greatly enlarged its scope and took it beyond the domain of simple, continuous manifolds where Riemann and Felix Klein had placed it.

6 AXIOMATIC MATHEMATICS

Hilbert's presentation came to have a decisive effect on many branches of mathematics. It was as if the pure mathematician's task was to provide axiom systems and check that they were self-consistent, which applied mathematicians, physicists and others could then use as they saw fit. This neatly defined a new relationship between pure and applied mathematics. Within pure mathematics, what was done for Euclidean geometry was done for other geometries. Hilbert showed the next year (1900) how non-Euclidean geometry can be obtained by changing just his version of the parallel postulate. Other mathematicians joined in, describing geometries which differed more and more in their nature from Euclidean geometry. Then other systems of mathematical ideas were given axiomatic treatments, starting with the theories of groups and fields. In time, there set in a reaction against a baroque profusion of axiom systems, serving less and less purpose; the present generation of mathematicians regards axiom systems as a good way to define or create mathematical objects, but prefers (with Hilbert) to concentrate on their intuitive meanings for the purposes of carrying out research. However, Hilbert's ideas had by then grown to embrace the hope that all of mathematics could be not only axiomatized, but given axioms that guaranteed every truth a proof expressible in a finite number of statements. This hope, partly forged in the battle against intuitionism, was shown to be futile by Kurt Gödel (§5.5).

7 THE CURRENT SITUATION

A further step away than non-Euclidean geometry lies the domain of special relativity (§9.12). With the rapid acceptance of Albert Einstein's ideas, space and time, to quote Hermann Minkowski, faded away to be replaced by spacetime. The geometry of spacetime, although linear (unlike that of non-Euclidean geometry), differs markedly from Euclid's in other ways, and so the dominant geometrical paradigm among physicists is no longer Euclidean. What, then, remains of Euclid's creation? It remains the naive formulation of space taught in schools. It is perfectly adequate for classical physics. The *Elements* proposed an ideal of rigorous deductive reasoning from clearly stated premisses that mathematicians continue to uphold. It is unlikely that anything like Kantian intuitionism will survive as an account of why we perceive space the way we do, but it is quite likely that we will learn that small, nearby regions of space are like patches of Euclidean space. Meanwhile, non-Euclidean geometry continues to arouse the greatest interest among mathematicians. Outstanding, but widely believed, conjectures by William Thurston imply, for example, that in some sense most three-dimensional manifolds are made up of patches of three-dimensional non-Euclidean space.

BIBLIOGRAPHY

Bonola, R. *1912, Non-Euclidean Geometry. A Critical and Historical Study of its Development*, Chicago, IL: Open Court. [English transl., with additional appendices, by H. S. Carslaw, repr. 1955, New York: Dover. Contains J. Bolyai, *Science of Absolute Space*, and N. I. Lobachevsky, *Geometrical Researches in the Theory of Parallels*, both transl. G. B. Halsted.]

Engel, F. and Stäckel, P. *1895, Die Theorie der Parallellinien von Euklid bis auf Gauss*, Leipzig: Teubner.

Euclid *1918, Elements* (ed. T. L. Heath), 3 vols, Cambridge: Cambridge University Press. [Repr. 1956, New York: Dover.]

Gray, J. J. *1989, Ideas of Space: Euclidean, non-Euclidean, and Relativistic*, 2nd edn, Oxford: Oxford University Press.

Hilbert, D. *1899, Grundlagen der Geometrie*, Leipzig: Teubner. [2nd edn transl. by L. Unger as *Foundations of Geometry*, 1902, La Salle, IL: Open Court; repr. 1971.]

Nagel, E. *1939*, 'The formation of modern conceptions of formal logic in the development of geometry', *Osiris*, **7**, 142–224.

Toepell, M.-M. *1986, Über die Entstehung von David Hilberts 'Grundlagen der Geometrie'*, Göttingen: Vandenhoeck & Ruprecht.

Torretti, R. *1978, Philosophy of Geometry from Riemann to Poincaré*, Dordrecht: Reidel.

7.5

Descriptive geometry

KIRSTI ANDERSEN AND

I. GRATTAN-GUINNESS

Today, 'descriptive geometry' denotes a discipline which deals with how to represent three-dimensional objects on a plane. The term was introduced by Gaspard Monge, and used by him in one sense more narrowly and in another more broadly. He required that knowledge about the shape and measures of an object could be obtained from its image, and restricted his considerations to a representation fulfilling this. However, not all his theory was directly connected with two-dimensional representations: some of it dealt more generally with synthetic geometrical methods; his book on descriptive geometry covers in particular topics which later became part of differential geometry. In treating the prehistory of the subject and its development after Monge, we leave out the history of differential geometry (on which see §3.4); see §12.6 for more details on the particular two-dimensional representation called 'perspective'.

1 PREHISTORY

What is now known as Monge's descriptive geometry developed out of a technique of representing a three-dimensional object on a plane by projecting it onto two suitable planes, perpendicular to each other, projections known as the plan and the elevation. The origin of this technique is unknown: plans have been used since very early days; elevations were also used in Antiquity, but apparently not together with plans. It is unclear when the two representations were first used simultaneously. But Renaissance architects knew the technique, and in the second part of the fifteenth century a perspective method relating directly to plans and elevations was described by the Italian painter Piero della Francesca in the book *De prospettiva pingendi* ('On Perspective in Painting') which, however, remained unpublished for more than four centuries. The German artist

Albrecht Dürer published a method similar to Piero's in his *Underweysung der Messung* in 1525 (Figure 1). Moreover, Dürer used the technique of plan-and-elevation for describing skew curves (Figure 2).

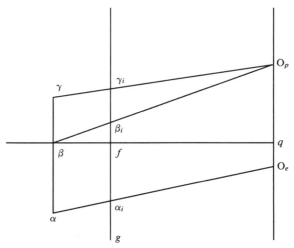

Figure 1 Dürer's perspective method based on plan and elevation; a redrawing of part of Dürer's figure 56 in *Underweysung der Messung* (1525) (the letters are not original). Below the line *fq* is the plan, and above it the elevation; the line *fg* represents the picture plane. The eye-point has been projected orthogonally onto the points O_p and O_e, the points α and β are the projections of a point that lies in the plan, and the points α and γ are the projections of a point lying a distance $\beta\gamma$ from the plan above α. By joining the projections of the eye and the projections of a point, and finding their intersections with *fg*, Dürer obtained the plan and elevation of the perspective image of the point; thus $f\alpha_i$ determines the horizontal position and $f\beta_i$ the vertical position of the image of the point that is projected onto (α, β)

Plans and elevations were handy tools for architects and engineers in planning buildings and fortifications, and they were required in the majority of perspective constructions. The method for producing them was therefore taught in many textbooks on civil and military architecture, and in textbooks on perspective; it was also included in some works on projections. In general the method was described rather briefly, and no new theoretical insights were added. A more comprehensive study of orthogonal projections of curved surfaces was published by the French engineer Amédé-François Frézier in the 1730s; like Monge later, Frézier combined his investigation with a kinematic description of curves and surfaces.

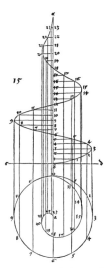

Figure 2 Dürer's plan and elevation of the cylindrical spiral and the helix (from Dürer, *Underweysung der Messung* (1525), figure 15)

2 MONGE'S NEW TRADITION OF DESCRIPTIVE GEOMETRY

Monge began to develop descriptive geometry during the late 1760s, when he was in his twenties and teaching at the military school at Mézières; he continued to practise and advocate its use throughout his life (Taton *1951*). Because it was applied to warlike needs, it was classified mathematics; for example, he used it to prepare the plates for his book *Description de l'art de fabriquer les canons* (1793), but he was not permitted to describe the manner of their preparation.

When the Parisian system of higher education was renovated in the 1790s, Monge found opportunities to publicize the details. Courses on descriptive geometry were given at the Ecole Normale from 1795, and soon after also at the Ecole Polytechnique; the lecture notes from the former were edited in a stenographic version in a special journal. In 1799 Monge's lectures on descriptive geometry were collected and published as the book *Géométrie descriptive* by J. N. P. Hachette, who helped Monge with the teaching. This book went through many editions, from 1820 enlarged with sections on shadows and perspective edited by Monge's student, B. Brisson.

The first chapter of *Géométrie descriptive* deals with projections of points, lines and planes. Among Monge's innovations to traditional plan-and-elevation methods were a systematization of the description of the images and an investigation of how some basic problems about determining lines and angles are solved when the constructions concern the images rather than the original objects in Euclidean space. In Figure 3, LMNO is the plan and LMPQ the elevation, and they intersect in the ground line LM; onto these two planes objects are projected by orthogonal projection. To obtain a plane configuration, the elevation LMPQ is rotated about LM into the plane of LMNO. It is handy to have a word for the two half-planes containing, respectively, plans and elevations of objects; hence the term 'descriptive plane' is used. The image of a point like A is the pair of points *a* and *a* "lying on a line perpendicular to LM intersecting it in C; the position of the original point A is determined by the distances C*a* and C*a*". The image of a line segment which cuts both the plan and the elevation is a pair of line segments determined by the images of the points where the line segment meets these two planes (Figure 4). Rather untraditionally, Monge also

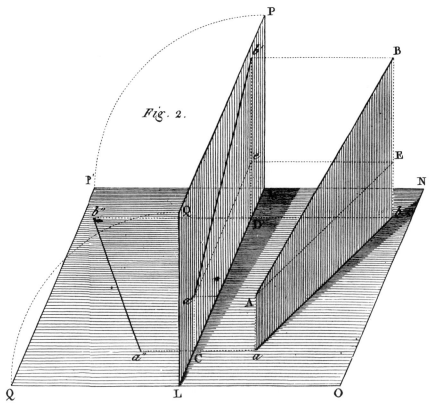

Figure 3 Monge's projections of points and lines onto the plan and the elevation (from Monge, *Géométrie descriptive* (1820), figure 2)

890

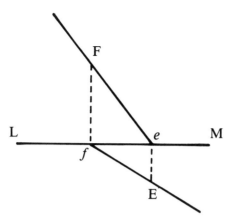

Figure 4 A line segment intersecting the plan in E and the elevation in F is represented by the lines (E*f*, *e*F); the points of intersection are represented as (E, *e*) and (F, *f*)

considered the images of planes; the image of a plane that meets the plan as well as the elevation he defined as the two lines of intersection. Thus, in general, a plane is represented in the descriptive plane by two half-lines intersecting each other on the ground line (Figure 5).

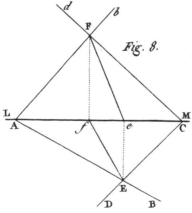

Figure 5 The two pairs of lines (BA, A*b*) and (DC, C*d*) represent two planes, their trace is represented by the lines (E*f*, *e*F), where (E, *e*) and (F, *f*) are the representations of the points that are common for both planes and, respectively, the plan and elevation (from Monge, *Géométrie descriptive* (1820), figure 8)

Incidences between points and lines are obviously preserved when they are mapped into the descriptive plane. Since a plane is represented by a pair of lines in the descriptive plane, the incidence relations involving planes are a little more complicated. It is, however, quite straightforward to construct, for instance, the image of a line which is the trace between two planes whose images in the descriptive plane are given (Figure 5). Monge described this construction; he also showed how the lengths of original line segments can be determined from their images, and similarly how the original angles between lines and planes can be determined from the images of these objects. Moreover, he demonstrated how a number of basic constructions can be performed in the descriptive plane. For instance, for a given point and a line (plane) not containing the point in the descriptive plane, he constructed the line (plane) through the given point parallel to the given line (plane), and the plane orthogonal (the normal) to the given line (plane) through the given point. This idea of investigating the geometry in the descriptive plane is analogous to an examination of direct constructions in the perspective plane carried out through the seventeenth century and the first part of the eighteenth.

Although some of the direct constructions in a descriptive plane are simple, the procedures are in general most interesting from a theoretical point of view, because the practical constructions involve so many steps that it becomes complicated to keep track of them. Figure 6 shows an example of a rather intricate construction. In the applications, the scheme was to work out various things on the descriptive representations to determine properties in space, and maybe even to modify the object by manipulating the projections and then projecting back again.

The largest part of Monge's *Géométrie descriptive* is devoted to curved surfaces and skew curves; in treating these Monge turned to kinematics. He considered in particular surfaces that can be generated by two moving curves, and through an intuitive interpretation of motion he found several results, for example a method for determining the tangent plane through a given point on a surface. His approach was a continuation of the line which had been followed by Gilles Personne de Roberval and Isaac Newton, among others. Monge referred to the former and generalized his method of tangents, or rather an incorrect application of it (§3.1), to three dimensions, the consequence being that Monge obtained a wrong result.

Besides tangents and tangent planes, Monge dealt with the problem of determining the curve of intersection of two surfaces, and with curvature; his treatment of the latter subject was not much related to descriptive representations.

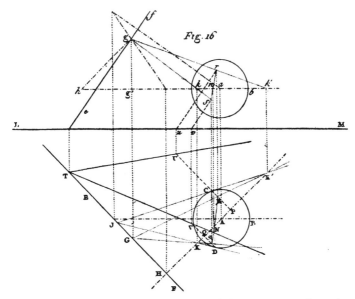

Figure 6 Illustration to a problem of determining a tangent plane through the line LM to a sphere of which the two projections (with centres A and *a*) are shown (from Monge, *Géométrie descriptive* (1820), figure 16)

3 MONGE'S FOLLOWERS

Monge's dream was that descriptive geometry should become an important part of all mathematical curricula. He constantly claimed that the discipline was very useful, and could be applied almost everywhere, but in illustrating this point he chose rather academic applications. Thus he seems to have been more interested in elevating the discipline into a recognized branch of mathematics than developing it into a valuable tool for engineering mathematicians. In the event, it never became a major topic in pure mathematics, but it was relevant for the creation of projective geometry (§7.6).

During the early years of the Ecole Polytechnique, Monge was very influential in setting the policy, and he saw to it that descriptive geometry was given much time on the syllabus. However, the critics (especially Laplace) gradually gained the upper hand in the new century, and the time devoted to the topic was substantially reduced (Paul *1980*).

The industrious textbook writer Sylvestre Lacroix who, like Hachette, assisted Monge, included elements of descriptive geometry in editions of his *Essais de géométrie* from as early as 1795; Lacroix's treatment was presumably inspired by, but not completely dependent on, Monge's work.

893

In 1822 Hachette wrote his own textbook on the discipline. Monge's other followers were mainly from his circle at the Ecole polytechnique. The most vocal was Théodore Olivier, who taught the topic to industrial and commercial engineers at the Ecole Centrale des Arts et Manufactures, which he helped to found in 1829, and published textbooks from the 1840s. However, as at the Ecole Polytechnique, descriptive geometry lost ground there and also at other French educational institutions.

The most interesting practitioner among Monge's followers was Charles Dupin, who applied descriptive geometry to the stability of non-symmetric floating bodies, and used it together with differential geometry to find many nice geometrical properties of surfaces. Thus, when descriptive geometry was allied to differential geometry valuable results could be found, but even then the mathematical scope was limited.

4 LATER DEVELOPMENTS IN DESCRIPTIVE GEOMETRY

Although descriptive geometry did not become as central in mathematical education as Monge had wished, it did spread and enjoyed a long life. Throughout the nineteenth century and in the first decades of the twentieth, descriptive geometry was taught in many countries at the various kinds of new institutions of engineering that were being created (Wiener *1884*, Papperitz *1910*, Loria *1921*, Booker *1963*). The preparation of scientific and technological patents was an important stimulus to making machine drawing an important discipline. Otherwise, the normal range of applications covered the construction of arches, stone-cutting, carpentry and topography; it was also used for determining shadows and perspective pictures, and later in photogrammetry.

The various textbooks on descriptive geometry reflect the ambiguity in Monge's programme; some focused on the theoretical aspects and others on the practical; moreover, each category contains rather varied approaches. Thus some authors derived their main results by analytical methods, whereas others built upon projective geometry. In the practical books, different traditions developed in various countries and caused much difficulty for international projects where drawing was important. For example, in Britain, a particular fashion developed (Figure 7). In the USA (where the subject became popular only from the mid-century onwards, with the growth of industrialization), some preference emerged for projection onto plan, section and elevation.

Theoretical books on descriptive geometry have become rarer, whereas there still is a market for textbooks on technical drawing.

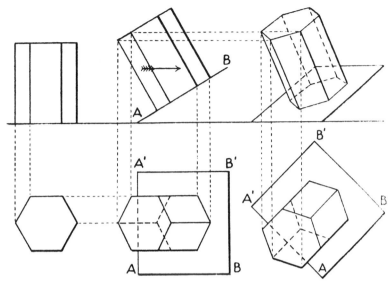

Figure 7 An example of British descriptive geometry. The image of an obliquely lying hexagonal prism is produced in the following way. First, the plan and elevation of the prism are drawn, then the ground line of the elevation is rotated into AB. At the same time, the plan is drawn twice corresponding to the upper and lower faces of the prism. In the third step the plan is rotated through the same angle as AB, and the image is constructed by determining the points of intersection of vertical lines through the vertices of the rotated plan, and horizontal lines through the vertices of the rotated elevation. (From Booker *1963*: 138)

BIBLIOGRAPHY

Booker, P. J. *1963*, *A History of Engineering Drawing*, London: Chatto & Windus. [Repr. 1979, London: Northgate.]

Loria, G. *1921*, *Storia della geometria descrittiva dalle origini sino ai giorni nostri*, Milan: Hoepli. [Good coverage of Europe. Worth learning Italian to appreciate his ironic remarks!]

Papperitz, E. *1910*, 'Darstellende Geometrie', in *Encyclopädie der mathematischen Wissenschaften*, Vol. 3, Part 1, 520–95 (article III AB 6).

Paul, M. *1980*, *Gaspard Monges 'Géométrie descriptive' und die Ecole Polytechnique* [. . .], Bielefeld: University.

Rowe, C. E. and McFarland, J. D. *1939*, *Engineering Descriptive Geometry* [. . .], 1st edn, Princeton, NJ: Princeton University Press. [2nd edn 1953. Not historical, but gives an excellent presentation of the different forms and uses of descriptive geometry.]

Schelling, F. *1904, Über die Anwendungen der darstellenden Geometrie* [...], Leipzig and Berlin: Teubner. [Includes art, photogrammetry and map-making.]

Taton, R. *1951, L'Oeuvre scientifique de Monge*, Paris: Presses Universitaires de France.

——*1954, L'Histoire de la géométrie descriptive*, Paris: Palais de la Découverte.

Wiener, C. *1884*, 'Geschichte der darstellende Geometrie', in his *Lehrbuch der darstellenden Geometrie*, Vol. 1, Leipzig: Teubner, 1–61.

7.6

Projective geometry

J. J. GRAY

1 INTRODUCTION

Projective geometry is the study of geometric figures and their transformations under projection. For example, the shadow of a figure in a plane cast onto a screen by a point source of light may be thought of as a projective transformation of the figure, called a 'perspectivity' (Figure 1(a)). The original figure and its image are said to be in perspective, or to be perspective images of each other; older books speak of 'transformation by projection and section'. A sequence of perspectivities is called a projective transformation. If the direction of the light is reversed, and the source of light replaced by an (idealized or simplified) eye, it is apparent that a figure and its perspective image can look exactly alike (Figure 1(b)). The precise appearance is dependent on the correct positioning of the eye, which compensates for some distortions but not others. For an account of the relationships between perspective theory, art and descriptive geometry, see §12.6 and §7.5.

A connection between optics and geometry had been appreciated by Greek mathematicians (§1.5). In Euclid's *Optics* the apparent sizes of objects is discussed; properties of reflection were discussed in his *Catoptrics*. Several theorems in the *Collection* of the later commentator and mathematician, Pappus, were devoted to reconstructing a lost book by Apollonius called *Determinate Section*. Pappus also discovered this result, later to become central in the theory of projective geometry (Figure 2): three points, A, B and C, lie on one line, and three more points, A', B' and C', lie on another. Let the lines AB' and A'B meet at R, the lines BC' and B'C at P, and the lines CA' and C'A at Q; then the points P, Q and R lie on a line. Modern commentators from Michel Chasles *1837* to B. L. van der Waerden *1961* have indicated how profitably these theorems can be seen as part of projective geometry; but it is unlikely that they were so regarded in Classical times.

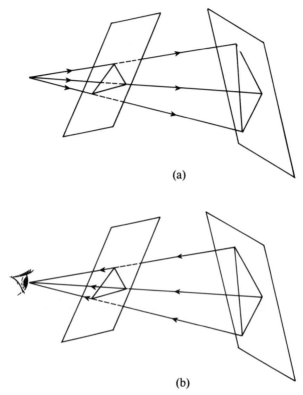

(a)

(b)

Figure 1 Projective geometry: (a) a 'perspectivity';
(b) similarity to the eye of figure and projection

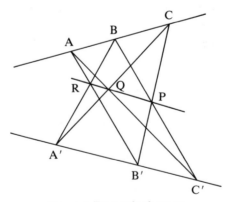

Figure 2 Pappus's theorem

2 DESARGUES

A mathematical theory of perspective and projective transformations requires two things: the discovery of interesting properties of figures that remain true under such transformations, and satisfactory proofs of the main theorems. Rigorous proofs of the known constructions in the theory of perspective were given from the late sixteenth century on. For the most part they were ingenious or subtle exercises in Euclidean geometry, aimed at showing that, although a perspectivity changes lengths and angles, one can rigorously show how to draw a perspective image correctly. The first to appreciate that it was worth while looking for properties of figures that are not altered was the architect and mathematician Girard Desargues, who wrote about it in a short pamphlet of 1639 called the 'Brouillon project', or 'Rough draft on conics'. This difficult work circulated for a time in Paris before almost disappearing; for a recent discussion of the origin of Desargues's ideas and their impact, see Field and Gray *1987*.

It is trivially true that the projective transformation of a straight line is another straight line. Desargues found that a certain ratio of six points on a line was equal to the same ratio calculated for the image points. A special case of this observation, when two pairs of points coincide, was also studied by him because it arises naturally in geometrical problems; it is called 'four points in involution'.

Four points, A, B, C and D, on a line are said to be in involution if the ratios AC/CB and AD/DB are equal and opposite: AC/CB = − AD/DB. The minus sign arises because one of the segments is measured in the opposite direction to the others (DB in Figure 3). Desargues showed that if a perspectivity from a point O casts the points A, B, C and D onto the four points A′, B′, C′ and D′, respectively, then the image points are also in involution. Later mathematicians expressed this by saying that the 'cross-ratio' (AC/CB)/(− AD/DB) is unaltered by the perspectivity:

$$(AC/CB)/(-AD/DB) = (A'C'/C'B')/(-A'D'/D'B'). \qquad (1)$$

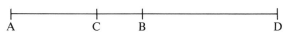

Figure 3 Four points in involution on a straight line

The configuration of four points in involution is important because it occurs frequently in the theory of conics. For example, given an ellipse and a point P outside it, one may draw the two tangents from P to the ellipse, touching it at the points R and S, say (Figure 4). Then any line through P

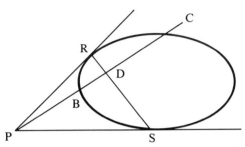

Figure 4 Four points in involution on a conic

cutting the ellipse in two points, B and C say, and the line RS at D provides four points in involution: PBDC. Since every conic arises as a plane section of a cone on a circular base, Desargues could regard Figure 4 as a perspective image of a figure involving tangents and chords in a circle. As a result, many theorems about conics, whether ellipses, hyperbolas or parabolas, could be proved by formulating them in terms of points in involution and reducing to the simple case of the circle. In this way, Desargues's 'Brouillon project' gave the first unified treatment of the conic sections, surpassing the classical theory due to Apollonius. Some time later he contributed some theorems to his friend Abraham Bosse's work on perspective, including the lovely result that bears his name: if two triangles ABC and A'B'C' are in perspective from a point O, and the lines AB and A'B', BC and B'C', and CA and C'A' meet at the points R, P and Q respectively, then the points R, P and Q lie on a straight line (Figure 5).

As Desargues clearly saw, any theory of the projective properties of figures requires radical revision of some of the most elementary concepts of geometry, not least that of the straight line. For, by suitably positioning the image plane it is possible to cast the image of two intersecting lines as two parallel lines, and vice versa. Projective geometry cannot therefore distinguish between intersecting and parallel lines. In order to make sense of the point of intersection of two parallel lines, Desargues spoke of their intersection at infinity. This unhappy locution satisfies and repels readers in equal numbers, and was only made rigorous in the nineteenth century. Similarly, the hyperbola and parabola are given points at infinity, which is how they can be projectively equivalent to the ellipse.

Points at infinity were not the only source of difficulty in Desargues's work. His original manuscript, which circulated in fifty copies, seems to have been written with a view to the eventual production of a more polished version. Indeed, a later book may have been written, but only its title, *Leçons sur ténèbres*, survives. The original was very poorly organized,

900

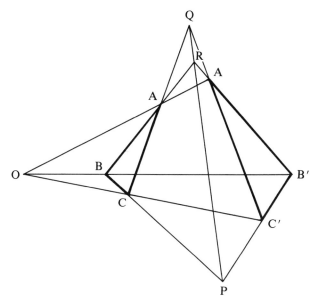

Figure 5 Desargues's theorem

written in an opaque style and marred by misprints. Even René Descartes, a friend of Desargues, found it hard to read, and its impact was probably less than its author hoped for. But impact it had. The gifted young Blaise Pascal read it when only 16, and produced his wonderful theorem about a hexagon inscribed in a conic: if A, B, C, D, E and F are six points on a conic, and the lines AB and DE, BC and EF, and CD and FA meet at the points P, Q and R, respectively, then the points P, Q and R lie on a straight line (Figure 6). In Pascal's original presentation the special case of a circle

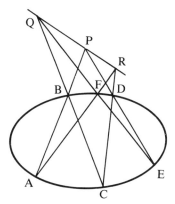

Figure 6 Pascal's theorem

901

immediately precedes the general case of a conic; presumably the one was reduced to the other by a projection, but unhappily his proof for the circular case has not come down to us. Another, later author who was inspired by Desargues was Phillipe De La Hire, who had come to some of the same ideas independently. La Hire's later treatment *1685* emphasized the role of four points in involution, and is much clearer and more systematic than Desargues's.

Another who came to projective geometry was Isaac Newton, whether by reading La Hire or independently it is not possible to determine. He described a projective transformation in Book I of the *Principia* (1687), and used it to solve all the problems of the form 'given k points and $5 - k$ lines, find the conic through those points and having those lines as tangents'. The most dramatic use of the idea of projective transformations, however, he reserved for his *Opticks* (1704), in which he summarized his analysis of the different curves, more than seventy of them, that can be represented by cubic equations in x and y by saying that the shadows of such curves were of only five different types. This cryptic remark towards the end of a difficult and lengthy analysis eventually prompted book-length explanations from J. P. de Gua de Malves *1740*, and from Patrick Murdoch *1746*.

At this juncture, projective geometry might have blossomed. A projective account of the conic sections written in Cartesian terms might have been produced, and the Newtonian projective theory of cubic curves acted as a spur to further research. But this did not happen. The influential treatises on geometry in the second half of the eighteenth century, those by Gabriel Cramer and Leonhard Euler, turned their backs on projective methods and the subject went into a fifty-year decline. When its study was resumed in the 1820s it was as an independent discovery, although French scholars did their best to make amends to Desargues, whose work had by then almost completely disappeared.

3 THE REDISCOVERY OF PROJECTIVE GEOMETRY

The stimulus for the rediscovery was the influential French geometer Gaspard Monge, a founder and initial organizer of the Ecole Polytechnique in Paris. Through his lectures he created a new school of geometers in France (§11.1). One of the most gifted of these, Jean Victor Poncelet, took the occasion of his capture by the Russians during the Napoleonic invasion of 1812 to rethink Monge's use of geometrical transformations. He generalized Monge's projection along parallel rays to projection along a pencil of intersecting rays, and thus rediscovered projective transformations. He also saw that in this way he could unify the study of the conics, and so obtain for geometry a level of generality that he felt had hitherto been the province of algebra (geometry being tied too closely to specific figures). His book *1822* gave a remarkably thorough treatment of the results of these insights. A problem for Poncelet, as it had been earlier for Desargues, was how to treat simultaneously those situations where a line does, and where it does not, cut a conic (Figure 7). Poncelet's solution was ingenious, but not easy to accept. Later geometers split into two camps: those who, like Poncelet, wanted to reason exclusively in geometrical terms (line, curve, intersection, involution), and those who were happier to pass into algebra. The former group, the 'synthetic geometers', following the example of C. G. C. von Staudt, came to unify Figures 7(a) and 7(b) around the concept of involution. The latter group, the 'algebraic geometers', invoked complex numbers and spoke instead of real and complex points of intersection; they were led by August Ferdinand Möbius and Julius Plücker. In particular, Plücker showed what a projective theory of cubic and quartic curves could accomplish. In a remarkable extension of these ideas towards the end of his life, he showed how one could study the geometry of all lines in space; this was a four-dimensional geometry which he and others, notably Ferdinand Lindemann, showed to be useful in the study of mechanics (§8.2).

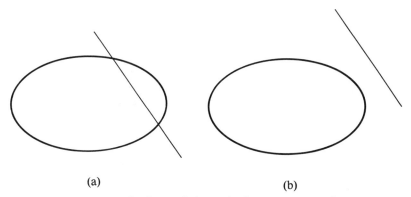

(a)   (b)

Figure 7 The line and the conic: in contact or not?

Synthetic projective geometry produced a distinguished line of adherents attracted by its methodological purity: Jacob Steiner, a Swiss who taught for most of his life in Berlin, von Staudt in Germany, Chasles in France, Arthur Cayley in England and Luigi Cremona in Italy. To those gifted with the ability to see projective transformations (in their mind's eye), it was a suggestive field full of rich results. But the problem of points of intersection bedevilled the theory, making it unappealing to read and difficult to master. For these reasons J. L. Coolidge, the American mathematician and historian of mathematics, lamented in *1940* that it had died in the late nineteenth century. Algebraic projective geometry, on the other hand, had only to win acceptance for points with complex coordinates and the way would be open for a dramatic extension of Cartesian methods for the simultaneous study of algebra and geometry. The generation of Plücker and Otto Hesse equivocated on the issue, but complex points were firmly accepted by Bernhard Riemann throughout the 1850s. Consequently, those who wished to understand his ideas and to apply them – as he had occasionally done – to the study of curves, could bring the rigour of algebra to their study of geometry. Prominent in this endeavour were the members of the school that formed around Alfred Clebsch: above all Paul Gordan, Alexander Brill, Max Noether and Felix Klein.

Of these, all but Klein sought to recast as much as possible of Riemann's theory of complex functions in terms of the projective geometry of curves, and in this they were successful. Klein stood alone for a while in feeling that Riemann's ideas had in fact struck deeper than that, and his view was satisfactorily vindicated somewhat later by Hermann Weyl. But Klein's own original contribution lay in another direction. As a young man he was struck by the profusion of geometries that had been created in the nineteenth century: not just Euclidean and projective, but also non-Euclidean geometry and affine geometries now existed. In his famous *Erlanger Programm* of 1872, Klein showed how all these geometries could be thought of as variants of projective geometry, thus reunifying a subject he felt had become regrettably broken up. He went on to show how other geometries, such as Plücker's line geometry, could be considered, and how they too fitted his scheme of things. Klein's unification proceeded by regarding any geometry as consisting of a space of points and a group of transformations that moved the figures in the space around while preserving the properties appropriate to that geometry. So plane Euclidean geometry consists of a plane of points and has for its group all the transformations that preserve length; plane projective geometry consists of a plane of points suitably augmented by a line at infinity, and has for its group all the continuous transformations that preserve the cross-ratio of four points along a line. By the 1890s, when the idea of groups acting on spaces had become central (because of the work of Henri Poincaré and Sophus Lie), Klein's presentation, which had originally aroused little interest, came to seem prescient, and was widely seen as a forceful statement of the view (held also for example by Cayley) that projective geometry was the most basic kind of geometry. So, from being an unorthodox way to transform figures in a

Euclidean setting, projective methods had become fundamental and Euclidean ones but special cases.

In the latter half of the nineteenth century, a start was also made on the study of surfaces using projective transformations (§7.9). That topic led Giuseppe Veronese in 1880 to describe *n*-dimensional projective geometry. He was able to show that a number of difficulties in the study of surfaces had arisen because the surfaces in question were confusing projections onto three-dimensional space of simple surfaces in some higher-dimensional space. In this way he inspired mathematicians to look for simplification by passing to higher dimensions, and so promoted the study of geometry beyond the familiar space of physical intuition.

4 THE FOUNDATIONS OF GEOMETRY

Although geometry had become extensively reformulated by the end of the nineteenth century, with the advent of non-Euclidean as well as *n*-dimensional geometries, it retained traces of its original Euclidean state. It was still possible to draw figures and to imagine them being moved around. But even this feature changed with the publication of David Hilbert's *Grundlagen der Geometrie* in *1899*. (Hilbert's route to his discoveries is well described in Toepell *1986*.) Hilbert had been led to his ideas by a book by Moritz Pasch. Pasch had sought to give a rigorous treatment of geometry, one that would be worthy of the esteem that had been heaped upon Euclid's *Elements* (so naively, it was now thought). He discovered rules of inference used – but never made explicit – in elementary geometry, for example the one now known as Pasch's axiom: if a line enters a triangle ABC through the side AB, and does not pass through the vertex C, then it must leave the triangle either between B and C or between C and A. He sought to free the deductive side of geometry from hidden appeals to intuition, not because such appeals were flawed, but because until they were made explicit one could not be sure of the validity of a geometric argument. But he equivocated about the role of intuition in the formation of geometrical concepts such as that of the straight line, arguing that in these respects geometry was akin to the natural sciences.

Hilbert took up Pasch's ideas and found, to his surprise, that they led him not just to a philosophical critique of geometry but also to the possibility of doing geometry in a new way. By emphasizing the rules of inference to the exclusion of any consideration of the objects of geometry, Hilbert saw that one could start with certain rules, which he called axioms, and deduce their consequences. Different initial axioms would give different sets of consequences; in particular, there were sets of axioms from which particular theorems could be derived, and others from which they could not. Ingenuity in proving theorems in this new setting appealed to Hilbert, as did the production of counter-examples to conjectures suggesting that one set of axioms implied another.

In his book *1899*, Hilbert gave a set of axioms for Euclidean geometry, grouped into axioms governing incidence, order, congruence, parallelism and continuity. He then investigated the connection between geometry as he presented it and coordinate geometry. He showed that his axioms, and even suitable subsets of them, yielded an arithmetic of segments. That is, it was possible to define the length of a segment, and to add, subtract, multiply and divide these lengths. Depending on the axioms chosen, such an arithmetic need not faithfully resemble ordinary arithmetic, but Hilbert showed that if Desargues's theorem is true in a geometry then the arithmetic of segments is a (possibly non-commutative) field, and if Pappus's theorem holds then the associated arithmetic is commutative. Moreover, if in a geometry Pappus's theorem holds, then so does Desargues's, but not conversely.

Hilbert soon incorporated non-Euclidean geometry into this new setting (§7.4), and the book by Veblen and Young *1910–13* showed how all known geometries, most notably projective geometry, could be described in this spirit. The result, as Hilbert was not slow to emphasize, was that for the first time geometrical concepts could be handled rigorously without the need for any definitions. 'Line', 'point' and 'plane' could be undefined terms in a set of axioms, the axioms being such statements as (to quote from the start of the *Grundlagen*) 'for every two points there exists a line that contains each of the points', and 'for every two points there exists no more than one line that contains each of the points'. The axioms implied certain consequences, and if those consequences were those of, say, Euclidean geometry, then specific undefined terms could be interpreted as corresponding to specific Euclidean concepts. But the interpretation was not necessary, and in particular it did not intervene in the process of deduction. It was as if applied mathematicians could take a ready-made system of axioms and interpret it as they wished, secure in the knowledge that the deductions were always valid. This radical approach struck at the heart of any view that the truth of mathematics rested on the nature of its basic definitions. Other mathematicians were quick to praise this achievement, and to see that it could be extended. A. Hurwitz, for example, wrote to Hilbert to say that he had created a new field of study, the mathematics of axioms, going far beyond the domain of geometry (see §5.5, on metamathematics).

The mathematics of axioms did indeed flourish for a while, only to abate when it became clear that some systems of axioms were of very little interest. It is now usual in most areas of pure mathematics to give an axiomatic description of the basic entities. As for projective geometry, the techniques of modern algebraic geometry have given it a new lease of life. There is much interest in the study of such objects as the space of all curves of a given genus, or of all surfaces of a given kind. Such collections often form subspaces of a projective space, and so invite one to parametrize them algebraically. While the techniques are formidable, the problems addressed are once again in the mainstream of nineteenth-century geometry, and many questions once treated only intuitively have recently yielded to modern investigations.

BIBLIOGRAPHY

Apollonius *1891–3*, *Apollonii Pergaei quae Graece existant cum commentariis antiquis*, 2 vols (ed. J. L. Heiberg), Leipzig: Teubner.
——*1896*, *Treatise on Conic Sections* (ed. T. L. Heath), Cambridge: Cambridge University Press.
Chasles, M. *1837*, *Aperçu historique sur l'origine et le développement des méthodes en géométrie*, Brussels: Académie des Sciences.
Coolidge, J. L. *1940*, *A History of Geometrical Methods*, Oxford: Oxford University Press. [Repr. 1963, New York: Dover.]
Field, J. V. and Gray, J. J. *1987*, *The Geometrical Work of Girard Desargues*, New York: Springer.
Gua de Malves, J. P. de *1740*, *Usages d'analyse* [. . .], Paris: Briasson.
Hilbert, D. *1899*, *Grundlagen der Geometrie*, Leipzig: Teubner [2nd edn, transl. by L. Unger as *Foundations of Geometry*, 1902, La Salle: IL: Open Court, repr. 1971.]
La Hire, P. de *1685*, *Sectiones conicae*, Paris: Michallet.
Murdoch, P. *1746*, *Neutoni genesis curvarum per umbras*, London: Miller.
Poncelet, J. V. *1822*, *Traité des propriétés projectives des figures*, Paris: Bachelier.
Toepell, M.-M. *1986*, *Über die Entstehung von David Hilberts 'Grundlagen der Geometrie'*, Göttingen: Vandenhoeck & Ruprecht.
van der Waerden, B. L. *1961*, *Science Awakening* (transl. A. Dresden), New York: Oxford University Press.
Veblen, O. and Young, J. W. *1910–18*, *Projective Geometry*, 2 vols, Boston, MA: Ginn.

7.7

Line geometry

DAVID E. ROWE

1 ORIGINS AND SOURCES

The origins of line geometry can be traced back to two sources: early nineteenth-century investigations at the crossroads of projective geometry and mechanics, and optics. Its actual inception as an independent field of research, however, came about only in 1865, the year in which Julius Plücker, returning from a twenty-year hiatus from geometry, introduced the fundamental idea of regarding the four-parameter collection of lines in space as the ground elements in a new system of geometry. Three years later, he laid the foundations for the subject of line geometry in the first volume of his *Neue Geometrie des Raumes gegründet auf die Betrachtung der geraden Linie als Raumelement* (Plücker *1868–9*). The appearance of the second volume was interrupted by Plücker's sudden death, but his young assistant, Felix Klein, completed the work and went on to undertake ground-breaking research in the field (Rowe *1989*).

Line geometry is concerned with the properties of line complexes, which are three-parameter families of lines in space satisfying an algebraic condition in the system of coordinates for the lines. Although Gaspard Monge had already found a coordinatization for the lines of space as early as 1770, and Arthur Cayley had applied these in 1860, it was Plücker who took the decisive step by investigating the complex geometrical configurations associated with line complexes. Nearly all the more fundamental notions are due to him, although much of his terminology was supplanted by others, mostly German synthetic geometers, who came after him. Plücker's fame as one of the most important contributors to modern analytic geometry and his rivalry with Jacob Steiner, for many years the leading synthetic geometer in Germany, have sometimes obscured the fact that his approach actually represented a mixture of the analytic and synthetic styles. His characteristic *modus operandi* was to strive to simplify the analytic apparatus whenever possible, typically by introducing elegant parameters that lead to a straightforward geometric interpretation.

2 ORIGINS IN PHYSICS

Intimately related to Plücker's line complexes are so-called ray systems, which are two-parameter collections of lines. These arise, for example, by taking the lines common to two line complexes, a type of ray system Plücker called a 'congruence'. The theory of ray systems arose in a natural way from problems in optics involving the convergence of light rays: the determination of focal points, caustic curves and caustic surfaces (in German: *Brenn-punkte*, *-kurven*, *-flächen*). Important contributions to this field came from such figures as E. Malus, William Rowan Hamilton and Ernst Kummer. It was in connection with his researches on ray systems (*Strahlensysteme*) that Kummer found that famous surface which bears his name and which contains the Fresnel wave surface as a special case (§7.1).

Just as the study of ray systems had its origins in problems that arose in optics (§2.9), so can one find a rich prehistory for the theory of line complexes in the field of mechanics. Here one should distinguish the tradition of analytical mechanics (Leonhard Euler, Joseph Louis Lagrange, Pierre Simon Laplace and others) from a geometrical tradition that included Louis Poinsot, J. P. M. Binet, Michel Chasles and August Ferdinand Möbius (Ziegler *1985*). Their collective contributions were largely concerned with the geometry of motion for rigid bodies independent of considerations of the time variable, the kinematical properties alone. In considering the motion of a rigid body not subject to net forces or torque, one can easily show that the centre of mass either remains at rest or moves uniformly. Thus, without loss of generality, one can confine the analysis to motions in which the centre of mass stays fixed. Euler gave a classical analytic treatment of this problem, deriving the equations of motion (Euler's equations) and proving that at each instant in time the body's motion consists of a rotation about a certain axis. With the exception of three special (inertial) axes, the axis of rotation will vary with time.

A similar situation arises if one holds some arbitrary point fixed: for each point P one obtains three inertial axes about which the body will rotate without wobbling. In 1811 Binet found that by varying the point P the collection of all such inertial axes forms a three-parameter family of lines – i.e. a line complex – consisting of the normals to a one-parameter system of second-degree confocal surfaces. By Dupin's theorem, exactly three mutually orthogonal confocal surfaces pass through an arbitrary point P, and these intersect one another along lines of curvature, which means that the three inertial axes through P are tangents to these lines of curvature. Considerations such as these led A. M. Ampère and others to further investigations of this kind.

3 POINSOT, CHASLES AND MÖBIUS

Although he appears to have made no direct references to line complexes as such, Poinsot, the veritable father of geometrical mechanics, deserves some passing mention. In his *Eléments de statique* (1803), Poinsot first showed how a system of forces acting on a rigid body in space could be uniquely resolved into a single force together with a couple whose axis is parallel to the force. His analysis revealed a close connection between systems of forces and lines in space. Later, in studying the torque-free motion of a rigid body, Poinsot held the angular momentum vector **L** fixed and noted that the motion associated with the inertial ellipsoid traces out two curves, the 'polhode' and 'herpolhode' (meaning 'snakelike'), that give a clear, intuitive picture of the motion. In effect, the polhode rolls without slipping along the herpolhode, which lies in a certain invariant plane. For a symmetrical body these curves reduce to two circles generated by the motion of the 'body cone' and 'space cone'. In the same context, Binet considered what would happen in a reference system in which **L** were no longer held fixed. This led to his discovery of the so-called 'Binet ellipsoid', which essentially describes the possible positions that **L** may assume. Its three axes are related to the stability of the body's motion (minor deviations from the largest and smallest of the axes are stable, but not those near the intermediate axis).

Systems of lines played a central role in Chasles's kinematical investigations. In 1861 he studied a direct connection between rigid-body mechanics and line geometry by noting that the collection of lines connecting corresponding points of a body in two distinct positions forms a line complex of the second degree. Another key theorem of his states that, by an appropriate choice of reference system, the general motion of a rigid body can be reduced to a rotation together with a translation along the axis of the rotation, a so-called 'screw' motion. The elaborate theory of screw motions and their intimate connection with linear line complexes was developed by Robert S. Ball, whose work was championed in Germany by Felix Klein (Ball *1900*, Klein *1892–3*). Chasles's theorem, coupled with Poinsot's characterization of force systems acting on a rigid body, forged an important connection between the analysis of such systems and infinitesimal motions. The latter were investigated by Sophus Lie in connection with non-Euclidean geometries and the Riemann–Helmholtz–Lie space problem, which deals with the geometrical properties of spaces that permit the free movement of rigid bodies.

The most probing of the numerous studies connecting systems of lines and mechanics was undertaken by August Ferdinand Möbius, whose name is largely remembered today in connection with the single-sided surface known as the Möbius strip or band (§7.10). Although he was a pioneering figure in geometry and mechanics, he spent most of his career working as an astronomer in Leipzig. In 1828 Möbius made the following key discovery: if one considers any system of forces and fixes a plane M, then the collection of lines in M with moment equal to zero all pass through a definite point, the null point of M. This led him to investigate so-called 'null systems' of lines, and these turned out to be identical to the first-degree line complexes of Plücker's theory.

4 THE APPROACH OF PLÜCKER AND KLEIN

Although Plücker was well aware of the numerous connections between line geometry and mechanics, he was principally concerned with the elaboration of the geometrical theory rather than its potential applications. There are several different ways to coordinatize the lines in space, but the most elegant approach represents the four-parameter family of lines in $P^3(\mathbb{C})$ by six homogeneous coordinates, p_{ij}. Following Hermann Grassmann, these p_{ij} can be obtained by taking any two points, (x_1, x_2, x_3, x_4) and (y_1, y_2, y_3, y_4), on a given line l and forming the six possible 2×2 determinants $\rho(p_{ij}) = x_i y_j - x_j y_i$. The p_{ij} thereby satisfy the fundamental relation

$$P = p_{12}p_{34} + p_{13}p_{42} + p_{14}p_{23} = 0. \tag{1}$$

One also has the possibility of a dual representation by taking any two planes, (u_1, u_2, u_3, u_4) and (v_1, v_2, v_3, v_4) that contain l and forming $\sigma(q_{ij}) = u_i v_j - u_j v_i$, where the q_{ij} satisfy the same identity:

$$Q = q_{12}q_{34} + q_{13}q_{42} + q_{14}q_{23} = 0. \tag{2}$$

The coordinates p_{ij} and q_{ij} represent the same line if

$$\rho p_{ij} = \frac{\partial Q}{\partial q_{ij}} \quad \text{or} \quad \sigma q_{ij} = \frac{\partial P}{\partial p_{ij}}. \tag{3}$$

An nth-degree line complex is determined by a homogeneous nth-degree equation $F(p_{ij}) = 0$. Its local behaviour can be studied by fixing a point $x = (x_1, x_2, x_3, x_4)$, so that $F(p_{ij}) = F(x_i y_j - x_j y_i) = 0$ results in an nth-degree equation in (y_1, y_2, y_3, y_4); the lines through x then determine a cone of the nth order. As mentioned above, the lines shared by two line complexes, $\Omega_m = 0$ and $\Omega_n = 0$, determine a 'congruence' with order and class equal to mn. This means that mn lines pass through a typical point, and mn lines lie in a typical plane. The lines of a congruence envelope a caustic surface, which generally has two components. If one adjoins a third line complex $\Omega_l = 0$, then the lines common to all three determine a ruled surface of order $2lmn$, and the intersection with a fourth complex Ω_o results in $2lmno$ lines in space.

The theory of first-degree or linear line complexes clearly illustrates the central role played by invariant theory in this field. Such complexes have the form $\Omega_1(p_{ij}) = \Sigma\, a_{ij}p_{ij} = 0$, which leads to a null system of lines. Following the approach taken by Plücker and Klein, the expression

$$A := a_{12}a_{34} + a_{13}a_{42} + a_{14}a_{23} \tag{4}$$

is called the invariant of the complex. Its geometric significance can be interpreted as follows: if $A = 0$, then the a_{ij} satisfy the identity $P = 0$, and therefore can themselves be viewed as line coordinates. By setting $a_{ij} = q'_{ij}$, one has $\Sigma\, a_{ij}p_{ij} = \Sigma\, q'_{ij}p_{ij} = 0$, and the complex consists of all lines p_{ij} that intersect a fixed line q'_{ij}, a so-called special linear complex.

After Plücker's death, Klein promoted both the theoretical development of line geometry and its applications to mechanics. In his doctoral dissertation of 1868, he introduced a classification scheme for second-degree line complexes based on G. B. Battaglini's earlier work, and utilizing Karl Weierstrass's new theory of elementary divisors (§6.6). Klein left it to his Erlangen student A. Weiler to work out the detailed analysis, which leads to 57 different types. Klein also established the connection between non-Euclidean line geometry, based on the Cayley metric, and non-Euclidean kinematics and statics. This became the dissertation topic of Ferdinand Lindemann, who also studied under Klein at Erlangen. Klein's own publications on line geometry had already culminated in the broad vision of geometry he set forth in his *Erlanger Programm* (Klein *1872*: see §7.6).

BIBLIOGRAPHY

Ball, R. S. *1900, Theory of Screws*, Cambridge: Cambridge University Press.

Klein, F. *1872, Vergleichende Betrachtungen über neuere geometrische Forschungen. Programm zum Eintritt in die philosophische Facultät und den Senat der Friedrich-Alexanders-Universität zu Erlangen*, Erlangen: Deichert. [Also in *Gesammelte mathematische Abhandlungen*, Vol. 1, 460–97.]

—— *1892–3, Vorlesungen über höhere Geometrie*, lithograph. [3rd edn (revised and ed. W. Blaschke), 1949, New York: Chelsea.]

Plücker, J. *1868–9, Neue Geometrie des Raumes gegründet auf die Betrachtung der geraden Linie als Raumelement*, Vol. 1 (ed. A. Clebsch), Vol. 2 (ed. F. Klein), Leipzig: Teubner.

Rowe, D. *1989*, 'The early geometrical works of Sophus Lie and Felix Klein', in D. Rowe and J. McCleary (eds), *The History of Modern Mathematics*, Vol. 1, Boston, MA: Academic Press, 209–73.

Ziegler, R. *1985, Die Geschichte der geometrischen Mechanik im 19. Jahrhundert*, Stuttgart: Steiner.

7.8

The philosophy of geometry to 1900

JOAN L. RICHARDS

1 EUCLIDEAN GEOMETRY

Throughout the nineteenth century, long after other classical works were rendered obsolete by the developments of the scientific revolution, Euclid's *Elements* remained the text from which many schoolchildren learned the subject. Even today, Euclidean theorems and proofs are readily accessible. In writing this remarkable text, Euclid seems to have been acting primarily as the compiler of a system of geometry which had been developing for years before he appeared on the scene. His major innovation was also the most controversial part of his book, the fifth or parallel postulate:

> That, if a straight line falling on two straight lines make the interior angles on the same side less than two right angles, the two straight lines, if produced indefinitely, meet on that side on which are the angles less than the two right angles. (Euclid *1926*: Vol. 1, 202)

Euclid accepted this postulate without proof, despite its being so crabbed and complex in comparison to the others (e.g. 'that all right angles are equal to one another'). Much of his posterity was unwilling to do so, and, according to the fifth-century commentator Proclus, unsuccessful attempts to prove it directly date back to the appearance of his system; all these attempts ended in failure, however (§7.4).

Thus Euclidean geometry, complete with its disputed but unrivalled theory of parallels, remained constant over several millennia. However, there was always a wide variety of interpretations of its meaning and significance. From the ancients come two radically different interpretations: that which saw geometry as the most essential subject, figuring the ultimate reality underlying all superficial impressions; and that which saw geometry as a highly abstract subject, far removed from the real world of experience. These two perspectives, which can be loosely labelled as Platonic and

Aristotelian, respectively, formed the bases upon which were developed a whole range of views of the nature of geometry in the Arab and medieval worlds.

2 NEWTON, KANT AND SPACE

The beginning of the modern discussion of the philosophy of geometry can be tied to the work of Isaac Newton. His formulation of dynamics entailed identifying the mathematical space so well described by Euclid with the physical space in which the Sun, Earth, planets and stars move. In the Scholium to the Definitions of the *Principia* (1687), he explicitly indicated the relationship between mathematical and physical space:

> Absolute space, in its own nature, without relation to anything external, remains always similar and immovable. Relative space is some movable dimension or measure of the absolute spaces; which our senses determine by its position [with respect] to bodies; and which is commonly taken for immovable space .... Absolute and relative space are the same in figure and magnitude; but they do not remain always numerically the same.

The essential linkage which Newton here maintained between relative, physical space and absolute, mathematical space allowed him to cloak his physics with the mantle of mathematical certainty. This meant that his system was remarkably solid but, at the same time, it raised the epistemological question of how such certain knowledge of a physical reality was possible. For many philosophers, from John Locke onward, the power of Newton's physics was indubitable, and the knowledge of space it rested upon exemplified the kind of certainty to which human knowledge could aspire.

The most notable example of the philosophical power of Newtonian space is found in the eighteenth-century philosophy of Immanuel Kant. For Kant, space was a form of our pure intuition, ontologically prior to the objects we perceive as existing in it. The unique combination of the absolute certainty of geometry and the substantial reality of physics made Newtonian spatial knowledge the quintessential exemplar of Kant's category of synthetic, *a priori* knowledge.

3 NON-EUCLIDEAN GEOMETRIES

From a geometrical standpoint, when he spoke of space Kant meant the space of Euclid's *Elements*. Even as he and others were incorporating the certainties of this geometry into their philosophies, however, mathematicians were re-examining the geometrical bases for such claims. The

914

attempt to clarify geometry's certainty led them again to consider the status of Euclid's fifth postulate.

Recognizing the futility of trying to prove the postulate outright, the Italian Jesuit Girolamo Saccheri spearheaded the eighteenth-century movement to prove it indirectly by indicating the depths of contradiction entailed in assuming it to be false. Focusing on a quadrilateral whose base angles were assumed to be equal and right, Saccheri divided the various alternatives to Euclid's geometry into clearly demarcated categories. He was able to prove quite easily that the two remaining angles were equal to each other. From there he went on to show that if they were both right angles, Euclid's theory of parallels would hold; if, on the other hand, the remaining angles were either acute or obtuse, then the theory would not hold. Thus, in order to prove the necessity of Euclidean geometry, Saccheri set out to demonstrate the impossibility of both the acute-angle and the obtuse-angle hypothesis (Bonola *1912*: 22–44; Gray *1979a*: 51–62).

Saccheri apparently believed he had succeeded in this endeavour; other mathematicians who followed him, like Johann Heinrich Lambert, seemed more ambivalent about the implications of their results. Whatever their conclusions, however, efforts to generate contradictions from alternative postulates led to a number of geometrical theorems developed from systems with non-Euclidean parallels.

In the early decades of the nineteenth century these scattered results began to bear fruit; Ferdinand Schweikart and Carl Friedrich Gauss privately, and Nikolai Lobachevsky and János Bolyai more publicly, developed self-consistent 'non-Euclidean' geometries (§7.4). These men recognized that their success in creating alternatives to Euclid's geometrical system raised significant questions about the exact certainty which had been ascribed to Newtonian space; whether physical space actually was Euclidean became for them an empirical issue as yet undetermined (Bonola *1912*: 84–121; Gray *1979a*: 96–116).

4 RIEMANN AND HELMHOLTZ

Despite their importance, these mathematical developments had remarkably little immediate impact. They were not widely discussed, either philosophically or mathematically, until the 1860s. In 1863 Gauss's correspondence about non-Euclidean geometry was published for the first time; the rest of the decade saw the publication of a French edition of Lobachevsky's work, and Bernhard Riemann's 'Über die Hypothesen, welche der Geometrie zu Grunde liegen' (*1867*).

This last was Riemann's *Habilitationsvortrag*, initially delivered in Göttingen in 1854. Whereas Lobachevsky, Bolyai and their contemporaries were 'synthetic geometers' who constructed alternative spaces based on non-Euclidean assumptions, in this work Riemann worked analytically. In his view the concept of space was a particular instance of a more general concept, that of multiply extended magnitude. The identifying mark of space within this broader category was that a metric is defined on it:

> Hence [it] follows as a necessary consequence that the propositions of geometry cannot be derived from general notions of magnitude, but that the properties which distinguish space from other conceivable triply extended magnitudes are only to be deduced from experience. (Riemann *1867*: 272–3).

He went on to establish that Euclidean space was characterized by the particular distance function, $ds = (dx^2 + dy^2)^{1/2}$, but that other functions would generate alternative spaces. He further speculated that space might not even have a single, constant distance function, but that the measure of distance might be different for the infinitely large or small.

Riemann did not try to specify the actual experiences which might ground the Euclidean spatial concept; he saw this as 'a problem which from the nature of the case is not completely determinate, since there may be several systems of matters of fact which suffice to determine the measure-relations of space' (*Ibid.*: p. 273). The issue seemed more determinate to the German physicist Hermann von Helmholtz, who was initially drawn into geometrical discussion by studies of vision. In an attempt to justify his conviction that visual space had its roots in childhood experiences, Helmholtz set out to identify them. He settled on the motions of rigid bodies, which all infants have experienced by the end of their first year, and showed how the basic structures of Euclidean space could be generated from them.

Before he had quite finished his first paper on this subject, however, Helmholtz read Riemann's newly published work. This was his first introduction to non-Euclidean geometries, and Helmholtz quickly realized that they could also be constructed from his rigid-body motions. In subsequent papers Helmholtz offered these motions as the 'matters of fact' on which Riemannian measure relations were based. At the same time, he used the possibility of non-Euclidean geometries as the basis for a frontal attack on the so-called 'nativist' physiologists who maintained that Euclidean space is a physiologically determined concept. The fact that we could have constructed several different geometries, and yet see only one, indicated to Helmholtz that we have a choice which we make based on experience. This argument was easily extended to challenge philosophies which, like Kant's, asserted that space is known intuitively and absolutely without contingent empirical input (Richards *1977*; Torretti *1978*: 155–71).

Helmholtz's argument and analysis were clear and persuasive; his writings served to introduce many mathematicians as well as others to non-Euclidean ideas. His empirical conclusions were not universally embraced, however. Philosophers who were committed to the necessity of geometrical truth, like the Dutchman J. P. N. Land, countered his arguments by focusing on the nature of conceivability. The ability to describe a space mathematically, they argued, was not enough to render it truly conceivable; in their view, despite the mathematical alternatives Helmholtz had constructed, Euclidean space remained uniquely certain because of its conceptual status. This line of argument led away from mathematics into discussions of the nature of conception.

5 PROJECTIVE GEOMETRY

A related development grew out of yet another mathematical approach to non-Euclidean geometries, through projective geometry (§7.6). Initially rooted in techniques of technical drawing developed after the French Revolution by Gaspard Monge, Jean Victor Poncelet, and others, projective geometry focused on those aspects of figures which remain constant throughout the transformations of projection and section (Daston *1986*). Metric relations are not invariant in this way and all lines intersect at infinity, so, initially, there was no relation between projective and non-Euclidean ideas. However, in his 'Sixth Memoir on Quantics' (1859), Arthur Cayley developed from projective properties a function which displayed the defining characteristics of a Euclidean distance function. Having shown in this way how metric Euclidean space may be generated from the less structured projective space, Cayley concluded: 'Metrical geometry is thus a part of [projective] geometry, and [projective] geometry is *all* geometry, and reciprocally' (Richards *1988*: 130). Cayley completed this work before he knew of non-Euclidean geometry; but in the wake of the non-Euclidean publications of the 1860s the German mathematician Felix Klein showed that non-Euclidean metrics could also be generated within projective spaces. With this insight, he opened a powerful new projective approach to non-Euclidean geometry.

Projective geometry held a great deal of mathematical interest in the final decades of the nineteenth century, and for some mathematicians who were committed to the necessity of geometrical truth it held out a solution to Helmholtz's empirical challenge. By emphasizing the conceptual primacy of projective space over any metric spaces, they were able to preserve the necessity of geometrical knowledge, albeit now of projective rather than Euclidean space; in this view the choice among metric geometries might be contingent, but knowledge of the projective space in which they were embedded was not. Bertrand Russell's book of 1897, *An Essay on the*

Foundations of Geometry, marks the culmination of this tradition (Richards *1988*: Chap. 5).

6 CONVENTIONALISM AND AXIOMATICS

In the last decade of the nineteenth century, the French mathematician Henri Poincaré developed yet a different approach to geometry. He focused on the structural equivalences among the various geometries, and claimed that choices among them could not be physically determined; rather, he emphasized, they were the result of decisions about which was the most convenient (*'la plus commode'*) (Torretti *1978*: 320–56).

In Poincaré's conventionalist view, the category of necessary truth is basically irrelevant to geometry; in this he was feeding into a movement towards an axiomatic or formal view of geometry which had been growing since the middle of the century. The 1844 *Ausdehnungslehre* of Hermann Grassmann was an early attempt to avoid the epistemological quagmire surrounding geometry by developing a geometrical structure which did not rely at all on intuitive content (§6.2). In 1882, Moritz Pasch published a series of 'Lectures on Modern Geometry' with a similar intent. Starting from a carefully circumscribed set of undefined concepts drawn from experience, Pasch tried to erect a purely deductive, non-conceptual geometrical structure. In his words:

> If geometry is to be truly deductive, the processes of inference must be independent in all its parts from the meaning of the geometrical concepts, just as it must be independent from the diagrams. All that need to be considered are the relations between the geometrical concepts, recorded in the propositions and definitions. (Torretti *1978*: 211)

The Italian Giuseppe Peano went a step further by not only proceeding from a strictly limited number of geometrical terms, but also creating an artificial language as a way of controlling the logical content of theory. It was under the influence of his ideas that Russell abandoned the kind of work found in his early *Essay*, and moved towards the logicism of *Principia mathematica* (1910–13; see §5.2).

Thus, by the end of the nineteenth century the mathematical understanding of geometry had changed considerably. With these developments came challenges to the claims of special, necessary truth which had underlain the subject for two centuries. The philosophical interest of the subject changed accordingly, and the twentieth century has opened a whole new chapter in its development.

BIBLIOGRAPHY

Bonola, R. *1912, Non-Euclidean Geometry. A Critical and Historical Study of its Development*, Chicago, IL: Open Court. [English transl., with additional appendices, by H. S. Carslaw, repr. 1955, New York: Dover.]

Daston, L. J. *1986*, 'The physicalist tradition in early nineteenth century French geometry', *Studies in the History and Philosophy of Science*, **17**, 269–95.

Euclid *1926, The Thirteen Books of Euclid's Elements* (introduction and commentary by T. L. Heath), 3 vols, Cambridge: Cambridge University Press. [Repr. 1956, New York: Dover.]

Gray, J. J. *1979a, Ideas of Space: Euclidean, Non-Euclidean, and Relativistic*, 1st edn, Oxford: Oxford University Press. [2nd edn 1989.]

—— *1979b*, 'Non-Euclidean geometry – A re-interpretation', *Historia mathematica*, **6**, 236–58.

Gray, J. J. and Tilling, L. *1978*, 'Johann Heinrich Lambert, mathematician and scientist, 1728–1777', *Historia mathematica*, **5**, 13–41.

Helmholtz, H. *1977, Epistemological writings* (transl. by M. F. Lowe, ed. with an introduction and bibliography by R. S. Cohen and Y. Elkana), Dordrecht: Reidel.

Nowak, G. *1989*, 'Riemann's *Habilitationsvortrag* and the synthetic *a priori* status of geometry', in D. Rowe and J. McCleary (eds), *The History of Modern Mathematics*, Boston, MA: Academic Press, Vol. 1, 17–46.

Richards, J. L. *1977*, 'The evolution of empiricism: Hermann von Helmholtz and the foundations of geometry', *British Journal for the Philosophy of Science*, **28**, 235–53.

—— *1988, Mathematical Visions: The Pursuit of Geometry in Victorian England*, Boston, MA: and San Diego, CA: Academic Press.

Riemann, G. F. B. *1867*, 'Über die Hypothesen, welche der Geometrie zu Grunde liegen', in *Gesammelte mathematische Werke*, 2nd edn, 1892, 272–87.

Rougier, L. *1920, La Philosophie géométrique de Henri Poincaré*, Paris: Alcan.

Scholz, E. *1982*, 'Herbart's influence on Bernhard Riemann', *Historia mathematica*, **9**, 413–40.

Torretti, R. *1978, Philosophy of Geometry from Riemann to Poincaré*, Dordrecht: Reidel.

7.9

Early modern algebraic geometry

J. J. GRAY

Algebraic geometry is the application of algebra to the study of geometry. As such, its origins lie in the invention of coordinate or analytic geometry by René Descartes in the 1630s (§7.1). His work began a tradition whereby the algebra of polynomial expressions was applied to the study of curves defined by equations relating their coordinates. Over the subsequent centuries both the algebraic and the geometrical sides developed in sophistication, but a crucial aspect of the modern period has been the development of newer kinds of algebra to solve geometrical problems, as is described with his customary briskness and mathematical clarity by Dieudonné *1974*. These new problems often involve three or more variables, and often arise in contexts which are not exclusively geometric, such as the study of complex functions. Most influentially, this was the case with the work of Bernhard Riemann.

1 RIEMANN

The history of algebraic geometry from the work of Riemann onwards can be regarded as a dialogue between two approaches, one emphasizing the geometry and the other the algebra. In two major papers of 1851 and 1857 (see his *1990*), Riemann showed how the ideas of complex-function theory could be extended and applied to study integrals on an algebraic curve (later also to be called a 'Riemann surface'). The central theorem in his presentation of the subject is Niels Abel's theorem concerning integrals on such a curve. This asserts that there is a number, q, determined by the curve, such that the sum of any number of integrals (with the same integrand), having a fixed initial point but variable end-points, is the sum of some rational functions and some logarithmic terms minus exactly q integrals having the same integrand. Moreover, the end-points of these integrals depend algebraically on the given variable end-points.

Riemann simplified this theorem by considering the integrands. If they have simple poles (if, that is, they look like dz/z near the origin), then logarithmic terms will arise if the path of integration winds around the pole. If there are poles of higher order (terms like dz/z^n, $n > 1$) then a rational function can be expected. But if the integrand is 'holomorphic' (i.e. has no poles) then neither of these terms can arise, and Abel's theorem becomes the statement that the sum of any number of integrals having a fixed initial point but variable end-points, is equal to minus exactly q integrals having specified integrands and whose end-points depend algebraically on the given variable end-points. The number q was given a novel interpretation by Riemann (it had been left unexplained by Abel): he showed that the original algebraic curve could be regarded as a real surface upon which a system of $2q$ (but not $2q + 1$) cuts can be made, leaving the dissected surface still in one piece. He then showed that, as a result, there are exactly q linearly independent holomorphic integrands.

Riemann's approach was a difficult mixture of the profound and the naive, and to make sense of it the next generation of German mathematicians sought to rewrite it in a language they better understood – that of algebra (Gray *1989*). Alfred Clebsch (the leader of this movement) and his colleague Paul Gordan found Riemann's topological arguments obscure, and the deduction about the number of integrands, resting as it did on Dirichlet's principle (§3.17), they found unacceptable. So they based their book of *1866* on the idea that a complex curve is a locus defined by a polynomial equation in two complex variables, x and y (such a curve is called an algebraic curve), and went back to Abel's explicit presentation of the holomorphic integrands. Abel had written them in the form $\int_0^z g\,dx/F_y$, where $F(x, y) = 0$ is the equation of the curve, F_y is the partial derivative of F with respect to y, and $g = g(x, y)$ is a polynomial of low enough degree for the quotient $g\,dx/F_y$ to be holomorphic. He had then obtained the number q by counting the number of possible terms in g. If the curve $F = 0$ has no singular points and is of degree n, then this number is $q = (n - 1)(n - 2)/2$.

Clebsch and Gordan used geometry to understand the way in which the end-points enter Abel's theorem. They considered families of algebraic curves depending linearly on some parameters, and cutting the curve in the end-points of the given integrals. It is then clear that the remaining points of intersection of the original curve and these variable ones are determined by algebraic conditions. Problems arose with this approach when the curve could have singular points, self-intersections or cusps. It was possible to show that each of d double points and r cusps lowers this number q by 1.

This meant that Plücker's formulas from projective geometry could be used to show that $\frac{1}{2}(n-1)(n-2) - d - r$ was invariant under any projective transformation of the curve, and could therefore be taken as the definition of q. But for curves with more complicated singularities the definition of q was not clear. Alexander Brill and Max Noether, following the work of Clebsch, then sought to show that birational transformations could always be found to reduce a curve to one having singularities which gave only a straightforward contribution to the number q. In this way they gave a definition of q which was indeed birationally invariant. This they did by insisting that the variable curves through the singular points have the right degree and passed through the singular points the right number of times ($n-1$ times through a singularity of order n). Riemann's topological approach was replaced with a geometrical definition of q (called the genus by Clebsch) which was, however, valid for both singular and non-singular curves. Then in 1880 Richard Dedekind and Heinrich Weber put forward a remarkable theory of Riemann surfaces which showed how a complex curve could be studied entirely in terms of the families of meromorphic functions it supported.

2 HIGHER DIMENSIONS

For a long time the path to studying geometrical objects other than curves (for example, surfaces) seemed blocked, until an approach was developed by Leopold Kronecker and extended by David Hilbert which provided a rich, uniform treatment and offered the securities of algebra in exchange for the insights of geometry. Kronecker's work was but a part of what he hoped would be a theory embracing algebraic number theory as well as geometry. In the event, he put forward the important definition of an algebraic variety as the set of points in $\mathbb{C}^n$ satisfying a family of polynomial equations, and showed that each such variety could be decomposed into a number of irreducible pieces, each having a definite dimension. He observed that the variety depends only on the ideal in the polynomial ring $C[x_1, x_2, \ldots, x_n]$ that the given polynomials define, and showed that if the ideal is prime then the variety it generates is irreducible. (An ideal a in a ring R is a set such that $u, v \in a$ implies that $u + v \in a$, and $u \in a$ and $r \in R$ imply that $ru \in a$; it is prime if $uv \in a$ implies either $u \in a$ or $v \in a$; see §6.4, Section 2.2.)

Hilbert's profound contributions were part of his reformulation of invariant theory in the late 1880s (§6.8). Above all, he investigated the polynomials that vanished at every point of a variety, and showed that if f has this property then some power of f belongs to the ideal defining the variety. This is Hilbert's *Nullstellensatz* ('theorem of the zeros'). Powers of f need

922

to be taken in case the variety has multiple points. It means that one may study either the variety or the so-called radical of the ideal defining it, where the radical of an ideal a is the set of all $u \in \mathbb{R}$ such that $u^k \in \mathbb{R}$ for some k.

3 MODERN ALGEBRA

Kronecker's work opened the way for the transcription of geometrical properties of varieties into a new algebraic language, that of commutative algebra. In this spirit, the type of ideal that an irreducible variety generates was first elucidated in 1905 by Emanuel Lasker (who later became the world chess champion). The important concept of a field was axiomatized by Ernst Steinitz, who in 1910 isolated the concepts that underlay a profusion of examples previously discussed by Heinrich Weber in his influential book on algebra (§6.4, Section 2). He also defined the concept of the transcendence degree of one field over a field that it contains, which counts the number of linearly independent elements of the larger field, and connected it to Kronecker's definition of the dimension of a variety. The connection is provided by the idea of the coordinate ring of a variety (loosely, the polynomials defined on the variety). It turns out that every geometric property of a variety is expressible as an algebraic property of its coordinate ring. In particular, the construction of the coordinate ring yields a field of transcendence degree d over $\mathbb{C}$, where d is the dimension of the variety.

4 GEOMETRICAL RESPONSES

A programme that aimed to solve problems in geometry by translating them so completely into the language of commutative algebra naturally did not appeal to geometers, nor did everyone welcome the drift away from the questions that had so profitably animated Abel and Riemann. It seemed to geometers of the flourishing Italian school in particular that at least algebraic surfaces could be tackled more geometrically, and concentrated effort was devoted to this problem from 1893, when two decisive papers were published. One, by E. Bertini, summarized the Brill–Noether approach to curves. The other, by Corrado Segre, reformulated that theory in a way that invited generalization to higher dimensions. What Bertini described via families of curves through certain points on a given curve, Segre sought to describe in terms of certain sets of points on the curve. In particular, Segre moved away from Bertini's description of a curve as a locus in the plane and towards Riemann's hazy appreciation of a curve as an object to be studied intrinsically, without reference to any particular embedding in an ambient space.

Segre's paper opened the way to a series of papers by Guido Castelnuovo and Federigo Enriques over the next twenty years that culminated in a classification of algebraic surfaces analogous to Riemann's classification of complex curves according to their genus. They also gave complete descriptions of several families of curves, analogous to the rich picture one has of curves of low genus. Meanwhile, Emile Picard had been working methodically on a generalization of Riemann's function-theoretic ideas to complex surfaces. The result was an analysis of surfaces according to the kinds of function they can support (as described by their singularities and their integrals) that exactly complemented the Italian theory of surfaces and the families of curves they can have. When Picard wrote up his findings in the definitive form of his book with F. Simart (*1897, 1906*) this overlap was often alluded to, and in a lengthy appendix Castelnuovo and Enriques set out the main features of their approach. The roles were reversed in their article on the algebraic geometry of surfaces in the *Encyklopädie der mathematischen Wissenschaften* (*1908*), in which the Italians showed how Picard's discoveries fitted into their approach.

5 THE NEED FOR RIGOUR

For good reasons, the Italian theory was soon felt to lack rigour. It was criticized from within most decisively by Francesco Severi for relying too heavily on intuition, and even Castelnuovo came to feel that progress required other methods. For this reason he encouraged his student O. Zariski in Rome, saying to him, however, 'You are here with us, but you are not of us'. Zariski's book *Algebraic Surfaces* (1935) was a turning point for him. He went as far as he felt he could to present the ideas underlying the Italian approach, but at the price, he later wrote, of feeling compelled to leave the geometric paradise for the greater rigour of algebra. The problem was the singular points and curves of an algebraic surface, which can be very complicated. He therefore turned to the ideas that E. Krull, and Emmy Noether and her student B. L. van der Waerden, were developing in the theory of rings.

Meanwhile, complex algebraic geometry developed under the forceful guidance of Solomon Lefschetz in the direction of algebraic topology. Picard and Simart's book had been a great source of inspiration for Lefschetz, but they had not been able to draw on Henri Poincaré's visionary ideas about topology (§7.10). It was Lefschetz's contribution to do just that, in forging an algebraic topology of algebraic varieties of higher dimension. This theory was consummated by W. V. D. Hodge's theory of harmonic forms, which represents the subtle generalization to higher dimensions of Riemann's insight into the intimate connection between holomorphic and harmonic functions.

Throughout the period, other questions also motivated algebraic geometers. It must suffice here to describe just two of these. A remarkable calculus (the enumerative calculus) had been developed for counting the number of curves touching a given set of curves. Proposed by Michel Chasles, and developed by H. Schubert and others, notably H. G. Zeuthen, in the later nineteenth century, the move towards rigour in geometry found this theory almost incomprehensible. Only recently has it been rescued from oblivion and several of Schubert's remarkable counts confirmed. The study of particular surfaces was a continuing source of discovery. The 27 lines on a cubic surface, discovered by Arthur Cayley and counted by George Salmon, have yielded many gems, including a connection with the Weyl groups of exceptional Lie groups, while Ernst Kummer's quartic surface with its 16 nodal points, which has its origins in Augustin Jean Fresnel's theory of optics (§9.1), has recently yielded the discovery of non-standard differentiable surfaces on $\mathbb{R}^4$.

6 A GLIMPSE OF THE MODERN SITUATION

Some of the problems in algebraic geometry today curiously resemble those that inspired Kronecker. In the hands of A. Grothendieck, a language has been developed specifically to move algebraic geometry beyond its reliance on field-theoretic concepts, to give rings equal weight. This amounts to passing from the study of the rational numbers to the integers, almost always a considerable step, and it has brought to the problems of algebraic number theory the resources of modern geometry, often with considerable success. Grothendieck shares with Zariski and others an unwillingness to accord the complex numbers any special preference, but some problems remain that have finally been solved only over the complex numbers. For example, in 1966 the Japanese mathematician H. Hironaka showed how singularities of any kind may be resolved, but only over the complex numbers and certain related fields (those of characteristic zero). This theory permits one to pass in a controllable way from a singular variety to a non-singular one related to it. Sometimes even this theory does not suffice; the recent successful classification of algebraic varieties of (complex) dimension 3 requires an analysis of certain mildly singular cases, unlike the Enriques classification in dimension 2.

Since the 1930s there has been a fruitful tension between the half of algebraic geometry that sought to stay near the intuitively accessible complex varieties and the half that, attracted by the generality of algebra, sought to deal with objects defined over any field (or ring). This tension is perhaps at its most productive in its analysis of the so-called problem of moduli. Riemann had observed that curves of a given genus $g > 1$ seemed to form a family of complex dimension $3g - 3$. In other words, a curve of genus g could be specified uniquely by choosing $3g - 3$ parameters or moduli. Despite some early attempts to describe the space of all parameters (for a given value of g), no progress was made on this problem until the work of K. Kodaira and D. C. Spencer in the 1950s. What they began for the case of complex curves and certain types of surface was extended by David Mumford for other examples and broadened to include fields of arbitrary characteristic. Most unexpectedly, the hope today is that these theories will play an essential part in rigorizing the theory of the Feynman integral in quantum-field theory, which has already begun in its turn to produce remarkable insights in the difficult territory of low-dimensional topology.

BIBLIOGRAPHY

Clebsch, A. and Gordan, P. *1866*, *Theorie der Abelschen Functionen*, Leipzig: Teubner. [Repr. 1967, Würzburg: Physica.]

Dieudonné, J. *1974*, *Cours de géométrie algébrique*, Paris: Presses Universitaires de France.

Enriques, F. and Castelnuovo, G. *1908*, 'Grundeigenschaften der algebraischen Flächen', in *Encyklopädie der mathematischen Wissenschaften*, Vol. 3, Part 2, 635–73 (article III C 6a).

Gray, J. J. *1989*, 'Algebraic geometry in the late nineteenth century', in D. Rowe and J. McCleary (eds), *The History of Modern Mathematics*, Boston, MA: Academic Press, Vol. 1, 361–85.

Picard, E. and Simart, F. *1897, 1906*, *Théorie des fonctions algébriques de deux variables indépendantes*, 2 vols, Paris: Gauthier-Villars. [Repr. 1971, New York: Chelsea.]

Riemann, B. *1990*, *Gesammelte mathematische Werke*, 3rd edn, New York: Springer.

Schubert, H. *1979*, *Kalkül der abzählenden Geometrie* (reprint of original 1879 edition, with an introductory essay by S. L. Kleiman), New York: Springer.

7.10

Topology:
Geometric, algebraic

E. SCHOLZ

The move from classical to modern geometry was linked with a reorganization of geometrical thought around new concepts which were much more general than the traditional concept of space. This change opened new reference fields for geometry within mathematics, and created a whole network of methods which gave rise to new geometrical subdisciplines. Most important in this respect, certainly for the nineteenth century and with its strong impact in the twentieth, was the new concept of manifold, introduced by Bernhard Riemann in his inaugural lecture of 1854 (Scholz *1980*: Chap. 2). One of the striking new features of Riemann's approach was the investigation of global properties of manifolds: *analysis situs*, in his terminology, now better known as topology.

1 RIEMANN'S APPROACH TO GEOMETRY

1.1 Manifolds

Riemann started his inaugural lecture with a semi-philosophical introduction of the concept of a multiply extended magnitude, or manifold. By this he meant a (somehow given) general concept, the individual instances of which allow continuous transitions between one another and a locally bijective correspondence to a finite (or in some cases even infinite) number of real quantities $x_1, x_2, \ldots, x_n$ which may serve as coordinates for the description of the points in the manifold (of dimension n).

Riemann initiated investigations of manifolds on different levels. He distinguished metric-free investigations from those using differential geometrical metrical concepts as the two main research areas for manifolds. Of the investigations which are free from metrical considerations, he emphasized particularly those of *analysis situs* (topology), although in his function-theoretic work he also studied the analytical and/or algebraic birational structures of surfaces (i.e. for dimension $n = 2$), which also are essentially free of metrical considerations.

1.2 Differential geometry

In his approach to differential geometry, Riemann used ideas from Carl Friedrich Gauss's theory of surfaces, but liberated them from the restriction of being embedded in (three-dimensional) Euclidean space (§3.4). He started from a determination of the length of a line element as a positive-definite quadratic differential form $ds^2 = \Sigma_{i,j} g_{ij} dx_i dx_j$ to derive further notions depending on metrics, in particular that of the geodesic line. Moreover, he introduced the sectional curvature of an infinitely small surface element, derived from the Gaussian curvature of the associated finite surface inside the manifold, which is generated by all geodesic lines starting in the surface element. He looked for geometrical properties, particularly of manifolds of constant (sectional) curvature, and identified them as the simplest cases generalizing Euclidean geometry (§7.4). In another context, while studying heat distribution, he introduced an analytical criterion for a metric ds^2 to reduce to the Euclidean case after proper choice of coordinates. This criterion led him to a 4-index quantity formed from the second partial derivatives of the metrical coefficients g_{ij}, later to be known as Riemann's curvature tensor (§3.4).

1.3 Topology

Riemann published only those parts of his topological ideas about manifolds which refer to the theory of surfaces, but his *Nachlass* shows that he started to generalize his topological methods to n dimensions. In 1851, in his doctoral dissertation, he had developed a topological classification of bounded (compact, orientable) surfaces that was based essentially on dissections of the surface by a finite number e of cross-cuts into f simply connected pieces. He defined the order of connectivity m of the surface F as $m = e - f$ ($= -\chi(F)$, the 'Euler characteristic' of F; see Section 2 below) and classified the (compact, orientable) surfaces by their number of boundary components and their order of connectivity. Later, in a function-theoretic memoir written in 1857, he introduced a second topological method of studying closed surfaces. He considered systems of closed curves (1-cycles), and whether or not they form together the complete boundary of a piece of the surface. He considered the maximal number of closed curves which do not completely bound a piece of the surface (and so are in this sense homologically independent), and showed that this number is

928

even, say $2p$, and gives a good classifying invariant for closed (orientable) surfaces.

2 TOPOLOGICAL THEORY OF SURFACES IN THE NINETEENTH CENTURY

In the 1860s and 1870s, knowledge from three sources flowed together to form the outline of a new geometrical subdiscipline, the topological theory of surfaces. One source of this knowledge was geometric function theory, where Riemann's dissection method was developed to a level where it could be used to determine the order of connectivity (or slight variants of it) of Riemann surfaces. Carl Neumann, Jakob Lüroth, Alfred Clebsch, William Kingdon Clifford and others studied the topology of Riemann surfaces, covering the complex-number sphere with n leaves and branching with total order w in a given number of points by the dissection method. Neumann found that in this case Riemann's invariant is $2p = w - 2n + 2$, and others studied more in detail how the covering surface may be constructed from the topological and branching data.

Other sources were the topological theory of polyhedra, originating from René Descartes and Leonhard Euler (§7.3), and the topological investigation of real projective surfaces. Main contributions in the theory of polyhedra came from August Ferdinand Möbius and Camille Jordan in the 1860s (Pont *1974*). In 1858 Möbius discovered the famous Möbius strip (or band), an example of a non-orientable surface (details were published in 1865). And in an unpublished prize paper of 1860 for the Paris Academy of Sciences he discussed oriented surfaces from a point of view which he called 'elementary kinship' (now, topological maps), and classified the surfaces according to the number of boundary components, using a dissection process of his own (governed by an early and special version of what today is called Morse theory). In the result he characterized normal forms of oriented surfaces which were very similar to the attachment of handles to a sphere.

Jordan, on the other hand, not knowing of Möbius's work but well aware of Riemann's results, introduced yet another invariant by counting the maximal number p of cuts along disjoint, boundary-independent cycles without self-intersection that do not separate the surface into different components. He showed that p and the number of boundary components classify the (orientable) surfaces, and remarked that for closed surfaces Riemann's cycle number is $n = 2p$. Moreover in 1866 Jordan introduced more clearly than anybody else before the idea of the homotopy of paths in a surface by deformation into one another, and characterized the homotopy group of a closed orientable surface constructively, although not using explicit group terminology.

Early in the 1870s L. Schläfli and Felix Klein discussed the topological characterization of non-orientable real projective surfaces, starting with the projective plane itself as the most simple case (§7.6). After initial difficulties in giving a correct value for the connectivity number Z of the projective plane, which Schläfli defined such that $Z = 2 - \chi(F)$, in 1875 he gave a consistent value for his version of the connectivity number ($Z = 1$). In 1888, Klein's former student W. Dyck introduced the 'Euler characteristic' of surfaces or higher manifolds constructed as finite cell complexes (see Section 3), and showed that the number of boundary components and the Euler characteristic give sufficient information to classify (orientable or non-orientable) surfaces up to topological equivalence. Thus, by the end of the 1880s a complete topological classification of compact surfaces of all possible types (orientable or non-orientable, bounded or unbounded) had been achieved.

3 FIRST EXPLORATIONS OF THE TOPOLOGY OF HIGHER MANIFOLDS

In his last years in the 1860s, Riemann had discussed mathematical questions, including the topology of manifolds, with the Italian mathematician Enrico Betti. So it was no coincidence that Betti was the first to publish on the topology of n-dimensional manifolds. Most important in this respect was an article of his, published in 1871, which gave a rough, intuitive introduction to 'higher connectivity numbers' (later to be known as Betti numbers) of an n-dimensional manifold. Betti used Riemann's method of the maximal number of closed k-dimensional submanifolds (k-cycles) which do not completely bound a $(k + 1)$-dimensional submanifold. Although Betti's argumentation was very vague, it became the first published heuristic approach to the problem (Scholz *1980*: Chap. 6).

It was not long before higher manifolds appeared in other mathematical subdisciplines, in particular function theory and algebraic geometry (Bollinger *1972*). Early in the 1880s, when Klein and Henri Poincaré created the theory of automorphic functions (§3.16), they tackled the problem of uniformization of multi-valued functions z on a Riemann surface F of genus p with predetermined branching properties in k given points $x_1, x_2, \ldots, x_k$. In more modern terminology, they were looking for the simply connected analytical covering surface of F with given branching properties in $x_1, x_2, \ldots, x_k$. From their point of view it was much more natural, however, to consider the covering map (presupposing its existence) as an automorphic function (Gray *1986*); the existence of such a covering then appeared as the question of whether there are 'sufficiently many' automorphic functions of given type (leading to Riemann surfaces of genus p and given branching behaviour). Klein (in 1882) and Poincaré (in 1884) tried to answer this question by studying all the Riemann surfaces of given type as a 'manifold' M_1, and all the automorphic functions of given type as a 'manifold' M_2 (in fact, M_1 and M_2 are not exactly manifolds, but rather more complicated spaces, the so-called 'module spaces'), and tried to prove that there is a bijective map between the two by using uniformization proof by the continuity method. It was relatively easy for them to show the injectivity and continuity of the map, but the essential question of surjectivity they discussed only vaguely, although it was addressed in more detail by Poincaré than by Klein (Scholz *1980*: Chap. 7).

In fact, Klein and Poincaré had unknowingly entered a difficult area. It seemed obvious to them that an injective continuous function from $\mathbb{R}^n$ to $\mathbb{R}^n$ maps open sets onto open sets, thus keeping the dimension constant. This soon became highly controversial, and was resolved only be L. E. J. Brouwer, who attacked the problem with new methods in 1911 (§7.11) and also discussed more carefully the problem of M_1 and M_2 not being manifolds in the proper sense.

Other aspects of higher manifolds came to the fore when, in 1888, Dyck approached n-dimensional manifolds by generating them from one or several n-cells by cutting along submanifolds and/or pasting together along submanifolds. He considered this as a recursive definition of higher manifolds (which was similar to the inverse process of building up a manifold as a finite cell complex), and introduced the Euler characteristic as the alternating sum of k-cells involved in the construction process $(0 \leqslant k \leqslant n)$. In 1889, Emile Picard started to analyse complex algebraic surfaces (i.e. real 4-dimensional manifolds), dealing particularly with the second Betti number. The situation was too complicated, however, to calculate the second Betti number on the basis of the rudimentary topological knowledge of the time (§7.9).

4 POINCARÉ'S *ANALYSIS SITUS*

4.1 *Analysis situs* and manifolds

The situation started to change with Poincaré, who came across manifolds in different contexts. One has already been mentioned – geometric function theory and automorphic functions. Two more fields in which he realized the importance of the topology of manifolds were his investigations in the qualitative theory of differential equations, which he initiated with a series of four memoirs in the 1880s (§3.14); and complex double integrals over

2-dimensional cycles in algebraic surfaces, which he undertook in the late 1880s. Early in the 1890s he began to study the topology of manifolds in its own right. He published the principles of his new and deep insights into the topology of manifolds in a memoir entitled '*Analysis situs*', in 1895, and went on in a famous series of five 'Complements to *analysis situs*', written between 1899 and 1904.

Poincaré worked with a constructive manifold concept, and used diverse methods to determine an n-dimensional manifold: characterization by $k - n$ equations (and possibly some inequalities) in $\mathbb{R}^k$, local parametrization of a set in $\mathbb{R}^k$ by n parameters, a combined method of parametrized equations, and finally construction as a generalized 'polyhedron' as a finite cell complex given by a list of all the k-dimensional cells (homeomorphic to the k-ball in $\mathbb{R}^k$) and, for $1 \leqslant k \leqslant n$, a set of incidence matrices determining whether a $(k - 1)$-cell coincides with the boundary of a k-cell (entry ± 1, dependent on orientation) or not (entry 0).

4.2 Betti numbers and homotopy

In his memoir of 1895, Poincaré introduced the Betti numbers of manifolds by the method already proposed by Riemann and Betti, but slightly improved. He studied linear combinations of closed submanifolds N_1, $N_2, \ldots, N_m$ of dimension k with integer coefficients (k-cycles of submanifolds). His shift towards algebraization thus led him to calculate with modules of closed submanifolds of dimension k (although he did not, of course, use the module concept explicitly, which at the time was still restricted to number theory). If all the N_i together ($1 \leqslant i \leqslant k$) form the complete boundary of a $(k + 1)$-dimensional submanifold, Poincaré said, then they satisfy a homology, expressing this fact symbolically by the relation

$$N_1 + N_2 + \cdots + N_k \sim 0, \tag{1}$$

which states that the $\{N_i\}$ are homologous to zero. This allowed him to define an equivalence concept of k-cycles

$$N_1 + N_2 + \cdots + N_k \sim N_{k+1} + \cdots + N_m$$
$$\Leftrightarrow k(N_1 + \cdots + N_k - N_{k+1} - \cdots - N_m) \sim 0 \tag{2}$$

for some integer k, which he complemented by the rule to allow 'division by common integer factors'. As a consequence, he also divided out torsion elements ($k(\Sigma \lambda_i N_i) = 0 \Rightarrow \Sigma \lambda_i N_i = 0$). He thus introduced a calculus of boundary homology classes of closed submanifolds of M in the sense of a free module with integer coefficients, and defined the kth Betti number P_k

as the maximal number of homologically independent k-cycles (i.e. the dimension of the free part of the underlying $\mathbb{Z}$-module).

Poincaré's method was soon criticized by the Danish mathematician P. Heegard as conceptually and technically too vague to admit well-determined results. This prompted Poincaré to develop a more manageable method in his 'First complement to *analysis situs*' (1899). He now characterized a manifold M as a finite cell complex given by incidence matrices (see Section 4.1). This approach was well adapted to linear algebraical calculations, as Poincaré now restricted his considerations to cycles of closed k-dimensional submanifolds built up from cells of the given cell division of M (combinatorial k-cycles). After a diagonalization procedure by elementary transformations of all the incidence matrices, he could easily read off the rank of the combinatorial k-cycles, as well as the number of linearly independent homology relations and their coefficients, and thus the kth Betti number (and even the torsion coefficients) in the combinatorial sense. Moreover, he showed that subdivisions of the cells do not influence the value of the Betti numbers. All in all, this new approach led to an ingenious device and enabled mathematicians for the first time to calculate the Betti numbers of rather complicated higher-dimensional examples. Cell complexes also allowed Poincaré to prove a generalization of Euler's polyhedron theorem for n-dimensional compact manifolds. In his previous, vaguer but − at first glance − more general approach, he had derived a duality theorem for orientable manifolds (Poincaré duality),

$$P_i = P_{n-i}, \quad \dim M = n, \quad 1 \leqslant i \leqslant \tfrac{1}{2}n, \tag{3}$$

which he now derived anew by the combinatorial method of 1899.

4.3 Poincaré's conjecture concerning the fundamental group

Also in 1895, Poincaré had introduced the concept of the fundamental group of an n-dimensional manifold, and improved his first vague (and, at times, even wrong) arguments in the 'Fifth complement to *analysis situs*', in 1904. The fundamental group became an important research instrument with which he investigated how well the topology of a manifold is characterized by homology properties. He gave an argument that there exist manifolds with equal Betti numbers greater than zero but different in their fundamental group, which therefore are not topologically equivalent. At first he was convinced that the same cannot happen for simply connected, 3-dimensional manifolds (i.e. $P_1 = P_2 = P_3 = 0$ in equation (3)). When he realized that even this is not the case (as a counter-example he cited the so-called 'spherical dodecahedron space'), he proposed instead as his final conjecture that a 3-dimensional closed manifold with trivial

933

fundamental group is equivalent to the 3-dimensional sphere S^3, whereas a 3-dimensional open manifold with trivial fundamental group is equivalent to the open 3-ball.

This is known as 'the Poincaré conjecture'. It was highly regarded by algebraic topologists well into the twentieth century, comparable in status only to Hilbert's famous problems in other mathematical fields. Its original version is still open, but a generalized version has been formulated and partially solved: is a closed connected manifold of dimension n with trivial fundamental group and trivial homology topologically equivalent to the n-sphere S^n? A positive answer for dimension $n > 4$ was given in the 1960s, with increasing generality of assumptions about the manifold M (differentiable, topological, and finally combinatorial) by S. Smale, M. H. A. Newman, J. R. Stallings and E. C. Zeeman. Recently, the case $n = 4$ has been solved positively by M. H. Freedman, whereas the original conjecture still remains unsolved.

4.4 Two more conjectures

Two other points appeared clear to Poincaré, which later turned out to be rather difficult conjectures, and – at least from the point of view of a more general concept of manifold – not always correct. First, he thought it would always be possible for a (compact differentiable) manifold to be decomposed into a finite cell complex which can be chosen, after a normalization of the cells, as simplexes (k-dimensional triangles/pyramids): the 'triangulation conjecture'. Second, he discussed subdivisions of cell complexes or combinatorial simplicial complexes and argued that, for any two representations of a (compact differentiable) manifold by such complexes, there exists a common subdivision (this, called the *Hauptvermutung* or 'central conjecture' by H. Kneser in 1925, had already been expressed as a conjecture by Heinrich Tietze and Ernst Steinitz in 1908). This would imply the independence of the topological invariants (Betti numbers, torsion coefficients, etc.) from the chosen cell or simplicial decomposition. Both conjectures have been investigated in detail, and have led to some surprising results (see Section 6).

5 SPECIFICATIONS OF THE CONCEPT OF MANIFOLD IN THE TWENTIETH CENTURY

5.1 Differentiable manifolds and differential geometry

With Poincaré manifolds became the object of a mathematical theory of their own: algebraic and differential topology. Thus the global characterization of these objects became more understandable and manageable for potential users of the concept. This enabled differential geometers, for example, to think more in terms of higher-dimensional manifolds than just in terms of coordinate systems and their transformations. In fact, the twentieth century saw an influx of manifold concepts into differential geometry, slowly but surely transforming the tensor calculus of Gregorio Ricci-Curbastro and Tullio Levi-Città (§3.4) from a purely local and coordinate-bound instrument into a characterization of differential-geometric structures on the tangent and cotangent bundles of the underlying manifold and their exterior and tensorial products. A first step into this type of approach was taken by Hermann Weyl, who in 1918 described affine connections, independent of a metric on or an embedding of the manifold. His work was part of the mathematical clarification of the foundations of relativity, which triggered a whole range of investigations into more general differential-geometric structures on manifolds (for more details see Thomas *1938*). It must be added, however, that the concept of the tangent space of a manifold continued until the 1940s to be formulated by differential geometers in the language of differentials and infinitesimal displacements. It seems that Claude Chevalley was the first to introduce tangent spaces of a manifold by the modern construction of appropriate bundles of linear spaces, in his famous book on Lie groups (1946).

5.2 Combinatorial manifolds

At that time the fundamentals of the concept of a differentiable manifold started to be understood quite well. That cannot be said for the first decades of the twentieth century. Poincaré's early successors became (rightly) suspicious of the master's easily pronounced but never proved identification between the different modes of defining a manifold. Most general among his characterizations, and most promising from the point of view of algebraic topology, was that of a cell (or simplicial) complex which satisfies certain conditions such that the local property of manifolds (to be topologically equivalent to a ball in $\mathbb{R}^n$) is secured. This was the approach taken by Brouwer, P. Heegard and M. Dehn, and most of the other topologists early in the century, although it has to be added that algebraic topology quickly liberated itself from the restriction of working only with manifolds. From the combinatorial point of view, cell or simplicial complexes without the local restriction of manifolds seemed at least as good.

5.3 Axiomatic definition of manifolds

In his book on Riemann surfaces (1913), Weyl had taken up an idea of

Hilbert's and introduced a (real) two-dimensional manifold as a point set F endowed with a collection of (topologically consistent) coordinate systems which define neighbourhood relations and thus a topology on F. Strangely enough, it was not until 1932 that manifolds in general were similarly defined, by Oswald Veblen and J. H. C. Whitehead, as point sets with (maximal) atlases of coordinate systems and continuous local transformation of coordinates (respectively differentiable transformations of any order). That was an axiomatization of the manifold concept, soon accepted in topology and differential geometry. It included a precise (even if at the time nearly only nominal) distinction between C^0-manifolds (continuous, not necessarily differentiable transformations), C^1-, C^2-, ..., C^∞-manifolds (once, twice, ..., infinitely often differentiable coordinate transformations).

6 SOME FUNDAMENTAL INSIGHTS OF THE TWENTIETH CENTURY INTO CONTINUOUS AND DIFFERENTIABLE MANIFOLDS

Given the abstract definition of topological and differential manifolds, the question of its relationship to the combinatorial approach and/or the description by submanifolds of a sufficiently high-dimensional $\mathbb{R}^m$ became even more important than before. American mathematicians were prominent in finding solutions to several of the basic questions (§11.10). In 1936 H. Whitney had essentially solved the embedding problem for differential C^r-manifolds by showing that, for any manifold of dimension n, there are C^r-diffeomorphisms onto a submanifold of $\mathbb{R}^{2n+1}$ (even $\mathbb{R}^{2n}$, as he showed later, in 1944). Also, between 1932 and 1935, S. Cairns had shown that triangulation is always possible for compact differentiable (C^1-) manifolds. As nearly all the known continuous manifolds can be endowed with a differentiable structure, Cairns's answer to the existence question was nearly exhaustive. Most recent results in the topology of 4-manifolds, however, seem to indicate that there may exist continuous manifolds without any differentiable atlas.

Much more difficult, and in the event surprising, was the answer to the question of the unicity of the combinatorial structure of manifolds. A positive answer to the *Hauptvermutung* was generally accepted for dimension $n = 2$ (although definitively proved perhaps only in the 1960s) and extended in 1952 to $n = 3$ by E. Moise. But all expectations were broken when J. Milnor found counter-examples of dimension 8 in 1963, and results obtained by R. D. Edwards and J. W. Cannon from 1979 led easily to counter-examples for dimensions $n \geqslant 5$, derived from Poincaré's dodecahedron space. So, at the moment the *Hauptvermutung* seems to be open only for $n = 4$, whereas it has turned out to be true for $n \leqslant 3$ and wrong, if taken in full generality, for $n \geqslant 5$ (Henn and Puppe *1990*).

Other very surprising insights about the topological and differentiable structures of manifolds arose from Milnor's observation of 1956 that on the sphere S^7 there exist different C^∞-structures (i.e. non-diffeomorphic differential atlases of class C^∞). These structures are not even particularly artificial, as was shown by E. Brieskorn's construction in 1966 of neighbourhood boundaries to singularities of classical hypersurfaces in algebraic geometry. M. Kervaire and J. Milnor showed that similar phenomena arise for all higher-dimensional spheres, and moreover that the differential structures of S^n, for $n > 3$, form a finite group.

7 CLOSING REMARKS

Whereas in the nineteenth century early algebraic topology was the theoretical medium in which manifolds came to be considered in their own right, the twentieth century saw a highly technical turn of algebraic topology combined with an increasing extension of the (topological) spaces it dealt with (Dieudonné *1989*). Thus the rising discipline of differential topology more or less substituted for algebraic topology as the mathematical vehicle for manifolds. This does not imply that manifolds became central in differential topology only; in fact, quite the opposite happened. The diversification of modern geometrical theories was accompanied by a similar diversification of the manifold concept. As a proper expression of this development, genuine algebraic concepts of (algebraic) varieties were formed between 1930 and 1950, and similarly in complex analysis the concepts of complex manifolds were shaped. These are developments in their own right which cannot be covered in this encyclopedia. Even the description of differential topology here has been restricted to the most fundamental insights into manifolds and their elementary structures. For more information see Dieudonné *1989* and Henn and Puppe *1990*.

BIBLIOGRAPHY

Bollinger, M. *1972*, 'Geschichtliche Entwicklung des Homologiebegriffs', *Archive for History of Exact Sciences*, 9, 94–166.

Dieudonné, J. *1974*, *Cours de géometrie algébrique*, Vol. 1, Paris: Presses Universitaires de France.

—— *1989*, *A History of Algebraic and Differential Topology, 1900–1960*, Boston, MA and Basel: Birkhäuser.

Gray, J. J. *1986*, *Linear Differential Equations and Group Theory from Riemann to Poincaré*, Basel: Birkhäuser.

Henn, H. and Puppe, D. *1990*, 'Algebraische Topologie', in G. Fischer *et al.* (eds), *Ein Jahrhundert Mathematik, 1890–1990. Festschrift zum Jubiläum der DMV*, Braunschweig: Vieweg, 674–716.

Pont, J.-C. *1974*, *La Topologie algébrique des origines à Poincaré*, Paris: Presses Universitaires de France.

Scholz, E. *1980*, *Geschichte des Mannigfaltigkeitsbegriffs von Riemann bis Poincaré*, Basel: Birkhäuser.

Thomas, T. Y. *1938*, 'Recent trends in geometry', in *Semicentennial Addresses of the American Mathematical Society*, Vol. 2, 98–135.

7.11

Topology: Invariance of dimension

JOSEPH W. DAUBEN

Although the subject of topological dimension theory was first raised seriously by Bernhard Riemann and Georg Cantor in the nineteenth century, it was only satisfactorily resolved in the early decades of the twentieth century through the efforts of Henri Poincaré, L. E. J. Brouwer, Pavel Urysohn and Karl Menger, who brought the subject to a satisfactory conclusion on rigorous terms by about 1930. Nevertheless, the first to appreciate the importance of dimension conceptually was the Czech mathematician-philosopher Bernard Bolzano, who (largely for philosophical reasons) was the first to explore seriously the problem in geometry of defining the differences between lines, surfaces, solids and continua in general.

1 BERNARD BOLZANO'S ANTICIPATIONS

In 1817 Bolzano published a pamphlet, *Die drey Probleme der Rectification, der Complanation und der Cubirung*, in which his first 'quasi-topological' definitions were given for lines, surfaces and solids in terms of set-theoretic, dimensional properties. Later, in the 1830s and 1840s, he made further refinements in his approach to dimension, all of which remained unpublished until this century, except for some brief comments on the subject which appeared in his (posthumous) *Paradoxien des Unendlichen*, published in 1851. Among many of Bolzano's insights, including distinctions between continua and discontinua, was his recognition of the need to prove what is today known as the Jordan curve theorem: that a simple closed curve divides the plane into two parts.

In the 1840s Bolzano also studied the problem of skew (non-planar) curves, which led him to consider a number of neighbourhood properties for both curves and surfaces. In still later work, he used 'distance' and 'neighbourhood' to distinguish continua from discontinua: a continuum or extension had no isolated points, whereas a discontinuum consisted only of isolated points (Bolzano *1948*: 145–9).

Unfortunately, few of these ideas were known to Bolzano's contemporaries. Most of his mathematical insights lay buried, incompletely described in his philosophical publications (like the *Wissenschaftslehre* of 1837, or *Paradoxien des Unendlichen*). Nevertheless, his interests were visionary in so far as he recognized that there was a significant mathematical problem in adequately defining lines, curves and solids in terms that required consideration of dimensional concepts (Johnson *1977*).

2 RIEMANN

One of the nineteenth-century milestones in the history of geometry was Riemann's *Habilitationsvortrag*, 'Über die Hypothesen, welche der Geometrie zu Grunde liegen' (written 1854, first published 1868). In this work he took up a number of essential subjects, including non-Euclidean geometry (§7.4) and the problem of defining n-dimensional manifolds topologically. In doing so he was led to introduce an informal theory of topological manifolds and dimension.

Only a few years earlier, Riemann's *Dissertation* (1851), 'Grundlagen für eine allgemeine Theorie der Functionen einer veränderlichen complexen Grösse', devoted to complex-function theory (§3.12), had already included a discussion of the topology of surfaces in order to handle mutli-valued analytic functions in terms of Riemann surfaces. This alone made clear the importance of treating the subject of manifolds topologically. In four papers on Abelian functions (§4.6) published as *1857*, he adopted the modern view that topology is concerned with the theory of continuous quantities, not with respect to their 'measure' relations (*Massverhältnissen*), but solely with respect to their local and regional properties.

3 CANTOR

Despite the important topological insights into dimension apparent in Riemann's papers, it was the remarkable discovery that dimension was not simply a matter of the number of coordinates needed to define a given n-dimensional manifold that opened a new chapter in the topological study of dimension. This startling find was due to a logical extension of questions raised by Georg Cantor's revolutionary discovery, in 1873, that the set of real numbers was non-denumerably infinite – of a greater order of infinity than the denumerably infinite set of integers (§3.6). Consequently, Cantor wondered if point sets of higher dimensions might also display increasingly higher degrees of non-denumerable infinities. Was the plane infinitely richer in points than the line, solids still richer in points than planes?

To his dismay, Cantor found that the points of any n-dimensional space could in fact be mapped in a 1–1 fashion onto a 1-dimensional line by means of a simple correspondence. For example, any point (x, y) in the plane can be corresponded to the single point p on the real line in terms of the infinite decimal expansions for x and y. If $x = x_1 x_2 x_3 \ldots$ and $y = y_1 y_2 y_3 \ldots$, p could be determined by $p = x_1 y_1 x_2 y_2 x_3 y_3 \ldots$. To avoid a difficulty that Richard Dedekind called to Cantor's attention (namely that there is a problem with the uniqueness of such decimal expansions, since $0.2000 \ldots$ and $0.1999 \ldots$ are equivalent, for example), Julius König found that if blocks of numbers were used whereby zeros were taken together with the first non-zero number following any group of zeros, Cantor's simple method was no longer susceptible to Dedekind's objection. In this case the point (x, y), if $x = 0.590\,600\,78 \ldots$ and $y = 0.200\,540\,5 \ldots$, would correspond to the point $p = 0.529\,005\,064\,007\,057 \ldots$. A similar rule governed untangling a point, say $p = 0.600\,705\,600\,074 \ldots$, to find its corresponding point (x, y) in the plane. In this case, p would correspond to the point (x, y) with $x = 0.605\,000\,7 \ldots$ and $y = 0.007\,64 \ldots$.

Cantor was so unprepared for his discovery of the 1–1 correspondence between lines and planes that it prompted him to exclaim, in a letter to his friend Dedekind, 'I see it, but I don't believe it!' (29 June 1877, in Cantor *1937*: 34). What was especially important about Cantor's result, however, was its explicit linkage of mappings and correspondences to the question of the dimensions of figures and spaces. This proved to be a direct challenge to the foundations of geometry, especially to the concept of dimension which geometers until then had been using intuitively but uncritically. After working out a number of improvements (and corrections needed to answer various criticisms pointed out by Dedekind), Cantor published his discovery in the paper *1878*.

Cantor believed that his discovery required a new look at the problem of dimension, that it was not simply a question of the number n of 'coordinates' needed to determine a point in n-dimensional space. Dedekind, however, was certain that 'the dimension number of a continuous manifold now as before is the first and most important invariant' (Cantor *1937*: 37). As he pointed out, it was significant that Cantor's correspondence was not continuous, and this fact eventually led to a variety of proofs of the invariance of dimension.

4 EARLY 'PROOFS' OF THE INVARIANCE OF DIMENSION

Over the next few years a number of attempts were made to establish the invariance of dimension using the condition of continuity with varying degrees of success. Jakob Lüroth gave several proofs limited to cases of dimension 3 or less, while Enno Jürgens presented a proof for the 2-dimensional case. Johannes Thomae attempted a general proof of the invariance of dimension that was soon found lacking, and in 1878 Eugen Netto also gave a general proof by induction, but this too failed to be completely satisfactory (Dauben *1975*, Johnson *1979*).

Meanwhile, Cantor was dissatisfied with all these attempts, and published a proof of his own for the invariance of dimension as *1879*. Like Netto's, his proof depended on an argument by induction. Cantor's proof remained the only generally accepted proof until 1899, when Jürgens found a counter-example. By then, many of the essential details of point set topology had been worked out, and many of the topological tools previously unavailable but necessary for a satisfactory resolution of the topological problem of dimension had been perfected (largely by Cantor).

5 PEANO AND POINCARÉ

In 1890 Giuseppe Peano provided an analytic description of a space-filling curve in an article entitled 'Sur une courbe, qui remplit toute une aire plane'. Peano's curve, which covers all the points of a square, again upset all previous geometric intuition about dimension. Shortly thereafter, David Hilbert conceived of the better-known version of Peano's idea, whereby a space-filling curve was envisaged as the limit of a sequence of polygonal paths (see §3.8, Figure 4).

Although Peano's use of ternary representations of points made his mapping continuous, it was not one-to-one. Nevertheless, Peano's curve soon led mathematicians to propose alternative definitions for curves in order to exclude the disturbingly non-intuitive case of his 1-dimensional curve that filled the 2-dimensional plane.

At the turn of the century, one of the mathematicians concerned with geometry and aspects of dimension was Poincaré, who said (in 1901), 'Every problem I attacked led me to *analysis situs*' (topology). He took two approaches to defining dimension, one algebraic and based on displacement groups, the second topological and based on 'cuts' (*coupure*; 1903). Even so, he never took up the invariance of dimension itself. Others at the time who did show an interest in the invariance problem included René Baire, Maurice Fréchet and Frigyes Riesz.

6 BROUWER

It was Brouwer who gave the first rigorous proof of the invariance of dimension, which he actually published in two versions (the first in 1911, the second in 1913). Because much of his early work on finite continuous groups depended on results obtained by Arthur Schönflies on the topology of the plane, when Brouwer discovered serious shortcomings in Schönflies's work, he was forced to undertake his own analysis of point set topology (see Brouwer's letter to Hilbert of 14 May 1909, in Johnson *1981*: 129).

This soon resulted in his important paper *1910*, 'Zur *analysis situs*'. In addition to a variety of counter-examples discrediting many of Schönflies's conclusions, Brouwer's article also gave examples of the first decomposable continuum. Above all, what Brouwer did was to construct point sets with properties that, according to Schönflies's theory, should have been impossible. Several especially clever examples were even published in two colours, red and black, to make clear the way in which rather complicated curves could be constructed to disprove a number of Schönflies's results. For example, he illustrated nowhere dense, connected sets dividing the plane into separate areas, each of which has the same boundary (Figure 1). As he noted, with suitable modifications of the figure it was even possible to show that the plane could be divided into infinitely many sub-domains, all with a common boundary. Moreover, in addition to the theorems of Schönflies that were wrong, Brouwer found that others were open to question, among them theorems directly related to the problem of dimension.

It was in trying to deal with the theory of functions of a complex variable and their Riemann surfaces that Brouwer was led to introduce the concept of the degree of a mapping, which immediately led him to the theory of the invariance of dimension, resulting in his first rigorous proof. Through his work on vector fields and mappings of spheres, Brouwer came to define the degree of a continuous mapping, which became one of his principal weapons for attacking topological problems. Henri Lebesgue, at virtually the same time, also considered the problem of the invariance of dimension, and devised a proof using a 'tiling principle', motivated by the example of Peano's space-filling curve and the attendant problem this raised for the concept of dimension.

Otto Blumenthal, then managing editor of the *Mathematische Annalen*, published both Brouwer's and Lebesgue's proofs simultaneously in the issue for February 1911. Despite the controversy which ensued between Lebesgue and Brouwer over how best to approach the problem of dimension, Lebesgue's early attempts were less than totally satisfactory, whereas Brouwer's proof was rigorous if difficult. His continued consideration of the dimension problem soon led to a second proof, by induction, published

Brouwer, Zur Analysis Situs.

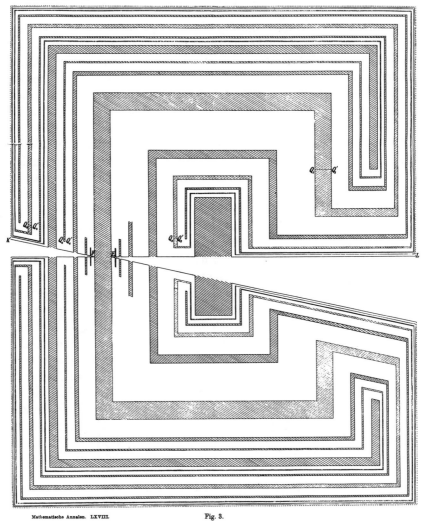

Mathematische Annalen. LXVIII.   Fig. 3.

Figure 1 Nowhere dense, connected sets dividing the plane into separate areas, each of which has the same boundary. In this diagram as it was originally printed, the central rectangle was black, the paths marked $Q_n Q'_n$ were indicated in red, and the rest was cross-hatched in black
(From Brouwer *1910*: figure 3)

in 1913, this one based on vector distributions over the n-dimensional cube, and simplexes ('Über den natürlichen Dimensionsbegriff').

7 URYSOHN AND MENGER

The task of drawing further consequences from Brouwer's work, especially from his definition of dimension, fell to Urysohn, Menger, P. S. Aleksandrov, W. Hurewicz and L. A. Tumarkin, among others, who between them served to outline the essence of a formal theory of dimension in the 1920s (Johnson *1981*).

Urysohn studied with N. N. Luzin and D. Egorov in Moscow. In 1921–2 he worked out the basic details of a topological theory of continua, which he presented in his course on topology at Moscow University, although it was published only after his tragic, premature death in 1924. Above all, Urysohn succeeded in defining curves and surfaces, as well as n-dimensional manifolds, insisting that all his definitions be intrinsic, limited only to the internal properties of the sets in question and not considering their situation in larger spaces. Thus he sought local rather than global definitions. He also located his theory within the framework of compact metric spaces, having found that compactness was crucial to most of his arguments. Urysohn often collaborated with Aleksandrov. In their work, the abstract theories of both metric and topological spaces were extended and deepened.

Menger, at about the same time, was a student at the University of Vienna, where he took part in a seminar offered by Hans Hahn on the problem of curves. Menger soon began to develop his own ideas on the subject, emphasizing that 1-dimensionality was the most important property of a curve. A hallmark of Meyer's approach was his exploration of concepts of neighbourhood and boundary. He went on to define 1-dimensional and n-dimensional sets recursively from his definitions of zero-dimensional sets. Soon he reached a definition of dimension that was to provide the basis for future topological research with respect to dimension: that n-dimensional spaces in Euclidean n-space are those sets which contain interior points. Over the years he proceeded to prove various theorems and draw further conclusions from his definition of the dimension concept.

In his major work, 'Über die Dimension von Punktmengen' (Part II), published in 1924, Menger set out his major results on the topological invariance of dimension, the structure of n-dimensional sets and the dimension of sets in Euclidean n-space. He also proved (among other results) his dimensional-structure theorem and the sum theorem.

After the initial publications of Urysohn and Menger, topological dimension theory grew rapidly thanks to continued work by Menger,

Aleksandrov, Hurewicz and Tumarkin. Hurewicz and Tumarkin independently extended earlier results from compact metric spaces to separable metric spaces. Meyer, in fact, published the first comprehensive textbook on the subject, *Dimensionstheorie* (1928). Thereafter, many mathematicians continued to add further refinements, and in 1941 Hurewicz and Wallman published their book *Dimension theory*, which brought the subject of separable metric spaces to a high level of completion.

BIBLIOGRAPHY

Aleksandrov, P. S. *1955*, 'The present status of the theory of dimension', *American Mathematical Society Translations*, **1**, 1–26.

Bolzano, B. *1948*, 'Über Haltung, Richtung, Krümmung und Schnörkelung bei Linien sowohl als Flächen sammt einigen verwandten Begriffen', manuscript, in J. Vojtech (ed.), *Spisy Bernarda Bolzana*, Vol. 5, Prague: Academy, 139–83.

Brouwer, L. E. J. *1910*, 'Zur Analysis Situs', *Mathematische Annalen*, **68**, 422–34. [Also in *Collected Works*, Vol. 2, 352–66.]

Cantor, G. *1878*, 'Ein Beitrag zur Mannigfaltigkeitslehre', *Journal für die reine und angewandte Mathematik*, **84**, 242–58. [Also in Cantor *1932*: 119–33.]

—— *1879*, 'Über einen Satz aus der Theorie der stetigen Mannigfaltigkeiten', *Nachrichten von der Königlichen Gesellschaft der Wissenschaften und der Georg-Augusts-Universität zu Göttingen*, 127–35. [Also in Cantor *1932*: 134–8.]

—— *1932*, *Gesammelte Abhandlungen mathematischen und philosophischen Inhalts* (ed. E. Zermelo), Berlin: Springer. [Repr. 1962, Hildesheim: Olms; and 1980, Berlin: Springer.]

—— *1937*, *Briefwechsel Cantor-Dedekind* (ed. E. Noether and J. Cavaillès), Paris: Hermann.

Dauben, J. W. *1975*, 'The invariance of dimension: Problems in the early development of set theory and topology', *Historia mathematica*, **2**, 273–88.

Johnson, D. M. *1977*, 'Prelude to dimension theory: The geometrical investigations of Bernard Bolzano', *Archive for History of Exact Sciences*, **17**, 261–95.

—— *1979*, *1981*, 'The problem of the invariance of dimension in the growth of modern topology', *Archive for History of Exact Sciences*, **20**, 97–188; **25**, 85–267.

Riemann, G. F. B. *1857*, 'Theorie der Abel'schen Functionen', *Journal für die reine und angewandte Mathematik*, **54**, 115–55. [Also in *Gesammelte mathematische Werke*, 2nd edn, 1892, 88–144.]

7.12

Finite-dimensional vector spaces

J. J. GRAY

1 SOURCES FOR THE CONCEPT OF A VECTOR SPACE

Vector spaces, like prose, were around long before the concept became explicit; they have an extensive prehistory, while their history lies in the shadow of the much more complicated idea of infinite-dimensional vector spaces. There were many occasions during the nineteenth century when mathematicians dealt with objects that could be added together and multiplied by numbers to yield other objects of the same kind. Since these simple properties form the intuitive core of the idea of a vector space, vector spaces can be found in many mathematical areas, and several basic theorems concerning them were developed in various specific domains. One often finds the idea that in such a situation the objects can all be expressed in a unique way as a sum of a basic set of them. This was recognized, for example, by Leonhard Euler in 1743 in his account of the solutions to a linear ordinary differential equation with constant coefficients, and then taken up by Joseph Louis Lagrange when he began the study of the variable-coefficient case. The culmination of this line of enquiry was L. I. Fuchs's study in 1865 of the nth-order linear ordinary differential equation with complex functions as coefficients (Gray *1986*).

Throughout the nineteenth century other investigations gave rise to vector spaces: complex numbers were presented by William Rowan Hamilton, lacking only the term itself, as a two-dimensional vector space over the reals, and the ensuing search for quaternions and hypercomplex numbers is naturally full of such objects. In the 1880s Leopold Kronecker, in his study of Galois theory, stressed the importance of what he called a *Rationalitätsbereich* ('domain of rationality'), made up of all finite linear combinations of a family of algebraic numbers taken with rational coefficients. This gave him a finite vector space over the rationals, and indeed a field (§6.4). The idea of field extensions occurred at the same time to

947

Richard Dedekind and Heinrich Weber in their study of complex functions, where the extension (called a *Modul*) is usually over the field of rational functions, and so is itself a field and a vector space.

The study of geometry is particularly rich, not just because the plane and 3-space are almost the paradigm examples themselves. For example, if $S = 0$ and $S' = 0$ are the equations of two conics, then the conics in the family of conics through their four points of intersection all have equations of the form $kS + k'S' = 0$. Similar statements can be made about curves of higher degree satisfying certain constraints, and the theory of these so-called linear systems of curves became and remains of considerable importance in algebraic geometry (§7.9). Strictly speaking, these systems form projective spaces, but they illustrate the ubiquity of the idea of forming linear combinations of basic objects.

Without doubt, geometry and mechanics are the main areas in which the concept of a vector space was forged, for in both subjects the idea of a directed magnitude or vector is naturally to hand. In plane geometry, one might say, of a triangle ABC, that the line segment AB is added to the line segment BC to give the line segment AC, and one may regard the line segments AB and BA as equal but opposite. In mechanics, the forces represented by the line segments AB and BC combine to give the force represented by AC. The idea of force in its modern sense had entered physics with Isaac Newton's *Principia* (*1687*), and investigations of the parallelogram law for their addition had occupied many subsequent authors. Less obviously, one may represent a couple (a pair of forces that would rotate a body about a fixed point) by a directed line segment in space, and find that couples combine vectorially. This was the discovery by Louis Poinsot in France at the start of the nineteenth century, and it led to the theory of graphical statics (§8.2). With the discovery of other vectorial quantities in physics, notably in electromagnetism, the stage was gradually set for the introduction of a simplifying notation.

2 THE FORMAL STUDY OF VECTORS AND VECTOR SPACES

The idea that directed line segments can profitably be represented and studied algebraically was slow to catch on. It was put forward in August Ferdinand Möbius's book on statics *1837*, where the idea of resolving a vectorial quantity along arbitrarily chosen axes was also clearly outlined. It was also part of the elaborate but difficult treatise of Hermann Grassmann, his *Ausdehnungslehre* (*1844*); but the treatment that caught on was Hamilton's (1843), in which spatial vectors appeared as the imaginary part of a quaternion (§6.2). Indeed, as Crowe *1985* has described, much of the history of vectors in the nineteenth century is the story of their historical background in quaternions. Quaternions were energetically advocated by Hamilton and Peter Tait, and employed by James Clerk Maxwell in his *Treatise on Electricity and Magnetism* (1873), but in a way that highlighted the utility of the vector part. Finally, Josiah Willard Gibbs successfully separated the vector part from the full quaternion and showed how elegantly vectors could be used in physics. In so doing he took support from his reading of Grassmann, whose work was only then becoming appreciated, and so brought together the thinking of mathematicians and physicists.

In common with all algebraic structures in mathematics, vector spaces became axiomatized about a century ago. The first to describe a vector space this way was Giuseppe Peano, in his study of differential equations. From this perspective, a vector space is a set of objects (the vectors) which themselves form an Abelian group (§6.4), together with a field (the field of scalars) such that, if $\mathbf{u}$ and $\mathbf{v}$ are vectors and a is a scalar, then $a(\mathbf{u} + \mathbf{v}) = a\mathbf{u} + a\mathbf{v}$. Peano, who had just read Grassmann, had a lifelong interest in freeing mathematics from any illusory dependence on intuition, and this was his motivation here. In the abstract setting, the essential theorem is that a vector space has a unique finite dimension, n say. This means that there is a set of n independent vectors forming a basis (a set such that any vector is a sum of vectors in the basis); indeed, any n independent vectors will do, and conversely any basis consists of the same number n of independent vectors. (There are infinite-dimensional vector spaces, but the appropriate concept of basis is harder to define for them.) The fact that any set of n independent vectors forms a basis enables one to change basis, a technique that simplifies the mathematics and forms part of the subject of the theory of matrices. It also enables one to describe vectors in a coordinate-free way, thus making them even easier to use.

Several familiar objects were recognized to be vector spaces, once the concept became available. For example, the complex numbers are a two-dimensional vector space over the reals, and the quaternions are a four-dimensional vector space over the reals.

3 THE IMPORTANCE OF VECTOR SPACES

The importance of vector spaces derives not from the power of their theory, which is elementary, but in the conceptual simplicity they bestow on everything they touch. The widespread belief is that linear problems are easy. This philosophy derives its appeal from its success in the calculus and the application of the calculus to mechanics. One of the unsung but significant achievements of Newton and Gottfried Wilhelm Leibniz in their invention of the calculus (§3.2) was their perception that a tangent (to a given curve at a given point) can be represented in a linear way, using the derivative of the function. Previously, a tangent had been expressed via its so-called sub-tangent (see Figure 1), which measures the length of the segment of the x-axis underneath the tangent. However, if a curve is obtained from two others by addition, the subtangent of the new curve is not easy to find, whereas it is easy to find the slope of the tangent. Let the curves be given by the equations $y = f_1(x)$ and $y = f_2(x)$. The derivative of the new curve is found additively:

$$f_1'(a) + f_2'(a) = (f_1 + f_2)'(a). \tag{1}$$

It is because differentiation is additive that finding derivatives, and hence tangents, is an almost automatic process in the calculus, which accounts for much of its power.

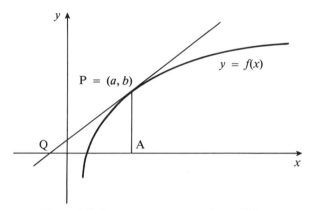

Figure 1 Subtangent to a curve: $QA = b/f'(a)$

Linear problems are simple in another important way, alluded to above. For a linear ordinary differential equation, the sum of two solutions is again a solution, and every solution is a sum of a finite number of basic ones. The elementary differential equation

$$d^2y/dt^2 = -k^2y \tag{2}$$

illustrates this. Two solutions are $y = \cos kt$ and $y = \sin kt$, every sum of these ($a \cos kt + b \sin kt$) is a solution, and every solution is a sum of these. This observation not only simplifies the mathematics; it is at the heart of any explanation of why, in some physical settings, two modes of behaviour can be added to yield a third, a fact of crucial importance in the theory of sound.

A guitar string can emit several notes at once, each oscillating independently and combining to give the resultant sound. The differential equation that describes the vibrations of a string is a linear partial differential equation called the wave equation (§3.15). Being linear, the sum of two solutions is again a solution, which is the basis of the mathematical explanation of the behaviour of the string. In more technical terms, the technique of separation of variables leads to the ordinary differential equation just studied, where (in suitable units) the constant k is an integer multiple of the length of the string. So solutions of the wave equation contain factors of the form $a \cos kt$ and $b \sin kt$ for integers k, thus once again explaining the fact about the string.

As this example indicates, the study of vector spaces generalizes naturally to the infinite-dimensional case. But infinite sums of vectors require a discussion of their convergence if they are to make numerical sense, and so involve questions of topology. For that reason they are not discussed here; see the articles on functional analysis, integral equations and harmonic analysis (§3.9–3.11).

Problems which are not linear are usually much harder to deal with precisely because the sum of two solutions is not a solution. But so powerful are linear methods, and so few and so specialized are the non-linear ones so far known, that passing to a linear simplification is often still not only a heuristic first step, but the only way known to generate solutions mathematically.

BIBLIOGRAPHY

Crowe, M. J. *1985, A History of Vector Analysis*, 2nd edn, New York: Dover.
Grassmann, H. G. *1844, Die lineale Ausdehnungslehre*, Leipzig: Wigand.
Gray, J. J. *1980*, 'The history of the concept of a finite-dimensional vector space', *Historia mathematica*, 7, 65–70.
——— *1986, Linear Differential Equations and Group Theory from Riemann to Poincaré*, Boston, MA and Basel: Birkhäuser.
Möbius, A. F. *1837, Lehrbuch der Statik*, 2 vols, Leipzig: Göschen.
Newton, I. *1687, Philosophiae naturalis principia mathematica*, 1st edn, London: Streater. [English transl. by A. Motte, 1729; rev. F. Cajori, 1934, as *The Mathematical Principles of Natural Philosophy* [. . .], 2 vols, Berkeley, CA: University of California Press.]

7.13

Combinatorics

ROBIN J. WILSON AND

E. KEITH LLOYD

In his *Dissertatio de arte combinatoria* (1666), Gottfried Wilhelm Leibniz described combinatorics as the study of placing, ordering and choosing a number of objects. Several books on the subject were published in the late eighteenth and early nineteenth centuries, including some in Latin and some in German by the school based around K. F. Hindenburg in Leipzig. In his *Essays on the Combinatorial Analysis* (1818), the Scottish architect and mathematician Peter Nicholson defined the subject as 'a branch of mathematics which teaches us to ascertain and exhibit all the possible ways in which a given number of things may be associated and mixed together'; later this definition was cited in various editions of the *Encyclopaedia Britannica*. By the end of the nineteenth century, designs were also included under the heading of combinatorics (see, for example, Netto *1901*, which includes good historical material). More recently the subject has further broadened in scope, and nowadays the term 'combinatorics' is often used to embrace the whole of finite or discrete mathematics.

Much of the material in Sections 1 and 2 is taken from Biggs *1979*; a fuller account of material in later sections may be found in Biggs *et al. 1994*. A more detailed account of the history of graph theory appears in Biggs *et al. 1976*, which includes extracts from several works cited here; in Sections 5 and 6, (Biggs *et al. 1976*: *n*X) refers to extract X of Chapter *n*.

1 COMBINATORICS IN ANTIQUITY

The basic rules of counting have been taken for granted since the early civilizations, and their application has been emphasized by seemingly nonsensical examples possessing the elusive property of memorability. For example, Problem 79 of the Egyptian Rhind Papyrus (*circa* 1650 BC), pictured in Figure 7 of §1.2, can be interpreted as follows:

> In each of seven houses are seven cats, each cat kills seven mice, each mouse would have eaten seven ears of *spelt*, each ear of *spelt* will produce seven *hekat* of grain; how many items altogether?

The scribe has drawn some hieroglyphs which may be roughly translated as: houses 7, cats 49, mice 343, spelt 2401, hekat 16807, with a total of 19607. A related problem occurs in Leonardo of Pisa's *Liber abaci* (1202). Such problems are discussed by Biggs *1979*, who compares them with the well-known nursery rhyme 'As I was going to St Ives, I met a man with seven wives ... '.

Another type of problem which has a long history is that of constructing magic squares. The earliest recorded example is the Chinese *luoshu*, pictured in Figure 1, which dates from the first century BC or earlier. Later appearances were in Baghdād, where Arab and Islamic scholars took a great interest in them − in particular, a group of scholars around AD 990 called 'the brethren of purity'. In their *Encyclopaedia*, they constructed magic squares of sides 3, 4, 5 and 6, and elsewhere of squares of certain higher values. Later, Aḥmed al-Būnī (died 1225) showed how to construct magic squares using a simple bordering technique, and Moschopoulos (*circa* 1315) provided the link between early work on magic squares and later European treatments of the subject.

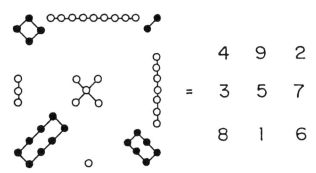

Figure 1 The Chinese *luoshu* magic square

During the thirteenth century there was some exchange of knowledge between scholars of China and Islam. Iron tablets bearing 6 × 6 magic squares inscribed in East Arabic numerals have been unearthed in China (Figure 2). Yang Hui (1275) wrote an account of magic squares of sides 3 to 10, and presented a 9 × 9 magic square constructed from nine 3 × 3

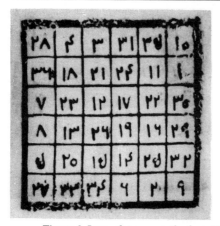

28	4	3	31	35	10
36	18	21	24	11	1
7	23	12	17	22	30
8	13	26	19	16	29
5	20	15	14	25	32
27	33	34	6	2	9

Figure 2 Iron plate unearthed at Xi'an, bearing a magic square in East Arabic numerals (from Li and Du *1987*: 172; with the permission of Oxford University Press)

luoshu-type magic squares. Subsequently the scene moved to Europe, where many methods of construction were proposed. In the seventeenth century the French mathematician B. Frénicle de Bessy found all 880 squares of order 4, and S. de la Loubère described a simple method of construction learnt from the people of Siam (Rouse Ball *1892*).

2 PERMUTATIONS AND COMBINATIONS

Combinatorics originated in China and India. Possibly the earliest example of combinations with repetitions occurs in the Chinese *Yijing* (Book of changes, *circa* seventh century BC), where the two symbols for the *yang* (—) and the *yin* (– –) were combined into hexagrams (strings of six symbols) in $2^6 = 64$ ways. Replacing — by 0 and – – by 1 gives the binary representations of the numbers 0 to 63, although there is no evidence that the Chinese thought of hexagrams in this way; nevertheless, Leibniz later credited the Chinese with inventing the binary number system.

The Hindu contribution included the medical treatise by Suśruta (possibly as old as the sixth century BC) which contains a systematic list of combinations of tastes obtained from six basic qualities – sweet, acid, saline, pungent, bitter and astringent: namely, 6 taken separately, 15 taken two at a time, 20 in threes, and so on. In modern notation, the results in Suśruta can be written as

$$\binom{6}{1} = 6, \quad \binom{6}{2} = 15, \quad \binom{6}{3} = 20, \quad \binom{6}{4} = 15, \quad \binom{6}{5} = 6, \quad \binom{6}{6} = 1. \quad (1)$$

Other examples are found in the work of Jáinas (second century BC) in India, involving combinations of philosophical categories, of sense, and of males, females and eunuchs. A later example occurs in the sixth-century *Bṛhatsaṃhitā* by Varāhamihira, where it clearly states that there are 1820 ways of choosing 4 ingredients out of 16 to make perfumes. In modern notation,

$$\binom{16}{4} = 1820. \tag{2}$$

It is unlikely that Varāhamihira listed all the cases, but he may have used the formula

$$\binom{n}{r} = \frac{n(n-1)\cdots(n-r+1)}{r(r-1)\cdots 1}. \tag{3}$$

In Bhāskara's *Līlāvatī* (eleventh century AD) there are two very significant passages. One relates to combinations, as above, and the other contains the *n*! rule for permutations, stated in words and including the following examples:

Example How many are the variations of form of the god Sambhu by the exchange of his ten attributes held reciprocally in his several hands: namely, the rope, the elephant's hook, the serpent, the tabor, the skull, the trident, the bedstead, the dagger, the arrow and the bow...
Statement Number of places 10. In the same mode as above shown, the variations of form are found to be 3 628 800...

According to 'Umar al-Khayyāmī (died 1122), the Hindus were also familiar with the binomial expansion for $(a + b)^n$, for small integer values of *n*. Related to this is the arithmetical triangle containing the binomial coefficients, now usually called Pascal's triangle (Figure 3). It is claimed that the triangle was familiar to the Hindu scholar Piṅgala around 200 BC. It certainly appeared in Jordanus de Nemore's *De arithmetica* (*circa* 1225), in the *Handbook of Arithmetic Using Board and Dust* (1265), by the Arab astronomer and mathematician Naṣīr al-Dīn al-Ṭūsī, and in a Chinese text by Zhu Shijie of 1303. It acquired its current name from Blaise Pascal's

Figure 3 Pascal's triangle

955

extensive discussion of it in his famous *Traité du triangle arithmétique* (1665). A fuller discussion of the triangle appears in Edwards *1987*.

3 ORIGINS OF MODERN COMBINATORICS

Modern combinatorics dates from the late seventeenth century. Within a few years, three notable books appeared: Pascal's *Traité* (1665), Leibniz's *Dissertatio de arte combinatoria* (1666) and Athanasius Kircher's *Ars magna sciendi sive combinatoria* (1669).

Kircher was a prolific polymath. His aim in the *Ars magna* was to develop work by the thirteenth-century philosopher Ramón Lull into a system of logic which could be applied to all branches of learning. In this way, he was following a tradition laid down by earlier sixteenth- and seventeenth-century scholars, including Paul Guldin, Marin Mersenne and Sebastian Izquierdo, whose books included extensive tables of combinations, permutations and factorials. Further details appear in Knobloch *1979*.

Abraham De Moivre (1697) generalized the binomial theorem (§4.1) to the multinomial theorem − this gives a rule for finding the coefficients in the expansion of $(x_1 + x_2 + \cdots + x_r)^n$. Subsequently (1718) he used a form of the principle of inclusion and exclusion to obtain an expression for the number of derangements (permutations in which no object remains in its original position) of n objects, although this result had been obtained earlier by Nicholas II Bernoulli and P. R. de Montmort.

There are, broadly speaking, four aspects to modern combinatorics: listing, counting, estimation and existence − most of which may be illustrated by the problem of arranging n distinguishable objects in a row. With three objects A, B and C, it is easy to list all possibilities: ABC, ACB, BAC, BCA, CAB, CBA − six arrangements. With n objects there are $n!$ arrangements, and Kircher lists factorials up to 64! In particular, he gives

$$50! = 1\,273\,726\,838\,815\,420\,399\,851\,343\,083\,767\,005\,515\,293\,749\,454\,795$$
$$473\,408\,000\,000\,000\,000. \quad (4)$$

In such a case it is impossible to list all the arrangements. Since $n!$ grows very rapidly with n, it soon becomes impossible even to write down the exact number of arrangements, so the emphasis shifts towards estimation. By 1730, both De Moivre and James Stirling had obtained results related to what is now usually called Stirling's formula; this states that, for large n,

$$n! \simeq (n/e)^n (2\pi n)^{1/2}. \quad (5)$$

This may be used to give the estimate $69! \simeq 1.7 \times 10^{98}$ for the number of ways of arranging 69 objects in a row.

In many counting problems the existence of solutions is evident, but sometimes this is not so. In 1782, Leonhard Euler considered the following problem:

> If there are n^2 officers, one of each of n ranks from each of n different regiments, can they be arranged in a square in such a way that each row and column contains exactly one officer of each rank and from each regiment?

Euler was particularly interested in the case $n = 6$, for which he failed to find a solution. He conjectured that no solution exists when $n \equiv 2 \pmod 4$; the resolution of this is discussed in Section 7. When it is hard to establish the existence of an arrangement, then to count them all is usually quite intractable.

4 PARTITIONS

If $\{\lambda_1, \lambda_2, \ldots, \lambda_k\}$ is a set of positive integers with sum n, then the λ_i are said to form a partition of n. For example, there are seven partitions of 5:

$$1 + 1 + 1 + 1 + 1, \quad 2 + 1 + 1 + 1, \quad 2 + 2 + 1, \quad 3 + 1 + 1, \quad 3 + 2, \quad 4 + 1, \quad 5.$$

An obvious problem is to find the number $p(n)$ of ways of partitioning n into parts, but there are many variations of this problem: for example, one can ask for the number of partitions in which all parts are odd numbers. Leibniz studied partitions, but published little on it (Knobloch *1974*); Euler made substantial contributions in his book *Introductio in analysin infinitorum* (1748). By manipulating power series, Euler established many theorems on partitions, including the fundamental result that

$$\sum_{n=0}^{\infty} p(n)x^n = \prod_{n=1}^{\infty} (1 - x^n)^{-1}. \tag{6}$$

He calculated $p(n)$ for small values of n, but did not obtain a general formula. There was little further progress until the 1840s; later, several people, including Arthur Cayley and J. J. Sylvester, used analytical methods, such as Augustin Louis Cauchy's theory of residues (§3.12), to seek a general expression for $p(n)$. A different approach was adopted by N. M. Ferrers, who represented partitions by rows of dots, now known as Ferrers's diagrams. For example, the diagram for $4 + 4 + 2 + 1$ is

$$
\begin{array}{cccc}
\cdot & \cdot & \cdot & \cdot \\
\cdot & \cdot & \cdot & \cdot \\
\cdot & \cdot & & \\
\cdot & & &
\end{array}
$$

This idea was taken up by Sylvester and others; rearranging some of the dots enabled them to establish correspondences between different sorts of partition and to find certain proofs which are much more transparent than Euler's proofs.

Further progress was made early in the twentieth century. P. A. MacMahon was able to extend the known values of $p(n)$ to $p(200) = 3\,972\,999\,029\,388$; and this value was included in a remarkable paper by G. H. Hardy and Srinivasa Ramanujan, in the London Mathematical Society's *Proceedings* (1918), where they gave an explicit expression for $p(n)$. Further refinements were made by H. Rademacher, but the final formula is still very complicated; it is in the form of an infinite sum and involves derivatives, square roots, exponentials and a 24th root of unity.

5 GRAPH THEORY

The subject of graph theory originated with Euler's 1735 presentation of the Königsberg bridges problem, which asks for a route crossing each of the seven bridges of Königsberg just once (Figure 4). Euler proved there was no

KONINGSBERGA

Figure 4 Königsberg in the seventeenth century, showing the seven bridges spanning the River Pregel (from Biggs *et al. 1976*: 2; with the permission of Oxford University Press)

route (Biggs *et al. 1976*: 1A), and showed how his method can be extended to any arrangement of islands and bridges. In particular, he formulated necessary and sufficient conditions for a route to exist, but he did not prove their sufficiency. Although his methods were essentially graph-theoretic, he did not use graphs as such, and the graph usually drawn to represent the problem (Figure 5) did not appear until 150 years later. Indeed, the problem did not become well known until Lucas *1882* and Rouse Ball *1892* presented it in their books on recreational mathematics.

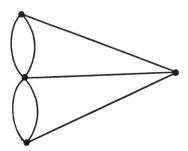

Figure 5 The Königsberg graph

Another type of traversal problem involves finding a cycle through every vertex of a graph, rather than a route passing along each edge. An example is the knight's tour problem, which asks for a sequence of knight's moves on a chessboard, visiting each square just once and returning to the starting-point; this problem was investigated mathematically in the eighteenth century by both Euler and Alexandre Vandermonde (Biggs *et al. 1976*: 2A). The first general discussion of vertex-traversal problems was given in 1855 by the English clergyman T. P. Kirkman (Biggs *et al. 1976*: 2B), who investigated which polyhedra have a cycle passing through all vertices. Graphs admitting such a cycle are now called Hamiltonian, after William Rowan Hamilton's 1856 investigation of cycles on a dodecahedron (Biggs *et al. 1976*: 2C), an offshoot of his work on non-commutative algebra. Unlike the Eulerian problem, the general Hamiltonian problem is hard to solve, although a number of sufficient conditions have been obtained.

One of the most fundamental results concerning polyhedra is Euler's polyhedral formula, $v - e + f = 2$, relating the number of vertices, edges and faces of a polyhedron or connected planar graph. This first appeared in a letter from Euler to Christian Goldbach in 1750 (Biggs *et al. 1976*: 5A), but Euler was unable to prove it. (The formula is sometimes incorrectly attributed to René Descartes, who obtained the sum of the angles of all the faces of a polyhedron; however, although Euler's formula can be deduced from this, Descartes does not seem to have done so.) A metrical proof was given in 1794 by Adrien Marie Legendre; Cauchy later gave a topological proof (Biggs *et al. 1976*: 5B). For a graph embedded on a sphere with *g* handles, S. A. J. Lhuilier (Biggs *et al. 1976*: 5C) obtained the corresponding formula $v - e + f = 2 - 2g$ in 1813. This was the starting-point for an extensive investigation by J. B. Listing (published in 1861–2) into simplicial complexes, thereby laying the groundwork for Henri Poincaré's development of algebraic topology at the end of the nineteenth century. An extensive discussion of the formula and its generalizations, together with many historical notes, is given by Lakatos *1976*.

It was known in the nineteenth century that certain graphs cannot be embedded in the plane: for example, the complete graph K_5 (Figure 6(a)), arising from a recreational problem posed by August Ferdinand Möbius, and the complete bipartite graph $K_{3,3}$ (Figure 6(b)), which corresponds to the 'gas–water–electricity puzzle', in which three houses are to be connected to three utilities without the cables/pipes crossing over one another. In 1930, K. Kuratowski proved that these are the 'basic' non-planar graphs, in the sense that every non-planar graph must contain a subdivision of at least one of them (Biggs *et al. 1976*: 8C). Recently, N. Robertson and P. D. Seymour have proved that there is a corresponding list of 'forbidden subgraphs' for surfaces of any genus.

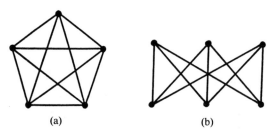

(a)　　　　　　　　　　(b)

Figure 6 (a) The graph K_5; (b) the graph $K_{3,3}$

The most celebrated problem in graph theory is the four-colour problem, which asks whether every map can be coloured with just four colours in such a way that neighbouring countries are differently coloured. This first occurred to Francis Guthrie in 1852, and was communicated via his brother Frederick to Augustus De Morgan, who in turn communicated it to other mathematicians (Biggs *et al. 1976*: 6A). In 1878, Cayley posed the problem at a London Mathematical Society meeting, and the next year A. B. Kempe produced his celebrated 'proof' for the newly founded *American Journal of Mathematics* (Biggs *et al. 1976*: 6B). In 1890, P. J. Heawood pointed out an error in Kempe's proof, but salvaged enough to prove that every map can be coloured with at most five colours. He also gave a formula for the number of colours needed for maps on other surfaces; but his proof was deficient, and the gap was not filled until 1968, by G. Ringel and J. W. T. Youngs. The four-colour theorem was eventually proved, by K. Appel and W. Haken, in 1976; the main ideas of their proof can be traced back to Kempe, but the details were very complicated, involving the analysis of almost 2000 configurations and the use of hundreds of hours of computing time.

Increasingly algorithmic methods have been used to attack problems in graph theory (§5.10). For example, there are algorithms which can be used to find the shortest path in a road network, or the longest path (or 'critical path') in an activity network. A celebrated problem of this type is the so-called travelling salesman problem, in which a salesman is required to make a cyclic tour of a number of cities covering the shortest possible distance. This problem can be traced back to 1831; although no efficient algorithms are known, heuristic methods have been developed which can cope with problems involving hundreds of cities.

6 COUNTING WITH SYMMETRY

Trees are connected graphs without cycles, and their mathematical study dates from the nineteenth century. The initial stimulus came from differential calculus, but chemistry provided further motivation. Both mathematicians and chemists became interested in counting various types of isomers of chemical compounds, including Cayley, who counted the alkanes (paraffins) C_nH_{2n+2} (Biggs *et al. 1976*: 4B).

In counting mathematical (or chemical) structures, it is important to formulate precisely the conditions under which two structures are to be considered different. Two identical structures may appear to be different, simply because they are being viewed from different angles. Many counting problems can be formulated and solved by defining an equivalence relation on a finite set of objects and counting the number of equivalence classes. Often this equivalence can be specified in terms of symmetry operations acting on the set of structures. Cauchy and Georg Frobenius expressed the number of equivalence classes as the average number of structures fixed by a symmetry; this is often erroneously known as Burnside's lemma (Neumann *1979*).

The development of enumeration under the action of a symmetry group has occupied many chemists and mathematicians, and there have been several instances of independent discovery. Most of the important steps in the development are detailed by R. C. Read (Pólya and Read *1987*).

Many counting problems are equivalent to placing objects in certain positions in a given framework. For example, the framework might consist of the edges of a cube, with a coloured ball to be placed at each corner of the cube. If all the balls have the same colour, then there is essentially only one structure, but if balls of several colours are available then many different structures may be formed. The symmetry group acting on the framework also induces an action on the set of structures, and the averaging process permits the number of structures to be calculated.

A refinement of the problem is to ask for the number of structures which use specified numbers of balls of various colours. The key idea for solving such problems was published independently by the American polymath J. Howard Redfield in 1927 and by the Hungarian mathematician George Pólya in the mid-1930s. Both introduced a polynomial which records the number of objects fixed by each symmetry, together with additional information about the way in which the various symmetries act on the framework. Pólya named it the 'cycle index', whilst Redfield called it the 'group reduction function'; both used it in conjunction with generating functions. Unfortunately, Redfield's paper was generally overlooked for about thirty years, and some of his ideas were rediscovered independently by Read in the 1950s.

7 DESIGNS

In 1839, the geometer Julius Plücker remarked that a system $S(n)$ of n points, arranged in triples so that any two points belong to just one triple, is possible only when $n \equiv 1$ or $3 \pmod 6$. This came to the attention of the editor of the *Lady's and Gentleman's Diary*, who proposed the following prize question in the 1844 issue:

> Determine the number of combinations that can be made out of n symbols, p symbols in each; with this limitation, that no combination of q symbols which may appear in any one of them shall be repeated in any other.

In modern terminology, the question asks for the number of blocks in a q-design; Plücker's system corresponds to the case $p = 3$, $q = 2$.

In 1846–7, Kirkman showed how to construct the system $S(n)$ whenever $n \equiv 1$ or $3 \pmod 6$ – a major achievement. Later he observed that there is a system $S(15)$ whose 35 triples may be partitioned into seven sets of 5 triples so that each symbol occurs just once in each set of 5. This is the Kirkman schoolgirls problem, which appeared in the *Lady's and Gentleman's Diary* for 1850:

> Fifteen young ladies in a school walk out three abreast for seven days in succession: it is required to arrange them daily, so that no two shall walk twice abreast.

Kirkman also constructed designs with parameters $n = r^2 + r + 1$, $p = r + 1$ and $q = 2$, when r is prime and when $r = 4$ and $r = 8$; this is the projective plane of order r. Unfortunately his work went almost unnoticed at the time. Indeed, Plücker's question was revived in 1853 by the geometer Jacob Steiner; this is why $S(n)$ is now usually called a Steiner triple system, even though Kirkman had discussed the matter six years earlier. A related problem is to prove that, for each value of $n \equiv 3 \pmod{6}$, there is a Steiner triple system that may be partitioned as in Kirkman's schoolgirls problem. This remained unsolved until 1971, when D. K. Ray-Chaudhuri and R. M. Wilson obtained a proof.

Both Cayley and Sylvester published in this area, and Cayley coined the term 'tactic' for the general area of configurations and designs. In his important *Tactical Memoranda* of 1896, the American E. H. Moore gave a systematic treatment of the numerical constraints on the existence of designs, and used finite fields to construct families of designs.

Around this time, geometrical ideas began to revitalize the subject of 'tactic'. The geometrical requirements that two points determine a unique line, and that two lines determine a point, provide the link between geometry and designs. Thus, in 1892, the Italian geometer G. Fano described several finite geometries, including the seven-point plane that now bears his name; this is just the projective plane of order 2, or the Steiner system $S(7)$. In 1906, O. Veblen and W. H. Bussey extended Fano's work by studying Desargues's theorem in finite projective geometries. In 1907, Veblen and J. M. Wedderburn constructed a non-Desarguesian plane of order 9.

A latin square is an arrangement of letters a, b, c, ..., such as in Figure 7(a), each letter occurring just once in each row and column. Two latin squares are said to be orthogonal if, when superimposed, each of the ordered pairs aa, ab, ba, ac, ca, ... appears just once, as in Figure 7(b). An early problem involving the finding of a pair of orthogonal latin squares occurs in J. Ozanam's *Récreations mathématiques* (1725): arrange the 16 court cards so that each row and each column contains one card of each suit and one of each value.

$$
\begin{array}{cccc}
a & b & c & d \\
b & c & d & a \\
c & d & a & b \\
d & a & b & c \\
\end{array}
\qquad
\begin{array}{cccc}
aa & bb & cc & dd \\
bc & ad & da & cb \\
cd & dc & ab & ba \\
db & ca & bd & ac \\
\end{array}
$$
$$\text{(a)} \qquad\qquad \text{(b)}$$

Figure 7 (a) A latin square; (b) two orthogonal latin squares superimposed

The existence of a finite projective plane of order n is equivalent to the existence of $n-1$ mutually orthogonal latin squares (MOLS) of order n. Euler's conjecture (see Section 3) implies that there are no projective planes of order $n \equiv 2 \pmod 4$. In 1900, G. Tarry proved that there is no pair of orthogonal latin squares of order 6, thereby showing that there is no projective plane of order 6, and answering the '36 officers' problem in the negative.

The first known use of latin squares in an experimental design was by Cretté de Palluel in 1788, but it was not until the 1920s that interest in latin squares and designs was renewed, mainly as a result of the work of R. A. Fisher and F. Yates on the design of agricultural experiments (see §10.11). In the years 1934–40 they completely classified the latin squares of order 6, introduced balanced incomplete block designs (BIBDs), and proved the necessary conditions

$$vr = bk, \qquad \lambda(v-1) = r(k-1) \qquad v \leqslant b \tag{7}$$

for the existence of BIBDs; their book *Statistical Tables for Biological Agricultural and Medical Research* (1938) lists many BIBDs.

Towards the end of the 1950s, the problem of constructing a pair of MOLS of order $n \equiv 2 \pmod 4$ was again revived. In 1959–60 R. C. Bose and S. S. Shrikhande constructed two MOLS of order 22, then E. T. Parker found two of order 10, and finally all three combined to show that Euler's conjecture is false for *all* $n \equiv 2 \pmod 4$, $n > 6$. Eventually in 1988, as a result of an extensive computer search, Clement Lam and his colleagues announced that a projective plane of order 10 does not exist.

BIBLIOGRAPHY

Ball, W. W. Rouse, *1892*, *Mathematical Recreations and Problems of Past and Present Times*, London: Macmillan. [Also later edns.]

Biggs, N. L. *1979*, 'The roots of combinatorics', *Historia mathematica*, **6**, 109–36.

Biggs, N. L., Lloyd, E. K. and Wilson, R. J. *1976*, *Graph Theory 1736–1936*, Oxford: Clarendon Press. [Rev. edn 1986.]

——*1994*, 'The history of combinatorics', in *Handbook of Combinatorics*, Amsterdam: North-Holland, to appear.

Edwards, A. W. F. *1987*, *Pascal's Arithmetical Triangle*, London: Griffin, and New York: Oxford University Press.

Knobloch, E. *1974*, 'Leibniz on combinatorics', *Historia mathematica*, **1**, 409–30.

——*1979*, 'Musurgia universalis: Unknown combinatorial studies in the age of Baroque absolutism', *History of Science*, **17**, 258–75.

Lakatos, I. *1976*, *Proofs and Refutations: The Logic of Mathematical Discovery* (ed. J. Worrall and E. Zahar), Cambridge: Cambridge University Press.

Li, Y. and Du, S. *1987*, *Chinese Mathematics: A Concise History*, Oxford: Clarendon Press.

Lucas, E. *1882*, *Récréations mathématiques*, Vol. 1, Paris: Gauthier-Villars.

Netto, E. *1901*, *Lehrbuch der Combinatorik*, Leipzig: Teubner. [2nd edn 1927; repr. 1958, New York: Chelsea.]

Neumann, P. M. *1979*, 'A lemma that is not Burnside's', *Mathematical Scientist*, **4**, 133–41.

Pólya, G. and Read, R. C. *1987*, *Combinatorial Enumeration of Groups, Graphs and Chemical Compounds*, New York: Springer.

Part 8
Mechanics and Mechanical Engineering

8.0

Introduction

For at least four centuries mechanics has been a major branch of mathematics, and in the eighteenth century it was perhaps the dominant one. This is reflected in the size of this Part.

Since the mid-eighteenth century, mechanics has fallen into five main areas (Grattan-Guinness *1990*), which has been of some guidance in designing this Part. The areas are, in the order in which they are handled here: 'corporeal' (as I call it), oriented around 'ordinary-sized' objects and in which the main principles are presented (§8.1–8.7); celestial, in which the heavenly bodies are treated as mass points (§8.8–8.10); planetary, in which the shapes are allowed for and indeed can constitute a major factor in the study (§8.11–8.16); engineering and technology, including structures, artefacts and instruments (prominent in §8.4–8.7 and §8.12, and with §8.14–8.18 relating closely to naval matters); and molecular, involving the study of the supposedly intimate constitution of matter (the smallest area, and relating more closely to the development of mathematical physics described in the next Part, but present here especially in aspects of hydrodynamics and elasticity theory in §8.5–8.6).

As is clear from this list of topics, there is overlap between areas, and more than one article could be used in conjunction. Furthermore, the order of the articles represents much less of a chronological sequence than in many other Parts; all these topics received much attention over the same long period. Several articles in the next Part (on physics and mathematical physics) are related, and complicate the chronology still further.

There have been a number of general histories of mechanics; some of the better ones are listed in the bibliography below. None fully demonstrates the range of the subject; the technological area has been particularly neglected. In addition, the different traditions within which it has been developed are not always fully stressed (Newton's law of motion, or variational principles, or considerations of energy and work); this range is considered explicitly in §8.1, and arises also in several other articles.

In compensation, let me underline the general praise uttered in §0 for the *Encyklopädie der mathematischen Wissenschaften*. Its survey of mechanics in Volumes 4 (mechanics) and 6 (astronomy), and parts of Volume 5 (physics, geodesy and geophysics) provide not only an extraordinary survey of mechanics as of the early years of the twentieth century, but also, in several articles, extensive historical surveys of particular branches and aspects.

BIBLIOGRAPHY

Dugas, R. *1950*, *Histoire de la mécanique*, Paris: Griffon. [English transl.: 1957, New York: Central.]

Grattan-Guinness, I. *1990*, 'The varieties of mechanics by 1800', *Historia mathematica*, **17**, 313–38.

Grigoryan, A. T. and Pogrebyssky, V. (eds) *1971–2*, *Istoriya mekhanika*, 2 vols, Moscow: Nauka.

Jouguet, E. *1908–9*, *Lectures de mécanique. La Mécanique enseignée par les auteurs originaux*, 2 vols, Paris: Gauthier-Villars.

Rühlmann, M. *1891–5*, *Vorträge zur Geschichte der theoretischen Maschinenlehre und der damit in Zusammenhang stehenden mathematischen Wissenschaften*, 2 parts, Braunschweig: Schwetschke.

Truesdell, C. *1968*, *Essays in the History of Mechanics*, Berlin: Springer.

8.1

Classical mechanics

CRAIG G. FRASER

1 INTRODUCTION

Classical mechanics may be defined most generally as the study of the equilibrium and motion of bodies based on the principle of inertia, and employing the mathematics of the differential and integral calculus. It refers to the range of theories in dynamics and material science that have developed historically from the seventeenth century to the present. It emphasizes formal abstraction and mathematically grounded concepts as tools in the investigation of physical phenomena, but also recognizes the essential role of experiment and engineering experience in achieving successful theories. The adjective 'classical' is used to distinguish the subject from two more recent developments: Albert Einstein's special theory of relativity of 1905, and quantum mechanics, invented by E. Schrödinger and W. Heisenberg in the 1920s (§9.13 and §9.15).

In the seventeenth century, several traditions of quantitative research emerged involving distinct methodological and physical precepts. The most prominent developments were connected with advances in astronomy, with the new theories of Nicolaus Copernicus and Johannes Kepler. The beginning of classical mechanics, conventionally taken to be Isaac Newton's *Philosophiae naturalis principia mathematica* (1687), provided an appropriate physics for a heliocentric astronomy. The celebrated 'Newtonian synthesis' was the brilliant culmination of the pioneering work of Galileo Galilei and others in inertial mechanics. Celestial dynamics and problems of stability continued to be a central area of investigation in the history of the subject.

Astronomy was by no means the only source of problems that led to classical mechanics. Historical writing over the past few decades has discredited the view, originating in the late nineteenth century, according to which the basic theoretical structure of the subject was supposed to have emerged in its entirety with the publication of the *Principia*. This work successfully treated only a limited range of phenomena – the central-force dynamics of freely moving particles – and employed mathematical methods that have since been discarded. Theoretical engineering and technology also contributed problems, concepts and techniques to major branches of the subject. More generally, although specific physical hypotheses put forward by René Descartes and Christiaan Huygens were rejected in the eighteenth century, Cartesianism as a scientific philosophy continued to exert a profound influence on mechanical thought.

2 NEWTON'S *PRINCIPIA*

The *Principia* was written by Newton from 1684 to 1687 in a period of intense intellectual activity that is probably unmatched in the history of science. At the time he was Lucasian Professor of Mathematics at Cambridge, and in his early forties. He presented the treatise in the neo-Euclidean style favoured by mathematical scientists of the early modern period. It begins with a preliminary set of definitions and 'axioms, or laws of nature', and is followed by three parts, or 'books'. Book I, which opens with eleven mathematical lemmas containing Newton's version of the calculus, a kind of geometrical theory of limits, is a systematic treatise on particle dynamics; Book II, the least successful of the three, investigates the motion of bodies in resisting media; and Book III introduces the universal law of gravitation and applies the mathematical theory of Book I to the Solar System.

The *Principia* grew from a draft essay on dynamics entitled 'De motu', written in 1684. This essay, which would be incorporated into the opening sections of Book I, contains Newton's significant theoretical innovations in dynamics. An indication of his approach is provided by the results published as Propositions 1, 6 and 11 of Book I. He considers a particle P acted upon by a force directed toward a fixed centre S (Figure 1). In Proposition 1, Newton uses geometrical–infinitesimal techniques to show that, as a consequence of the action of the force and the inertial motion of P, the line connecting S and P sweeps out equal areas in equal times. In Proposition 6 he introduces a measure for the force, and in the corollaries uses the area law to express this measure entirely in terms of spatial quantities. With reference to Figure 1, he shows that the force acting on P is proportional to $(QR)/(SP)^2(QT)^2$. This result prepares the way for the initial problem of his dynamics, to calculate the force from a knowledge of the orbit or trajectory of P. The inspiration for the entire treatise developed from Newton's discovery that this purely technical–mathematical question led to a coherent and substantial body of results. Thus, in Proposition 11 he shows mathematically that if the trajectory is an ellipse with S at one focus, then the force is inversely proportional to the square of SP. In

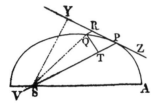

Figure 1 The equal-areas law, from Newton's *Principia* (1687)

subsequent propositions he extends this result to cases in which the trajectory is a parabola and hyperbola.

Although Proposition 11 was presented by Newton as a mathematical theorem, it had an immediate application to planetary motion. The planets were known by Kepler's laws to move in ellipses with the Sun at one focus. If one assumed they moved as a consequence of an attractive force directed towards the Sun, then it was clear from Proposition 11 that the force would vary inversely as the square of the distance from the Sun. Proposition 45 provided further confirmation of this law. For a particle moving about a centre in a closed trajectory, the line of apsides is defined as the axis joining the closest and most distant points of approach. Consider a force law of the form k/r^n. Newton showed mathematically that if n is any number other than 2, then the line of apsides will experience a regular rotation. Since such a rotation was not observed in planetary motion, it followed once again that the planets were governed by an inverse-square law.

The *Principia* was a large, difficult work containing many completed results and many suggestions for future research. It became the cornerstone of classical inertial physics and provided the paradigm for subsequent research in physical astronomy. Newton was fortunate in having chosen to investigate a range of physical phenomena – the dynamics of particles – that was so entirely amenable to systematic analysis in terms of the physical concepts and mathematical techniques of the period. The scientific significance of his theory was potentially far-reaching. If one assumed that all phenomena were derived from the interaction by forces among corpuscles or atoms, then it followed that Newtonian particle dynamics was in principle the ultimate foundation for all of physical science.

3 ANALYTICAL MECHANICS

Newton had invented an analytical calculus in his study of Cartesian algebraic geometry in the 1660s. In the next two decades he developed a strong aversion to all aspects of Descartes's philosophical and scientific thought. In the *Principia* he avoided analytical techniques, preferring instead a thoroughly geometrical formulation of dynamical theory. The archaic appearance of his treatise to a modern reader is due in no small part to the rather peculiar mathematical idiom he adopted.

The eighteenth-century progress of theoretical mechanics occurred almost entirely on the Continent, using the mathematics of the Leibnizian calculus (§3.2). The *Principia* became historically influential once its theory had been absorbed by researchers working in the scientific academies of Paris, Berlin and St Petersburg. The French priest Pierre Varignon wrote a series of memoirs beginning in 1700 on orbital dynamics that employed Leibnizian analytical methods. He gave an elegant demonstration of *Principia*, Book I, Proposition 11 that replaced Newton's involved geometric reasoning by a simple algorithmic procedure. Varignon's contemporaries John I Bernoulli and Jakob Hermann also contributed to his subject, providing a uniform mathematical treatment of the so-called 'inverse problem' in which it was required to determine the orbit from a knowledge of the force law.

The analytical development of dynamics is illustrated in the differential-equation form of what is known today as Newton's second law:

$$M\frac{d^2x}{dt^2} = F_x, \qquad M\frac{d^2y}{dt^2} = F_y, \qquad M\frac{d^2z}{dt^2} = F_z, \tag{1}$$

where M is the mass of the body, x, y and z are its spatial coordinates, and F_x, F_y and F_z are the components of the total force acting on it. Newton's original statement of the law was verbal and referred to increments of velocity rather than to acceleration. Only slowly, in the first half of the eighteenth century, did equation (1) come to be recognized as a fundamental principle. Special cases of the law had appeared in the orbital dynamics of Varignon and John I Bernoulli, where it had constituted the mathematical foundation of the investigation. In more complicated mechanical systems researchers tended to analyse the phenomena from first principles in ways that obscured the possibility of recognizing (1) as the operative physical law. Part of the difficulty was that sophistication in using the calculus developed gradually, and progress in the formation of new physical concepts were even slower. In his *Traité de dynamique* (1743), the young French scientist Jean d'Alembert employed a relation connecting force and time derivatives to investigate the motion of a heavy hanging chain. Leonhard Euler, perhaps the greatest theorist of eighteenth-century exact science, introduced equation (1) as an explicit general law in his 1750 memoir 'Découverte d'un nouveau principe de mécanique'; the title itself suggests the novelty and originality of these equations, introduced over sixty years after the appearance of Newton's *Principia*. In this paper he used it to find the 'Euler equations' for the rotation of a rigid continuous body.

Mathematical mechanics developed more broadly in the eighteenth and nineteenth centuries in association with advances in the theory of infinite series, the calculus of variations, the theory of ordinary and partial differentiation, and differential geometry. In his investigation of the vibrating string in 1747, d'Alembert derived a second-order partial differential equation, the wave equation, and integrated it in terms of arbitrary functions (§3.15). His research started an interesting debate within mathematics on the foundations of analysis and the function concept. Conversely, results in the theory of partial differentiation contributed to an understanding of the behaviour of fluids, elastic beams and continuous media in general. As the eighteenth century progressed, links between the two subjects became deeper and more diverse.

4 THE ESTABLISHMENT OF VARIATIONAL MECHANICS

The establishment of variational mechanics was largely the work of Euler and Joseph Louis Lagrange. Although Lagrange's *Méchanique analitique* (1788) is usually cited as the definitive presentation of the subject, the theory was developed earlier, by Euler between 1740 and 1750 and by Lagrange between 1760 and 1780. Euler provided some of the essential ideas, while the systematic mathematical elaboration of the theory was Lagrange's achievement.

Variational mechanics had its origins in the rule for equilibrium known as the principle of virtual velocities. This principle was a basic axiom in the medieval statics of Jordanus de Nemore, and was discussed by both Descartes and Galileo in the seventeenth century (§2.6). The formulation of the principle common in the eighteenth century appeared in a letter of 1717 from John I Bernoulli to Varignon. Consider a constrained system of bodies in equilibrium with respect to a set of applied forces, and suppose that the system is subjected to a small disturbance which imparts to each mass m a 'virtual' velocity w. This velocity must necessarily be compatible with the constrains in the system. The principle asserts that the sum of the product of the forces and the virtual velocities will be zero:

$$\sum Fw \cos \theta = 0, \qquad (2)$$

where F is the applied force acting on m, and θ is the angle between the directions of F and w.

The principle of live force was also important in the establishment of variational mechanics. This principle stipulated that the motion of a dynamical system satisfies the equation

$$\sum mv^2 + 2\Phi(x, y, z, \ldots) = \text{constant}, \tag{3}$$

where v is the speed of a typical mass m, and the variables x, y and z denote its position. The quantity $\Sigma\, mv^2$ was known as the 'live force' of the system, and the expression $\Phi(x, y, z, \ldots)$ was obtained by integrating the forces over the spatial variables. In later mechanics, Φ would be called the potential function of the system.

Equation (3) was derived for particle dynamics, elastic collision and the motion of constrained bodies. It was obtained for a particle acted upon by a central force in Propositions 39–41 in Book I of Newton's *Principia*, and was a standard relation in the orbital dynamics of Varignon and John I Bernoulli. For constrained systems the principle took on a special significance. The forces of constraint do not contribute to the quantity Φ in equation (3); only applied forces need to be considered in calculating this integral. Assume now that, in a given configuration of the system, the velocity of each body is zero. If it is further supposed that in this configuration the quantity Φ is a minimum, then it follows (because $\Sigma\, mv^2$ is always positive) that the system is in static equilibrium with respect to the given applied forces.

It was evident that in particular examples the condition that Φ be a minimum yields precisely the same relation for equilibrium as does the principle of virtual velocities. Since the latter was a fundamental principle, it seemed that the science of statics could be derived from a single variational law asserting that in equilibrium a determined quantity – the integral Φ – is a minimum.

Euler introduced the term 'effort' to denote Φ and called the statement that Φ be a minimum in static equilibrium his 'law of rest'. Following his contemporary Pierre de Maupertuis, he generalized the law of rest to dynamics, thereby obtaining the celebrated principle of least action. This principle asserts that among all curves joining two points in a plane, a particle follows the curve for which the integral $\int mv\,ds$ is a minimum. Euler first introduced the principle in an appendix to his treatise of 1744 on the calculus of variations. He considered a particle of unit mass with Cartesian coordinates x and y moving freely under the action of a central force. Using the equation of live force, $\frac{1}{2}v^2 + \Phi(x, y) = c$, he wrote the action integral $\int v\,ds$ as

$$\int [2(c - \Phi)]^{1/2}(1 + y'^2)^{1/2}\,dx, \quad y' := dy/dx, \tag{4}$$

a definite integral that is to be evaluated between the initial and final values of the variable x. By means of mathematical variational methods, he showed that the problem of extremalizing this integral leads to the same trajectory as does a calculation based on direct methods employing forces and accelerations.

Although Euler had made a substantial beginning, he did not continue with his research on variational principles; his ideas remained promising suggestions rather than fully developed concepts. The elaboration of the theory was achieved in the decades that followed by his younger contemporary Lagrange. In his first contribution to the subject in 1762, Lagrange employed the new δ-process that he had introduced into the calculus of variations (§3.5). Beginning with a more general form of Euler's principle of least action, he showed how it led to a uniform procedure for generating the equations of motion of an arbitrary dynamical system (Fraser *1983*).

Despite his considerable success with this principle, Lagrange proceeded in the next few years to replace it with another law, derived from a dynamical generalization of the principle of virtual velocities. His shift in approach was influenced by technical considerations, by a desire for a unified treatment of both statics and dynamics, and by an aversion to the metaphysical associations of least action. The new formulation became the basis of his *Méchanique analitique*, published in 1788 in Paris when he was 52 years old. He began with the fundamental axiom

$$\sum m \frac{d\mathbf{v}}{dt} \cdot \delta\mathbf{r} = \sum \mathbf{F} \cdot \delta\mathbf{r}, \tag{5}$$

where m is the mass of a typical body, $\mathbf{v}$ is its velocity, $\delta\mathbf{r}$ is its virtual displacement and $\mathbf{F}$ is the applied force acting on it. Given that the system is described in terms of a set of 'generalized' coordinates $(q_1, q_2, q_3, \ldots, q_n)$, he derived the 'Lagrangian' equations of motion

$$\frac{\partial T}{\partial q_i} - \frac{d}{dt}\left(\frac{\partial T}{\partial \dot{q}_i}\right) - \frac{\partial \Phi}{\partial q_i} = 0, \tag{6}$$

where T denotes half the live force, and Φ is the potential.

The *Méchanique analitique*, published a century after Newton's *Principia*, was the most developed expression of the eighteenth-century abstract analytical tendency in mechanics. Lagrange's variational approach was most important in advancing new methods, in mathematicizing the study of dynamical systems, rather than in enlarging our understanding of physical phenomena. He provided a uniform procedure for generating the equations of motion that was independent of the coordinatization employed or of any assumption concerning the material constitution of bodies. A conspicuous feature of his approach was his use of constraints to simplify mathematically the description of the system. The notion of a constraint enabled one to idealize the analysis in a way that avoided detailed assumptions concerning the physical basis of the phenomena. Formal mathematical development replaced physical hypothesis and experimental verification.

5 HAMILTON–JACOBI THEORY

The legacy of Lagrange was continued in the nineteenth century in the development of what is known in modern physics as 'Hamilton–Jacobi theory', the set of methods for formulating and integrating the differential equations of motion of a general dynamical system. The theory was established by the young Irish scientist William Rowan Hamilton in two papers published in the Royal Society's *Philosophical Transactions* in 1834 and 1835. These researches originated in his investigation of questions in mathematical optics and in his study of the three-body problem in celestial dynamics (§8.9). Two years later, Carl Jacobi introduced important modifications and additions to Hamilton's work in a paper that appeared in Crelle's *Journal*.

By means of variational methods, Hamilton showed that Lagrange's system (6), a set of n second-order differential equations, could be replaced by a new system consisting of $2n$ first-order equations, the so-called 'canonical differential equations of motion'. He did so by expressing (6) in terms of the q_i and an additional set of variables, the 'conjugate momenta' $p_i = \partial T / \partial \dot{q}_i$. In his further study of a problem of perturbation, he employed a 'contact transformation' – that is, a change of variables that preserved the form of the canonical equations. Jacobi subsequently developed a systematic theory of contact transformations and reduced the problem of integration to the solution of a single partial differential equation containing a principal or generating function (Prange *1933*).

In the later nineteenth century the methods of Hamilton and Jacobi were investigated by such people as Joseph Liouville and Rudolph Lipschitz as part of the further development of mathematical mechanics and differential geometry. In addition to their purely theoretical interest, these methods were important in applications to celestial mechanics (§8.8). In the 1920s, Schrödinger and Louis de Broglie showed that they also provided an appropriate formalism for expressing the new quantum-wave physics. As a consequence of its role in quantum mechanics, the theory of Hamilton and Jacobi occupies a rather more prominent place today than its historical origins and position in the classical subject (as well perhaps as its intrinsic interest) might otherwise indicate.

6 CONTINUOUS MEDIA

Newton achieved only special (albeit significant) results in his investigation of fluids in Book II of the *Principia*, and he contributed little to the study of flexible and elastic bodies. Much effort in the period 1700–1830 was devoted to constructing an adequate mathematical theory for these materials. These investigations had their origins in the work of such early mechanicians as Huygens, James I Bernoulli and Antoine Parent, and were continued and brought to maturity by mathematicians Euler, d'Alembert, Alexis Clairaut, Lagrange, Augustin Louis Cauchy, and many others. The resulting body of results makes up a substantial part of modern civil and mechanical engineering.

Theories of continuous media developed from the study of such special problems as the shape of a suspended cable, the deflection of an elastic beam, the strength of struts and columns, the vibration of a taut string and the resistance of bodies in fluids. Each of these problems required its own concepts and techniques of solution. The progress of the subject abounded with false starts, long detours and brilliant successes.

The analysis of elastic beams provides an illustrative study in the history of continuum mechanics (see §8.7 on the theory of structures). Between 1695 and 1705, James I Bernoulli wrote three seminal papers analysing the deformation of an elastic blade or lamina subject to forces acting at its ends. He considered a lamina built into a support at one end and loaded at the other end by a weight (Figure 2); the problem was to determine the curve of deflection. By considering the structure of the deformed lamina at a cross-section, he concluded that the tensile force on the fibres was inversely proportional to the radius of curvature of the curve of deflection at its intersection with the cross-section. The force would exert an internal resisting moment that in static equilibrium would just balance the moment due to the external load. Let the lamina be described in a Cartesian (x, y) coordinate system in which the load is situated at the origin, the y-axis coincides with the line of action of the load, and the coordinates of an arbitrary point of the lamina are x and y. The calculus, for Bernoulli a very new and exciting mathematical tool, gave the expression $r = -\mathrm{d}^2 y / \mathrm{d}x\,\mathrm{d}s$ (where $\mathrm{d}s^2 = \mathrm{d}x^2 + \mathrm{d}y^2$) for the radius of curvature. Using the condition that the moments balance, he obtained a differential equation to describe the static configuration of the elastica (the function representing the elastic line curved in the plane):

$$-k\frac{\mathrm{d}^2 y}{\mathrm{d}x\,\mathrm{d}s} = Px.\tag{7}$$

He considered the case where the load acts perpendicularly to the lamina,

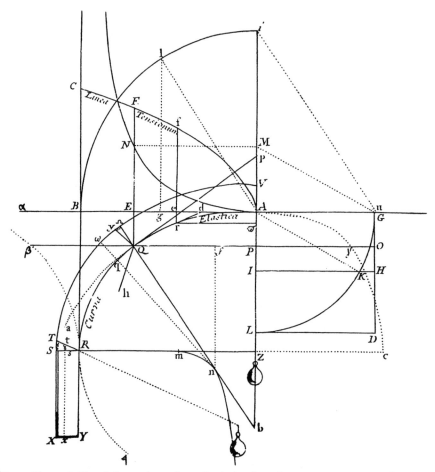

Figure 2 The deformation of an elastic lamina, from James I Bernoulli's
Curvatura laminae elasticae (1694)

the so-called 'rectangular elastica', and integrated equation (7) once to
obtain a first-order differential equation connecting x and y. (In later
mathematics its solution would be called an elliptic integral; see §4.5.)
James I Bernoulli introduced infinite series to investigate this equation, and
discussed some of its properties.

In 1744 Euler, then a 37-year-old academician in Berlin, published an
important mathematical analysis of the elastica. He took equation (7) and
completed James I Bernoulli's investigation by enumerating all the forms
assumed by an elastic lamina loaded at its ends. He also considered the
application of the theory to the study of the strength of columns. An

elastica that is subject to external terminal forces making small angles with its long axis provides a suitable model for a column undergoing compression. Euler derived an expression (known later as 'Euler's buckling formula') giving the minimal or critical load necessary to bend the column. In 1779 Lagrange extended his results using elliptic integrals, and showed that higher-order configurations are associated with a larger value for the critical load and therefore lead to a substantial strengthening of the column.

The work of James I Bernoulli, Euler and Lagrange was carried out in an atmosphere of intellectual excitement created by the power of the new analytical methods to solve physical problems. At this time, mathematical sophistication often ran far ahead of physical understanding. A weak point in everyone's treatment of the elastica concerned the constant k in equation (7). This quantity incorporated all the information about the individual structure and material elasticity of the lamina, and was simply introduced – on no certain grounds – in order to derive equation (7). A satisfactory theory would need a detailed analysis of the three-dimensional structure of the lamina in terms of the concepts of neutral axis, moment of inertia, and tensile and compressive elastic stress. It was only very slowly and with great difficulty that such a theory was forthcoming.

Although one can find during the eighteenth century significant special anticipations of modern stress analysis, the general concept of elastic stress itself never emerged. A developed mathematical theory of continuous media first appeared in the 1820s, in the writings of Cauchy and Claude Navier. Working independently of Cauchy, Navier employed the idea of elastic stress to construct a complete analysis of the elastica, published in 1826 in his *Resumé des leçons données à l'Ecole des Ponts et Chaussées*. Beginning with the results of the late-eighteenth-century scientist Charles Coulomb, he obtained the modern formula, EI/R, for the resisting moment of a section of a loaded beam, where E is the modulus of elasticity, I is the moment of inertia of the section about the neutral axis, and R is the radius of curvature of the curve of deflection at the point where it passes through the section.

In 1829 Cauchy stipulated generally that, given any surface S on or within a body that bounds a part V of this body, we may assume that the matter exterior to V exerts a field of stresses or forces per unit area on S. Using this seminal idea, together with the principles of linear momentum and moment of momentum, he developed an analytical theory applicable to elastic materials and fluids. Modern continuum mechanics was born with his treatises of the 1820s.

For a more detailed survey of elasticity theory, see §8.6.

7 MECHANICS IN FRANCE, 1788–1830

Navier worked in mechanics at a time when Paris dominated world science. Along with fellow engineers Jean Victor Poncelet and Gustave Coriolis, he graduated from the Ecole Polytechnique, founded in 1795, and with them participated in the extraordinary scientific vitality of this school in its first decades. The three men built on the researches of such earlier figures as Bernard Bélidor, Coulomb, Lazare Carnot and Gaspard de Prony. Their distinctive achievement was to combine the highly theoretical programme of mathematical mechanics of Euler and Lagrange with the more practical tradition of civil and mechanical engineering long established in France. The Ecole Polytechnique and the Ecole des Ponts et Chaussées played a prominent role in focusing and coordinating their efforts, in which research concerns had strong links to engineering education. As historian I. Grattan-Guinness has noted (*1984*: 31), the story here is 'an extraordinary *potage* of mathematics, mechanics, engineering, education and social change'.

The development of the concept of mechanical work illustrates the achievements of the French engineers. The principle of live force (equation (3)) was a well-established relation of theoretical mechanics by the middle of the eighteenth century. Although in later physics it would become known as the conservation of mechanical energy, there was at this time little recognition of the concepts of work and kinetic energy, and of their interchangeability. The principle served largely as a mathematical relation that imposed an analytical condition on the motion of the system.

In 1783 Lazare Carnot, then an unknown 30-year-old French engineer, published a treatise on efficiency in machines. He introduced the term 'moment of activity' for force times distance, and identified it as the determinative quantity in analysing the performance of weight-driven machines. Beginning with the principle of virtual velocities, he derived a relation connecting moment of activity and live force. He concluded that to obtain the maximum 'effect' or work from a machine it was necessary to avoid inelastic shocks in its parts. In a first primitive recognition of the interconvertibility of kinetic energy and work, he observed that such shocks would reduce the total live force and hence also the work available (Gillispie *1971*).

In 1819 Navier edited a reprinting of Bélidor's *Architecture hydraulique* (1737) in which he called attention to Carnot's researches on machine efficiency. With the rapid development of industrial technology in France during this period, there was a strong interest in the theoretical analysis of machines. In 1829 Coriolis and Poncelet published books in which they introduced the term 'work' for the product of force and distance. Coriolis explicitly redefined the live force of a mechanical system as $\frac{1}{2}\sum mv^2$ (sometimes called the 'reformed live force'), where the factor $\frac{1}{2}$ constituted explicit theoretical recognition of the primary place of the concept of work.

8 PHILOSOPHICAL CURRENTS

Since Newton, philosophical questions about the nature and meaning of classical mechanics have intruded in the development of the subject. Although these discussions have often concerned general methodological assumptions unconnected to technical work in the field, no historical survey would be entirely complete without some consideration of them.

Throughout the history of mechanics there has been an opposition between an empirical, phenomenalist approach to the world, and an approach that emphasizes theory and going beyond the appearances. Newton wrote in the *Principia* that he 'frames no hypotheses' to defend his refusal to seek a physical basis for gravitational interaction. Whereas he believed that systematic mathematical analysis was sufficient to account for the observable phenomena, the Cartesians sought a more fundamental explanation that would interpret the world from first principles in terms of primitive concepts of extended matter in motion. The Cartesians accepted Newtonian dynamics, but denied that it provided an ultimate explanation of physical reality.

This tension reappears in a different form in the early nineteenth century in the opposition between Lagrange's analytical mechanics and Siméon Denis Poisson's 'physical mechanics' (as he called it). Poisson believed that Lagrange's theory involved too severe an idealization and abstraction of physical reality, that it was necessary to approach nature at a deeper level in terms of molecular models of the structure of matter. He wanted to replace the notion of constraint and the assumption of perfectly hard bodies by definite molecular mechanisms. Although his attitude to physical theory placed him within the contemporary programme of Laplacian physics, it set him at odds with the influential mathematical positivism of his countryman Joseph Fourier, and led to his progressive isolation within French science.

The dialectic between phenomenology and realism arises at the end of the nineteenth century in the writings of Ernst Mach. His *Die Mechanik in ihrer Entwicklung historisch–kritisch dargestellt* (1883) was one of several treatises of the period devoted to a critical study of the foundations of mechanics. Along with John Stallo and Pierre Duhem, he was a leading *fin-de-siècle* proponent of positivistic physics, and criticized what he regarded as the excessive employment of atomic–mechanical hypotheses. He believed that phenomenological concepts such as energy were more likely to reveal the inner character and unity of physical theory. His philosophy of science led him to oppose Ludwig Boltzmann's statistical mechanics (§9.14) and increasingly placed him at odds with prominent trends in contemporary physics.

Positivistic philosophy can result in a sterile and limited conception of physical theory and its possibilities. There are nevertheless two respects in which it has had a beneficial influence in the history of science. In granting each theory autonomy on the basis of its individual phenomenological domain, it avoids a rigid mechanistic reductionism and leads to a principle of tolerance in the development of new theories.

In emphasizing observation, empiricism necessarily takes note of the role of the observer in the description of a physical system. Newton's insistence on the absolute character of space and time contradicted his own empiricist refusal to frame hypotheses, and it was natural of Mach to have rejected this part of Newton's mechanical philosophy. Mach's positivism prepared the way for the advent of relativistic conceptions in the work of Henri Poincaré and Einstein. Einstein had read Mach, and the latter's influence is apparent in Einstein's fundamental recognition in special relativity that it is necessary to incorporate the observer into the description of a physical system.

With the establishment of special relativity (§9.13), it became necessary to introduce the adjective 'classical' to delineate the vast range of mechanical doctrines from Newton to Einstein. Classical theories retain their validity and continue to be cultivated extensively today in mathematical engineering. Nevertheless, since Einstein, the classical viewpoint has lost its epistemological primacy as a final description of material motion in space and time.

9 POSTSCRIPT: ON THE TERM 'RATIONAL MECHANICS'

In the preface to the *Principia*, Newton wrote of 'rational mechanics' to refer to the study of motions engendered by forces and, conversely, the study of forces that correspond to given motions. The adjective 'rational' was used to distinguish this subject from practical or common mechanics, and served to emphasize the abstract, general character of his investigation.

The term 'rational mechanics' was not employed as a formal category by the leading mathematical scientists of the eighteenth century. Although d'Alembert mentioned Newton's terminology in his article 'Méchanique' in Volume 10 of the *Encyclopédie* (1765), *mécanique rationelle* (sic) was not included as a subject classification and did not appear in the famous 'Système figurée des connoissances humaines' at the beginning of the work. Neither was it in his *Traité de dynamique* (1743).

984

August Comte adopted the term *mécanique rationnelle* in his *Cours de philosophie positive* (1830 onwards). Comte recognized the basis of mechanics in experience, but he also wished to emphasize the theoretical character of the subject and felt that the adjective 'rational' was well suited to do that. He would in fact have preferred the term *phoronomie*, originally introduced by Jacob Hermann in 1716.

As a result of Comte's influence, the designation *mécanique rationnelle* was commonly used in France in the nineteenth century to refer to theoretical mechanics. It seems, however, not to have gained much currency in Britain, the USA or Germany. It is not in the major English, American and German dictionaries. In Harrap's French–English dictionary, *mécanique rationnelle* is translated as 'theoretic mechanics, pure mechanics'. The term has been introduced more recently by Clifford Truesdell, who uses it in his historical and scientific writings to identify a mathematically rigorous, deductive approach to mechanics.

In discussing the eighteenth century, the term 'rational mechanics' as used by Newton should be distinguished from Cartesian rationalism. Hankins *1970*, Gaukroger *1982* and Pulte *1989* have documented the influence of Cartesianism on the physical science of the period, referring however to a definite set of philosophical attitudes with specific historical origins. While one can recognize Cartesian echoes in Newton's usage, he was not expressing a formal philosophical doctrine and it would be premature to infer a deeper meaning in his terminology. In view of the lack of usage by scientists of the period, its rareness in modern writing outside France, and the possibly misleading associations with Cartesian rationalism, it seems preferable not to employ the term 'rational mechanics' for theoretical mechanics in the eighteenth century.

BIBLIOGRAPHY

Aiton, E. J. *1972, The Vortex Theory of Planetary Motions*, London: Macdonald.
Dugas, R. *1957, A History of Mechanics*, London: Routledge & Kegan Paul.
Fraser, C. G. *1983*, 'J. L. Lagrange's early contributions to the principles and methods of mechanics', *Archive for History of Exact Sciences*, **28**, 197–241.
—— *1985*, 'D'Alembert's principle: The original formulation and application in d'Alembert's *Traité de dynamique*', *Centaurus*, **28**, 31–61, 145–59.
Gaukroger, S. *1982*, 'The metaphysics of impenetrability: Euler's conception of force', *British Journal for the History of Science*, **15**, 132–54.
Gillispie, C. C. *1971*, 'The Carnot approach and the mechanics of work and power, 1803–1829', in his *Lazare Carnot Savant*, Princeton, NJ: Princeton University Press, 101–20.
Gillmor, C. S. *1971, Coulomb and the Evolution of Physics and Engineering in Eighteenth-Century France*, Princeton, NJ: Princeton University Press.

Grattan-Guinness, I. *1984*, 'Work for the workers: Advances in engineering mechanics and instruction in France, 1808–1830', *Annals of Science*, **41**, 1–33.
—— *1990*, 'The varieties of mechanics by 1800', *Historia mathematica*, **17**, 313–38.
Hankins, T. L. *1970*, *Jean d'Alembert: Science and the Enlightenment*, Oxford: Clarendon Press.
Heyman, J. *1972*, *Coulomb's Memoir on Statics. An Essay in the History of Civil Engineering*, Cambridge: Cambridge University Press.
Jouguet, E. *1908–9*, *Lectures de mécanique. La Mécanique enseignée par les auteurs originaux*, 2 vols, Paris: Gauthier-Villars.
Kuhn, T. S. *1969*, 'Energy conservation as an example of simultaneous discovery', in M. Clagett (ed.), *Critical Problems in the History of Science*, Madison, WI: University of Wisconsin Press, 321–56.
Mach, E. *1883*, *Die Mechanik in ihrer Entwicklung historisch–kritisch dargestellt*, Prague. [English transl. by T. J. McCormack as *The Science of Mechanics*, 1893, Chicago, IL: Open Court. Latest (6th) English edn from the 9th German edn with a new introduction by Karl Menger, 1960, La Salle, IL: Open Court.]
Prange, H. F. W. G. *1933*, 'Die allgemeine Integrationsmethoden der analytischen Mechanik', in *Encyklopädie der mathematischen Wissenschaften*, Vol. 4, Part 2, 505–804 (article IV 12–13).
Pulte, H. *1989*, *Das Prinzip der kleinsten Wirkung und die Kraftkonzeptionen der rationellen Mechanik* [...], Stuttgart: Steiner (*Studia Leibnitiana*, Vol. 19).
Stäckel, P. G. *1905*, 'Elementare Dynamik der Punktsysteme und starren Körper', in *Encyklopädie der mathematischen Wissenschaften*, Vol. 4, Part 1, 435–684 (article IV 6).
Todhunter, I. *1886–93*, *A History of the Theory of Elasticity and of the Strength of Materials from Galilei to the Present Time*, 2 vols (ed. and completed by K. Pearson), Cambridge: Cambridge University Press. [Repr. 1960, New York: Dover.]
Truesdell, C. A. *1960*, 'The rational mechanics of flexible or elastic bodies 1638–1788', in L. Euler, *Opera omnia*, Series 2, Vol. 11, Part 2, Zurich: Orell Füssli.
—— *1968*, *Essays in the History of Mechanics*, Berlin: Springer.

8.2

Graphical statics

E. SCHOLZ

1 ORIGINS

For a long time graphical methods have been used not only to represent the geometrical shape and data of an engineering construction, but also to analyse the stability of its underlying structure. In 1687, Pierre Varignon proposed to analyse the stability of vaults by considering the structure to be inverted, and wondering whether a fictitious rope polygon with weights at the vertices comparable to the loads on the vault would, if hanging in equilibrium, take on roughly the shape of a polygon inside the vault. If so,

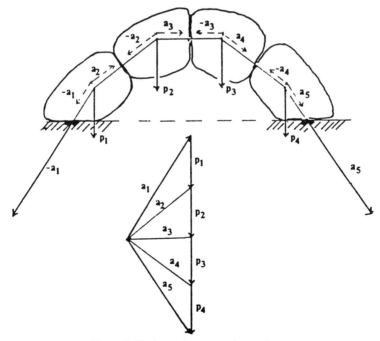

Figure 1 Varignon's supporting polygon

987

the original vault should also be in equilibrium, because when inverted the 'funicular polygon' transforms into a fictitious supporting polygon for the forces (see Figure 1; all figures are taken from Scholz *1989*).

Varignon's idea was generalized and introduced into the Ecole Polytechnique tradition as the so-called 'method of the funicular polygon' for combining plane systems of forces, and checking the balance of forces and the stability of simple structures. Given a system of forces in a plane, a fictitious funicular polygon was constructed with vertices on the lines of action of the given forces; the decomposition of forces along the funicular edges (including the directions of the latter) was constructed by means of a 'force diagram', in the later sense of vectorial force decomposition (Figure 2).

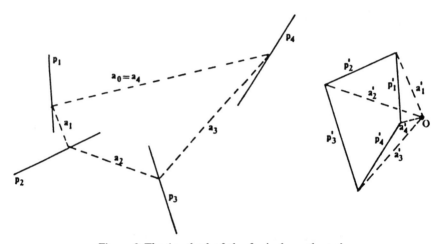

Figure 2 The 'method of the funicular polygon'

2 DETERMINATE FRAMEWORKS

With the rise of the engineering sciences during the nineteenth century and the spread of descriptive geometry (§7.5), graphical methods of statics were further developed and systematized. During the 1850s and 1860s the iron structures of railway stations, in particular bridges and halls, were analysed and constructed as determinate frames; that is, as (ideal) linkage structures which, by their geometrical properties, admit a unique decomposition of the load forces at each node in the directions of the rods meeting there. The problem of frame stability could then be reduced by graphical means to the question of the stability of the rods under well-determined longitudinal load forces (Figure 3).

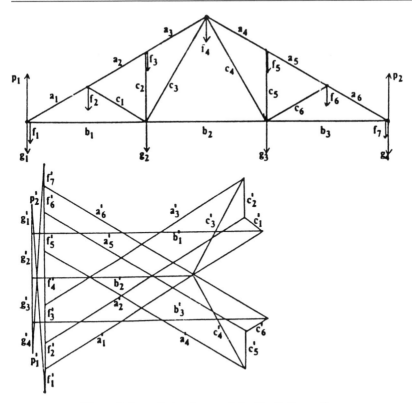

Figure 3 A roof structure and its (dual) force diagram

Graphical methods were also used for other problems in structural engineering like vault theory, and earth pressure, but it was particularly useful for determinate frame structures. In the last third of the nineteenth century engineers turned to more complicated structures; the use of iron frames with 'redundant' rods (from the point of view of determinate frameworks) would lead to questions about whether elasticity was more accessible to algebraic and analytical than to graphical methods, not to mention the problems arising from the use of reinforced-concrete structures early in the twentieth century (Straub *1964*; compare §8.7).

3 CULMANN'S CONCEPT OF GRAPHICAL STATICS

In the 1860s several engineering scientists developed systematizations of graphical methods in statics, among them Carl Culmann in Switzerland, S. Whipple in the USA and D. J. Zhouravsky in Russia (Scholz *1989*: Chap. 2). Culmann, who had been called to the newly founded

Eidgenössisches Polytechnikum (later the ETH) in Zürich after several years' experience of railway construction, wrote an influential textbook on graphical statics (*1866*), and used the idea to found the graphical principles on concepts of projective geometry. He looked, for example, for projective relationships between the funicular and the force polygons, and realized that there is a projective correlation transforming one polygon into the other if and only if the external forces of the funicular polygon intersect in one point (the 'pole' of the funicular polygon).

Culmann applied this duality theorem to achieve striking simplifications of the graphical constructions of supporting polygons in elliptical vaults. This remained, however, an application on the theoretical level, as this type of structure was no longer of particular importance to engineering practice during the second phase of industrialization. Other attempts by Culmann to introduce projective concepts to graphical statics in 1866 were even less convincing. But the programme of foundation was made public, and had a certain attraction for several more theoretically minded engineers of the time.

4 RANKINE–MAXWELL–CREMONA DUALITY

The Scottish engineer W. J. M. Rankine was confronted with problems similar to those that faced Culmann, and developed methods of his own to solve the force distribution in determinate frame structures. He used the observation that a rod polygon connected in joints is in equilibrium under external forces applied at each vertex if the polygon of external forces (in the vectorial sense) can be extended by a set of edges parallel to the rods and converging at one point, the pole of the force polygon. In this case the straight-line segments between the pole and the vertices of the force polygon represent the stresses inside the corresponding rods of the rod polygon. By repeatedly applying this observation to the analysis of determinate frame structures, Rankine hit upon the correspondence between the polygons of the rod diagram and the points (the respective poles) of a system of force diagrams. Moreover, he generalized this observation to the three-dimensional case in a paper published in 1864.

When James Clerk Maxwell became aware of Rankine's observation, he immediately abstracted from it the underlying geometrical conception, which he characterized as 'reciprocal figures' in two papers of 1864. A 'figure' in this sense of Maxwell can be understood as a two-dimensional cell complex with polygons as two-dimensional cells and straight-line segments as one-dimensional cells. Maxwell called two figures F and F' 'reciprocal' if there is a system of bijective correspondences between the polygons of F and the vertices of F', between the edges of F and the edges of F', and between the vertices of F and the polygons of F', such that corresponding edges of F and F' are parallel or, more generally, meet at a fixed angle (Figure 4). Maxwell remarked that the forces in F', when applied along the edges of F, were in equilibrium. Therefore the equilibrium and stability of frame structures could possibly be analysed by looking for a reciprocal net of force diagrams to a given frame structure and a system of external forces. For a modern version of the analysis, see §8.7.

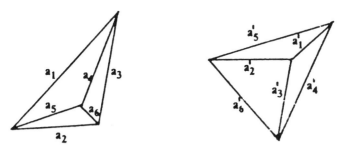

Figure 4 Maxwell's 'reciprocal figures'

Whereas his 1864 papers were of a rather theoretical character, Maxwell returned to frame structures in 1867, explaining in an engineering journal how to use his idea of reciprocal figures for a particularly effective and clear analysis of force distribution in determinate frame structures with parallel external forces (the example shown in Figure 4). This method was close enough to engineering practice and of sufficient practical advantage that it was adopted by engineers over the next few years.

In addition, Maxwell's method had some impact on Culmann's programme of projective theorization of graphical statics. Luigi Cremona, one of the Italian mathematicians who took part in the reform of academic education in Italy and taught in engineering institutions for much of the 1860s and 1870s (§11.8), took Maxwell's concept of reciprocal figures in 1872 and interpreted it from the point of view of duality in projective 3-space, the figures being plane projections of spatial polygonal nets. Using a clever method of amending the diagram of rods and external forces, he succeeded also in extending the range of applicability of the method to the case of non-parallel external forces. Thus Cremona achieved not only an integration of Maxwell's duality into Culmann's programme, but also a generalization of at least some potential practical importance.

5 THE DECLINE OF CULMANN'S PROGRAMME

When Culmann reorganized his textbook on graphical statics for its second edition of 1875, he put even more emphasis on projective conceptions than he had in 1866. He could not, however, include the larger successes of his theorization programme. The most striking example of projective ideas having consequences for the practice of graphical calculation was still the case of Maxwell–Cremona duality; his other examples remained of purely theoretical value. Culmann did not view this restriction as a severe disadvantage, because he had in mind an educational goal for engineering students which was not too far from the educational ideal (*Bildungsideal*) of the German universities (§11.2), even if concentrated on 'geometrical–statical formation', as he once explained.

This conception of theory oriented around *Bildung* became highly controversial in the 1890s in German technical institutions (Hensel *1989*). In the 1870s and 1880s a reform of the training of the young generation of German-speaking engineers had begun, resulting in a redefinition of the goals of the technical sciences, differently from those of the classical university sciences. This contextual shift, combined with relatively weak technical developments, resulted in a rapid decline of interest in Culmann's theorization programme of graphical statics. What remained in the engineering textbooks of the early twentieth century was a collection of graphical methods which, with the rising complexity of construction tasks, lost more and more of their direct practical importance and were reduced to didactical material comparable to a rather pragmatic version of geometrical–statical formation – a far cry from what Culmann had dreamt of in the 1860s and 1870s.

BIBLIOGRAPHY

Culmann, C. *1866*, *Die graphische Statik*, Zürich: Meyer and Zeller. [2nd edn 1875.]

Henneberg, E. L. *1903*, 'Die graphische Statik der starren Körper', in *Encyklopädie der mathematischen Wissenschaften*, Vol. 4, Part 1, 345–435 (article IV 5).

Hensel, S. *1989*, 'Die Auseinandersetzung um die mathematische Ausbildung der Ingenieure an den Technischen Hochschulen in Deutschland Ende des 19. Jahrhunderts', in Hensel *et al.*, *Mathematik und Technik im 19. Jahrhundert in Deutschland. Soziale Auseinandersetzung und philosophische Problematik*, Göttingen: Vandenhoek & Ruprecht, 1–111.

Scholz, E. *1989*, *Symmetrie – Gruppe – Dualität. Zur Beziehung zwischen theoretischer Mathematik und Anwendungen in Kristallographie und Baustatik des 19. Jahrhunderts*, Basel: Birkhäuser.

Straub, H. *1949, Die Geschichte der Bauingenieurkunst*, Basel: Birkhäuser. [2nd edn 1964 (revised); 3rd edn 1975.]

Timoshenko, S. P. *1953, History of Strength of Materials*, New York: McGraw-Hill.

8.3

Kinematics

TEUN KOETSIER

1 INTRODUCTION

In his *Essai sur la philosophie des sciences* (1834), in which he introduced a new classification of the sciences, André-Marie Ampère proposed the word 'kinematics' (*cinématique*) for a field of mechanics that would be concerned with motion independent of its causes. The time was ripe for such a proposal. In geometry and theoretical mechanics interesting kinematical results concerning motion had been obtained, but it was developments in mechanical engineering that were particularly striking. At the time of the foundation of the Ecole Polytechnique in 1794, Gaspard Monge had decided that a course on machine elements had to be included in the curriculum, and Jean Hachette was given the task of preparing a text. Unfortunately, changes of curriculum delayed introduction of the course until 1806, and not until 1811 was Hachette's textbook, the *Traité élémentaire des machines*, ready (Ferguson *1962*). Although Hachette's book also contained dynamical considerations, an important part of it concerned the geometry of motion of different mechanisms.

In this article the kinematics of mechanisms, which deals with the kinematic properties of particular mechanisms, is distinguished from theoretical kinematics, which deals with the kinematic properties of motion in general. Soon after Ampère had coined the word 'kinematics', these became independent areas of investigation. Although the French dominated on the theoretical level, the first book entirely devoted to the kinematics of mechanisms was by an Englishman: R. Willis's *Principles of Mechanism* (1841). The first book entirely devoted to theoretical kinematics was H. Résal's *Traité de cinématique pure* (1862).

2 KINEMATICS OF MECHANISMS: WATT'S PARALLEL MOTION

A famous mechanism that played an important role in the development of

kinematics in the nineteenth century was the so-called 'parallel motion'
invented by James Watt in 1784. Watt's problem was to connect the piston-
rod of a double-acting steam engine pushing and pulling perpendicularly up
and down, with the engine's working beam rocking back and forth. The
existing solutions were very unsatisfactory. Watt had the brilliant idea of
guiding the upper end of the piston rod 'by only fixing it', as he wrote to
his partner Matthew Boulton, 'to a piece of iron upon the beam without
chains, or perpendicular guides, or untowardly frictions, archheads, or
other pieces of clumsiness'. In practice, the mechanism worked perfectly
and, wrote Watt, did 'not make the shadow of a noise'. The success of
the mechanism is based upon the property that the point D (Figure 1(a)),
connected to the piston rod, ascends and descends over a long distance
in an approximately straight line. Watt explained it by saying that the
'convexities' described by the two points E (end of beam) and C (rocking
about F, attached to the engine frame), lying in opposite directions,
compensate each other.

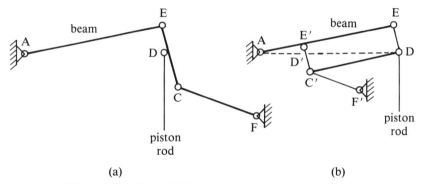

Figure 1 Watt's parallel motion: (a) in its original form;
(b) compact mechanism utilizing parallelogram extension E'C'DE.
The triangles indicate fixed pivots

Later in the same year, Watt succeeded in making the mechanism considerably more compact by means of a parallelogram (Figure 1(b)). Because
E'C'DE is a parallelogram, D and D' describe similar curves. Since the
bars AE', E'C' and C'F' form a smaller version of the mechanism of
Figure 1(a), the dimensions can be chosen such that D in both figures
describes exactly the same approximately straight line. The parallel motion
achieved with the parallellogram was very successfully applied in many
steam engines and greatly admired. Dominique François Arago wrote about
it in 1834: 'At each double oscillation it unfolds itself and closes itself, with

the softness, I would almost say with the grace that fascinates us in the gestures of an accomplished actor.'

Linkages – that is, mechanisms consisting of links or bars connected by pins – had been used before 1784. However, with his parallel motion Watt applied a linkage in a completely new way, and this application decisively influenced the development of the theory of linkages (Koetsier *1983*). Under the influence of Watt's invention, linkages became the object of serious investigations that led to many interesting results, particularly in the second half of the nineteenth century.

3 THEORETICAL KINEMATICS: CHASLES

Watt's parallel motion belongs to the kinematics of mechanisms. At the time of Ampère, there were also interesting results belonging to theoretical kinematics in theoretical mechanics and in geometry. A simple but very fundamental result had already been proved in the first half of the eighteenth century by John Bernoulli: that planar motion is instantaneously always a rotation or a translation. Bernoulli's argument was dynamical, based upon the occurrence of forces. Another proof of the same result was given by Augustin Louis Cauchy in 1827. In the same paper Cauchy also proved another important theorem: that, generally speaking, all planar motion can be generated by means of a curve rolling without slipping along another fixed curve. Cauchy argued as follows. Excluding instantaneous translations, the path during the motion of the instantaneous rotation centre (also frequently called the pole) is a curve, or rather two curves – one in the fixed system, the fixed 'polhode', and one in the moving system, the moving 'polhode'. At each instant the two polhodes touch at the instantaneous rotation centre. Cauchy showed that during the motion the moving polhode rolls without slipping on the fixed polhode.

Cauchy's proofs were based upon the existence of forces. However, in 1829 Michel Chasles showed that such theorems could be proved easily by geometrical means alone. His proof of the existence of the instantaneous rotation centre is extremely simple. The two congruent triangles ABC and A'B'C' (Figure 2) represent two positions of a moving plane. It is easy to see that the perpendicular bisectors of AA', BB' and CC' have a common point of intersection, P, and that by rotation about P, triangle ABC can be made to coincide with triangle A'B'C'. If the two triangles are infinitesimally close, the perpendiculars are the normals to the trajectories of A, B and C, and P is the instantaneous centre of rotation. Chasles applied these results in different ways: he showed, for example, how the instantaneous centre of rotation can be used to draw tangents to curves, if those curves can be defined by a suitable motion.

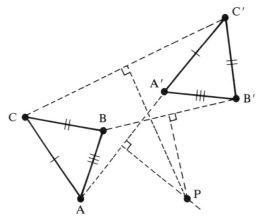

Figure 2 Chasles's 1829 proof of the existence of the instantanéous centre of rotation

Not only the first-order kinematical properties of motion (tangents, velocities) were studied, but also second-order properties (centres of curvature, accelerations). For example, in the 1830s, E. Bobillier gave a well-known construction to determine, for arbitrary instantaneous planar motion, the centre of curvature of the trajectory of an arbitrary point in the moving plane, if the centres of curvature of the trajectories of two other points are given (Koetsier *1986*).

Another interesting result from Ampère's time concerns the Coriolis acceleration. Gustave de Coriolis showed in 1835 that the ordinary laws of motion may be used in a rotating frame of reference if what is now called the Coriolis acceleration is included in the equations of motion. The acceleration appeared in Coriolis's work as an extra force that had to be taken into consideration, but soon the purely kinematical nature of the result became clear. The Coriolis acceleration is, for example, very important in studies of the dynamics of the atmosphere and in ballistics (§9.6, §8.11).

4 WORK ON LINKAGES

Watt was aware that his 'parallel motion' produced motion (of point D in Figure 1(a)) in only approximately a straight line. The question must have arisen of whether a similar mechanism describing an exact straight line could be made. At the beginning of the 1850s the Russian mathematician Pafnuty Chebyshev became fascinated by Watt's linkages. He did not believe that an exact straight-line linkage could be constructed, and directed his attention to the problem of how to choose the dimensions of a given mechanism so as to minimize the maximum deviation from a straight line over a given range. In this context Chebyshev discovered the well-known Chebyshev polynomials (§4.4).

At the beginning of the 1860s the Frenchman J.-N. Peaucellier discovered an exact straight-line linkage (Figure 3). The mechanism consists of seven links. The links PA and PC are of equal length, and are both pivoted at the fixed point P. Their other ends, A and C, are attached to the opposite angles of a rhombus composed of four equal links. The seventh link, BQ, is pivoted at the fixed point Q in such a way that QB equals QP. When QB rotates about Q, the hinge D describe a straight line (perpendicular to PQ).

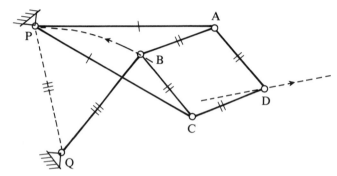

Figure 3 Peaucellier's exact straight-line linkage. The triangles indicate fixed pivots

Peaucellier's invention did not attract much attention. Then, in 1873, at an international exhibition in Vienna, the Russians showed precisely the same straight-line mechanism, independently invented in 1870 by one of Chebyshev's students, L. Lipkin. Immediately the French woke up; Peaucellier received an important prize for his invention, and internationally the mechanism became the subject of much interest. In England, the mathematician J. J. Sylvester, who became acquainted with the mechanism through Chebyshev, was immensely impressed by it. When Sylvester said that 'it would be difficult to quote any other discovery which opens out such vast and varied horizons as this of Peaucellier', he was exaggerating, but he stimulated young mathematicians like A. B. Kempe and H. Hart to study linkages.

In the 1870s the investigations on linkages by these English mathematicians and two others, Arthur Cayley and S. Roberts, led to several interesting results. The text of a series of lectures on exact straight-line linkages given in 1876 by Kempe wonderfully illustrates the kind of research that was involved (Kempe *1877*). New linkages were invented and, for example, in 1875 Roberts derived the main properties of the peculiar sixth-degree curves that are described by 'three-bar linkages' like Watt's parallel motion in its simple form (Figure 1(a)): they are tri-circular, sixth-degree curves with three finite double points that lie on the circumscribed circle of the triangle formed by the three singular foci the curves possess (Koetsier *1983*).

5 THE NEW SCHOOL OF KINEMATICS OF MECHANISMS: REULEAUX

Especially in the second half of the nineteenth century, kinematics enjoyed great popularity among both mechanical engineers and mathematicians. A figure of major importance was the German engineer Franz Reuleaux. In 1864 he started lecturing on his ideas, and in 1875 he published the book that made him famous: *Theoretische Kinematik* (a theoretical treatise on kinematics of mechanisms, in spite of its title).

We owe the way in which we look at mechanisms today to Reuleaux. In the course of the nineteenth century many new mechanisms had been invented, and it was difficult to classify them in a sensible way. Reuleaux introduced a new abstract point of view. He recognized that the pedestal of a mechanism has to be considered as an element of the mechanism, just like the other elements. He realized that one ought to classify mechanisms by considering them as consisting of a chain of links, counting also the pedestal as a link. Fixing different links of the same kinematic chain creates mechanisms that look quite different but are in a sense kinematically identical. Reuleaux convincingly proved the fertility of his ideas. In his book he gave several examples of rotary steam engines that had been patented as different but, after his analysis, appeared to be based upon the same kinematic chain (Ferguson *1962*).

6 THE MATHEMATIZATION OF THE KINEMATICS OF MECHANISMS: BURMESTER

Although, for example, in 1830 Hachette used the instantaneous centre of rotation to determine, in Watt's parallel motion, the tangents of the curve described by the end of the piston-rod, for a long time the kinematics of mechanisms did not profit very much from theoretical kinematics. In this respect the German mathematician L. E. H. Burmester brought about an important change. In 1888 his *Lehrbuch der Kinematik* appeared, partially based upon earlier work. In its 941 pages he gave an extensive treatment of planar theoretical kinematics, and applied the results thus obtained to a wealth of examples from the kinematics of mechanisms. The book contains graphical methods for determining the instantaneous velocities and accelerations for a given mechanism in any position. Similar methods for the 'kinematical analysis' of mechanisms were developed in Britain by the engineers Alexander B. W. Kennedy (who translated Reuleaux's book into English) and R. H. Smith.

Burmester's work, however, was much more influential. A very original contribution of his was the so-called 'Burmester theory', which deals with three, four or five given discrete positions of a 'moving plane' in a fixed plane. For three discrete positions there will obviously be points in the moving plane that are in the three positions on a straight line in the fixed plane. Burmester showed that all such points lie on a circle in the moving plane. He also showed that for four positions the points of the moving plane that are in these four positions on a circle in the fixed plane all lie on a third-degree curve in the moving plane, the so-called circle-point curve. The theory can be used in 'kinematical synthesis' – to design mechanisms. Points in the moving plane that are in a number of given positions on a circle can become the end-points of a link pivoted at the centre of that circle. Two such links plus a connecting coupler constitute a mechanism that can, in principle, move the moving plane through the given positions (Koetsier *1989*).

7 CONCLUDING REMARKS

The roots of kinematics lie in geometry, in theoretical mechanics and in mechanical engineering. Although in the nineteenth century kinematics became an independent area of research, the institutionalization of kinematics as an independent discipline remained limited, and interest in kinematics was usually a derived interest. For example, mathematicians' interest in kinematics was connected with their interest in geometry (Schönflies and Grübler *1902* contains an excellent survey of the nineteenth-century results). In the course of the twentieth century mathematicians in general turned their attention away from (classical) geometry, other areas of mathematics seen as promising more (or more general) results; the interest of mathematicians in kinematics as a research topic almost completely disappeared.

In mechanical engineering, the situation was different. In the 1870s and 1880s the engineering interest in kinematics was considerable. This was particularly so in Germany, where Reuleaux's ideas were very influential. Unfortunately, there was a considerable gap between the work of Reuleaux, Burmester and others, on the one hand, and engineering practice on the other hand. The designers and builders of machines often viewed Reuleaux's kinematical theories as far too theoretical, and in the course of time he had to face serious opposition. In the 1890s the so-called 'engineer movement' succeeded in removing most of the mathematics, and also Reuleaux's kinematics, from the programme of the German technological universities (Hensel *et al. 1989*). It was only after the First World War that mechanical engineers began to appreciate kinematics again. Since then, within mechanical engineering, the kinematics of mechanisms has been a small but fertile area of research. As for applications, before the Second World War graphical methods were in particular quite popular, but in the second half of the twentieth century, with the ever more extensive use of computers, analytical methods have prevailed.

BIBLIOGRAPHY

Ferguson, E. S. *1962*, 'Kinematics of mechanisms from the time of Watt', *Contributions from the Museum of History and Technology, United States National Museum*, Bulletin 228, Washington, DC: Smithsonian Institute.

Hensel, S., Ihmig, K. N. and Otte, M. *1989, Mathematik und Technik im 19. Jahrhundert in Deutschland. Soziale Auseinandersetzung und philosophische Problematik*, Göttingen: Vandenhoeck & Ruprecht.

Kempe, A. B. *1877, How to Draw a Straight Line: A Lecture on Linkages*, London: Macmillan. [Repr. as Vol. 6 of the series Classics in Mathematics Education, The National Council of Teachers of America, 1977, Albion, MI: Pentagon.]

Koetsier, T. *1983*, 'A contribution to the history of kinematics', *Mechanism and Machine Theory*, **18**, 37–48. [On French and English work.]

—— *1986*, 'From kinematically generated curves to instantaneous invariants: Episodes in the history of instantaneous planar kinematics', *Mechanism and Machine Theory*, **21**, 489–98.

—— *1989*, 'The centenary of Ludwig Burmester's "Lehrbuch der Kinematik"', *Mechanism and Machine Theory*, **24**, 37–8.

Schönflies, A. and Grubler, M. *1902*, 'Kinematik', in *Encyklopädie der mathematischen Wissenschaften*, Vol. 4, Part 3, 190–278 (article IV 3).

8.4

Feedback control systems

A. T. FULLER

1 THE NOTION OF FEEDBACK CONTROL

The basic idea of feedback control is quite simple. Suppose there is some quantity, for example the temperature of a room, which is required to be kept close to a certain reference value even when there are disturbances, such as changes in the weather. One can measure the temperature, and arrange for more fuel to be supplied if the room temperature drops, or less fuel if it rises. All this can be done automatically, and the resulting system is a feedback control system. It comprises a feedback loop because an output signal dependent on an effect (temperature) is fed back to change an input or cause (fuel supply).

A naively designed control system can often exhibit untoward behaviour. The output variable can approach and reach the reference value, but may tend to keep on moving and thus overshoot. Such overshoots can build up into a sustained or increasing oscillation, which may prevent the system from operating effectively or, worse, may result in catastrophic failure. The designer is thus faced with the problem of making the system respond sufficiently quickly while at the same time avoiding instability, a problem entailing considerable mathematical difficulties.

2 EARLY DEVELOPMENT OF CONTROL THEORY IN BRITAIN

Automatic control was brought to public attention in a dramatic way in 1788 when James Watt began to fit fly-ball governors to his steam engines. The invention was not new, however, as similar contrivances had previously been used in watermills and windmills. In the Watt governor the two fly-balls act as conical pendulums (§8.13). When the engine speed changes, the pendula move inwards or outwards due to change of centrifugal force, and this inward or outward motion is exploited to increase or decrease the steam supply, thus tending to restore the engine speed to its normal value.

Watt's governor was sometimes found to create instability. The first person to make a theoretical study of the instability problem in speed governors was George Biddell Airy, who was Astronomer Royal at Greenwich from 1835 to 1881. He encountered the problem in the design of clockwork drives for large telescopes (Airy *1840*). The clockwork turned the telescope so that it compensated for the rotation of the Earth, and a centrifugal governor was fitted in an attempt to keep the speed constant.

Astronomers were familiar with the study of the stability of dynamical systems — in particular Joseph Louis Lagrange and Pierre Simon Laplace had discussed at length the stability of the Solar System (§8.8). Thus Airy was well placed to initiate the investigation of stability problems associated with governors. He set up the differential equations for a simple system, found first integrals corresponding to expressions for energy and angular momentum, and arrived at a non-linear ordinary differential equation for the outward displacement of the fly-balls. He then considered the evolution of small deviations from the normal operating conditions. Unfortunately Airy's equation, while of first order, was of second degree, and this meant that linearization about the operating point was not straightforward. He nevertheless succeeded in showing that, for the system he studied, small oscillations would build up exponentially; that is, he was able to account mathematically for the instability phenomenon.

The next major contribution to control theory was by James Clerk Maxwell, in a paper *1868* entitled 'On governors'. Although aware of steam-engine governors, Maxwell was primarily motivated by the need to govern the speed of a spinning electrical coil, which he was using in connection with experiments to determine the ratio of electromagnetic and electrostatic units of electricity. He hoped to show that this ratio was equal to the velocity of light — a crucial point for the field theory he was developing (§9.10). Like Airy, Maxwell was already familiar with stability problems in astronomy, as he had previously made a major investigation of the stability of Saturn's rings in an essay for which he obtained the Adams Prize at Cambridge in 1859.

Maxwell's study of governors was more general and more systematic than Airy's. He set up differential equations and immediately linearized them, rather than using the awkward technique of working with first integrals. He pointed out that, for stability, the characteristic equation of the resulting linear differential equations had to have all its roots with negative real parts. There then arose the interesting question of how to interpret this condition in terms of the coefficients of the characteristic equation. Maxwell succeeded in answering the question for systems of third order, gave partial results for a system of fifth order, and posed the problem of finding similar stability criteria for systems of general order.

Maxwell's paper on governors was written when he was aged about 35 and many of his ideas on other subjects were coming to fruition. It bears signs of hasty preparation, and the consequent cryptic and fragmentary style prevented the paper from having a direct effect on the subsequent development of control theory. It did, however, have an indirect influence through the work of E. J. Routh. In the second edition of 'the advanced part' of his book on rigid dynamics (*1868*: 291–6), Routh wrote a new section on governors that was based essentially on the work of Airy and Maxwell. The book was gradually expanded in successive editions, and eventually included Routh's own solution of Maxwell's problem of stability criteria. Routh's solution was originally given in an essay which won for him the Adams Prize in 1877, 18 years after Maxwell had obtained the award.

3 SOME CONTINENTAL CONTRIBUTIONS

Meanwhile, related work was being done in Continental Europe. J. J. L. Farcot invented position control systems, which he called 'servo-motors', around 1873; these were mainly for naval applications. G. Marié in 1878 drew attention to the common features of speed, pressure and temperature control systems. F. Lincke indicated the analogy with biological control systems in 1879.

J. Wischnegradski *1876*, unaware of Maxwell's work, studied the dynamics of the governing of steam engines, and although his treatment in this and subsequent papers was less deep and less correct than Maxwell's, it was more clearly expressed and consequently had greater impact. A. Stodola, working on the control of hydro-electric turbines in Switzerland, was influenced by Wischnegradski's treatment, and asked his mathematician colleague A. Hurwitz to solve the problem of finding stability criteria for systems of general order. Hurwitz did not know about Routh's work, and duly came up with his own version of the stability criteria. These took the form of elegant determinantal inequalities: for example, for the real equation

$$a_0 z^4 + a_1 z^3 + a_2 z^2 + a_3 z + a_4 = 0, \quad a_0 > 0, \tag{1}$$

his necessary and sufficient conditions for every root to have a negative real part are (Hurwitz *1895*):

$$a_1 > 0, \quad \begin{vmatrix} a_1 & a_3 \\ a_0 & a_2 \end{vmatrix} > 0, \quad \begin{vmatrix} a_1 & a_3 & 0 \\ a_0 & a_2 & a_4 \\ 0 & a_1 & a_3 \end{vmatrix} > 0, \quad \begin{vmatrix} a_1 & a_3 & 0 & 0 \\ a_0 & a_2 & a_4 & 0 \\ 0 & a_1 & a_3 & 0 \\ 0 & a_0 & a_2 & a_4 \end{vmatrix} > 0.$$

$$\tag{2}$$

Aleksander Lyapunov *1892* raised stability theory to a new level of generality and rigour. He had astronomical rather than control problems in mind, but his work would be used in control studies in the mid-twentieth century. For non-linear systems he gave a stability criterion which is roughly as follows. Suppose that the state of the system can be represented by a finite number of variables $x_1, x_2, \ldots, x_n$ which evolve with time, and that there is a positive-definite function $f(x_1, x_2, \ldots, x_n)$ which decreases as time increases whenever f is non-zero; then the system is stable. Lyapunov also gave conditions under which the method of linearization, as used by Maxwell and others, yields a valid assessment of stability for non-linear systems.

Detailed discussions of the history of control have been given by Mayr *1970*, Fuller *1975, 1976, 1979* and Bennett *1979*. For an introduction to modern control theory see Takahashi *et al. 1970*; and for an extensive guide to the literature see the *Systems and Control Encyclopedia* edited by Singh *1987*.

BIBLIOGRAPHY

Airy, G. B. *1840*, 'On the regulator of the clock-work [...]', *Memoirs of the Royal Astronomical Society*, **11**, 249–67.
Bennett, S. *1979*, *A History of Control Engineering 1800–1930*, London: Institution of Electrical Engineers.
Fuller, A. T. (ed.) *1975*, *Stability of Motion*, London: Taylor & Francis.
—— *1976*, 'The early development of control theory', *Transactions of the American Society of Mechanical Engineers*, **98G**, 109–18, 224–35.
—— *1979*, 'Clerk Maxwell's London notebooks [...]', *International Journal of Control*, **30**, 729–44.
Hurwitz, A. *1895*, 'Über die Bedingungen [...]', *Mathematische Annalen*, **46**, 273–84. [Also in *Mathematische Werke*, Vol. 2, 533–45.]
Lyapunov, A. *1892*, *Obshchaya zadacha ob ustoichivosti dvizheniya*, Kharkov: Kharkov Mathematical Society. [French transl. 1949 as *Problème générale de la stabilité du mouvement*, Princeton, NJ: Princeton University Press. English transl. 1992 as *The General Problem of the Stability of Motion*, London: Taylor & Francis.]
Maxwell, J. C. *1868*, 'On governors', *Proceedings of the Royal Society of London*, **16**, 270–83. [Also in *Scientific Papers*, Vol. 2, 105–20.]
Mayr, O. *1970*, *The Origins of Feedback Control*, Cambridge, MA: MIT Press.
Routh, E. J. *1868*, *Dynamics of a System of Rigid Bodies*, 2nd edn, Part 2, London: Macmillan.
—— *1877*, *Stability of a Given State of Motion*, London: Macmillan.
Singh, M. G. (ed.) *1987*, *Systems and Control Encyclopedia*, Oxford: Pergamon.
Takahashi, Y., Rabins, M. J. and Auslander, D. M. *1970*, *Control and Dynamic Systems*, Reading, MA: Addison-Wesley.
Wischnegradski, J. *1876*, 'Sur la théorie générale des régulateurs', *Comptes Rendus Hebdomadaires des Séances de l'Académie des Sciences*, **83**, 318–21.

8.5

Hydrodynamics and hydraulics

G. K. MIKHAILOV

1 INTRODUCTION

A reasonably clear distinction between the sciences of hydraulics and hydrodynamics was made only in the nineteenth century. The ninth edition of the *Encyclopaedia Britannica* (1880) treats hydromechanics as 'the mechanics of water and fluids in general' and divides it 'into three branches: hydrostatics, which deals with the equilibrium of fluids; hydrodynamics, which deals with the mathematical theory of the motion of fluids, neglecting the viscosity; and hydraulics, in which the motion of water in pipes and canals is considered, and hydrodynamical questions of practical application are investigated'. Published half a century later, the 14th edition of the *Encyclopaedia* (1929) gives no definition of hydromechanics, but divides it into hydrostatics and hydrodynamics, considering hydraulics as 'that branch of engineering science which deals with the practical application of the laws of hydrodynamics'.

The term 'hydraulics' appeared in the period of the early Renaissance and came from the conjunction of two Greek words: ὕδωρ ('water') and αὐλός ('pipe'). However, its etymological origin goes back, most likely, to ὕδραυλος, the name of the Greek water organ invented about the second century BC. 'Hydraulics' initially covered all practical aspects of and rules for handling water-works; in ancient times the Greeks seemed to use for it the word ὑδρᾰγωγία.

According to a modern point of view, theoretical hydraulics is the study of quasi-one-dimensional (or even all vertically averaged) flow of an incompressible or a weakly compressible liquid, and engineering hydraulics deals with practical applications connected with the flow of such a liquid.

The term 'hydrodynamics' was first used by Daniel Bernoulli as the title of his book *Hydrodynamica* on the forces and motions of fluids (*1738*). He explained its content as a junction of the sciences of hydrostatics ('considering the pressures and various equilibria of stagnant liquids') and hydraulics ('considering the motion of fluids'). In the modern sense, hydrodynamics begins with the work of Jean d'Alembert and Leonhard Euler, published in the middle of the eighteenth century, and treats the motion of fluids defined by some physico-mechanical models.

2 HYDRAULICS BEFORE THE NINETEENTH CENTURY

2.1 General sources

There is little information about water-works and hydraulic devices of ancient times. The early centres of civilization in Egypt, Mesopotamia, India and China systematically used various machines for irrigation and water supply. Water-screws and simple water-wheels seem to have been used to raise water at least as far back as the third century BC. The invention of the two-piston water pump has been dated to the second century BC. The earliest use of water power is connected also with water-wheels (undershot, primitive overshot and horizontal water-wheels) developed for this purpose during the last two centuries BC (Usher *1954*).

The earliest extant treatise devoted to practical hydraulics is a manual on urban water supply written by Roman soldier and engineer Sextus Julius Frontinus (flourished first century AD), superintendent of the aqueducts of Rome. His book was republished by the end of the fifteenth century.

It is impossible not to mention the artistic observations of open, flowing water drawn by Leonardo da Vinci and his sketches of devices for studying flow (including a prototype of Hele-Shaw's cell). However, his influence on the development of science is rather dim, as his notebooks (written in a mirror image) remained forgotten until the end of the eighteenth century.

The new engineering hydraulics begins in the seventeenth century. The development of hydraulic machinery at that time is vividly reflected in G. Branca's *Le macchine* (1629), a large collection of pictures of various machines, including pumps and water-wheels.

The hydraulic principle of continuity was explicitly stated by a pupil of Galileo, in 1628. The first properly quantitative and essentially correct relation for the velocity of water efflux from orifices seems to be given by Evangelista Torricelli and Marin Mersenne, in 1644.

2.2 Theoretical hydraulics

Elements of a water-motion theory can be perceived in Book II of Isaac Newton's *Principia* (1687), where one finds the first mathematical analysis of water efflux from vessels through orifices. The analysis was improved in the second edition (1713).

The subject of theoretical hydraulics was created by Daniel and John Bernoulli, and Euler. Following the principle of live forces (§8.1), actively popularized by John Bernoulli, his son Daniel gave in 1727 the first complete hydraulic analysis of the efflux of an ideal liquid from vessels, based on the principle of live forces and the use of parallel cross-sections with velocities averaged over them. At the same time, Euler found a similar solution in the same way, but he did not publish it (Mikhailov *1983*).

A general hydraulic theory of the flow through, and efflux from, compound vessels, founded on the same basis, was presented by Daniel Bernoulli in his skilfully elaborated *Hydrodynamica* (*1738*), in which he analysed mean liquid velocities and wall pressures (without having any general concept of hydrodynamic pressure). The work contains many deep physical insights. By a sophisticated method he came, in particular, to a relation between velocities and (wall) pressures for steady flow, equivalent to that which is now generally known as the Bernoulli equation. This relation was given in analytical form (for the general case of non-steady flow) in John Bernoulli's *Hydraulica* (*1742*). He systematically used a method corresponding, in essence, to Newton's method of accelerating forces (Newton's second law), introducing the idea of the internal pressure of the flow, necessary for the future formulation of the equations of hydrodynamics. John's *Hydraulica* was, to a considerable extent, an exposition and development of his son Daniel's results in a more general dynamic language, without recourse to the principle of live forces he had formerly supported so ardently. The Bernoullis' fundamental works determined the progress of theoretical hydraulics for more than a century.

Among other theoretical results was an explanation of hydraulic-head loss at sudden expansions of a flow, given by Jean Charles Borda (1776) by means of the principle of live forces. With the help of momentum analysis, he also found the contraction factor (0.5) for a jet entering the so-called 'Borda mouthpiece' before leaving an orifice (Mascart *1919*).

2.3 Applied hydraulics

A number of experiments were carried out during the eighteenth century, mainly in France, to investigate the resistance of fluids to moving bodies, and in flow past bodies or through pipes and channels. The main result of those experiments was to confirm the law of the proportionality of the resistance to the square of the flow velocity V, but many later authors proposed the form $aV + bV^2$ that was first used for analytical calculations in Newton's *Principia*. New experimental devices were invented for measuring flow: Pitot's velocity tube (1732), the rotating arm for resistance tests and Woltman's turbine current meter (1790).

Only a few results were achieved by the engineering hydraulics of open channels. The well-known formula $v = C(RI)^{1/2}$ for the velocity of a uniform flow of water in open channels goes back to an analysis carried out by the French engineer A. Chézy in about 1770 (but not duly published), based on the hypothesis of a square-law flow resistance and balancing the effects of gravity and resistance (Mouret *1921*).

There were attempts to find engineering applications for the theory: the design of a reaction water-jet ship (Daniel Bernoulli *1738*) and a reaction water turbine (Euler, 1754). However, both proposals were realized only in the next century, and they had to be designed anew.

3 HYDRODYNAMICS DURING THE EIGHTEENTH CENTURY

3.1 The beginning

Some rudimentary elements of hydrodynamics appear in Newton's *Principia*, where he proposed a formal 'molecular' model for studying the resistance of bodies moving in rarefied and dense fluids, introduced the idea of a viscous (Newtonian) liquid and discussed the propagation of surface water waves. An essentially continuum approach was developed by Newton only in studying the propagation of sound through air, but it was too complicated and vague for his contemporaries (Cannon and Dostrovsky *1981*).

The real foundation of hydrodynamics was laid by d'Alembert and Euler in the 1750s. Studies in this field were connected with attempts both to create a reasonable theory of fluid resistance (which was not achieved at the time) and to apply general dynamic principles to continuous media. Some traces of fluid velocity analysed as a continuum can be found in Euler's translation of Benjamin Robins's *New Principles of Gunnery* (1745), where Euler also gave a proof (later forgotten by him) of the so-called 'd'Alembert's paradox' (the absence of drag in the flow of an ideal fluid past a body).

An important step towards constructing the general equations of motion for continuous media consisted in using fixed Cartesian spatial coordinates, first introduced in a clear manner by Colin Maclaurin in his *Treatise on Fluxions* (1742), and independently by Euler in a paper published in the Berlin Academy's *Mémoires* in 1752. There Euler formulated 'the general and fundamental principle of the whole mechanics', comprising a systematic use of the Newtonian laws of dynamics, treated as projections onto fixed coordinate axes, for each particle of the body (§8.1). Euler commented on his principle thus (Truesdell *1954*: xlii):

> This sole principle shall be the basis of all other principles, both those already obtained in Mechanics and Hydraulics... and also those that are not yet known and that we need for developing both... the cases of solid bodies and many other concerning fluid bodies.

3.2 Equations of motion

The first step to formulating the equations of fluid motion was taken by d'Alembert in his *Essai d'une nouvelle théorie de la résistance des fluides* (1752) in which he developed, in typically tortuous manner, a sophisticated approach to deriving differential equations of motion, and obtained hydrodynamic equations for some special cases.

The general equations of motion for an ideal fluid were given by Euler in his paper 'Principes généraux du mouvement des fluides', presented in the Berlin Academy of Sciences in 1755 and published in its *Mémoires* in 1757. Using the modern concept of the internal pressure of a fluid, and considering a fixed infinitesimal rectangular parallelepiped in the flow domain, Euler obtained three equations of motion and the continuity equation. Also necessary, he noted, is a relation between fluid density q, pressure p and another suitable physical parameter (e.g. temperature), an equation we now call the constitutive equation. Euler cast his equations in quite an up-to-date form (except for his notation '$(\mathrm{d}p/\mathrm{d}x)$' for partial derivatives). For a particle moving under the action of force (P, Q, R) per unit mass with velocity (u, v, w),

$$P - \frac{1}{q}\left(\frac{\mathrm{d}p}{\mathrm{d}x}\right) = \left(\frac{\mathrm{d}u}{\mathrm{d}t}\right) + u\left(\frac{\mathrm{d}u}{\mathrm{d}x}\right) + v\left(\frac{\mathrm{d}u}{\mathrm{d}y}\right) + w\left(\frac{\mathrm{d}u}{\mathrm{d}z}\right), \qquad (1)$$

with two similar equations for the y- and z-axes; and

$$\left(\frac{\mathrm{d}q}{\mathrm{d}t}\right) + \left(\frac{\mathrm{d}.qu}{\mathrm{d}x}\right) + \left(\frac{\mathrm{d}.qv}{\mathrm{d}y}\right) + \left(\frac{\mathrm{d}.qw}{\mathrm{d}z}\right) = 0. \qquad (2)$$

His paper was set out so clearly that there is no difficulty in understanding it, even for a student today.

In the same paper Euler introduced force and velocity potentials and obtained some equations that are now usually called Lagrange–Cauchy integrals. (For further details about his and his contemporaries' contributions to hydrodynamics, see the comprehensive surveys by Truesdell *1954*, *1955*.) He systematically used the spatial ('Eulerian') description, but also introduced the material ('Lagrangian') description. Later, Joseph Louis

Lagrange used the latter widely, including in his *Méchanique analitique* (1788); hence the description 'Lagrangian'.

3.3 Dimensional units

When reading old original works in mechanics, it is necessary to bear in mind that there was no general physical theory of dimensions (Macagno *1971*). The systems of primary dimensional units (and even the quantity of units) used by various scientists in different periods were not identical. In his earlier works, Euler often used a system with only two primary units: L(ength) and F(orce). The systems of primary units L(ength)–M(ass)–T(ime) and L(ength)–F(orce)–T(ime) were introduced in the following century. An essential point is that the form of equations in mechanics can depend on the chosen system of primary dimensional units. Thus, in his paper of 1752 Euler wrote Newton's second law in the form $2M \, dd x = P \, dt^2$. The factor 2 appeared here because Euler used L and F as primary units, so that masses are measured in units of force, accelerations are dimensionless, and time is actually measured by the square root of length. The fact is that Euler introduced velocity and time independently: velocity was measured as the square root of the height defined by the condition that a body, falling from a state of rest, would attain the given velocity during its fall through this height, and time was measured by the ratio of the path described by the body to the determined velocity. The original Euler equations of that type can be converted into their modern form by conserving lengths and forces and changing other quantities according to the following scheme:

$$\text{mass} \to mg, \qquad \text{velocity} \to v/(2g)^{1/2}, \qquad \text{time} \to t(2g)^{1/2}. \qquad (3)$$

4 HYDRAULICS DURING THE NINETEENTH CENTURY

4.1 Introduction

Throughout the nineteenth century, hydraulics, as a part of fluid dynamics, was in a peculiar state. Water supply, water-energy utilization, and the construction of canals and various water-works needed more and more reliable design methods for water-flow in pipes, open channels and special engineering devices. However, the theory provided no answer to the main question: how to explain and to estimate the basic characteristic – the hydraulic resistance of real flows. The principal aim of hydraulics in the nineteenth century was therefore the experimental study of flow resistance under various conditions. In practice, the design of water flows and water-works was based on elementary dynamic and equilibrium equations, and on a number of empirical coefficients. Theoretical studies dealt mainly with quasi-one-dimensional analysis of water flow in open channels (Rouse and Ince *1957*).

4.2 Applied hydraulics

The empirical binomial relation for hydraulic resistance mentioned in Section 2.3 was widely used by engineers at the beginning of the century. Fundamental experimental studies of the water flow through pipes were undertaken in the second quarter of the century in Germany by G. Hagen, but the most important practical investigations of hydraulic resistance were carried out in the middle of the century by the French engineers H. Darcy and H. Bazin. Many empirical equations for the friction factor, including those used today, were proposed in the second half of the century (P. G. Gauckler, E. O. Ganguillet and W. R. Kutter, R. Manning and Bazin). Bazin's formula included explicitly a wall-roughness factor.

Flow in thin (capillary) tubes was investigated experimentally by Hagen in the 1830s, and especially thoroughly by French physicist Jean Poiseuille (1840–41), who established an empirical formula for the flow rate, brilliantly confirming the later theoretical solution for laminar motion (§9.9).

Various industrial hydraulic devices were in use, and an enormous amount of information was required about their hydraulic characteristics. The basic reference book for hydraulic design in mechanical engineering in the middle of the century was the *Lehrbuch der Ingenieur- und Maschinen-Mechanik* compiled by the German engineer J. Weisbach, and re-edited and translated many times and in several languages from 1845 until 1891. With this book, Weisbach essentially shaped the subsequent development of hydraulics; particularly important was the introduction of dimensionless coefficients in semi-empirical equations for hydraulic-head losses.

During the whole of the nineteenth century, engineering hydraulics had a wide field of practical applications in water-wheel and turbine design. The first practical outward-flow reaction turbines were constructed by B. Fourneyron in France in the second quarter of the century. In the decades that followed, J. B. Francis improved inward-flow runners in the USA. An essential improvement of the bucket form for impulse wheels made in the 1880s by L. A. Pelton allowed them to be used more effectively for high-head power-plants. Propeller turbines were introduced for the low-head field only in the 1910s (Kaplan's turbine with adjustable blades).

The methods of hydraulic analysis of reaction water turbines remained basically on the level of the Euler theory until the final decades of the nineteenth century (but were not always associated in the literature with Euler's name).

4.3 Theoretical hydraulics

The first attempt to analyse steady, non-uniform (gradually varying) water flows in open channels, including hydraulic jump, seems to have been undertaken in Italy in the 1820s, but advanced hydraulic studies of the whole problem were carried out by French engineers in the second quarter of the century, first by J. B. Belanger. He also proposed the principle of maximum discharge for broad-crested weirs. A general analysis of water-profile types for channels with inclined bottoms was carried out, under the simplest assumptions, by J. Dupuit (1848). The equations of non-steady water flow in open channels, being widely studied nowadays, were derived by Barré de Saint-Venant (1871). The problem of sediment motion in rivers was attacked from the 1870s, first by P. F. D. du Boys.

A fundamental contribution to hydraulics was made by J. Boussinesq. His large *Essai sur la théorie des eaux courantes* (1872) deals with all aspects of water flow, especially with analysis and establishing general equations for quasi-one-dimensional water flow in open channels with accurate averaging of velocities.

By the end of the century, transient processes in pipelines turned out to be very important in the design of water-distribution systems because of water hammer arising at fast valve operations, resulting sometimes in the destruction of pipes. The problem of pressure-wave propagation in elastic pipes bearing a weakly compressible liquid was studied by N. E. Zhukovsky (1899) and L. Allievi (1909).

4.4 Groundwater flow

Groundwater-flow theory is based on the linear law of hydraulic resistance for fluid filtering through porous media, which was found experimentally by Darcy (1856). The hydraulic theory of steady groundwater motion was developed by Dupuit (1863), similar to the general hydraulic theory of water flow in pipes and open channels. P. Forchheimer generalized the Darcy law for three-dimensional flow and reduced planar problems of confined and non-confined groundwater motion in homogeneous media to the integration of the Laplace equation. The hydraulic theory of non-steady groundwater flow was developed at the very beginning of the twentieth century by Boussinesq.

5 HYDRODYNAMICS OF A VISCOUS FLUID DURING THE NINETEENTH CENTURY

5.1 General equations

The equations of fluid motion were studied in the first half of the nineteenth century with a view to incorporating the flow resistance. Claude Navier developed a molecular approach (1821) in which, following Newton, he assumed the proportionality of the additional forces of interaction between ultimate molecules in their motion to the velocity of their divergence. Using the principle of virtual displacements and limiting the analysis to the case of an incompressible liquid, he obtained equations of motion in quite a modern form (he used the symbol d rather than ∂):

$$P - \frac{dp}{dx} = \rho\left(\frac{du}{dt} + u\frac{du}{dx} + v\frac{du}{dy} + w\frac{du}{dz}\right) - \varepsilon\left(\frac{d^2u}{dx^2} + \frac{d^2u}{dy^2} + \frac{d^2u}{dz^2}\right), \qquad (4)$$

and two similar equations for the y- and z-axes; here P is the component of the external force per unit volume, and ρ and ε are density and viscosity. His equations were generalized for a compressible fluid by Siméon-Denis Poisson (1829), who based his analysis essentially on molecular models, but used some continuous ones too. He began also to study heat distributions in flows.

A fully continuous approach to the flow of a viscous fluid was developed first by Saint-Venant in a short note published in the Paris *Comptes Rendus* in 1843. He explicitly formulated a generalized hypothesis on linear relations between tangential stresses and shear-strain rates, and indicated that the equations of motion arose from the Cauchy dynamic equations after the substitution of linear relations for stress components and properly taking into account the pressure in the fluid.

The hydrodynamics of a viscous fluid was comprehensively treated by George Stokes in a paper presented to the Cambridge Philosophical Society in 1845. He gave, independently of Poisson and Saint-Venant, a rigorous derivation of the dynamic equations on the basis of linear relations between stress and strain-rate components (Stokes *1847*).

5.2 Particular problems

Only a few rigorous solutions of the Navier–Stokes equations were obtained in the nineteenth century. An analytical solution for the steady laminar flow in a cylindrical pipe was obtained by Stokes (1845), who also solved the problem of viscous flow between rotating coaxial cylinders (and discovered an error committed by Newton in his approach to the plane problem of the axisymmetric rotation of a viscous liquid in an infinite region). The first solutions for pipe flow were based on the assumption that fluid particles slide along pipe walls, but Stokes had already come to the conclusion that fluid particles normally adhere to walls.

The further development of the theory of viscous flow went in the direction of simplifying the equations of motion. Stokes proposed a simple linearization of the equations by omitting their convective terms (1851). Under such conditions, he studied the uniform descent of a sphere through an unbounded fluid. A series of problems on the slow oscillation of bodies in a viscous liquid were solved in the 1850s, and their solutions were used for experimental verification of the equations of motion.

It was noticed that the general equations for non-steady viscous flow are in special cases analogous to the heat equation. This direction of investigation was connected with studies of the diffusion of vortices and energy dissipation in viscous liquids, and of the stability of laminar flow. Lord Rayleigh did fundamental work in this area.

Another system of simplified Navier–Stokes equations have found some practical application in the hydrodynamic theory of lubrication laid down by N. Petrov (1883) and O. Reynolds (1886).

General equations of the hydrodynamics of porous media were analysed by Zhukovsky (1889), who substituted viscous effects in flow by a volume force defined according to the Darcy law (see Section 4.4). The theory of groundwater flow became an independent field of hydrodynamics in the work of the American hydrogeologist C. Slichter (1899).

5.3 Flow regimes

Hagen seemed to be the first to observe quite accurately the progressive disruption of laminar flow with increasing velocity. But a turning-point in the experimental study of flow regimes was achieved by Osborne Reynolds (1883), who connected the dimensionless quantity $\rho VL/\mu$ with the flow regime, and studied the threshold value of this parameter as laminar flow gives way to turbulent ('sinuous') flow (ρ is the density, V the mean velocity, L a characterizing measure of the regime and μ the coefficient of viscosity). This quantity was later named the Reynolds number (Rott *1990*).

The first attempt to analyse turbulent flow was made by Boussinesq in his memoir of 1872 (Section 4.3). He resolved real velocities in turbulent flow into mean velocities and pulsation components, and evaluated the effect of pulsations by introducing a variable 'effective' viscosity coefficient.

A general approach to turbulent flow was developed by Reynolds (1894), who constructed equations for mean velocities which included mean values of squares and products of pulsation components. As a result, the system of Reynolds equations turned out to be, generally speaking, non-closed.

5.4 Similarity in hydrodynamics

Some of the simplest problems of similarity (or analogy) for the strength of materials had already been considered by Galileo (1638), and a more general approach to the principles of dynamic similarity was developed by Newton (1687). Joseph Fourier widely used similarity in his theory of heat (§9.4). However, the first general similarity analysis of the hydrodynamic equations was by Stokes (1851); a more detailed analysis was carried out by Hermann von Helmholtz (1873), who used some of his theoretical results to reinterpret various experiments in hydrodynamics in terms of similarity. A lively discussion of the general backgrounds of the similarity and dimensionality methods took place in the 1910s (Macagno *1971*, Görtler *1975*). As a result, the real practical assimilation of the similarity methods in engineering design (even a wide introduction of the Reynolds number) was completed by the end of the first quarter of the twentieth century.

6 HYDRODYNAMICS OF AN IDEAL FLUID DURING THE NINETEENTH CENTURY

6.1 Potential flow

As far back as 1781 Lagrange formulated an important theorem stating that if there exists a velocity potential in a flowing ideal barotropic fluid, then that velocity potential will remain unchanged provided the forces acting on the fluid are conservative. Potential flows were thereby regarded as important: for example, all flows of an ideal fluid, beginning from a state of equilibrium, were potentials.

The theory of potential flows attracted the special attention of scientists during the nineteenth century. In particular, the method of sources and sinks, going back to George Green's *Essay* of 1828 (§3.17), was widely developed for the construction of such flows. The corresponding singularity reflection principle was introduced into hydrodynamics by Stokes (1843).

An important place in hydrodynamics during the last third of the nineteenth century belongs to studies of the motions of a rigid body in an ideal fluid. Fundamental work in this field was carried out in the 1860s by William Thomson and Gustav Kirchhoff, who reduced the body and the fluid motion to a unified dynamic system. Difficult problems of integrating the differential equations were closely connected with problems of rigid-body dynamics, and this attracted the attention of many mathematicians at the end of the nineteenth century and the beginning of the twentieth. The concept of an apparent mass, so important for applications, had been introduced for particular cases of a moving sphere as far back as the 1830s and 1840s by Green and Stokes.

The most remarkable hydrodynamic results were obtained in the theory of the plane potential flow of an incompressible liquid. The concept of stream function for plane flow was introduced by Lagrange (1781), and its clearest kinematic interpretation was given by W. J. M. Rankine (1864). The stream function, as well as the velocity potential, satisfies the Laplace equation, and they are both connected by the Cauchy–Riemann equations (which are to be found in the earlier works of d'Alembert and Euler; §3.12). Thus, plane potential flows have an intrinsic connection with the theory of analytic functions. The rapid development of the latter in the second half of the nineteenth century was spurred on by the increasing field of its applications in hydrodynamics and electrostatics. The first elementary hydrodynamic problems were investigated with the use of complex variables in the 1860s (Hicks *1882*, *1883*).

6.2 Flow with free surfaces

A new class of flow, having free (isobaric) streamlines, was discovered by Helmholtz (1868). Interest in such flows arose in connection with attempts to analyse drag, and more realistic patterns of flow past bodies, within the framework of the ideal-fluid model. A solution for the free-streamline efflux from Borda's mouthpiece was given by Helmholtz, and Kirchhoff proposed (1869) a method of solving free-streamline problems based on the use of the reciprocal complex velocity hodograph $\zeta \equiv dz/dw$ and the complex velocity potential w in the plane z of the flow. Kirchhoff found solutions for a free-streamline flow past a plate and efflux from a sharp-edged orifice. In particular, for the contraction factor of a free jet, he obtained a value close to the experimental one. Important contributions to the theory of free-streamline flow were made later, simultaneously by J. Michell and Zhukovsky (1890–91). They proposed an effective modification of the Kirchhoff method by introducing the function $\Omega \equiv \log \zeta$ and connecting flow domains on the planes w and Ω with a definite domain on the plane of an auxiliary complex variable.

A large number of hydrodynamical investigations in the nineteenth century dealt with water waves. To a unique extent among the various branches of the hydrodynamics of an ideal liquid, this one gave results corroborated by experiment. As far back as 1815–16, Augustin Louis Cauchy and Poisson had laid down the theory of infinitesimal water waves based on a linearization of boundary conditions on the undisturbed water surface (Burkhardt *1912*). During the course of the century, water-wave theory was developed mainly by English scientists, and was used widely in the study of tides (§8.16) and the wave resistance of ships (§8.17). The first experimental investigation of solitons for solitary surface water waves was carried out by J. Scott Russell in the 1830s.

6.3 Vortex flow

Studies of the kinematics of continuous media by Cauchy and Stokes led to the mathematical concept of vortices and to analysis of vortex flow. A series of elegant and important theorems on vortex lines and vortex tubes were published by Helmholtz (1858), who brought vortex flows to the attention of scientists. The notion of velocity circulation, connected with vortex flux, was introduced at about the same time.

A specific form of the equations of motion, containing the velocity vortex vector only in a vector product with the velocity vector itself, was given by Horace Lamb (1879). It made easily distinguishable the class of so-called helical flow, in which velocity vortex and velocity vectors are collinear. This class of flow, studied in the nineteenth century by I. S. Gromeka and Eugenio Beltrami, is remarkable in that the Bernoulli integral remains constant throughout any steady flow. Thus, helical flows are to some extent a natural generalization of potential flows.

A very important problem in fluid dynamics is how to explain the nature of lifting forces, and estimate their value. It does not necessarily require the explicit introduction of the fluid viscosity, and its first solution for plane profiles was given within the framework of the ideal incompressible fluid, by introducing velocity circulation around profiles (see §8.12 on aerodynamics; for some details, see Mikhailov *1981*).

7 HYDRODYNAMICS AND HYDRAULICS DURING THE FIRST HALF OF THE TWENTIETH CENTURY

7.1 Introduction

Problems of fluid dynamics became prominent in mechanics during the first half of the twentieth century (Goldstein *1969*), owing to the rapid rise of aviation. The development of fluid dynamics was characterized by an intensive *rapprochement* of theoretical and engineering investigations. Large research bodies were set up, combining the efforts of scientists in solving complex problems.

Before passing on to details, let us note several remarkable scientific schools originating around this time: the Moscow school of Zhukovsky and S. A. Chaplygin, the Göttingen school of Ludwig Prandtl, and the slightly younger school of Theodore von Kármán in the USA, connected with

Prandtl's traditions. G. I. Taylor was a more solitary, but equally influential, figure in England. Many investigations in hydrodynamics were concentrated between the two World Wars around such institutions as the Zhukovsky Central Aero-Hydrodynamic Institute in Moscow, and the Kaiser-Wilhelm-Institut für Strömungsforschung und Aerodynamische Versuchsanstalt in Göttingen.

7.2 Viscous flow

Difficult problems of the stability of viscous flows were attacked for small disturbances by W. M. Orr and Arnold Sommerfeld in the first decade of the century, and by Taylor and Werner Heisenberg in the 1920s. The von Kármán vortex sheets should also be mentioned here (see §8.12 on aerodynamics).

A fundamental achievement was the idea of a boundary layer (BL) in a viscous flow, proposed by Prandtl in 1904. He obtained the laminar BL equations first with the help of not very rigorous but clear intuitive considerations. An effective approximate method of solving these equations by means of integral equations was proposed by von Kármán and K. Pohlhausen (1921). Prandtl explained (1914) an experimentally discovered peculiarity, the dependence of drag on Reynolds number ('drag crisis') in terms of BL turbulization; this stimulated the development of turbulent BL theory (for details see Tani *1977* and Loitsiansky *1970*).

A formal closure procedure for the Reynolds equations of turbulent flow was proposed by L. V. Keller and A. A. Friedmann (1924), who considered a system of equations of motion with a consecutive increase of order of correlation moments for velocity pulsations. The equally fruitful idea of describing turbulence by a hierarchical cascade of diminishing vortices was proposed by L. Richardson in the 1920s. The main concern of the 1930s was the theory of isotropic turbulence laid down by Taylor; the concept of a locally isotropic turbulence appeared in the 1940s.

The practical demands of analysing turbulent flow were met in the 1930s by semi-empirical theories of turbulence based on the concept of a 'mixing length' (Prandtl and Taylor). Developing this idea, von Kármán obtained the logarithmic velocity distribution near walls, and studied the turbulent core of flows.

7.3 Ideal fluids

The problem of the flow of an ideal fluid presents chiefly purely analytic difficulties. Therefore new mathematical tools, such as integro-differential equations and functional analysis (§3.9–3.10) were used in this field. The most interesting problems concern flows with free boundaries, particularly free-streamline flows past obstacles. A general problem of a plane free-streamline flow past a curved obstacle was reduced to an integro-differential equation first by Tullio Levi-Città (1906) and H. Villat (1911). Some general existence and uniqueness theorems for such flows were proved by J. Léray and M. A. Lavrentiev in the 1930s. The theory of free jets found an unexpected military application in the 1940s in connection with the study of cumulative charge effects.

A special class of flow is presented by surface waves. General investigations of steady, finite-amplitude surface waves were begun in the 1920s by A. I. Nekrasov and Levi-Città, who reduced the problem to an integral equation. An existence theorem for solitary waves was demonstrated in the 1940s by M. A. Lavrentiev. A paper by D. Korteweg and J. de Vries (1895) deserves special mention, as their 'KdV' equation is now considered a milestone in the development of general non-linear wave theory.

Interaction of a body with free surface (impact and water entry) presents another class of problems, investigated in the 1930s; in a paper by H. Wagner (1932) an appropriate new complex function was introduced.

7.4 Some modern trends

There is now a necessity to construct new models of media and processes, as the classical ones (ideal and Newtonian fluids) are no longer sufficient to describe reality. A characteristic feature of modern fluid dynamics is a deep interest in essentially non-linear problems in general, and particularly in the problems of stability (bifurcations) and turbulence, internal waves and stratified flows, multiphase flows and non-Newtonian fluids (rheology), heat transfer with chemical reactions in flows, and geophysical fluid dynamics, including both meteorology and oceanology. Refined numerical methods and large-scale (as well as micro-scale) experiments have become indispensable. Modern hydraulics is characterized by the assimilation of all formerly traditionally hydrodynamic methods, and particularly by the use of various semi-empirical theories of turbulence.

As a result, yesterday's clear distinction between fluid dynamics and related branches of physics and Earth sciences is being washed away. Meanwhile, stochastic and fully non-classical concepts (such as coherent structures, strange attractors and fractals (§3.8)) are invading the respectably deterministic Newtonian mechanics.

BIBLIOGRAPHY

Bateman, H. *1915*, 'Some recent researches on the motion of fluids', *Monthly Weather Review*, 163–170. [Short survey of the development of general hydrodynamics and its geophysical applications.]

Bernoulli, D. *1738, Hydrodynamica sive de viribus et motibus fluidorum commentarii*, Strassburg: Dulsecker. [Russian transl. 1959, Leningrad: Academy of Sciences. English transl. 1968, New York: Dover, with next item.]

Bernoulli, J. *1742, Hydraulica*, in *Opera omnia*, Vol. 4, 387–488. [Repr. 1968, Hildesheim: Olms. English transl. 1968, New York: Dover, with previous item.]

Binnie, A. M. *1978*, 'Some notes on the study of fluid mechanics in Cambridge, England', *Annual Review of Fluid Mechanics*, **10**, 1–10.

Birkhoff, G. *1983*, 'Numerical fluid dynamics', *SIAM Review*, **25**, 1–34. [Short survey of analytical (1740–1940) and numerical methods in fluid dynamics.]

Burkhardt, H. *1912*, 'Die Untersuchungen von Cauchy und Poisson über Wasserwellen', *Sitzungsberichte der mathematisch–physikalischen Klasse der Bayerischen Akademie der Wissenschaften*, 97–120.

Cannon, J. T. and Dostrovsky, S. *1981, The Evolution of Dynamics: Vibration Theory from 1687 to 1742*, New York: Springer.

Challis, J. *1834*, 'Report on the present state of the analytical theory of hydrostatics and hydrodynamics', *Report of the British Association for the Advancement of Science* (1833), 131–51. ['Supplementary report', (1836), 225–53.]

Cisotti, U. *1924*, 'Über den Anteil Italiens an dem Fortschritt der klassischen Hydrodynamik in den letzten 15 Jahren', in *Vorträge aus dem Gebiete der Hydro- und Aerodynamik*, Berlin: Springer, 1–17.

Dryden, H. L., Murnaghan, F. D. and Bateman, H. *1956, Hydrodynamics*, New York: Dover. [Large bibliography, including many older sources.]

Garbrecht, G. (ed.) *1987, Hydraulics and Hydraulic Research. A Historical Review*, Rotterdam: Balkema. [Thirty surveys, from the early history to the mid-twentieth century.]

Görtler, H. *1975*, 'Zur Geschichte des π-Theorems', *Zeitschrift für angewandte Mathematik und Mechanik*, **55**, 3–8.

Goldstein, S. (ed.) *1938, Modern Developments in Fluid Dynamics*, 2 vols, Oxford: Clarendon Press. [A detailed account of theory and experiment relating to boundary layers, turbulent motion and wakes between the two World Wars. Corrected repr. 1965, New York: Dover.]

—— *1969*, 'Fluid mechanics in the first half of this century', *Annual Review of Fluid Mechanics*, **1**, 1–28. [Each of the next volumes of this annual review series contains a short introductory historical survey.]

Hicks, W. M. *1882, 1883*, 'Report on recent progress in hydrodynamics', *Report of the British Association for the Advancement of Science*, (1881), 57–88; (1882), 39–70.

Jouguet, E. *1925*, 'Histoire des études hydrodynamique en France', *Revue Scientifique*, **63**, 345–60. [History up to the beginning of the nineteenth century.]

Lamb, H. *1932, Hydrodynamics*, 6th edn, Cambridge: Cambridge University Press. [1st edn 1879. Influential textbook; contains large bibliography and historical notes.]

Loitsiansky, L. G. *1970*, 'The development of boundary-layer theory in the USSR', *Annual Review of Fluid Mechanics*, **2**, 1–14.

Love, A. E. H. *1901*, 'Hydrodynamik', in *Encyklopädie der mathematischen Wissenschaften*, Vol. 4, Part C, 48–147 (articles IV 15–16). [See also articles IV 19 and 20, by G. Zemplén and P. Forchheimer.]

Macagno, E. O. *1971*, 'Historico-critical review of dimensional analysis', *Journal of the Franklin Institute*, **292**, 391–402.

Mascart, J. M. *1919*, *La Vie et les travaux du Chevalier Jean-Charles Borda (1733–1799)*, Lyon: Croissant (*Annales de l'Université de Lyon*, n.s., Section 2 (*droit, lettres*), fasc. 33).

Mikhailov, G. K. *1981*, 'History of mechanics: Present state and problems', in *Advances in Theoretical and Applied Mechanics*, Moscow: Mir, 148–65.

—— *1983*, 'Leonhard Euler und die Entwicklung der theoretischen Hydraulik im zweiten Viertel des 18. Jahrhunderts', in *Leonhard Euler*, Basel: Birkhäuser, 229–41.

Mouret, G. *1921*, 'Antoine de Chézy, histoire d'une formule d'hydraulique', *Annales des Ponts et Chaussées*, Vol. 2, 165–269.

Neményi, P. F. *1962*, 'The main concepts and ideas of fluid dynamics in their historical development', *Archive for History of Exact Sciences*, **2**, 52–86.

Nordon, M. *1991–2*, *L'Histoire de l'hydraulique*, 2 vols, Paris: Masson.

Oseen, C. W. *1916*, 'En blick på hydrodynamikens utvecklingshistoria', *Nyt tidsskrift for matematik*, **27B**, 2–18. [From Euler to von Kármán.]

Rennie, G. *1834, 1835*, 'Report on the progress and present state of our knowledge of hydraulics as a branch of engineering', *Report of the British Association for the Advancement of Science*, (1833), 153–84; (1834), 415–513.

Rott, N. *1990*, 'Note on the history of the Reynolds number', *Annual Review of Fluid Mechanics*, **22**, 1–11.

Rouse, H. *1976*, *Hydraulics in the United States, 1776–1976*, Iowa City, IA: Institute of Hydraulic Research.

Rouse, H. and Ince, S. *1957*, *History of Hydraulics*, 1st edn, Iowa City, IA: Institute of Hydraulic Research. [Repr. 1963, New York: Dover. Also includes history of hydrodynamics.]

Stokes, G. G. *1847*, 'Report on recent researches in hydrodynamics', *Report of the British Association for the Advancement of Science*, (1846), 1–20. [Also in *Papers*, Vol. 1, 157–87.]

Tani, I. *1977*, 'History of boundary-layer theory', *Annual Review of Fluid Mechanics*, **9**, 87–111.

Truesdell, C. *1954*, 'Rational fluid mechanics, 1687–1765', in L. Euler, *Opera omnia*, Series 2, Vol. 12, vii–cxxv. [The most comprehensive survey of the matter.]

—— *1955*, 'Editor's introduction', in L. Euler, *Opera omnia*, Series 2, Vol. 13, ix–cv. [History of the theory of aerial sound (1687–1788) and rational fluid mechanics (1765–88).]

Usher, A. P. *1954*, *A History of Mechanical Inventions*, rev. edn, Cambridge, MA: Harvard University Press. [Repr. n.d., New York: Dover. From the early history to the mid-nineteenth century.]

8.6

Theories of elasticity

J. J. CROSS

1 EARLY THEORY

The use and study of elastic materials does not begin with Robert Hooke in 1678, but the formulation of relations between forces and geometry for elastic ('springy') materials certainly does. He stated that the restoring force is proportional to the change in length from the natural position, by elongation or compression. If F is the force and e the change in length, then $F = ke$ for some constant k.

Galileo, Edmé Mariotte, Gottfried Wilhelm Leibniz and Pierre Varignon had considered loaded beams, and concluded that the beams could be regarded as being made up of fibres running lengthwise, which extend under the action of the load at the free end, so producing a moment at the fixed end as the beam bends or curves under the load. James Bernoulli (1705) showed that compression also occurred, and that the so-called neutral line or surface, which was neither extended nor compressed, need not be at the bottom surface of the beam. In modern terms, he derived the differential equation for the deflection y of this line along the length x of the beam in the linear case as

$$dy = \frac{(x^2 + ab)\,dx}{[a^4 - (x^2 + ab)^2]^{1/2}}.$$
(1)

While he made an error in the moments by overlooking their independence of the position of the neutral line (Truesdell *1960*), he did treat the problem of finding the curvature of an elastic line or of a bent elastic lamina, in general, where Hooke's law is a special case of the basic law of nature linking physics (forces) and geometry (change of place and shape).

The second major source of theory for elastic bodies is sound (§9.8). After many had commented on the link between the vibrations of strings and elasticity, it was Brook Taylor (1713) who first applied the momentum principle to an element of the string and deduced its fundamental frequency. John Bernoulli (1727) did the same for loaded strings; the work was taken up by his son Daniel (1733) who found the nodes (stationary points) and frequencies of the sinusoidal motion, and also by Leonhard Euler (1733). The three of them followed this by studying the transverse oscillation of bars, using the moment equation, with frequencies proportional to the square roots of the absolute elasticities (those related to extension). Jean d'Alembert (1746) obtained the second-order partial differential equation governing the motion of the string, and its general solution in terms of two arbitrary functions. Daniel Bernoulli (1753) and Joseph Louis Lagrange (1759) started the long trek to the Dirichlet–Cantor proof of the relation of the 'Fourier' series given by Daniel Bernoulli to the general functions of d'Alembert's solutions (§3.15). The relation between the physics and the geometry was in the same confused state as in the previous half-century, and not at the forefront of their thinking. Too many other problems were involved: which functions could be solutions, what functions were, how one could take limits to go from a discrete to a continuous medium (§3.11).

The next step was taken by Charles Coulomb (1777), in investigating the torsion of thin rods (wires) by a force perpendicular to their length, giving a twisting moment. He stated that the moment of torsion is proportional to the angle through which the point of application turns. He derived the equation of motion for torsional vibrations, involving the geometry of the wire through its moments of inertia (cross-section perpendicular to its length) and its length, and a physical coefficient which is constant for the particular material the wire is made of. The law

$$I\ddot{\theta} = T \equiv -n\theta, \tag{2}$$

where θ is the angle through which the end-plane of the body is rotated, T is the torsional moment (independent of the amount of tension in the wire for moderate tensions), I is the moment of inertia of the wire and n is the elastic constant, was stated as the result of experiment and not as derived from Hooke's law or any other model (see also §8.7 on structures, for Coulomb).

2 TRANSITION: GERMAIN

Ernst Chladni (1787) published pictures showing the nodal lines on circular and square plates vibrating under various conditions. These pictures created a sensation in both Germany and France on their republication in 1802, but first attempts to explain the patterns produced in the sand heaped along the stationary lines foundered on inadequate physical models for the elasticity of the plates. So too did Sophie Germain's two attempts (1811, 1815) which purported to give the equation of motion of the neutral surface $z = z(x, y, t)$ of the plate. Both her equations were wrong, both being based on the mean curvature (the sum of the principal curvatures) associated with the surface, and one drawing on incorrect physics. Lagrange (1811) obtained the generally accepted equation from Germain's first attempt:

$$\frac{d^2 z}{dt^2} = k^2 \left(\frac{d^4 z}{dx^4} + 2\frac{d^4 z}{dx^2\,dy^2} + \frac{d^4 z}{dy^4} \right), \tag{3}$$

and this is now easily derived from the general equations of elasticity. That Germain wrongly derived Siméon-Denis Poisson's (1814) incorrect equations is palliation but no excuse, although she was awarded a prize for her work by the Paris Academy of Sciences. The molecular attractions explaining the coherence of matter passed into elasticity theory through Poisson (Grattan-Guinness *1990*: Chap. 7; Belhoste *1991*: Chap. 6; Bucciarelli and Dworsky *1980*).

3 ESTABLISHMENT? NAVIER, CAUCHY AND POISSON

On 14 May 1821 Claude Navier presented the general equations of equilibrium and motion for a body of elastic material, equations which were to hold at all points inside the body and on its surface. His ideas were based on a theory of molecular attraction, and therefore subject to the theory's constraints and faults, as Poisson had already found and Augustin Louis Cauchy was about to. He used a calculus-of-variations approach to derive the equations from an integral of the force function for the attraction expressed as a function of the intermolecular distance; there seem to be errors in the derivation. Only one elastic constant was introduced, but the equations coincide with those of George Green (1837), Gabriel Lamé (1852) and George Stokes (1845) when the constants are properly identified. Navier used these ideas in 1822 to develop the equations of motion for viscous liquids, giving equations again with only one parameter ε:

$$P - \frac{\partial p}{\partial x} = \rho\left(\frac{\partial U}{\partial t} + U\frac{\partial U}{\partial x} + V\frac{\partial U}{\partial y} + W\frac{\partial U}{\partial z} \right) - \varepsilon\left(\frac{\partial^2 U}{\partial x^2} + \frac{\partial^2 U}{\partial y^2} + \frac{\partial^2 U}{\partial z^2} \right), \tag{4}$$

$$Q - \frac{\partial p}{\partial y} = \rho\left(\frac{\partial V}{\partial t} + U\frac{\partial V}{\partial x} + V\frac{\partial V}{\partial y} + W\frac{\partial V}{\partial z} \right) - \varepsilon\left(\frac{\partial^2 V}{\partial x^2} + \frac{\partial^2 V}{\partial y^2} + \frac{\partial^2 V}{\partial z^2} \right), \tag{5}$$

$$R - \frac{\partial p}{\partial z} = \rho\left(\frac{\partial W}{\partial t} + U\frac{\partial W}{\partial x} + V\frac{\partial W}{\partial y} + W\frac{\partial W}{\partial z} \right) - \varepsilon\left(\frac{\partial^2 W}{\partial x^2} + \frac{\partial^2 W}{\partial y^2} + \frac{\partial^2 W}{\partial z^2} \right), \tag{6}$$

where P, Q and R are the forces acting through space on the body (such as gravity), U, V and W are the velocities of the fluid molecule, and p is the hydrostatic pressure.

Poisson (1814) had already made one failed attempt at elasticity when Cauchy's first paper (1822) was read to the Paris Academy. Cauchy saw Navier's paper (1822), and reported the results of his own researches soon after. The stress could be perpendicular (like hydrostatic pressure) to the surface considered, or could act at any angle to the surface, even tangentially to it (as with friction), and it could act inwards (compression) or outwards (extension). Hence, with his now immortal tetrahedron of small faces, three parallel to the coordinate planes and the fourth cutting obliquely across them, he showed that the stress on the oblique face depends linearly on the normal to this face and is derived from it by six stress components. He also expressed his results in terms of a certain ellipsoid, giving principal tractions. Poisson and Cauchy now batted this theory back and forth, with interference from Navier, up to 1830. In essence they adopted the molecular-theory approach, confused the properties of the elastic constants and the relations between them, and ended up with equations of motion involving a single constant (unlike Cauchy).

Cauchy adopted several positions: first the general theory of linear stress and, later, strain (1822–1827), then a two-parameter isotropic elastic theory (1828). Poisson (1827) wrote a paper giving the ratio of torsional compression to longitudinal extension as $\frac{1}{2}$ for wires, associated with Poisson's ratio, which experimental evidence soon showed not to be $\frac{1}{2}$ for all materials. He (1828) then redeveloped the molecular theory, provoking a dispute with Navier about its nature and details. Cauchy (1828) used the molecular theory to derive his (faulty) relations between the elastic constants. It is interesting that all three developed very similar equations, often identical in form, from very differential physical theories (Grattan-Guinness *1990*: Chap. 15).

4 INTERMISSION: LINEAR THEORIES

We begin by looking at geometry: we take a body – a rod, a shell, a solid rubber ball, say – and label the original positions of each particle in the body by Cartesian coordinates x, y, z. When this particle moves to another position, $X = x + u$, $Y = y + v$, $Z = z + w$, we say that the particle has undergone a displacement u, v, w and that the body is in general deformed.

Note that u, v, w may depend on x, y, z and t. We call the matrix $\mathbf{F}$ the deformation matrix:

$$\mathbf{F} \equiv \begin{pmatrix} \dfrac{\partial X}{\partial x} & \dfrac{\partial X}{\partial y} & \dfrac{\partial X}{\partial z} \\[2mm] \dfrac{\partial Y}{\partial x} & \dfrac{\partial Y}{\partial y} & \dfrac{\partial Y}{\partial z} \\[2mm] \dfrac{\partial Z}{\partial x} & \dfrac{\partial Z}{\partial y} & \dfrac{\partial Z}{\partial z} \end{pmatrix} = \begin{pmatrix} 1+\dfrac{\partial u}{\partial x} & \dfrac{\partial u}{\partial y} & \dfrac{\partial u}{\partial z} \\[2mm] \dfrac{\partial v}{\partial x} & 1+\dfrac{\partial v}{\partial y} & \dfrac{\partial v}{\partial z} \\[2mm] \dfrac{\partial w}{\partial x} & \dfrac{\partial w}{\partial y} & 1+\dfrac{\partial w}{\partial z} \end{pmatrix} \tag{7}$$

The strain $\mathbf{E}$ is defined as

$$\mathbf{E} \equiv \tfrac{1}{2}(\mathbf{F}\mathbf{F}^{\mathrm{T}} - \mathbf{I}) \tag{8}$$

(where $\mathbf{F}^{\mathrm{T}}$ denotes the transpose of $\mathbf{F}$); it is linearly approximated by neglecting the squares and products of the derivatives of the displacements u, v, w:

$$\mathbf{E} = \begin{pmatrix} \dfrac{\partial u}{\partial x} & \dfrac{1}{2}\left(\dfrac{\partial u}{\partial y}+\dfrac{\partial v}{\partial x}\right) & \dfrac{1}{2}\left(\dfrac{\partial u}{\partial z}+\dfrac{\partial w}{\partial x}\right) \\[3mm] \dfrac{1}{2}\left(\dfrac{\partial u}{\partial y}+\dfrac{\partial v}{\partial x}\right) & \dfrac{\partial v}{\partial y} & \dfrac{1}{2}\left(\dfrac{\partial v}{\partial z}+\dfrac{\partial w}{\partial y}\right) \\[3mm] \dfrac{1}{2}\left(\dfrac{\partial u}{\partial z}+\dfrac{\partial w}{\partial x}\right) & \dfrac{1}{2}\left(\dfrac{\partial v}{\partial z}+\dfrac{\partial w}{\partial y}\right) & \dfrac{\partial w}{\partial z} \end{pmatrix} \equiv \begin{pmatrix} e_{xx} & e_{xy} & e_{xz} \\ e_{xy} & e_{yy} & e_{yz} \\ e_{xz} & e_{yz} & e_{zz} \end{pmatrix}. \tag{9}$$

The extension is reckoned as the change in length per unit length; Hooke's original law, *ut tensio sic vis*, or in modern terms, 'the force per unit area of a cross-section (stress) S is proportional to the extension e (perpendicular to the cross-section)', gives a relation $S = Ee$. The constant E (not to be confused with $\mathbf{E}$ in equation (8)) is called the 'Young's modulus' of the material of which the body is made. Perpendicular to the axis of extension there is usually a shortening or shrinkage s in length per unit length, given by $s = \sigma e$. The constant σ is called 'Poisson's ratio'. In Poisson's time it was thought that this ratio always was $\tfrac{1}{4}$, but W. Wertheim gave experimental evidence to the contrary in the 1840s.

The torsion of a body about an axis is the angular twisting of the body, measured as the change in angle (in radians) per unit length, with the total angular change θ at a position x from the end held fixed according to Coulomb's formula (1784) $\theta = \alpha x$, where α is a constant for the material of the body. When the stress is perpendicular to the axis of torsion, $S = \mu\theta$, where μ is a material constant called the shear modulus. It is a result of linear elasticity that

$$\mu = \frac{E}{2(1 + \sigma)}.$$ (10)

The material constants E, μ and σ all depend on the material of which the body is made, and are coefficients in the theory of linear elasticity founded by Hooke and Coulomb.

We began by looking at the geometry of the body, and have now introduced the other half of the theory of elasticity – the material of the body, which is characterized by the response of the body to the forces acting upon it. The forces in the body, usually calculated as force per unit area, or stress, are assumed to depend on the deformation, and therefore there is a mathematical function linking the stress vector **t** to the deformation matrix **F**: if we can write $\mathbf{t} = \mathbf{t(F)}$, we have an elastic material. For other materials one uses more than just the present value of the deformation **F**: for example, for a visco-elastic fluid one uses the deformation and one or more of its time derivatives; for a liquid one restricts the deformations by requiring that $\det \mathbf{F} \equiv 1$; for an elasto-plastic material one uses **F**, but assumes that there is a bound for the stress **t** beyond which the response is no longer elastic but obeys some other law. Still other materials, such as liquid crystals or fibre-reinforced materials, have a local structure separate from the deformation, a set of one, two or three 'director vectors' on which the stress also depends.

Cauchy (1822) found that (in modern terms) the stress vector **t** at a point on a surface is the product of the stress tensor T and the normal **n** to the surface at that point: $\mathbf{t} = T\mathbf{n}$, and that for elastic materials the matrix **T** is symmetric:

$$\mathbf{T} = \mathbf{T(F)} = \mathbf{T}^{\mathrm{T}}$$ (11)

(a second-order tensor being expressible as a matrix). The function $\mathbf{T(F)}$ is now called the material response function, or the stress function. The properties of this function reflect the properties of the material.

The first law of thermodynamics (conservation of energy) was used by William Thomson (1855) to justify Green's use of a strain potential function W for the stress:

$$W = W(e_{xx}, e_{xy}, e_{xz}, e_{xy}, e_{yy}, e_{yz}, e_{xz}, e_{yz}, e_{zz}).$$ (12)

From this work-energy function, the stress T is obtained by differentiation:

$$T_{ab} = \frac{\partial W}{\partial e_{ab}}.$$ (13)

1028

Green arranged the function W as the sum of terms homogeneous of degree
$0, 1, 2, 3, \ldots$ in the strains, and then asserted that the function of degree 1
was zero for static equilibrium at some natural position, and that the terms
of degree 3 or higher could be assumed to be zero because of the smallness
of the strains (Cross 1985).

A material is called 'linear elastic' if the stress tensor T is a linear function
of the strain E:

$$T = \mathcal{M}E \quad \text{or} \quad T_{ab} = \mathcal{M}_{abcd}E_{cd}, \tag{14}$$

where $\mathcal{M}$ is a fourth-order tensor of material constants (such as E, μ and
σ in equation (10)). The symmetry of the material – crystal structure
(aeolotropy) or lack of it (isotropy) – is reflected in the stress function.
In linear elasticity this is shown in the constants $\mathcal{M}$. Ostensibly there
are $3^4 = 81$ constants, but by the symmetry of T and E there are only
$6^2 = 36$. However, since in linear elasticity the stress comes from a
potential, then

$$T_{ab} = \frac{\partial W}{\partial e_{ab}}, \quad \frac{\partial T_{ab}}{\partial e_{cd}} = \frac{\partial T_{cd}}{\partial e_{ab}} \quad \text{or} \quad \mathcal{M}_{abcd} = \mathcal{M}_{cdab}. \tag{15}$$

These are similar to, but quite different from, the incorrect Cauchy relations. There are 15 off-diagonal coefficients affected by these relations, so
there are 21 independent coefficients in the most general case. Cauchy
manufactured six relations based on the distance law for forces, but these
relations have been shown to be false in general.

Green (1837) now proceeded to reduce the coefficients. By enforcing
symmetry with respect to the x, y-, y, z- and z, x-planes, the number of
coefficients was reduced to 9. By making the material indifferent to
symmetry about the z-axis, the number of coefficients was further reduced
to 5; by making the material indifferent to symmetry about all three
axes, the number is finally reduced to 2. Green's method involved
the infinitesimal generators of the orthogonal group, and if all these
symmetries are present in the material the material is called isotropic. So
finally we have the stress–strain relation for linear-elastic isotropic
materials:

$$T_{ab} = \lambda(e_{xx} + e_{yy} + e_{zz})\delta_{ab} + 3\mu e_{ab}, \tag{16}$$

where δ_{ab} is Kronecker's delta (equal to 1 if $a = b$ and 0 otherwise). These
coefficients are called the Lamé constants, and they are related to Young's
modulus and Poisson's ratio:

$$\sigma = \frac{\lambda}{2(\lambda + \mu)}, \quad E = \frac{\mu(3\lambda + 2\mu)}{\lambda + \mu}; \quad \lambda = \frac{E\sigma}{(1 + \sigma)(1 - 2\sigma)}, \quad \mu = \frac{E}{2(1 + \sigma)}. \tag{17}$$

5 CLARITY: GREEN

In his course at the Ecole Polytechnique (1836), Lamé continued with Navier's one-constant elasticity in the face of the experimental evidence, of which he was aware. In his *Leçons* (1852) he adopted a totally different approach. What had happened in between? The simplest reply is Green. Green's contributions were few but fruitful, and known only in Germany up to 1845; his works are collected in Green *1871*.

Green read Laplace and Lacroix, Poisson on electricity and magnetism, and probably Fourier, Lagrange and Coulomb. From his reading he saw the central problems in gravity, electricity and magnetism; and he saw that these problems were similar and could be treated in the same way, as he stated in the *Essay* of 1828 on electricity. His background gave him an interest in waves and fluid flow of all kinds, from the winds needed to run the mill which he owned and the canals which supplied it.

Green applied his techniques to derive the equations of motion of (light in) an elastic medium as well as the boundary conditions that apply. The two motions (of light and of the medium) were identified. If the displacement of a particle is specified by $\mathbf{u} = (u, v, w)$ as functions of the coordinates x, y, z and the time t, then the motion of the medium derives from an energy balance under a virtual displacement $\delta\mathbf{u}$, and the corresponding variation $\delta\phi$ of an energy function ϕ representing the actions of the particles on each other (ρ is the density, and δ here is the Lagrange variable operator):

$$\iiint \rho \, dx \, dy \, dz \left(\frac{d^2u}{dt^2} \delta u + \frac{d^2v}{dt^2} \delta v + \frac{d^2w}{dt^2} \delta w \right) = \iiint dx \, dy \, dz \, \delta\phi. \quad (18)$$

He made several assumptions about the nature of the motion defined by u, v, w and the properties of the energy function ϕ, and used Lagrange's calculus-of-variations technique to obtain the equations of motion in the absence of external forces:

$$\rho \frac{\partial^2 u}{\partial t^2} = -2G \frac{\partial}{\partial x} \left(\frac{\partial u}{\partial x} + \frac{\partial v}{\partial y} + \frac{\partial w}{\partial z} \right) - 2L \left[\frac{\partial^2 u}{\partial y^2} + \frac{\partial^2 u}{\partial z^2} - \frac{\partial}{\partial x} \left(\frac{\partial v}{\partial y} + \frac{\partial w}{\partial z} \right) \right], \quad (19)$$

$$\rho \frac{\partial^2 v}{\partial t^2} = -2G \frac{\partial}{\partial y} \left(\frac{\partial u}{\partial x} + \frac{\partial v}{\partial y} + \frac{\partial w}{\partial z} \right) - 2L \left[\frac{\partial^2 u}{\partial y^2} + \frac{\partial^2 u}{\partial z^2} - \frac{\partial}{\partial x} \left(\frac{\partial v}{\partial y} + \frac{\partial w}{\partial z} \right) \right], \quad (20)$$

$$\rho \frac{\partial^2 w}{\partial t^2} = -2G \frac{\partial}{\partial z} \left(\frac{\partial u}{\partial x} + \frac{\partial v}{\partial y} + \frac{\partial w}{\partial z} \right) - 2L \left[\frac{\partial^2 u}{\partial y^2} + \frac{\partial^2 u}{\partial z^2} - \frac{\partial}{\partial x} \left(\frac{\partial v}{\partial y} + \frac{\partial w}{\partial z} \right) \right], \quad (21)$$

or the vector equation (with $\lambda \equiv -G$, $\mu \equiv -2L$)

$$\rho \frac{d^2\mathbf{u}}{dt^2} = (\lambda + \mu) \, \text{grad}(\text{div } \mathbf{u}) + \mu \, \nabla^2 \mathbf{u}. \quad (22)$$

The variational process yields surface integrals which give the conditions at the boundary to be imposed on the displacements u, v, w and their derivatives.

Green acknowledged that his methods were in part due to Cauchy. His method eliminated the effects of the molecular model from the equations, since the function ϕ was originally chosen to be the most general and then simplified by geometric (small displacements and derivatives) and algebraic (rotations and reflections) constraints rather than physical laws. He was read by Stokes, Thomson and James Clerk Maxwell after 1845, so that the sphere of his influence widened in Britain. However, Barré de Saint-Venant criticized the British school unmercifully (Navier *1864*: 712, 732–3, 738): they produced no arguments against Cauchy's relations, their own arguments were specious and plausible, their theories did not apply to practical materials and processes, and generally would please analysts since they reduced the theory of elasticity to pure analysis (mathematics) without invoking any physical law, let alone a molecular hypothesis.

6 DEVELOPMENT: STOKES

Stokes's first great paper *1845* fully established the equations of motion for both viscous liquids and elastic materials. In it he derived the relation between the stress and the rate of strain (or the time derivative of the strain tensor) from two axioms, together with the use of principal axes and the rules of changes of axes. Instead of the energy potential of Green and the molecular models of Navier, Cauchy and Poisson, Stokes used the axioms (art. 1):

> That the difference between the pressure on a plane in a given direction passing through any point P of a fluid in motion and the pressure which would exist in all directions about P if the fluid in its neighbourhood were in a state of equilibrium depends only on the relative motion of the fluid immediately about P; and that the relative motion due to any motion of rotation may be eliminated without affecting the differences of the pressures above mentioned.

This means that the relative motion reduces to a rate of strain, and the relation between the force or stress and this rate of strain is invariant under rotation, just as Cauchy had shown for other models. In modern terminology, the stress tensor is invariant under the special orthogonal group and is thus an isotropic tensor whose special form allows only two arbitrary constants dependent on the material. The equations Stokes obtained for fluid motion were those of Navier and Poisson, where the assumption of liquidity introduced the Stokes relation $3\lambda = 2\mu$. But for the equilibrium of elastic materials his work led him to different conclusions: the equations of motion have two constants, as did Green's.

7 DISSEMINATION

Saint-Venant (in his introduction to Navier *1864*) proudly boasted that Carl Neumann, S. Houghton, Rudolf Clausius, Wilhelm Weber, and all the savants he cites in his 1200 pages agree 'without exception' that only one isotropic elastic constant is needed: French engineers from the Ecole Polytechnique had one theory for extension and another for torsion. Unfortunately for him, Lamé (1852) and Alfred Clebsch (1862) already disagreed, and followed Green and Stokes. Franz Ernst Neumann tried to preserve the one-parameter theory by imposing inequalities on the elastic coefficients, but without conviction. But his influence on the German school is immense, both directly and through his students and his son Carl. The German university education system required one or two theses to graduate and to teach, so a wealth of solutions for particular problems in elasticity (and other disciplines) flowed from them: Gustav Kirchhoff, Ludwig Boltzmann, O. E. Meyer, W. Voigt and J. Finger. There were others: Stokes, Thomson and Maxwell in Britain; Saint-Venant, J. Boussinesq, P. Duhem, M. Brillouin and the Cosserat brothers (Eugene and Félix) in France; G. Piola, Enrico Betti and Vito Volterra in Italy; and a host of other minor players to 1914.

Developments in the theory of partial differential equations, in the Fourier–Sturm–Liouville theory of eigenfunction expansions and special functions, in differential geometry and vector analysis, meant that a rich tapestry of special problems and structures – bars, membranes, plates – could be and were attacked, leading to impressive practical structures: for example, thin concrete shells shaped as parabolic hyperboloids and forming spectacular roofs for buildings, reinforced rubber wheat storage bags attached to poles, thin metal membrane roofs for outdoor theatres.

8 NEW MATERIALS

In the twentieth century, many new materials have been discovered. Some have an internal structure superimposed on the elastic material: liquid crystals, polar elastic materials. Others show a response to forces and deformations which depends on higher-order derivatives of the displacement: rubbers and gels. Fluids have been found which respond in peculiar ways: they climb the sticks with which they are stirred, or flow down circular pipes in helical flows rather than in straight lines; their response depends on second and higher time derivatives of deformation, as did the original visco-elastic fluids. And there are engineered materials such as plastics and metals whose behaviour depends on the whole history of their deformation, and so require the theories of integration (§3.7) and functional analysis (§3.9) to make any progress.

BIBLIOGRAPHY

There are several 'histories' of elasticity: Grattan-Guinness *1990 passim*, Truesdell *1960*, Todhunter *1886–93*, the short introduction to Love *1927*, and the long introduction and notes by Saint-Venant to Navier *1864*. Truesdell is probably the most reliable.

Belhoste, B. *1991*, *Augustin-Louis Cauchy: A Biography* (transl. F. Ragland), New York and Berlin: Springer.

Bucciarelli, L. L. and Dworsky, N. *1980*, *Sophie Germain: An Essay in the History of the Theory of Elasticity*, Dordrecht: Reidel.

Cross, J. J. *1985*, 'Integral theorems in Cambridge mathematical physics, 1830–1855', in P. M. Harman (ed.), *Wranglers and Physicists: Studies on Cambridge Mathematical Physics in the Nineteenth Century*, Manchester: Manchester University Press, 112–48.

Grattan-Guinness, I. *1990*, *Convolutions in French Mathematics, 1800–1840* [. . .], 3 vols, Basel: Birkhäuser.

Green, G. *1871*, *Mathematical Papers* (ed. N. M. Ferrers), Cambridge: Cambridge University Press. [Repr. 1970, New York: Chelsea.]

Love, A. E. H. *1927*, *A Treatise on the Mathematical Theory of Elasticity*, 4th edn, Cambridge: Cambridge University Press. [1st edn 1892–3; Repr. 1944, New York: Dover.]

Navier, C. L. M. H. *1864*, *Résumé des leçons données à l'Ecole des Ponts et Chausées, sur l'application de la mécanique à l'établissement des constructions et des machines*, 3rd edn (ed. B. de Saint-Venant), Paris: Dunod.

Stokes, G. G. *1845*, 'On the theories of the internal friction of fluids [. . .]', *Transactions of the Cambridge Philosophical Society*, **8**, 287–319. [Also in *Mathematical and Physical Papers*, Vol. 1, 75–129.]

Tedone, O. *1906*, 'Allgemeine Theoreme der mathematischen Elastizitätslehre (Integrationstheorie)' and 'Spezieller Ausführungen zur Statik elastischer Körper', in *Encyklopädie der mathematischen Wissenschaften*, Vol. 4, Part D, 55–124, 125–214 (articles IV 24–25). [Second article written with A. Timpe; see also A. Timpe and C. Müller in article IV 23.]

Todhunter, I. *1886–93*, *A History of the Theory of Elasticity and of the Strength of Materials from Galilei to the Present Time*, 2 vols (ed. and completed by K. Pearson), Cambridge: Cambridge University Press. [Repr. 1960, New York: Dover.]

Truesdell, C. *1960*, 'The rational mechanics of flexible or elastic bodies: 1638–1788', in L. Euler, *Opera omnia*, Series 2, Vol. 11, Part 2, Zürich: Orell Füssli.

8.7

The theory of structures

J. HEYMAN

A structure (from the Latin *struere*) is anything built: say a cathedral or arched bridge from stone; a ship or a roof (and perhaps a spire) from timber; an earth dam, or an excavation in soil for a fortification; or (as isolated usages) iron bars (in China) or vegetable ropes to form suspension chains in bridges. Before the Renaissance all these structures were built without calculation, but not without 'theory', or what would today be called a 'code of practice'. Mignot's statement in 1400 that *ars sine scientia nihil est* ('practice is nothing without theory') testifies to the existence of a medieval rule-book for the construction of cathedrals (the rules were numerical rules of proportion and, as such, effectively correct: see §2.8).

The introduction two centuries ago of iron, and then steel, as a structural material generated new structural forms (the large-span lattice girder, or truss, the high-rise framed building), and the advent of reinforced concrete made possible such new forms as the large-span shell roof. With this increase in complexity of structural form came the need for a structural theory to assess whether or not a particular building was fulfilling its design criteria.

1 STRUCTURAL CRITERIA

A structure is deemed successful if it satisfies the three main criteria of strength, stiffness and stability. (There are other minor criteria that the modern designer will also take into account.) The homely example of a four-legged table may make clear the three aspects of performance that are being examined. The legs of the table must not break when a (normal) weight is placed on top, and the table-top itself must not deflect unduly. (Both these criteria will usually be easily satisfied by the demands imposed by a dinner party.) Finally, the stability criterion may be manifest locally, or overall. If the legs of the table are slender, they may buckle when the overall load on the table is increased. Alternatively, if the legs are not at the four corners, but situated so that the top overhangs them, then placing a heavy weight near an edge may result in the whole table overturning.

Stresses in medieval structures are low (and this has helped to ensure their survival). Thus the stone in a cathedral, or in the arch ring of a masonry bridge, is working at a level one or two orders of magnitude below its crushing strength. Similarly, deflections due to loading (but not those imposed by slow movements of the foundations) are negligibly small. The structural criterion for such structures is that of stability (the four-legged table must not overturn), and stability will be ensured by imposing the right form on the construction (a flying buttress must be of the right shape; an arch ring must have a certain depth; a river pier must have a minimum width). Correct form is a matter of correct geometry, and medieval rules of proportion (which can be traced back to the biblical Book of Ezekiel, *circa* 600 BC) were established empirically to give satisfactory design rules.

'Modern' structures, from the eighteenth century on, work their materials harder. A slender steel truss will work at stresses which are a sensible proportion of the yield stress (say one half or more), and strength must be checked; the push/pull bar forces will cause elastic compressions/ extensions of those bars, leading to overall deformation of the structure, and stiffness must be checked. Finally, individual members must be checked for stability.

2 THE MASTER EQUATION OF VIRTUAL WORK

The simple theory of structures is based upon the mechanics of slightly deformable bodies. By 'slight' is implied that, viewed overall, the geometry of the structure does not appear to alter. Thus, when overall equilibrium equations are written for slightly deformable structures they are identical to those obtained by rigid-body statics. The use of the equation of virtual work to obtain such equations for rigid bodies can be traced back through Aristotle to Archimedes (3rd century BC) with his formulation of the laws of the lever and, in more recent times, to Jordanus de Nemore in the thirteenth century AD (§2.6). The insight obtained by the use of virtual work in the study of rigid bodies is deep, but the application is simple. The clarifying concept of an energy balance is used to obtain relations between external forces by the study of a small rigid-body displacement (which need not be a possible displacement for the structure in question – hence 'virtual').

The equation of virtual work is altered profoundly when the body being studied suffers small deformations. Not only will external forces be involved; account must somehow be taken of the internal forces in equilibrium with those external forces. (The words 'in equilibrium' imply that any such set of forces may be used; in general, there will be (infinitely) many states of internal forces in equilibrium with one given set of external loads.) The essential feature of the equation of virtual work is that it relates two completely independent statements about a given structure, one to do with the equilibrium of forces and one with the compatibility of deformation.

External loads **p** act on a body, and the equilibrium equations are satisfied by internal forces **t** (which could be bar forces in a truss, bending moments from section to section of a frame, and so on). As a completely separate matter, the same body suffers internal deformations **e** (bar elongations, changes of curvature, and so on) which lead to surface displacements **d**.

The external virtual work of the loads **p** moving through distances **d** is $\mathbf{p}^T\mathbf{d}$, and the internal virtual work is similarly $\mathbf{t}^T\mathbf{e}$. The equation of virtual work states that these two quantities are equal.

There is no necessary connection between the equilibrium set $(\mathbf{p}, \mathbf{t})$ and the deformation set $(\mathbf{d}, \mathbf{e})$ – the loads **p** do not necessarily cause the displacements **d**. However, the equation allows the calculation of a set of internal forces in the structure which are in equilibrium with the given loading. This set of internal forces may then be examined, for all portions of the structure, to determine whether or not each section has sufficient strength, to compute corresponding deformations to determine whether or not the structure has sufficient stiffness, and to check local and overall stability.

3 THE STRENGTH OF BEAMS TO 1826

Galileo *1638* was concerned with the fracture strength of beams (in modern terms, with the calculation of the maximum bending moment that can be imposed on a beam made from a material with a known tensile stress at fracture). He deduced correctly that the strength of a beam of rectangular cross-section is proportional to the breadth and the square of the depth of the beam; he assumed that, at fracture, the whole of the cross-section was in a state of uniform tension, so that the 'neutral axis' lay in the surface of the beam. Edmé Mariotte *1686* carried out experiments, and extended the theory by arguing that the extensions (and hence the stresses) should vary linearly with distance from the neutral axis. There were numerical mistakes in this analysis, repeated by James Bernoulli in 1705, and it fell to Antoine Parent *1713* to derive the first proper account of the position of the neutral axis in elastic bending. Charles Coulomb *1776* is often credited with the solution of the elastic problem, and he does use explicitly a linear stress–strain relationship to derive a linearly varying stress distribution over a cross-section in flexure. However, he, like Galileo, was interested in the problem of fracture, and a full grasp of the proper basis of structural analysis is only glimpsed in his work. Coulomb's memoir is concerned with each of the four main problems of structural analysis in the eighteenth century; in addition to the fracture of beams, he discussed the fracture of columns, the thrust of arches, and the thrust of soil (i.e. one of the basic problems of modern soil mechanics).

Claude Navier (*1826*, with some correction and much expansion by Barré de Saint-Venant in 1864) gives a full and correct account of the problem of elastic bending of beams; that is, of the determination of elastic stresses in a cross-section resulting from the flexure of that section by a specified bending moment. This part, and it is only a part, of the general structural problem had therefore been solved – given an internal force (the bending moment), the corresponding elastic stresses could be computed and compared with the known strength of the material.

4 THE STIFFNESS AND STABILITY OF BEAMS TO 1826

James Bernoulli *1691* made a start in the determination of the shape of a bent elastic member, with his statement that the curvature at any point of an initially straight beam would be proportional to the bending moment at that point. Daniel Bernoulli *1751* demonstrated that the resulting elastic curve gave minimum strain energy in bending, and he proposed to Leonhard Euler that the calculus of variations should be applied to the inverse problem of finding the shape of the curve of given length, satisfying given end-conditions of position and direction, so that the strain energy was minimized.

Euler made the analysis in *1744*, and his calculus of variations quickly yielded the governing fourth-order differential equation (§3.5). He integrated this equation three times, and finally tabulated numerical solutions (i.e. values of the elliptic integrals). Before this, however, he discussed graphically the first-order equation, and sketched nine different classes of solution. The work is of the utmost brilliance, and the solution to the first class is of great practical importance. The class deals with very small excursions from the linear form, and Euler showed that those excursions are sinusoidal, and can be maintained only in the presence of a calculable value of the axial load – in practical terms, the elastica (§8.1) buckles in the presence of the 'Euler buckling load'.

Small lateral deflexions of a cantilever had been treated directly by Daniel Bernoulli *1751* ($EI\,\mathrm{d}^2y/\mathrm{d}x^2 = Wx$), and Euler in turn made a direct study of the buckling problem in *1757* ($EI\,\mathrm{d}^2y/\mathrm{d}x^2 = -Py$). Joseph Louis Lagrange *1776* gave the first satisfactory account of higher buckling modes.

Navier, just before his collected *Leçons* (*1826*), had studied the elastic bending of plates, and he gave a recognizably modern treatment of the small deflection of beams.

5 ELASTIC ANALYSIS FROM 1826

Navier also treated statically indeterminate beams. He found that the first master equation of structural analysis, that of equilibrium, may not be sufficient by itself to determine the internal forces. In order to calculate these internal forces for a hyperstatic structure, the two remaining master equations must be used: a statement of compatibility of deformation must be made, and some material law (stress–strain relationship) must be postulated. Thus, for a loaded beam resting on two simple supports, the reactions on those supports may be calculated by statics. If the same beam rests on three supports, however, only two equations are available to determine three reactions. In order to proceed with the analysis, the (very small) elastic displacements of the beam must be examined, and – most importantly – boundary conditions must be specified (that the supports are rigid, for example). All this is necessary, not to compute displacements (which may or may not be of interest), but in order to solve the main structural problem, that of finding the internal forces in the (hyperstatic) structure resulting from given external loading.

Indeed, it was Navier who formulated clearly the elastic method of design, which is based on the actual working values of the internal forces; the resulting internal stresses should not exceed a proportion of the limiting stress for the material. For the four-legged table, the load in each leg is determined for a given load on the table-top, and the corresponding stresses are calculated. Now only three equations of overall equilibrium can be written; the forces in the legs of a tripod can be found from statics, but not those in the legs of a four-legged table. To calculate these, the flexure of the table-top and the elastic compressions of the legs must be taken into account. Finally, the engineer will assume (unthinkingly) that all legs are of the same length and that the floor is completely rigid (compare §6.11 on optimization).

A real table has legs of unequal length and stands on an uneven floor. Its 'actual' observable state is an accidental product of its history – it will be standing on three of its four legs, and the fourth will be unloaded. At any moment, however, a small shift of loading will rock the table so that the unloaded leg now carries weight, while one of the other three legs is relieved. All that can truly be said of the table is that the load in a leg lies somewhere between zero and a value calculable by simple statics, but to ask for the value of the 'actual' load in a leg is meaningless; there are an infinite number of equilibrium states for a hyperstatic structure subjected to a given loading. All this is ignored by the elastic (Navier) designer, who pretends to calculate the 'real' state of a structure. The situation was not recognized explicitly until the 1930s, when the so-called plastic theory was developed.

6 PLASTIC THEORY

Plastic theory has an advantage not foreseen in the 1930s – the elastic equations for a complex structure might number hundreds, to be solved simultaneously, and this was a formidable task (circumvented by ingenious mathematical techniques, e.g. moment distribution, relaxation methods, and so on) until the invention of the electronic computer, whereas plastic methods operate with far fewer equations. However, the main virtue of plastic theory is that it is concerned with a real structural attribute (rather than the calculation of hypothetical stresses), and that it has led to new, powerful, fundamental theorems for structures made of any 'sensible' structural material (e.g. steel, aluminium or reinforced concrete, but not cast iron or glass). These theorems appear to have been formulated first in Russia, by S. M. Feinberg *1948*; Drucker *et al. 1951a* translated them and provided proofs. The four-legged table will collapse when two adjacent legs reach a stable crushing state (the plastic theorems are strictly not applicable to unstable behaviour). The 'upper-bound theorem' states that, of all possible mechanisms of collapse (e.g. of all combinations of adjacent table legs), the one that is critical is the one that corresponds to the least value of the applied load. The 'uniqueness theorem' guarantees that there is a calculable value of actual collapse load.

The most powerful of the theorems is the 'safe' or 'lower-bound theorem'. If any equilibrium solution of internal forces is established, and the structure checked to be serviceable under those forces (e.g. the yield stress is not exceeded in a steel structure), then the structure cannot collapse under the corresponding load, whatever other equilibrium state the structure might happen to find itself in. (Navier's elastic method is therefore, by virtue of the plastic theorems, a safe method, since it operates with one particular set of equilibrium forces.) Of all the possible states of equilibrium, the one that is critical is the one that corresponds to the greatest value of the applied load.

7 THE MASONRY ARCH

The plastic theorems may be applied to masonry, provided simplifying assumptions are made. Of these the chief is that stresses are so low (as indeed in general they are) that there is no danger of the material failing.

The loading theorems then translate into geometrical statements – the structure must be designed so that there are paths to transmit forces from one portion of the fabric to the next.

The thrust of arches was one of Coulomb's four problems. Phillipe de La Hire *1695* made a substantial contribution (to the problem of the design of the piers and abutments of an arch) with his invention of the funicular and force polygons. Hooke had solved the problem in 1675 (conceptually and experimentally, but not in closed form) with his statement that the shape of a flexible hanging cord will, inverted, give the shape of the ideal arch to carry the same vertical loads. David Gregory *1697* did the mathematics (the shape of the catenary) and, moreover, had a full though unproved understanding of the safe theorem; for a virtually unstressed material, stability of the arch would be assured if the inverted catenary lay (anywhere) within the thickness of the masonry. P. Couplet *1729–30* made a full and essentially correct analysis of the arch, discussing safe equilibrium states on the one hand and mechanisms of collapse on the other. A. A. H. Danyzy carried out experiments in 1732, and G. Poleni *1748* reviewed all known work on masonry during the course of his study of the dome of St Peter's in Rome. He also stated (explicitly but without proof) the safe theorem, and applied it in a remarkable way to the three-dimensional dome.

Coulomb was in his late thirties when he wrote his memoir, and probably ignorant of the work of Couplet and Poleni (but he knew of that of La Hire). He was clear in the formulation of his assumptions, and arrived at the single condition necessary to confirm the stability of an arch, namely that the line of thrust should lie everywhere within the masonry. In broad terms, nothing has been added to the theory of arches since Coulomb's time, except to confirm formally the embracement of the modern plastic theorems.

8 THE THRUST OF SOIL

The stability of slopes, including that of vertical cuts in cohesive soil, is a problem noted as early as Cetius Faventinus (*circa* AD *300*); the sides of a ditch must be prevented from falling in. First attempts to apply mechanics to a theory of earth pressure (e.g. by P. Bullet *1691*) treated soil as a granular material. Couplet *1728–30* advanced this work, which was mainly geometrical, but it fell to Coulomb *1776* to introduce two parameters, cohesion and friction, in the discussion of the failure of soil structures. Two parameters greatly complicate the analysis, but Coulomb started successfully by discussing the fracture of columns in compression, and he correctly established the plane of fracture. He was equally successful with the simpler soil-mechanics problems – the thrust against a vertical retaining wall, and the greatest depth of a vertically sided ditch. He failed, however, to develop a general theory, being defeated (effectively) by his lack of knowledge of the stress tensor (Cauchy *1823*; see §8.6). With a general theory of stress analysis available, nineteenth-century work concentrated on the cohesionless problem. The final exponent of purely frictional behaviour is probably V. V. Sokolovsky *1960*.

Coulomb's problem involving both cohesion and friction may well repay further study, but the results will not contribute markedly to the advancement of soil mechanics. K. Terzaghi *1943* introduced water content as a third and major parameter to be taken into account in the analysis of soil, and it fell to Roscoe *et al. 1958* to establish the concept of critical-state theory as the proper way forward.

9 MAXWELL AND MATRIX ALGEBRA

James Clerk Maxwell *1864* studied the truss made of straight pin-ended members (§8.2). He showed that such a space truss, simply stiff and supported by six external reactions, will be statically determinate if $b = 3j - 6$, where b is the number of bars the forces $\mathbf{t}$ in which are to be found, and j is the number of joints at which loads $\mathbf{p}$ (having $3j$ components) can be applied. (Maxwell foresaw exceptions to this rule for exceptional structures.) In matrix notation, for the general truss, the equilibrium matrix $\mathbf{H}$, with $n = 3j$ rows and $m = b + 6$ columns, relates $\mathbf{p}$ and $\mathbf{t}$ by

$$\mathbf{Ht} = \mathbf{p}. \tag{1}$$

Similarly, the vector $\mathbf{d}$ of joint displacements is related to the bar extensions $\mathbf{e}$ by the compatibility matrix $\mathbf{C}$:

$$\mathbf{Cd} = \mathbf{e}. \tag{2}$$

The principle of virtual work states that

$$\mathbf{p}^{\mathrm{T}}\mathbf{d} = \mathbf{t}^{\mathrm{T}}\mathbf{e} \tag{3}$$

for any $\mathbf{p}$, $\mathbf{t}$ satisfying equation (1), and any $\mathbf{d}$, $\mathbf{e}$ satisfying equation (2), and it is then easy to show that $\mathbf{C} = \mathbf{H}^{\mathrm{T}}$.

If $m = n$ (i.e. if Maxwell's rule is satisfied), the rank of $\mathbf{H}$ is (except for unusual structures) equal to n. $\mathbf{H}$ may be inverted to give, from equation (1),

$$\mathbf{t} = \mathbf{H}^{-1}\mathbf{p}. \tag{4}$$

The frame is thus statically determinate; if $\mathbf{p} = \mathbf{0}$ then $\mathbf{t} = \mathbf{0}$, and no states of self-stress are possible. Similarly, from equation (2), if $\mathbf{e} = \mathbf{0}$ then $\mathbf{d} = \mathbf{0}$ and the truss is rigid.

If the same frame has one extra bar, then **H** will have n rows and $n + 1$ columns. The rank is unchanged, so that the dimension of the null space of the solution vector **t** is 1; there is a single state of self-stress. Similarly, if one bar is removed from the original frame the frame will display a mechanism of one degree of freedom.

Calladine *1978* has explored such matters, including unusual structures which appear to have too few bars by Maxwell's rule, but which nevertheless are of practical importance. Pellegrino and Calladine *1986* have developed the work, and devised computational schemes suitable for modern high-speed computers.

BIBLIOGRAPHY

Baker, J. F., Horne, M. R. and Heyman, J. *1956*, *The Steel Skeleton*, Vol. 2: *Plastic Behaviour and Design*, Cambridge: Cambridge University Press.

Bernoulli, D. *1751*, 'De vibrationibus et sono laminarum elasticarum', *Commentarii Academiae Scientiarum Imperialis Petropolitanae*, **13** (1741–3), 105–20.

Bernoulli, James *1691*, 'Specimen alterum calculi differentialis', *Acta eruditorum*, 13–23. [Also in *Opera omnia*, 1st edn, Vol. 1 (1744), 431–42.]

—— *1705*, 'Véritable hypothèse de la résistance des solides, avec la démonstration de la courbure des corps qui font ressort', *Mémoires de l'Académie des Sciences* (Paris), quarto edn, 176–86. [Also in *Opera omnia*, 1st edn, Vol. 2 (1744), 976–89.]

Bullet, P. *1691*, *L'Architecture pratique* [...], Paris: Michallet.

Calladine, C. R. *1978*, 'Buckminster Fuller's "tensegrity" structures and Clerk Maxwell's rules for the construction of stiff frames', *International Journal of Solids and Structures*, **14**, 161–72.

Cauchy, A. L. *1823*, 'Recherches sur l'équilibre et le mouvement intérieur des corps solides ou fluides, élastiques ou non élastiques', *Bulletin de la Société Philomathique de Paris*, 9–13. [Also in *Oeuvres complètes*, Series 2, Vol. 2, 300–304.]

Coulomb, C. A. *1776*, 'Essai sur une application des règles *de maximis & minimis* à quelques problèmes de statique, relatifs à l'architecture', *Mémoires de Mathématique & de Physique, présentés à l'Académie Royale des Sciences par divers Savans, & lûs dans ses Assemblées*, **7** (1773), 343–82. [Also in his *Théorie des machines simples* [...], 1821, Paris: Bachelier, 318–63.]

Couplet, P. *1728–30*, 'De la poussée des terres contre leurs revestements, et de la force des revestements qu'on leur doit opposer', *Histoire de l'Académie Royale des Sciences* (1726, published 1728), 106–64; (1727, published 1729), 139–78; (1728, published 1730), 113–38.

—— *1731–2*, 'De la poussée des voûtes', *Histoire de l'Académie Royale des Sciences*, (1729, published 1731), 79–117; (1730, published 1732), 117–41.

Danyzy, A.-A.-H. *1778*, 'Méthode générale pour déterminer la résistance qu'il faut opposer à la poussée des voûtes' (27 February 1732), *Histoire de la Société Royale des Sciences établie à Montpellier*, **2**, 40*ff*.

Drucker, D. C., Greenberg, H. J. and Prager, W. *1951a*, 'The safety factor of an elastic plastic body in plane stress', *Journal of Applied Mechanics*, **18**, 371–8.

Drucker, D. C., Prager, W. and Greenberg, H. J. *1951b*, 'Extended limit design theorems for continuous media', *Quarterly Journal of Applied Mathematics*, **9**, 381–9.

Euler, L. *1744, Methodus inveniendi lineas curvas maximi minimive proprietate gaudentes, sive solutio problematis isoperimetrici latissimo sensu accepti*, Lausanne and Geneva: Bousquet. [Also *Opera omnia*, Series 1, Vol. 24.]

—— *1757*, 'Sur la force des colonnes', *Mémoires de l'Académie Royale des Sciences de Berlin*, **13**, 252–82. [Also in *Opera omnia*, Series 2, Vol. 17, 89–118.]

Faventinus, M. Cetius *circa 300, De diversis fabricis architectonicae.* [Editions include: in Vitruvius, *De architectura* (ed. A. Choisy), 1909, Paris: Lahure.]

Feinberg, S. M. *1948*, 'The principle of limiting stress', *Prikladnaia matematika i mekhanika*, **12**, 63–8. [In Russian.]

Galileo Galilei *1638, Dialogues Concerning Two New Sciences.* [Edition used: transl. H. Crew and A. de Salvio, 1914, New York: Macmillan.]

Gregory, D. *1697*, 'Catenaria', *Philosophical Transactions of the Royal Society of London*, **19**, 637–52. [Also in *Acta eruditorum*, 1698, 305–21.]

Hooke, R. *1675, A Description of Helioscopes, and Some Other Instruments*, London: 'T. R. for John Martyn'.

Lagrange, J. L. *1770*, 'Sur la figure des colonnes', *Miscellanea Taurinensia*, **5**, 123–66. [Also in *Oeuvres complètes*, Vol. 4, 125–69.]

La Hire, P. de *1695, Traité de mécanique*, Paris: Anisson.

Mariotte, E. *1686, Traité du mouvement des eaux*, Paris: Michallet.

Maxwell, J. C. *1864*, 'On the calculation of the equilibrium and stiffness of frames', *Philosophical Magazine*, Series 4, **27**, 294–9. [Also in *Scientific Papers*, Vol. 1, 598–604.]

Navier, C. L. M. H. *1826, Résumé des leçons données à l'Ecole des Ponts et Chausées, sur l'application de la mécanique à l'établissement des constructions et des machines*, Paris: Didot. [3rd edn, with notes and appendices by B. de Saint-Venant, 1864, Paris: Dunod.]

Parent, A. *1713, Essais et recherches de mathématique et de physique*, 3 vols, Paris: Nully.

Pellegrino, S. and Calladine, C. R. *1986*, 'Matrix analysis of statically and kinematically indeterminate frameworks', *International Journal of Solids and Structures*, **22**, 409–28.

Poleni, G. *1748, Memorie istoriche della gran cupola del Tempio Vaticano*, Padova: Libreria di Venezia.

Roscoe, K. H., Schofield, A. N. and Wroth, C. P. *1958*, 'On the yielding of soils', *Géotechnique* (London), **8**, 22–53.

Sokolovsky, V. V. *1960, Statics of Soil Media* (transl. D. H. Jones and A. N. Schofield), London: Butterworths.

Terzaghi, K. *1943, Theoretical Soil Mechanics*, New York: Wiley.

8.8

The dynamics of the Solar System

C. WILSON

1 NEWTON'S INSIGHTS

It was Robert Hooke who, in 1679, first confronted Newton with the problem of central forces. Straight away, Newton showed that conic-section orbits with centre of force at a focus imply an inverse-square force. Assuming the uniqueness of the solution for given initial conditions, he claimed that an inverse-square force implies a conic-section orbit. Beginning in late 1684, at Edmond Halley's urging, he proceeded to derive implications of this idea in a treatise which expanded into his *Principia* (1687). In the process, he discovered the proportionality of gravitation to the masses of the attracted and attracting bodies – the constant association of gravitation with mass, hence its 'universality'.

The *Principia* contained the chief ideas and problems that would be pursued in 'physical astronomy' (later dubbed 'celestial mechanics' by Pierre Simon Laplace) for two centuries. Its qualitative arguments were impressive, but many of its quantitative conclusions were questionable. Newton accounted qualitatively for the known inequalities in the orbit of the Moon, but his lunar theory yielded errors as high as 8' of arc. He suggested the true cause of the precession of the equinoxes, but his explanation was flawed by a faulty dynamics of rotation. His mathematical method – geometry plus limiting procedures – did not lend itself to successful imitation. (His claim to have derived the propositions of the *Principia* 'analytically', then to have translated them into geometrical form, is, on the available evidence and on the later understanding of the word 'analytical', implausible.)

Analytical celestial mechanics began in the 1740s, its emergence made possible by an emerging calculus of the trigonometric functions (§4.2). Roger Cotes had given the derivatives of the sine, tangent and secant in a work published posthumously in 1722; but it was Leonhard Euler, in 1739, in studies of the integration of linear differential equations with constant coefficients, who codified this calculus and made it known.

2 EULER ON PERTURBATION THEORY

Euler was also the first to apply the calculus of trigonometric functions to the three-body problem (§8.9). By 1744 he had constructed lunar tables incorporating perturbations. The first published account of his procedures was an essay on the perturbations of Saturn which won a prize offered by the Paris Academy of Sciences in 1748.

The anomalies in the motions of Saturn and Jupiter had troubled astronomers since Johannes Kepler's time. Euler, although failing to achieve a very successful theory, introduced techniques that would prove important for future investigators. He obtained two second-order equations in r and ϕ for the motions of the perturbed planet as projected onto the plane of the perturbing planet's orbit, and two first-order equations giving the inclination and ascending node of the orbital plane of the perturbed planet.

In these equations it was necessary to express v^{-3}, the inverse cube of the distance between perturbing and perturbed planets, as a function of the angle θ between the radius vectors of the two planets. For the planets, as opposed to the Moon, v^{-3} varies wildly in value. To put this quantity into integrable form, Euler transformed a Taylor series in such a way as to obtain something quite new, a trigonometric series:

$$A + B\sin\theta + C\sin 2\theta + \cdots. \tag{1}$$

The coefficients $A, B, C, \ldots$ were themselves infinite series, but Euler showed that each coefficient after the first two could be calculated from the two preceding ones. To approximate A and B he provided two different procedures, one of them a summation whose limit is a Fourier integral.

To integrate his equations Euler used the method of undetermined coefficients, carrying the approximation as far as the first power of the eccentricities. Owing to mistakes in sign, his coefficients for certain large terms were wrong. The existence of sizable higher-order terms he did not even suspect.

In an attempt to correct the orbital elements, and determine the coefficients of doubtful perturbational terms, Euler introduced the use of multiple equations of condition – a first step towards bringing statistics into astronomy (compare §10.13).

In solving his third and fourth differential equations, he found that the nodes of Saturn's orbit are steadily regressing; an analogous result must apply to all the planets. In a memoir completed in 1754 he deduced from this the fact that the obliquity of the ecliptic was diminishing by some 47".5 per century (Wilson *1980*).

The Paris Academy again posed the problem of Jupiter and Saturn in 1750 and 1752, and in 1752 Euler again won the prize. His essay of 1752 showed that the orbital eccentricities were partly 'proper' and partly 'adventitious', induced by the attraction of the second planet. Unfortunately, his calculation of the secular changes in eccentricities, apsides and mean motions was riddled with algebraic error. This essay was not published until 1769; but, carrying the ideas of this essay further in an appendix to his *Theoria motus lunae* (1753), Euler proposed calculating perturbations by the variation of orbital parameters.

3 CLAIRAUT AND THE LUNAR THEORY

Beginning in 1746, both Alexis Clairaut and Jean d'Alembert undertook to formulate perturbational theories, and to apply them to the Moon. In November 1747 Clairaut announced that the motion of the lunar apse deducible from the inverse-square law was only half as great as the observed motion (a conclusion reached also by Euler and d'Alembert). To resolve the discrepancy, he proposed adding an inverse fourth-power term to the gravitational law. By December 1748, however, he had discovered that most or all of the motion of the lunar apsides was deducible from the unmodified inverse-square law, provided the approximation was carried far enough. The new deduction was published in 1752 in Clairaut's *Théorie de la lune*. It was this result that first led to widespread acceptance outside England of Newton's law of gravitation.

Clairaut began by assuming that the motion of the perturbed body lay in a single plane. By a twofold integration of his differential equations he eliminated time, and so arrived at an expression for the radius vector r in terms of the true longitude ϕ in the perturbed orbit:

$$f^2/Mr = 1 - g \sin \phi - c \cos \phi + \sin \phi \int \Omega \cos \phi \, d\phi - \cos \phi \int \Omega \sin \phi \, d\phi, \quad (2)$$

where f, g and c are constants, and Ω is a function of the radial and transverse components of the perturbing force.

If the terms involving Ω are deleted from (2), it becomes the equation for a conic section. The integrals in the last two terms of equation (2) can be evaluated if Ω can be expressed as a sum of terms of the form $A \cos p\phi + B \cos q\phi + \cdots$, where $A, p, B, q, \ldots$ are constants. For this it is necessary that r be replaced by an expression in terms of ϕ. To substitute the expression for the fixed, unperturbed ellipse, Clairaut urged, would be a mistake since observations teach us that the orbital apsides of the Moon and planets advance. He therefore proposed using the formula

$$k/r = 1 - e \cos m\phi,\qquad(3)$$

which represents a rotating ellipse; this is justified, Clairaut explained, if once the substitutions are made, the larger terms of equation (2) prove to have the form of (3), permitting k, e and m to be evaluated in terms of the other constants of the theory. This turns out to be the case.

But substituting (3) in (2) led to a value of m giving only half the observed motion of the lunar apsides. Clairaut knew that he could generalize the new formula for r, replacing the numerical coefficients of the sinusoidal terms with indeterminate ones, then substituting this more elaborate formula into equation (2) to obtain a closer approximation; but it was not until the second half of 1748 that he attempted this tedious second-stage calculation. To his surprise, it gave a value of m implying an apsidal motion nearly equal to the observed motion (Waff *1976*).

D'Alembert objected to Clairaut's assumption of an apsidal motion on the basis of observation, instead of deriving this fact from the differential equations. In his own treatise on the subject, published in 1754, he carried the algebraic articulation of the theory further than Clairaut had. Calculating the first four terms in the motion of the apogee, he found close agreement with the observed motion. But, he warned, the validity of the result remained uncertain: further iterations might fail to converge.

4 D'ALEMBERT ON THE PRECESSION AND NUTATION OF THE EARTH'S AXIS

Nutation is a small wobble in the precession of the Earth's axis, induced by the changing orientation of the Moon's orbit as its nodes regress on the ecliptic. James Bradley discovered it in the early 1730s, but announced it only in January 1748 after following it through a full 18.6-year cycle of the lunar nodes. D'Alembert, learning of Bradley's discovery in the summer, completed by May 1749 a derivation of the precession and nutation from Newton's inverse-square law – a triumph for the theory.

The important innovation in d'Alembert's treatise was to set the external torque about each coordinate axis, due to the Sun's and Moon's attraction of the equatorial bulge, equal to the integral of the moments about the same axis of the products of mass and acceleration for each mass point in the body of the Earth. This is the principle of moment of momentum. D'Alembert justified the innovation by a confused appeal to 'd'Alembert's principle' (§8.1). In successive memoirs during the 1750s, Euler clarified this deduction, and developed a coherent mechanics of rigid bodies (Wilson *1987*).

5 THE WORK OF LAGRANGE TO 1774

Joseph Louis Lagrange's first foray into celestial mechanics was an essay on the libration of the Moon which won the Paris Academy's contest of 1664. Libration produces an apparent oscillation of the side of the Moon facing the Earth, so that regions near the edge of the Moon's disc are sometimes visible, sometimes not. Using a combination of d'Alembert's dynamical principle with the principle of virtual work (it would become the basis of his *Méchanique analitique*, 1788), he obtained three second-order equations for the Moon's motion about its centre of gravity. Integration of the first showed that, in addition to the 'optical' libration (the difference between the Moon's uniform axial rotation and its variable speed in orbit) there can be the physical libration that Newton had hypothesized: a slight, slow oscillation of the Moon's body about its polar axis, owing to its being elongated in the direction of the Earth. The exact equality now obtaining between the Moon's axial rotation and its mean orbital speed would follow from this oscillation, had the initial values not been too unequal.

In integrating the other two equations, Lagrange neglected the second-order differentials and could not account for the observationally verified coincidence between the nodes of the lunar orbit and the nodes of the Moon's equator. In 1780 he returned to the problem, substituting new variables to integrate the second-order equations, and so deduced the alignment.

In 1766 Lagrange won the Paris Academy's contest concerning Jupiter's satellites. In his essay he introduced an elegant algorithm for determining the coefficients of the trigonometric series for v^{-3}, using the relation

$$(\cos\theta + \mathrm{i}\sin\theta)^n = \cos n\theta + \mathrm{i}\sin n\theta. \tag{4}$$

He was able to account very exactly for the inequalities that P. Wargentin had observed in a long sequence of eclipses of the satellites. The first three satellites have mean motions, n_1, n_2 and n_3, very nearly in the ratio $4:2:1$; this near-commensurability leads to small divisors in certain terms of the integrals. The time taken for these inequalities to return to the same phase in eclipses proved to be 437.6 days, just as Wargentin had found. For the second satellite two perturbational terms are involved, with arguments $(n_1 - n_2)t$ and $2(n_3 - n_2)t$, but these combine into a single term since, as Lagrange deduced from the observations,

$$2(n_3 - n_2)t = 180° - (n_1 - n_2)t, \tag{5}$$

so that the sines of the two arguments are equal.

The second part of the essay dealt with a serious difficulty in the method of successive approximations used by both Euler and Lagrange. In the first approximation they had assumed the orbits to be circular, except for the perturbations. When the first-order solutions were substituted back into the differential equations and the latter again integrated, among the terms emerging were some belonging to the first-order solution and others with zero denominators.

To overcome this difficulty, Lagrange engaged in an elaborate algebraic process, using integration by parts to obtain expressions for the radius vector and equation of the centre embodying all the first-order terms. His two expressions each contained four terms – one 'proper' to the satellite, the other three contributed by the other satellites. Because the apsides of the orbits were in motion, the four terms added together to form a slowly changing sum. Lagrange had rediscovered the essential insight of Euler's still unpublished essay of 1752.

Lagrange proceeded forthwith to apply this analysis to the mutual perturbations of Jupiter and Saturn, thereby obtaining values for the secular variations of the eccentricities, aphelia, inclinations, nodes and mean motions. These were correct except for the mean motions, as Laplace was soon to prove. Lagrange showed the secular variations to be periodic, with periods in the tens of thousands of years; their essential difference from the ordinary periodic inequalities being that they do not depend on planetary positions.

Lagrange shared with Euler the prize offered at the Paris contest of 1772 concerning the lunar theory. Euler's new lunar theory introduced a coordinate system rotating with the mean angular speed of the Moon; Lagrange's memoir was an analysis of the three-body problem (§8.9).

The subject of the Paris contest of 1774 was the secular acceleration of the Moon. Halley had discovered it from a comparison of ancient, Arabic and modern observations of eclipses. Lagrange's winning essay showed that the non-spherical shapes of the Earth and Moon could not account for it, and he introduced the potential in the calculation of the forces.

In a memoir sent to Paris in October 1774, Lagrange showed how to formulate first-order linear differential equations for the secular variations of the nodes and inclinations of planetary orbits. The crucial step was the introduction of new variables:

$$s = \lambda \sin \chi, \qquad u = \lambda \cos \chi, \qquad (6)$$

where χ is the longitude of the node and λ the tangent of the inclination. The integrals were sums of slowly changing sinusoidal terms. The same procedure was applicable to the aphelia and eccentricities of the orbits, but here Laplace beat Lagrange by publishing first.

6 THE WORK OF LAPLACE AND LAGRANGE TO 1785

Laplace's first coup was to show, in a memoir completed by early 1774, that the secular variations of the mean motions came out identically to zero, if the approximations were carried to the order of the squares of the eccentricities and inclinations and the first power of the perturbing mass. In late 1774 he applied Lagrange's new method for secular variations to the aphelia and eccentricities. But then, deterred from further work on planetary perturbations by imponderables (the unknown masses of the inner planets, the indeterminate role of comets), he turned to other topics – the tides, precession and the attraction of spheroids.

Meanwhile, in two memoirs of 1776 and 1777, Lagrange introduced the perturbing function, a potential function from which the forces on a perturbed planet were obtainable by partial differentiation. The first memoir was a new proof that the mean motions were, to any order in the eccentricities and inclinations, and to the first powers of the perturbing masses, immune to secular variation provided the mean motions were incommensurable (as Lagrange assumed they were). The second memoir derived the known integrals of the equations of motion.

In two long memoirs published in 1783 and 1784, Lagrange derived the secular variations systematically. He then turned to the periodic variations, using the same method of variation of orbital elements. At this point he discovered that his earlier proof of the immunity of the mean motions to secular variation needed qualification: the mean motions *did* vary secularly, insofar as they depended on the secular variations of the other orbital elements. To symbolize this, he expressed the variation of the mean motion as $d(p + \Sigma)$, where dp represents periodic inequalities, and $d\Sigma$ represents the newly discovered secular variations; $d\Sigma$, he proposed, was to be viewed as a variation of the epoch.

Lagrange devoted a special memoir, published in 1785, to computing $d\Sigma$ for Jupiter and Saturn; he found it to be negligible. To save labour in computing the higher-order terms, he suggested calculating only those terms that would be greatly enlarged by integration – terms of the form $A \sin(\nu t + B)$, where ν is small relative to the mean motion. But ν is a linear combination of the mean motions of the planets involved. By late 1785, deploying this very selection procedure, Laplace had discovered the source of the Jovian and Saturnian anomalies – in dp, not in $d\Sigma$.

For a detailed account of these developments, see Gautier *1817*.

7 LAPLACE'S TRIUMPH

On 23 November 1785 Laplace read to the Paris Academy a memoir on the secular inequalities of planets and satellites, the first of four memoirs in which he accounted for the chief remaining anomalies in the theory of the Solar System. In it he announced his discovery of hitherto unknown inequalities of Jupiter and Saturn, explained the permanence of pattern in the dance of Jupiter's first three satellites, and gave an *a priori* proof of the stability of the Solar System.

The 'great inequality' of Jupiter and Saturn was due to terms of the third degree in the eccentricities and inclinations, with the argument $(5n' - 2n)t$, where n and n' are the Jovian and Saturnian mean motions. Since $n : n'$ is very nearly $5 : 2$, $5n' - 2n$ is small relative to n or n', and with two integrations the terms are multiplied by $1/(5n' - 2n)^2$. The resulting coefficients were nearly 50 arc minutes for Saturn, and 20 arc minutes for Jupiter; the period of the inequality was about 900 years. In addition to these large terms, Laplace located sizable second-order terms in the eccentricities with coefficients containing the first power of the small divisor $5n' - 2n$.

In his excursus on the satellites of Jupiter, Laplace showed that a relation indicated by the observations,

$$n_1 - 3n_2 + 2n_3 = 0, \qquad (7)$$

where n_1, n_2 and n_3 are the mean motions of the first three satellites, was exact, maintained as the mean of a theoretically derived oscillation. The same was true of the relation Lagrange had found, the adding of the arguments of the two main inequalities of the second satellite to $180°$ (equation (5)).

The final section of the memoir concerned the stability of the Solar System as a whole. Starting from the conservation of angular momentum and proceeding by *reductio ad absurdum*, Laplace argued that the orbital eccentricities and inclinations of the planets are oscillatory and do not contain terms proportional to the time or exponentials dependent on the time.

Laplace's 'Théorie de Jupiter et de Saturne' was presented to the Academy in two parts in 1786. Here, by means of the perturbing function and the integrals of motion, he derived his fundamental equations for computing perturbations, as he would give them in the first volume of his *Mécanique céleste* (1799). As always, he used Lagrange's variational operator 'δ' to express the perturbations, while the other quantities in the equations referred to the unperturbed elliptical motion. In fitting his theories to observational data, he used multiple equations of condition (Wilson *1985*).

One major unexplained anomaly in the theory of the Solar System remained: the secular equation of the Moon. After several attempts Laplace succeeded in discovering a cause for it: the action of the Sun on the Moon, combined with the secular diminution of the eccentricity of the Earth's orbit. His memoir on the subject was submitted on 19 December 1787. He took account only of the radial component in the Sun's perturbing force. In 1854 John Couch Adams pointed out that since the terrestrial eccentricity diminishes continuously, the transverse components of the perturbing force before and after syzygy (that is, when the three bodies are in close alignment) do not exactly balance. The more exact computation gave only half the observed value of the secular acceleration. It was eventually realized that the remaining half was due to tidal retardation of the Earth's rotation.

8 CELESTIAL MECHANICS IN THE NINETEENTH AND TWENTIETH CENTURIES

Nineteenth-century work in celestial mechanics was mainly a skilful exploitation of the earlier achievements. William Rowan Hamilton and Carl Jacobi developed the analytical mechanics that Euler and Lagrange had founded (§8.1); Peter Andreas Hansen, Charles Delaunay, Urbain Le Verrier, Simon Newcomb, G. W. Hill and others calculated perturbations by methods that derived from Euler, Lagrange and Laplace. Hamilton's introduction of the 'Hamiltonian', and Jacobi's introduction of canonical equations with conjugated variables facilitated the transformation of equations into systems that could be treated more advantageously.

The chief preoccupation of astronomers was the construction of accurate theories of the planets (including the minor planets, the first of which was discovered on 1 January 1801), satellites and comets. Much effort went into the development of the perturbing function; by 1855 Le Verrier had pushed it to seventh-order terms. In publications of 1829 and 1859, Hansen showed how all perturbations could be applied to the mean longitude, and in his lunar theory he used as intermediary orbit an ellipse of fixed dimensions with uniformly turning perigee. Delaunay developed the more precise literal theory of the Moon, accurate to 1 arc second (1″). Hill's lunar theory as completed by E. W. Brown achieved a like accuracy.

Anomalies in the motions of Uranus, the planet discovered by William Herschel in 1781, led to the discovery of Neptune. Both Adams and Le Verrier, in calculating a position for the perturbing planet, followed the empirical Titius–Bode law, which implied a mean solar distance twice that of Uranus. The planet, actually located by Johann Galle on 23 September 1846, proved to have a smaller mean solar distance, mass and orbital eccentricity than had been assumed by either Adams or Le Verrier, but the differences were mutually compensatory. Their prediction was regarded as a triumph for the Newtonian theory (Grosser *1962*).

In 1859 Le Verrier published his theory of Mercury, which pointed out a 38″ discrepancy between the predicted motion of the perihelion, 527″ per century, and the observed motion, 565″. (Newcomb later put the discrepancy at 43″.) Le Verrier looked for a planet or ring of matter within the orbit of Mercury to explain the discrepancy, but by 1896 François Félix Tisserand had concluded that no such body or bodies existed. Newcomb, in his theories of the inner planets, assumed a tiny departure from the inverse-square law to account for the excess motions of their perihelia. In 1915, Albert Einstein showed that the extra advance of Mercury's perihelion followed from his general theory of relativity (§9.13).

Le Verrier's planetary theories were used for ephemerides in France until 1960. One flaw in them was the lack of a consistent set of planetary masses. This flaw was corrected in Newcomb's and Hill's theories. Except for certain corrections to Newcomb's theory of Mars by Frank E. Ross, these were used for ephemerides in the USA and UK until 1960, when Wallis Eckert's numerical integration of the orbits of the five outer planets was adopted. Improved values of the constants used in the theories have come from increased accuracy in the measurement of the fluctuations in the Earth's rate of rotation, more accurate mapping of the Earth's gravitational potential by means of artificial satellites, and improved knowledge of planetary distances given by radar and laser observing techniques. The accuracy of prediction for the three innermost planets is rated at about ±0.1″, and it is expected that the errors will be at least halved by new theories adopted from the 1980s onwards. For Mars and, especially, the outer planets, the accuracy is somewhat less.

The deepest probing into the theory of orbits was initiated by Poincaré, who studied the problem of stability topologically. This arose from work on the three-body problem, to which we now turn.

BIBLIOGRAPHY

Gautier, A. *1817, Essai historique sur le problème des trois corps*, Paris: Veuve Courcier.

Grant, R. *1852, History of Physical Astronomy*, London: Baldwin. [Repr. 1966, New York and London: Johnson.]

Grosser, M. *1962, The Discovery of Neptune*, Cambridge, MA: Harvard University Press.

Tisserand, F. *1889–96, Traité de mécanique céleste*, 4 vols, Paris: Gauthier-Villars.

Waff, C. *1976*, 'Universal gravitation and the motion of the Moon's apogee', Ph.D. Dissertation, Johns Hopkins University, Baltimore, MD.

Wilson, C. *1980*, 'Perturbations and solar tables from Lacaille to Delambre', *Archive for History of Exact Sciences*, **22**, 53–304.

—— *1985*, 'The great inequality of Jupiter and Saturn', *Archive for History of Exact Sciences*, **33**, 15–290.

—— *1987*, 'D'Alembert *versus* Euler on the precession of the equinoxes and the mechanics of rigid bodies', *Archive for History of Exact Sciences*, **37**, 233–73.

8.9

The three-body problem

C. WILSON

1 BEGINNINGS: NEWTON, EULER, LAGRANGE

The two-body problem – to determine the motions of two mutually attracting bodies according to the inverse-square law – is solvable in closed form. Isaac Newton solved it in Propositions 1–17 and 57–60 of Book I of his *Principia* (1687). In Propositions 65 and 66 he went on to treat the three-body problem qualitatively. An exact solution, he speculated elsewhere, 'exceeds, if I am not mistaken, the force of any human mind'.

Post-Newtonian dynamical astronomy has had the task of deriving ephemerides of the celestial bodies by approximate resolution of the three-body problem. The theory of the problem, meanwhile, has evoked extensive discussion, and it is with the history of the theory that this article is concerned.

Leonhard Euler was the first to investigate restricted forms of the three-body problem with a view to obtaining exact integrals. Beginning in 1760, he wrote three memoirs on the motion of a body attracted to two fixed centres (Euler *1766a*, *1767a*, *1767b*). Using the available first integrals and with several changes of variables, he derived the general solution as an equation between two elliptic integrals with separated variables. In the third memoir he extended the solution to the three-dimensional case.

In a memoir presented in 1762 (Euler *1766b*), Euler derived differential equations for a 'Moon' of negligible mass attracted by an 'Earth' and a 'Sun' circling each other; it is what Henri Poincaré was to call 'the restricted problem of three bodies'. He showed that there were collinear solutions, with the 'Moon' in constant conjunction or opposition at a distance from the 'Earth' (as given by the root of a quintic equation) equal to about four times the distance of the actual Moon. For slightly different initial conditions, Euler proposed that the 'Moon' would oscillate about the point of conjunction or opposition. In an essay presented in 1763, Euler *1767c* studied the possible rectilinear motions of three interacting bodies, without any restriction as to their masses, and arrived at the same quintic equation as before.

Joseph Louis Lagrange, stimulated by Euler's work on the motion of a body attracted by two fixed centres, wrote two memoirs on the subject in the late 1760s. In the first he achieved the Eulerian results with improved elegance. In the second he investigated the integrability of the equations under hypotheses of attraction other than the inverse square; in a final paragraph he adapted the analysis to the lunar problem by adding to inverse-square forces a force on the Moon equal to the Sun's force on the Earth, but of opposite direction. The problem in this form, however, proved unsolvable.

The problem of two fixed force centres was treated by Carl Charlier early in the twentieth century (*Die Mechanik des Himmels*, Volume 1, 1903). It has had indirect applications in more recent work, for instance in supplying reference orbits for artificial satellites.

In his prize-winning 'Essai sur le problème des trois corps', Lagrange *1772* set out to determine the motions of the bodies relative to each other solely in terms of the distances between them. (This is now called the 'general' three-body problem.) Using the integral of *forces vives* (the expression of the day for kinetic energy) and the three integrals for conservation of angular momentum, he reduced the problem to one of the seventh order, involving the solution of differential equations in terms of the mutual distances of the bodies and their masses, and certain auxiliary variables. (According to Whittaker *1900*, the elimination of the nodes with consequent reduction to the sixth order, later carried out explicitly by Carl Jacobi (see Section 2), is implicit in Lagrange's essay.) Assuming these equations solved, Lagrange showed that the coordinates of the bodies in a fixed frame of reference could be found without further integration.

Seeking the closed solutions of which his equations admitted, he found two. One was the collinear solution given earlier by Euler; in the other, the bodies were located at the vertices of an equilateral triangle. In both solutions the motions were confined to a single plane, and the distances could be either constant or in fixed ratios to one another (in the latter case the orbits were conic sections). Lagrange thought these solutions inapplicable in 'the system of the world', but it is now known that they are exemplified in the Trojan asteroids, which form with Jupiter and the Sun two equilateral triangles.

2 THE WORK OF JACOBI AND HAMILTON

In 1836, Jacobi announced an integral for what is now called the 'three-dimensional' restricted problem of three bodies. A massless point orbits the Sun, its motion perturbed by a planet moving in a circular orbit round the Sun. The x- and y-coordinates of the point are taken in the plane of the perturbing planet; z designates the departure from this plane. Jacobi employed sidereal coordinates rather than synodic coordinates, which were introduced later and which simplify the expression of the integral; they are defined relative to the Earth's rotation rather than the Moon's orbit.

Jacobi's 'Sur l'élimination des noeuds dans le problème des trois corps' appeared as *1843*. In it he showed how to reduce a general three-body problem – that of the motion of two planets about the Sun – to a problem of the motion of two fictive bodies. He subjected the masses, positions and velocities of these fictive bodies to the restrictions that their centre of gravity should be identical to that of the original system, and their accelerations derivable from the force function for that system; the total *force vive* was the same, and the conservation of areas still held. Then, following a general Lagrangian formula, he obtained the equations

$$\mu_1 x_{1i} = \partial U / \partial x_{1i}, \quad \mu_2 x_{2i} = \partial U / \partial x_{2i}, \quad i = 1, 2, 3. \tag{1}$$

The fictive masses μ_1 and μ_2 were dependent in value on arbitrary constants, which Jacobi now chose so as to make the μ_1 and μ_2 nearly equal to the two planetary masses, and their coordinates nearly equal to those of the two planets. The previously derived equations, with all but a single term deleted from their right-hand terms, became solvable by the known formulas for elliptical motion, so that, to a first approximation, the fictive masses (and hence the planets) could be regarded as moving in ellipses about the centre of gravity of the original system as a focus. The perturbations could then be introduced by varying the orbital elements in the Lagrangian manner.

But the varying orbital elements were subject to the three integrals expressing conservation of areas. From these, Jacobi deduced that the common intersection of the orbital planes of the two fictive bodies remained constantly in a fixed plane. The six equations of motion, expressed in terms of (a) the radius vectors, (b) their angular distances from the common ascending node and (c) the inclinations of the two planes, were of the first order (except for one of the second order), and they did not contain the node, which was given directly by an integration.

The possibility, left open in Jacobi's memoir, of choosing other values for the fictive masses was followed up by later investigators. Thus in 1852 Joseph Bertrand analysed a system often employed thereafter: the first fictive mass, placed at the centre of gravity of m_1 and m_2, is equal to $m_1 m_2 / (m_1 + m_2)$; the second fictive mass, placed at the original location of m_3, is equal to $m_3(m_1 + m_2) / (m_1 + m_2 + m_3)$. Like Jacobi, Bertrand was able to reduce the system to the sixth order.

In the further history of the three-body problem, Hamilton–Jacobi dynamics played a dominant role (§8.1). Its roots lay partly in Lagrangian dynamics. In 1808 Lagrange discovered a relation giving the partial derivatives of the perturbing function with respect to the orbital parameters as linear functions of the time derivatives of these parameters:

$$\frac{\partial H}{\partial a_i} = \sum_{j \neq i} [a_i, a_j] \frac{da_j}{dt}, \qquad (2)$$

where H is the perturbing function, and the term in square brackets consists of certain time-independent functions now known as 'Lagrange brackets'. In 1809, Siméon-Denis Poisson derived an inverse relation,

$$\frac{da_j}{dt} = \sum_{j \neq i} (a_j, a_i) \frac{\partial H}{\partial a_i}, \qquad (3)$$

in which the term in parentheses consists of time-independent functions now known as 'Poisson brackets'. Lagrange also showed that, for a certain choice of the orbital parameters, the right-hand side of equation (2) reduces to a single time derivative.

In two essays published in 1834 and 1835, William Rowan Hamilton gave these formulas an extraordinary extension. Previously, the force function had always been defined in terms of the coordinates alone, but Hamilton allowed it to be an arbitrary function H of the momenta p_i as well as the generalized coordinates q_i. With this function, taken to be time-independent, and applying the principle of least action, he derived the equations of motion for any system of n degrees of freedom as n pairs of first-order partial differential equations:

$$\dot{p}_i = -\partial H/\partial q_i, \qquad \dot{q}_i = \partial H/\partial p_i. \qquad (4)$$

(It was Jacobi who later gave the designation 'canonical' to the equations of motion written in this form.) Hamilton then went on to show that the integrals of these equations could be represented by the partial derivatives of a function S which he called the 'principal function'. In general, S, defined by partial differential equations, could be found only by successive approximations.

Jacobi simplified and generalized Hamilton's theory. He showed that the restriction to conservative systems was unnecessary: H could be a function of the time t as well as of the q_i and p_i. He focused on the transforming of one canonical system into another by change of variables; H, enjoying a minimum property to begin with, retained it under the transformation, and necessary and sufficient conditions on the new variables could be given for the maintenance of canonical form. (The first published proof of these conditions was by A. H. Desboves, in 1848.) These general rules of transformation facilitated choices of variables that would simplify the dynamical problem, frequently making possible a reduction in the number of degrees of freedom.

3 DELAUNAY'S LUNAR THEORY, AND THE AVOIDANCE OF SECULAR TERMS

The various reductions of the general and restricted three-body problems during the later nineteenth century are too numerous to review here; for details see Whittaker *1900* and Marcolongo *1919*.

The lunar theory of Charles Delaunay, published in two volumes in 1860 and 1867, applied canonical transformations to the 'elliptic' restricted problem in three dimensions (in which the two primaries move in ellipses). The variables were six orbital elements, in three canonically related pairs. The Hamiltonian H was expanded as an infinite series, each term consisting of a function of the 'momenta', multiplied by the cosine of a linear function of the 'coordinates' and the time. Delaunay's procedure was to select an important term in H, and to carry out a canonical transformation so as to yield a new perturbing function H' in which the selected term was no longer present. This operation was then repeated – Delaunay carried it out 57 times – to obtain a Hamiltonian function from which all the important periodic terms were eliminated; and the equations were then integrated. The longitude, latitude and parallax of the Moon thus came to be expressed as infinite series in which the time t occurred only in the arguments of periodic terms.

Delaunay's theory, the product of twenty years of labour, attained an accuracy of 1 arc second. Its practicality was limited by the slow convergence of its series, but it showed that secular terms – ones in which t appeared outside the arguments of periodic terms – could be avoided. In 1874, Simon Newcomb gave a demonstration that the problem of three bodies may be formally solved by infinite series of purely periodic terms. In 1883, A. Lindstedt, starting from Lagrange's equations for the three-body problem, showed again that such a series solution was obtainable. But were these series convergent? In 1882–4, Poincaré proved that a function represented by a series that is not uniformly convergent can assume arbitrarily large values. H. Bruns proved in 1884 that the series can fluctuate between convergence and divergence when, in the terms of long period, the constants on which the coefficients of the time depend are varied by small amounts.

4 THE INTERMEDIATE ORBIT OF HILL

Meanwhile, G. W. Hill published the first account *1878* of his lunar theory; its ideas were to be seminal. Earlier astronomers had taken as a first approximation the ellipse the Moon would follow if the solar perturbation were absent; Hill chose instead the 'variation ellipse' produced by a Sun in circular orbit about the Earth. His first step was thus to address the restricted problem of three bodies, using a synodic system of coordinates rotating with the Sun; the solution was periodic, the Sun, Earth and massless Moon periodically returning to the same relative positions. He then went on to vary this solution in order to take account of the eccentricity of the lunar orbit. Hill also studied a range of periodic orbits by analytic continuation, and by means of the Jacobian integral he established curves of zero velocity delimiting the regions of space accessible to the Moon.

5 THE WORK OF POINCARÉ

In the late 1880s, Bruns and Poincaré once again produced closely associated proofs. In 1887, Bruns proved that in the three-body problem, apart from the ten classical integrals (the six of the centre of gravity, the three of angular momentum and the energy integral), no other algebraic integrals exist. In 1889 Poincaré proved that in the restricted three-body problem no integral exists besides the Jacobian.

Poincaré's interest now came to focus on periodic solutions of the restricted three-body problem. In his prize-winning essay *1890*, he first proved a recurrence theorem: that in any region r_0 of the phase space, however small, there exist trajectories that pass through this region an infinite number of times. Moreover, the points in r_0 which do not give rise to such trajectories have a volume infinitely smaller than r_0; recurrent trajectories are thus infinitely more numerous than non-recurrent ones.

Poincaré next considered solutions differing only slightly from a given periodic solution. Suppose that the Hamiltonian is developable in terms of a parameter μ:

$$F = F_0 + \mu F_1 + \mu^2 F_2 + \cdots. \tag{5}$$

Suppose also that when $\mu = 0$, the solution

$$x_i = \phi_i(t), \qquad y_i = \psi_i(t) \tag{6}$$

is periodic. Poincaré showed that, for small values of μ, there exist periodic solutions

$$x_i = \phi_i(t) + \xi_i, \qquad y_i = \psi_i(t) + \eta_i, \tag{7}$$

where

$$\xi_i = S_i \exp(\alpha_k t), \qquad \eta_i = T_i \exp(\alpha_k t); \tag{8}$$

here the S_i and T_i are periodic functions of t, and the α_k, called 'characteristic exponents', are the roots of a certain algebraic equation, and must be imaginary for the solution to be stable.

Poincaré's essay also includes a theory of 'asymptotic solutions', solutions such that the ξ_i diminish indefinitely as t goes to $+\infty$ or to $-\infty$, or to both. Thus these solutions approach periodicity.

The three volumes of Poincaré's *Les Méthodes nouvelles de la mécanique celeste* (*1892–9*) carried further the discussion of these topics. The thrust was towards characterizing completely the totality of motions of dynamical systems by their qualitative properties. As in his recurrence theorem, he invoked a hydrodynamic analogy in which the orbits appear as streamlines of a three-dimensional incompressible fluid of finite volume in steady motion. A moving particle of such a fluid must indefinitely often partially reoccupy its original position. Another analogy introduced by Poincaré (and developed by Jacques Hadamard, E. T. Whittaker, and G. D. Birkhoff) was that of geodesics on a surface of revolution. Here closed geodesics correspond to periodic orbits, and the existence of certain closed geodesics of minimum length can be used as a basis for deductions.

In *Les Méthodes nouvelles*, Poincaré also discussed the convergence of the series used by astronomers in treating the three-body problem, in particular Lindstedt's series and the series in an elaborate theory that H. Gyldén had developed with a view to ensuring convergence. Poincaré showed that in general these series were not uniformly convergent, but that they represented the dynamical coordinates in an asymptotic sense. This result showed Lagrange's and Laplace's demonstrations of stability to be inconclusive.

In his last year (1912), Poincaré introduced another topological analogy for the theory of orbits: the variables of the restricted problem are interpreted as rectangular coordinates of a moving point, and the successive intersections of this point with a stationary surface are studied as transformations of the surface into itself. A conjectured theorem, later proved by Birkhoff, implied that periodic solutions in the restricted problem are infinite in number.

6 AFTER POINCARÉ

Poincaré's *Les Méthodes nouvelles* stimulated further work on periodic orbits. In 1901 Tullio Levi-Cività showed that, in the restricted problem, if the mean motions of the massless point and the other two bodies are commensurable, then the motion is unstable. George Darwin, Forest Ray Moulton, E. W. Brown and Elis Strömgren studied various classes of periodic orbit in the restricted problem. In 1912, Birkhoff showed that every stable orbit in the restricted problem has certain properties of recurrence, or asymptotically approaches and recedes from orbits with such properties. Also in 1912, Karl Sundmann showed that the problem was solvable, albeit by means of series that converge too slowly to be of practical use.

T. N. Thiele (1895), Levi-Città (1906, 1919) and Sundmann (1912) showed that singularities in the motion of a system can often be removed by a change of variables (regularization), so that it becomes possible to follow the motion through collision. Sundmann showed that in the general three-body problem the sum of the three mutual distances always exceeds a definite positive quantity if the motions are not confined to a single plane. J. Chazy (1918–32) and Birkhoff (1927) studied further the tendency to instability in the general problem: escape of one of the three bodies from the other two appeared to be the predominant outcome.

The advent of artificial satellites and the emergence of the electronic computer have produced a resurgence in studies of the various three-body problems, restricted and general. Regularization, stability and instability are the major foci of interest.

BIBLIOGRAPHY

Euler, L. *1766a*, 'De motu corporis ad duo centra virium fixa attracti', *Novi commentarii Academiae Scientiarum Petropolitanae*, **10** (1764, publ. 1766), 207–42. [Also in *Opera omnia*, Series 2, Vol. 6, 209–46. Journal listed as '*NC*' hereafter.]

—— *1766b*, 'Considerationes de motu corporum coelestium', *NC*, **10**, 544–58. [Also in *Opera omnia*, Series 2, Vol. 25, 246–57.]

—— *1767a*, 'Problème: Un corps étant attiré en raison réciproque quarrée des distances vers deux points fixés donnés trouver des cas où le courbe décrite par ce corps sera algébrique', *Mémoires de l'Académie des Sciences de Berlin*, **16** (1760, publ. 1767), 228–49. [Also in *Opera omnia*, Series 2, Vol. 6, 274–93.]

—— *1767b*, 'De motu corporis ad duo centra virium fixa attracti', *NC*, **11**, (1765, publ. 1767), 152–84. [Also in *Opera omnia*, Series 2, Vol. 6, 247–73.]

—— *1767c*, 'De motu rectilineo trium corporum se mutuo attrahentium', *NC*, **11** (1765, publ. 1767), 144–51. [Also in *Opera omnia*, Series 2, Vol. 25, 281–9.]

Gautier, A. *1817*, *Essai historique sur le problème des trois corps*, Paris: Veuve Courcier.

Hill, G. W. *1878*, 'Researches in the lunar theory', *American Journal of Mathematics*, **1**, 5–26, 129–47, 245–61.

Jacobi, C. G. J. *1843*, 'Sur l'élimination des noeuds dans le problème des trois corps', *Comptes rendus de l'Académie des Sciences de Paris*, **15**, 236–55. [Also in *Gesammelte Werke*, Vol. 4, 297–314.]

—— *1866*, *Vorlesungen über Dynamik* (ed. A. Clebsch), Leipzig; Teubner. [Also in *Gesammelte Werke*, Supplementband. Lectures originally delivered in 1842–3.]

Lagrange, J. L. *1772*, 'Essai sur le problème des trois corps', *Prix de l'Académie Royale des Sciences de Paris*, **9**, Part 9. [Also in *Oeuvres*, Vol. 6, 229–324. See also pp. 67–121.]

Marcolongo, R. *1919*, *Il problema dei tre corpi da Newton ai nostri giorni*, Milan: Hoepli. [Detailed bibliographical study of the literature between 1870 and 1910.]

Poincaré, H. *1890*, 'Sur le problème des trois corps et les équations de la dynamique', *Acta mathematica*, **13**, 1–270. [Also in *Oeuvres*, Vol. 7, 262–479.]

—— *1892, 1893, 1899*, *Les Méthodes nouvelles de la mécanique céleste*, 3 vols, Paris: Gauthier-Villars.

Szebehely, V. *1967*, *Theory of Orbits*, New York and London: Academic Press.

Whittaker, E. T. *1900*, 'Report on the progress of the solution of the problem of three bodies', *Report of the British Association for the Advancement of Science*, (1899), 121–59.

8.10

Astrophysics and cosmology

C. W. KILMISTER

1 INTRODUCTION

The link between astrophysics and cosmology is close: for the study of the nature and constitution of the stars one needs to know their distribution in order to assess the empirical evidence, while in order to assess the evidence for the distribution one needs to know the mechanism of the radiation sources. The only empirical material is the light from the stars. In the mid-eighteenth century the speculations of many (including Immanuel Kant in 1755) that the stars might be arranged in 'island universes' were ignored by astronomers until William Herschel in the early nineteenth century. But despite his careful observations, his work was hampered by false assumptions; and the realization that many of the observed nebulae are outside the Milky Way galaxy dates only from Cleveland Abbe's paper of 1887. Even then, another fifty years were needed before Abbe's views were accepted. Meanwhile, the resolution of various globular clusters into individual stars was upstaged by the increasing knowledge of stellar constitution. In 1859 Gustav Kirchhoff and Robert Bunsen correctly interpreted the light and dark lines in the Sun's spectrum, noticed nearly half a century earlier by Joseph von Fraunhofer, as emission and absorption lines of chemical elements, and over the next decade the spectra of the Sun and some stars were analysed. However, spectral evidence alone did not suffice to determine the nature of the nebulae, and for some time they were all regarded as lesser entities than the Milky Way; but everything changed rapidly in the 1920s. For a detailed history, see North *1965*.

2 STELLAR STRUCTURE

The problem of stellar constitution was tackled then, principally by Arthur Eddington, whose *The Internal Constitution of the Stars* was published as his *1926*. Auguste Comte had cited the chemical nature of stars as something that could never be known, but this is far from the case: a star's interior is revealed to us through its gravitational field and also through the radiation that streams out from it. These two effects determine the nature of the outer layer, which can be observed. Most stars are in a steady state, the atmosphere adjusting itself to let out the radiation from the interior. One can therefore describe the surface by two parameters: g, the gravitational field, and T, the effective surface temperature. From 1914 this was reflected in two classifications of stars: the Harvard sequence of spectral types (differing in T) and the absolute magnitude scale (with g varying).

The problem in establishing stellar constitution is to relate these classifications to those in terms of M and R, the mass and radius. In particular, how is T determined by M and R? Walther Nernst and James Jeans held that nothing could be done without knowing the actual heat output; the heat, it was guessed, came from a nuclear reaction, about which nothing was known. Eddington saw that this impasse could be avoided by simply studying the outflow. He built on the earlier studies of distribution in polytropic gas spheres by Homer Lane in 1870 and the complete solution of that problem by R. Emden. These results had no effect on theories of stellar evolution while astronomers continued to believe that the hotter a star, the younger it was. However, Ejnar Hertzsprung (1905) and Henry Norris Russell (1913) suggested that stars start as cool red 'giants' of low density and high mass, then contract as they rise in temperature until they are very dense, when they would cease to be a perfect gas. They then cool, to become dwarfs. Eddington retained this picture, but modified it by assuming that stars remain a perfect gas even when very dense.

The equations for mechanical equilibrium and for thermal equilibrium, and the perfect gas law, together determine a star's pressure, density and temperature. Eddington saw that the general integration of these equations was not necessary to an understanding of the problem; he integrated them only in two cases, that in which energy generation took place entirely at one point, the centre, and that in which the energy generated per unit mass was uniform. This enabled him to find the mass–luminosity relation, because his two solutions gave very consistent results, and also to evaluate the central temperature, which came to the same value for all main-sequence stars. This temperature, therefore, must be that needed for the production of radiation. So, although nothing was known about the mechanism of generation, it was evidently extremely temperature-dependent. Eddington's book revolutionized the subject, but he was not totally successful; he greatly overestimated the effect of radiation pressure, and expressed his results in terms of it, which was not helpful. He could not discuss opacity, because this would have needed modern quantum theory; and he spent a great deal of time on variable stars, a topic too difficult for him to make progress.

Eddington's interest in variable stars was related to the importance of the role of the type known as Cepheid variables in determining distance on the large scale. These stars had been found to have periods that were a function of their luminosity; at least, this was known to be true for those close enough for their distances to be determined by a parallax method. So, under the assumption that this was always true, Cepheids could be used to determine greater distances. These distances turned out to be consistent with those found by another method based on the apparent brightness of novae, all of which were assumed to be similar. The search for Cepheids in the various nebulae began in 1922, and from then on it became apparent that many nebulae were indeed separate systems – other galaxies – some of them as big as the Milky Way, distributed in an apparently random manner and at very great distances indeed. This was the success of what may be called the astronomical approach: careful observation, with the hope of some systematic pattern emerging.

3 COSMOLOGICAL MODELS IN RELATIVITY

Any mathematical advance beyond this point depended critically on the existence of the observational evidence, but also on the theoretical approach made possible by Albert Einstein's general theory of relativity (§9.13). Once this had replaced Newton's gravitational theory, it was natural to hope that it might be useful in resolving the paradox of the infinite nature of the Newtonian universe: that in an infinite universe with a distribution of matter which is on average uniform and of non-zero density, no value can be given to the gravitational potential at a point. Einstein's field equations for empty space were expressed in terms of the Ricci tensor, R_{ij}, as $R_{ij} - \frac{1}{2}g_{ij}R = 0$; here, g_{ij} is the metric tensor and R (not to be confused with the radius R) the 'contraction' tensor of R_{ij}, defined by the sum of its diagonal elements (see §3.4 on tensors). For a smeared-out model of the universe, it was natural to replace these by the equations $R_{ij} - \frac{1}{2}g_{ij}R = -T_{ij}$, where T_{ij} is the momentum–energy tensor, which specifies the distribution and motion of the matter.

Einstein, who first tried to find cosmological solutions in 1917, argued in a more complex manner. He combined a 'cosmological principle', that the averaged-out matter should be the same at all points and in all directions (with which there would be general agreement) with a further requirement which he saw in Ernst Mach's writings: 'there can be no inertia relative to space, but only an inertia of masses relative to each other'. This would have the form, in the actual universe, that the local inertial frames would be determined in some way by the position and motion of matter, and, since the more distant matter predominates, by that. By using a Newtonian analogy he argued for modifying the field equations to the form

$$R_{ij} - \tfrac{1}{2}g_{ij}(R - 2\lambda) = -T_{ij} \tag{1}$$

for a small constant λ. Taking a T_{ij} appropriate for uniform dust at rest and setting $\lambda R^2 = 1$ gives the first cosmological solution (the 'Einstein universe') for a point in the gravitational field with spherical polar co-ordinates (r, θ, ϕ) and path variable s:

$$ds^2 = dt^2 - (1 - r^2/R^2)^{-1}dr^2 - r^2(d\theta^2 + \sin^2\theta \, d\phi^2) \tag{2}$$

(with $\lambda R^2 = 1$). This removed Einstein's worries over boundary conditions, for the universe was unbounded, though finite in extent. In the same year, 1917, Willem de Sitter found another solution of the new field equations which corresponded to zero density and pressure everywhere, and yet was not the special-relativity solution corresponding to empty space. By 1919 Einstein was convinced that the extra term λ in the field equations had been a mistake, but it is still often used.

The equations of motion for the de Sitter universe predict that if a test-particle (i.e. one which does not disturb the solution) is introduced, it begins to accelerate away from the origin, and this is independent of what point is chosen as the origin because the velocity that results is proportional to distance. The de Sitter solution suggested that astronomers should seek a systematic recession of distant matter. C. Wirtz found some evidence of this in 1924, using spectroscopic evidence for the velocities, but the finished form of the hypothesis was due to Edwin Hubble in 1929. He based his 'velocity–distance law' on the measured velocities of 46 objects, of which he knew the distances of 18. In subsequent observations, determinations of velocities tended to be used to define distances, but this is not a circular process, for many conditions of consistency (brightness and so on) have to be satisfied.

4 THE EXPANDING UNIVERSE

Hubble was guided in his search by the behaviour of the de Sitter solution, but was unaware of the theoretical work by the Russian Alexander Friedmann in 1922. Friedmann had considered time-dependent solutions, but ones for which at any instant the spatial part of the metric was a space of constant curvature; the curvature was then a function of time. His work was ignored, and much of it was done again in 1927 and 1931 by Georges Lemaître. This was the foundation of the 'expanding universe' models which predict the required velocity–distance relation. Not all the solutions of the equations expand indefinitely: it is possible for an expanding phase to be followed by a contracting one or to be preceded by one, but the most popular version is that of continual expansion. Such a model suggests that the universe originated from a singularity in the 'big bang', which has sometimes been thought a difficulty. Other explanations of the velocity–distance law have been given. Edward Milne suggested that it might simply be the result of looking at a cloud of objects moving at arbitrary speeds, for even if the speeds were constant, and whatever their direction, after sufficient time they will all be moving away from any fixed origin and the velocity–distance law will result because the fastest will have gone the farthest. In 1948, Hermann Bondi, Thomas Gold and Fred Hoyle suggested the steady-state theory in which, to avoid the single creation of everything at the first instant of time, a continual creation is postulated of just enough matter for the total observed universe to stay unchanged, replacing matter receding out of observational distance as its velocity approaches that of light. But at present some form of evolution from an initial singularity (events at which quantum mechanics is required for their careful description) is widely accepted.

Observational work in the 1930s was directed towards establishing one of the expanding models, but it was inconclusive. At the same time a number of theoretical discussions were in progress, one of which concerned the so-called 'cosmological principle'. The models described above were of a smeared-out distribution of matter and were based on the assumption that the universe was homogeneous, so that the average density of matter was the same in all directions. But although this makes good sense in the model, it is difficult to reconcile with the 'clumping' observed in the actual universe, first in the stars, then in the clumping of stars into galaxies, galaxies into clusters of galaxies, and larger structures. It is usual to say that the homogeneity is there provided that the average density is determined over a sufficiently large volume; but the criterion for 'sufficiently large' is not known independently, so that the cosmological principle is in danger of being a tautology.

5 DIMENSIONLESS CONSTANTS

A great deal of attention was paid to certain numerical coincidences. These are significant because they link numerical constants from different fields – quantum mechanics, gravitation and cosmology. The two cosmological constants are the mean density of matter in the universe, ρ, and the Hubble time T, the time for which the universe has been expanding, defined by Hubble's velocity–distance law $v = r/T$ for the velocity of recession v at distance r. From quantum mechanics come the electron charge e, and the masses of the proton M and the electron m. Finally, from gravitation come the constant of gravitation G in the Newtonian inverse-square law for the force G/r^2 between unit masses at distance r, and the speed of light, c. One striking coincidence is that the ratio of the electrical and gravitational forces in a hydrogen atom, e^2/GMm, is about 10^{39}, the same order of magnitude as the ratio between the size of the observable universe cT, as defined by the Hubble time, and the size of the electron, as defined by its Compton wavelength e^2/mc^2. The first ratio compares quantum-mechanical and gravitational numbers, the second cosmological and quantum-mechanical. This suggested some deep connection between the two theories, and was supported by the wholly independent relation $G\rho T^2 = 1$.

Paul Dirac argued in 1937 that coincidences between such large numbers could be explained if their largeness were due to a single factor. He proposed that all fundamental physical constants could be seen as falling into three groups – those of order unity, those of order 10^{39}, and those of order 10^{78} – and he believed that the factor 10^{39} was the 'age of the universe', so that only the constants of order 1 were truly constant and the others increased with time. A different line was taken by Eddington from 1929 onwards: he believed that the fundamental constants were determined by the structure of the algebras which Dirac had introduced into quantum mechanics to express the wave equation of the electron in Lorentz-invariant form. Neither of these approaches gained wide acceptance, and the problem of the numerical coincidences remains unsolved.

BIBLIOGRAPHY

Bondi, H. *1960, Cosmology*, 2nd edn, Cambridge: Cambridge University Press.
Eddington, A. S. *1926, The Internal Constitution of the Stars*, Cambridge: Cambridge University Press. [Repr. 1959, New York: Dover.]
Hawking, S. W. and Israel, W. *1979, General Relativity – An Einstein Centenary Survey*, Cambridge: Cambridge University Press.
North, J. D. *1965, The Measure of the Universe*, Oxford: Clarendon Press.

8.11

Ballistics and projectiles

A. R. HALL AND

I. GRATTAN-GUINNESS

Ballistics divides into two parts: in the terms long used, 'internal' ballistics dealt with the dynamics and effects of explosions inside the launching instrument, while 'external' ballistics analysed the path followed by the object fired. The practice corresponding to external ballistics is called 'gunnery'; and it links with projectile theory, where the method of launch can be arbitrary but where, as in ballistics, air resistance and density, and the rotation of the Earth, are likely to be considered.

The treatment here is chronological; authorship switches at Section 3, around the late eighteenth century. Of the items in the bibliography, Mandrika *1964* is especially recommended for its attention to mathematical details.

1 EARLY THEORIES OF GUNNERY

Treatises on gunnery (*Buchsenmeisterei*) date from the early fifteenth century; the earliest attempts to consider projectile motion theoretically were by Leonardo da Vinci, in about 1500 (§2.8). Leonardo drew pictures of lofty curving trajectories and knew that the descending curve is steeper than the ascending; like many successors he supposed high shots to return to the ground vertically. His 'ballistics' was intuitive and visual; the first mathematician to tackle its problems was Niccolò Tartaglia (*La nova scientia*, 1537). His discussion, still philosophical rather than geometrical, sprang from the medieval theory of impetus, dividing the trajectory into three segments: rectilinear point-blank (i.e. along the axis of the gun-barrel) and rectilinear, vertical line of descent, coupled by an arc of a circle to which the straight segments are tangential. From symmetry, he guessed that the maximum range would be achieved with an elevation (projection angle) of 45°. Later, he recognized that no part of the trajectory is geometrically rectilinear.

Like many successors, including the English military expert Thomas Digges (*circa* 1570–90), Tartaglia found no theory connecting velocity, elevation and range in shooting. Writers on gunnery offered various arbitrary, rule-of-thumb tables. There could be no satisfactory ballistics until the way of compounding motions following different lines was understood, and the acceleration of freely-falling bodies had been satisfactorily defined. In *Two New Sciences* (1638) Galileo accomplished both these steps and at the same time postulated the symmetry of the ascending and descending segments of the trajectory (§2.6).

We now know that experimenting with projectiles (accelerated by descent down a sloping plane) was essential to Galileo's discovery of the law of fall ($s = \frac{1}{2}at^2$), before 1609. By this same date he had also realized that in consequence of this law the trajectory of a projectile must be a parabola (postulating the horizontal component of motion to be uniform, and therefore the absence of any resistance to motion). Galileo imagined that air resistance had negligible effect upon a slow-moving body, but thought that a musket-ball acquired a 'supernatural' velocity. In *Two New Sciences* he printed a table relating range, r, to angle of elevation, d (from $0°$ to $90°$), actually derivable from a table of sines, since $r = R \sin 2\alpha$, where R is the extreme ($45°$) range.

The parabolic theory was first published by Bonaventura Cavalieri, to Galileo's chagrin (*Lo specchio ustorio*, 1632). The first to conceive of it was Thomas Harriot, in the first years of the century (Lohne *1979*). However, where Galileo used an upright parabola, Harriot preferred one *inclined*, giving a decided asymmetry to the trajectory, the descending segment being (correctly) both shorter and steeper (Figure 1). It is not clear what reasons Harriot had for this choice, which indeed accorded with the ideas of gunners (e.g. Bourne, 1587) and was supported by his own experiments. Lohne opines: 'If Harriot had prepared his results properly and published them, we might say that with him ballistics emerged as a science' (*1979*: 239). In fact, his study remained completely unknown at the time.

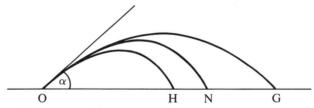

Figure 1 Three comparable trajectories, velocity constant, angle of elevation α fixed: OG, Galileo's upright parabola; OH, Harriot's tilted parabola; ON, Newton's curve for $r \propto v$ (redrawn from Lohne *1979*: 241)

Given the concept of the upright parabola, development of the geometry (e.g. for non-horizontal ranges) and tabulation was straightforward; it was accomplished by Evangelista Torricelli (1644) and, with practice in mind, by François Blondel (1683). Mathematicians possessing physical intuition were not long satisfied with the upright parabola, however. To them it was clear that both components of motion must be resisted by the air; that this resistance must be proportional to the square of the velocity was first guessed by Christiaan Huygens. In 1672 James Gregory, restricting resistance to the horizontal component of motion, proposed the tilted parabola (Figure 1) as a suitable model (knowing nothing of Harriot).

2 FROM NEWTON TO ROBINS

At about this time various investigators attempted to measure the effect of air resistance on the speed of falling bodies, from which to infer its influence on the trajectory of a projectile. Among them was Isaac Newton, who tackled ballistics as a problem of rational fluid mechanics in Book II of his *Principia* (1687). Postulating a horizontal range, and neglecting the sphericity of the Earth and other physical complications, Newton accepted the Galilean parabola as true for the vacuum, but showed that for air resistance directly proportional to the projectile's velocity v the trajectory is an asymmetric curve related to the logarithmic. (Huygens verified this by his own methods leading to the equivalent hyperbolic curve.) Newton also explored the physically more likely cases of $r \propto v^2$ and $r \propto v(v + 1)$, without, however, being able to derive general expressions for the trajectories. He employed the inverse method − taking the ratio of the resistance to the velocity as the unknown variable − to demonstrate that the hyperbola offered the best obvious approximation to the v^2 trajectory: 'That line is certainly of the hyperbolic kind, but about the vertex is more distant from the asymptotes and in the parts remote from the vertex draws nearer to them than these hyperbolas here described.' Newton was also the first mathematician to examine the effect of the body's shape upon its resistance to motion in a fluid, proposing (Book II, Proposition 34) a curve defining the solid of least resistance; the demonstration of this solution − not given in the *Principia* − is the unique example of the use of fluxions in the preparation of that work.

Gottfried Wilhelm Leibniz too undertook post-*Principia* investigations parallel to those of Newton and Huygens, starting from his own principles; it was the involvement of ballistics in the fluxions−calculus dispute (§3.2) that led Leibniz's lieutenant, John Bernoulli, to provide general equations for defining the trajectory for $r \propto v^n$. He proved that if $n = 1$ the curve is logarithmic; if $n = 2$, it is related to the hyperbola, as Newton had stated.

Bernoulli's equations are not always capable of numerical solution. In any case the parabola, with the theory further refined by Halley and others, was still the only curve offered to artillerists for practical use.

Benjamin Robins (1742) was the founder of modern ballistics research, in that he refuted the dogma 'that no considerable variation could arise from the resistance of the air, in the flight of shells or cannon-shot'; and he devised the first instrument, the ballistic pendulum, relying on the collision equality $MV^2 = mv^2$ of bodies with masses and velocities m, v and M, V to measure the speed of projectiles. Robins was convinced that, at velocities above 1000 feet per second (300 m/s), air resistance increased far more than as the square of the velocity. While he produced a number of calculations of *ad hoc* comparative trajectories, he could find no satisfactory general theory for the flight of projectiles in air. In 1745 Leonhard Euler translated Robins's book into German, adding far more new material of his own.

3 FRENCH CONTRASTS IN EXTERNAL BALLISTICS: BORDA AND LAPLACE

The French applied their mathematical skills to this area, with interesting differences of approach. Among figures of the late eighteenth century, Jean Charles Borda wrote a widely admired paper. Ignoring the rotation of the Earth but assuming air resistance to be a quadratic function of velocity (the normal view by then), he found ordinary differential equations for horizontal (x) and vertical (y) motion, and by ingenious manipulations of x, y and arc length s he obtained parametric solutions with dy/dx as the parameter. Much of the later work centred on the military school at Metz: F. J. Français tried exact solutions in the early 1800s that were too lengthy to be useful, and his successors of the mid-century, I. Didion and G. Piobert, preferred more modest desimplifications, such as modifying the law of air resistance to include a cubic term in velocity (Charbonnier *1927*).

Meanwhile, Pierre Simon Laplace had carried out a similar but different analysis of projectiles in a paper of 1803 which was extended in the fourth volume of his *Mécanique céleste* (1805). By a brilliant deployment of fixed and moving frames of reference, he produced three ordinary differential equations for latitude, longitude and altitude of the projectile, and proceeded to calculations concerning, for example, the well-known westward 'drift'. He included the so-called Coriolis force in appropriate directional components: this name for the force commemorates the *savant* who realized (in *1835*) its importance for dynamics in general (§8.1).

Laplace's disciple Siméon-Denis Poisson made several different contributions. In his textbook of 1811 on mechanics he presented Borda's treatment as more suitable than Laplace's for students, but in research of the late 1830s he extended Laplace's analysis to allow also for the rotation of the projectile while in motion; it was these two pieces of work which helped Léon Foucault in mid-century to conceive of his pendulum to demonstrate the rotation of the Earth (Acloque *1981*). Poisson also applied his growing interest in probability theory (§10.4) to analyse the distribution of shots in target-shooting; and on the 'internal' side of ballistics he extended some incomplete and unpublished research by Joseph Louis Lagrange in the mid-1790s on the differential equations to represent the firing of cannon. In addition, he republished an analysis of the mid-1820s on the recoil of cannon, treated as an exercise in several-variable dynamics.

4 PROGRESS UP TO THE EARLY TWENTIETH CENTURY

Military concerns kept ballistics in the limelight, in both its external and internal aspects. New propellants and rifled barrels in steel made it worth while to lay guns for ranges of many kilometres, which encouraged the development of numerical techniques to produce range tables for gunners. Interestingly, while many workers were involved on both sides (Cranz *1903*), few were major mathematicians or scientists.

On the external side, in addition to allowing for the rotation of the projectile, the effect of winds was entertained. Approximative methods became the norm: they included choosing forms of the differential equations which represented the apparent effects in a 'reasonable' way and yet allowed closed-form solutions to be obtained. The Italian F. Sciacci was influential; working out partly from Didion, but allowing for a more general function of air resistance of the form $av^m + bv^n$ (with v as velocity, and a, b, m and n as determinable constants), he defined various indefinite integrals (now named after him) having this function in their integrands, from which relatively compact expressions for the various parameters of space and time could be determined.

On the internal side, the developing subject of thermodynamics (§9.5), along with some relevant chemistry, was applied to analyse the behaviour of gunpowder by means of appropriate versions of the energy equation. The analyses involved not only the exit velocity but also the maximal temperature that might be generated during and after ignition.

By the beginning of the twentieth century, ballistics and projectile theory were merging as the distances of projections of missiles was increasing; in addition, the problems posed came to include the correct point of launch of bombs from aircraft. During and after the First World War, these topics themselves shaded into aerodynamics (which comes next). In particular, in

Britain G. I. Taylor worked with distinction in both areas almost at once, as in a paper of 1929 (in Taylor *1963*) on 'The flow of air at high speeds past curved surfaces', while in Germany Carl Cranz enhanced the experimental side (Nelson *1951*).

Since then ballistics has taken in new problems connected with the development of long-range rockets and the use of radar and servo-mechanical guidance systems, and has expanded to compute the paths of space probes across the Solar System. The original interest in stone, bullet and shot has been extended to take account of missile, satellite and spaceship.

BIBLIOGRAPHY

Acloque, P. *1981*, *Oscillations et stabilité selon Foucault* [. . .], Paris: CNRS.

Charbonnier, P. J. *1927*, 'Essais sur l'histoire de la balistique', *Mémorial de l'Artillerie Française*, **6**, 955–1251. [Also published as a book, 1928, Paris: Société d'Editions Géographiques [. . .]. Valuable, although contains little mathematics. Main text goes up to 1800.]

Cranz, C. *1903*, 'Ballistik', in *Encyklopädie der mathematischen Wissenschaften*, Vol. 4, Part C, 185–279 (article IV 18).

Hall, A. R. *1952*, *Ballistics in the 17th Century*, Cambridge: Cambridge University Press.

Lohne, J. A. *1979*, 'Harriot studies II. Ballistic parabolas', *Archive for History of Exact Sciences*, **20**, 230–64.

Mandrika, A. P. *1964*, *Istoriya ballistika (go seredini XIX b.)*, Moscow and Leningrad: Nauka. [The best history for mathematical purposes. Goes up to 1850.]

Nelson, W. C. (ed.) *1951*, *Selected Topics in Ballistics*, London: Pergamon. [In honour of Cranz; some historical articles.]

Seeger, R. J. *1951*, 'On aerophysics research', *American Journal of Physics*, **19**, 459–69. [On aeroballistics and the use of wind-tunnels.]

Tallqvist, H. J. *1931*, 'Oversikt av ballistikens historia', *Svenska Tekniska Vetenskapsacademien i Finland historia, Acta*, **9** [the full volume].

Taylor, G. I. *1963*, *Scientific Papers*, Vol. 3 (ed. K. Batchelor), Cambridge: Cambridge University Press.

8.12

Mathematics and flight

J. A. BAGLEY AND J. H. B. SMITH

1 INTRODUCTION

Practical flying-machines are a twentieth-century invention, and in the second half of this century a considerable volume of mathematics has been generated in connection with the business of designing, building and flying aeroplanes. The roots of this work lie in the theories of classical hydrodynamics developed in the eighteenth and nineteenth centuries, but these theories had little influence on the pioneers who brought the practical aeroplanes into existence. In fact, even a casual survey of the six-volume encyclopedia entitled *Aerodynamic Theory* (Durand *1934*) shows that the science of aerodynamics still relied very heavily at this time on experiment. The primary purpose of these volumes was to encourage the aircraft industry to make more use of mathematical methods and not rely on purely empirical results, but it is clear that for most purposes the existing mathematical theories could provide little more than a framework for the better understanding of experimental measurements obtained from flight and wind-tunnel tests.

A useful survey of the history of the subject is given by R. Giacomelli in Durand (*1934*: Section D), and the historical development of classical hydrodynamics has been comprehensively reviewed in Szabó (*1987*: Section 3). A much briefer (but more entertaining) survey is by the pioneer aerodynamicist Theodore von Kármán *1954*; and another useful book, Tokaty *1971*, pays particular attention to Russian work.

There are two principal areas in which mathematics has been applied to flight: the design of aeroplanes to have good performance, and the development of satisfactory flying qualities. The first requires the use of aerodynamic theories to generate shapes with (in general) low drag and adequate lift; in practice, the constraints are such that a direct solution of the design problem is too difficult, and the designer seeks rather to predict the aerodynamic forces on a defined shape, which is then modified if necessary.

In this article only the wing, which generates most of the lift and an important part of the drag, is considered.

The second problem is concerned with the motion of the complete aeroplane, and the assessment of its stability and controllability. Again, the designer can only expect to proceed by successive modification of an initially chosen shape.

2 THE LIFT AND DRAG OF WINGS

The first mathematical investigation of the forces on a body immersed in a fluid stems from Isaac Newton's *Principia* (*1726*: Book II, Proposition 34); he imagined the fluid as a stream of discrete particles impinging on the body and exchanging momentum with it. He deduced that the force would be proportional to the area of the body, the density of the fluid and the square of the velocity. Applying his principles to a flat plate of area S inclined at an angle α to a stream of speed U of a fluid of density ρ leads to the so-called 'Newton's law'

$$F = \rho S U^2 \sin^2\alpha, \tag{1}$$

where the force F acts normal to the plate.

It has to be stressed that this 'law' was never stated as such by Newton himself, but was deduced later. The forces predicted are so small that the flight of birds is inexplicable, and it has been argued that this result was sufficient in the nineteenth century to discourage serious attempts to make aeroplanes capable of carrying people. In fact, it seems clear that the aerodynamic forces on plates were well known to be much larger than this theory predicts, and that no would-be fliers were seriously deterred by the anomaly.

The Newtonian model was in any case obviously flawed, inasmuch as it ignored the flow over the upper side of the plate; but a fundamental difficulty emerged in subsequent attempts to predict the force on a closed body immersed in a continuous fluid. This was clearly stated by Jean d'Alembert *1768*, who published three different attempts to solve this problem. He summed up the result in his famous paradox:

> I do not see then, I admit, how one can explain the resistance of fluids by the theory in a satisfactory manner. It seems to me on the contrary that this theory, dealt with and studied with profound attention, gives, at least in most cases, resistance absolutely zero: a singular paradox which I leave to geometers to explain.

'The theory' is here the theory of inviscid, incompressible, irrotational steady flow, and 'resistance' is the total force on the body (see §8.5, on hydrodynamics).

Fortunately, practical men take little account of paradoxes. The first to set out a rational basis for a practical aeroplane was George Cayley, in his paper 'On aerial navigation' (*1809*): 'The whole problem is confined within these limits, viz. To make a surface support a given weight by the application of power to the resistance of the air.' In modern terminology, this is equivalent to the resolution of forces along and perpendicular to the direction of travel. For level, unaccelerated flight in a straight line:

$$\text{thrust} = \text{drag} \quad \text{and} \quad \text{weight} = \text{lift}. \tag{2}$$

Cayley observed that the 'Newton's law' (1) was incorrect, and used his own and other experimental measurements of the aerodynamic forces on flat plates (using whirling-arm apparatus) to deduce the dimensions of a wing which would support a person. His experiments culminated, towards the end of his life, in the construction of two gliders – the 'boy-carrier' of 1849 and the 'man-carrier' of 1853 (Gibbs-Smith *1962*). Although the evidence for successful flights is little more than anecdotal, there is sufficient solid information in Cayley's published and private papers for a reproduction of the larger machine to have been built and flown for a television film in 1972.

After Cayley, the next significant aviator was Otto Lilienthal, whose several hundred gliding flights in Germany from 1891 to 1896 were widely publicized and directly inspired other pioneers. The most notable of these were the Wright brothers, whose successful powered flights on 17 December 1903 are recognized as the world's first. Before embarking on his flying experiments, Lilienthal published as *1889* the results of his tests on flat-plate and circular-arc model wings, showing that the latter developed much more lift at a given angle for a similar amount of drag.

The attempt to explain Lilienthal's results finally led to a partial resolution of d'Alembert's paradox by the German mathematician Wilhelm Kutta *1902* and the Russian Nikolai Zhukovsky (more familiar in Western Europe as Joukowsky) *1906*. Both were well-acquainted with Lilienthal: indeed, Zhukovsky bought one of Lilienthal's later gliders, though it does not seem that he tried to fly it himself. They independently arrived at the concept of a circulation around the wing section, whose value can be fixed by assuming that the flow leaves the trailing edge smoothly. The lift on the flat plate is found to be proportional to $\sin \alpha$, as in experiment.

Kutta did little further work in aerodynamics, but Zhukovsky became the leader of a major research institute in Moscow where experimental work was supported by theoretical studies. His major contribution was to show that, in two-dimensional flow, a lift force $L = \rho U \Gamma$ is produced on a body, where Γ is the circulation around it. He went on to describe a family of aerofoil shapes generated by simple conformal mapping of a circle, and to show how their lift, pitching moment and pressure distribution could be calculated for potential flow. Zhukovsky's aerofoils were not adopted by aircraft builders, but they provided useful test cases for supporting the validity of wind-tunnel testing, and for later calculation methods which could also be used for more practical aerofoil shapes.

On a wing of finite span, the circulation varies from section to section across the span, giving rise to a vortex wake. This was first recognized by Frederick Lanchester in Britain and Ludwig Prandtl in Germany, but it was Prandtl whose work led to a useful mathematical model. He visualized the wing as a lifting line, represented by a bound vortex varying in strength across the span, from which trailing vortices extend downstream. The wing and its wake are thus represented by a set of coplanar horseshoe vortices whose individual strengths correspond to the spanwise increments in the lift distribution. The vertical velocity induced by each vortex element is calculated by the Biot–Savart law (§9.10) and summed to match the boundary condition at each point of the span. This leads to an integral equation which takes the form

$$\frac{\Gamma(y)}{\pi U c(y)} = \alpha(y) - \frac{1}{4\pi U} \int_{-s}^{s} \frac{d\Gamma}{dy'} \frac{1}{y - y'} \, dy' \tag{3}$$

connecting the vortex strength $\Gamma(y)$, wing chord $c(y)$ and incidence $\alpha(y)$, on a wing of span $2s$.

This theory was developed at Göttingen under Prandtl's direction during the First World War (Prandtl *1920*); a brief summary of the numerous methods of solving the equation in practice has been given by Küchemann (*1978*: 127*ff*), who also discusses how the basic methods have been extended to cover swept wings and other topics. In particular, Küchemann shows how the effects of compressibility in subsonic flow may be represented by using an analogy first developed by Glauert *1927*, relating the flow at a Mach number M_0 on a body to the incompressible flow on a body whose lateral dimensions are reduced by a factor $(1 - M_0^2)^{1/2}$.

When the stream speed is a large fraction of the speed of sound, at say $M_0 > 0.8$, very complex flows occur around practical aeroplane shapes; these have become susceptible to mathematical treatment only since about 1950. However, some mathematical solutions were obtained for supersonic flow (with $M_0 > 1$) over simple two-dimensional shapes, first by Meyer *1908*. The extensions of this work are dealt with by Taylor and Maccoll in Durand (*1934*: Vol. 3, division H.4).

Prandtl was also responsible for the other great step needed to reconcile theoretical fluid mechanics with reality: the concept of the boundary layer. This he saw as a thin layer of fluid adjacent to the surface of a wing or body, within which viscous forces are important and the total vorticity is concentrated. Downstream of the body, the boundary layers merge into a wake, which is the physical equivalent of the dividing surface postulated by Kutta and Zhukovsky.

Prandtl *1905* obtained a solution for a flat plate exposed to a uniform parallel stream, assuming that the flow remains laminar (i.e. steady, even on the smallest scale), with the velocity increasing from zero on the surface of the plate itself. He found that the boundary layer thickness increased with $\sqrt{x}$ (where x is distance along the plate), and the local skin friction decreased as $1/\sqrt{x}$.

In most cases of aeronautical importance, the flow in the boundary layer becomes turbulent in a short distance. This transition to a flow which is unsteady on the smallest scale results in an increase in local skin friction and important changes in the development of the boundary layer. In the period under review, despite lively controversy, understanding of these matters remained at the phenomenological level. Early attempts to deal with these problems are reviewed by Prandtl (in Durand *1934*: Vol. 3, Division G), and less formally by von Kármán *1954*.

3 DYNAMIC STABILITY

Remarkably early on in the practical development of aviation, Bryan (*1904, 1911*) tackled the mathematical analysis of the stability of aeroplane flight, using the principles of E. J. Routh's treatment of rigid-body dynamics. He considered an aeroplane slightly disturbed from uniform level flight. The six equations for translation in three perpendicular directions and rotation about the three axes separate into two groups when the shape of the aeroplane is taken to be symmetrical about a vertical plane through the centre of gravity. One group is concerned with motion (commonly called longitudinal) in the plane of symmetry, and the other with asymmetrical (lateral) motion. Assuming small disturbances from equilibrium, varying with time like $e^{\lambda t}$, each set of equations reduces to a biquadratic in λ. The coefficients of these equations involve a set of quantities representing the aerodynamic forces due to linear and rotary motions in the three directions, which Bryan assumed could be represented by linearized approximations.

The conditions for stability are given by Routh; when these are satisfied the four roots of the biquadratic are all complex numbers with negative real parts. The motions all have the form of damped oscillations, so that the aeroplane eventually returns to equilibrium (steady flight).

To make use of Bryan's analysis, values of the various aerodynamic-force coefficients (usually referred to as 'derivatives') are required, most of which could only be determined by experiment. The early work to devise suitable experimental methods, mainly by testing models in wind-tunnels, was done by Bairstow *et al. 1913* at the UK's National Physical Laboratory, Teddington. At Farnborough, E. T. Busk used Bairstow's analysis to modify an aeroplane to fly stably without the pilot's intervention (O'Gorman *et al. 1914*). Bairstow extended the theory to deal with more complex cases (*1920*: 447–520), and much experimental and theoretical work has been done since; however, Bryan's original work still provides the basis for most practical work in the field today.

BIBLIOGRAPHY

Bairstow, L. *1920, Applied Aerodynamics*, London: Longman, Green.

Bairstow, L., Jones, B. M. and Thompson, B. A. *1913*, 'Investigation into the stability of an aeroplane', *(British) Advisory Committee for Aeronautics, Reports and Memoranda*, No. 77.

Bryan, G. H. *1911, Stability in Aviation*, London: Macmillan.

Bryan, G. H. and Williams, W. E. *1904*, 'The longitudinal stability of aerial gliders', *Proceedings of the Royal Society of London*, Series A, **73**, 100–16.

Cayley, G. *1809*, 'On aerial navigation (part I)', *Nicholson's Journal of Natural Philosophy*, **24**, 164–74.

d'Alembert, J. le R. *1768, Opuscules mathématiques*, Vol. 5, Paris: Briasson, 132–8.

Durand, W. F. (ed.) *1934, Aerodynamic Theory*, 6 vols, Berlin: Springer. [Repr. 1943, Pasadena, CA: California Institute of Technology.]

Finsterwalder, S. *1902*, 'Aërodynamik', in *Encyklopädie der mathematischen Wissenschaften*, Vol. 4, Part C, 149–84 (article IV 17).

Gibbs-Smith, C. H. *1962, Sir George Cayley's Aeronautics, 1796–1855*, London: Her Majesty's Stationery Office.

Glauert, H. *1927*, 'The effect of compressibility on the lift of an aerofoil', *(British) Aeronautical Research Council, Reports and Memoranda*, No. 1135. [Also in *Proceedings of the Royal Society of London*, Series A, **118**, 113–19.]

Kármán, T. von *1954, Aerodynamics: Selected Topics in the Light of their Historical Development*, Ithica, NY: Cornell University Press.

Küchemann, D. *1978, The Aerodynamic Design of Aircraft*, London: Pergamon.

Kutta, M. W. *1902*, 'Auftriebskrafte in strömenden Flüssigkeiten', *Illustrierte Aeronautische Mitteilungen*, **6**, 133–5.

Lilienthal, O. *1889, Der Vogelflug als Grundlage der Fliegekunst*, Berlin: Gärtner.

Meyer, T. *1908*, 'Über zweidimensionale Bewegungsvorgänge in einem Gas, das mit Überschallgeschwindigkeit strömt', *Forschungsarbeiten auf dem Gebiete des Ingenieurwesens, VDI*, Heft 62.

Newton, I. *1726, Philosophiae naturalis principia mathematica*, 3rd edn, London: Innys. [Edition used: (transl. A. Motte, ed. F. Cajori), 1934, Berkeley, CA: University of California Press.]

O'Gorman, M., Mayo, R. H. and Busk, E. T. *1914*, 'Longitudinal stability', *(British) Advisory Committee for Aeronautics, Reports and Memoranda*, No. 134.

Prandtl, L. *1905*, 'Grenzschichten in Flüssigkeiten mit sehr kleiner Reibung', in A. Krazer (ed.), *Verhandlungen des dritten Internationalen Mathematiker-Kongresses*, Leipzig: Teubner, 484–91.

—— *1920*, 'Tragflächenauftrieb und -widerstand in der Theorie', *Jahrbuch der Wissenschaftlichen Gesellschaft für Luftfahrt*, 37–65.

Szabó, I. *1987*, *Geschichte der mechanischen Prinzipien*, 3rd edn, Basel: Birkhäuser.

Tokaty, G. A. *1971*, *A History and Philosophy of Fluidmechanics* [sic], Henley-on-Thames: Foulis.

Zhukovsky, N. E. *1906*, 'O prisoedinennikh vikhryakh', in *Obshchestvo liubitelei estestvoznaniya*, Vol. 13, Part 2, Moscow: Trudi otdeleniya fizicheskikh nauk, 12–25.

8.13

The pendulum:
Theory, and its use in clocks

PAUL FOULKES

1 INTRODUCTION

A pendulum is a mass suspended from a fixed point through a join (thread, string, rod) and set into a swinging motion. The swing is used as a measure of time, and therefore as a way of regulating it.

It is only in relatively modern times that the pendulum was recognized as a means of recording time, some unsubstantiated mention of its use by Arab astronomers notwithstanding. A sketch by Leonardo da Vinci from about 1500 shows a pendulum, as a kind of flywheel. The first to state that the pendulum could be used to record time was Galileo Galilei. However, he was wrong in holding that the time of swing of a pendulum swinging in a circular arc was independent of the amplitude: this is true only of small amplitudes. That he thought otherwise shows how inaccurate his clocks must have been. Indeed, if the pendulum is started at $90°$ from the vertical, the time of swing is about 18% greater than for very small swings (see equation (13) below). There is a curve of swing under gravity that does yield swings of equal duration whatever the initial deviation from rest: this feature of isochrony belongs to the cycloid, and was discovered by Christiaan Huygens. He proved this late in 1659, and made the necessary adjustments to the pendulum clock he had invented in 1656–7, details of which were published in his *Horologium* (1658). The clock of 1657 was the first time-keeper to use the pendulum as a regulator. The results on the cycloid were not published until 1673, when his *Horologium oscillatorium* appeared in Paris (Yoder *1988*).

Huygens was a mathematician rather than an experimenter, and tended to lack patience with his clock-makers, who had to cope with practical problems not directly amenable to theoretical treatment, at least at that stage. Indeed, even as regards theory, Huygens could not give a general account of pendular motion under gravity; the mathematics and the physics needed were lacking. It would take until the mid-eighteenth century for calculus to grow far enough to settle that problem. Later physicists did produce elaborate theoretical investigations on various aspects that complicate the elementary theory of the pendulum (see Section 5). Meanwhile, the physicists of the seventeenth century used geometrical methods that had to be adapted to each particular case. Huygens performed truly heroic feats of geometrical insight and construction. Even Newton's *Principia* (1687) still contains complicated geometrical arguments that are not easy to follow for moderns used to calculus. It would be difficult to present such demonstrations here, so the theory is presented in modern garb.

2 MOTION OF A BODY UNDER GRAVITY ALONG A CURVE

Consider a body (taken to be concentrated at its centre of mass) moving under gravity along a curve from $y = y_0$ down to $y = 0$. The equation of motion is easily read from Figure 1 as:

$$(d^2s/dt^2) = -g(dy/ds), \tag{1}$$

which integrates into

$$\tfrac{1}{2}(ds/dt)^2 = g(y_0 - y). \tag{2}$$

Integrating again,

$$t - t_0 = (1/2g)^{1/2} \int_{y_0}^{0} (ds/dy)(y_0 - y)^{-1/2} \, dy, \tag{3}$$

where t_0 is the initial time and y_0 the initial height. Let $t_0 = 0$, and take

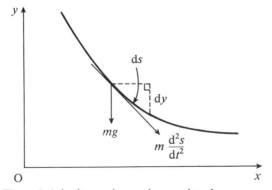

Figure 1 A body moving under gravity along a curve

$y = y_0 \sin^2 \psi$, so that when $y = y_0$, $\psi = \frac{1}{2}\pi$; $y = 0$, $\psi = 0$. Now,

$$(ds/dy)(y_0 - y)^{-1/2} = y^{1/2}(ds/dy)(y_0 y - y^2)^{-1/2}. \qquad (4)$$

On substituting for y, the integrand becomes

$$y^{1/2}(ds/dy)/y_0 \sin \psi \cos \psi, \qquad dy = 2y_0 \sin \psi \cos \psi \, d\psi. \qquad (5)$$

Putting $-y^{1/2}(ds/dy) = u$, the time of descent T from release to the bottom is

$$T = (2/g)^{1/2} \int_0^{\pi/2} u \, d\psi. \qquad (6)$$

This is the general formula for the time of descent under gravity. We proceed to examine two special cases.

3 THE CIRCULAR ARC

Consider motion along a circular path, as in Figure 2. The parametric equation of that circle is $x = a + a\cos\theta$, $y = a + a\sin\theta$. Hence

$$(ds/dy)^2 = 1 + (dx/dy)^2 = 1 + \tan^2\theta = 1/\cos^2\theta = 1/[1 - (y-a)^2/a^2]$$
$$= a^2/(2ay - y^2); \qquad (7)$$

so

$$u^2 = y(ds/dy)^2 = a^2/(2a - y). \qquad (8)$$

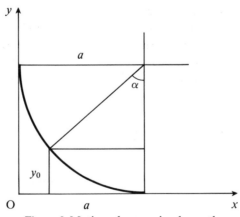

Figure 2 Motion along a circular path

From Figure 2,

$$y_0 = a(1 - \cos \alpha) = 2a \sin^2(\tfrac{1}{2}\alpha), \tag{9}$$

where α is the amplitude angle. Hence

$$u^2 = a/2[1 - \sin^2(\tfrac{1}{2}\alpha)\sin^2\psi], \tag{10}$$

$$T = (a/g)^{1/2} \int_0^{\pi/2} \{1 - \sin^2(\tfrac{1}{2}\alpha)\sin^2\psi\}^{-1/2}\, d\psi. \tag{11}$$

The integral is $K(\sin(\tfrac{1}{2}\alpha))$, with $K(k)$ the complete elliptic integral of the first kind (§4.5):

$$T = (a/g)^{1/2}K(\sin(\tfrac{1}{2}\alpha)). \tag{12}$$

For $k < 1$, $K(k)$ expands into an infinite series integrable term by term, and

$$K(k) = \frac{\pi}{2}(1 + \tfrac{1}{4}k^2 + \tfrac{9}{64}k^4 + \tfrac{25}{256}k^6 + \cdots). \tag{13}$$

If $\alpha = \tfrac{1}{2}\pi$, $k^2 = \sin^2(\pi/4) = \tfrac{1}{2}$, and equation (13) now becomes $K(1/\sqrt{2}) \simeq 1.18\pi/2$. Thus a swing from the horizontal position takes 18% longer than for very small angles: $k^2 \simeq 0$ and $K(0) = \tfrac{1}{2}\pi$.

Pendulums in modern regulator clocks use amplitudes of $2°$ (i.e. $\alpha = \pi/90$), for which, from equation (12), the period deviates from that with $K(0)$ by less than 1 part in 10 000. Since regulators have constant amplitude, this is no error. The period for one oscillation is

$$\tau = 4T = 4(a/g)^{1/2}K[\sin(\tfrac{1}{2}\alpha)], \quad \alpha = 0; \quad \tau_0 = 2\pi(a/g)^{1/2}. \tag{14}$$

The circular swing is often treated by approximating before integration. The equation of motion is $a(d^2\phi/dt^2) = -g \sin\phi$, where ϕ is the angle of deviation. Since $\sin\phi \simeq \phi$ for small ϕ, this becomes $(d^2\phi/dt^2) = -(g/a)\phi$, and describes a harmonic oscillator of period $\tau = 2\pi(a/g)^{1/2}$. Here we cannot tell how far the actual circular pendulum departs from a harmonic oscillator.

4 ISOCHRONIC DESCENT: THE CYCLOIDAL ARC

If the descent is to be isochronic, T must be independent of y_0. Therefore $dT/dy_0 = 0$. This means that

$$\int_0^{\pi/2} \partial u/\partial y_0\, d\psi = 0. \tag{15}$$

Now, $\partial u/\partial y_0 = du/dy \cdot \partial y/\partial y_0 = du/dy \cdot \sin^2\psi$. Hence $du/dy = 0$; therefore u is constant; put $u^2 = 2a$. The cycloid in Figure 3 has the parametric form

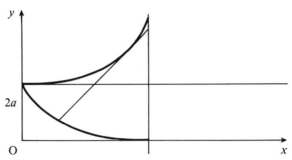

Figure 3 Isochronic descent along a cycloidal arc

$x = a\theta - a\sin\theta$, $y = a + a\cos\theta$, with a the radius of the generating circle. Then

$$(ds/dy)^2 = 1 + (dx/dy)^2 = 1 + (a - a\cos\theta)^2/a^2\sin^2\theta$$
$$= 2(1 - \cos\theta)/(1 - \cos^2\theta) = 2/(1 + \cos\theta) = 2a/y; \qquad (16)$$

hence

$$u^2 = y(ds/dy)^2 = 2a. \qquad (17)$$

Therefore isochronic motion occurs along a cycloidal arc. Taking the length $l = 4a$, the size of the arc from cusp to trough, we can view the thread as unwinding from another, similar and equal cycloid which is the evolute of the curve of swing: its equation is $\xi = a\theta + a\sin\theta$, $\eta = 3a - a\cos\theta$. Hence

$$T = \tfrac{1}{2}\pi(l/g)^{1/2}, \qquad \tau = 2\pi(l/g)^{1/2}. \qquad (18)$$

5 CORRECTIONS

Huygens' suspension, with cycloidal 'cheeks' for isochrony, was soon abandoned, since other factors cause rather larger discrepancies than isochrony solves. In his case it was the thread changing its length with humidity. However, the foremost complication is that g varies with geographic latitude, a fact first discovered by Jean Richer in 1672. Indeed, it was through observing shorter periods at higher latitudes that the shape of the Earth was found to be a spheroid flattened at the poles. Thus the pendulum has to be lengthened somewhat to maintain the same period.

Interest in the pendulum for measuring g and for geodesy became widespread, with contributions by physicists (Pierre de Maupertuis, 1738), mathematicians (Siméon-Denis Poisson and Pierre Simon Laplace, 1816), engineers (Jean Charles Borda, 1792) and astronomers (Jacques Dominique Cassini, 1810), to name but a few. These pendulums were compound and increasingly complicated (see Wolf *1889* for discussion and bibliography). Such measurements continued to be made until the 1930s, after which the pendulum was replaced by the balance-spring, which gives quicker and more accurate results. In seismometry, pendulums are still used today.

Because its total mass is not concentrated at the mass centre of the bob, any real pendulum is compound. Its effective length is from suspension to the actual mass centre, taking into account the mass of the rod. That point is also called the 'centre of oscillation', and Huygens was well aware of the need for making the appropriate adjustments in the formula.

The original Huygens clock had a verge escapement. The later deadbeat anchor escapements interfere less with the free motion of the pendulum. Various other types have been invented over the years. Early eighteenth-century makers discovered that temperature changes produce distortions, affecting the length. To counter this, one can use wood, which does not change in length with temperature, or a metal alloy such as Invar with negligible temperature coefficient of expansion. Meanwhile, elaborate mathematical work was done in the nineteenth century on different physical effects that influence the performance of the pendulum as regulator. For example, Friedrich Wilhelm Bessel considered elasticity and friction in the suspension, while George Stokes analysed the damping caused by the air through which the pendulum moves.

For domestic uses, pendulum clocks remain quite accurate enough. Even as regulator clocks, with suitable adjustments for the effects mentioned, they are still used. However, the accuracy of the best pendulum clocks and even better is obtained more easily with quartz oscillators. Since the mid-1950s, atomic clocks consisting of quartz oscillators regulated by the vibrations of the caesium atom have displaced all other standards of time measurement. This is rather more accurate than any previous standard.

ACKNOWLEDGEMENT

Thanks are due to John Leopold, British Museum (London), for valuable horological advice.

BIBLIOGRAPHY

Ariotti, P. E. *1972*, 'Aspects of the conception and development of the pendulum in the 17th century', *Archive for History of Exact Sciences*, **8**, 329–410.

Huygens, C. *1932–4*, *Oeuvres*, Vols 17–18, The Hague: Royal Dutch Academy. [Contain his main papers on the pendulum.]

Wolf, C. J. E. (ed.) *1889, Mémoires sur le pendule* [. . .], 2 parts, Paris: Gauthier-Villars. [Full bibliography on the pendulum and its applications for the period 1629–1885, and commentary.]

Yoder, J. G. *1988, Unravelling Time* [. . .], Cambridge: Cambridge University Press. [On Huygens' mathematics.]

8.14

Geodesy

The late SEYMOUR L. CHAPIN

In its etymological sense of 'dividing' the Earth, geodesy, among its Greek originators, was that area of applied mathematics dealing with the establishment of property and political boundaries and other surveying matters. It is also true, however, that the Greeks further concerned themselves with the 'higher' subject of determining the size and shape of the Earth as a whole. And, after the latter question had been scientifically and philosophically resolved in favour of a sphere, the former could be – and was – determined by estimating distances separating two points on the same meridian whose angular separation, seen from the centre of the sphere, was established by celestial observations. The details of these and subsequent efforts, continuing from the third century BC through to the sixteenth century AD, need not detain us, except to point out that at the end of this period there were proposals for a new method of measuring such arcs with greater certitude than by pacing, counting wheel revolutions or employing mariners' guesses.

1 TRIANGULATION AND MATHEMATICAL GEODESY

It was the Netherlander Reiner Gemma Frisius who, in the sixteenth century, was the first to suggest the method of triangulation by which such arcs would be measured by means of terrestrial observation of angles between established sites, followed by appropriate geometrical and trigonometrical procedures. It remained for others in the following centuries, however, to develop and apply that technique; this would again raise the question of the sphericity of the Earth and provide solutions to the new queries of its deviation from that form.

Frisius's countryman Wilibrord Snel, a professor at the University of Leiden, greatly matured this proposed approach. Starting the necessary baseline from his house in Leiden, he used the spires of three churches in the town as reference sites, and then established a network of triangles that enabled him to compute the distance between two Dutch towns lying on the same meridian and separated by an arc whose amplitude was determined by celestial observations. Snel's work not only established precise geographical details about his newly independent country, but also enabled the perfection of sea charts by making available the essential knowledge of the length of a degree. Unfortunately, his and other efforts in the first half of the seventeenth century were not as accurate as might have been desired.

That was to change in the second half of the century, with the famous measure carried out by Jean Picard. He carried through a revolution in observational astronomy made possible by the filar micrometer, the astronomical pendulum clock and the application of telescopes to large-scale graduated instruments appropriate for the measurement of small angles. With this equipment Picard executed a measure on the meridian of Paris that has been said to have been marked by a precision thirty to forty times greater than any previously achieved. The work provided the basis on which the desired rectification of French cartography could be accomplished, as well as a model for all later undertakings of this kind of 'geometrical' geodesy. Subsequent improvements were to depend far more upon instrumental perfection than upon changes in the techniques themselves (Perrier *1939*).

Picard's measure also, of course, found application in Isaac Newton's elaboration of the law of universal gravitation in his *Principia* (1687), a work destined to have a profound impact upon geodesy because of its insistence that the Earth cannot have a perfectly spherical form. Rather, Newton postulated (without demonstration, and assuming that the Earth could originally have been a fluid homogeneous mass) that the form of equilibrium of such a mass, subject to the law of attraction and rotating about an axis, is an ellipsoid of revolution around that axis, flattened at the poles. Employing a technique that anticipated what was to be called the potential function (§3.17), he determined the attraction that would cause the fluid particles in two hypothetical canals leading from the centre of the Earth to its surface (one in the equatorial plane, the other coinciding with the polar axis) to be in balance, and concluded that the Earth has the shape of an oblate spheroid in which the ratio of the equatorial to the polar axis is 230 to 229 (Chapin *1956*).

Newton also undertook to compare effective weights at different latitudes on the Earth's surface, claiming, on the principle of 'balanced canals', that the weights should vary inversely with their distance to the centre. It followed that the increase in effective weight from the equator to the poles varied as the square of the sine of the latitude. From this result, he deduced that a pendulum beating seconds would have to be shortened by about one-twelfth of an inch (2 millimetres) when taken from Paris to the equator (Chapin *1994*).

Newton's point of departure for this was the experience of one of the young 'student' members of the Paris Academy of Sciences, Jean Richer, on being sent to Cayenne in 1672 to conduct a series of astronomical observations. Having taken with him a pendulum that had been regulated to beat seconds in Paris, Richer found it necessary to shorten that apparatus in Cayenne in order to make it continue to do so. In his classic work on the pendulum (§8.13), Christiaan Huygens had found that the period of a simple pendulum could be expressed by the formula $T = 2\pi (l/g)^{1/2}$, where T is the time for a complete back-and-forth swing, l is the pendulum's length and g is the acceleration due to gravity. But since he did not imagine g to be variable, Huygens was inclined to attribute Richer's necessitated shortening simply to carelessness. Newton, however, made it clear that Richer's experience – which had been confirmed by others prior to the publication of the *Principia* – was a consequence of the variation in the acceleration due to gravity over the surface of an oblate spheroid. Newton also invoked in his support the astronomical observations of the flattening of Jupiter towards its poles.

Newton's postulates had two broad outcomes for the study of the shape of the Earth. The first was to provide the underpinning for an entirely new branch of geodesy which was originally of uniquely theoretical interest, and which has come to be known as 'physical' or 'dynamic' geodesy. The second was to alter the thrust of the older mathematical geodesy by imposing upon it the necessity to determine not simply the single unknown of the terrestrial diameter, but also the two unknowns of either the two axes of the ellipsoid or its major axis and the amount of flattening. In order to obtain two equations furnishing the two unknowns, it became necessary to measure two arcs of the meridian instead of the single one required by the hypothesis of a spherical Earth (Perrier *1937*). In this article the latter approach will be pursued to its logical termination, after which the former will receive brief consideration.

2 ARC MEASURES, DISPUTES AND EXPEDITIONS

The first proposal to measure an arc capable of being compared directly to Picard's actually predated Newton's work and was tied closely to cartographical concerns. This was the suggestion put forward as early as 1682 by Giovanni Domenico Cassini, an Italian astronomer lured to Paris by a large pension from Louis XIV in order to participate in the works of the Academy's planned astronomical adjunct, the Observatoire Royal in Paris (where, incidentally, he subsequently made the observations of Jupiter that served Newton's cause). With backing for a plan that called for a southward extension of Picard's earlier measure, several of the Academy's astronomers were dispatched to that end; but unfortunately they were recalled when the death of Colbert, Louis XIV's finance minister, brought about the project's cancellation. It was resumed only in 1700, although then in a northerly direction as well. These efforts were carried out by Jacques Cassini, the second in a dynasty which was to dominate the Observatoire Royal for more than a hundred years. The outcome of his labours was published in 1720 as a work dealing with the size and shape of the Earth.

If the Earth has the shape of an oblate spheroid, the length of one degree of latitude – and the force of gravity on its surface – ought to increase as one moves from the equator toward the poles. In the new measure, therefore, a northern extension of the same amplitude as Picard's arc ought to have been slightly longer than the original, which, in turn, ought to have exceeded in length a southern extension. The figures published by Cassini, however, indicated the opposite relationship. Having found that the most southerly portion of the arc had the greatest length, he and a group of followers maintained that the Earth had the shape of a prolate spheroid – elongated rather than flattened at the poles.

The dispute between the proponents of oblate and prolate spheroids (known respectively as Newtonians and Cassinians) was largely the result of a fundamental split between theories based on assumptions, and those based on observation: between the newer and the older geodesy. That the evidence from observation was suspect was first pointed out in 1720 by Joseph Nicolas Delisle, when he expressed reservations about what one could infer about the Earth's shape from any local measurement, even those associated with his own suggestion for substituting a measure of a degree of longitude along a parallel to the equator for the measurement of latitudinal length. Since Delisle subsequently left France without making public these considerations (which he had published only in 1737), it fell to others to publicize his suggestion. They, particularly Pierre de Maupertuis, did so in 1733 on the basis of a small book first published in Italy in 1724 by Giovanni, the Marquis Poleni and Professor of Mathematics at the University of Padua (Greenberg *1983*), which also urged the utilization of longitudinal measures. Maupertuis presented this idea to the Academy when Cassini was engaged in the field work necessary to trace across France the arc of the great circle perpendicular to the meridian through Paris. Reporting on this to the Academy a few months later, Cassini not only insinuated that he had himself arrived at the idea of using longitudinal lengths to determine the Earth's shape, but also that the work already accomplished had again indicated a prolate spheroid (Greenberg *1984*).

The real problem with the evidence of the extended measure, as Newton himself pointed out in the third (1726) edition of his *Principia*, was that the small differences in length involved could have resulted simply from errors in the observations. One proposed solution to this was simply to measure degrees of latitude considerably distant from one another. So, in 1735 the Academy resolved to send an expedition to measure a degree of latitude near the equator and another to do the same near the pole (Jones *1967*). Shortly thereafter, two groups of academicians left France equipped with the best available examples of the instruments made basic by Picard: carefully divided zenith sectors and quadrants, both with good telescopes, as well as wooden rulers 2 *toises* (approximately a fathom) in length that had been carefully compared to the *toise du Chatelet*, the official standard of length then only recently established (Hallock and Wade *1906*). One group went to Peru, the other to Lapland.

The expedition to Lapland actually did not leave until 1736, almost a year after the other. It returned, however, in 1737 − seven years before its Peruvian counterpart. An account of this northern undertaking was published in 1737 by Maupertuis, who had both promoted and headed it. The arc that it measured was found to be considerably longer than the Picard standard.

In addition to substantiating the flattening of the Earth at its poles, the value of the Lapland degree invalidated the results proclaimed after the prolongation of Picard's arc, and necessitated a remeasure of the meridian of Paris. This was undertaken in 1739 and 1740 under the auspices of the Academy, and featured the work of Nicolas Louis de Lacaille and César François Cassini, the third member of the Cassini dynasty to be involved in that project. The *méridienne vérifiée*, as it was appropriately called in the title of Cassini's book describing the operations, reversed the earlier findings. By thus removing any doubts which might have remained, the question of the shape of the Earth had been definitely decided in favour of the Newtonian theory, although the ratio of axes as determined by the Lapland arc and the 'verified meridian' was 178 to 177, an ellipticity greater than the theory called for. For some, the latter outcome was reversed by the results brought back from Peru, although three different accounts of that venture produced three slightly differing outcomes.

3 MORE MEASURES, FROM ONE INTERNATIONAL AGREEMENT TO ANOTHER

The 1750s and 1760s saw several new arc measures in various parts of the world. Thus, for example, Lacaille undertook a measurement at the Cape of Good Hope, Charles Mason and Jeremiah J. Dixon did the same in Pennsylvania and Maryland, and J. X. Lieseganig provided data from measurements in Austria and Hungary. More important than these, however, was the Italian effort by the Jesuits Roger Boscovich and C. Maire (1770) which, while closely approximating the latitudes of the measures in

France, produced such differing outcomes for the length of a degree as to cause Boscovich to conclude his account of it with the observation that, the more one measures degrees, the more uncertain the shape of the Earth becomes (Lacombe and Costabel *1988*).

It was partly such discrepancies and uncertainties that led French scientists in the early 1790s to propose the new measure of a long arc between Dunkirk in the north of France and Barcelona in the north of Spain. With new instrumentation promising hitherto impossible degrees of accuracy, the results would provide a necessary foundation for the establishment of a new standard of length. The new measure, by using a figure for ellipticity drawn from a comparison of it with the degree length established by the Peruvian measure, was to enable the determination of the length of one quarter of the terrestrial meridian, the ten-millionth part of which became the new metre.

The authorized operations were begun in 1792 by two young academicians, Jean-Baptiste Delambre and Pierre Méchain, and their aides. They were suspended at the end of the following year, but resumed in 1795, and were completed by 1799. In this last year the French government invited other states to send representatives to Paris for the purpose of examining the completed work. The result was the adoption of an ellipticity, based upon the stated comparison, of 1/334. Interestingly, Delambre later proposed, in the *Base du système métrique* – the full three-volume account of the operations that he and Méchain later published in 1806–10 – that a figure of 1/309 would more accurately represent the Earth's deviation from sphericity (Delambre *1912*).

The arc that had been measured by Méchain and Delambre was soon extended southward to the Balearic Islands so that it would centre around a latitude of 45°. This extension was begun by Méchain, who, however, died during the operations. It was continued by two new young academicians, François Arago and Jean-Baptiste Biot, who published in 1821 an account of their work under the title of *Recueil d'observations géodésiques*.

The operations involved in the establishment of the metric system did not, of course, finally resolve the question of the exact shape of the Earth. Still, they were of great importance. New instruments and techniques were employed in this measure which, for a long time thereafter, was considered a model. The example thus set gave rise to a whole new series of arc measures, of which there is space here to provide only the sketchiest of illustrations. Moreover, it brought together for the study of that fundamental question what may be considered the first ever international congress of scientists; 125 years later, an international meeting of geodetical specialists would arrive at a new decision.

Nineteenth-century efforts of this sort began with J. Svanberg's 1801–03 remeasurement of the Lapland arc which brought it into far better relation with the metric ellipticity, and ended with a reconnaissance mission to Ecuador with a view to studying the possibility for remeasuring the arc at the equator, a project that was actually carried out between 1901 and 1906. In between these operations there were many many others, in England, Russia, Austria, various German states and India. They all benefited from continued refinement of methods and instruments. The most notable example of better methods was the ever more frequent employment of the method of least squares (§10.5) – first published by Adrien Marie Legendre in 1805, although used in practice by Carl Friedrich Gauss more than ten years before – for the reduction of personal error and the finding of the best equation fitting a group of observations. Perhaps the most significant example of the application of new instruments was the use of, initially, the telegraph, and later, the radio for the easy determination of longitudinal differences (Perrier *1939*).

For all of their improved accuracy, however, arc measures undertaken over a limited part of the globe's surface can yield at best an ellipsoid representing the Earth's figure in that region. What is needed for a more truly international standard ellipsoid – a concern for the establishment of which became an early goal of the International Geodesic Association, created in 1885 after a gestation period of almost twenty years – is a measure based not simply on arc measures but on an area measure of a large land mass. The US Coast and Geodetic Survey provided the opportunity for such, and J. F. Hayford, chief of its calculating division, made the first extended application of the method of areas in 1909. The result was an ellipsoid with a flattening of 1/297, which was accepted by the Association in 1924 as the international reference surface: that is, the world's standardized geoid, a term coined in the second half of the nineteenth century to mean the spheroid of the Earth taken as its mean sea level conceived as extending continuously through all the continents (Perrier *1939*). That development and definition serves to return us to the question of the shape of the Earth as determined by theory, or, in keeping with the distinction made in Section 1, to the subject of dynamical geodesy, an approach capable of providing only the form and not the dimensions of the terrestrial ellipsoid.

4 THEORIES OF SHAPES AFTER NEWTON

The first successor to Newton was Huygens, whose pendulum work had been so crucial to the *Principia*. In his *Discours de la cause de la pesanteur* (1690), he did not admit the reciprocal attraction of all particles of matter. Rather, he considered that each particle of a fluid, homogeneous mass was attracted only by the centre of gravity of the mass, thus greatly simplifying the problem of the equilibrium form of the mass under rotation. Huygens also arrived at the shape of an oblate spheroid. Owing to his initial assumption, however, he found the polar radius to be shorter than the equatorial radius by only 1 part in 578.

Huygens' conclusion in favour of the flattening of the Earth at its poles was extremely important, in view of the then almost universal opposition to Newtonian gravitation on the Continent. Nowhere was that truer than in France, where the Academy's Secrétaire Perpétuel, the illustrious Bernard de Fontenelle, charged J. B. Dortous de Mairan with attempting a theoretical conciliation between the polar elongation proposed by Jacques Cassini and the undeniable flattening effect stemming from the centrifugal forces due to the rotation of the Earth. Mairan's 1720 solution was arrived at by denying the primitive sphericity of the Earth which Huygens had assumed as a postulate. Mairan argued instead that the Earth was originally an oblong spheroid, which would have made it depart even further from its spherical form (i.e. more elongated) in the absence of rotation (Lacombe and Costabel *1988*).

Mairan subsequently found himself looking more and more to Newton's *Principia*; but it was Maupertuis who became the real champion of Newtonianism in France, although the problem of the shape of the Earth was a major difficulty for him. In readying his work on Newtonian principles, the 1732 'Discours sur les différentes figures des astres', he realized that he understood nothing of the method that Newton had followed in establishing his oblate spheroid, since, like Huygens, he had obtained a value for the flattening of the globe that differed substantially from Newton's (Greenberg *1987*). Indeed, he could not even understand why Newton had chosen an ellipsoid of revolution since homogeneous solids of revolution acted on by a central force whose strength varies directly according to r^n (where r is the distance to the centre of the force), are not ellipsoids except if $n = 1$. It was because of his problems here that Maupertuis turned to Poleni's suggestion with the results mentioned in Section 2.

It was while the measurements called for by Maupertuis were still being carried out that the question of the relation of the empirically determined shape and variation of gravity to Newton's proposed explanation was given prominence by new theoretical investigations. The first of these were the researches by Colin Maclaurin, presented in his *A Treatise of Fluxions* (1742) (Todhunter *1873*). Maclaurin claimed more than he was actually justified in doing when he maintained that he had demonstrated that, if the Earth were a fluid homogeneous mass, it would assume the shape of an oblate spheroid in consequence of its diurnal rotation. In fact, he had not proved that the planet *would* assume that form, but he had shown for the first time that the oblate spheroid was *a* form of equilibrium. His value for the ratio of the axes was practically the same as Newton's, namely 230 to 229. Since that value now disagreed with the amount of flattening derived from the measurements in Lapland and in France, Maclaurin next proposed to treat the Earth as non-uniform in density, with either greater or lesser density at the centre. His investigation of these hypotheses being unsatisfactory, however, it fell to the French mathematician Alexis Clairaut to continue the investigation.

Clairaut's epochal *Théorie de la figure de la terre* appeared in 1743, and attacked that problem by purely geometrical methods (Greenberg *1988*). As he began by treating the Earth as a homogeneous fluid, his results were essentially the same as Maclaurin's, so that he also demonstrated Newton's postulate. He then went on to consider the Earth as heterogeneous in order to test Newton's remark that, if the Earth were denser towards the centre, it would be more flattened than if it were homogeneous, and that the contrary would be true. In so doing, he generalized the problem and considered a spheroid composed of concentric layers of variable densities. His results may be expressed as the following simple formulas:

$$g_L = g_e(1 + \beta \sin L), \quad \beta := (g_p - g_e)/g_e, \tag{1}$$

where g_L is the gravitational acceleration at latitude L, and g_e and g_p the values at the equator and poles, respectively; and

$$a = (5Y/2) - \beta, \tag{2}$$

where a is the flattening and Y the ratio of the centrifugal force to the intensity of gravity at the equator.

Clairaut's work was later placed by Pierre Simon Laplace 'in the class of the most beautiful mathematical productions'. Some contemporaries, however, claimed to have brought in needed improvements. This position was maintained, for example, by Jean d'Alembert, whose most extensive treatment was contained in his *Recherches sur différents points importans du système du monde*, the first two volumes of which appeared in 1754 and the third two years later. D'Alembert did introduce a valuable method for estimating the attraction of a spheroid by resolving the body into a sphere and a thin additional shell, but, again according to Laplace, his investigations 'lack the clarity so necessary in complicated calculations'.

5 LAPLACE

Although Clairaut's theory did not produce a value of the ellipticity that was borne out by comparison, initially, of the recent measures in Lapland and in France (or even, subsequently, of the former with that from Peru), later work undertaken by Legendre and Laplace demonstrated that it was less in error than his eccentricity value would have implied. Of Legendre and Laplace, the researches of the latter are far more important, although their initiation appears to owe something to the work of the former (Puissant *1819*). Indeed, it has recently been said that, in his studies on attraction, 'Laplace played leapfrog with Legendre'.

Since the latter's work is noted in §4.4 on special functions, only that of Laplace will be looked at here. Although his first offering in this area, a memoir on the attraction of elliptical spheroids, was read to the Academy in the spring of 1783, it was not published in the Academy's *Mémoires*. Rather, Laplace chose to develop it further as Part II of the *Théorie du mouvement et de la figure elliptique des planètes*, a book brought out in 1784 with the personal subsidy of the honorary academician, Bochard de Saron. It was notable for its clearer use of the concept, earlier utilized by Clairaut, of what came in 1828 to be called potential (§3.17). Laplace subsequently read to the Academy two more memoirs dealing with this subject, one a general theory on the attractions of spheroids and the shape of planets, the other a more specific study of the shape of the Earth (Todhunter *1873*). The essence of both these efforts, as well as his 1784 *Théorie*, were later repeated – with some revision – in the second volume of his monumental *Traité de mécanique céleste* (1799), of which the *Théorie* had been the 'veritable embryo'. From both pendulum observations and the phenomena of the precession and nutation of the Earth's axis, Laplace concluded that the flattening of the Earth had to be less than 1/230, and that the limits between which this fraction must fall, as set by precession and nutation, were 1/304 and 1/578.

Laplace frequently showed a tendency to dismiss complicated mathematical argument with the remark that 'it is easy to see' (Chapin *1956*). In translating his great *Traité* into English, the American astronomer Nathaniel Bowditch not only supplied these missing arguments, but also corrected numerous arithmetical and other errors which had appeared in the original. In addition, he took into account Laplace's own subsequent revisions, stating that the upper limit on ellipticity was 1/279, not 1/304, when consideration was given to Laplace's later alteration of constants.

On the basis of pendulum experiments and the assumption of ellipticity, Laplace found the most probable ellipsoid to have an ellipticity of 1/336, a fraction that Bowditch later changed to 1/315 by correcting errors in his calculations. Since the total variation in the length of the seconds-pendulum from equator to pole amounts to only 5.24 millimetres, to obtain a reliable value of the ellipticity requires that a large number of carefully performed measurements be brought to bear on the formation of a statistical mean. Laplace had only 15 such measurements at his disposal. By 1832, when he brought out his translation of Laplace's second volume, Bowditch had assembled 52 measurements, to which he applied the method of least squares; neglecting the eight that differed most from the rest, he obtained an ellipticity of 1/297, precisely the amount adopted internationally 92 years later (Chapin *1991*).

Laplace also examined the implications of the arc measures at different latitudes, employing the seven from Lapland, Peru, the Cape of Good Hope, Pennsylvania, Italy, Austria and the French metric result. Determining the ellipticity of an ellipsoid that would reduce the largest absolute error to a minimum (§6.11), he found that minimum to be 189 metres and to occur in the measures from Lapland, the Cape and Pennsylvania; on this assumption, the ellipticity turned out to be 1/277. Since Laplace believed that these measures could not be so greatly in error, he concluded that the shape of the Earth must differ sensibly from the ellipsoidal. When he, nevertheless, calculated the most probable ellipsoid, based on the same seven measures, he found an ellipticity of 1/312, and the error in the Lapland measure to be 336 metres. Certain that the error could not be nearly so large (as in fact Svanberg's remeasure was to show it to be), he again concluded that the departure from ellipticity must be sizeable. Bowditch, however, found later measurements fitting the elliptic hypothesis more closely. From five measures from Peru, India, France, England and Sweden, he obtained an ellipticity of 1/310.

In the second volume of the *Mécanique céleste*, Laplace could claim with assurance that the flattening of the Earth was less than 1/230, the Newtonian value implied by the supposition of homogeneity; the result was confirmed by pendulum experiments, arc measures, and precession and nutation. In the next volume of that work, which appeared in 1802, he drew added support for the same conclusion from lunar theory. In lunar tables published in 1780 by Mason (of the Pennsylvania measure), and also in those currently being prepared by J. T. Burgh, there appeared in the Moon's longitude a term proportional to the sine of the longitude of the lunar node; it had been determined empirically, Mason putting the coefficient at 7".7, while Burg gave it as 6".8. From gravitational theory, Laplace showed the coefficient to be proportional to the flattening of the Earth, and that Burg's value thus implied an ellipticity of 1/305.

Although the nineteenth century was to witness the elaboration of various other geoids and the clear emergence of the problem of the 'deflection of the vertical' – the departure of a plumb line from perpendicularity to the reference ellipsoid of revolution, a departure due to anomalies in the distribution of the mass of the Earth, first noted during the Peruvian expedition – it seems well to end this survey with the mention of Laplace's use of lunar theory. The reason for its appropriateness in this regard is that the launching in 1957 of Sputnik, the first artificial satellite, led to a resurgence of geodesy (Lacombe and Costabel *1988*). The use of artificial moons, containing photographic equipment or electronic transmitters enabling the further use of the Doppler–Fizeau effect, has generated a new era of ever more precise studies. Still, such developments are really only new instances of instrumental improvement, and the mathematical tools forged by Newton, Clairaut, Legendre, Laplace and their many successors remain basic to the determination of the exact shape of the globe on which we live.

BIBLIOGRAPHY

Chapin, S. L. *1956*, 'The size and shape of the world: A catalogue of an exhibition from the collection of Robert B. Honeyman, Jr', *UCLA Library Occasional Papers*, No. 6.

—— *1994*, 'The shape of the Earth', in *The General History of Astronomy*, Vol. 2B, *Planetary Astronomy from the Renaissance to the Rise of Astrophysics*, Cambridge: Cambridge University Press, Chap. 15, to appear.

Delambre, J. B. J. *1912*, *Grandeur et figure de la terre* (ed. G. Bigourdan), Paris: Gauthier-Villars.

Greenberg, J. L. *1983*, 'Geodesy in Paris in the 1730s and the Paduan connection', *Historical Studies in the Physical Sciences*, **13**, 239–60.

—— *1984*, 'Degrees of longitude and the Earth's shape: The diffusion of a scientific idea in Paris in the 1730s', *Annals of Science*, **41**, 151–8.

—— *1987*, 'Isaac Newton et la théorie de la figure de la Terre', *Revue d'Histoire des Sciences*, **40**, 357–66.

—— *1988*, 'Breaking a "vicious circle": Unscrambling A.-C. Clairaut's iterative method of 1743', *Historia mathematica*, **15**, 228–39.

Hallock, W. and Wade, H. T. *1906*, *Outlines of the Evolution of Weights and Measures and the Metric System*, London: Macmillan.

Jones, T. B. *1967*, *The Figure of the Earth*, Lawrence, KS: Coronado Press.

Lacombe, H. and Costabel, P. (eds) *1988*, *La Figure de la Terre du XVIIIe siècle à l'ère spatiale*, Paris: Gauthier-Villars.

Perrier, G. *1937*, 'Le développement de la géodésie de ses origines à nos jours', *The Rice Institute Pamphlets*, **24**, 168–88.

—— *1939*, *Petite Histoire de la géodésie* [. . .], Paris: Presses Universitaires de France.

Puissant, L. *1819*, *Traité de géodésie* [. . .] 2nd edn, 2 vols, Paris: Veuve Courcier.

Todhunter, I. *1873*, *A History of the Mathematical Theories of Attraction and of the Figure of the Earth*, London: Constable. [Repr. 1962, New York: Dover.]

8.15

Cartography

H. WALLIS AND M. H. EDNEY

1 PREHISTORY

Discoveries of rock art in Europe and elsewhere show that people were making maps in prehistoric times. The earliest known town plan is a neolithic wall painting found in a settlement at Çatal Hüyük, Turkey, dated around 6200 BC. The petroglyph from Bedolina, Valcamonica, in northern Italy depicts a village settlement forming a topographical map from around 1500 BC. From more recent years, the birchbark maps of North American Indians and the carved coastal charts of the Inuit (Eskimos) are intricate topological and topographical depictions. Although not of planimetric accuracy, they are based on geometrical perceptions of the relationship of objects and places in space. As C. Delano Smith points out, the idea of showing images in plan representation dates from the Upper Palaeolithic age, 40 000 to 10 000 years ago (Harley and Woodward *1987*: 504).

The science of cartography, on the other hand, originated in the early civilizations of Mesopotamia, Greece, Egypt and China. People in these societies made maps for practical purposes, as part of elaborate processes of building, taxation, travel and waging war. Surveying was one of the oldest professions, and served the demands of a ruling class of officials. Position, distance, direction and area were the basic mathematical concepts. One of the statues representing Gudea, ruler of the Sumerian city-state of Lagash (*circa* 2200 BC), shows the surveyor holding a scale plan of the Ningirsu temple, with a measuring rule and a writing tool. This is the earliest extant plan drawn to scale. Plans on cuneiform tablets recorded surveys of an area, and measurements of fields. In Babylonia in the second millennium BC, *kudurrus* (carved stones) represented grants of private and public lands and were inscribed with the cardinal points and details of boundaries.

In addition to such local plans, a map of northern Mesopotamia, on a clay tablet from Nuzi dating from between 2500 and 2300 BC, is an example of regional mapping. It shows rivers and mountains, and is the

1101

earliest-known example of a map on which cardinal directions are marked. The Babylonian world map, from the seventh or sixth century BC, displays on a clay tablet the Earth (i.e. the Babylonian world), as a circular disc, and the surrounding cosmos. It illustrates the predilection of early mapmakers for showing the world in a circular form, with the important places at the centre. The Greek philosopher Eudoxus of Cnidus (*circa* 408–355 BC) was to develop the concept of geocentric circles for the mapping of the sky and Earth, and was the first to draw stars on a globe.

In ancient Egypt, described in the fifth century BC by the Greek historian Herodotus as the birthplace of geometry, the annual flooding of the Nile made it necessary to resurvey land at regular intervals to establish property boundaries. The Rhind Papyrus of around 1650 BC (British Museum) gives instructions in surveying, with diagrams (§1.2). The 'Fields of the Dead' (*circa* 1400 BC), drawn on papyrus, depict plots of land and waterways in the afterlife. These may be seen as idealized pictorial examples of surveys of real estates. Wood and stone measuring-rods used for such surveys survive as votive offerings. Surveying teams measuring fields for taxation purposes used ropes knotted at regular intervals.

Local Chinese topographies dating back to the second century BC include regional maps showing mountains, rivers, towns and roads. The Han maps (*circa* 200 BC) found in Ming tombs mark military posts and numbers of households, and distances between settlements. As early as 227 BC, the Chinese were drawing maps on silk. According to tradition, the warp and weft of the silk inspired the idea of using grid squares as a reference system; this became the standard Chinese reference system, and remained in use from about 200 BC until the sixteenth century. Although such maps do not explicitly reveal knowledge of a spherical Earth, Chinese astronomers conveyed this concept in their analogies of the Earth to the yolk of an egg or a cross-bow bullet (*circa* 200 BC).

2 GREEK AND ROMAN PRACTICES

In ancient Greece, land measuring (geometry) and philosophy developed as two distinct professions. Hippodamus of Miletus (flourished 473–443 BC) is said to have originated surveys in which squares and rectangles were used as a framework for surveying towns and rural areas. The Romans carried on the Greek tradition with their surveyors (*mensores*) and land surveyors (*agrimensores*). Their textbook, the *Corpus agrimensorum* (first century AD onwards) included plans of land laid out by 'centuriation' (i.e. in bands of 100 units) at Orange in southern France from AD 77. At Rome, the 'Forma urbis Romae', a plan of the city from the third century AD, was inscribed on marble tablets fixed to a wall. The principal surveying instrument was a *groma*, a surveyor's cross with a staff and cross-arms. The mathematics required accurate linear measurement, the laying out of lines at right angles to each other, and an adequate numbering system. The unit of measurement for a survey was the *actus* of 120 Roman feet, originally the distance oxen would plough before turning. 'Centuries' (i.e. successive hundreds) were named according to a system of coordinates, starting from the main roads of a centuriated area. An instruction in the *Corpus agrimensorum* apparently referring to a scale of 1:5000 (one Roman foot to a Roman mile) has been identified as an explicit indication that surveyors had a clear concept of ratio (Dilke, in Harley and Woodward *1987*: 276).

The philosophers of ancient Greece were concerned with explaining the shape and size of the world. The Pythagoreans of the sixth century BC are regarded as the first to express nature in quantitative terms. The astronomer Anaximander of Miletus is reputed to be the first Greek to have made a world map. The construction of such a map involves the concepts of a central feature, distance, direction and form. The Greeks inherited from the Babylonians knowledge of the division of the circle into 360 degrees, which derived from the sexagesimal system used in Babylonian calculations. On a framework representing the conceived shape and size of the Earth, the mapmaker would lay down countries, seas and coastlines as a compilation from the itineraries and the reports of travellers.

The idea of a spherical Earth is said to have derived from Pythagoras, an Ionian of South Italy, although even in Antiquity this was disputed (Kahn *1960*: 115–18). Parmenides of Elea in southern Italy (born *circa* 515 BC) is reputed to have been the first to divide the sphere into five zones, one hot, two temperate and two cold, and he probably illustrated these divisions on a map or globe. How to depict the *oecumene* (the inhabited world) was a problem for mapmakers. The Homeric scholar Crates of Mallos (*circa* 150 BC) showed the *oecumene* in the form of a hemisphere ('Crates' Orb'). The world was divided into four symmetrical land-masses: Europe, Asia and the known part of Africa formed three, while the fourth comprised to the south the Antoikoi, 'dwellers opposite', to the west the Perioikoi, 'dwellers round', and to the south-west the Antipodes (Dilke, in Harley and Woodward *1987*: 36). Macrobius developed the system, which was then followed by some medieval mapmakers.

Eratosthenes, the head librarian at Alexandria from 240 BC, was the first to make a scientific map of the world. It was based on a method of measuring the size of the Earth and displayed parallels and meridians with an irregular grid. The inhabited world was divided into areas called *sphragides*. Although he was graded as 'beta' by his contemporaries, his method was correct in principle, and his estimate of the circumference of the Earth – at 252 000 *stades* – was the most accurate in Antiquity. It may have been within 300 kilometres of the correct figure (Thrower *1972*: 180); however, the *stade*, consisting of 600 Greek feet, has been calculated by some historians to be around 185 metres.

Claudius Ptolemy was the most influential mapmaker of the ancient world. His *Geography* (*circa* AD 150) was the dominating cartographic work for a period of some fourteen centuries. As Dilke remarks (Harley and Woodward *1987*: 183), the *Geography* was planned as a manual for mapmakers. It dealt first with the size of the inhabited world, and criticized Marinus of Tyre (flourished AD 100) for extending the inhabited world too far. Ptolemy used the smaller value of the circumference of the world proposed by the Greek astronomer Posidonius (186–135 BC), 180 000 *stades*, about three-quarters of the true figure. Ptolemy then set out the different forms of map projection, which has proved to be his most important contribution to the mathematics of mapmaking (*Ibid.*: 185).

3 EARLY GRAPHICAL METHODS

Beyond the small area covered by a plane survey, the cartographer must take into account the curvature of the Earth. The transformation from the spherical (or spheroidal) Earth to the plane map is achieved through a 'map projection'. The heart of mathematical cartographic theory for centuries, the description of projections has naturally depended on the mathematical sophistication of their creators. While modern descriptions define at least two functions,

$$x = f(\phi, \lambda) \quad \text{and} \quad y = g(\phi, \lambda), \tag{1}$$

which relate a point's position on the Earth (latitude ϕ, longitude λ) to its position on the map (x, y), the earliest projections were constructed graphically. In the most general sense, a projection is any network (or 'graticule') of meridians and parallels on a map which corresponds to the lines of longitude and latitude on the Earth.

The projection of the curved Earth onto the plane cannot be achieved without distortion. The fundamental issue in map projections is therefore to minimize distortion in angles (and shapes), areas and certain distances. Projections which completely eliminate one of these three types of distortion are called, respectively, conformal (or 'orthomorphic'), equal-area (or 'equivalent') and equidistant. But minimizing one set of distortions will increase other distortions.

The earliest-known projection for geographical maps is equidistant in nature. Comprising an equirectangular grid of straight meridians and parallels, the parallels were spaced equally for equal changes in latitude, so that scale was preserved along the meridians; the spacing of the meridians was varied to preserve the scale along a specific parallel. Basing his map on the parallel of Rhodes (about 36°N), Marinus of Tyre (second century AD) spaced his meridians at about $\frac{4}{5}$ (about $\cos 36°$) of the spacing of the parallels. Precise knowledge of the Earth's size is not necessary for this type of projection, just basic trigonometrical functions. In his *Geography*, Ptolemy recommended that the projection be used for regional maps. For maps of the inhabited world he presented three other projections, one to give a perspective view of the spherical world, the other two to be equidistant but with parallels formed by circular arcs.

Knowledge of Ptolemy's astronomical and geographical works became lost in Europe in the early Middle Ages, but was preserved and developed by the Arabs. In Europe, world maps developed according to various traditions derived from Roman and Greek prototypes. One was the tripartite world map: the schematic 'T–O map', in which a T-shape divides the O representing the Earth into three parts. The encyclopedist Isidore of Seville (*circa* 560–636), working in a Roman and early Christian tradition, provided most of the schematic T–O maps. Another type was zonal, and based upon the ideas of Theodosius Macrobius (fifth century AD).

The T–O maps were developed into works of great complexity. Thirteenth-century England was notable for the projection of three of the largest and most important *mappae mundi*. The Ebstorf by Gervais of Tilbury (1229), now destroyed but known in facsimile, was the largest (3.56 × 3.58 metres). The Hereford map (preserved in Hereford Cathedral) made by Richard de Bello (Richard of Haldingham) around 1285, at 1.32 metres in diameter, is the largest extant. All places in the Roman Empire are marked on the map, which is the last in a long series of post-Roman maps.

These maps performed a symbolic function, showing Christ in judgement presiding over the universe. Woodward has suggested that medieval world maps were drawn with a frame from a number of geometric shapes, circular, oval, rectangular or mandorla (in the shape of a mandolin), each with symbolic significance. Since later medieval authorities agreed that the Earth was a sphere, it has been suggested by Waldo Tobler and others that such world maps were constructed according to a type of projection approaching the azimuthal logarithmic (Harley and Woodward *1987*: 322).

At the same time that the *mappae mundi* were developing to a high degree of cartographic art, a very different type of map made its appearance in the Mediterranean region in the later years of the thirteenth century, namely the nautical portolan chart. Beazley *1904* has described such charts as 'the first true maps'; Armando Cortesão considered the advent of the portolan chart to be one of the most important turning points in the history of cartography (*1969*: 215–16).

4 SOME EARLY CHARTS

From now on we follow convention in distinguishing maps (for land) from charts (for water-covered areas). The term 'portulan' (from the Italian *portolano*) refers to written sailing directions, and it was extended in the 1890s to maritime atlases and charts which some commentators thought owed their origin to written directions. The features generally present are a network of interconnecting lines ('rhumb lines') radiating from 16 equidistant points around the circumference of one (or sometimes two) 'hidden circles'. Place-names are written inland at right angles to the coast.

The standard area covered by the charts during the next three centuries consisted of the Mediterranean and Black Seas, with the Atlantic coasts from Denmark to Morocco and including the British Isles. The main centres of production were Venice, Genoa and Majorca. The earliest surviving portulan chart is the 'Carte Pisane', made at Genoa around 1300 and so named because it belonged to a family from Pisa. The earliest dated portulan chart is that of Pietro Vesconte of Genoa (1311); the earliest atlas, consisting of a standard chart arranged in sheets, is also by Vesconte, dating from 1313.

The origin of the portulan chart is still unknown. The charts present a remarkably accurate portrayal of the Mediterranean. Theories of Greek, Roman and Arab origins have been put forward, but are not generally accepted, and it is considered likely that the chart originated in the thirteenth century. Another question concerns the role of the magnetic compass, which came into use in the early thirteenth century at almost the same time the first charts were made, but this connection has not been generally accepted and remains conjectural.

Controversy has also centred upon whether the charts are based on a projection. The general view is that they are projectionless, or that any projection is accidental. While each chart carries its own scale, the lack of any statement of value raises the question of the unit involved. Although the basic form of the charts did not change from 1300 to 1450, Abraham Cresques of Minorca did introduce a major innovation with his world map in portolan style in 1375.

The revival of Ptolemy's *Geography* in Europe in the late fourteenth and fifteenth centuries was a major factor in the geographical renaissance. Although the geographical features showed the world as it was known around AD 150, the framework of latitude (*climata*, or zones) and longitude gave the map a scientific form which could be adapted to contemporary knowledge, and stimulated the growth of projective geometries. Mapmakers now introduced a new type of world map, which was Ptolemaic in form (e.g. with north at the top), but revised to take account of new discoveries; Henricus Martellus Germanus included such a manuscript map in his 'Isolario' made in Florence in about 1490 (British Library, Add. Ms. 15760, fols 68b–69). His larger world map, slightly later in date, is in the Yale University Library. These are the last world maps of the *oecumene*.

The discovery of America stimulated the evolution of a new type of world map, the planisphere. Those of Diego Ribeiro, a Portuguese working for Spain, dating from 1525, 1527 and 1529, are the first to show the whole circuit of the globe between the polar circles. As manuscript maps they were of limited influence. The printed world map dating from 1472 onwards was a much more powerful document of geographical intelligence.

5 PTOLEMY'S INFLUENCE: MERCATOR

Ptolemy's *Geography* was the vehicle for some of the most influential new world and regional maps. The work had circulated in manuscript from 1410 onwards, and was first printed with maps at Bologna in 1477. The successive editions were enlarged with *tabulae novae* based on new information and new projections. The cartographers of the Renaissance also applied the classical astronomical projections to terrestrial data. Some of the new projections were designed for their aesthetics, such as the popular oval world projections first used by Francesco Rosselli for his world map of around 1508, and made popular by Benedetto Bordone through publication of his world map in his *Libro* [...] *de tutte l'Isole del mondo* (1528). Most projections, however, had mathematical purposes. Thus Henricus Martellus Germanus replaced the circular parallels on Ptolemy's second world projection with non-circular parallels in order to preserve scale along all, instead of just three, lines of longitude. Projections were still defined graphically.

The great geographer Gerardus Mercator of the Netherlands, described by his friend Abraham Ortelius as 'the Ptolemy of our Age', was the first to invent a projection to aid navigation. The famous projection named after him, which he designed for his world map of 1569, shows lines of constant bearing ('rhumb lines' or 'loxodromes') as straight lines. The projection is also conformal. In contrast, all equal-area projections destroy angles and shapes.

Mercator most probably derived the idea of increasing the spacing of the parallels on his 1569 world projection not from a mathematical formula (he lacked the necessary secant tables), but by graphically transferring loxodromes from a globe to the plane. An initial step in the truly mathematical consideration of projections was Edward Wright's *Certaine Errors in Navigation* (1599), which explained the mathematics involved in Mercator's projection and presented a table of parallel spacing. Wright's world map on Mercator's projection, published in Richard Hakluyt's *Principal*

Navigations (1599), was to be the basis of many later maps and charts (see §8.18, on navigation).

Topographical mapping also made great progress in the fifteenth and sixteenth centuries as a result of the development of astronomical and mathematical studies at the European universities, especially Vienna, Nuremberg and Freiburg. The Flemish mathematician Reiner Gemma Frisius (1508–55), a student and then lecturer at the University of Louvain, described and presumably invented in 1533 the technique of triangulation. He described and illustrated the principles in his *Libellus de locorum describendorum ratione*, which formed the appendix to his edition of Peter Apian's *Cosmographicus liber* (1533). The technique was essential for large-scale mapping, but only slowly came into use. Philipp Apian (son of Peter) used it for his map of Bavaria, published at Ingolstadt in 1568. The astronomer Tycho Brahe used it for his map of the island of Hveen, Denmark, in 1596, and for a map of Baden around 1600.

6 ENTER THE CALCULUS

The first topographical survey over a whole country was achieved in France. A regional survey of Paris and its environs was undertaken in 1668 under the direction of the Académie Royale des Sciences in Paris, to test the feasibility of the project. The first map of national production in multisheet form was begun in France in 1747, based on the national triangulation 1733–44 by Jacques Cassini and his son César François. Known officially as the 'Carte géométrique de la France' but more popularly as the 'Carte de Cassini', it was finally finished in 1818 (Pelletier *1990*).

For such mapping to match the increasing quality of geographic data (through improved optics and technology) and the increasingly sophisticated knowledge of the Earth's shape (§8.14), new projections were required. These were created through the application of the calculus (§3.2), in particular through the consideration of infinitesimal scale factors (a scale factor is the ratio between distances on the map and on the ground). When calculated along the meridian and parallel at a given point, they define the distortions inherent in the projection in use. Most importantly, integral calculus can transform a desired pattern of distortion into a set of formulas for a new projection.

Consider Johann Heinrich Lambert's derivation of the 'conformal conic' projection, one of seven different projections he described in 1772 and which is still in common use. The term 'conic' indicates that the Earth (taken as a sphere for simplicity) is to be projected onto a cone, which can then be opened out to create the plane of the map; the cone is tangential to the globe along the 'standard parallel' (latitude ϕ_0).

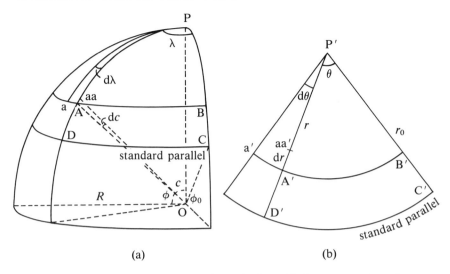

Figure 1 Lambert's conformal conic projection: (a) shows part of a sphere (the Earth), which is projected onto a cone; (b) the cone, rolled out flat, forms the map. P'C' = r_0, P'A' = r and aa'A' = dr

The geometry of a conic projection is illustrated in Figure 1. Point A (ϕ, λ) is projected to point A' (r, θ); small variations in c (colatitude) and λ give rise to small variations in r and θ on the map. The geometry of cones means that the ratio θ/λ is a constant (n), equal to the sine of the latitude of the standard parallel:

$$\theta/\lambda = n = \sin \phi_0; \qquad \therefore \ \mathrm{d}\theta/\mathrm{d}\lambda = n. \qquad (2)$$

From Figure 1, the scale factors along the parallel (k_p) and the meridian (k_m) are

$$k_p = \frac{\text{map distance}}{\text{ground distance}} = \frac{r\,\mathrm{d}\theta}{R \sin \mathrm{d}c\,\mathrm{d}\lambda} = \frac{rn}{R \sin c}, \qquad k_m = \frac{\mathrm{d}r}{R\,\mathrm{d}c}. \qquad (3)$$

For conformality, scalar distortion is constant in each direction, so that angles are not distorted; thus $k_m = k_p$, and so

$$\frac{\mathrm{d}r}{r} = \frac{n\,\mathrm{d}c}{\sin c}. \qquad (4)$$

Integrating (4) from initial values r_0 and c_0 of r and c yields

$$\ln r - \ln r_0 = n\left[(\ln \tan \tfrac{1}{2}c) - (\ln \tan \tfrac{1}{2}c_0)\right] \tag{5}$$

$$\therefore\ r = r_0 \cot^n \tfrac{1}{2}c_0 \tan^n \tfrac{1}{2}c = k_0 \tan^n \tfrac{1}{2}c, \tag{6}$$

with k_0 a constant of the given projection. Also, from equation (2), $\theta = n\lambda$. These polar coordinates (r, θ) are either transformed into Cartesian coordinates or are plotted directly.

Lambert's concerns were taken up by Leonhard Euler. For example, Euler showed in 1778, by a clever proof by *reductio ad absurdum*, that a spherical surface could not be mapped onto a plane in a manner which preserved area (d'Avezac *1863*). He, along with Joseph Louis Lagrange, also studied conformal mappings in ways which relate via the (misnamed) 'Cauchy–Riemann equations' to the later development of complex-variable analysis (§3.12).

7 METHODS IN THE NINETEENTH AND TWENTIETH CENTURIES

The nineteenth and twentieth centuries have witnessed the description of hundreds of new and variant projections (Bourgeois and Furtwängler *1909*), often designed for a specific purpose or a specific ellipsoidal shape of the Earth (building upon Carl Friedrich Gauss's work on spheroidal geometries). Gauss himself described the first ellipsoidal transverse Mercator projection in 1822. This projection, a Mercator projection reoriented so that the line of zero distortion is not the equator but a meridian, is one of the standard projections for topographic mapping today. In contrast, the projection as devised by Mercator is notorious for its extreme distortion of area in medium to high latitudes.

More importantly, the development of statistical analysis and the modelling of error in the nineteenth century has furthered the analysis of map distortions. In 1859 August Tissot presented his 'indicatrix', a graphical means of portraying distortion. An infinitesimal circle on the Earth is projected to an ellipse on the map; the shape of the ellipse indicates the amount of angular distortion at a point, while its size indicates the degree of areal distortion. George Biddell Airy (in 1861) used least-square analysis (§10.5) to produce the first 'minimum-error' projection as a compromise between the large distortions associated with conformal, equal-area and equidistant projections.

Beyond the niceties of the theory of map projections, the cartographer has always had to face certain practical demands. Projections which feature straight lines or circular arcs have always been preferred since they are easily constructed. There is also the issue of how to project geographical features in addition to the meridians and parallels. The basic technique has been first to draw the map's graticule, and then to plot all the points known by latitude and longitude; other data are then inserted by a process of judicious interpolation.

By the eighteenth century this situation began to change with respect to larger-scale maps owing to refinements in the process of triangulation and its growing adoption for surveys of extensive areas. Treatment of a triangulation across a spheroid as opposed to a sphere is a complex procedure. Originally, the triangles were solved on the spheroid, so that the latitude and longitude of each vertex would be calculated. A simpler method was provided by Adrien Marie Legendre (1794), who proved that a spherical triangle may be treated as a plane triangle if a third of its spherical excess is subtracted from each interior angle. This correction converts a spheroidal triangle (assumed to be spherical when small) from a set of arcs in different planes to a plane triangle with the chords of the spheroidal arcs as its sides. Further sophistication came with the modelling of observation errors through triangulation networks and their reduction via least-square analysis.

The extensive triangulations were used for the dual purpose of geodesy and as the basis of the great national topographic surveys. Following the surveys of France by the famous Cassini family (in 1680–1744 and 1747–89), such surveys were rapidly adopted throughout Europe: in Britain (the Ordnance Survey, 1791 onwards), Austria-Hungary (1806–69), the Netherlands (1801–23), Sweden (1805 onwards) and Bavaria (1801–67). For each of these, a projection was chosen (César François Cassini created his own projection for the second survey of France, which was later adopted by the British), and the principal triangulation stations mapped to the plane of projection. All subsequent survey operations, both the lesser triangulation and the detailed survey, were then undertaken on the plane of the map.

In France, cartography developed especially well from the mathematical point of view, and also professionally with the military Corps des Ingénieurs Géographes (Bret *1991*). The leading figure after Cassini (Section 6) was Louis Puissant. He produced widely used textbooks (*Traité de géodésie* and *Traité de topographie*, first editions 1805 and 1807, later editions around 1820) and directed an Ecole de Géographes in Paris from 1809 to 1833. Above all, he invented the projection for the new map of France to replace Cassini's, and was heavily involved in its direction (Berthaut *1898–9*). The journal *Mémorial du Dépôt Général de la Guerre* devoted three volumes to the details (1832, 1840, 1853, the first two edited by Puissant himself). He also chose a conical projection which gave good areal preservation over the range of latitudes covering France. He set the apex along the terrestrial pole, the one mean latitude at 45°, and the rectified longitude through the Paris Observatoire. The map was prepared with a great deal of cartographic and triangulative detail; the projection, especially the calculation of small features when allowing for the spheroidicity of the Earth, drew upon spheroidal trigonometry (helped by Legendre's theorems; Section 6), truncated power-series expansions of variables to the desired level of accuracy, and much differential geometry (§3.5).

The phenomenal growth in the amount of geographic data now available to cartographers has led to increased interest in portraying those data and their errors mathematically. Computed map projections are enjoying a revival, since the computer removes the drudgery of calculation. The modern cartographer needs to be more numerate than ever before. But at the same time, the non-numerical side of mapmaking remains significant. The newest compromise map projection, described by Arthur Robinson in 1974, is based upon aesthetic rather than scientific criteria: its parameters were chosen so as to make the configuration and shape of the continents pleasing to the human eye.

8 THEMATIC MAPS

An important class of maps designed to show distributions was developed in the later part of the seventeenth century, for which N. Creutzberg devised the term 'thematic maps' in 1953. The earliest are the map of the ocean currents by Athanasius Kircher published in his *Mundus subterraneus* (1665), the world map of winds by Edmond Halley (1686), and Halley's map of compass variation in the Atlantic (1701). For this map Halley invented isogones, which he described as 'curve lines' – lines of equal value, along which some mathematical relationship remains constant.

The logical implications of Halley's lines of equal value were not appreciated. Only in 1817 was the concept developed when Alexander von Humboldt employed the technique for isotherms, lines of equal values of temperature. The principle was then rapidly taken up and extended to a wide range of phenomena. The technique was seen as an ideal tool for the developing geographical sciences. Heinrich Berghaus's *Physikalischer Atlas* (1845) and the English version, *The Physical Atlas* by A. K. Johnston (1848), rank as the first comprehensive thematic atlases.

Another type of thematic mapping, developed in the 1820s under the title 'moral statistics', but only recently recognized as a cartographic genre (Wallis and Robinson *1982*: xv), deals with the mapping of social phenomena. Thus Charles Dupin in his *Forces productives et commerciales de la France* (1827) included a map of the distribution of literacy in France.

The French statistician A. M. Guerry (1802–66) in his *Statistique morale de l'Angleterre comparée avec celle de la France* (1864) set out in his introduction a brilliant survey of the development of geographical statistics and their application to mapping.

Thematic maps may be seen as providing a new cartographic language. The 'chloropleth' technique of showing density, for example, was developed into the 'dasymetric' map, in which density values are shown unevenly distributed in an area. Henry Drury Harness showed rural population in a shaded dasymetric style on his population map of Ireland in 1837, made for the *Atlas to Accompany the Second Report of the Commissioners Appointed to Consider and Recommend a General System of Railways for Ireland* (1838).

Thematic cartography in all its varied aspects may be regarded as illustrating the truth of Humboldt's dictum set out in his *Political Essay on the Kingdom of New Spain* (1811): 'whatever relates to extent and quantity may be represented by geometrical figures; and statistical projections which speak to the senses without fatiguing the mind, possess the advantage of fixing the attention on a great number of important facts'. The map in all its varied ramifications is one of the most remarkable coded documents devised by man.

BIBLIOGRAPHY

d'Avezac(-Macaya), M. A. P. *1863*, 'Coup d'oeil historique sur la projection des cartes de géographie', *Bulletin de la Société Géographique de France*, Series 5, 5, 257–361, 437–85. [Also publ. in book form, Paris: Martinet.]

Bagrow, L. *1964*, *History of Cartography* (rev. R. A. Skelton), London: Watts, and Cambridge, MA: Harvard University Press.

Beazley, C. *1904*, 'The first true maps', *Nature*, 71, 159–61.

Berthaut, H. M. A. *1898–9*, *La Carte de France, 1750–1898. Etude historique*, 2 vols, Paris: Service Géographique.

Bourgeois, R. and Furtwängler, P. *1909*, 'Kartographie', in *Encyklopädie der mathematischen Wissenschaften*, Vol. 6, Part 1, 250–300 (article VI 1,4).

Bret, P. *1991*, 'Le Dépôt Général de la Guerre et la formation scientifique des ingénieurs-géographes militaires en France (1789–1830)', *Annals of Science*, 48, 113–57.

British National Committee for Geography *1966*, *Glossary of Technical Terms in Cartography*, London: The Royal Society. [Invaluable list, with historical remarks.]

Brown, L. A. *1949*, *The Story of Maps*, Boston, MA: Little, Brown.

Cortesão, A. F. Zuzarte *1969*, *História da Cartografia Portugesa*, Vol. 1, Lisbon: Junta de Investigaçoes do Ultramar.

Crone, G. R. *1978*, *Maps and their Makers*, Folkestone, Kent: Dawson, and Hamden, CT: Archon Books.

Dilke, O. A. W. *1987, Mathematics and Measurement*, London: British Museum Publications.

Harley, J. B. and Woodward, D. (eds) *1987, The History of Cartography*, Vol. 1: *Cartography in Prehistoric, Ancient and Medieval Europe and the Mediterranean*, Chicago, IL, and London: University of Chicago Press.

Harvey, P. D. A. *1980, The History of Topographical Maps*, London: Thames & Hudson.

International Cartographic Association *1973, Multilingual Dictionary of Technical Terms in Cartography*, Wiesbaden: Steiner.

Kahn, C. H. *1960, Anaximander and the Origins of Greek Cosmology*, New York: Columbia University Press.

Pelletier, M. *1990, La Carte de Cassini*, Paris: Presses de l'Ecole Nationale des Ponts et Chaussées.

Robinson, A. H. *1982, Early Thematic Mapping in the History of Cartography*, Chicago, IL: University of Chicago Press.

Thrower, N. J. W. *1972, Maps and Man*, Englewood Cliffs, NJ: Prentice-Hall.

Wallis, H. M. and Robinson, A. H. (eds) *1982, Cartographical Innovations: An International Handbook of Mapping Terms to 1900*, Tring: Map Collector Publications, in association with the International Cartographic Association.

8.16

The tides

The late E. J. AITON

1 EARLY THEORIES

Pytheas of Massilia (*circa* 325 BC), who sailed possibly as far north as Norway and Iceland, observed the connection between the tides and the Moon, noting the alternation of the spring and neap tides with the phases of the Moon. Posidonius, whose work is known through the quotations from it in Strabo's *Geography*, reported the annual inequality in the tides depending on the Sun's position in the zodiac, though his informants were not very reliable. His report on the diurnal inequality (i.e. the inequality between the two tides on the same day) was more accurate. As a result of the expansion of the Roman Empire beyond the Mediterranean area, the tides were common knowledge among the Romans. Pliny the Elder (AD 23–79), who attributed the tides to the action of the Sun and the Moon, described the principal phenomena in his *Historia naturalis*.

The precise nature of the connection between the tides and the Moon had long been sought when, in the sixteenth century, a revival of interest in the magnetic force provided a clue for Julius Caesar Scaliger, who hinted at the existence of some influence analogous to that of the magnet. Among those in the seventeenth century who turned their minds to the problem of the tides before Isaac Newton, only Johannes Kepler pursued Scaliger's suggestion of some kind of attraction, analogous to that of the magnet, between the Moon and the ocean. In the introduction to his *Astronomia nova* (1609), Kepler explained the tides as an oscillation maintained by the attractive force of the Moon.

Galileo Galilei believed that the tides could be explained by the motions of the Earth on its axis and around the Sun, and in no other way (Riccioli *1651*). So he held that his theory of the tides demonstrated the Earth's double motion and hence also the Copernican system. The theory was improved by John Wallis and by G. B. Baliani to give a better representation of the phenomena. It was finally disproved in an essay on the tides published by Daniel Bernoulli in 1740, who showed that, from the point of view of Newtonian mechanics, the motion of the Earth around the Sun could not give rise to a motion of the sea relative to the Earth. Another influential theory was that of René Descartes, based on his theory of vortices. This theory, with modifications, was defended up to the time of Antoine Cavalleri's prize-winning essay of 1740.

2 NEWTON

The first moderately successful attempt to relate the tides to the forces of the Sun and the Moon causing them was made by Isaac Newton in his *Principia* (1687) (Aiton *1955*, Proudman *1927*). He began by considering a canal encircling the Earth along the equator and applying to the particles of water in it the laws he had deduced for the motion of a satellite disturbed by a third body. At this point, however, Newton made the transition to the equilibrium theory, according to which the water was supposed to be in equilibrium with the disturbing forces. Assuming the figure of the Earth to be that adopted by a fluid of uniform density in equilibrium under the forces of attraction and rotation, and making use of his calculation of the attraction of a spheroid, Newton deduced that the Earth is larger at the equator than at the poles by 85 472 feet (26 052 metres). As this difference is the result of the centrifugal force at the equator arising from the Earth's rotation, $g/289$ (expressed in foot-poundals), he concluded that the force of the Sun to move the sea, $g/12\,868\,200$, would account for less than two feet (0.6 metres). The ratio of the masses of the Earth and the Moon being unknown, Newton could not calculate directly the force of the Moon to move the sea. So, to estimate the lunar tide-generating force he used the ratio of the solar and lunar tides, which he estimated from the systematic observations by Samuel Sturmy at Bristol and published in the *Philosophical Transactions of the Royal Society* in 1668. Newton's estimate of the ratio of the lunar and solar tide-generating forces was much too large. The correct value is about 2.3 : 1, approximately half Newton's estimate.

The anomalous tides at the Bar of Tunquin (now called the Gulf of Tonkin) in China, reported by Edmond Halley in the *Philosophical Transactions* for 1684, presented a problem to which Newton offered an interesting solution. According to the report, there was just one high water in 24 hours, and in each month there were two intermissions, about 14 days apart, when no tide was observed. This happened about the time when the Moon passed near one of the equinoxes. Newton explained that two equal tides flow towards the Bar of Tunquin through equal channels, one preceding the other by six hours. Consequently the semi-diurnal oscillation is destroyed according to the principle of interference of waves, leaving only the diurnal oscillation, which also vanishes when the Moon is near the equator, so that the intermissions occur at these times.

3 THE PRIZE ESSAYS OF 1740

In the early years of the eighteenth century the Academy of Sciences in Paris promoted systematic observations of the tides at several Continental ports. Then, in 1738, the Academy proposed the causes of the tides as the subject for a prize essay. In 1740 the prize was divided between Cavalleri, Colin Maclaurin, Daniel Bernoulli and Leonhard Euler (*Recueil des pièces 1752*). Although Cavalleri's essay was based on Cartesian principles, it followed a recently established tradition of assimilating Newton's mathematical results. The other essays were based firmly on the principle of universal gravitation. Maclaurin's included an elegant geometrical demonstration of the theorem that, under the force of the Sun or the Moon, a homogeneous fluid sphere would assume the form of a prolate spheroid having its axis directed towards the disturbing body. He was also the first to describe and explain the deflection towards the right in ocean currents along a meridian in the northern hemisphere of a rotating Earth. Daniel Bernoulli developed the equilibrium theory to give it a practical value in predicting the tides at a given port. The tables in the essay give the time and relative heights for the various phases and distances of the Moon. Observing that the period (but not the amplitude) of the oscillations of a pendulum responds immediately to a change in the force causing the motion, he recommended that observations of the duration and intervals of the tides provided a better basis for the comparison of the lunar and solar tides than observations of the heights employed by Newton. Using this principle, he estimated the mean lunar tide to be two and a half times the mean solar tide, a much more accurate value than that obtained by Newton.

Euler's prize essay contains two significant original contributions to the theory of the tides. The first was the recognition that the horizontal components of the disturbing forces are the effective tide-generating forces; the second was a direct analytical approach in place of Newton's argument by analogy with the problem of the figure of the Earth. Euler determined the equilibrium figure of the ocean under the action of the Sun by composing the condition that the resultant force is perpendicular to the surface of the water. He neglected the attraction of the water in the two high-water regions, an approximation that greatly simplifies the calculation; however, the effect is not negligible, as Bernoulli pointed out to him.

Besides his formation of the equilibrium theory, Euler also attempted what he described as a mathematical theory, though in this he was less successful, not having yet worked out his theory of the motion of fluids.

Neglecting the horizontal motion and considering only the vertical motion, he supposed that each particle of water tends to return to the vertical position of equilibrium with a force proportional to the vertical displacement from this position. Combining this force with the disturbing force of the Sun or Moon, he established and solved the second-order differential equation for this forced harmonic oscillation. Euler also demonstrated that the solar and lunar tides could be added together, a result that Newton had simply assumed, and he emphasized the need for an acceptable theory of the tides to be able to account in detail for the observations that had been collected and systematized, especially those recorded in the first decades of the eighteenth century. The success of his own theory in this respect, he claimed, confirmed its validity.

Euler made a further contribution to the theory of the tides in a memoir presented to the St Petersburg Academy of Sciences in 1775. Using a principle he had established in a memoir presented to the Berlin Academy of Sciences in 1755, he derived the equilibrium state of the ocean from the condition that the potential is constant over the surface of the water.

4 LAPLACE

Pierre Simon Laplace succeeded in arriving at the equations for the tidal oscillations in memoirs presented in the Paris Academy of Sciences in 1775 (Harris *1898*). The first equation is the equation of continuity, expressing the constancy of the total mass of water in the ocean. The second equation is the equation of motion, expressing the relation between the horizontal components of the disturbing forces and the accelerations of the water produced by these components. The disturbing forces are the forces of the Sun and Moon and the attraction of the heaped-up water. Laplace analysed the tidal oscillation into its three principal harmonic constituents, taking account of the effect on the tides of the depth of the ocean and the rotation of the Earth.

The general integration of the equations of motion and continuity presents many difficulties, and Laplace confined his attention to the case in which the depth of the ocean is assumed to be constant along parallels of latitude. He analysed the total effect into a sum of three separate terms, giving rise to three different oscillations which coexist without mutual interference. The oscillations of the first kind produce the annual and monthly inequalities due to the apparent motion of each of the disturbing bodies in its orbit. The oscillations of the second kind have a period of about a day, so that, superimposed on the chief semi-diurnal oscillation, they give rise to the diurnal inequality in the tides. The theory predicts the reversal of the diurnal inequality which is observed to take place when the disturbing body crosses the equator. With the assumption that the ocean has a depth of 1.4 miles (2.25 kilometres), the theory predicts an inversion of the tide at the eighteenth degree of latitude, as is observed to take place.

As one of Laplace's greatest achievements was his proof of the stability of the orbits of the planets (§8.8), it is not surprising that he devoted some attention to the problem of the stability of the oscillations of the oceans. He succeeded in showing that, whatever the depth of the sea and the speed of rotation of the Earth, the sea is stable provided its density is less than the mean density of the Earth.

The irregularities of the depth of the ocean and of the local coastline caused modifications that Laplace could not derive by calculation. Consequently, to complete the theory for particular ports, he had to combine theory with observation. For this purpose he used the general principle of dynamics that the state of a system in which the initial conditions of movement have been destroyed by friction has the same period as the forces that disturb it. Therefore the amplitudes and phases of the tides had to be determined by observation.

5 THE NINETEENTH AND TWENTIETH CENTURIES

In the first part of the nineteenth century, J. W. Lubbock and William Whewell made important contributions to the practical problem of the tides by coordinating and analysing enormous masses of data for various ports, and using the results for the compilation of tide tables. Whewell also advanced a progressive wave theory according to which a forced wave originating in the southern ocean dominates the tides of the world, giving rise to progressive waves moving northwards through the various oceans. George Biddell Airy (1842) studied waves in estuaries and considered the effects of frictional resistances on the progress of tidal and other waves, while George Darwin (1879–82) made extensive researches into the theory of tidal friction.

The interesting phenomenon of the oscillations in Lake Geneva were studied in great detail by the Swiss naturalist F. A. Forel. These oscillations consist of stationary waves.

At the turn of the century, R. A. Harris *1898* advanced the hypothesis that the genuinely tidal waves were also stationary oscillations. This theory replaces the idea of the tides as a single worldwide phenomenon by the idea of regional oscillatory areas in various parts of the oceans as the origin of the dominant tides, the oscillations being maintained by the periodic tidal forces of the Sun and the Moon. A region is divided into a continuous series of basins whose natural periods correspond to the tidal period. Owing to the rotation of the Earth, the nodal lines become points about which the vertical oscillations appear to circulate. North of the equator the rotation is generally clockwise, with anticlockwise rotation in the southern hemisphere. The charts produced by Harris were much in advance of any existing at the time, and his work was a great stimulus to others. For the first time the predominance of the oceanic circulatory systems just described, or 'amphidromic systems' as they are called, was made evident. In 1920–21, charts for the semi-diurnal tide were given by R. Sterneck for all oceans of the world (Doodson *1958*). For many regions his charts were in broad agreement with those of Harris. For example, both agreed that in the North Atlantic semi-diurnal tides could be represented by an amphidromic system in the main part of the basin, with small systems on the outskirts; but they differed over the assignment of the centres or amphidromic points, and over the distribution of the co-tidal lines. Co-tidal charts for all the oceans were published in 1944 and 1951 by G. Dietrich and C. Villain, both of whom had more material than had been available to their predecessors.

In recent years, finite-difference methods have been used in solving the differential equations. Such methods were first applied to small seas for which the tides were well known. W. Hansen applied them to the North Sea with considerable success, and then applied the same methods to the North Atlantic, finally extending his analysis to the whole Atlantic.

BIBLIOGRAPHY

Aiton, E. J. *1955*, 'The contributions of Newton, Bernoulli and Euler to the theory of the tides', *Annals of Science*, **11**, 206–23.

Doodson, A. T. *1958*, 'Oceanic tides', *Advances in Geophysics*, **5**, 117–52.

Harris, R. A. *1898*, *Manual of Tides*, Part 1, Washington, DC: United States Coast and Geodetic Survey.

Proudman, J. *1927*, 'Newton's work on the theory of the tides', in W. J. Greenstreet (ed.), *Isaac Newton 1642–1727*, London: Bell, 87–95.

Recueil des pièces qui ont remporté les prix de l'Académie Royale des Sciences, *1752*, Vol. 4, Paris.

Riccioli, G. B. *1651*, *Almagestum novum*, Bologna: Benatij.

8.17

Shipbuilding and ship operation

I. GRATTAN-GUINNESS

1 TERMS OF REFERENCE

A wide range of mathematical and scientific topics and problems has fallen under this heading. One can mention the basic design of the lines of the ship; the design and placing of masts and sails; the flexural properties of wood (and later, iron and steel) that have to be taken into account when laying down the keel and building the frame and superstructure; the estimation of the stresses and strains that they would sustain when the ship is in motion; quantity surveying to measure and cost the amounts of material required; the measure of strength required to withstand military attack; the volume of the hull; the functioning of pulleys; the flexure and strength of ropes, and even the proto-topology of special knots; methods of safe launching; the resistance of water to the motion of the vessel; theories of pitch, roll and yaw; the effects of two of the most unpredictable of forces, the wind and the motion of the sea; the safe loading of a ship with military or civil cargo; the action of oars and rudders and, later, of engine-driven propeller or paddle-wheel (and their efficiency); the effects of vibration caused by engines; the use and role of scale models; the design of ports, docks and quays...

Following common practice, here this ensemble of topics is simply called 'shipbuilding' (the expression 'naval architecture' has also often been used), and vessels in general are referred to as 'ships'. Clearly, a ship is a scientifically complicated object, even before it enters the water: many variables obtain, and the relationships between them are not clear. Thus simplified theories and/or approximate methods have always had to be used.

The history of shipbuilding is very ancient (see e.g. Abell *1948*); but much of it is hard to detect. For ancient civilizations the developments are as obscure as any other aspect, and as denuded of records. But even within the Western traditions since the seventeenth century (the period upon which this article concentrates), early records are often scanty. Various procedures were followed, but the masters did not publish them – maybe for reasons of professional secrecy or fear of commercial exploitation (e.g. with the design of sails).

Thus, especially for the beginning of our period, it is hard to appraise the extent to which mathematical *theory* based on proportions was used; that is, something beyond empirical rules with a mathematical ring to them. For periods for which evidence is available, however, it appears that mathematical (and scientific) theories were usually being applied to shipbuilding (as in most trades) after being developed elsewhere. This is clear especially from the eighteenth century onwards, when marine schools were established in many leading countries and some mathematical shipbuilding was taught (together with topics treated in neighbouring articles in this encyclopedia, such as cartography and navigation).

Whether used explicitly or intuitively, the principal, pertinent branches of mathematics were in mechanics, including hydrostatics, hydrodynamics, elasticity theory and structures. Various aspects of geometry were also used, especially the descriptive and differential modes (§7.5, §3.4), the latter as part of the general use of the calculus. In addition, a variety of numerical methods were used, including even the compilation of tables of various kinds, and topics such as the approximation of the roots of equations (§4.10), together with graphical presentations of information *à la* nomography (§4.12).

2 FROM HARRIOT TO EULER

From the fifteenth century onwards, the main countries for shipbuilding were Britain, Holland, Spain and Portugal. However, despite the importance of the industry, many of the best technical accounts remained in manuscript, and some designs are only partially known (e.g. Christopher Columbus's caravelle).

One of the first comprehensive mathematical studies was prepared around 1610 by the English polymath Thomas Harriot (Pepper *1978*: Chap. 9). It was not published, but its details may have circulated among contemporary shipbuilders (such as Matthew Baker). Harriot laid out the midship elevation (or 'mould') of the ship as a sequence of arcs of circles, in various radial proportions; but he allowed for cubic and quartic curves in designing this and other moulds, and also in laying out the lines of breadth of the ship and of the floor of the hold (vertical longitudinal elevation) and lines of beam and floor (horizontal section). With these procedures he built upon known practices; but he appears to have introduced the idea of relating the breadth of the ship to the height of the mainmast by a hyperbola.

In general, seventeenth-century manuals on shipbuilding continued to use empirical rules or elementary geometrical routines for the basic design; P. Hoste was a well-known author in this tradition. The question of stability was more demanding, but a line of reasoning developed from Archimedes' insight about displacement (§1.3). Starting with a suggestion by Christiaan Huygens in 1695 that stability could be likened to the motion of a compound pendulum, through contributions from John Bernoulli and others to Pierre Bouguer's *Traité du navire* (1746), some essential distinctions emerged. Whereas some held that the centre of gravity G of the ship should lie below the centre B of buoyancy of the displaced water for stability to obtain, Bouguer made it clear the role of the metacentre (his word) M, the point of intersection of the axis of symmetry of the ship and the limiting position of the vertical through B as the angular displacement of the ship, tended to zero; if it fell below G, the ship would be unstable when floating upright (Gille *1958*).

Bouguer also gave extensive accounts of the design of ships, the effects of wind upon sails, the distribution of loads, and so on; but the most significant contributor in the eighteenth century was Leonhard Euler. In his two-volume treatise *Scientia navalis* (1749) and the more elementary *Théorie complette de la construction et de la manoeuvre des vaisseaux* (1773), he treated most of the problems listed at the head of this article (Habicht *1974, 1978*). He gave these matters further attention in a number of papers, some of which were written in response to prize problems set by the Paris Academy of Sciences, in which another participant was Daniel Bernoulli.

In the *Scientia navalis*, Euler expounded the general principles of the stability of floating bodies, and applied them to ships. Among his main contributions three may be stressed. First, he used the calculus of variations, then a rather new subject (§3.5), to determine the optimal profile of the ship according to an assumed law of water resistance. Second, he used the basic equations of hydrodynamics (§8.5) to analyse the basic motion of the ship and the effects of recoil. And third, his analyses gave him insights into the basic dynamics of rigid bodies which would culminate in 1752 in the 'Euler equations' for the rotation of a rigid body. In addition, in particular cases he made great use of the composition of forces, in order (for example) to determine kinematically the direction and manner in which a ship would proceed under the action of forces of steerage, wind and water. He also studied the consequences of different shapes of the rudder, and the use of more than one, mainly by forming equations of moments.

Euler's contributions form a massive and impressive achievement. However, they seem not to have become general knowledge among shipbuilders or marine scientists, although in the late 1790s the Englishman G. Atwood extended them to the stability of ships with large angles of roll (where the notion of metacentre is less effective).

3 THE NINETEENTH CENTURY

In the eighteenth century, France created various marine and hydrographic schools (see F. Russo and P. Gille in Taton *1964*). In this context C. Bossut was an influential teacher, and the much reprinted *Cours des mathématiques* by Etienne Bézout was originally published (in 1764–69) for use in marine schools.

With the expansion of education after the French Revolution (§11.1), marine engineering was favoured, and specialist schools in marine engineering and ship construction were (re)established, usually in port towns. However, few major mathematicians came from the marine corps. The most prominent was Charles Dupin. In his books *Développements de géométrie* [...] (1813), *Mémoires sur la marine* [...] (1818) and *Applications de géométrie et de méchanique à la marine* [...] (1822), he applied descriptive geometry (of which he was an ardent advocate) to study in detail the relationships between the metacentre and the centre of buoyancy, including for ships with a non-symmetric profile. (From this time descriptive geometry became a standard guide for the preparation of design drawings for ships.) Dupin also found that the flexure of timber varied non-linearly with the loading, reviewed the current practices in British shipbuilding, summarized the needs of 'military naval architecture', and (perhaps from noting the creation of a School of Naval Architecture at Portsmouth in 1811) fought, albeit unsuccessfully, for the restoration in France of the eighteenth-century Académie de la Marine.

The introduction of metal ships and ship engines during the first half of the nineteenth century changed the character of some of the problems, and even introduced new ones, such as the use of kinematic stress diagrams (§8.3) to analyse the strength of designs (see Harvey *1836* for a typical 'transitional' survey). Lloyd's Register of British and Foreign Shipping, founded in 1834, played an important role worldwide in promoting iron ships and in relating safety and operation to the fundamental principles of shipbuilding.

The growth of energy physics in the mid-nineteenth century (§9.5) led to a somewhat greater use of energy-conservation equations. A typical author is W. J. M. Rankine, with his textbook *Shipbuilding Theoretical and Practical* (1866) written with colleagues; he was also an active early member of the Institution of Naval Architects, a typical example of a professional body for shipbuilding and related topics, which was set up in 1860 and at once published a survey on current mathematical theory (Woolley *1860*).

Among other British workers, from the 1850s W. Froude carried out important mathematical and experimental investigations of the effects of water resistance on the stability, pitching and rolling of large ships; his work was partly stimulated by problems posed in the design of the *Great Eastern* built by Isambard Kingdom Brunel. His work led him to pioneer the construction of special tanks in which small-scale models could be tested, and to propose laws relating the results determined from these models to effects found with real ships, especially concerning the properties of resistance. Among his theoretical contributions, he analysed rolling by making certain assumptions about the position of stability of the ship in the water, approximating to the wave surface by a finite trigonometric series, taking rolling to occur with pendular regularity, and thereby finding variants of the wave equation to represent the inclinations of the masts to the vertical. By considering laws of water resistance, he also analysed the form of waves which a moving ship created by its motion.

A rather different perspective was cast by J. Scott Russell, who had somewhat difficult relations with Brunel. His *The Modern System of Naval Architecture* (three gigantic volumes, 1865) gave small place to mathematical theory, but was distinguished by many fine illustrations, some showing good use of descriptive geometry. In France at this time, F. Reech made extensive studies of the effect on the metacentre of different loadings of a ship. In the USA, the nature of the Mississippi river forced attention upon the design and operation of shallow-draught paddle-steamers.

4 THE EARLY TWENTIETH CENTURY

By the end of the nineteenth century the power of engines increased concern about the effects of vibrations, in all three axial directions of the ship. Interpreting them as quasi-acoustic effects, cousins to the wave equation or the fourth-order equation for the ringing of bells could be deployed and solved by non-harmonic trigonometric series (§9.4).

At this time the field became well established internationally (Kriloff and Müller *1907*, Watts *1911*). Compendia such as J. Pollard and A. Dudebout's *Théorie du navire* (four volumes, 1890–94) provided comprehensive up-to-date surveys (including of mathematical aspects), and also a fine bibliography; and *Jane's All the World's Fighting Ships* began to appear in 1898. The First World War extended the shipbuilding industry considerably, although the main novelties concerned the scale and speed of production, and the further and rapid development of special kinds of ship (the submarine, to take an important example) and special purposes such as the firing of torpedos. After the war, both civil and military concerns went for ever bigger ships, the luxury of the one genre contrasting with the power of the other.

BIBLIOGRAPHY

Abell, W. *1948*, *The Shipwright's Trade*, Cambridge: Cambridge University Press.

Gille, P. *1958*, 'Les mathématiques et la construction navale', in *Actes du VIIIe Congrès International d'Histoire des Sciences*, Paris: Hermann, 57–63.

Habicht, W. *1974*, 'Einleitung', in L. Euler, *Opera omnia*, Series 2, Vol. 20, Zürich: Orell Füssli, vii–lx.

—— *1978*, 'Leonhard Eulers Schiffstheorie' and 'Einleitung [. . .]', in L. Euler, *Opera omnia*, Series 2, Vol. 21, Zürich: Orell Füssli, vii–cxcvi, cxcvii–ccxlii.

Harvey, G. *1836*, 'Naval architecture', in *Encyclopaedia Metropolitana*, Vol. 6, 329–424. [Dating conjectured. Mostly British material.]

Kriloff, A. and Müller, C. *1907*, 'Die Theorie des Schiffes', in *Encyklopädie der mathematischen Wissenschaften*, Vol. 4, Part 3, 517–93 (article IV 22).

Meyer, H. B. (compiler) *1919*, *List of References on Shipping and Shipbuilding*, Washington, DC: United States Government Printing Office. [Valuable list, including historical works.]

Pepper, J. V. *1978*, 'Studies of some of Thomas Harriot's unpublished mathematical and scientific manuscripts', Ph.D. Thesis, University of London.

Taton, R. (ed.) *1964*, *Enseignement et diffusion des sciences en France au XVIII siècle*, Paris: Hermann. [Repr. 1986.]

Watts, P. *1911*, 'Ship' and 'Shipbuilding', in *Encyclopaedia Britannica*, 11th edn, Vol. 24, 860–922, 922–81.

Woolley, J. *1860*, 'On the present state of the mathematical theory of naval architecture', *Transactions of the Institution of Naval Architects*, 1, 10–38.

8.18

Astronomical navigation

DEREK HOWSE

1 INTRODUCTION

Italian merchants and seamen were the first professional groups to use mathematics in their everyday work, from the early thirteenth century. For seamen, courses and distances sailed (as measured by magnetic sea compass and sandglass) were written down in Hindu–Arabic numerals, and geometry and trigonometry were used to calculate courses and distances to be sailed to reach the intended landfall when winds were adverse. In the virtually tideless Mediterranean, such calculations were sufficiently reliable for use with a geometrically constructed sea chart (a 'portulan chart'), drawn to scale from magnetic courses and estimated distances, with directional lines and distance scales in miles (Taylor *1959*).

In his popular *Practical Navigation*, John Seller (*1694*) linked navigation and mathematics:

Navigation (that useful part of the Mathematicks) is a Science which has been highly valued by the Ancients, especially by our ancestors of this island...
... the Art of Navigation... which guides the Ship in her Course through the Immense Ocean, to any part of the known World, which cannot be done unless it be determined in what place the Ship is at all times, both in respect to Latitude and Longitude; this being the principal Care of the Navigator, and the Master-Piece of Nautical Science.
To the commendable Accomplishment of which Knowledge, these four Things are subordinate Requisites:

VIZ. $\begin{cases} \text{Arithmetick} \\ \text{Geometry} \\ \text{Trigonometry, and} \\ \text{The Doctrine of the Spheres.} \end{cases}$

The 'Doctrine of the Spheres' concerned astronomical navigation, first developed through the initiative of Portugal's Prince Henry the Navigator to enable his seamen to find more exactly their position in the ocean, where currents of unknown velocity falsified their 'dead reckoning' – the navigator's best estimate of his position from courses steered and estimated distances run, taking into account prevailing wind and weather and experience of the ship's performance. By the 1420s, the Portuguese had developed 'altitude navigation', correcting their north–south position by measuring the altitude of the north celestial pole. At first, they observed with a quadrant the angular height of the pole star, Polaris, above the horizon and applied rules to allow for the fact that it was at that time some $3\frac{1}{2}°$ from the celestial pole itself. However, as they approached the equator (reached in 1474), they saw Polaris sink lower in the sky until, near Sierra Leone (latitude 9°N), it became navigationally useless. From 1485, now exploring south of the equator, they learned from astronomers how to find their north–south position at sea by observing the Sun (as described in Section 3), and using the sea astrolabe (*circa* 1485) and cross-staff (*circa* 1515). But there was to be no practical method of measuring longitude at sea until the 1760s.

European seamen learned the art of oceanic navigation primarily from Portuguese-derived Spanish manuals, translated into English in the 1550s. Thenceforth, for more than two hundred years, it was English mathematicians and seamen who made many fundamental navigational and hydrographic improvements, particularly after the foundation of Gresham College in London (1598) and the Royal Observatory at Greenwich (1675).

By the 1570s, the English had devised the log and log-line to measure a ship's speed (they would remain in use into the twentieth century); and Leonard and Thomas Digges, with John Dee, had initiated mathematical navigation, for which Captain John Davis developed the columnar logbook (1580s), still in use, which systematized navigational entries and facilitated calculation. Davis's quadrant or backstaff (1595) improved the accuracy of solar observations, displacing the sea astrolabe and continuing in use until superseded in turn by John Hadley's reflecting quadrant (1731) and Captain Campbell's sextant (1756).

For compasses, Robert Norman's discovery of magnetic dip and William Borough's magnetic-variation studies (1581) inspired William Gilbert's *De magnete* (1600). Henry Gellibrand's discovery of the secular change of variation (1631) improved the accuracy of course-setting.

In 1594, Robert Hues published the six fundamental navigational propositions involved in solving what came to be known as the 'nautical triangle' (Figure 1), involving course, distance, latitude and departure (or longitude). In 1614, Ralph Handson published the trigonometrical solutions to the triangle, and the mid-latitude formula from which the triangle could be solved arithmetically. John Tapp's concurrently published quarto tables of natural sines, tangents and secants, with examples of navigational use, made 'arithmetical navigation' generally practicable at sea. John Napier's

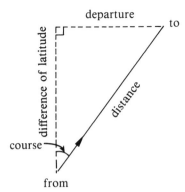

Figure 1 The nautical triangle. Its components are:

Plane sailing		*Other sailings*	
Course	Co.	Difference of longitude	D.long.
Distance	Dist.	Difference of	
Difference of latitude	D.lat	meridional parts	D.mer.parts
Departure	Dep.		

logarithms for numerals (1614), which Edward Wright adapted to base 10 for the use of seamen (1615), together with Edmund Gunter's logarithms for trigonometrical functions (1623), revolutionized calculations by reducing them to simple addition and subtraction (§2.5). Already, instruments such as Thomas Hood's sector (1598), John Speidell's plain scale (1607) and Edward Gunter's sector (*circa* 1607, 'the Gunter'; still in use in the late eighteenth century) made rapid navigational calculation practicable. Gunter's traverse tables and wooden logarithmic scale (1623) and Richard Delmaine's (1630) and William Oughtred's (1632) slide rules – only superseded by the electronic calculator – made calculation even easier and faster.

Wright's introduction of the Mercator chart (§8.15) in 1599 was of fundamental importance for navigation over long distances and in the higher latitudes because it takes into account the spherical shape of the Earth. The earlier plane (or portulan) chart did not do this, so that the distances between the lines of latitude were correctly spaced but those between lines of longitude – which are shown as parallel whereas in fact they converge – are progressively increased in the proportion of the secant of the latitude: by $60°$ latitude the distance between meridians on the chart was twice as large as it should be (since $\sec 60° = 2$). This did not matter when only a small area was concerned, when near the equator, or when voyages were primarily east–west, as in the Mediterranean. But in higher latitudes and over larger areas like the North Atlantic, the consequence for the navigator was that, on the plane chart, the proportional relation of northing (latitude) to easting (longitude) was everywhere falsified, and any course laid off (except due north, south, east or west) was incorrect.

Mercator's remedy was to falsify the spacing of the lines of latitude in exactly the same proportion as that of the plane chart's parallel lines of longitude, making all angles correct and courses laid off consequently true. However, the scale differs from latitude to latitude, increasing progressively as the table of secants, which many simple sailors found difficult to understand (Taylor *1956*: 222–3). In the event, despite the advice of the writers of navigation manuals, the plane chart continued to be used at sea until well into the nineteenth century.

2 LATITUDE-FINDING BY ASTRONOMY

At any place, the altitude of the celestial pole above the horizon is precisely equal to the latitude of the place (Figure 2). As we saw in Section 1, the earliest measurement of latitude at sea made use of observations of Polaris, but this cannot be done as the equator is approached. From 1485, Portuguese astronomers therefore taught their pilots to find their latitude by measuring the Sun's altitude at noon (i.e. at meridian passage), and then to use declination tables, tabulating for each day of the year the Sun's angular distance north or south of the celestial equator.

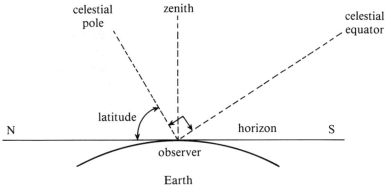

Figure 2 The observer's latitude equals the altitude of the celestial pole. The diagram is in the plane of the observer's meridian

As can be seen from Figure 3, the formula

$$\text{latitude N}(+) \text{ or S}(-) = \text{zenith distance} \pm \text{declination} \qquad (1)$$

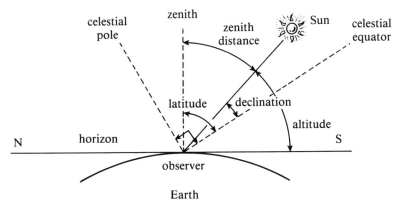

Figure 3 The altitude of a celestial body at its meridian passage

(where zenith distance is 90° − altitude), requires only simple addition or subtraction, and care in applying the signs, for which simple rules were used.

By the mid-eighteenth century, refinements such as tables allowing for terrestrial refraction and dip (correcting observations to sea level) were produced (Cotter *1968*: 97−117); and methods were evolved whereby observations for latitude could be made at times other than at meridian passage (*Ibid.*: 165−79). Latitude is measured in much the same way today.

3 THE SAILINGS AND DEAD RECKONING

The 'sailings' were the different ways in which the path of a ship at sea could be represented on paper, either graphically or by calculation. In essence, what navigators did, in various degrees of sophistication, was to solve the problem, 'given the present position and the courses and distances sailed thereafter, find the final position', or its converse, 'given the present position and the destination, find the course and distance'. Today, these problems would generally be solved by plotting the tracks on a chart, but in the seventeenth and eighteenth centuries this was often done by trigonometry, using either tables or 'the Gunter'.

3.1 Plane sailing

Plane (or, erroneously, 'plain') sailing resolves the ship's actual track into difference of latitude, and departure (the linear distance in miles sailed east and west). The assumption is made that, over the area concerned, the Earth is flat, so that meridians were assumed to be parallel (whereas in fact they

1131

converge). The plane right-angled 'nautical triangle' (Figure 1) comprises course, distance, difference of latitude and departure. Given any two of these, the other two can be found by plotting or calculation using tables of sines, tangents and secants.

3.2 Traverse sailing

Except when running before the wind, the ship has to sail a zigzag course to make good a given track (Figure 4). The resultant course and distance

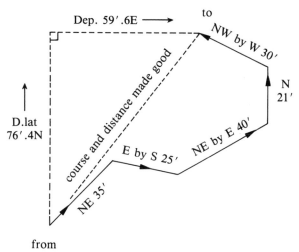

Figure 4 Traverse sailing. A day's run of traverses resolved into northings/ southings and eastings/westings from published traverse tables (nautical miles indicated by primes in the diagram):

Course	Dist. (nautical miles)	D.lat N	S	Dep. E	W
NE	35	24.7	–	24.7	–
E by S	25	–	4.9	24.5	–
NE by E	40	18.9	–	35.3	–
N	21	21.0	–	–	–
NW by W	30	16.7	–	–	24.9
		81.3N	4.9S	84.3E	24.9W
	less	4.9S		24.9W	
Resultant day's traverse		76.4N		59.6E	

1132

'made good' are found by 'resolving a traverse', by plotting or calculation. 'Traverse Tables' (Gunter, 1623) tabulated difference of latitude and departure for various courses and distances.

3.3 Middle-latitude sailing

Departure can be converted into difference of longitude for use on a Mercator chart by the middle-latitude formula

$$D.long. = Dep. \times \sec(mid.\text{-}lat.), \tag{2}$$

where mid.-lat. is the mean of the latitudes of the start and finish positions.

3.4 Mercator sailing

By using the principles of the construction of the Mercator chart, Mercator sailing gave a more accurate answer than middle-latitude sailing when applied to problems of plane or traverse sailing. 'Meridional parts' (mer.parts), tabulated in nautical tables, are the lengths of the arc of the meridian between the equator and a given parallel on a Mercator chart, expressed in units of 1′ of longitude on the equator (Figure 5). The method makes use of the formula

$$D.long = D.mer.parts \times \tan(course). \tag{3}$$

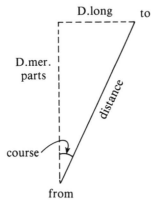

Figure 5 Mercator sailing

4 LONGITUDE-FINDING BY ASTRONOMY

The theory of finding longitude astronomically was known in Classical times, the concept being inexorably bound up with the rotation of the Earth on its axis – one complete revolution ($360°$) in a day, $15°$ in an hour. To find the difference of longitude between any place and some reference or 'prime' meridian (say Greenwich, used as such since the 1760s), all that is needed is the local time of the place (which can be measured by a handy sundial) and the local time on the prime meridian *at that very same moment*. The difference between these two times (say 1 hour) is the difference of longitude ($15°$).

Finding local time astronomically is not difficult, though something rather more refined than a sundial is needed. The difficult part is knowing Greenwich time at that same instant. One method (proposed by Gemma Frisius in 1530) is for a ship to have a timekeeper which keeps Greenwich time, but producing such a machine – capable of keeping accurate time for months, even years, in every climate and in a ship which rolls, heaves and pitches – long defied contrivance.

In the seventeenth and early eighteenth centuries, the astronomical longitude-finding method which seemed to offer most promise at sea depended upon the fact that the Moon moves comparatively rapidly against the background stars – approximately its own diameter ($1°$) in 1 hour. Thus the Moon may be considered as the hand of a clock, and the Sun – or stars along the zodiac – as time-markers on the clock dial. As long ago as 1514, Johann Werner had suggested that predicted angular distances between the Moon and these time-markers could be used to give in effect Greenwich time. But this would be useful at sea only if these lunar distances, as they were called, could be predicted accurately for years ahead, so that the navigator could have them when he started his voyage.

However, the going of that Moon clock is highly irregular and, in the seventeenth century, neither the laws governing the Moon's apparent motion nor the relative positions of the star time-markers with respect to one another were known to the required accuracy. Indeed, it was to provide these very data that the Royal Greenwich Observatory was founded in 1675. Then, in 1714, the British government offered enormous rewards to anyone, of any nationality, who could discover a method for determining longitude: £10,000 if accurate to 60 miles, £15,000 to 40 miles and £20,000 to 30 miles, with smaller rewards also.

These enormous prizes caused mathematicians such as Jean d'Alembert, Alexis Clairaut and Leonhard Euler to turn their minds to the laws governing the Moon's motion (using observational data provided by astronomers such as James Bradley, Pierre Charles Lemonnier and

César-François Cassini de Thury) so that the navigator might have predictions of the Moon's position in lunar-distance form (§8.9). The person who finally produced lunar tables of the required accuracy was the German astronomer Tobias Mayer, using equations produced by Euler, the former's heirs receiving in 1765 a reward of £3000, the latter, £300. (Eight years later, John Harrison was to receive from Parliament the last instalment of the £18,750 reward for his time-keeper – in 1773, when he was 80.) This led to the publication in 1767 of the first annual *Nautical Almanac*, edited by Astronomer Royal Nevil Maskelyne (himself a distinguished mathematician and practical navigator), in which were included lunar distance tables for every three hours throughout the year. At the same time, Maskelyne published the *Tables Requisite* [. . .], containing data which did not change from year to year, or (like star positions) changed only very slowly.

Precomputed lunar distance tables had originally been suggested by the French astronomer Nicolas Louis de Lacaille, and the idea was enthusiastically put into execution by Maskelyne, who gave full credit to Lacaille (who had died in 1762). The many wars between France and England failed to stop the closest cooperation between astronomers and mathematicians of the two countries in these matters. The French official almanac *Connaissance des temps* for the years 1774 to 1789 included Maskelyne's lunar distance tables, unchanged except for column headings in French. Despite the American War of Independence, Maskelyne always managed to get the year's tables to Paris in time for publication in the *Connaissance*, until the issue for 1790 (published in 1788), which included for the first time tables based on the meridian of Paris, calculated in France.

French mathematicians used the observational results of Maskelyne and his predecessors at Greenwich to improve the equations describing the motion of the various bodies in the Solar System; Maskelyne in turn used the French equations to improve the predictions in the British almanacs for 1805 onwards; the French then used many of these British predictions in the *Connaissance*. A fine example of international scientific cooperation! (Howse *1989*: 85–96, 120–21, 210–11).

5 TYPICAL LATE-EIGHTEENTH-CENTURY OBSERVATIONAL PROCEDURES

5.1 Latitude by meridian altitude, the 'Noon Sight'

Observation Measure the altitude of the Sun at meridian passage, when it is at culmination (the highest point of its daily circuit).

Reduction

1 Correct observed altitude for dip, refraction and semi-diameter (from the Requisite Tables) to obtain true altitude.
2 Subtract from 90° to obtain zenith distance.
3 Apply the Sun's declination (from the *Nautical Almanac*) to obtain latitude (Cotter *1968*: 137–9).

This takes a few minutes only.

5.2 Longitude by lunar distance, the 'Lunar'

Observation Make three more-or-less simultaneous observations (preferably with at least two observers) – one lunar distance and one altitude of each of two bodies (Figure 6), all timed with a pocket-watch, not necessarily a chronometer. For the Sun, observations were taken in daylight when both Sun and Moon were at least 10° above the horizon. For a star, morning or evening twilight (when the horizon is easily visible) is preferable, otherwise a separate observation to find local times has to be taken when the horizon *is* visible, and a 'running fix' used.

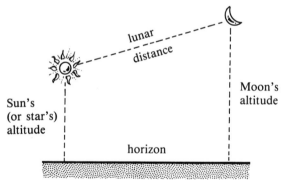

Figure 6 Lunar-distance observation

Reduction

1 Find the local apparent time of observation from the measured altitude of the Sun or star, which must be at least two hours from the meridian. Find the Sun's place from the almanac, or the star's from the Requisite Tables, and solve a spherical triangle by trigonometry (Cotter *1968*: 51–2).
2 Clear the measured lunar distance of the effects of refraction and

parallax, using Lyons's or Dunthorne's tables in the Requisite Tables, to obtain true lunar distance (*Ibid.*: 225–32).

3 From this true lunar distance, find the Greenwich apparent time of the observation by interpolation in the lunar distance tables in the almanac (*Ibid.*: 237–42).

4 The difference between local apparent time in step 1 and Greenwich apparent time in step 3 gives the longitude.

Before Maskelyne published the first *Nautical Almanac* in 1767, probably not more than a score of navigators of any nationality had succeeded in measuring their longitude when out of sight of land, so daunting was the spherical trigonometry involved in the reduction. With the precomputed lunar distance tables provided by Maskelyne, the time taken was reduced to an hour, or even less.

Howse (*1989*: 95) reproduces a contemporary printed form completed in ink, showing the reduction of an actual observation taken 500 miles west of the Canary Isles in 1772. The corresponding 1772 almanac pages are reproduced on pp. 88–9.

5.3 Longitude by chronometer

Observation   Measure the altitude of the Sun or star at least two hours from the meridian, timing by chronometer, or more usually, on deck by a pocket-watch which is immediately compared with the chronometer, so that the latter does not have to be moved from its stowage below.

Reduction

1 Follow step 1 for the lunar distance reduction method, to obtain the local apparent time of the observation.

2 Convert local apparent time to local mean time by applying the equation of time, taken from the almanac.

3 Find the Greenwich mean time of observation by applying the known (or, more usually, estimated) error of the chronometer.

4 The difference between local mean time in step 2 and Greenwich mean time in step 3 gives the longitude (Cotter *1968*: 243–54).

Reduction takes 15 minutes or so.

It was not until about 1800 that the marine chronometer became a practical and economic possibility for many ships, but by the middle of the nineteenth century enough were available and cheap enough for almost all ocean-going ships to have one or more. 'Lunars' soon fell out of use.

6 SOME LATER DEVELOPMENTS

Before the 1840s, latitude and longitude were found in two distinct operations. Then the 'Sumner line' (Cotter *1968*: 275–8) yielded integrated positions, though just as much spherical trigonometry was required. It was not until 1873, when Marcq St Hilaire, a French naval captain, proposed the 'Intercept method' (or cosine–haversine method), that at last the observed position was given as a *point*. Called the 'new navigation' in Britain, where it rapidly became popular, it widely replaced the Sumner method, though the noon sight for latitude and longitude-by-chronometer were never completely superseded (Cotter *1968*: 293–308).

ACKNOWLEDGEMENT

For much valuable advice, I express thanks to Commander D. W. Waters.

BIBLIOGRAPHY

Cotter, C. H. *1968*, *A History of Nautical Astronomy*, London: Hollis & Carter. [This gives a very full account of the mathematical aspects of astronomical position-finding from Classical times to the present day.]

Howse, D. *1980*, *Greenwich Time and the Discovery of the Longitude*, Oxford: Oxford University Press.

—— *1989*, *Nevil Maskelyne, the Seaman's Astronomer*, Cambridge: Cambridge University Press.

Plumley, N. *1976*, 'The Royal Mathematical School within Christ's Hospital, the early years', *Vistas in Astronomy*, **20**, 51–9.

Seller, J. *1694*, *Practical Navigation, or an Introduction to the Whole Art*, 7th edn, London: 'J. D. for the Author'. [1st edn, 1669.]

Taylor, E. G. R. *1956*, *The Haven-Finding Art. A History of Navigation from Odysseus to Captain Cook*, London: Hollis & Carter.

—— *1959*, 'Mathematics and the navigator in the thirteenth century', *The Journal of the Institute of Navigation*, **13** (1), 3–14.

Waters, D. W. *1958*, *The Art of Navigation in England in Elizabethan and Early Stuart Times*, London: Hollis & Carter.

Part 9
Physics and Mathematical Physics, and Electrical Engineering

9.0

Introduction

In tandem with the development of mechanics in its five areas (§8.0), mathematical physics grew out of mechanics in the early nineteenth century as mathematicization was introduced into heat theory, physical optics, and electricity and magnetism (and into their interconnections, after the discovery of electromagnetism in 1820). Prior to 1800 some mathematics was used in these disciplines, but it was extremely modest in scope, and physics in general – very much in contrast with mechanics – was viewed as an experimentally rather than a mathematically driven field. Incidentally, physics had a fairly low reputation among the sciences; optics was its strongest branch.

The opening articles of this Part are ordered somewhat according to the chronology of these innovations, although thereafter they developed simultaneously, and with interactions with each other and with mechanics. Optics occupies §9.1–9.3 (with the last article going back a little into the seventeenth century); heat theory dominates §9.4–9.9 (including acoustics, where consideration of the thermal properties of air advanced studies of central questions such as the velocity of sound; and capillarity, with its utility in aiding the careful use of thermometers and barometers); and electricity and magnetism occupy §9.10–9.12, including the (slow) arrival of electrical engineering among the technological sciences in the late nineteenth century.

The next three topics (§9.13–9.15) link somewhat more specifically to physics, and gained momentum especially in the late nineteenth century. To some extent they involve probability theory, which is treated in general in the next Part. The final three articles treat of chemistry and the life sciences.

9.1

Physical optics

N. KIPNIS

The term 'physical optics' was introduced by Thomas Young in 1802, and initially it meant all optics except for geometrical optics and vision (Young *1802*), with vision accommodated afterwards (Young *1807*). Nowadays, it usually refers to such phenomena as interference, diffraction, double refraction, polarization and dispersion (see also §9.2 on the velocity of light). This article looks at how physical optics became a mathematical science, using selected examples from its history. By 'mathematical science' is meant a science quantified by *any* mathematical means so as to allow a comparison with a quantitative experiment.

1 HUYGENS AND NEWTON

The mathematical approach to physical optics was pioneered by Christiaan Huygens and Isaac Newton. Huygens focused on a theory of double refraction, while Newton tried to build a theory of all phenomena of colours. In his *Traité de la lumière* (1690), Huygens assumed that light was a wave process in a special medium, the ether, which permeated all bodies. He described the propagation of a wavefront as follows: to locate a wavefront at any moment t given the wavefront at the preceding moment, $t - \Delta t$, consider each point on the old wavefront to be a centre of secondary spherical waves of radius $R = c \times \Delta t$ spreading forward, and draw a surface tangential to them all. This hypothesis later became known as the Huygens principle.

With this principle, Huygens explained double refraction in Iceland spar by the propagation of two different waves. The ordinary wavefront was a sphere, and propagated at the same velocity in all directions, while the extraordinary wavefront was an ellipsoid of revolution, the velocity of which depended on the direction of propagation at each point and could be represented by a vector drawn from the centre of the wave to the chosen point. To determine the ratio of the two unequal axes of the ellipsoid, Huygens used the indices of refraction of the two rays measured in the principal plane. He did not provide any experimental data, but stated instead that the experiment fully confirmed his theory.

Newton developed the first version of his theory of light and colours around 1668, when he was 26 years old, and it was supposed to show 'how valuable mathematics is in natural philosophy'. In his *Opticks*, first published in 1704, he did not mention mathematics in his opening statement: 'my design in this book is not to explain the properties of light by hypotheses, but to propose and prove them by reason and experiments'. However, mathematics might have been included in the 'reason' because it is clear from other passages that he did try to mathematicize physical optics using geometrical optics as a model (Newton *1730*: 131, 240, 244).

Actually, Newton's mathematical approach affected not only the form of his theory, but also its content. Indeed, he claimed that: (a) there were many kinds of light, which differed in their refraction; (b) light of each kind produced a specific simple colour; (c) simple colours mixed together produce compound light, and in particular, white light; and (d) a prism (or another device) produced colours not by modifying white light, but by separating different rays from one another. This part of the theory could be demonstrated by qualitative experiments. This was not true of his claim that colorific ability is an innate and immutable property of light.

Proving this was equivalent to proving that a colour which appears to be the same when observed in different phenomena is produced by the same kind of light. To achieve this, Newton had to describe light quantitatively. For this purpose he introduced two parameters, the 'refrangibility' and the 'reflexibility', which accounted for a difference in refraction (by a prism) or reflection (by a thin film), respectively, given the same incidence. Newton then demonstrated a constant relation between colour and either parameter. For the colours of thin films, he found that dark and bright rings were produced by specific thicknesses of a film which formed an arithmetical progression. To explain this he postulated the existence in light particles of periodical changes ('fits'), which made them susceptible to either reflection or refraction. He characterized this spatial periodicity by a separate parameter, the 'interval of fits', which was different for different colours (it also depended on the angle of incidence, which means that Newton's concept of periodicity differed from the modern one). Later, Young found that the interval of fits at normal incidence of light was equal to half the wavelength.

Newton extended the concept of intervals of fits to colours of thick plates, and obtained very good agreement with experiment. For prismatic colours, he showed that the distances from each part of a spectrum to a specific point were in a constant proportion taken from an optico-acoustical analogy. In doing so he drew an analogy with acoustics, designing the theory so that the proportions of prismatic colours were the same as those of periodic colours in thin films or thick plates (Newton *1730*: 127, 212, 225–7, 284, 295, 305).

The conclusion about a constant proportion for prismatic colours turned out to be erroneous, and it delayed the invention of achromatic lenses. On the other hand, without it Newton could not have supported his theory of colours. Demonstrating that a quantitative relation between any two colours is the same in different optical phenomena was the next thing to finding a universal 'measure' for coloured light, which was provided in the nineteenth century by the wavelength. Whether Newton's proof was valid is beyond the scope of this article, but at least he understood the importance of such a quantitative parameter in a theory of colour.

Thus, Huygens and Newton demonstrated that optical phenomena could be mathematicized in either the wave or the corpuscular theory of light. However, neither their contemporaries nor the following generations of physicists shared their concern with quantifying physical optics. Consequently, they rejected Huygens' theory of double refraction and Newton's theory of fits as mechanically unsound, but offered nothing in their place. As a result, throughout the eighteenth century physical optics remained a qualitative science. The new era of mathematical physical optics began early in the nineteenth century, and it started with the rediscovery of Huygens' and Newton's theories.

2 YOUNG AND FRESNEL

In 1801, at the age of 28, Thomas Young revitalized the old wave theory of light by adding to it the principle of interference. According to this principle, under certain conditions two rays of light can destroy one another. The mathematical concept behind it was the principle of superposition of waves, which Young obtained by generalizing the concept of superposition of forces and vibrations. He considered a superposition of only two vibrations, which had either the same or opposite phase. Initially, he applied the superposition of waves to acoustics and explained beats of sound (1799), then he modified it so as to make it applicable to light (1801). Young believed that every phenomenon of alternate colours required the principle of interference (together with other hypotheses) for its explanation, and that the difficulty was only in selecting a suitable pair of interfering waves (Kipnis *1991*: Chap. 5). He found that using only two interfering waves simplified the mathematical part, reducing it to simple geometry, while still providing sufficiently precise explanations of the coloured fringes produced by thin films, thick plates and 'mixed plates', additional rainbows sometimes observed inside the primary or outside the secondary bow, and diffraction (Young *1802*: 108–14; *1807*: 434–46).

In France, mathematical physicists led by Pierre Simon Laplace preferred analytical methods to geometry. Thus, when Etienne Malus decided to check William Hyde Wollaston's (1802) claim about verifying Huygens' theory, he began by rewriting it in an analytical form (1810). When Malus concluded that his experiments supported it, Laplace suggested using the theory but replacing its wave foundation with a corpuscular one (Chappert *1977*). Huygens' theory was compatible with Fermat's principle, which could be replaced with the principle of least action by substituting the velocity of light with its inverse. According to Laplace, the principle of least action implied the existence in Iceland spar of short-range attractive and repulsive forces, which acted on particles of light. In Young's view (1809), this conclusion was totally unfounded.

Another important piece of work done on the basis of the corpuscular theory was Jean-Baptiste Biot's theory of chromatic polarization (1812–14). In 1811, François Arago discovered that a thin plate of mica displayed different colours when viewed through Iceland spar. By using Newton's theory of fits, Biot developed a quantitative theory in which the colour of a crystal plate depends on its thickness and the orientation of its optical axis about the plane of polarization of polarizer and analyser. Biot's precise measurements confirmed his theoretical predictions of colours, and his theory received a favourable response.

The next breakthrough in physical optics was linked with the wave theory. In 1815, at the age of 27, Augustin Fresnel rediscovered the principle of interference and offered a theory of diffraction very similar to Young's. In 1818 he improved the theory and presented it in the mathematical contest announced by the Paris Academy of Sciences. By that time Fresnel had discovered how to add two vibrations with an arbitrary phase difference and how to add more than two vibrations. This enabled him to consider a diffraction fringe as the result of interference of secondary waves coming from all points on an open wavefront AMI (Figure 1). Thus, the intensity of light at the point of observations P is

$$\left[\int dz \cos\left(\frac{\pi z^2(a+b)}{ab\lambda}\right)\right]^2 + \left[\int dz \sin\left(\frac{\pi z^2(a+b)}{ab\lambda}\right)\right]^2, \tag{1}$$

where a and b are the distances of the diffractor from the source C and the screen DB, respectively, and λ is the wavelength. This integral could be evaluated only numerically.

To Fresnel, mathematics was a tool rather than an end in itself. Wherever possible he used simple mathematical means, as, for instance, in his theories of chromatic polarization and of the reflection and refraction of polarized

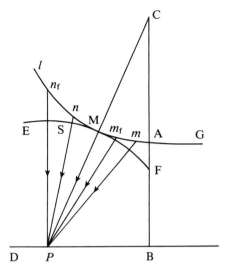

Figure 1 Fresnel's model of a diffraction fringe

light. There he introduced the very important concept of the transversality of light waves. He conceived of this idea in 1816 after discovering, together with Arago, the non-interference of light beams polarized in perpendicular directions; however, only in 1821 did he become convinced that light waves do not have a longitudinal component. Assuming that all refracted and reflected waves were transverse, Fresnel derived the laws for their intensity and confirmed them experimentally (Kipnis *1991*: Chap. 7).

The transversality of light waves became the physical foundation of Fresnel's theory of double refraction. To explain it mechanically, Fresnel assumed that the aether's resistance to compression is much greater than to distortion; the resistance to distortion implied that the aether somehow resembled a solid body. Incidentally, in this analogy Fresnel could not draw upon the theory of elasticity, which had yet to be developed (§8.6); in contrast, Augustin Louis Cauchy, one of the founders of the theory of elasticity, was to be influenced in some of his ideas by Fresnel. Fresnel showed that in a solid body there are three orthogonal directions ('elasticity axes') in which the displacement of a particle produces a force parallel to the displacement, and that an arbitrary displacement produces a force which has its components along the elasticity axes. He proposed to determine the velocity of light in a crystal by means of the elasticity ellipsoid, the axes of which coincided with x, y, z (Figure 2). If the ellipsoid is cut through its centre perpendicularly to the direction of light, the section will be an ellipse, the main axes of which are the directions of vibrations in the two waves, the semi-axes representing the magnitudes of their ray velocities.

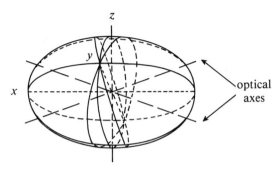

Figure 2 Fresnel's 'elasticity ellipsoid'

If the section is a circle, both velocities are equal, which means that the direction of propagation of light coincides with an optical axis of the crystal. From such considerations Fresnel concluded that in a biaxial crystal neither wave obeys Snel's law (Buchwald *1989*).

3 MATHEMATICAL THEORIES OF THE AETHER

Fresnel's death in 1827 ended the era in which new advances in optics began with physical discoveries, and mathematics was subordinate to physics. Now, mathematicians had taken the lead, and they were more concerned with the generality of their equations and solutions than with physical limitations of the results. While adopting Fresnel's laws for double refraction, reflection and refraction, and other phenomena, mathematicians rejected their derivation as lacking generality and being mechanically inconsistent. Cauchy suggested basing physical optics on the wave equation borrowed from the theory of elasticity (Whittaker *1910*: Chap. 5). In this way he obtained in 1830 his first theory of double refraction and that of reflection. Franz Ernst Neumann, James MacCullagh and George Green followed his lead and produced a number of theories of different optical phenomena. The new approach to physical optics had its own difficulties, as is illustrated by the theory of the reflection and refraction of polarized light.

Although all theories started from the same wave equation and aimed to reproduce Fresnel's laws of reflection, they differed in several points. First, it had to be decided whether to assume the aether vibrations to be perpendicular to the plane of polarization or parallel to it. The former assumption was consistent with the hypothesis that the distortional elasticity of the aether was constant and its density was variable, while the latter assumption was consistent with constant density of the aether and variable elasticity. Cauchy (1830, 1836) and Green (1837) used constant elasticity, while Carl Neumann (1835) and MacCullagh (1835) preferred constant density. The second decision to be made was what to do with longitudinal waves, which coexisted with transverse waves in a solid body. MacCullagh (1835) and Neumann (1835) avoided them, while Cauchy (1830, 1836, 1839) and Green (1837) tried to deal with them. The final decision was how closely the models of the aether should imitate real solids. Cauchy, for instance, was satisfied to select any boundary conditions which led to Fresnel's experimental laws. On the other hand, Neumann and MacCullagh wanted some similarity with the boundary conditions of elastic solids, so they kept the continuity of three components of velocity; another condition being the conservation of *vis viva* (or energy). To achieve this they had to opt for the hypothesis of constant density. Since this hypothesis was also used in the theory of double refraction, choosing it had the advantage of a unified approach to two different phenomena. Yet, their aether was not a true solid (Whittaker *1910*).

Green (1837) went even further and took *all* the boundary conditions from elastic solids: continuity of three components of displacement and of three components of stress. He abandoned Cauchy's molecular aether and replaced it with a continuous model, which he represented with a suitable potential function. This 'truly dynamic' theory, however, had its own difficulties, in particular the longitudinal waves. Neumann and MacCullagh had only four boundary conditions, which could be satisfied by transverse waves alone: thus they *postulated* that all waves in the aether were transverse. Green, on the other hand, had six boundary conditions, and he had to retain longitudinal waves. To prevent them from carrying away any energy, he postulated their velocity to be much higher than that of transverse waves. Yet the theory contradicted experiment, and it was rejected.

Green's failure to make the aether similar to an ordinary solid body gave rise to some unusual models. MacCullagh (1839) invented an aether in which elasticity was caused solely by rotation of its volume elements, which eliminated longitudinal waves. The same year Cauchy offered another way to ban these waves: he attributed to them zero velocity. This led to a negative compressibility which made the aether unstable. Although both theories provided 'natural' boundary conditions and agreed with Fresnel's laws, they were too distant from reality to be accepted. In the second half of the nineteenth century, many mathematicians, including George Stokes, William Thomson (Lord Kelvin) and Lord Rayleigh, tried their hand at aether theories. Success eluded them, however, and they resolved neither the problem of the direction of the vibrations nor that of the existence of longitudinal waves. Those whose aether imitated ordinary solids came to conflicting theories of different phenomena, while those who managed a more-or-less unified theory found their model of the aether to be too different from real bodies.

James Clerk Maxwell was the first to develop mathematically the idea that optical, electrical and magnetic phenomena all result from disturbances of the same aether. This enabled him to give a new explanation of the transversality of light and of magneto-optical and electro-optical phenomena (Everitt *1974*). However, the electromagnetic theory of light did not resolve all the problems of wave optics because modelling the electromagnetic aether turned out to be no easier than modelling the solid aether. This does not mean that all the efforts of mathematicians were in vain. Apart from modelling the aether, they were successful: they made advances in explaining dispersion, selective absorption, diffraction and other phenomena. It is important to note that mathematical theories confirmed by experiment retained their validity even after their interpretation changed.

The first change was when the solid aether gave way to the electromagnetic aether. For instance, in 1878, George Fitzgerald discovered that by identifying **e** with magnetic force, where **e** is the displacement vector in MacCullagh's theory, and curl **e** with dielectric displacement, he could obtain the same expressions for kinetic and potential energy in Maxwell's theory as in MacCullagh's, which made MacCullagh's theory of reflection and refraction of light correct in the electromagnetic field free of charges and conduction currents. Cauchy's 1839 theory of the unstable aether was also resurrected: Josiah Willard Gibbs showed that its boundary conditions could be transferred into electromagnetic theory (Stein *1981*). Another change occurred with the removal from physics of the aether: the equations of physical optics continued to describe phenomena correctly, even though the electromagnetic waves were no longer connected with any carrier.

4 THE ORIGIN OF RELATIVITY AND QUANTUM THEORY

Besides difficulties with the mechanical modelling of the aether, there was another powerful reason for abandoning it: the impossibility of justifying its existence as an absolute reference system. For two centuries physicists had tried to discover some evidence of a uniform motion of the Earth relative to the aether, first by optical and later by electromagnetic means, and had failed. By 1900, the experiments had improved so much that there was no longer any doubt about their result. To explain the negative outcome, Joseph Larmor, Hendrik Lorentz and Henri Poincaré developed between 1900 and 1905 linear transformations of coordinates and time which made electromagnetic phenomena in a system independent of the system's uniform motion. Poincaré treated these transformations as a group of rotations in a four-dimensional space. In 1905, Albert Einstein brought forth his theory of relativity, which asserted that 'Lorentz transformations' were not limited to Maxwell's theory, but represented general properties of space and time (Miller *1981*; see also §9.13).

Another revolutionary theory of the twentieth century – the quantum theory – originated in the study of black-body radiation (§9.15). Max Planck applied the concept of entropy and Ludwig Boltzmann's statistics to a set of electromagnetic resonators. To obtain a spectral distribution compatible with experimental laws, he assumed (1900) that the energy of radiation is made up of discrete 'bundles'. However, it was only after 1905 that physicists began to interpret this mathematical hypothesis as expressing a physical discontinuity of energy (Kuhn *1978*). Einstein's discovery (1905) that the entropy of radiation in a given frequency interval is the same function of volume as that of gas was instrumental in this change. Following Planck, Arnold Sommerfeld (1912) suggested that the fundamental concept in the theory must be the quantum of action rather than the energy quantum, and that at every act of absorption or emission the time integral of the Lagrangian of the system must be equal to Planck's constant.

While such optical phenomena as the photoelectric effect, characteristic X-rays and the Compton effect lent their support to the quantum theory of radiation, spectroscopy provided necessary data for the first quantum theory of the hydrogen atom, proposed by Niels Bohr (1913) and improved by Sommerfeld (1915).

The rise of the quantum theory of radiation did not mean rejection of the wave theory; it was just that the latter was inapplicable for radiation of very short wavelength and extremely low intensity (§9.15). Interestingly, Einstein's 1916 paper on quantum radiation and the Bohr atom, in which he introduced the concept of stimulated emission, eventually led to the invention of lasers, which in turn stimulated the development of wave optics (holography).

BIBLIOGRAPHY

Buchwald, J. Z. *1980*, 'Optics and the theory of the punctiform ether', *Archive for History of Exact Sciences*, **21**, 245–78. [On the wave theory of dispersion.]
——*1989*, *The Rise of the Wave Theory of Light*, Chicago, IL: University of Chicago Press. [An excellent mathematical treatment of optical theories from Malus to Fresnel.]
Chappert, A. *1977*, *Etienne Louis Malus (1775–1812) et la théorie corpusculaire de la lumière*, Paris: Vrin. [On the theories of double refraction and polarization.]
Everitt, C. W. F. *1974*, 'Maxwell, James Clerk', in *Dictionary of Scientific Biography*, Vol, 9, New York: Scribner's, 198–230. [Especially pp. 209–14 on the electromagnetic theory of light.]
Garding, L. *1989*, 'History of the mathematics of double refraction', *Archive for History of Exact Sciences*, **40**, 355–85.

Grattan-Guinness, I. *1990, Convolutions in French Mathematics* [...], 3 vols, Basel: Birkhäuser, Chaps 7, 15, 17.

Kipnis, N. *1991, History of the Principle of Interference of Light*, Basel: Birkhäuser.

Kuhn, T. *1978, Black-Body Theory and the Quantum Discontinuity, 1894–1912*, New York: Oxford University Press.

Miller, A. *1981, Albert Einstein's Special Theory of Relativity*, Reading, MA: Addison-Wesley. [See Chap. 1 on optical background.]

Newton, I. *1730, Opticks*, 4th edn, London: Innys. [Edition used: 1952, New York: Dover.]

——*1984, The Optical Papers of Isaac Newton*, Vol. 1, *The Optical Lectures, 1670–1672* (ed. A. E. Shapiro), Cambridge: Cambridge University Press.

Shapiro, A. E. *1980*, 'Newton's "achromatic" dispersion law [...]', *Archive for History of Exact Sciences*, **21**, 92–128.

Stein, H. *1981*, '"Subtler forms of matter" [...]', in G. N. Cantor and J. S. Hodge (eds), *Conceptions of Aether: Studies in the History of Aether Theories, 1740–1900*, Cambridge: Cambridge University Press, 309–40. [Electromagnetic models of the aether.]

Stuewer, R. *1975, The Compton Effect*, New York: Science History Publications. [The debate between the wave and corpuscular theories of light, 1897–1925.]

Whittaker, E. T. *1910, A History of the Theories of Aether and Electricity*, 1st edn, London: Longmans, Green. [2nd edn 1951; repr. 1960, New York: Harper.]

Young, T. *1802, A Syllabus of a Course of Lectures*, London: The Royal Institution.

——*1807, A Course of Lectures on Natural Philosophy* [...], Vol. 1, London: Johnson.

9.2

The velocity of light

GEOFFREY CANTOR

The history on this subject encompasses both experimental determination of the velocity of light, c, and the role of c in physical theories. While the empirical aspect provides a commentary on the progressive refinement of metrological techniques, the theoretical dimension connects the major theories of physics and the understanding of measurement. These historically interwoven topics will, however, be discussed separately below.

1 DETERMINATIONS OF THE VELOCITY OF LIGHT

Before the seventeenth century, most writers followed Aristotle in considering that light operates instantaneously. Only with the rise of the mechanical philosophy did the speed of light become a meaningful question and one open to empirical solution. Both Galileo Galilei and René Descartes sought values for c, but they were only able to set crude lower limits. However, in 1676 Ole Römer, observing the eclipses of Jupiter's inner moon in order to provide an international time standard, found that the interval between eclipses varied by some 22 minutes over a six-month period, and argued that this time difference was due to the transit of light across the diameter of the Earth's orbit. Since the size of the Earth's orbit was in dispute in the seventeenth century and estimates differed significantly from present-day values, it is not legitimate to translate Römer's assessment of the speed of light into modern units. (However, in his *Traité de la lumière* (1690) Christiaan Huygens used the data to estimate a value equivalent to 2×10^8 m/s.)

A similar problem arises with James Bradley's determination in 1728. Bradley measured the aberration of starlight, and argued that the tangent of the angle of aberration is equal to v/c, where v is the Earth's velocity perpendicular to the line of sight. Such astronomical methods expressed the speed of light in terms of inaccurately known components of the Earth's motion (Boyer *1941*).

To some extent this difficulty engendered a lack of consistency in the published estimates of the velocity of light until the middle of the nineteenth century. However, its determination by Hippolyte Fizeau in 1849 marks the advent of sophisticated terrestrial methods and the demand for accuracy. Fizeau's method used a carefully measured baseline (of approximately 8.6 km) and a rotating toothed wheel to determine time. The speed of rotation was adjusted so that light which passed through the gap between one pair of teeth traversed the baseline and was reflected back again so as to pass through the neighbouring gap. From the rate of rotation, the time for this round-trip could readily be calculated, and hence too the speed of light.

There were many subsequent attempts to improve on Fizeau's apparatus, including Jean Foucault's substitution of a rotating mirror for the toothed wheel. In 1862 Foucault obtained the value of 298 000 km/s. Moreover, one advantage of Foucault's method was that, since it employed a much shorter baseline, it could be used in the laboratory to measure the speed of light not only in air but also in other media (Dorsey *1944*).

Albert Michelson devoted much of his professional life to this metrological project. With zeal and dogged resolution he repeatedly refined Foucault's apparatus, making ever more accurate determinations between 1878 and his death in 1931. In his most celebrated series of experiments, published in 1927, he utilized a 35 km baseline between Mount Wilson and Mount St Antonio, and obtained a value of 299 798 km/s. In another extensive series of experiments beginning in 1881, Michelson sought to determine whether the velocity of light varied with direction of propagation and, in particular, whether its velocity was affected by the Earth's motion. This problem arose from realist interpretations of the wave theory of light that postulated a ubiquitous aether (§9.1). While other assumptions were discussed, one prominent hypothesis required the aether to be stationary with respect to the Sun, and the Earth to be moving freely through it, thus producing an 'aether wind' on the Earth's surface. Michelson devised a method for detecting this 'wind' using a sensitive split-beam interferometer with light paths set parallel and perpendicular to the Earth's motion. In an impressive series of experiments with Edward Morley, he reported in 1887 that 'the relative velocity of the Earth and the ether is ... certainly less than one fourth' of the Earth's orbital velocity (Michelson and Morley *1887*: 341). If there was a 'wind', it had far less effect on the velocity of light than predicted by the aether drift hypothesis (Jaffe *1960*, Livingstone *1973*, Swenson *1972*).

More recent developments have included the introduction of a Kerr cell in place of Fizeau's toothed wheel. Moreover, much of the experimental work has concentrated on electromagnetic radiation outside the visible range. Thus K. D. Froome obtained the widely accepted standard of 299 792.5 ($\pm$ 0.1) km/s *in vacuo* at the UK's National Physical Laboratory in 1957 using a microwave interferometer. While much effort has been expended in measuring the value of c, the velocity of light itself has become a metric for measuring astronomical distances.

2 THEORETICAL IMPLICATIONS

The following examples illustrate ways in which the velocity of light has entered into many of the key theories in the history of physics, astronomy and cosmology.

The theories of light developed in the seventeenth and eighteenth centuries offered various predictions for the ratio between the velocities of light in different media. According to many corpuscular theories, such as the one propounded in Isaac Newton's *Opticks* (1704), a straightforward geometrical analysis of refraction implied that the velocity of light must increase when passing from air into, say, water, such that

$$\frac{\sin i}{\sin r} = \frac{v_w}{v_a} = \mu, \tag{1}$$

where v_a is the velocity of light in air, v_w is the velocity of light in water, and the refractive index $\mu > 1$. By contrast, some other writers, such as Pierre de Fermat, who utilized the principle of least time, considered the velocity to decrease such that the velocity of light is greater in air than in water and, for a ray refracted towards the normal on entering the water,

$$\frac{\sin i}{\sin r} = \frac{v_a}{v_w} = \mu. \tag{2}$$

This equation also formed part of the wave theory as developed by Thomas Young and, especially, Augustin Fresnel in the early nineteenth century. While the work of both Young and Fresnel indirectly confirmed that the velocity of light decreases on entering the water, Foucault succeeded in showing in 1850 that light travels faster in air than in water. While this result impressively confirmed the new wave theory, it had already been accepted by most scientists.

The velocity of light also entered crucially into mid-nineteenth century electromagnetism (§9.10). In his paper 'Thoughts on ray-vibrations' (1846), Michael Faraday noted that electricity is transmitted at approximately the speed of light. Moreover, he speculated that light is propagated by the vibration of lines of force which he considered to permeate space. James Clerk Maxwell developed and mathematicized this speculation in his paper 'On physical lines of force' (1861–2), in which he explored the analogy between Faraday's electromagnetic medium and the behaviour of an elastic fluid composed of small molecular vortices. Having calculated the density and elasticity of this fluid, he inferred that the velocity of propagation of transverse vibrations is 310 740 km/s. This figure, he noted,

> agrees so exactly with the velocity of light calculated from the optical experiments of M. Fizeau, that we can scarcely avoid the inference that *light consists in the transverse undulations of the same medium which is the cause of electric and magnetic phenomena.* (Maxwell *1890*: Vol. 1, 500).

In the fourth and final section of the paper he used his theory of molecular vortices to account for the magneto-optical (or Faraday) effect.

Maxwell subsequently dispensed with the molecular vortices (while retaining the notion of an elastic aether of finite density), and developed his electromagnetic field equations in a paper entitled 'A dynamical theory of the electromagnetic field' (1865). He also derived the behaviour of light waves from these electromagnetic equations, hence articulating what he called the electromagnetic theory of light. Furthermore, he explicitly recognized that the velocity of light is equal to the number of electrostatic units in one electromagnetic unit of electricity. Thus he connected the velocity of light to the units of electromagnetism (Everitt *1974*, Buchwald *1985*).

The velocity of light also entered crucially into Albert Einstein's special theory of relativity (§9.12). The foundations of the theory were laid in his 1905 paper, 'Zur Elektrodynamik bewegter Korper', which opened with a critique of Maxwell's electrodynamics. Einstein then postulated that 'light is always propagated in empty space with a definite velocity c which is independent of the state of motion of the emitting body' (Lorentz *et al. 1923*: 112). Taken with his principle of relativity, this postulate implied that the velocity of light is the same in all inertial frames, and it yielded new relativistic definitions for the measurement of times, lengths and masses which displaced the 'absolute' conceptions articulated in Newton's *Principia*. Einstein applied Lorentz transformations to correlate positions and times between two frames of reference, such that both time (t') and length (x', along the direction of motion) are functions of velocity (v):

$$x' = (x - vt)\left(1 - \frac{v^2}{c^2}\right)^{-1/2} \quad \text{and} \quad t' = \left(t - \frac{vx}{c^2}\right)\left(1 - \frac{v^2}{c^2}\right)^{-1/2}. \quad (3)$$

Implications of this analysis are that mass is also a function of velocity, and that no material body can travel at or above the speed of light. Moreover, Einstein's theory challenged the classical view that distinguished sharply between mass and energy — the energy of a body depending on its velocity or its position in a force field. Instead, Einstein showed the equivalence of mass and energy according to the celebrated equation $E = mc^2$, in which these two concepts are interrelated via the square of the velocity of light.

BIBLIOGRAPHY

Boyer, C. E. *1941*, 'Early estimates of the velocity of light', *Isis*, **33**, 24–40.

Buchwald, J. Z. *1985*, *From Maxwell to Microphysics. Aspects of Electromagnetic Theory in the Last Quarter of the Nineteenth Century*, Chicago, IL: University of Chicago Press.

Dorsey, N. E. *1944*, 'The velocity of light', *Transactions of the American Philosophical Society*, **34**, 1–110.

Everitt, C. W. F. *1974*, 'Maxwell, James Clerk', in *Dictionary of Scientific Biography*, Vol. 9, New York: Scribner's, 198–230.

Jaffe, B. *1960*, *Michelson and the Speed of Light*, New York: Doubleday.

Livingston, D. M. *1973*, *The Master of Light. A Biography of Albert A. Michelson*, Chicago, IL: University of Chicago Press.

Lorentz, H. A., Einstein, A., Minkowski, H. and Weyl, H. *1923*, *The Principle of Relativity*, London: Methuen.

Maxwell, J. C. *1890*, *Scientific papers*, Vol. 1 (ed. W. D. Niven), Cambridge: Cambridge University Press.

Michelson, A. A. and Morley, E. W. *1887*, 'On the relative motion of the earth and the luminiferous ether', *American Journal of Science*, **34**, 333–45.

Swenson, L. S. *1972*, *The Ethereal Aether. A History of the Michelson–Morley–Miller Aether-Drift Experiments, 1880–1930*, Austin, TX: University of Texas Press.

9.3

Optics and optical instruments, 1600–1800

J. C. DEIMAN

Optics is usually divided into physical optics, physiological optics and geometrical optics. Physical optics deals with the nature and properties of light (§2.9, §9.1); physiological optics deals with vision. Geometrical optics, the subject of this article, is that part of optics which deals with optical instruments. In this article are sketched the historical development of geometrical optics and its interaction with the development of two optical instruments, the telescope and the microscope.

1 THE GEOMETRICAL MODEL AND OPTICAL INSTRUMENTS

One of the most fundamental concepts of geometrical optics is the rectilinear propagation of light; derived from this are the ray and pencil of rays. This concept of the ray is very similar to that of the straight line in Euclidean geometry, which is no coincidence as it was described by Euclid. Thanks to this circumstance, geometrical optics was able to develop comparatively fast as soon as the need arose, since a suitable mathematical model was available. Also part of geometrical optics are the phenomena of reflection and refraction. The optical elements which correspond to these two are the mirror and the lens. Mirrors were known in Antiquity; lenses were used in spectacles from medieval times.

Optical instruments are those instruments in which prisms, mirrors, lenses or combinations of these elements form a major part. Their aim is to extend the capabilities of the human eye. Need for their development arose when people started observing nature more closely. This 'scientific revolution' is situated in the sixteenth and seventeenth centuries, with important developments in astronomy, mathematics and the natural sciences. As vision is one of our most important senses, it is not surprising that optical instruments played an important role in this revolution. The human eye has some severe limitations, one of which is its limited resolving power: we generally cannot see details less than 0.1 mm apart at a distance of 250 mm.

2 THE INVENTION OF THE TELESCOPE

The earliest known telescopes date from around the beginning of the seventeenth century. These 'Dutch' or 'Galilean' telescopes consist of a positive object lens and a negative eye lens, giving an erect image; their magnification was small, at most two to five diameters. A patent for this instrument was applied for by the Dutch spectacle-maker Hans Lipperhey of Middelburg in 1608. This does not mean that Lipperhey invented the telescope; it was probably known before in Italy, by G. della Porta amongst others. The Italian scientists were looking for a much more powerful device, more like a modern telescope, but they did not realize that the simple combination of two spectacle lenses, which they knew about, could lead to the telescope that they were looking for. Galileo Galilei heard rumours of this Dutch invention in May 1609, and a day later he assembled his first telescope – obviously the idea was not alien to him! His telescope started a chain reaction, and from that moment on the telescope spread rapidly over Europe (van Helden *1977*: 5–9).

3 THE TELESCOPE AND GEOMETRICAL OPTICS

The first telescopes were assembled by people who had a very limited knowledge of optics. The quality of the instruments did not suffer too much because of this, the bad quality of the glass and the technical problems of making perfectly spherical lenses of known dimensions being much stronger impediments.

Geometrical optics was virtually unchanged since Euclid's *Optics*, in which the rectilinear propagation of light and the theory of the flat, concave and convex mirrors were treated (§9.1). Although the spectacle lens had been in use since the end of the thirteenth century, it was still believed that light proceeded from the eye to the object seen, and the phenomenon of refraction was understood only to a very limited extent.

However, very shortly after 1610, when he received his first telescope, Johannes Kepler in Prague started studying the telescope, the human eye and the fundamental properties of lenses. In his *Dioptrice* (1611), he construed light as proceeding from the objects seen to the retina of the eye, where an inverted image was formed by the eye's lens. He also showed that a lens with spherical surfaces gives rise to spherical aberration, in which rays

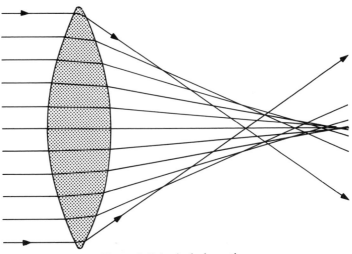

Figure 1 Spherical aberration:
near-axis rays focus farther from lens than do off-axis rays

entering the lens near the axis are focused at a point further from the lens
than are rays entering near the rim of the lens (Figure 1). The other defect
of lenses which Kepler discovered was chromatic aberration, caused by the
focus for the blue rays lying closer to the lens than that for the red rays
(Figure 2) (King *1955*). From this moment on, geometrical optics developed
very successfully: together with Newtonian mechanics it was at the forefront of mathematicized physics. Important contributions were made by
René Descartes, Christiaan Huygens and Isaac Newton. Later in the
eighteenth century, French mathematicians such as Alexis Clairaut and
Jean d'Alembert and the Swiss Leonhard Euler extended the theory of
systems of lenses, and the theory of aberrations and colours.

The impact of this highly developed mathematical theory on instrument-
making, however, was rather limited, though there were improvements –
like the reflecting telescope, in which a mirror is used instead of an objective
lens. In a reflecting telescope chromatic aberration is much less as there is
no refraction (except at the eye lens) and consequently no dispersion.
Newton experimented with these telescopes, as he erroneously deduced
from his experiments that an achromatic combination of lenses was impossible. A drawback of the reflecting telescope was the low reflectivity of the
speculum-metal mirrors, a technical problem which was solved in the
middle of the nineteenth century when it became possible to coat a glass
mirror with a highly reflective layer of silver.

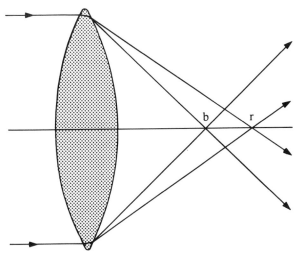

Figure 2 Chromatic aberration:
red rays focus farther from lens than do blue rays

This slow dissemination of knowledge was caused by social factors; mathematical scientists and optical instrument-makers belonged to different social classes, so that interaction of ideas was limited. In the middle of the eighteenth century the social status of instrument-makers became gradually much higher. John Dollond, the famous London maker, corresponded with Euler and was elected a Fellow of the Royal Society in 1761. Also, scientists became more interested in solving practical problems.

Another reason was the enormous technical complexity of the problem. Glass of optical quality did not exist before around 1820, and even then it was difficult to obtain. Also, the optical properties of glass (refractive index and dispersion) were not accurately known. For this reason instrument-makers preferred to construct their telescopes and microscopes according to simple rules of thumb.

The accurate mathematical design of an optical system is not easy, as the only way is by successive approximation. One method uses ray tracing, which is based on simple geometry, tracing the ray from surface to surface, repeatedly applying the laws of reflection and refraction. This method is very time-consuming as a great number of different rays (at least fifty) have to be traced to gain sufficient insight into the performance of the system. After one tracing an experienced designer can make small adjustments to curvatures, distances or refractive indices, and start a second tracing. To circumvent this problem special mathematical techniques were developed in the nineteenth century which gave direct approximations for aberrations in terms of the curvatures, distances and refractive indices of the system. It is only since the Second World War with the availability of fast digital computers that there has been a return to the more accurate method of ray tracing.

This situation changed gradually from 1757 onwards, when John and Peter Dollond started selling their achromatic telescopes. In an achromatic doublet objective there are two lenses: a positive one of ordinary crown glass with low refractive index (1.5–1.55) and low dispersion (less than 0.01), and a negative one of flint or lead glass with high refractive index (1.57–1.7) and higher dispersion (more than 0.01). By correctly choosing the four curvatures of the two lenses, it is possible to reduce the overall spherical and chromatic aberration of the system, the aberrations of one lens counteracting those of the other. As a result the aperture of the telescope can be made much larger before the image quality starts to deteriorate. A non-achromatic telescope made by Jonathan Sisson in 1735 had an aperture of 1 inch (25 mm); around 1775 Peter Dollond was making such instruments with an aperture of 3 inches (75 mm).

The design of these telescopes required more knowledge of geometrical optics to calculate the curvatures of the lenses. The assembled achromatic doublets had to be corrected for differences in the optical properties of the glass and differences in the curvatures by zoning. An experienced instrument-maker could see by observing the image of a star in which way the doublet should be corrected in order to make it perfect. As a result these telescopes were very expensive, and took a long time to make.

In Germany, Joseph Fraunhofer definitively solved the scientific and technical problems associated with the telescope. He succeeded in making optical-quality glass, and devised ways of accurately measuring the refractive index and the dispersion for precisely defined wavelengths, and of calculating and manufacturing objectives according to specifications. He discovered the great complexity of the solar spectrum, starting a line of research leading to atomic spectroscopy and astrophysics (§8.10).

4 THE MICROSCOPE

Although lenses were used as magnifiers before 1600, the compound microscope, consisting of an object lens and an eye lens, was probably invented when a 'Galilean' telescope was inverted. Unlike telescopes, which could be used for astronomical, nautical and military purposes, there was no immediate use for the microscope, so developments in microscopy proceeded at a slower pace. Another important reason is that microscopes were used by a different class of people. Astronomers were well versed in mathematics and were accustomed to improving their instruments. The users of microscopes however, formed a less well-defined group of people: some were medical men, some just amateurs. As a result, improvements were mainly mechanical, with the microscope becoming much easier to use and better adapted to the needs of users (Turner *1980*).

Natural scientists started using microscopes in the later part of the seventeenth century. In 1665 Robert Hooke's *Micrographia* was published, the first great account of microscopical observations. Hooke's ideas were picked up and elaborated by Anton van Leeuwenhoek, a linen-draper from Delft in the Netherlands. The home-made simple microscopes which van Leeuwenhoek used for his researches he constructed with no more than an elementary knowledge of optics. It is still surprising that the best one, which is still in existence and kept at Utrecht University's museum, has a resolution of about 1.4 μm (micrometres); this would not be bettered by a compound microscope until 1830. (The optical properties of the microscopes made by van Leeuwenhoek are treated in detail by van Zuylen *1981*.) The spherical and chromatic aberrations of simple microscopes are relatively small; their main drawback is that they are tedious to use. Compound microscopes with a simple biconvex objective lens were made up to 1830; their resolving power was at most 2 μm.

The Dutch telescope maker Jan van Deijl realized in about 1770 that it would be possible to improve the microscope in the same way as the telescope. However, the small lenses needed for a microscope were very difficult to make; as there were no interested customers, he preferred making telescopes. In 1807, however, he made a few microscopes with an achromatic doublet lens; recent research has shown that the doublet was constructed in the same way as the ones he used in his telescopes (van Zuylen *1987*). In France, J. G. Chevalier, an instrument-maker, combined two or three of these doublets; he started selling them in about 1825. In England, the amateur scientist Joseph Jackson Lister studied these lenses, and on the basis of his experiments and his knowledge of optics he succeeded in developing an improved type of objective lens. Lister also knew the relation between the aperture of the objective and the resulting resolving power. Instrument-makers like A. Ross, J. Smith and H. Powell, and P. H. Lealand made objectives based on Lister's designs from 1837 onwards. Incomplete knowledge of optics prevented the successful execution of these designs; they were just calculated roughly and assembled by trial-and-error.

Carl Zeiss from Jena, who also made microscopes, realized in about 1865 that he could improve his objectives no further himself, so he asked the Jena University professor Ernst Abbe to help him develop a scientific way of designing and manufacturing objectives for microscopes. Abbe developed a general theory of the microscope in which diffraction phenomena were also taken into account; furthermore, the production of the objectives was organized by the division of labour and strict quality control, so that it became possible to make microscope objectives exactly according to specification (Abbe *1904*). Geometrical optics was enormously stimulated by this work, for lengthy design procedures became worth while. Theory and practice went together from this period on as technical optics.

BIBLIOGRAPHY

Abbe, E. *1904*, *Gesammelte Abhandlungen*, Jena: Fischer.

Born, M. and Wolf, E. *1959*, *Principles of Optics*, 1st edn, Oxford: Pergamon.

Bradbury, S. and Turner, G. l'E. (eds) *1967*, *Historical Aspects of Microscopy*, Cambridge: Heffer.

Harting, P. *1866*, *Das Mikroskop*, Braunschweig: Vieweg.

Helden, A. van *1977*, 'The invention of the telescope', *Transactions of the American Philosophical Society*, **67**, 1–67.

King, H. C. *1955*, *The History of the Telescope*, London: Griffin.

Kingslake, R. *1989*, *A History of the Photographic Lens*, London: Academic Press.

Turner, G. l'E. *1980*, *Essays on the History of the Microscope*, Oxford: Senecio.

Twyman, F. *1988*, *Prism and Lens Making*, Bristol: Adam Hilger.

Welford, W. T. *1986*, *Aberrations of Optical Systems*, Bristol: Adam Hilger.

Zuylen, J. van *1981*, 'The microscopes of Antoni van Leeuwenhoek', *Journal of Microscopy*, **121**, 309–28.

—— *1987*, 'Jan en Harmanus van Deijl [. . .]', *Tijdschrift voor de geschiedenis der geneeskunde, natuurwetenschappen, wiskunde en techniek*, **10**, 208–28.

9.4

Heat diffusion

I. GRATTAN-GUINNESS

1 FOURIER'S CONTRIBUTIONS: THE DIFFUSION EQUATION AND FOURIER ANALYSIS

The principal founder of this topic was Joseph Fourier in the early 1800s, when he was in his mid-thirties. His problem was: given an initial distribution of heat in a body, to calculate the temperature of any point at any time when it is placed in an external cooling environment. His results constitute the first important advance in mathematical physics outside the range of mechanics.

Fourier's first attempt, made around 1802, used discrete models in which n equal and equally spaced objects were analysed; but a conceptual error in specifying internal heat conductivity gave a false constant temperature solution in the limiting case $n \to \infty$ of a continuous body (e.g. n objects in a line becoming a bar). But in 1804 his interest in heat diffusion was rekindled by a short paper recently published by Jean-Baptiste Biot. Biot's mathematical ability was limited (like Fourier before him, he seems to have misdefined conductivity), but he had indicated a fresh approach: form directly the differential equation representing the phenomenon, and then solve it by mathematical methods.

Fourier took Biot's example of heat diffusion in a straight bar. He considered heat diffusing within the bar, of conductivity K, and into the environment (assumed to be at zero temperature) with conductivity h; y was the temperature at the point a distance x along the bar. Using Leonhard Euler's version of the differential calculus (§3.2), he divided the bar into infinitesimal slices of constant thickness dx. Assuming Newton's law of cooling (that the rate of cooling of a body is proportional to its current difference of temperature over that of the environment), he showed that $-K d(dy/dx) dt$ of heat flowed through the slice at x in time dt; balancing this with the external heat flow $(-hy \, dx \, dt)$ and the calorimetric gain of heat

($CD\,dx\,dy$, with C the specific heat and D the linear density), he found the (literally) differential equation

$$K(d^2 y/dx^2) - hy = CD\,dy/dt, \quad \text{or} \quad Ky_{xx} - hy = CDy_t, \tag{1}$$

where the suffices denote partial differential coefficients.

This is one version of the partial differential equation known as the 'diffusion equation'. Fourier also studied diffusion in the lamina, annulus, sphere, cylinder and cube, giving the equation each time in a suitable coordinate system. For the last three of these bodies, he also used the important additional idea of a separate equation to represent the external diffusion of heat from its surface, for example

$$ky_x + hy = 0, \quad x = a \text{ (a constant).} \tag{2}$$

Mindful of the earlier error with internal conductivity, he reintroduced dimensional analysis into physics by determining the units of his various parameters.

The form of solution to the diffusion equation that Fourier developed was similar to, and probably inspired by, that in the unsuccessful n-body analysis. He used the method of separation of variables, setting $y = x(x)T(t)$ and splitting equation (1)$_2$ into two ordinary differential equations; already known as a method (§3.15), this was now raised to a new level of prominence by his work. For interior diffusion (no h), it led to the solution

$$y = a_0 + \sum_{r=1}^{\infty} (a_r \cos rx + b_r \sin rx) \exp(-Kr^2 t); \tag{3}$$

the 'Fourier series' followed from it by putting $y = f(x)$ and $t = 0$.

Fourier found for himself the now standard method of term-by-term integration (already known in fact to Euler) for evaluating the coefficients a_0, a_r and b_r as integrals, for both continuous and discontinuous functions. He anticipated by nearly a century the discovery of the 'completeness' aspects of the series (§3.9), to which that method is insensitive. He first calculated the coefficients by another method, which is of significance in its own right for it marked a first step in the theory of infinite matrices.

Fourier understood that the periodicity of the trigonometric functions allowed the series to represent a function over only a finite range of values of x: outside it, series and function part company. Thereby he resolved one aspect of the eighteenth-century controversy over solving the wave equation (§3.15). He also made a notable study of the convergence of his series (§3.11). The various extensions of equation (1) that Fourier found included multiple-series solutions; the so-called 'non-harmonic' series (in which the integer r in rx was replaced by n_r, the roots of the equation obtained from the surface equation (2)); and a beautiful treatment of the cooling cylinder, for which he developed the first comprehensive theory of the 'Bessel function' $J_0(x)$ (§4.4).

Fourier presented his work in December 1807 to the Institut de France, and a controversy developed with Joseph Louis Lagrange (on mathematical aspects, including periodicity) and with Pierre Simon Laplace (on the physical modelling: see Section 3); so his large manuscript remained unpublished until Grattan-Guinness and Ravetz *1972*. But with Laplace's eventual help, a prize problem in heat diffusion was set for early in 1812. A revised version of the 1807 paper won the prize in 1812, but it also remained in manuscript; so Fourier prepared a third version, which was published in 1822 as the book *Théorie analytique de la chaleur*.

Laplace had also helped Fourier with the following lacuna: in 1807 Fourier could not solve the diffusion equation with a boundary condition over an infinite interval. One reason was the periodicity of the trigonometric functions; but with an infinite body no surface diffusion in the sense of equation (2) could take place anyway, so the methodology of the solution by series broke down. Laplace solved this problem in 1809, by producing the integral solution

$$y = \frac{1}{\sqrt{\pi}} \int_0^\infty \exp(-u^2) f(x + 2u(Kt)^{1/2}) \, du, \tag{4}$$

of equation (1) with $h = 0$ over $[0, \infty]$ of x, with the initial temperature distribution $f(x)$ encased under the integral sign. This result gave Fourier the clue that enabled him to find soon afterwards his own integral solution form (via the contemporary rarity of an integral equation; see §3.10) and thence the 'Fourier integral equation' for $t = 0$, which he wrote in forms such as

$$f(x) = \frac{1}{\pi} \int_0^\infty f(t) \, dt \int_0^\infty \cos \, [q(x - t)] \, dq \tag{5}$$

(Grattan-Guinness *1990*: Chaps 9 and 10; see also §3.11–3.12 on mathematical methods).

2 FOURIER'S CONTRIBUTIONS: HEAT THEORY

One important problem concerned the values of the n_r in the non-harmonic series: it was essential for the validity of the solution (3) as heat theory that all these values were real (and infinite in number); for if any were complex, then the corresponding t-term would become trigonometric, and thereby inadmissibly periodic. Fourier used his knowledge of the theory of equations (§4.10) on this problem, but he could produce only plausible, but not rigorous, arguments for their reality.

As this work was applied mathematics, Fourier had to confront it with experimental evidence. Although he worked diligently, corroboration was very slight; a defect of Fourier series is that they are not very number-friendly. Among his successors, C. Despretz carried out experiments on heat diffusion in both solid bodies and fluids which gave some encouragement.

Regarding the nature of heat itself, the prevailing view around 1800 was that thermal phenomena were ascribable to a fluid called 'caloric', although a rival wave theory was developing (Brush *1976*: Chap. 9; see also §9.14). In particular, Laplace used caloric in his molecular interpretation of 'all' physical phenomena (the successes of Etienne Malus and Biot in optics, recorded in §9.1, belong to this movement); so he rederived equation (1) in molecularist terms. Fourier preferred not to address such questions, taking a positivistic view of heat (and its opposite, cold) as basic effects; he was to be an important influence in the 1820s on the emergence of the positivistic philosophy of Auguste Comte. However, he could not be so aloof from the criticism of Newton's law of cooling for high temperatures made by Laplace's followers Pierre Louis Dulong and Alexis Petit in the 1810s on experimental grounds (Fox *1974*).

Despite Fourier's non-conversion to the Laplacian faith, Laplace publicly praised his work around 1820 when, in concurrent work, he and Fourier analysed the Earth as a cooling sphere (Laplace by surface harmonics, Fourier by Fourier integrals – see Grattan-Guinness *1990*: Chap. 12). They came up with estimates for its age of at least 15 million years.

3 FOURIER, POISSON AND THEIR SUCCESSORS

More determined opposition of a Laplacian kind was to come in the 1820s from Laplace's principal mathematico-physical disciple, Siméon-Denis Poisson (Grattan-Guinness *1990*: Chap. 12). Some years earlier he and Fourier had had a rather sarcastic exchange on methods of analysing heat radiation and the conditions for thermal equilibrium. Now, in a suite of papers, Poisson reworked the principles of heat diffusion (such as equations (1) and (2)) for 'any' body, in Laplacian molecular terms; he also analysed a number of desimplified Fourierian cases (such as a bar in two parts of different materials). For solutions he often used Laplace's (4) rather than Fourier's (3). The results are not of major significance, but he made two valuable contributions: the diffusion of heat in a sphere with arbitrary initial distribution, with Bessel and Legendre functions both in place (1823); and a good argument for the reality of the roots of non-harmonic series (1826). He continued this approach, and added his own study of the cooling of the Earth in his treatise *Théorie mathématique de la chaleur* (1835–7).

By the mid-1820s, both men were well in print; in addition, a new generation of French mathematicians was beginning to emerge, and heat theory was a favoured topic (Grattan-Guinness *1990*: 1165–80; Bachelard *1928*). J. M. C. Duhamel took versions of equation (1) generalized to linear second-order differential equations to represent crystalline media, and produced analyses mathematically similar to Augustin Fresnel's recent work on double refraction in optics (§9.1); he also applied Poisson's very recent introduction of divergence theorems (§3.17) in magnetism (§9.10) to justify the surface condition (2), as did the Russian visitor Mikhail Ostrogradsky. On his return from Russia (§11.6), Gabriel Lamé also studied diffusion in crystalline media, but by working in various coordinate systems. Meanwhile, the Swiss visitor J. C. F. Sturm studied diffusion 'in vases' (i.e. in separate vessels, analysed by means of difference equations); his work was not published, but consideration of the reality of roots seems to have led him to his famous theorem of 1829 on the number of zeros of a polynomial $f(x)$ within an interval of values of x (§4.11).

4 SOME LATER WORK

After the 1820s, heat diffusion did not capture the same measure of attention but became absorbed into the growing empire of mathematical physics alongside optics, electromagnetism and elasticity theory (including aether modelling). In addition, in the emergence of energy physics in the 1840s (Planck *1887*) heat theory was a prominent component, with the recognition of the importance of thermodynamics (§9.5); and gas dynamics and geophysics grew rapidly in significance (§9.6). But a variety of studies continued within diffusion itself, often based upon the new ideas described in this article (for excellent surveys, see Hobson and Diesselhorst *1904*; and Burkhardt *1908*: Section 14).

William Thomson (later Lord Kelvin) used Laplace's solution (5) to put forward, around 1850, a theory of 'point-sources' at which heat was suddenly generated. He also played a prominent role in the discussion of the age of the Earth; this became an important interdisciplinary issue in physics, and was a major objection to Charles Darwin's theory of evolution (Burchfield *1975*). Thomson also showed that the diffusion equation could represent other phenomena, such as the flow of electricity in wires; this work had implications for long-distance telegraphy and Brownian motion (see §9.11 and §9.14).

Joseph Bertrand and others applied to questions in heat diffusion the Sturm–Liouville theory of differential equations (§3.15), which itself had been partially inspired by Fourier's treatment of the cylinder. Lamé's interest in coordinate systems extended into his important contributions to differential geometry (§3.4). Possibly at his prompting, the Paris Academy of Sciences proposed in 1858 a prize problem of determining isothermal surfaces; for the solution by Bernhard Riemann of 1861 (published in 1868) contained the analytical presentation of his views on the foundations of geometry (§7.4), which was significant in paving the way for the emergence of tensor calculus (§3.4).

Some proposals drew on contemporary developments in elasticity theory. Crystalline media attracted much interest, partly for their bearing upon crystallography (§9.17). Again, in the 1890s Gustav Kirchhoff imitated the basic equations to analyse the heat generated by the 'inner friction' of a moving fluid. At that time J. Boussinesq used potential theory to propose an aetherian model of 'caloric propagation' in his interestingly entitled treatise *Théorie analytique de la chaleur, mise en harmonie avec la thermodynamique et avec la théorie mécanique de la chaleur* (1901).

With the advent of quantum mechanics (§9.15) a new era was opened, for the basic processes of heat and radiation were examined afresh. Before that, however, heat theory had been enriched by the advent of thermodynamics, to which we now turn.

BIBLIOGRAPHY

Bachelard, G. *1928, Etude sur l'évolution d'un problème de physique. La Propagation thermique dans les solides*, Paris: Vrin. [Repr. 1973.]

Brush, S. G. *1976, The Kind of Motion We Call Heat: A History of the Kinetic Theory of Gases in the 19th Century*, 2 vols, Amsterdam: North-Holland.

Burchfield, J. D. *1975, Lord Kelvin and the Age of the Earth*, New York: Science History.

Burkhardt, H. K. F. L. *1908*, 'Entwicklungen nach oscillirenden Functionen und Integration der Differentialgleichungen der mathematischen Physik', *Jahresbericht der Deutschen Mathematiker-Vereinigung*, **10**, Part 2, xii + 1804 pp.

Fox, R. *1974*, 'The rise and fall of Laplacian physics', *Historical Studies in the Physical Sciences*, **4**, 81–136.

Grattan-Guinness, I. *1990, Convolutions in French Mathematics, 1800–1840* [. . .], 3 vols, Basel: Birkhäuser, and Berlin: Deutscher Verlag der Wissenschaften.

Grattan-Guinness, I. and Ravetz, J. R. *1972, Joseph Fourier 1768–1830. A Survey of his Life and Work, Based on a Critical Edition of his Monograph on the Propagation of Heat, Presented to the Institut de France in 1807*, Cambridge, MA: MIT Press.

Hobson, E. W. and Diesselhorst, H. *1904*, 'Wärmeleitung', in *Encyklopädie der mathematischen Wissenschaften*, Vol. 5, Parts A–B, 161–231 (article V 5).

Planck, M. *1887, Das Princip der Erhaltung der Energie*, Leipzig and Berlin: Teubner. [2nd edn 1908.]

9.5

Thermodynamics

ERI YAGI

1 INTRODUCTION

The first and second laws of thermodynamics were stated by Rudolf Clausius in a paper published in 1865, when he was 43 years old:

1. The energy of the universe is constant.
2. The entropy of the universe tends to a maximum.

The corresponding mathematical equations for an irreversible process were written as

$$dQ = dU + A\,dW \quad \text{or} \quad dQ = dU + dW, \tag{1}$$

$$\int \frac{dQ}{T} < 0. \tag{2}$$

These equations were also written without the inequality sign, as

$$Q = U - U_0 - W, \tag{3}$$

$$N = S - S_0 - \int \frac{dQ}{T}. \tag{4}$$

For a reversible process, the equations were given as

$$dU = dQ - dW, \tag{5}$$

$$dS = \frac{dQ}{T}. \tag{6}$$

In equations (1)–(6), Q indicates heat, W work, A Clausius's constant (the thermal equivalence of work, corresponding to $1/J$, where J is Joule's constant, the mechanical equivalent of heat), U energy, T absolute temperature, N uncompensated transformation and S entropy. The suffix 0 denotes initial values of the system (gas).

It is useful to show how a modern textbook of thermodynamics writes these laws. Along the lines of the German tradition, Arnold Sommerfeld's

1171

Thermodynamik und Statik (3rd edn, 1964), for example, has the following expressions:

$$dU = dQ - dW', \quad \oint dU = 0. \tag{7}$$

$$\oint \frac{dQ_{rev}}{T} = 0, \quad dS = \frac{dQ_{rev}}{T} \quad \text{for a reversible process;} \tag{8}$$

$$\oint \frac{dQ'}{T} < 0 \quad \text{for an irreversible process.} \tag{9}$$

Sommerfeld's and Clausius's notations are similar except for Clausius's use of the integral sign '∫' instead of '∮' for a circular process; the latter was not used in Clausius's time.

The main difference is with the first law. Clausius wrote the equation with dU on the left-hand side only for the reversible process in connection with the second law, for which dS is also on the left-hand side. This difference is very important in the development of the theory of heat. In Clausius's eight famous papers, compiled in his *Abhandlungen* (1864), he always writes the first law with dQ on the left-hand side to show its structural content. It was in the last (ninth) paper, written in 1865, that he expressed the first law with dU on the left-hand side for a reversible process in connection with the second law. This is well explained in that the entropy change dS was proposed by Clausius primarily as a complete differential corresponding to the energy change dU which he had already introduced as a complete differential in his first paper, published by the Berlin Academy in 1850.

Let us now consider some prehistory of these laws (Bryan *1903*).

2 THE NATURE OF HEAT

There were two trains of thought concerning the nature of heat in the eighteenth and early nineteenth centuries. The older one was the material theory of heat (caloric theory); the other was the mechanical theory of heat. Antoine Lavoisier, known as the founder of the law of conservation of mass in chemical reactions, placed 'caloric' at the head of his table of chemical elements in the eighteenth century. N. L. Sadi Carnot's important work on the motive power of fire, *Réflexions sur la puissance motrice du feu* (1824), which introduced the ideal (Carnot) cycle, was written on the basis of caloric theory.

Ten years later, Sadi Carnot's work was expressed analytically by Emile Clapeyron, with the help of a graphic representation (Section 4). This memoir (in the *Journal de l'Ecole Polytechnique* in 1834) had a great influence on both William Thomson (later Lord Kelvin) and Clausius, who wrote their first papers on the motive power of heat based only on Clapeyron's work rather than on Sadi Carnot's.

That heat is a mechanical effect, not a substance (caloric), was shown through experiments on the friction of solids by Benjamin Thomson (Count Rumford) and Humphry Davy in, respectively, the late eighteenth and early nineteenth centuries. The mechanical equivalence of heat was established by James Prescott Joule through his experimental research between 1843 and 1850. His final measurement on the mechanical equivalent of heat, *J*, was published in *Philosophical Transactions* in 1878. Through these experiments, Joule used two kinds of heat: heat evolved by the friction of fluids, and heat generated by an electric current. For further details see Cardwell *1989*: 29–184; on the power technology and background of Carnot's work, see Fox *1986*: 1–57.

3 THE FIRST LAW OF THERMODYNAMICS (CONSERVATION OF ENERGY)

As mentioned above, the mechanical equivalence of heat was proposed by Joule on the basis of his experiments. In 1842, the 42-year old Robert Mayer, a German physician on a Dutch ship sailing to Southeast Asia, proposed on the basis of physiological measurements made in Java the conservation of energy and the mechanical equivalence of heat. Hermann von Helmholtz published a theoretical approach to the conservation of energy in his *Über die Erhaltung der Kraft* (1847). His expression for the first law was

$$\tfrac{1}{2}mV^2 - \tfrac{1}{2}mv^2 = -\int_r^R \phi\,dr. \tag{10}$$

The left-hand side of equation (10) indicates the change in kinetic energy (of a mass under a central force) which was called *lebendige Kraft*, and the right-hand side the work done by an elastic force ϕ. Helmholtz demonstrated the conservation of various kinds of energy such as mechanical, thermal, electrical and electromagnetic.

Along the lines of Clapeyron, Clausius succeeded in obtaining an analytical expression of the first law of thermodynamics in his first paper. He arrived at the ratio equation (14) below from his understanding that both work produced and heat expended must be the second-order differentials in the infinitesimal Carnot cycle (see Figure 2). Since

$$\text{heat expended} = \left[\frac{d}{dt}\left(\frac{dQ}{dv}\right) - \frac{d}{dv}\left(\frac{dQ}{dt}\right)\right] dv \, dt \tag{11}$$

and

$$\text{work produced} = \frac{R \, dv \, dt}{v}, \tag{12}$$

where R is a gas constant, then

$$\text{heat expended/work produced} = A, \tag{13}$$

or

$$\left[\frac{d}{dt}\left(\frac{dQ}{dv}\right) - \frac{d}{dv}\left(\frac{dQ}{dt}\right)\right] dv \, dt \Big/ \frac{R \, dv \, dt}{v} = A. \tag{14}$$

(Clausius knew Carl Jacobi's notation for partials, $\partial Q/\partial v$ and $\partial Q/\partial t$, but he did not use them.) Equation (14) became

$$\frac{d}{dt}\left(\frac{dQ}{dv}\right) - \frac{d}{dv}\left(\frac{dQ}{dt}\right) = \frac{AR}{v}. \tag{15}$$

Then the following expression, the first analytical form for the first law of thermodynamics, was deduced in 1850:

$$dQ = dU + AR(a + t) \, dv/v \tag{16}$$

(or $dQ = dU + A \, dW$, which is the same as equation (1)). Here U was primarily (originally) regarded as an arbitrary function of the volume v and temperature t of a gas (Yagi *1984*: 182–6).

Judging from equation (15), Clausius knew that dQ is not a complete differential as we use it today. His confusing usage of the term 'complete differential' is discussed in Section 9.

4 THE CARNOT CYCLE

The first graphical presentation of the Carnot cycle was made by Clapeyron. The original cycle has the following four processes. During the first process a gas is expanded (in contact with a heat reservoir) at a constant temperature, shown by the curve AB in Figure 1. During the second, the gas is continuously expanded adiabatically (the curve BC). During the third, the gas is contracted (in contact with another heat reservoir) at a constant temperature lower than the first (the curve CD). During the fourth, the gas is continuously contracted (also adiabatically) and returned to its initial

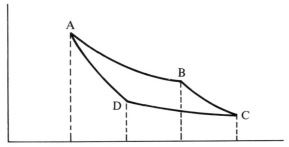

Figure 1 Clapeyron's diagram of the Carnot cycle

condition (the curve DA). The ordinate is pressure p, and the abscissa volume v.

The infinitesimal Carnot cycle was also presented by Clapeyron (Figure 2). The quadrilateral abcd was regarded as a parallelogram of area bn × eg (or fh) = $dp\,dv$. By the law of Mariotte and Gay-Lussac, $pv = R(a + t)$ and bn = $dp = R\,dt/v$, so Clapeyron obtained $R\,dv\,dt/v$ for the work (action) produced during the whole process.

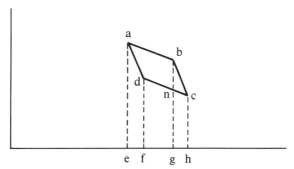

Figure 2 Clapeyron's diagram of the infinitesimal Carnot cycle

5 THE SECOND LAW OF THERMODYNAMICS FOR A REVERSIBLE PROCESS

The concept of the second law, first presented in Sadi Carnot's *Réflexions*, aimed to answer an important technological issue of those days: fuel economy in heat engines. Carnot expressed his principle in terms of caloric theory:

Whenever there is a difference in temperature, motive power can be produced. The converse is also true: whenever there is power which can be expended, it is possible to bring about a difference in temperature and disturb the equilibrium of caloric.... The motive power of heat is independent of the working substances that are used to develop it. The quantity is determined exclusively by the temperatures of the bodies between which, at the end of the process, the passage of caloric has taken place. (Quoted in Fox *1986*: 67, 76–7)

The analytical expression of Carnot's principle was provided by Clapeyron, as mentioned in Section 2. He treated the quantity of supposedly conserved heat dQ as a complete differential of two variables, volume v and pressure p. He derived the following ratio equations by the use of the infinitesimal Carnot cycle: since the heat (absorbed) is given by

$$dQ = \left(\frac{dQ}{dv}\right) dv + \left(\frac{dQ}{dp}\right) dp \quad \text{or} \quad dQ = \left[\frac{dQ}{dv} - \frac{p}{v}\left(\frac{dQ}{dp}\right)\right] dv, \quad (17)$$

where $dp = -(p/v)\, dv$, from the law of Mariotte and Gay-Lussac for a unit weight of a gas:

$$pv = R(a + t) \quad \text{and} \quad \text{work (action)} = R\, dv\, dt/v, \quad (18)$$

then

$$\text{action (work)/heat absorbed} = \text{maximum effect} \quad (19)$$

and

$$\text{maximum effect} = dt/C(t), \quad (20)$$

where $C(t)$ is Carnot's function of temperature. Here the concept of absolute temperature appeared in an analytical expression.

The first proposal of an absolute temperature scale (all degrees specified relative to an absolute zero) based on the Carnot principle and drawing upon the mechanical theory of heat was made by William Thomson in a paper published in 1848.

6 CLAUSIUS'S VERSION OF THE SECOND LAW

The first analytical expression based on the mechanical theory of heat was presented by Clausius, who accepted Clapeyron's expression as the first-order approximation in a mathematical sense for the infinitesimal Carnot cycle (Yagi *1984*: 186–9). Having recognized that the above approach for the second law could not result in anything new, Clausius came up with another new functional approach in his fourth paper, published in 1854, to find a quantity which he called the 'equivalence value' for two kinds of transformation. The first kind meant a heat transformation between work and heat, the second kind a heat transmission between higher and lower temperatures (or lower and higher ones), which was also regarded as a kind of transformation.

The concept of entropy, Q/T, for a reversible process was first proposed by Clausius as the equivalence value through these two kinds of transformation, explained below. Assuming that the equivalence value from work to heat Q during the first kind of transformation must be proportional to the heat produced, and that it can depend only on the temperature t, he expressed the value as $Qf(t)$, where $f(t)$ is a function of temperature. In the second kind of transformation it was also assumed that the equivalence value for heat transmission Q from temperature t_1 to t_2 must be proportional to the heat transmitted, and can depend only on these two temperatures. The equivalence value for the second kind of transformation was expressed as $QF(t_1, t_2)$, where $F(t_1, t_2)$ is a function of both temperatures t_1 and t_2.

Clausius himself defined these two functions in the following way. The two kinds of transformation were treated at the same time during the process of changes by the use of the Carnot cycle. First, during the ordinary Carnot cycle, heat Q at temperature t is changed to work, while heat Q_1 is transmitted from temperature t_1 to t_2. The total equivalence value for this case is assumed to be zero:

$$-Qf(t) + Q_1 F(t_1, t_2) = 0. \tag{21}$$

Second, during the reversed Carnot cycle, heat Q' at temperature t' is produced by work, while heat Q_1 is transmitted from temperature t_2 to t_1. The total equivalence value is also assumed to be zero:

$$Q'f(t') + Q_1 F(t_2, t_1) = 0. \tag{22}$$

Adding these two equations gives

$$-Qf(t) + Q'f(t') = 0, \tag{23}$$

since, on the basis of physical reality, the following property is assumed of F:

$$F(t_1, t_2) = -F(t_2, t_1). \tag{24}$$

In order to reconstruct equation (23), which has two terms of the first kinds of transformation, as a newly combined type of Carnot cycle, Clausius proposed the following two kinds of transformation: heat $Q' - Q$ at temperature t' is produced by work, while heat Q is transmitted from temperature t to t', where Q' and t' are larger than Q and t respectively. The equation of the total equivalence value for this newly combined cycle is

$$(Q' - Q)f(t') + QF(t, t') = 0. \tag{25}$$

From equations (23) and (25), the form of the functions was obtained as

$$F(t, t') = f(t') - f(t). \tag{26}$$

Finally, $f(t) = 1/T$ was given, where T was chosen as the absolute temperature $(= 273 + t)$. Therefore the functional form of the equivalence value, $Qf(t)$, was Q/T. Hence the transmission of heat Q from temperature t_1 to t_2 has the equivalence value

$$Q\left(\frac{1}{T_2} - \frac{1}{T_1}\right) = \frac{Q}{T_2} - \frac{Q}{T_1}. \tag{27}$$

If bodies K_1, K_2, K_3, ..., K_n serve as heat reservoirs at temperatures t_1, t_2, t_3, ..., t_n, respectively, and that heat Q_1, Q_2, Q_3, ..., Q_n is received by each of these bodies, then the total equivalence value N for all the transformations is

$$N = \sum_i \frac{Q_i}{T_i} = 0. \tag{28}$$

If now Q_i is replaced by dQ, the expression for a circular reversible process, whose irreversible process is given by equation (2), is obtained as

$$\int \frac{dQ}{T} = 0, \tag{29}$$

which was given as equation (6), in Section 1, namely, $dS = dQ/T$, where the entropy change dS was introduced by Clausius as a complete differential corresponding to the energy dU of the system.

7 TOMONAGA'S RECONSTRUCTION OF CLAUSIUS'S METHOD

The above method that Clausius used to define the form of functions in connection with entropy has been considered difficult to follow. A few historians and physicists have criticized its unclearness, among them Truesdell *1980*: 310–16. Shin'ichiro Tomonaga introduced another way to define the form of function in his popular book *Butsurigaku-towa-nandarouka?* ('What is Physics?', 1979).

Three ordinary kinds of Carnot cycle are used: the first one for temperatures between t_h and t_1, the second for between t_1 and t_0, and the third for between t_h and t_0. From a heat reservoir at the higher temperature t_h, heat Q_h is taken, while heat Q_1 is given to a reservoir at lower temperature, t_1. The functional equation is

$$Q_1/Q_h = F(t_h, t_1). \tag{30}$$

Using the second and third ordinary kinds of Carnot cycle, similar expressions are written (where t_0 is the lowest temperature):

$$F(t_h, t_l) = \frac{F(t_h, t_0)}{F(t_l, t_0)}.$$ (31)

The functional relation is then

$$F(t_h, t_l) = \frac{f(t_l)}{f(t_h)},$$ (32)

where $F(t, t_0) = 1/f(t)$ is assumed because of the arbitrary nature of t_0, and $f(t) = T$ (absolute temperature) is introduced. It is interesting to note that Tomonaga mentioned that this treatment was taken from a handwritten note (which might have been made by Carl Runge, the previous owner) on the copy of Clausius's book *Abhandlungen über mechanische Wärmetheorie* (1864) kept at the Library of the Institute of Physical and Chemical Research in Japan.

Enrico Fermi also gave a treatment of this problem (which seems the clearest) in his book *Thermodynamics* (1937). Comparing these treatments (Clausius, Tomonaga and Fermi), we may notice that there were two essential viewpoints for the second law of thermodynamics, proposed by the originator Clausius: (a) consideration of the first kind of transformation between work and heat, whose ratio was defined as the first law of thermodynamics; and (b) consideration of the integrating factor $1/T$, in order to change heat (which itself is not a complete differential) to a complete differential, namely to entropy. However, both Tomonaga and Fermi treated these functions F and f separately from the final form of entropy, whose existence had been familiar to them.

8 THE SECOND LAW OF THERMODYNAMICS FOR AN IRREVERSIBLE PROCESS

As mentioned in Section 5, the second law with the concept of entropy was proposed for the reversible process. The case of an irreversible process began to be discussed by Clausius in his fourth paper on the basis of such actual thermal phenomena as thermal conduction and the production of heat by friction. The fact that the total equivalence value N for the transformations tends to be positive was recognized through these thermal phenomena: namely, that no work is done by the thermal conduction while heat flows directly from a higher to a lower temperature; and there is no heat transmission during the production of heat by friction where work is changed to heat.

These two kinds of transformation have positive equivalence values, as was mentioned in connection with equation (25). Clausius therefore presented the following inequality, in his sixth paper:

$$\int \frac{\mathrm{d}Q}{T} > 0. \tag{33}$$

If dQ is regarded as positive when it represents heat received by the system, namely heat taken from a reservoir, this inequality must change its sign; it then becomes the same as equation (2):

$$\int \frac{\mathrm{d}Q}{T} < 0. \tag{34}$$

This derivation was influenced by William Thomson's way of thinking (Smith *1976*: 304–15).

9 MATHEMATICS IN THERMODYNAMICS

Generally speaking, the striving to understand relations among the quantities of physics as functional equations by Clapeyron, Clausius, Josiah Willard Gibbs and others played the most important role in the development of thermodynamics together with the accumulation of experimental results on heat by men such as Joule and Victor Regnault.

The heat change dQ was generally regarded by Clausius as a non-complete differential (as we would call it today). Therefore it was changed to a complete differential dS by the division of absolute temperature T, called the 'integral denominator'. The way in which mathematics was used in thermodynamics by Clausius has been a source of confusion to scientists, students and historians alike. What is most confusing is his modified definition of a complete differential, which he defined in terms of integrability. For example, dz was expressed by the following equation for two independent variables x and y:

$$\mathrm{d}z = \left(\frac{\mathrm{d}z}{\mathrm{d}x}\right)\mathrm{d}x + \left(\frac{\mathrm{d}z}{\mathrm{d}y}\right)\mathrm{d}y, \tag{35}$$

where dz became integrable when the following condition was fulfilled:

$$\frac{\mathrm{d}}{\mathrm{d}y}\left(\frac{\mathrm{d}z}{\mathrm{d}x}\right) = \frac{\mathrm{d}}{\mathrm{d}x}\left(\frac{\mathrm{d}z}{\mathrm{d}y}\right). \tag{36}$$

Up to this point everything is clear. The confusion begins with another case also called integrable by Clausius. He meant that dz can be integrable through the existence of a functional relation between the two variables x and y, where the order of integration was completely given by the relation. Applying it to thermodynamics, the heat change dQ became integrable by the use of such a functional relation as the law of Mariott and Gay-Lussac; and the following equation was generally called the complete differential equation of the first order:

$$dQ = \left(\frac{dQ}{dv}\right) dv + \left(\frac{dQ}{dt}\right) dt. \qquad (37)$$

An axiomatic approach to the foundations of thermodynamics was taken by Constantin Carathéodory in a paper of 1909. He expressed the first and second laws as his first and second axioms. The following two choices (different from the traditional approach to thermodynamics) were adopted: first, heat was reduced to mechanical work; and second, the system was treated mainly under the adiabatically isolated state instead of the traditional circular process. The first axiom was: $\bar{\varepsilon} - \varepsilon + A = 0$, where $\bar{\varepsilon} - \varepsilon$ indicates the difference between the final and initial energies, and A the external work done by the system. The second axiom was: in each neighbourhood of any given initial state, there is another state that cannot be arrived at by any adiabatic change. Carathéodory dealt with the complete differential equation with n variables (dimensional cases) as Pfaff's problem (§3.14; see Truesdell *1983*: 49–57). For entropy from the viewpoint of statistical mechanics, see §9.14.

BIBLIOGRAPHY

Brush, S. G. *1976, The Kind of Motion We Call Heat: A History of the Kinetic Theory of Gases in the 19th Century*, 2 vols, Amsterdam: North-Holland. [Extensive bibliographies.]

Bryan, G. H. *1903*, 'Allgemeine Grundlegung der Thermodynamik', in *Encyklopädie der mathematischen Wissenschaften*, Vol. 5, Part B, 71–160 (article V 3). [See also article V 5, on 'Technische Thermodynamik' (pp. 232–319).]

Cardwell, D. S. L. *1970, From Watt to Clausius*, London: Heinemann. [Useful information on heat and air engines.]

——*1989, James Joule*, Manchester: Manchester University Press.

Fox, R. (ed.) *1986, S. Carnot, Reflections on the Motive Power of Fire*, Manchester: Manchester University Press, and New York: Lilian Barber. [With an informative introduction.]

Gibbs, J. W. *1889*, 'R. Clausius', *Proceedings of the American Academy of Arts and Science at Boston*, **16**, 458–65. [Also in *Scientific Papers*, Vol. 2, 261–7.]

Hutchison, K. *1973*, 'Der Ursprung der Entropiefunktion bei Rankine und Clausius', *Annals of Science*, **30**, 341–64.

Kim, Y. S. *1983*, 'Clausius's endeavour to generalize the second law of thermodynamics', *Archives Internationales d'Histoire des Sciences*, **33**, 256–73.

Klein, M. J. *1969*, 'Clausius on Gibbs', *Historical Studies in the Physical Sciences*, **1**, 127–49.

Lervig, P. *1982*, 'What is heat? C. Truesdell's view of thermodynamics. A critical discussion', *Centaurus*, **26**, 85–122.

Mach, E. *1986*, *Principles of the Theory of Heat* (transl. from the 2nd German edn of 1900), Dordrecht: Reidel.

Smith, C. *1976*, 'Natural philosophy and thermodynamics', *British Journal for the History of Science*, **9**, 293–319.

Truesdell, C. *1980*, *The Tragicomical History of Thermodynamics*, New York: Springer. [Extensive critique from a modern physicist's viewpoint.]

—— *1983*, *Rational Thermodynamics*, 2nd edn, New York: Springer. [Critiques of Carathéodory.]

Yagi, E. *1981*, 'Analytical approach to Clausius's first memoir on the mechanical theory of heat', *Historia scientiarum*, (20), 77–94.

—— *1984*, 'Clausius's mathematical method and the mechanical theory of heat', *Historical Studies in the Physical Sciences*, **15**, 177–95.

9.6

Geophysics

STEPHEN G. BRUSH

Geophysics employs the concepts and methods of physics together with the resources of mathematics to investigate the present state and past history of the Earth and its fluid envelopes. Since its domain includes not only the Earth's solid crust but all of its interior as well as the oceans, atmosphere and ionosphere, it should not be considered a subfield of geology (in the sense that astrophysics is sometimes considered a subfield of astronomy). Nor should geophysics be regarded as merely an application to the Earth of physical principles established in the laboratory; geophysical researches have also led to major advances in the development of theories such as gravity, heat conduction and fluid dynamics (Brush *1978*). Geophysics has thus played an important role in the history of physical science since the Renaissance (Hall *1976*, Gillmor *1984–*, Brush and Landsberg *1985*).

1 THE MOTION AND SHAPE OF THE EARTH

In the scientific revolution of the seventeenth century, the Copernican hypothesis raised an important geophysical question: does the Earth's postulated motion produce observable effects?

Galileo Galilei believed that the pattern of ocean tides furnished direct evidence of the Earth's motion. Tidal theories developed by Isaac Newton, Daniel Bernoulli, Leonhard Euler and Pierre Simon Laplace incorporated the gravitational action of the Moon and Sun on the ocean and solid Earth along with the effects of rotation, and provided an opportunity to test the law of gravity for extended bodies (see references cited in Brush and Landsberg *1985*: 274–77; and §8.16).

In the eighteenth century, George Hadley and Colin Maclaurin showed that the global pattern of winds and ocean currents (westward in the equatorial regions, eastward at higher latitudes) is a consequence of the Earth's rotation. A detailed quantitative theory of this effect was developed by William Ferrel in the mid-nineteenth century.

Newton, in his *Principia* (1687), showed that the Earth's rotation changes its equilibrium shape from a sphere to an oblate spheroid. The centrifugal force due to rotation of a mass m at speed v in a circle of radius r is mv^2/r, a formula also derived earlier by Christiaan Huygens. For a rotating solid this force is greatest at the equator, where it cancels part of the gravitational force. If the solid is not perfectly rigid it will adjust its shape, flattening at the poles and bulging at the equator. The gravitational force exerted on the bulge by an external body such as the Moon can exert a torque, leading to precession of the axis of rotation. In this way Newton was able to explain the precession of the equinoxes, a phenomenon known to astronomers since Hipparchus (second century BC) (§8.8).

In the eighteenth century, measurements on the shape of the Earth provided an important test of Newtonian physics. If the Earth is flattened at the poles and bulges at the equator, the distance along the surface corresponding to a degree of latitude (as determined, for example, by the altitude of the north celestial pole) will be greater at the poles than at the equator. On the other hand, some theoretical deductions from Cartesian physics, and observations over a limited range of latitudes, suggested that the length of a degree is greater near the equator, implying that the Earth is a prolate spheroid.

To resolve the disagreement, the Paris Academy of Sciences sponsored expeditions to the Arctic regions (led by Pierre de Maupertuis) and to the equatorial regions (led by Charles Marie de la Condamine). The results, though now considered somewhat inaccurate, indicated that the length of a degree is greater near the poles, thus supporting Newton's dynamical and gravitational theories (§8.14).

In the eighteenth and early nineteenth centuries, theoretical analysis of the shape of the Earth and the gravitational attraction between it and external bodies stimulated the development of potential theory and the study of partial differential equations by Maclaurin, Alexis Clairaut, Jean d'Alembert, Adrien Marie Legendre, Laplace and Siméon-Denis Poisson (Todhunter *1873*). The results enriched the resources of mathematical physics, applicable in subsequent decades to theories of electricity, magnetism and hydrodynamics.

2 HEAT CONDUCTION AND THE GEOLOGICAL TIMESCALE

Another geophysical problem played a major role in the history of modern geology, biology and mathematical physics. Scientists in the eighteenth and nineteenth centuries generally believed that the Earth was originally formed as a hot fluid ball, which gradually cooled down and solidified at the outside while retaining a molten interior. In the middle of the eighteenth century,

George Louis Leclerc, Comte de Buffon, proposed to estimate a minimum value for the age of the Earth by investigating how long it would take for it to cool to its present temperature. On the basis of experiments with iron balls of different sizes Buffon obtained a value of 75 000 years, contrary to the biblical chronologies which allowed only 4000 to 6000 years since the Creation.

At the beginning of the nineteenth century, Joseph Fourier developed a mathematical theory of heat conduction, motivated in part by the need for a better method of analysing the Earth's cooling (§9.4). He derived a differential equation for the temperature as a function of space and time. To solve his heat-conduction equation, Fourier postulated that any such function could be represented as a sum of an infinite series (or, in some cases, an integral) of trigonometric functions, whose coefficients could be determined so as to satisfy the boundary conditions and initial conditions (§3.11). Fourier's method became an extremely useful technique in theoretical physics. It also raised difficult mathematical questions about what kinds of functions could actually be precisely represented by Fourier series, and the answers to those questions led to a broadening of the concept of function and stimulated the development of set theory (§3.3 and §3.6).

Applied to the Earth, Fourier's theory implied a cooling period of the order of a hundred million years. But by the mid-nineteenth century, geologists following Charles Lyell's 'uniformitarian' philosophy were assuming that processes such as erosion and weathering had been working on the Earth's surface for even longer periods. Charles Darwin suggested (*Origin of Species*, 1859) that periods such as 300 million years or more might be available for biological evolution. The influential British physicist William Thomson (later Lord Kelvin) used Fourier's theory to estimate that the Earth's surface has been solid for less than 100 million years, and perhaps as little as 20 million years, contrary to the assumptions of uniformitarian geology. Thomson's calculation thus provided, indirectly, an argument against Darwin, by suggesting that there has not been enough time for evolution to have taken place by the slow process of natural selection (Burchfield *1975*). (Thomson himself did not reject evolution, but argued that it had proceeded with divine guidance.)

3 THE INTERNAL STRUCTURE OF THE EARTH

In addition to casting doubt on the geological timescale accepted by British geologists, Thomson also used geophysical arguments to refute the prevalent assumption that the Earth has only a thin solid crust (about 80 km thick) surrounding a molten interior. The same gravitational forces (exerted by the Moon and Sun) that raise tides in the Earth's oceans would produce similar effects on the hypothetical liquid under the crust, causing the crust to rise and fall, with the result that no net ocean tides would be observable – or, more likely, the crust itself would break up and sink to the centre. Thomson concluded that the entire Earth is solid, 'as rigid as steel'. This geophysical inference undermined the most popular geological explanations of mountain-building and other processes, based on the assumption of a liquid region not far below the surface (Brush *1979*).

Around 1900, research on the age and physical state of the Earth was transformed by two developments. First, the analysis of wave motions inside the Earth produced by earthquakes (i.e. the science of seismology) made it possible to infer physical properties such as compressibility and rigidity at various depths. This was done by observing the time taken for waves of two different types – transverse and longitudinal – to travel along postulated paths within the Earth, and then constructing a model for the variation of wave velocity with depth. (The analysis employs an integral equation (§3.10) studied by Niels Abel in 1826.) In this way Beno Gutenberg was able to detect the boundary between mantle and core at a depth of 2900 km (identified as a discontinuity in wave velocity). In 1926, Harold Jeffreys showed that the core must have essentially zero rigidity (so that it does not propagate transverse waves), and so must be fluid. Ten years later Inge Lehmann detected the Earth's small inner core, subsequently found to be solid (Brush *1980*).

4 REVISION OF THE TIMESCALE

The second development at the beginning of the twentieth century was radiometric dating, based on the discovery of radioactivity by Henri Becquerel and the isolation of radium by Marie and Pierre Curie. Radium in the Earth's crust could generate enough heat to compensate for that radiated into space, thus invalidating the assumptions used in Thomson's estimate of the age of the Earth. Moreover, measurements of the proportions of uranium, radium and lead in rocks could be used to estimate how long those rocks had been solid, assuming that radioactive decay has been proceeding at a fixed rate. By 1920, estimates of 2000 or 3000 million years were generally accepted. More refined analyses, using ratios of different isotopes of lead, allowed C. C. Patterson in 1953 to determine the presently accepted age of the Earth, 4500 million years (Brush *1989*).

A third geophysical method, widely used in the mid-twentieth century, is the determination of the magnetization of rocks that solidified and cooled below the Curie point in past epochs when the Earth's magnetic field may have been reversed. In the 1960s this approach, known as palaeomagnetism, led to the construction of a chronology of geomagnetic reversals extending back several million years. The chronology, which depended on radiometric dating, was found to agree with patterns of magnetism found on the ocean floors, if one assumed that material had oozed up from mid-ocean ridges and spread outwards along the ocean floor at a rate of about 20 mm per year (Frankel *1982*).

The most spectacular advance in geophysics has been the 'revolution in the Earth sciences' that led to the establishment of continental drift and plate tectonics in the 1960s. Combining the results of seismology, radiometric dating and palaeomagnetism with new analyses of heat conduction through the Earth's crust, geophysicists created a new model of a dynamic Earth evolving through time, and persuaded the scientific community to accept this model (Glen *1982*; Frankel *1987, 1990*; Le Grand *1988*).

BIBLIOGRAPHY

Brush, S. G. *1978*, 'Planetary science: From underground to underdog', *Scientia*, **113**, 771–87.

—— *1979*, 'Nineteenth-century debates about the inside of the Earth: Solid, liquid or gas?' *Annals of Science*, **36**, 225–54.

—— *1980*, 'Discovery of the Earth's core', *American Journal of Physics*, **48**, 705–24.

—— *1989*, 'The age of the Earth in the twentieth century', *Earth Sciences History*, **8**, 170–82.

Brush, S. G. and Landsberg, H. E. *1985*, *The History of Geophysics and Meteorology: An Annotated Bibliography*, New York and London: Garland.

Burchfield, J. D. *1975*, *Lord Kelvin and the Age of the Earth*, New York: Science History. [Repr. with a new 'Afterword', 1990, Chicago, IL: University of Chicago Press.]

Frankel, H. *1982*, 'The development, reception, and acceptance of the Vine–Matthews–Morley hypothesis', *Historical Studies in the Physical Sciences*, **13**, 1–39.

—— *1987*, 'The continental drift debate', in H. T. Engelhardt, Jr and A. L. Caplan (eds), *Scientific Controversies*, New York: Cambridge University Press, 203–48.

—— *1990*, 'The development of plate tectonics by J. Morgan and D. McKenzie', *Terra Review*, **2**, 202–14.

Gillmor, C. S. (ed.) *1984–*, *History of Geophysics*, Washington, DC: American Geophysical Union. [A continuing series.]

Glen, W. *1982*, *The Road to Jaramillo: Critical Years of the Revolution in Earth Science*, Stanford, CA: Stanford University Press.

Hall, D. H. *1976*, *History of the Earth Sciences During the Scientific and Industrial Revolutions, with Special Emphasis on the Physical Geosciences*, Amsterdam: Elsevier.

Le Grand, H. E. *1988*, *Drifting Continents and Shifting Theories*, Cambridge: Cambridge University Press.

Todhunter, I. *1873*, *A History of the Mathematical Theories of Attraction and the Figure of the Earth*, London: Macmillan. [Repr. 1962, New York: Dover.]

9.7

Meteorology

ELIZABETH GARBER

1 INTRODUCTION

Studies of the behaviour of the atmosphere include attempts at understanding large- and small-scale physical processes as well as predicting their outcome. Understanding physical processes and forecasting were two separate activities until the early twentieth century, the cost of the latter activity being underwritten by the state that did not always appreciate the necessity for the former. Currently, forecasting depends on the numerical manipulation of models of physical processes made available through the computer. Understanding the physical processes depends on complex physical models developed through the coordinated collection and exchange of weather data around the globe, and the numerical calculation of particular solutions to the mathematical equations expressing the physical conditions.

Modelling the atmosphere is complex. Weather depends ultimately on the unequal heating of the atmosphere and the transformation of that heat into kinetic energy on a spinning, moving Earth; theory requires joining together hydrodynamics and thermodynamics.

Modern meteorology began with the idea of the atmosphere. We live at the bottom of an ocean of fluid that exerts pressure. This was expressed as Boyle's (Mariotte's) law in the seventeenth century and connected to temperature in the gas laws in the early nineteenth century. The importance of the moisture content and water vapour in the atmosphere was explored by John Dalton, again in the early nineteenth century (Frisinger *1977*).

While a great deal of weather data were collected during the latter half of the eighteenth century (recently meteorologists have reconstructed the weather of the 1780s across Europe using these data), no useful physical imagery of processes, or mathematical expressions for those processes, existed until the following century. The piecemeal assembly of the mathematical equations proceeded as more and better data revealed the complexity of weather patterns and their three-dimensional qualities. It was only after the thermal processes that drove storms were considered separately from the overlapping problem of the dynamics of their motion through the atmosphere that both processes were clarified. This was achieved after a generational conflict in the mid-nineteenth century.

Thermodynamics was the key to solving the dynamics of storms (Garber *1976*). William Thomson (later Lord Kelvin) was the first to use the first law of thermodynamics to explore the thermal changes in the atmosphere. But he stopped short at an expression for the heat released on condensation. What was needed was an expression for the change in temperature of a rising parcel of moist, warm air through a colder, drier atmosphere, and the subsequent release of condensed vapour and large quantities of heat convertible into kinetic energy. What happened to all this energy was left to the French meteorologist Theodor Reye and engineer H. Peslin to explore in their investigations of the energy budgets of storms.

The study of the hydrodynamical problem of the motion of the atmosphere on a rotating Earth (general circulation) was begun in the 1858 studies of vortices in an incompressible fluid by Hermann von Helmholtz (§8.5, Section 5). He used Euler's hydrodynamics and the equation of continuity, which led to stable vortex motion. For the atmosphere Helmholtz studied the propagation of small, wave-like disturbances of this equilibrium. The American meteorologist William Ferrel joined the two aspects, hydrodynamic and thermodynamic, of the problem in the 1870s (Kutzbach *1979*). From the 1860s, Francis Galton used statistical methods to prepare weather charts for newspapers.

2 EARLY ATTEMPTS TO MODEL THE ATMOSPHERE MATHEMATICALLY

To render the mathematics of the atmosphere tractable, Helmholtz treated his wave-like disturbances as perturbations using the perturbation techniques of astronomy. The intractable non-linear equations were thereby reduced to linear ones.

This abstract solution did not help the meteorologist, who was regarded as a forecaster rather than a scientist. Forecasters face the complexities of daily weather changes and prognostication. Nation states began to fund meteorology and forecasting as the weather affected both international trade and agriculture, both of which were enmeshed in policy. And in more traditional state concerns, weather could be crucial. The impact of weather on warfare was graphically illustrated in the Crimean War. On 14 November 1854 the British and French fleets were badly damaged in the Black Sea; a subsequent enquiry revealed that the storm had passed through

Europe the week before the disaster. Within a decade most European states had meteorological bureaucracies, cooperatively sharing data through the telegraph services.

Forecasting became critical in the First World War with the development of military, then civilian, air power. In this era forecasting depended on a good library of synoptic charts accumulated since the establishment of international cooperation. The forecaster juxtaposed past weather patterns with data gathered from across Europe, the USA and elsewhere. The meteorologist constructed his forecast with synoptic charts mapping the movement of wind etc., comparisons of present with past weather data, and knowledge of local weather patterns. Another method was to trace the weather through past decades and construct cycles of temperature and other parameters using Fourier analysis. The coefficients in the terms usually represented monthly means of temperature. The major problem was curve fitting (§10.4); a regression equation and the use of correlation coefficients usually defined the best fit (see Sheynin *1984* on statistical methods). However, none of these methods was satisfactory, especially for air-travel purposes.

These graphical and arithmetic methods were independent of physical processes. In 1902 the Norwegian meteorologist Vilhelm Bjerknes expressed the rash intention of generating accurate forecasts from mathematical equations that described the physical processes of the atmosphere. He also recognized that the atmosphere was heterogeneous and compressible, so that Helmholtz's assumptions were invalid. He also stated the general form of his theory that can be expressed in the three equations of motion,

$$\left.\begin{aligned}
\frac{\partial u}{\partial t} &= -u\frac{\partial u}{\partial x} - v\frac{\partial u}{\partial y} - w\frac{\partial u}{\partial z} + fv - \frac{1}{\rho}\frac{\partial p}{\partial x}, \\
\frac{\partial v}{\partial t} &= -u\frac{\partial v}{\partial x} - v\frac{\partial v}{\partial y} - w\frac{\partial v}{\partial z} - fu - \frac{1}{\rho}\frac{\partial p}{\partial y}, \\
\frac{\partial w}{\partial t} &= -u\frac{\partial w}{\partial x} - v\frac{\partial w}{\partial y} - w\frac{\partial w}{\partial z} - g - \frac{1}{\rho}\frac{\partial p}{\partial z},
\end{aligned}\right\} \tag{1}$$

the equation of continuity,

$$\frac{\partial \rho}{\partial t} = -u\frac{\partial \rho}{\partial x} - v\frac{\partial \rho}{\partial y} - w\frac{\partial \rho}{\partial z} - \rho\left(\frac{\partial u}{\partial x} + \frac{\partial v}{\partial y} + \frac{\partial w}{\partial z}\right), \tag{2}$$

and two equations expressing the changes in pressure and temperature

$$
\left.
\begin{aligned}
\frac{\partial p}{\partial t} &= -u\frac{\partial p}{\partial x} - v\frac{\partial p}{\partial y} - w\frac{\partial p}{\partial z} - \frac{c_p p}{c_V}\left(\frac{\partial u}{\partial x} + \frac{\partial v}{\partial y} + \frac{\partial w}{\partial z}\right), \\
\frac{\partial T}{\partial t} &= -u\frac{\partial T}{\partial x} - v\frac{\partial T}{\partial y} - w\frac{\partial T}{\partial z} - \frac{RT}{c_V}\left(\frac{\partial u}{\partial x} + \frac{\partial v}{\partial y} + \frac{\partial w}{\partial z}\right),
\end{aligned}
\right\}
\tag{3}
$$

where u, v and w are the components of velocity in two horizontal directions and one vertical direction, respectively; ρ is the local density; p is the pressure; f is the Coriolis parameter; g is the acceleration due to gravity; c_p and c_V are the specific heat capacities of air at constant pressure and volume, respectively; T is the temperature; and R is the gas constant (see §8.5, Section 3 on hydrodynamics).

Stating the equations is one problem; solving them another. In principle these equations can be used to predict the state of the atmosphere at some future time by using a set of values for the variables u, v, w, ρ, p and T to obtain the initial time derivatives. These can then be extrapolated to give new values for the variables after a short time period (explicit method). The new values then become the initial values for the variables, ready for the process to be repeated. Bjerknes was vague as to how his equations could be used for predictions. His prognostication methods depended on careful synoptic mapping over time. His equations remained a suggestive, yet abstract exercise (Friedman *1989*).

The British meteorologist Lewis Fry Richardson began the numerical solution of these equations in his book *1922* with the calculus of finite differences, using the method of central differences. He also included equations for radiation and turbulence for sets of seven equations for five different layers in the atmosphere. He added the moisture content of the air as one of his variables: the other independent variables were time, height of the layer and latitude and longitude. After eliminating the temperature and solving for the vertical velocity, there are five equations in five unknowns.

Richardson divided the surface of the Earth into quadrilaterals by latitude and longitude, along with his layers. The equations are integrated with respect to height across these layers. The centres of the squares are alternately observing points for pressure, temperature and humidity in one, density and wind velocity in the next. The observations lead to initial values for the variables which are then operated on to yield entities that are finally fed into the terms of the dynamical equations, yielding the velocity components, and so on. Richardson devised forms for each computational step that fed into the dynamical equations. He ended on a whimsical note: he envisaged a hall of calculators, each with a copy of his form, busily calculating his assigned step, forms passing from one calculator to the next for each square on the map; the overseer and master-calculator put together the forecast. He had invented a form of parallel computing, except that his calculators were human. Richardson estimated that he needed 64 000 of them to keep pace with the weather (Ashford *1985*). This was a gross underestimate: his retrospective prediction of the temperature, pressure and wind on one day in 1910 was a failure when compared with the synoptic charts of Bjerknes for that date.

3 COMPUTERS ENTER THE SCENE

Richardson's method was seen as a noble attempt to solve the problem at the heart of meteorology. In the 1930s, to make the problem tractable meteorologists tailored it to fit their mathematics. They used perturbation methods on linear versions of the equations.

In 1946 the US Office of Naval Research in Washington received from John von Neumann at Princeton a proposal to investigate dynamical meteorology and make it accessible to digital computing using the ENIAC (Electronic Numerical Integrator and Computer), which came on line in December 1945 (§5.12). Von Neumann regarded the problem as the most complex and interactive ever conceived and thought it would challenge stored-program machines for years, which it did. He had realized that Richardson's prediction failed because of unreliable initial data, and because of instability, discovered by Richard Courant in 1928, in finite-difference solutions.

However, the ENIAC could not handle Richardson's problem. Jule Charney modified the form of the hydrodynamical equations. On physical grounds he excluded the factors leading to the computational instability. Eventually the problem reduced to the solution of one equation in a single variable. Despite the clumsiness of the hardware and the simplifications, the first numerically calculated predictions of the atmosphere's large-scale motion were as accurate as those of existing methods.

Generalizations of the model on later computers could predict the development of new storms, and were used routinely in daily weather forecasting. Much effort went into understanding the instabilities of the numerical methods and the effects of truncation on energy propagation; higher-order truncations did not lead to greater accuracy. Other numerical methods were developed. One can represent the state of a fluid as a truncated series of superimposed waves whose amplitudes alone depend on time. Changes in the atmosphere are described by changes in the amplitudes, determined from a system of coupled ordinary differential equations derived from the original partial differential equations. These solutions became attractive with the development of the fast Fourier transforms, use of which reduces storage-space requirements and processing time.

Following labelled elements of the fluid moving with the flow, where the spin of each element is conserved, minimizes approximations and is amenable to exact analytical methods. The velocity at time t is deduced from the spin and the position of the element at that time. The size and shape of the elements can be chosen to model the atmosphere where there are localized structures distributed at irregular intervals.

Recent mathematical developments revolve around the realization that the present and future states of the atmosphere are known only probabilistically. Computing the evolution of the probability distribution for a sample of independent predictions using methods applied to the statistical theory of turbulence joins meteorology to the statistical analysis of an ensemble of non-linear systems.

BIBLIOGRAPHY

Ashford, O. M. *1985, Prophet or Professor: The Life of Lewis Fry Richardson*, Bristol: Adam Hilger.

Brush, S. G., Landsberg H. E. and Collins, M. *1989, The History of Geophysics and Meteorology: A Bibliographical Guide*, New York: Garland.

Darwin, G. H. *1908*, 'Bewegung der Hydrosphäre', in *Encyklopädie der mathematischen Wissenschaften*, Vol. 6, Part 1, 5–83 (article VI 1, 6).

Exner, F. M. *1912*, 'Dynamische Meteorologie', in *Encyklopädie der mathematischen Wissenschaften*, Vol. 6, Part 1, 180–233 (article VI 1, 8).

Friedman, M. R. *1989, Appropriating the Weather: Vilhelm Bjerknes and the Construction of a Modern Meteorology*, Ithaca, NY: Cornell University Press.

Frisinger, H. H. *1977, The History of Meteorology to 1800*, New York: Science History. [Repr. 1983, Boston, MA: American Meteorological Society.]

Garber, E. *1976*, 'Thermodynamics and meteorology', *Annals of Science*, **33**, 51–65.

Kutzbach, G. *1979, The Thermal Theory of Cyclones: A History of Meteorology in the Nineteenth Century*, Boston, MA: American Meteorological Society.

Richardson, L. F. *1922, Weather Prediction by Numerical Processes*, Cambridge: Cambridge University Press.

Shaw, W. N. *1926–31, Manual of Meteorology*, 4 vols, Cambridge: Cambridge University Press. [See especially Vol. 4, *Meteorological Calculus*.]

Sheynin, O. B. *1984*, 'On the history of the statistical method in meteorology', *Archive for History of Exact Sciences*, **31**, 53–95.

Thompson, F. D. *1978*, 'A history of numerical weather prediction in the United States', in *History of Meteorology in the United States, 1776–1976*, Boston, MA: American Meteorological Society, 125–52.

9.8

Acoustics

I. GRATTAN-GUINNESS

1 GENERATION, PROPAGATION, VIBRATION, RECEPTION

Acoustics is a very ancient science. Unusually, many of its principles have been known for a long time: the generation of sound by the vibration of components of the sounding source, its transmission by vibration of the air (or, more precisely, of whatever elements were supposed to constitute air) and its detection by processes in the ear. The great advances have come from the discovery of more properties, and from more sophisticated modelling of the processes involved. Mathematics is concerned mainly in the second context, but it has always been intertwined with aspects of physics, technology and instruments, music theory (§2.10) and phonetics; and experiment has come to play a major role. This article naturally focuses on the mathematical aspects: but even then the full panoply of interactions with vibrational aspects of dynamics (§8.1), potential theory (§3.17), pendulum theory (§8.13) and elasticity (§8.6) is too complex to be fully covered here (or in the articles just cited).

The topics treated here are, in order, the velocity of sound, the formulation and solution of the wave equation, the adoption of superposition theories, and some trends in this century. Lindsay *1972* is a valuable reader of primary sources, including several mentioned below; it also contains his own historical article.

2 CALCULATING THE VELOCITY OF SOUND

One of Isaac Newton's important contributions to acoustics was his argument in Book II of the *Principia* that the velocity v of sound in air is given by

$$v = \sqrt{(\text{pressure}/\text{density})} \text{ of the air.} \tag{1}$$

However, while his argument was clever, it led to a substantially inaccurate prediction of the velocity, and unfortunately he fudged his argument to obtain an exact correspondence with the currently accepted value. Various figures tried to rethink the argument, make analogies with the (apparent) propagation of light, and so on (Truesdell *1955*); but a major breakthrough came only around 1800 when Pierre Simon Laplace applied his growing interest in heat theory to suggest that, as the air vibrated as sound passed through it, the effects of extension and contraction caused heating and cooling. In 1802 Jean-Baptiste Biot modified the known equation for the propagation of sound (described in Section 3) to accommodate this idea and get an extra factor into equation (1); five years later Siméon-Denis Poisson wrote an influential paper, also involving recent work on the proportional increase in volume of a gas with temperature, and came up with a version of equation (1) in which all the quantities could be determined by experiment. A decade later, mathematicized models of molecular systems and the understanding of the thermal properties of gases had improved to the point where the distinction between specific heat capacities c_p at constant pressure and c_V at constant volume was recognized, and Laplace and Poisson developed their ideas so that equation (1) was replaced by

$$v = \sqrt{(c_p \times \text{pressure}/c_V \times \text{density})}; \qquad (2)$$

this gave good correspondence with experimental evidence, which had itself improved in quality in the meantime (Grattan-Guinness *1990*: 452–8, 805–18). Furthermore, c_p/c_V related to the distribution of energy among the different modes of action in the molecules. While many further studies were made of this problem and related ones (such as the velocities in various fluids), this work by Laplace and his followers remained classic – and conversely, the problem of the velocity helped the growth of heat theory itself.

3 THE FAMOUS VIBRATING STRING PROBLEM

As was noted in §2.10, the Pythagoreans had been fascinated with the properties of the vibrating string, and its status helped music to become regarded as one of the classical sciences. Mathematicization was difficult, however, for it involved thoughts on the means by which vibrations passed from sounding source to the air, analogies with the motion of the pendulum, the superposition of tones, and so on; and major advance came only with and after Newton. In the mid-1710s his follower Brook Taylor followed some predecessors in arguing that each element in the string behaved like a simple pendulum in that it executed simple harmonic motion, so that the sine function was essential to the analysis. Meanwhile, in France Joseph Sauveur stressed the analogy with pendulums in 'acoustics' (his word) to study the oscillations of the string from its 'fundamental' vibration to the 'harmonic' tones, and noted the 'nodes' at which the string was always at rest, in contrast to the 'loops' where vibration was greatest. All the quoted words are his.

The measure of mathematics in these considerations was modest. Various successors analysed small vibrations for both continuous and discrete cases, finding some differential equations for the motions, and fragments of the theory of various special functions for their solutions (Cannon and Dostrovsky *1981*). They included the young Leonhard Euler, who speculated upon the vibration of bells (considered as a stack of rings with centres lying on the axis), attempted to correct Newton's formula (1) for the velocity of sound, and studied the compound pendulum and the suspended chain. His friend Daniel Bernoulli also considered the last two cases, as well as the suspended rod. Then in the 1750s he proposed an idea which in his earlier work he had skirted: that all the constituents of a sound were provided by its harmonics on the fundamental, so that the shape $f(x)$ of the vibrating string was given by a series of sines:

$$f(x) = \sum a_r \sin rx, \quad r \geqslant 0, \tag{3}$$

and the constants $\{a_r\}$ were to be determined (somehow).

By this time mathematics had advanced far enough into the problem for the partial differential equation for transverse motions to be formulated. The 'wave equation' was the first to be so created (§3.15). It stated that if y is the displacement from equilibrium at time t of the element a distance x from one end, then

$$y_{xx} = y_{tt} \tag{4}$$

(where certain physical parameters are set to unity). This equation, obtained in the mid-1740s by Jean d'Alembert and Euler, was not doubted, but the general solution was disputed. One reason was that partial differential equations were themselves a new topic; but in addition these men gave different interpretations of their preferred functional solution

$$y = f(x + t) + g(x - t), \tag{5}$$

and rejected equation (3) when Bernoulli came to offer it, although it was also a candidate solution (when supplemented with the appropriate trigonometric t-terms, which he did not mention).

The significance of equation (3) for acoustics did not emerge for nearly a century; so the story is postponed until Section 4. But the wave equation (4) was a major concern at once, with its solution (5) (interpreted as wavefronts, for example), and it occupied most of the major mathematicians for the rest of the century (Truesdell *1960*). The wave equation was used in other axis frames, when Bessel and Legendre functions (§4.4) appeared in the solutions. There was a regular interest in discrete models using n linked masses, *à la* compound pendulum: Joseph Louis Lagrange's first essay on sound (1759) contained a (faulty) solution for the vibrating string obtained by taking n to infinity. D'Alembert's contributions included use of potential theory, especially the velocity potential (i.e. the function whose gradient is the velocity) as the dependent variable in the differential equation.

Interest was increased in the 1800s, when the German physicist Ernst Chladni visited Paris to publicize his work on the nodal patterns, or lines of repose, that he found on vibrating membranes. Sophie Germain and Poisson tried different but competing models of the behaviour; their work is more significant for elasticity theory than for acoustics as such (§8.6), but it helped clarify the differential equation for this type of body (namely, equation (4) with y_{x^4} for the spatial variable x), and Germain used the form (3) among her solutions.

By 1830 a significant mass of knowledge on various properties of sound had been accumulated; and it gained an impressive summary in a long article John Herschel *1830* wrote for the *Encyclopaedia Metropolitana*. The first part covered 'Propagation in general', including the differential equations and the calculation of the velocity of sound. The other two parts dealt with 'musical sound' and vibrations in general, and incorporated the motions of bars and plates and vibrational communication. He also noted the recent experimental work on many aspects of sound carried out in Paris by Félix Savart, which were significant for many later students of acoustics. However, there was little in Herschel's piece on equation (3) or its interpretability as the superposition of frequencies.

4 FOURIER SERIES AND SUPERPOSITION MODELS

Daniel Bernoulli had offered equation (3) on purely physical grounds, following his predecessors: he had no idea how to calculate the coefficients. This was one of Joseph Fourier's contributions to the series which we now name after him (§3.11). When he developed his theory in the 1800s, he was well aware that his work on equation (3) (and various related forms) improved its candidacy for the vibrating string problem (Grattan-Guinness and Ravetz *1972*: especially Chaps 7 and 9). But his concern was with heat theory (§9.4), and the interpretation of equation (3) in terms of superposition came only in a paper of 1843 by the physicist Georg Ohm (of the law in electromagnetism). Thereafter, however, it became a standard position in acoustical theory; it was particularly useful when considering both free and forced vibrations (i.e. those which are, respectively, integral to the system or imposed from outside).

Two major works of the mid-century exemplify the acceptance particularly well. One was Hermann von Helmholtz's major study *1863* of physiological acoustics, which appeared in editions until 1877 and in two editions of an English translation. As his main concern was with the consequences of his subject for musical theory and the perception of sound, he confined his mathematical details to appendices; but as an excellent mathematician he appreciated the consequences of his advocacy of Fourier series.

Similarly, at Oxford W. F. Donkin stressed superposition in his posthumous and incomplete *Acoustics* (1870); but it was his Cambridge contemporary Lord Rayleigh, then in his mid-thirties, who produced his treatise *The Theory of Sound* (*1877–8*), which has sustained its authority on the subject to a degree unusual in a developed science. In this work he covered all known aspects of the subject, and also projected ideas for later work, much of which was recorded in the second edition of 1894–6.

For Rayleigh the mathematical treatment was central. To obtain the basic equations he deployed in particular energy-conservation and variational methods; the latter also involved potential theory, with the newest major results (such as Green's and Stokes's theorems) in place. To solve these equations he not only used Fourier series but also cast the wave and related equations in various axis systems and so found solutions involving the special functions. He treated vibrational systems in general; vibrations of fluids, especially of air in its open state and when propagated down tubes or in chambers; resonators; and the vibration of strings, rods and plates. The additions made in the second edition included the noise produced by flames and by jets of fluid; vibration in shells; 'facts and theories of audition', where he drew much on Helmholtz; and some 'electrical vibrations', where he noted some of the effects currently being found in instruments and telephone cables.

Rayleigh also played historian when he had the Royal Society publish in 1892 a paper of 1845 by J. J. Waterston. It had not been appreciated at the time that Waterston had replaced Laplace's molecular model by a kinetic model, showing that the velocity of sound was of the order of magnitude of the average velocity of a molecule, and proportional to it.

One notable aspect of Rayleigh's and others' work of this time was the continuing interest in analogies from optics, as that science itself advanced (§9.1); thus much attention was paid to the reflection, refraction and diffraction of sound. However, the analogy had to be treated carefully, since opticians adopted Augustin Fresnel's model of vibration transverse to the direction of propagation in the aether, whereas acousticians frequently treated longitudinal transmission through body, fluid or gas.

A source of limitation for this analogy was the constraint to linear models (a usual difficulty with Fourier series). In 1858, pioneer studies of non-linear acoustics were made when Bernhard Riemann and the English mathematician Samuel Earnshaw studied cases where the vibration was of large amplitude. Non-linearity was to grow in importance in later concerns.

5 TO THE CENTURY OF SOUND

Many later workers started out from Rayleigh's and/or Helmholtz's treatises. One interesting application was to the design of concert halls, especially reverberation, where the American Wallace Sabine was an important pioneer; apart from exponential-decay laws, however, the mathematical component was modest.

During the early decades of the twentieth century acoustics grew in importance to an astonishing degree. We remember mostly the development of the telephone, and then of radio and the talking cinema; but there were many scientific and technological applications. Some of these were military; for example, underwater acoustics (see Hackmann *1984* for British activity in this area) and shock-waves, where non-linear modelling was essential (Courant and Friedrichs *1948*). The rise of the field is exemplified by the forming of The Acoustic Society of America in 1928.

The mathematical methods already in place were used and extended, and numerical techniques came more to the fore as more sophisticated methods of approximation were developed and matrix theory rose to prominence (§6.7). This was especially the case in the era of analogue sound; now we have the digital phase. The technology of acoustics produces an ever more verisimilar product, while music degenerates into 'sounds'.

BIBLIOGRAPHY

Cannon, J. T. and Dostrovsky, S. *1981, The Evolution of Dynamics. Vibration Theory from 1687 to 1742*, New York: Springer.

Courant, R. and Friedrichs, K. *1948, Supersonic Flow and Shock Waves*, New York: Interscience. [Reprinted 1976, New York: Springer. Useful bibliography.]

Grattan-Guinness, I. *1990, Convolutions in French Mathematics, 1800–1840* [...], 3 vols, Basel: Birkhäuser, and Berlin: Deutscher Verlag der Wissenschaften.

Grattan-Guinness, I. and Ravetz, J. R. *1972, Joseph Fourier 1768–1830* [...], Cambridge, MA: MIT Press.

Hackmann, W. *1984, Seek and Strike: Sonar, Anti-Submarine Warfare and the Royal Navy 1914–54*, London: Her Majesty's Stationery Office.

Helmholtz, H. von *1863, Die Lehre von den Tonempfindungen* [...], Braunschweig: Vieweg. [Edns to 4th, 1877. English transl.: 1st edn 1875, 2nd edn 1885.]

Herschel, J. F. W. *1830*, 'Sound', in *Encyclopaedia Metropolitana*, Vol. 4, 747–824. [Volume reprinted 1845.]

Lamb, H. *1906*, 'Schwingungen elastischer Systeme, insbesondere Akustik', in *Encyklopädie der mathematischen Wissenschaften*, Vol. 4, Part D, 215–310 (article IV 26).

Lindsay, R. B. (ed.) *1972, Acoustics: Historical and Philosophical Developments*, Stroudsburg, PA: Dowden, Hutchinson & Ross.

Miller, D. C. *1935, Anecdotal History of the Science of Sound to the Beginning of the 20th Century*, New York: Macmillan. [Anecdotal indeed, but good coverage and bibliography.]

Rayleigh, Lord (J. W. Strutt) *1877–8, The Theory of Sound*, 1st edn, London: Macmillan. [2nd edn 1894–6; various reprints.]

Sadie, S. (ed.) *1980, The New Grove's Dictionary of Music and Musicians*, 20 vols, London: Macmillan. [See especially the articles 'Acoustics', 'Physics of sound' and 'Sound'.]

Sauveur, J. *1984, Collected Writings on Musical Acoustics. Paris 1700–1713* (ed. R. Rasch), Utrecht: Diapason Press.

Truesdell, C. A. *1955*, 'Editor's introduction' [on sound and fluid mechanics], in L. Euler, *Opera omnia*, Series 2, Vol. 13, Zurich: Orell Füssli, vii–cxviii.

—— *1960, The Rational Mechanics of Flexible or Elastic Bodies 1638–1788*, Zurich: Orell Füssli [as L. Euler, *Opera omnia*, Series 2, Vol. 11, Part 2].

9.9

Capillarity

ALEXANDER RÜGER

Capillary phenomena and their explanation have undergone a considerable change of status in the physical sciences. Whereas the rise or fall of liquids in tubes or the apparent attraction or repulsion of small floating bodies figure only in subordinate chapters in twentieth-century textbooks or in specialized monographs, these phenomena claimed a prominent place in general physics from the early eighteenth well into the late nineteenth century. For James Clerk Maxwell in 1870, for instance, the study of capillary attraction helped to smooth 'the path which leads to the development of molecular physics'. The theory of capillarity was indeed, after the kinetic theory of gases, the most intensively cultivated field of nineteenth-century molecular physics.

1 CAPILLARITY IN NEWTON'S PROGRAMME

An account of capillarity phenomena formed part of Newton's programme, contained in the last Query in the second English edition of his *Opticks* (1718), to explain chemical, optical and other phenomena related to cohesion in terms of strong, short-range forces between the microscopic constituents of bodies. The importance of the phenomena as manifestations of the microscopic forces stimulated experimental work in the early eighteenth century. Building and improving upon earlier observations, F. Hauksbee (1706–12) and J. Jurin (1718–19) empirically established that the height of capillary rise is inversely related to the diameter of the tube; and, furthermore, that this height is independent of air pressure and of the kind and amount of material forming the tube, thus confirming the extremely short-range character of the forces. The extensive but futile attempts in the first half of the eighteenth century to infer the distance laws governing the non-gravitational forces (magnetism, chemical affinities and cohesion) did not, however, contribute to the execution of the Newtonian programme of deducing the phenomena from calculations with microscopic forces.

Alexis Clairaut, in his development of hydrostatics in *Théorie de la figure de la Terre* (1743), was instrumentalist enough to attempt such calculations without assuming a particular force law. A major obstacle in these calculations was, however, his assumption that the cohesive force acts over a small, but 'sensible' range. This led to, for instance, the erroneous inclusion of effects of the tube walls on the bulk of the liquid. The only other serious eighteenth-century attempt to analyse capillary phenomena mathematically, J. A. Segner's of 1751, was not much more successful. Neither Clairaut nor Segner was able to account, for example, for the inverse relation between capillary rise and tube diameter (Hardy *1922*, Millington *1945*, Bikerman *1978*).

2 LAPLACE'S THEORY AND ITS EARLY CRITICS

Under the influence of the Comte de Buffon's version of the Newtonian programme, capillary phenomena were studied in France in the second half of the eighteenth century (Millington *1947*). It is in this context that, at the end of the century, Pierre Simon Laplace developed his programme of reducing 'the phenomena of nature... ultimately to action *ad distans* between molecules' (1808). Optical refraction and capillary action figured as the first applications of 'Laplacian physics'; elasticity, heat, electricity, magnetism and chemical affinities were to follow (Fox *1974*, *1978*; see §9.4 on heat diffusion).

In the first supplement to book 10 of his *Mécanique céleste* (1806), Laplace calculated the pressure (force per unit surface) that a curved surface exerts on the column of liquid in a capillary tube. This capillary pressure is proportional to the mean curvature of the meniscus. According to whether the curvature is negative (concave) or positive (convex), the pressure will be smaller or greater than the pressure under a plane liquid surface, and this pressure difference will cause the liquid in the tube to rise above or sink below the level of the plane surface. Laplace's main result was the (second-order) differential equation for the free surface of a liquid. Integrating this equation in cases of cylindrical symmetry, which leads to the inverse relation of capillary rise and tube diameter, requires a fixed angle of contact between liquid and solid as a boundary condition (Dhombres *1989*).

Essential for Laplace's calculations was that no precise intermolecular force law was needed. It was sufficient to assume that these forces decrease fast enough with increasing distance from the source so that, for instance, the combined effect on a given molecule (at $r = 0$) of all the intermolecular forces $u(r)$, a distance r away, represented by $\int_0^\infty u(r)\,dr$, could be assumed to be constant but 'insensible' in value at any 'sensible' distance r.

The molecular theory of capillarity, however, was incomplete in several respects. First, Laplace had not considered repulsive forces between molecules – forces that were manifest, for example, in thermal expansion or in evaporation (he had assumed the liquid to be incompressible). Thus, he had chosen a quite special molecular model that was in conflict with generally accepted belief. Second, the boundary condition, the constancy of the angle of contact between liquid and solid body, was not deducible from the theory but had to be postulated independently.

Both defects loomed large in Thomas Young's polemic against Laplace's theory. About a year before Laplace, Young had devised his own theory in terms of surface tension (1805). He derived the equation for the capillary surface from considering the surface as analogous to an elastic membrane. These results were independent, Young claimed, from the special molecular models of the French. Only after he had obtained his results from more phenomenological premises did Young attempt to give a (qualitative) model in terms of attractive and repulsive molecular forces.

3 POISSON'S MOLECULAR THEORY

Laplace's theory survived criticisms of this sort unimpaired, and remained in excellent agreement with observations. It was only in 1831 that Siméon-Denis Poisson, in his attempt to radicalize the Laplacian programme into a truly molecular mechanics (*mécanique physique*), cast a dark shadow on the theory of his teacher. The principal criticism was that Laplace had treated matter as a homogeneous distribution of force centres, whereas from a strictly molecular point of view one had to consider aggregates of discrete particles. Laplace's integrals over sources thus needed to be expressed as sums over sources, and one had to investigate whether the approximation of the sum by an integral was admissible (Arnold *1984*). Poisson furthermore claimed that Laplace's assumption of uniform density (incompressibility) of the liquid led to serious difficulties; instead, one had to allow for a rapid variation of the density over the surface of the liquid.

Poisson's complex arguments led to expressions for the cohesive and the capillary pressure in terms of intermolecular forces that differed from Laplace's. But since the exact force law remained unknown, there seemed to be no observable differences between the two approaches: Poisson had simply suggested a different mechanism for capillary phenomena, which led to the same observable consequences as the Laplacian mechanism.

This 'mathematical scandal' (as Dominique François Arago called it) in the French school of theoretical physics supplied its opponents – in particular some German and many British physicists – with welcome arguments against the molecular approach in general (Rüger *1985*). The moral these physicists drew from this case, as well as from other problems in optics and the theory of heat, was that derivations from specific microscopic models (as opposed to phenomenological premises) were generally of very dubious merit since the premises were, to a great extent, arbitrary.

4 GAUSS'S ALTERNATIVE APPROACH

Carl Friedrich Gauss had in fact already succeeded, in 1830, in formulating a theory of capillarity that, although it used Laplace's assumptions about the range of molecular forces, was much less vulnerable to criticism directed against speculative models of the microscopic constitution of matter. The essential ingredients of Gauss's theory were the potential energies associated with the acting forces (gravity, cohesion and adhesion). Applying the calculus of variations to this problem, Gauss required the sum of all potential energies to become extremal, which led to the equation of the capillary surface and to the rule about the constant angle of contact. One of the advantages of this 'energeticist' approach was that no detailed considerations about how the capillary pressure arises from the combined force effects of the molecules were necessary. Gauss's method soon became the standard treatment in lectures and textbooks on theoretical physics in Germany and England. The energeticist approach was further developed by A. Dupré (1869), and finally purged of all recourse to molecular notions, only thermodynamic concepts being required, in the hands of Josiah Willard Gibbs (1878) (Bakker *1928*: Chaps 1–3; Rowlinson and Widom *1982*: Chap 2).

5 THE ROLE OF CAPILLARY PHENOMENA IN THE DEVELOPMENT OF NINETEENTH-CENTURY MOLECULAR PHYSICS

The fact that a strictly molecular treatment of capillary phenomena was not necessary and could be replaced by a phenomenological one did not, however, render these phenomena unimportant in the development of molecular physics. Laplace's theory employed what later became known as a 'mean field approximation': it is assumed that the liquid may be divided into elements of volume which are small compared with the sphere of molecular action, but large enough for them to contain so many molecules that the particles may be regarded as uniformly distributed throughout the volume. For any inhomogeneities, any molecular structures in the substratum, to manifest themselves, one therefore had to find situations where the mean field approximation was not valid. In the 1890s it became clear that the approximation cannot be valid, even for bulk liquids (Rowlinson and Widom *1982*: Chap. 1).

The main field of study here was liquid films, thin enough not to contain 'many' molecules in the vicinity of any given particle. The pioneer of such experiments was Joseph Plateau who, starting in 1861, made soap bubbles and other minimal surfaces beautiful instruments of research (Rosenberger *1890*: 444–7). One could, for instance, optically determine the thickness of a liquid film just before it disintegrated. According to the theory, the tension in the film is constant as long as there are several layers of molecules. The observed decrease in the tension immediately preceding the destruction of the film indicated that it was now reduced to approximately one layer of molecules, thus providing an estimate of the extension of the molecular sphere of action.

This line of research—finding evidence for the discrete constitution of matter from studying the limitations of continuous descriptions of matter – proved especially interesting to the British 'dynamists' (William Thomson, James Clerk Maxwell). It is also noteworthy that Johannes Diderik van der Waals's work (1873) on the equation of state for real gases had close connections with the theory of capillarity (Klein *1974*, Rüger *1985*).

BIBLIOGRAPHY

Arnold, D. H. *1984*, 'The Mécanique Physique of S. D. Poisson [...] Part VIII', *Archive for History of Exact Sciences*, **29**, 53–72.

Bakker, G. *1928*, *Kapillarität und Oberflächenspannung*, Leipzig: Akademische Verlagsgesellschaft (*Handbuch der Experimentalphysik*, Vol. 6).

Bikerman, J. *1975*, 'Theories of capillary attraction', *Centaurus*, **19**, 182–206.

——*1978*, 'Capillarity before Laplace: Clairaut, Segner, Monge, Young', *Archive for History of Exact Sciences*, **18**, 103–22.

Challis, J. *1834*, 'Report on the theory of capillary attraction', *Report of the British Association for the Advancement of Science*, (1833), 253–94.

Dhombres, J. *1989*, 'La théorie de la capillarité selon Laplace: Mathématisation superficielle ou étendue?', *Revue d'Histoire des Sciences*, **42**, 43–77.

Fox, R. *1974*, 'The rise and fall of Laplacian physics', *Historical Studies in the Physical Sciences*, **4**, 89–136.

——*1978*, 'Laplace', in *Dictionary of Scientific Biography*, Vol. 15, Suppl. 1, New York: Scribner's, 358–60.

Hardy, W. B. *1922*, 'Historical notes upon surface energy and forces of short range', *Nature*, **109**, 375–8.

Klein, M. J. *1974*, 'The historical origins of the Van der Waals equation', *Physica*, **73**, 28–47.

Millington, E. C. *1945*, 'Theories of cohesion in the 17th century', *Annals of Science*, **5**, 253–69.

——*1947*, 'Studies in capillarity and cohesion in the 18th century', *Annals of Science*, **6**, 352–69.

Minkowski, H. *1906*, 'Kapillarität', in *Encyklopädie der mathematischen Wissenschaften*, Vol. 5, Part 1, 558–613 (article V 9). [Also in *Gesammelte Abhandlungen*, Vol. 2, 298–351.]

Rosenberger, F. *1890, Die Geschichte der Physik*, Vol. 3, Braunschweig: Vieweg. [Repr. 1965, Hildesheim: Olms.]

Rowlinson, J. S. and Widom, B. *1982, Molecular Theory of Capillarity*, Oxford: Clarendon Press.

Rüger, A. *1985*, 'Die Molekularhypothese in der Theorie der Kapillarerscheinungen', *Centaurus*, **28**, 244–76.

9.10

Mathematical theories of electricity and magnetism to 1900

THOMAS ARCHIBALD

1 INTRODUCTION

Attempts to give mathematical descriptions or theories of electromagnetic phenomena have led quite directly to the creation and diffusion of much important mathematics, and often the theories themselves are of interest to the historian of mathematics because of the insights gained by mathematicians who have worked on them. This article outlines the main developments in electromagnetic theory, prior to the advent of relativity theory, which have been important for the history of mathematics: in essence, developments between 1750 and 1900.

There are of course many rather special theorems, objects and techniques which make their first appearance in papers on electromagnetic questions. Examples are the basic theorems of vector analysis now named after George Green and Carl Friedrich Gauss (§6.2), the method of electric images (which uses Möbius inversions to solve boundary-value problems), and the use of the Heaviside step function and the Dirac delta function (§4.8). In addition, the activities of some mathematicians were shaped in important ways by their studies of electrical and magnetic questions, at least for part of their careers: J. P. G. Lejeune Dirichlet, Bernhard Riemann and Rudolf Lipschitz, and many more. In some instances, entire fields have been substantially transformed by an attempt to deal with electromagnetic questions: this happened with the theory of boundary-value problems, potential theory and, arguably, the theory of Abelian functions. However, this article concentrates on the electrical and magnetic theories that inspired these developments; there are many omissions, and the reader is referred to the bibliography for fuller accounts.

2 THE EIGHTEENTH CENTURY

Eighteenth-century developments have entirely to do with electrostatics, since the production of current in conductors by a battery was not achieved until the end of the century. The key experimental fact which permitted the development of a mathematical theory was the discovery of the existence of two different kinds of charge, achieved by Charles Dufay in 1732 and 1733. It was possible to explain this in terms of various simple models: one assuming that two distinct 'fluids' existed, one termed 'positive' and the other 'negative'; one in which there was only one kind of electric substance, which could be present in excess or defect; and those that explained attraction and repulsion due to electrification by proposing modifications of an all-pervasive subtle aether. Associated with these were a host of physical explanations for these models (for details see Heilbron *1979*). The two-fluid theory was the most common during of the eighteenth century, largely because of its elaboration by Benjamin Franklin in the late 1740s and early 1750s; Franklin's theory won much support because of its success in qualitatively explaining many phenomena besides the electrostatic attraction and repulsion of charged bodies, among them the capacity of the Leyden jar.

The earliest attempt at providing a mathematical model of electrostatics (and also of magnetism) was by F. U. T. Aepinus (1724–1802), a German from Rostock who worked in St Petersburg from 1757. The mathematical theory he proposed in his 1759 Latin *Essay on the Theory of Electricity and Magnetism* (introduced and translated by Home and Connor *1979*) is rather rudimentary by the standards of later mathematical physics. Its aim is not so much to produce results which will be compared with the results of experiment, but to formulate a qualitative theory using mathematical symbolism in a way which might be fruitful for a general assessment of the model.

In Aepinus's theory, the dependence of the force between two charged (or magnetic) bodies on distance and charge was assumed to be unknown. These were experimentally established as Newtonian inverse-square forces by Charles Coulomb in memoirs written in 1785 and 1787 (Société Française de Physique *1884*). (Joseph Priestley had earlier expressed the same view, though he published neither mathematical proof nor experimental verification.) This in turn opened the way to a treatment of the electrical and magnetic forces in a fashion analogous to the treatment of gravitational forces in the work of Joseph Louis Lagrange, Pierre Simon Laplace and many others (Fox *1974*).

3 THE EARLY NINETEENTH CENTURY

Such a Laplacian treatment was provided by Siméon-Denis Poisson for electrostatics in 1812, substantially aided by Laplace and based on approaches in the latter's *Mécanique céleste* of the previous decade. Poisson supposed electricity to consist of two fluids, present in all bodies, behaving in such a way that like particles repel one another, and unlike particles attract, with an inverse-square force. The bodies in the physical world are divided into conductors and insulators; in the former, electricity is free (completely free, in the view of Poisson) to move. The phenomenon of charge results when an excess of one kind of electricity persists in a conductor bounded by insulators. The kinds of question Poisson addressed have to do with the equilibrium distribution of such excess electricity on charged conductors, and with the forces exerted by extended conductors on other bodies at a distance. This view of electrostatics and its fundamental problems would alter little until the work of James Clerk Maxwell.

The key feature of Poisson's electrostatics was the observation that, because of the repulsion of like electricities and the absence of resistance to the motion of electricity in the interior of the conductor, all free charge must migrate to the surface. Furthermore, the surface layer must exert no force on the interior of the body, otherwise there would be an additional separation of electricities in the interior, leading to a change in the body's electric state. Hence, to determine the equilibrium distribution on the surface, it sufficed to find the thickness of the layer at each point such that the resulting force on the interior of the body is zero.

Poisson considered a spheroid coated with an infinitesimal, yet variable layer of electric fluid. He used a potential function V, so that the problem of determining the force exerted by the layer becomes the problem of determining the values of V at any point in space. (The partial derivatives of V give the force components.) Using spherical coordinates, he obtained

$$V = \iint \frac{y'(r')^2 \sin\theta' \, d\theta \, d\omega'}{\{(r')^2 - 2r'x[\cos\theta\cos\theta' + \sin\theta\sin\theta'\cos(\omega - \omega')] + x^2\}^{1/2}}, \quad (1)$$

where the numerator in the integrand is the volume of an infinitesimal portion of the fluid layer, of thickness y' (here the density is assumed to be uniform and equal to unity), and the denominator is the distance between a point (r', θ', ω') on the surface and an arbitrary point in space with coordinates (x, θ, ω). The integration extends over the surface of the sphere. We can just as easily think of y as the mass of the fluid, incorporating the density, as Poisson pointed out.

In the remainder of the memoir, Poisson worked directly with this definition of V, rather than with its differential equation. In the following year he wrote a differential equation for V, the now-familiar 'Poisson equation':

$$\frac{\partial^2 V}{\partial x^2} + \frac{\partial^2 V}{\partial y^2} + \frac{\partial^2 V}{\partial z^2} = -4\pi\rho, \tag{2}$$

where ρ is the charge density. The result was known to Laplace for points outside the mass where the density is zero, and follows by direct calculation. For non-zero density, a problem arises because the distance between the acting mass and the acted-on mass becomes zero in the course of the integration, so that $1/r'$ becomes infinite. Poisson decomposed the region by placing a small sphere around the origin, and easily arrived at the result, though he ignored the necessity of imposing conditions on the density to ensure that the required derivatives exist. The differential formulation was perhaps less important for Poisson than it would be once the work of Joseph Fourier, Dirichlet, Green and others made an approach via boundary-value problems standard (see §3.17 on potential theory).

Important though the works of Poisson were for his contemporaries and for subsequent developments, even by the standards of the time they lacked completeness, not only in their failure to deal with anything besides attraction and repulsion, but also in their silence about the electric current, discovered by Luigi Galvani in the 1790s and then produced using piles of metal plates separated by moistened cardboard by Alessandro Volta. The relationship of current to electrostatic effects was problematic, as was the effect, discovered in 1820 by Hans Christian Oersted, of electromagnetism. These discoveries called for a unified theory of electricity and magnetism, and attempts to provide such a theory (omitting electrostatics) were made by Jean-Baptiste Biot and André Marie Ampère in the early 1820s.

Ampère founded his theory on the interactions between two infinitesimal line-elements, or *éléments de courant*. Arguing that the interaction between two such elements of the circuit could be taken to depend inversely on the square of the distance separating the centres of the elements, on the angle between them in a bilinear fashion, and on the current each element bore, he formulated his circuit-element force law:

$$\mathbf{F} = ii'\left\{\frac{ds \cdot ds'}{r^2} - \frac{3}{2r^4}(r \cdot ds)(r \cdot ds')\right\}\frac{r}{r} \tag{3}$$

(here we use vector notation for compactness of exposition; Ampère used Cartesian notation). Magnetism was treated as a derived phenomenon by postulating the existence of microscopic closed circuits (*courants moléculaires*) inside the magnetized substance (Blondel *1982*). Writing after Ampère published his first efforts at such a theory, Biot and his collaborator Félix Savart instead took magnetism as the fundamental phenomenon, retaining the Laplacian style of Poisson and avoiding the dependence on direction which Ampère had taken considerable rhetorical pains to characterize as Newtonian.

Ampère's force law, when integrated around closed circuits, yielded results in good agreement with observation, and avoided the multiplication of entities associated with the Biot–Savart approach. For this and other reasons, the Ampère law eventually became a fundamental of nineteenth-century electromagnetic theory, though until mid-century magnetic fluids were commonly employed in discussions of magnetism, most notably by Poisson himself in 1824. This approach was mathematically simpler than Ampère's. Indeed, the calculation of macroscopic effects from Ampère's law required the extensive use of line and surface integrals (§3.7), and Ampère's work was the first important mathematico-physical treatise to employ these concepts extensively.

It should be noted that infinitely many other such 'microscopic' laws, differing from that of Ampère by an exact differential, will yield the same result for closed circuits. This fact was important for subsequent developments, since other such microscopic laws were frequently proposed on the basis of physical models of electricity. Ampère himself had avoided a specific physical model.

Shortly after Ampère's memoir, a mathematical description of conduction in filamentary circuits was formulated between 1825 and 1827 by Georg Ohm, at the time a *Gymnasium* teacher in Köln, whose work was modelled on Fourier's study of heat conduction. Ohm's experimental determination of the linear relation between current (as measured by a galvanometer) and potential difference (which he kept roughly constant using a thermocouple) inspired him to formulate a mathematical model which has the same form as that we use today. Ohm's conceptual apparatus was quite different, however, and it was not until his work was reinterpreted by Gustav Kirchhoff in 1849 that Ohm's law took its present form.

4 THE MID-NINETEENTH CENTURY

With the discovery of electromagnetic induction by Michael Faraday in 1831, the agenda for those seeking to formulate a mathematical theory of electrical phenomena was set. A unified theory, whether or not based on a hypothetical model, would have to (a) yield Poisson's laws for electrostatics; (b) explain conduction in terms consistent with electrostatics in such a way as to yield Ohm's law for filamentary conductors; (c) yield Ampère's law (or something macroscopically equivalent to it, like the Biot–Savart law) for the 'electrodynamic' force between electric circuits; and (d) produce a quantitative description of electromagnetic induction, consistent with the foregoing, which was in accord with experimental results. These were minimal requirements, and a host of other phenomena and effects (such as those having to do with capacitance and spark discharge) would also ideally be accurately described by such a theory. Furthermore, general physical principles had to be respected by these models, a fact that became of critical importance with the formulation (by Hermann von Helmholtz, Rudolf Clausius, James Prescott Joule, William Thomson – later Lord Kelvin – and others in the late 1840s and early 1850s) of the principle of energy conservation, which was generally accepted by the late 1850s. Two principal types of theory, each satisfying some of these requirements, emerged as dominant: one, using an action-at-a-distance model, was developed in Germany in the 1840s and 1850s by Franz Ernst Neumann of Königsberg, Wilhelm Weber of Göttingen, and their colleagues and students; the other, the field theory of Maxwell, was elaborated at first almost entirely in Britain in the 1860s and 1870s.

The works of Neumann and Weber may conveniently be grouped together, since together they formed the standard picture of electromagnetic theory on the Continent during the middle years of the century, and defined the problem area and the direction of research (Sommerfeld and Reiff *1902*, Kaiser *1981*, Archibald *1987*). In the period between 1844 and 1848, Neumann found mathematical expressions for the current induced in filamentary circuits by variations in the current or in their relative position. In 1845 he published a derivation of a law which expressed the induced electromotive force (e.m.f.) acting on a circuit (of fixed form) due to a nearby moving filamentary circuit or magnet. He expressed the induced e.m.f. as a function of the velocity of the moving circuit and of the motive force (as given by Ampère's law) exerted on the moving circuit by an adjacent closed circuit. Neumann's work thus provided a basis for the examination of induced currents in a fashion which provided a link with Ampère's theory.

Neumann began with the derivation of an equation which specified the e.m.f. induced in an element of a circuit which is moving in a magnetic field generated by another, stationary circuit. His analysis rested on three basic facts: the proportionality of the motion-induced e.m.f. to the velocity of the motion, Ohm's law and Lenz's law. The latter, formulated on empirical grounds by Emil Lenz in 1834, stated that the induced current would flow in such a direction that the resulting Ampère force would oppose the inducing motion. If the circuit in which current is induced moves rigidly with constant velocity v, and if the induced current in an element of the circuit is i, then the force component in question may be written as $F_v i \, ds$, where F_v is the component of the Ampère force in the direction of motion. Using Ohm's law, integrating around the circuit, and making the simplest assumption consistent with Lenz's law yielded

$$E = iR = -eRv \int F_v \, ds \tag{4}$$

for the electromotive force induced, where R is the resistance. Neumann argued that eR must be a constant, which he denoted by ε.

More important for later developments, Neumann treated induction as the result of a change in a potential between two circuits. This potential was arrived at on mechanical grounds, using the *vis viva* (or energy) principle and the principle of virtual work (§8.1). Neumann employed it to generalize his approach to other cases of induction; in this picture, the induced e.m.f. was the rate of change of the potential. This approach gave his theory the possibility of being linked with the idea of energy conservation. By establishing a phenomenological theory of induced currents which allowed quantitative prediction, Neumann provided a possible test of physical hypotheses about the ultimate nature of electricity and electrical phenomena. No theory which gave results seriously at odds with Neumann's could be admitted without modification. This proved to be a critical test for Weber's theory, published shortly after Neumann's paper.

Weber, who had collaborated with Gauss intermittently since the early 1830s on electromagnetic questions, had been dismissed from Göttingen for political reasons in 1837, and had eventually moved to Leipzig. It was there that he produced his *Elektrodynamische Maassbestimmungen* (1846), which was to be the foundation of most subsequent electromagnetic research in Germany until the late 1860s. Weber sought to unify Poisson-type electrostatics with the theories of Ampère and Neumann by providing a single law which could permit all these to be derived. In so doing, he adapted a model of the electric current proposed by Gustav Theodor Fechner, which posited that when current flows in an electric circuit it consists of two oppositely directed streams of electric particles, one positive and one negative. (The pairing explains the absence of electrostatic effects, since the circuit is locally neutral.) Electromagnetism and induced currents must result from some sort of unbalanced force – in the former case ponderomotive force, in the latter case electromotive. Weber sought an interparticulate force law of the type that Newton had proposed, but to produce the required results his law was made to depend on the relative velocity (actually, the scalar rate of change of the distance between a pair of particles) and the acceleration (likewise scalar) of the particles. Written in modern electromagnetic units, the law Weber proposed was

$$F = \frac{ee'}{r^2} \left[1 - \frac{1}{2c^2} \left(\frac{dr}{dt} \right)^2 + \frac{r}{c^2} \left(\frac{d^2 r}{dt^2} \right) \right], \tag{5}$$

where e and e' are the charges of the particles, r is their separation, and c is the ratio of electrostatic to electromagnetic units – now known to be the speed of light *in vacuo*.

This force is electromotive; electricity was still seen by Weber as distinct from ordinary matter. The Weber law yields an electromagnetic effect by virtue of the fact that electricity cannot move outside conductors, so that when it attempts to do so the conductor is dragged along. The validation of this force law came largely because it yielded Ampère's law for the electromagnetic force between circuit elements, and is therefore subject to the same degree of uncertainty that Ampère's law is. However, Weber's law had the further merit of producing results in agreement with Neumann's induction law (except in a few problematic instances). By 1850, the Weber law and the Neumann induction theory were standard in Germany, and formed the backdrop for research in the area for the next three decades.

The Weber law was problematic on many grounds, however. Its dependence on a highly structured hypothetical model, its use of actions at a distance, and its dependence on velocity were all potential difficulties. Velocity dependence, in particular, made it seem likely that it was not a conservative force, a problem that came sharply into focus following the work of Helmholtz, Clausius and others in the period between 1847 and 1855. These men eventually proposed their own laws, modelled to some extent on that of Weber in that they attempted to give expressions for the force or potential between hypothetical electric substances. Both also submitted Weber's law to searching criticism in the 1870s and early 1880s, leading to an extended and inconclusive debate, which involved them, Weber and others, on the nature of a conservation law.

The dependence of the model on actions at a distance also led to a search for alternatives. In Germany, these took the form of enquiries into models where the electric force or potential was propagated with finite velocity. The earliest such efforts to reach completion were due to Riemann, who wrote a paper on the matter in 1858 (which was published only posthumously in 1867) and who lectured on it in the early 1860s. Such models were vexingly difficult, even in the 'simple' case of a pair of particles. Consider, for example, the case of two moving particles, each of which generates a potential depending on its velocity at time $t = 0$. This potential, received at time $t = t_1$, depends on what is known as the retarded distance and velocity, both of which vary continuously. To calculate how a system evolves with time involves integrating along retarded trajectories, a problem fraught with analytical difficulties – on which Riemann's early paper foundered. Others, most notably Carl Neumann, pursued this line of investigation during the 1860s and 1870s; once certain difficulties are cleared up it remains useful for certain problems today.

5 MAXWELL

While the theories discussed here so far were based on the fairly simple intuitive notion of electricity as a pair of fluids which by their imbalance define charge, and by their motions constitute current, the most successful nineteenth-century theory of electromagnetism was quite differently based. Maxwell's theory, as expressed in his paper 'A dynamical theory of the electromagnetic field' (*1865*) and the *Treatise on Electricity and Magnetism* (1873), differs so profoundly in its approach from those employed by the German writers just discussed that his work was incomprehensible to them (though some, most notably Helmholtz, were influenced by it). The following brief discussion of Maxwell's theory is largely based on the work of Buchwald *1985*, to which the reader is referred for references and further clarification.

In Maxwell's theory, charge and current are 'epiphenomena' (secondary appearances) of underlying processes in what he termed, following Faraday, the electric and magnetic fields. Indeed, Maxwell's mature theory stays completely away from the microstructure of matter and from any consideration of 'electric substance'. Instead, he proposed that certain quantities should be defined at every point in space, such that relations between them (the Maxwell equations) and functions of them (such as energy functions) determine phenomena. These quantities (the fields) may depend on microphysical events, and indeed Maxwell did expend some effort in his early papers on attempting to explain qualitatively how their relations could result from mechanical motions. However, the theory explains only large-scale phenomena, and it is not necessary to have the microscopic model in mind in order to work successfully with it.

Essential to Maxwell's notion is the idea that electrical and magnetic interactions are propagated by means of disturbances in a space-filling aether, a medium 'capable of being set in motion and of transmitting that motion from one part to another, and of communicating that motion to gross matter so as to heat it and to affect it in various ways' (Maxwell *1865*: art. 4). Unlike the Cartesian plenum, however, this medium is capable of elastic yielding. This may be seen by the fact that motion (for example, light propagation) in the aether is not instantaneous. Because of this elasticity, the medium can store kinetic energy as well as potential energy.

Different kinds of ordinary matter, as well as the aether, possess characteristics which may be thought of as scalar variables defined at every point in space. These are customarily denoted by ε, μ and σ; they characterize the specific inductive capacity, magnetic permeability and conductivity of the medium at various points. Of key concern to Maxwell was the action of e.m.f. in regions where conductivity is low − that is, in dielectrics.

The e.m.f. is a 'force called into play during the communication of motion from one part of the medium to another' (Maxwell *1865*: art. 10). When it acts on a dielectric, it produces a state of polarization, the result of an electric displacement. Such displacements, according to Maxwell, are an elastic yielding of the medium in question under the action of the e.m.f. Variations in this displacement are themselves currents. The polarization will be detectable at the boundaries of the dielectric – that is, where the value of ε changes – though not inside the dielectric. For a dielectric bounded by a conductor, the charge resides on the surface of the dielectric rather than on the adjacent conductor, and the sign of the charge is determined with reference to the component of displacement along the inward normal from the dielectric to the conductor. This approach allowed Maxwell to identify charge with discontinuities in the displacement. For him all charge arises in this way; it is exactly the reverse of the Poisson concept, or for that matter of the modern view. Instead of charged particles residing in conductors and causing 'fields', it is the fields that are causative. Conduction currents, which must be carefully distinguished from displacement currents, may be conceived of as a continual series of polarization chargings and dischargings in the bounding dielectric; it is a process of continual decay of induced charge due to the high ratio of σ to ε. Partly because charge and current were epiphenomena of polarization and displacement, the interpretation just outlined is only implicit in Maxwell's writings, rendering these writings exceedingly difficult to comprehend for the Continental theorists (or, for that matter, for the present-day reader).

More familiar to readers today are Maxwell's equations relating the field quantities. These took on various forms, and there is a large literature discussing their evolution in Maxwell's work. In his 1865 'dynamical theory' paper, Maxwell presented 20 equations – reducible now to eight by using vector notation. Some of these, the constitutive equations, express the results in the medium when the field acts; in present-day notation, these are

$$\mathbf{B} = \mu\mathbf{H}, \qquad \mathbf{D} = \varepsilon\mathbf{E}, \qquad \mathbf{J} = \sigma\mathbf{E}, \qquad (6)$$

where $\mathbf{J}$ consists of conduction and displacement current. Others express the relationships between the field vectors which bring about induction and electromagnetism, for example

$$\nabla \times \mathbf{H} = \mathbf{J}. \qquad (7)$$

A striking feature of Maxwell's *Treatise* is its occasional use of quaternion notation. This is generally accompanied by the corresponding Cartesian formulas, however, and so should not have consitituted a major obstacle to understanding. In the hands of Josiah Willard Gibbs and Oliver Heaviside it was transformed into the vector notation familiar today (§6.2).

The power of Maxwell's theory showed itself in its ability to handle problems which from the Continental point of view were intractable or which could not be dealt with in a unified way. Furthermore, it could treat dielectrics and diamagnetics with the same conceptual and mathematical apparatus as conductors, electromagnets and so on – a large advance over the Weber–Neumann style and those of its competitors. Of course it had the novel advantage of treating optics as a branch of electromagnetic theory (§9.1), and while immediate experimental evidence was lacking, it provided a stimulus for research. Maxwell's theory also solved the energy problems of velocity-dependent forces, since by allowing energy to be stored in the field it made it possible to retain the standard definition of energy. Maxwell appears to have regarded energy as being localized, and as flowing continuously through the system. However, the details of energy localization were not worked out by Maxwell himself, but by his students, notably John Poynting.

6 CONCLUSION

The final stage in the pre-relativistic, pre-quantum mathematical theory of electromagnetism was the transition to the electron-theoretic viewpoint, retaining as much of Maxwell's theory as possible. This is a complex story, the main lines of which depend, on the one hand, on the difficulties faced by the Continental writers (such as Hendrik Lorentz and Heinrich Hertz) in understanding Maxwell's concept of charge, and, on the other hand, on the discovery and analysis of a variety of physical phenomena (the Hall effect and the Faraday effect, among others) from the Continental standpoint as informed by an imperfect understanding of Maxwell. The outcome was a theory, familiar today, in which Maxwell's fields are arrived at as the averages over space of associated microscopic quantities based on the electron. The details of this classical theory were being worked out at the same time as relativity theory and quantum mechanics, and often by the same researchers. Hence it is not easy for the historian to distinguish the classical developments from the modern.

BIBLIOGRAPHY

Archibald, W. T. *1987*, 'Eine sinnreiche Hypothese: Aspects of action-at-a-distance electromagnetic theory, 1820–1880', Ph.D. Dissertation, University of Toronto.

Blondel, C. *1982*, *Ampère et la création de l'électrodynamique*, Paris: Bibliothèque Nationale (Ministère de l'Education Nationale, Comité des Travaux Historiques et Scientifiques, Mémoire de la Section des Sciences, No. 10).

Buchwald, J. Z. *1985*, *From Maxwell to Microphysics*, Chicago, IL: University of Chicago Press.

Everitt, C. W. *1974*, 'Maxwell', in *Dictionary of Scientific Biography*, Vol. 9, New York: Scribner's, 198–230.

Fox, R. *1974*, 'The rise and fall of Laplacian physics', *Historical Studies in the Physical Sciences*, **4**, 89–136.

Heilbron, J. L. *1979*, *Electricity in the 17th and 18th Centuries*, Berkeley, CA: University of California Press.

Home, R. W. and Connor, P. J. *1979*, *Aepinus's Essay on the Theory of Electricity and Magnetism*, Princeton, NJ: Princeton University Press.

Kaiser, W. *1981*, *Theorien der Elektrodynamik im 19. Jahrhundert*, Hildesheim: Gerstenberg.

Maxwell, J. C. *1865*, 'A dynamical theory of the electromagnetic field', in *Scientific Papers*, Vol. 1, 1890, Cambridge: Cambridge University Press, 526–97.

Société Française de Physique *1884*, *Collection de mémoires relatifs à la physique*, Vol. 1, *Mémoires de Coulomb*, Paris: Gauthier-Villars.

Sommerfeld, A. and Reiff, R. *1902*, 'Elektrizität und Optik. Standpunkt der Fernwirkung: die Elementärgesetze', in *Encyklopädie der mathematischen Wissenschaften*, Vol. 5, Part 2, 3–62 (article II A 7c).

Whittaker, E. T. *1910*, *A History of the Theories of Aether and Electricity: From the Age of Descartes to the Close of the Nineteenth Century*, London: Longmans, Green. [2nd edn 1951; repr. 1960, New York: Harper.]

9.11

Telecommunication theory

D. W. JORDAN

1 TELEGRAPHY AND TELEPHONY

The first commercial telegraphs date from the 1830s. By 1880 there was a worldwide network, brought into being by ingenious entrepreneur-inventors who assembled batteries, linear circuits, sending and receiving instruments, and relays into complex and powerful systems, but who on the whole made use of only elementary theoretical ideas. However, the growing demands of long-distance transmission, and the recent appearance of the telephone, indicated the need for a sophisticated theory of the behaviour of signals on long lines. The only study widely known in the early 1880s had been made by William Thomson (later Lord Kelvin) and George Stokes in 1855, in connection with the first Atlantic cable project. By solving the differential equations for a long line dominated by resistance and capacitance, they predicted how a signal such as a Morse-code dot would become attenuated over a great distance, and so were able to estimate how much traffic the line could carry. The theory, though correct, was limited strictly to long undersea cables, but it was adopted undiscriminatingly by practitioners during the mid-1880s, provoking a far-reaching controversy over the merits of theory as against practice.

James Clerk Maxwell's *Treatise on Electricity and Magnetism* (1873) transformed circuit theory, taken in a broad sense (§9.10). The most accomplished exponent of the new outlook was Oliver Heaviside, a one-time telegraphist, self-taught and remote from scientific society. He recast Maxwell's theory into a pure and general macroscopic form, interpreted it at many levels, and produced a great body of theory relating to telecommunications. His work in this field, largely carried out between 1884 and 1887, included general linear circuits and bridges, transmission lines and the low-distortion line, terminal apparatus, waveguides and the skin effect; in all cases he considered the effects of a periodic and an impulsive signal. This extraordinary body of work was at first generally ignored, but an attempt in 1887 to suppress his theory of the importance of self-induction in lines and the possibility of a low-distortion line became a scientific *cause célèbre*. Heaviside also established vector analysis as the natural formalism for electrodynamic theory, and greatly extended the scope of operational methods (see §6.2, §4.7–4.8).

2 OPERATIONAL METHODS

The main elements of Heaviside's operational calculus were developed between 1880 and 1887. To illustrate the technique, consider the differential equation $\mathscr{L}(x)(t) = H(t)$, where $\mathscr{L}$ is a linear ordinary differential operator with constant coefficients, and the 'Heaviside unit function' $H(t)$ is defined by

$$H(t) = 1, \quad t > 0; \qquad H(t) = 0, \quad t \leqslant 0. \tag{1}$$

By putting p^n in place of d^n/dt^n for $n \geqslant 0$, and thereafter manipulating p like an ordinary variable, the equation takes the form

$$Z(p)x(t) = H(t), \quad \text{or} \quad x(t) = Z^{-1}(p)H(t), \tag{2}$$

where $Z^{-1}(p)$ is regarded as an operator mapping Heaviside's standard forcing term $H(t)$ into the solution function. For electrical networks the expressions ('impedance operators') corresponding to Z for each branch combine like resistances. The system of simultaneous operational equations corresponding to the differential equations for the currents in the network can be solved algebraically, the resulting operators being rational functions of p. Partial differential equations, which arise from continuum problems, lead to operators which are non-rational functions.

Heaviside's methods for interpreting the operational forms were based on expanding the operators in positive or negative powers of p (with p^{-1} meaning $\int_0^t \ldots d\tau$), or as infinite partial fractions of Mittag–Leffler type, and interpreting them term by term. He also introduced new families such as those based on $\sqrt{p}$. Solutions obtained in this way correspond to zero initial conditions, implying initial quiescence of the system.

These symbolic methods were in Heaviside's view sufficiently justified by the results. He had learned the elements of the technique from the 1872 edition of George Boole's *Differential Equations*, in which symbolic methods (using D for p) were regarded as representing 'the highest abstraction'. Boole's view was that 'All special instances . . . seem to indicate that the mere processes of symbolic reasoning are independent of the conditions of their interpretation' (see §4.7 on operator methods). There are many processes in the book which may have led Heaviside to operate freely with the symbolism, as for example Boole's formal expansion of expressions such as $\exp(d/dx)$ and, in connection with a partial differential equation, the interpretation of an operator of the form $\sin(f(D))$. It was Heaviside's imaginative use of such imagined freedoms which enabled him to solve problems of great complexity, as during his careful passage from Maxwell's equations to the case of three-dimensional impulsive current flow in a cable.

However, during the 1890s he attempted to systematize his technique, and entered into an elaborate and purely mathematical exposition involving fractional differentiation, impulsive inputs, asymptotic theory, and generalizations of exponential and Bessel functions (Heaviside *1950*). In this way he collided with the contemporary mathematical preoccupation with analytical precision. He had been elected a Fellow of the Royal Society in 1891, and could have expected privileged treatment, but a paper on operators presented to the Society in 1894 (the third in a sequence, with more to follow) was rejected by the referee, William Burnside, as being partly unoriginal, containing errors of substance, and having irredeemable inadequacies in proof. Nevertheless, Edmund Whittaker, in a famous judgement, *1928*, rated Heaviside's operational calculus as one of the three most fruitful discoveries of the late nineteenth century, and fresh extensions and rationalizations have continued to appear to the present day.

Alternative approaches and retrospective rationalizations emerged only very gradually at first, and usually in the context of applications. Thomas Bromwich, in a sequence of papers written between 1916 and 1928, began by considering simultaneous systems arising from mechanical oscillations. He assumed that solutions took the form

$$\frac{1}{2\pi i} \int_\Gamma e^{pt} \xi(p)\,dp, \tag{3}$$

where Γ encloses all the poles of the unknown (rational) functions $\xi(p)$. In 1927 he modified the contour to the modern form

$$\frac{1}{2\pi i} \int_{c-i\infty}^{c+i\infty} e^{pt} \xi(p)\,dp, \tag{4}$$

where the point $(c, 0)$ lies to the right of all singularities. This enabled transcendental integrands arising from partial differential equations to be treated.

In 1916 John R. Carson, of the American Telephone and Telegraph Company, began a series of publications in which solutions became the subject of real integral equations of the form

$$\frac{Z^{-1}(p)}{p} = \int_0^\infty e^{-pt} x(t)\,dt. \tag{5}$$

A theorem proved by Matýaš Lerch *1903* enabled uniqueness of the solution to be demonstrated (under restrictions), and so the use of dictionaries of inversions was legitimatized, although in applications to partial differential equations such methods were still not necessarily rigorous since the analytic behaviour of the solution was not generally known in advance.

The essentials of a transform and its inverse are latent in the Bromwich and Carson expressions (3) and (5), and these were formalized in 1929 by H. W. March as

$$F(p) = p \int_0^\infty e^{-pt} f(t) \, dt, \qquad f(t) = \frac{1}{2\pi i} \int_{c-i\infty}^{c+i\infty} p^{-1} e^{pt} F(p) \, dp. \qquad (6)$$

The standard 'Laplace transform', as taught today, detaches the subject from its history to some extent by defining a slightly different transform pair, associated with a delta-function rather than Heaviside's step-function input (§4.8).

Later advances such as those made by Paul Lévy in 1926 and Jan Mikusiński in 1950 sought to accommodate broader function classes. Another developmental strand was the construction of Abelian and Tauberian theorems (§3.3) to exploit the capability of operational expressions to deliver asymptotic forms directly (Bleistein and Handelsman *1975*).

For most early electrical-engineering purposes, harmonic inputs to differential equations gave enough information to characterize systems, and complex impedances (complex functions of circular frequency) are natural to these problems. Calculations made in terms of these have vector-like interpretations ('phasor diagrams') which could be used by those knowing only a little geometry and having only rudimentary calculating facilities; but at the same time such methods gave rise to a comparatively sophisticated system for considering complex circuits abstractly. The initiative in developing these methods passed to electrical engineers in the USA in the late 1880s. Prominent exponents were Charles Proteus Steinmetz and Arthur E. Kennelly.

3 VECTOR ANALYSIS

The Cartesian form of Maxwell's equations conceals their structure and complicates transformations. The natural language for the directed quantities of physics remained a matter of debate for nearly fifty years. Vector analysis was to displace both Cartesian and quaternionic formulations (§6.2) in electrodynamics, but only against strong opposition: William Thomson regarded both vectors and quaternions as 'an unmixed evil', and Peter Guthrie Tait, defending quaternions, saw the vector form as 'cumbrous and unnatural' and containing 'fragments of the Cartesian shell'. In 1867 Tait wrote an elementary treatise on William Rowan Hamilton's quaternionic system of 1843 which was widely studied, though Heaviside claimed it was incomprehensible. Maxwell made some limited use of quaternionic notation in the 1873 *Treatise* and earlier work, but he relied largely on Cartesian representation (Crowe *1967*).

The simplicity and physical directness of vector as against quaternion analysis was strongly emphasized by Heaviside, who began using his own vectorial system in papers published in the magazine *The Electrician* in 1883. Josiah Willard Gibbs had developed a vectorial system, and produced a limited-circulation pamphlet on the subject in 1881; in this work he followed the quaternionic notational style, but in other ways the system was similar to Heaviside's. The honours for inventing a working system are considered to be shared between Heaviside and Gibbs.

4 RADIOCOMMUNICATIONS

In 1902 Guglielmo Marconi transmitted radio signals across the Atlantic, and so presented a problem described at the time as being one of the most difficult facing the mathematical and physical sciences: to account quantitatively for the transmission of electromagnetic waves to great distances beyond the horizon of the transmitter.

Attention was at first directed towards solving the following problem. A perfect sphere having in general imperfect conductivity (the Earth) is embedded in a homogeneous dielectric (the atmosphere) of infinite extent, and a vertical, harmonically oscillating dipole (the antenna) is the source of excitation. The fields near the surface of the sphere are to be found. There were a dozen or more contributors to this problem between 1903 and 1918, including Lord Rayleigh, Henri Poincaré, H. M. Macdonald, Arnold Sommerfeld, J. W. Nicholson, Augustus Love and G. H. Watson.

Before 1915 only limiting cases were considered, such as that of a perfectly conducting sphere, by Macdonald (first in 1903, and again in 1904, after correction by Lord Rayleigh and Poincaré of a subtle error); the diffraction of plane waves over an imperfectly conducting plane, by J. Zenneck in 1907; and of dipole-generated waves over an imperfectly conducting plane, by Sommerfeld in 1910 (Love *1915*, Bateman *1915*). These and other solutions of simplified problems led to discrepant conclusions, and to inconclusive attempts to combine them in order to make estimates for the complete problem.

In a critical review of the existing literature, Love *1915* concluded that to resolve the questions raised it was necessary to work through the theory of the imperfectly conducting sphere without simplification. Apart from the physical uncertainties in modelling the problem, the difficulty and the sources of error in previous treatments had lain not so much in producing formal solutions to boundary-value problems representing various physical hypotheses, as in transforming these solutions to a form in which it was practicable to carry out numerical calculations. Solutions typically took the form of infinite series, the nth terms of which was complicated combinations of Legendre functions of order n and Bessel functions of order $n + \frac{1}{2}$. They were of slowly diminishing magnitude and varying sign, producing very slow convergence. Love's predecessors, at a loss even for the simpler problems to sum such series numerically, had employed asymptotic estimates for apparently dominant groups of terms, and reformulations in terms of integrals, to deliver arithmetically manageable expressions. Some of these processes were erroneous and others of uncertain validity.

Love followed earlier writers by working in terms of the (scalar) Hertz potential Π, which satisfies an equation of the form

$$(\nabla^2 + k^2)\Pi = 0. \tag{7}$$

The fields are expressible in terms of its derivatives. However, instead of transforming the resulting series into another, approximate series, he carried out several direct arithmetical summations, considering a single wavelength of 5 km, a perfectly conducting sphere and a sphere having the conductivity of seawater, and he covered a range of distances up to about 3000 km from the antenna. This enabled him to dismiss certain earlier approximate formulas and physical speculations, to ratify Macdonald's results for infinite conductivity, and to obtain some agreement with the meagre available measured data.

In 1918 G. N. Watson undertook a fresh investigation of the same problem, at the request of Balthasar van der Pol. The object was to find a form of solution that would be computable, and would represent the solution uniformly over the Earth's surface. Starting from Love's series, he constructed a complex function, with variables s derived from the index n of the terms, which had two infinite sets of poles. The residues from one set delivered individually all the terms of Love's series. Also, the integral of the function taken around a suitable infinite contour enclosing both sets of poles was zero. Therefore the sum of the original series was the negative of the series obtained by summing over the residues of the second set of poles. The new series converged very rapidly, and this device is now known as Watson's transformation. The principal technical problem was to estimate, by asymptotic means, the general locations of the poles belonging to the second set. The technique was claimed by H. Bremmer to be the only thing that made it possible to master the problems of radio-wave propagation numerically.

Observations suggested that in practice the range attainable in radio transmission was far greater than Love's calculations indicated. Watson *1919*, with physical advice from van der Pol, inferred that simple diffraction might not be the only means by which waves were guided around the curve of the Earth, and that the possibility of an atmospheric reflecting layer, first suggested by Heaviside in 1902, should be explored. He calculated the solution of the original problem, with the addition of a conducting atmospheric layer but using the same techniques as before, and concluded that observed transmission was indeed consistent with the existence of a conducting layer at a height of about 100 km.

The theory of radio-wave transmission is related to acoustical theory, ray theory, optical scattering, the theory of the rainbow and elastic waves, and interpretations of complex wave structures have often been interpreted in terms borrowed from these other fields. As also with those branches of wave physics, the theory of partial differential equations, special functions (such as Bessel and Airy functions), complex-variable theory (as in Watson's transformation), transform theory and asymptotic theory played prominent roles (see e.g. Bremmer *1949*). Conversely, the demands of radio theory raised the practical significance of these areas of mathematics, and advanced their development by generating new procedures, approximation techniques and identities.

BIBLIOGRAPHY

Bateman, H. *1915*, *Electrical and Optical Wave Motion*, Cambridge: Cambridge University Press.

Bleistein, N. and Handelsman, R. A. *1975*, *Asymptotic Expansion of Integrals*, London: Holt, Reinhart & Winston. [Contains a large bibliography.]

Boole, G. *1872*, *A Treatise on Differential Equations*, 3rd edn, London: Macmillan.

Bremmer, H. *1949*, *Terrestrial Radio Waves*, London: Elsevier.

Crowe, M. J. *1967*, *A History of Vector Analysis: The Evolution of the Idea of a Vectorial System*, Notre Dame, IN, and London: University of Notre Dame Press. [Repr. with corrections and new bibliographical preface, 1985, New York: Dover.]

Heaviside, O. *1892*, *Electrical Papers*, London: Macmillan. [Heaviside's collected papers up to 1891.]

—— *1893*, *1899*, *1912*, *Electromagnetic Theory*, 3 vols, London: The Electrician. [Heaviside's collected papers, edited, from 1891 onward. Repr. 1950, New York: Dover; 1951, London: Spon.]

—— *1950*, *Heaviside Centenary Volume*, London: Institution of Electrical Engineers.

Jordan, D. W. *1982*, 'The adoption of self-induction by telephony', *Annals of Science*, **39**, 433–61.

King, R. W. P. *1955*, *Transmission Line Theory*, New York: McGraw-Hill. [Gives a comprehensive list of early texts on transmission-line theory.]

Lerch, M. *1903*, 'Sur un point de la théorie génératrices d'Abel', *Acta mathematica*, **27**, 339–51.

Love, A. E. H. *1915*, 'The transmission of electric waves over the surface of the earth', *Philosophical Transactions of the Royal Society of London*, **215**, 105–31.

Lützen, J. *1985*, 'Heaviside's operational calculus and attempts to rigorize it', *Archive for History of Exact Sciences*, **21**, 161–200.

Petrova, S. S. *1989*, 'Heaviside and the development of the symbolic calculus', *Archive for History of Exact Sciences*, **37**, 1–23.

Thomson, W. *1855*, 'On the theory of the electric telegraph', *Proceedings of the Royal Society*, **7**, 382–99. [Also in *Mathematical and Physical Papers*, Vol. 2, 61–76.]

Watson, G. H. *1918*, 'The diffraction of electric waves round the earth', *Proceedings of the Royal Society of London*, **95**, 83–99.

—— *1919*, 'The transmission of electric waves round the earth', *Proceedings of the Royal Society of London*, **95**, 546–63.

Whittaker, E. *1928*, 'Oliver Heaviside', *Bulletin of the Calcutta Mathematical Society*, **20**, 199–220.

Windred, G. *1930*, 'Early developments in AC circuit theory', *Philosophical Magazine*, Series 7, **10**, 905–16.

9.12

Electrical machines: Tensors and topology

M. C. DUFFY

1 INTRODUCTION

The development of electrical networks in supply systems for lighting, power and traction between 1880 and 1890 stimulated the analysis of transients, stability, and the interaction between one element of a network with other elements and with itself. It proved necessary to devise analogues for predicting the behaviour of systems at the planning stage. The simplest network analysis used Maxwell's 'loop current method', which assigned continuous paths to the currents of the circuits comprising the mesh or network. These mesh currents do not split at a junction into component currents, and because the algebraic sum of currents at each junction is zero, Kirchhoff's current law is not needed; the problem is simplified to applying Kirchhoff's voltage law, and the number of independent equations is reduced. This simple analysis (still used) led to matrix methods, and the use of models built up from ideal elements and equivalent components.

In this, the procedure resembled the network analysis of mechanical systems, but the existence of the electromagnetic field in electrical systems gave them a unique character which suggested the use of geometrized techniques in problem-solving.

In very simple cases, time-varying direct-current and alternating-current systems were described using equivalent circuits and differential equations based on the Kirchhoff voltage law, which were solved using the 'classical approach', or by Laplace transforms (§4.8), or by numerical methods such as the Euler or Runge–Kutta techniques (§4.13).

This article focuses on the geometrical and vectorial aspects of machine theory. The industrial importance of alternating-current systems, especially those using three-phase generation and transmission, encouraged the use of complex-number and complex-variable analysis. Charles Proteus Steinmetz used rotating vectors, later called 'phasors', in his text *Theory and Calculation of Alternating Current Phenomena* (1897), which developed methods first introduced in 1893 at the Chicago International Electrical Congress. Steinmetz's work was carried further by W. E. Sumpner in 1897 and 1901; major contributions were made in 1903 by E. Orlich and by G. T. Hanchett. In 1914 T. Wall used the concept of rotating fields to analyse three-phase machines running at the same speed as the fields. The next year, N. Shuttleworth extended this method to motors using commutators. Then, in 1919 M. Walker introduced vector methods, called 'phasor algebra', into general alternating-current network analysis. Many of these papers appeared in the *Journal of the Institution of Electrical Engineers*.

Steinmetz laid the foundations of the steady-state analysis of alternating-current machines, and developments of Oliver Heaviside's operational calculus (§9.11) enabled some incorporation of analysis of transients to be effected, but there were very great practical difficulties in solving equations for actual, complex networks made up of many components, each undergoing variations. A unified theory was needed which replaced all machine varieties by an idealized equivalent which could be included in analogues of power networks undergoing changes.

In the 1920s, headway was made in the USA by R. H. Park. Similarities between his 'two-axis' equations (magnetic north–south and perpendicular axes) and those for induction motors and certain direct-current motors were noted and used to advance unification.

2 THE UNIFIED THEORY OF ELECTRICAL MACHINES AND NETWORKS

In the late 1920s, the Hungarian engineer Gabriel Kron (1901–68), working in the USA, proposed a theory for unifying the analysis of all electrical and dynamical systems. He argued that, because an electromagnetic field surrounded any electrical machine or network, they should be analysed using tensors and geometrized formulations following the methods of Albert Einstein and Hermann Minkowski. His original invention of the connection tensor was the crucial, enabling innovation with which he realized his objective. Though much was derived from Einstein and Minkowski, Kron developed tensor methods pioneered by others: Hermann Grassmann, Bernhard Riemann and Gregorio Ricci-Curbastro. Tullio Levi-Civita's publications on the absolute differential calculus (§3.4) were particularly influential.

Between 1930 and 1938, Kron introduced the first phase of development of an ultimately very ambitious and comprehensive scheme. Crucial to this programme were the papers 'Generalized theory of electrical machines' (*1930*), 'Tensor analysis of rotating machinery' (1932), 'Non-Riemannian dynamics of rotating electrical machinery' (*1934*) and the serially issued (1935–8) 'Application of tensors to the analysis of rotating electrical machinery', published in the *General Electric Review* (for references see Kron *1939*).

In the face of some adverse criticism, Kron was encouraged not only by General Electric but also by the Institute for Advanced Study at Princeton, and by academics such as Oswald Veblen, Hermann Weyl, John von Neumann, Paul Langevin and Banesh Hoffmann. Einstein was greatly interested in Kron's use of geometrized techniques and mathematical models developed for the unified field theory, but applied by Kron to solve practical problems with actual motors and power lines.

The publication in 1936, by the American Institution of Electrical Engineers, of papers by A. Boyajian, A. P. Sah, L. V. Bewley and Kron gave tensor analysis a pronounced impetus (Stigant *1964*). The advantage of tensor methods was that a tensor equation was valid for all relevant reference axes, whereas a matrix equation was restricted to the one system for which it had been set up. However, matrix methods were adequate for many systems, such as static electrical networks, and were extensively developed in the late 1930s and after.

Between 1939 and 1942, the methods developed for rotating machines were extended to electrical networks, and after 1950 techniques were framed for dealing with general networks and systems under the name 'Diakoptiks' (meaning 'to gain insight through division'). This term was used by Kron to describe the method of analysis which split a system into separate, simpler units, which were solved for individually, and these individual solutions were then interconnected in such a way as to provide much simpler equations for the whole system. It became Kron's goal to express these solutions and equations through topology. In Japan, Kron's Diakoptiks was extended and generalized by a special group set up in 1951 to unify the study of fundamental problems in the engineering sciences through geometry. This group, called the Research Association of Applied Geometry after 1954, developed Kron's work under the name of 'Codiakoptiks', much of the research being directed by K. Kondo at Tokyo University.

3 DIAKOPTIKS, TOPOLOGY AND POLYHEDRAL NETWORKS

Kron's tensor methods overcame many difficulties in solving network and machine problems. Previously, if machine design was modified, connections were changed or the network was varied, the describing equations had to be set up again, and the complexity of actual systems made it impossible, in the 1930s, to solve the resulting equations. Only approximate solutions were available.

Kron's use of the 'primitive machine', the 'primitive network' and the 'connection tensor' greatly reduced the effort needed to solve problems otherwise practically beyond solution. Through the growth of Diakoptiks and Codiakoptiks these methods, devised for motors and transformers, were extended to complex physical systems of diverse natures, from crystalline arrays to nuclear reactors. Kron gave 1933 as the date by which he had invented the concept of tearing and interconnecting both 'dead' material wires and 'live' electric filaments in the form of a square, non-singular connection tensor and its inverse. Early analysis was founded on the notion of an imaginary, generalized rotating electrical machine, and by replacing this component by an imaginary, generalized magnetohydrodynamic generator, Kron extended his techniques to produce a theory with remarkable analytical potential, one which could analyse many systems including crystals and self-organizing automata. This was the theory of 'polyhedral networks', later developed as the theory of 'dynamic polyhedral networks', which was created by adding a large number of mechanical and thermodynamical parameters to the electromagnetic parameters of the early theory. He claimed that the structures described by the equations could assume self-organizing capabilities.

Kron stated that when the equations of rotating electrical machinery were cast in tensor form, they were formally analogous to Einstein's equations in contemporary electromagnetic theory, and he condemned much engineering analysis as methodologically erroneous because it failed to recognize the unique nature of electrical machines and networks, consisting as they did of both mechanical structures and electromagnetic fields. He argued that because the analyst needed to deal with both mechanical and electromagnetic entities, problem-solving in electrical engineering was akin to that in mathematical physics, and therefore similar methods should be used. Minkowski geometrized from a starting-point of the Lorentz theory of electrons, as is clear from his 1908 book *Raum and Zeit*; Kron geometrized from the rotating electromagnetic elements of electric machines, using tensors.

Kron further argued that electric-network theory needed a new type of tensor analysis because it required a discrete approach using global tensors in the large. He regarded the topology of differentiable manifolds (§7.10) as of crucial importance, and he incorporated contributions by Elie Cartan, G. De Rham, W. V. Hodge, L. Lichnerowicz, H. Whitney and others, on harmonic integrals, differential forms, fibre bundles and Grassmann algebra. He pointed out that his electrical-circuit models of partial differential equations, and his representations of Maxwell's equations and Schrödinger's equations, could be expressed using modern topology, and this became a crucial feature of polyhedral network theory. Electrified polyhedra modelled relativistic phenomena, and energized polyhedra provided quantum-dynamical models. In the mid-1960s, Kron believed that analogue and digital crystal computers and artificial brains would be built using the hierarchy of self-organizing polyhedral waves.

The evolution of unified machine theory, through the work of A. E. Blondel, Park and Kron, demanded that imaginary entities be interpreted in engineering terms, and the links delineated between actual working apparatus, and multi-dimensional geometrized formulations of their actions. The interpretations proposed and the status of such imaginary entities were the subject of controversy, as was the suggested ability of mathematical theory to lead engineering design.

Gibbs *1967* typifies the mathematically minded electrical engineer. He admitted that motor designers often came up with different interpretations of imaginary entities – such as fictitious reactance – which resulted in different design philosophies, and hence different actual machines, but he stressed that theorists should not change their outlook and so perpetuate discrepancies, because the main purpose of the unified theory was to bring out common features rather than direct attention to the particular differences. He claimed that the essence of a physical theory lay in the mathematical development of the theory. Many engineers, however, used tensor analysis without geometrizing the concepts, and there was controversy over the assumption that every use of tensors in electrical engineering must have a parallel meaning in geometry.

4 DIRECT-CURRENT MOTORS AND METADYNE ANALYSIS

Although the work of Kron and his associates resulted in the most ambitious and comprehensive theory, there were independent attempts to unify and generalize machine theory, sometimes limited to one class of electric motors. A unified theory for all electric motors, direct-current and alternating-current, was eventually framed by Kron's group and by others; but a unified theory for direct-current machines had been developed earlier, in the 1920s, though it was not given the tensor and geometrized formulation of the later, more comprehensive theory.

After twenty years of stagnation, direct-current analysis was greatly stimulated by the work of J. M. Pestarini, who treated the generalized direct-current machine as a form of electromechanical vacuum tube, and developed metadyne statics and metadyne dynamics (*1952*). The term 'metadyne' was applied by Pestarini to a class of direct-current machines developed by him after 1928 which used a special form of commutator, and which were found in three main types – the transformer metadyne, the generator metadyne and the motor metadyne. They enjoyed considerable vogue until the 1960s, when technical developments rendered the metadyne class of machines largely obsolete. Certain pioneers of the metadyne, such as Blondel and P. L. Alger, were later to make contributions to the body of theory introduced by Kron.

Pestarini's *Montefiore Manuscript* (1928) marks the beginning of metadyne analysis. He took a generalizing approach, noting that general theorems could be found which applied to all variants of that type of direct-current machine which he termed 'metadyne', while the steady-state characteristics of metadynes were similar to those of direct-current dynamo machines, and other characteristics were similar to those of alternating-current machines. Under rapid transient currents, metadyne operation resembled that of electronic tubes. Steady-state analysis gave metadyne statics; transient-state analysis gave metadyne dynamics. His work, though very widely known and respected, does not seem to have influenced Kron directly when the latter developed his more general theory.

5 SUMMARY

Kron's general theory originated in problems about electric motors and developed into a comprehensive system, expressed in the language of topology, using concepts paralleling those evolving in contemporary Einstein–Minkowski general relativity. It could be applied to all systems and networks reducible to mechanical and electrical elements and electromagnetic fields, and was later extended to include thermodynamics. Engineering considerations guided Kron throughout his work, and much of it was published in engineering journals. Its ability to relate actual mechanical devices, such as motors, mechanisms and media, to non-classical geometrized descriptions, which are equivalent, was perhaps its most interesting feature.

However, for routine work the whole analytical apparatus was rendered largely unnecessary by computers, which enable the equations, once all but unsolvable, to be handled without need of the Kron technique. It is now seldom taught in undergraduate courses, though it would be understood and perhaps used in research establishments, and for those engaged in doctoral work.

ACKNOWLEDGEMENTS

For advice and assistance I am indebted to my colleagues H. W. Bishop and R. H. Kitchin.

BIBLIOGRAPHY

Alger, P. A. *1965, The Nature of Induction Machines*, New York and London: Gordon & Breach. [Contains outline of Kron theory, with a biographical note on pp. 393–425.]

Gibbs, W. J. *1962, Electrical Machine Analysis using Matrices*, London: Pitman.

—— *1967, Electrical Machine Analysis using Tensors*, London: Pitman.

Hoffman, B. *1938*, 'What is tensor analysis?', *Electrical Engineering*, **57**, 3–9, 61–6, 108–9, 137, 360.

—— *1955*, 'Nature of the primitive system in Kron's theory', *American Journal of Physics*, **23**, 341–55.

Kron, G. *1930*, 'Generalized theory of electrical machines', *Transactions of the American Institution of Electrical Engineers*, **49**, 666–83.

—— *1934*, 'Non-Riemannian dynamics of rotating electrical machinery', *Journal of Mathematics and Physics*, **13**, 103–94.

—— 1939, *Tensor Analysis of Networks*, New York: John Wiley. [Repr. 1965, London: Macdonald. Contains useful bibliography (pp. 621–25), listing works which influenced Kron, and fundamental papers in Kron's theory.]

—— *1963, Diakoptics*, London: Macdonald. [Publication in book form of papers published serially in *The Electrical Journal* between 7 June 1957 and 13 February 1959.]

Lynn, J. W. *1963, Tensors in Electrical Engineering*, London: Edward Arnold.

Park, R. H. *1929*, 'The two-reaction theory of synchronous machines', *Transactions of the American Institution of Electrical Engineers*, **48**, 716–27.

Pestarini, J. M. *1952, Metadyne Statics*, New York: John Wiley/MIT Press, and London: Chapman & Hall.

Stigant, S. A. *1964, Applied Tensor Analysis for Electrical Students*, 2 vols, London: Macdonald. [Contains historical note on pp. 247–51.]

9.13

Relativity

C. W. KILMISTER

1 PREHISTORY

Notions of the relativity of motion developed in the seventeenth century. Long before, Aristotle had distinguished natural from violent motions: the fall of a stone or the motion of the planets was natural because each body was seeking its own place where it will be at rest; only violent motions, like pushing a cart, needed forces. Galileo Galilei widened this classification by admitting mixed motions (as a stone, thrown upwards but then falling naturally to the ground). He considered the thought-experiment in which a stone is dropped from the mast of a moving ship and falls to the deck by the combination of its natural downward movement and the forward movement of the ship. Galileo could admit mixed motions because he knew the general structure of a mechanical problem – general laws of motion applied along with specific starting conditions.

Galileo's physics was not without difficulties, but the mixed motion suggests an anti-Aristotelian position, of subsuming the natural motions under the violent ones by inventing an appropriate gravitational force. This was the position taken by Isaac Newton. Galileo also pointed to the possibility of a (modified) Aristotelian position, of treating motion under gravity as motion under certain acceleration, reserving the notion of force for the non-gravitational motion.

This was the position taken by Albert Einstein in the twentieth century (Barbour *1989*). Whereas Aristotle's original distinction required an absolute reference frame, provided for him by the Earth or the Solar System, the anti-Aristotelian position sees only acceleration as important, so that two descriptions of mechanical phenomena relative to two frames of reference in uniform motion with respect to each other will be equivalent. This was realized by Galileo at least as early as 1632. In modern notation, if two frames of reference are connected by the relations

$$\mathbf{r} \to \mathbf{r}' = \mathbf{r} - \mathbf{v}t, \qquad t \to t' = t, \tag{1}$$

a so-called Galilean transformation, then the same description of mechanics is given in each.

2 NEWTONIAN MECHANICS

By adding to a generalization of Galileo's mechanics the inverse-square law of gravitation, Newton was able to give a highly satisfactory explanation of the motion of the planets around the Sun (§8.8); but the lack of any mechanism for transmitting the gravitational force from one body to another had led to the rival vortex theory of René Descartes, which lingered on with Gottfried Wilhelm Leibniz, John Bernoulli and Giovanni Domenico Cassini. By the late eighteenth century the manifest superiority in calculating power of the Newtonian system had damped down the disquiet over the lack of a mechanism until the discovery of electromagnetism and wave optics. Thomas Young (1800) and Augustin Fresnel (1816) proposed a model for light as a transverse oscillation in a medium, the 'luminiferous aether', an elastic solid whose properties were to be found (§9.1). It was in terms of this theory, implicitly, that James Clerk Maxwell was able to formulate his electromagnetic theory and to identify the light waves with the electromagnetic ones detected by Heinrich Hertz in 1887 (§9.10). Such a success suggested the existence of a corresponding aether for gravitation; but Maxwell's attempt to demonstrate this failed, as did Max Abraham's in 1911, because the fact that gravitation is always attractive leads to difficulties over energy.

By the beginning of the twentieth century, the mystery over the mechanism of gravitation would have seemed more serious but for more practical difficulties which were also accumulating. One of these was the discovery, originally by Urbain Le Verrier in 1845, that the axis of Mercury's orbit was rotating by more than it should be if the effect were caused by the gravitational influence of the other planets. In the 1911 *Encyclopaedia Britannica*, Simon Newcomb had given this residual inequality as 43 arc seconds per century. This matter remained mysterious until 1915.

Another set of difficulties goes back to Leonhard Euler in 1739. He considered the problem of the abberation of starlight (the change in the measured position of a star caused by the Earth's orbital speed in combination with the speed of light) by two methods, and found discordant answers. In one method he used a change of reference frame by a Galilean transformation, and so this was one of the first in a series of investigations of the mechanics–optics interface. A later investigation in this series was by Hippolyte Fizeau, who in 1859 measured the speed of light in water contained in a long pipe. With the water at rest, he found it to be c/n, where c is the speed of light *in vacuo* and n is the refractive index of water; with the water flowing with speed v, he verified the formula, suggested by an analogy with water resistance, $(c/n) + [1 - (1/n^2)]v$, instead of $(c/n) + v$, as the Galilean transformation would give. Apparently the water dragged the aether along, but not completely (§9.2).

This raised the question of the motion of the Earth through the aether. In 1881 Albert Michelson carried out an experiment, suggested by Maxwell two years earlier, in which a ray of light is split by a half-silvered mirror, the two beams moving at right angles and reflected back so as to interfere. The interference fringes are watched as the apparatus is rotated through a right angle; any difference in the times of travel of the light would cause a shift in the fringes. No shifts occurred, nor did they when the experiment was repeated six months later to check that the first null result was not caused by the Earth happening to be stationary in the aether at the time (Kilmister *1970*).

3 SPECIAL RELATIVITY

Einstein's reformulation of the electrodynamics of moving media in 1905 also provided an explanation of all these experiments. This came by directing attention away from particular details and towards the need for an operational definition of simultaneity for distant events. This definition was in terms of light (or electromagnetic) signals between the event and the observer. The resultant special theory of relativity therefore operates by introducing a major ontological change: the Newtonian time (a one-place function of events), $t = f(E)$, is replaced by a two-place function, $t = f(E, O)$ where O is an observer (or coordinate system). The transformation between two coordinate systems (called inertial frames, in which both the 'law of inertia', Newton's first law, and Maxwell's equations hold, so that, among other consequences, light travels in straight lines), one of which is moving at speed v, but there is no rotation along the x-axis of the other, is

$$x \to x' = \beta(x - vt) \quad \text{and} \quad t \to t' = \beta\left(t - \frac{vx}{c^2}\right), \quad \beta := \left(1 - \frac{v^2}{c^2}\right)^{-1/2}. \quad (2)$$

These 'Lorentz transformations' had been found by Hendrik Lorentz, though with a different meaning for the time variable, in 1904. They leave invariant the differential form

$$ds^2 = c^2 dt^2 - dx^2 - dy^2 - dz^2, \quad (3)$$

which is also left invariant by the rotation group; conversely the set of transformations leaving the form invariant, the Lorentz group, consists essentially of Lorentz transformations and rotations. This much was made clear by Hermann Minkowski in 1909, though in a form in which ict is taken as a new (imaginary) variable so as to assimilate the Lorentz group into the four-dimensional rotation group; this trick impeded the progress of the subject for a number of years, since the structure of the two groups is different. Although Maxwell's equations are invariant under the Lorentz group, a small modification is needed to Newton's laws of motion to render them so. This results in an increase of mass with speed, which is in agreement with that found experimentally by Walter Kaufmann in 1901.

Since the Lorentz transformations show at once that

$$\frac{dx'}{dt'} = \left(\frac{dx}{dt} - v\right) \Big/ \left[\left(1 - \frac{v}{c^2}\right)\frac{dx}{dt}\right], \tag{4}$$

the speed in Fresnel's experiment should be $[v + (c/n)]/[1 + (v/nc)]$, which, when account is taken of the largeness of c, closely approximates to Fresnel's result. Similar arguments apply to the other experiments.

4 THE PROBLEM OF GRAVITATION

The question of Newton's gravitation remains. A Lorentz-invariant theory of gravitation was written down as early as 1905 by Henri Poincaré, in the form of an action-at-a-distance in which the effect was not simultaneous but depended on the position of the gravitating mass at an earlier time, so that the effect could be transmitted at the speed of light. Other theories followed, but there was usually disagreement with experiment as well as other defects. During the next ten years Einstein came slowly to the solution of this problem. A first step was to realize that the search for Lorentz-invariant gravitation was following the wrong problem, because a uniform gravitational field (in Newtonian mechanics) can be removed by an accelerated coordinate transformation. Thus, if $\ddot{\mathbf{r}} = \mathbf{g}$, then the transformation

$$\mathbf{r} \to \mathbf{r}' = \mathbf{r} - \tfrac{1}{2}\mathbf{g}t^2 \tag{5}$$

implies that $\ddot{\mathbf{r}} = \mathbf{0}$. A consequence is that, even in Newtonian theory, light is bent in an accelerated frame, and if one attributes this to the gravitational field and supposes that the result is also true for the irreducible field round a heavy body, then a ray passing a distance R from the centre of a mass M deviates by $2GM/Rc^2$, where G is the constant of gravitation. This could be sought experimentally by observing the positions of stars near the Sun's rim (at the time of a solar eclipse). The gravitational field is alone in being able to be treated in this way because it is an acceleration field, the acceleration imparted to a body being independent of its mass. Newton had verified this, to one part in a thousand, and in 1889 Roland von Eötvös improved this accuracy significantly. It is now verified to at least one part in 10^{12}. The importance of this became clearer to Einstein in 1907, and by 1912 he had finally reached the point (though he did not put it in this form) when he saw that it was useful – if not essential – to abandon the anti-Aristotelian position initiated by Galileo and Newton and take up instead the modified Aristotelian position also offered by Galileo.

The means of doing so came with the help of 'my friend, the mathematician, Marcel Grossman'. It seems as if Grossman explained to Einstein some aspects of Carl Friedrich Gauss's theory of surfaces (§3.4), thus showing him how to proceed. In the Gauss theory a surface is defined in terms of two parameters $\mathbf{r} = \mathbf{r}(u^1, u^2)$, where the parameters serve as coordinates on the surface, but are 'devoid of metrical meaning' (a situation which mirrored Einstein's conclusion about the coordinates on a rotating rigid disc). The metrical meaning is given by the 'second quadratic form':

$$ds^2 = d\mathbf{r} \cdot d\mathbf{r} = g_{ij} du^i du^j, \quad g_{ij} := \frac{\partial \mathbf{r}}{\partial u^i} \cdot \frac{\partial \mathbf{r}}{\partial u^j}, \tag{6}$$

where the repeated literal suffices conventionally imply a summation over appropriate values (here, 1 and 2). This form represents a property of the surface independent of how it is embedded in the ambient three-dimensional space. Other expressions (like the 'first quadratic form') represent, at least in part, 'external properties' which are not intrinsic to the geometry of the surface, but depend on the embedding.

Gauss pursued this distinction between intrinsic and external properties far enough to show that the measure of the curvature of the surface had one part, R, involving second derivatives of the g_{ij} as well as quadratic expressions in the first derivatives, which was intrinsic. Grossman was also able to tell Einstein that the Italian differential geometers, especially Gregorio Ricci-Curbastro and Tullio Levi-Cività, had generalized Gauss's work to any number of dimensions, when the generalization of R was an array R_{ijkl}, the Riemann–Christoffel tensor (§3.4). Intrinsic properties of the four-dimensional world were what had to be treated, and the correlation with gravitation came because the accelerated transformation turned the metric form into something that was no longer a sum and difference of squares; the g_{ij} (now 10 in number) were therefore to serve as gravitational potentials. In his joint paper with Grossman in 1913, quite general coordinate transformations

$$x^i \to x^{i'} = f^i(x^j) \tag{7}$$

produce a metric form like Gauss's, though with the 20 conditions $R_{ijkl} = 0$. The paths of particles in the space, originally straight lines, are expressed by the invariant form $\delta \int ds = 0$ (Torretti 1983: Chap. 5). The great leap forward was to realize that such a transformation, locally, could transform away any local gravitation, though not over a region. Such local transformations are not part of a single global one, so one is committed to the same metrical form but without the subsidiary condition on the Riemann–Christoffel tensor. The paths of particles were assumed to be given by the same equation; but what was lacking was a means to determine the g_{ij} for the gravitational field of a given set of masses. Something analogous to Poisson's equation was needed; but this analogue would be a set of 10 differential equations corresponding to the 10 g_{ij}.

It took Einstein two more years to see how to find the field equations. The 10 equations must in any case be satisfied if the 20 conditions $R_{ijkl} = 0$ are satisfied. But there is a standard procedure in differential geometry of deriving the 10 quantities $R_{il} = g^{jk} R_{ijkl}$, where the new g^{ij} are defined as the elements of the matrix inverse to g_{ij}. So Einstein was prompted to postulate, as the required generalization of Laplace's equation, $R_{ij} = 0$, and this was enough to allow him to find to a first approximation the 'three crucial tests': a rotation of the orbit of Mercury (as a single planet orbiting a fixed Sun) of 43 arc seconds per century, a deviation of light passing near the Sun by exactly twice the Newtonian value, and the prediction that the spectral lines of stars should be shifted towards the red end of the spectrum in a strong gravitational field. The initial confirmation of the second and third of these effects was enthusiastically received, but was in fact observationally fairly worthless; but later observations have confirmed the theory very well.

5 LATER DEVELOPMENTS

The first exact solution of the field equations, given by Karl Schwarzschild in 1916, corresponded to a spherically symmetric singularity. Many solutions are now known, but that given by Hermann Weyl in 1917, for the axially symmetric case of two point masses at rest, is of particular interest because it involves a non-massive singularity along the line joining the two masses (a 'light rigid rod' holding them apart). Unlike the Newtonian theory, the non-linear nature of the field equations makes the static, two-point-mass solution impossible without the 'light strut', because two such masses would attract each other. This suggests that the field equations could determine the equations of motion of a general set of particles as well. Weyl tried to show this in 1923, and Einstein and Jakob Grommer did so in 1927; but it was not until the joint paper by Einstein, Leopold Infeld and Banesh Hoffmann in 1938 that the demonstration was satisfactory. The conceptual difficulty lies in the fact that it is the motion of the singularities which has to be determined; Riemannian geometry works with a manifold free of singularities, so that the singularities have to be outside the spacetime manifold. The true nature of such singularities is only now becoming clear (on these and other developments, see Kilmister *1970*).

A full reconciliation between the two highly successful theories of relativity and quantum mechanics remains for the future. A first step was taken by Paul Dirac in 1928, when he showed how to write a wave equation for the electron which was Lorentz-invariant (although not of the tensor form which was thought at the time to be necessary). This led to a satisfactory theory of quantum electrodynamics, but the next step, to a quantum theory of the gravitational field, has not yet been successfully taken.

BIBLIOGRAPHY

Barbour, J. B. *1989*, *Absolute or Relative Motion?*, Vol. 1, *The Discovery of Dynamics*, Cambridge: Cambridge University Press.

Kilmister, C. W. (ed.) *1970*, *Special Theory of Relativity*, Oxford: Pergamon. [English translations of extracts of original papers, with editorial comments.]

——(ed.) *1973*, *General Theory of Relativity*, Oxford: Pergamon. [Sequel to *1970*.]

Kottler, F. *1920*, 'Gravitation und Relativitätstheorie', in *Encyklopädie der mathematischen Wissenschaften*, Vol. 6, part B, *159–237* (article VI 2, 22a). [Extensive literature survey.]

Ray, C. *1987*, *The Evolution of Relativity*, Bristol: Adam Hilger.

Torretti, R. *1983*, *Relativity and Geometry*, Oxford: Pergamon.

9.14

Statistical mechanics

STEPHEN G. BRUSH

Statistical mechanics is a branch of theoretical physics which attempts to deduce the properties of macroscopic matter and radiation from models based on large numbers of mathematical variables pertaining to atomic particles and their physical properties. It acquired its present name and formal mathematical structure from the work of the American physicist Josiah Willard Gibbs (*1902*); but its roots lie in the kinetic theory of gases, proposed by the Swiss mathematician Daniel Bernoulli in 1738 and developed during the second half of the nineteenth century by Rudolf Clausius in Germany, James Clerk Maxwell in Britain and Ludwig Boltzmann in Austria (Brush *1965*, *1966*, *1976*; Sheynin *1985*).

1 HISTORICAL ORIGINS, 1738–1850: A MODEL FOR GASES

In its original (non-statistical) form, the kinetic theory postulates that a gas consists of many (N) point particles, each having mass m and speed v, moving through empty space in a container of volume V. (Modern notation is used throughout this article.) Daniel Bernoulli showed that the pressure exerted by the gas on the container is

$$p = \tfrac{1}{3} Nmv^2 / V. \tag{1}$$

Thus, if the speed of the particles is fixed, the pressure is inversely proportional to the volume. This property of real gases was discovered by Henry Power, Richard Towneley, Robert Boyle and Edme Mariotte in the second half of the seventeenth century. Thus the theory achieved its first goal, to deduce a macroscopic property of matter (now known as 'Boyle's law') from a simple atomic model.

But the model itself was not consistent with generally accepted ideas about the atomic structure of gases in the eighteenth and early nineteenth centuries. Instead, it was believed that gases consist of particles that repel nearby particles with a force that decreases as the distance between them

1242

increases. Isaac Newton had shown in his *Principia* (1687) that Boyle's law could be deduced from such a model. Later, the repulsive force became associated with heat, thought to be a material fluid called 'caloric' (§9.4).

By the middle of the nineteenth century the caloric theory of heat had been replaced by the qualitative idea that heat is a form of energy, interconvertible with mechanical and other forms of energy (Cardwell *1971*). The law of conservation of energy, applied to heat, was expressed as the first law of thermodynamics by Clausius and by the British physicist William Thomson (later Lord Kelvin) (§9.5). The recognition that heat can be transformed into mechanical energy (and conversely) suggested, though it did not require, that heat *is* the mechanical energy of atomic motion. In particular, as physicists came to accept an absolute temperature scale they could assume it to be proportional to the average kinetic energy of molecular motion, $\langle \frac{1}{2}mv^2 \rangle$. The pressure given by equation (1) would then be directly proportional to the absolute temperature T, in agreement with the empirical law found earlier by Joseph Louis Gay-Lussac, John Dalton and Jacques Charles. Thus the modern 'equation of state' for a standard quantity of an ideal gas,

$$pV = RT, \qquad (2)$$

is explained in terms of the motion of atomic particles (1).

In the 1840s the kinetic theory of gases was revived independently by two British physicists (Brush *1976*, Mendoza *1983*). James Prescott Joule, whose experiments helped to establish the law of energy conservation (§9.5), pointed out that equation (1) could be used to compute the particle speed v from macroscopic data; this had been done earlier by John Herapath. J. J. Waterston, in a paper denied publication by the Royal Society and buried in its archives for nearly fifty years, presented an extensive development of the theory. He did succeed in publishing a short note which included the first statement of a basic principle of statistical mechanics for systems obeying Newtonian mechanics: in thermal equilibrium, the average kinetic energy of particles of different masses is the same:

$$\langle \tfrac{1}{2}m_1v_1^2 \rangle = \langle \tfrac{1}{2}mv_2^2 \rangle. \qquad (3)$$

This is now called the 'equipartition theorem'.

2 KINETIC THEORY OF INTERACTING PARTICLES, 1857–1872

Beginning in 1857, Clausius published a series of papers developing the kinetic theory and showing how it could explain, at least qualitatively, the properties of gases, liquids and solids. Statistical methods began to appear in his discussion of the 'mean free path' L travelled by a particle of finite diameter d before it collides with another particle. The mean free path L is inversely proportional to Nd^2, which measures the cross-sectional area occupied by other particles in a thin layer through which the first particle moves, and thus determines the probability of it encountering another particle in the layer. If L is less than 0.1 mm, the gas will diffuse rather slowly into a larger space even though the particle speed v may be several hundred metres per second. (Three of his papers are reprinted in Brush *1965*; see also Brush *1976*: Chap. 4; and Garber *1970*.)

Maxwell made more extensive use of statistical methods by introducing a Gaussian ('normal') distribution (§10.4) for the particle speeds. He was apparently influenced in this direction by the tradition of social statistics developed by the Belgian scientist Adolphe Quetelet. Maxwell read a review of Quetelet's works by John Herschel, who gave a derivation of the normal distribution formula identical to that subsequently presented for gas particles by Maxwell (Gillispie *1963*, Garber *1972*, Porter *1986*).

In 1868, Boltzmann generalized Maxwell's distribution law by including the effects of forces (external, such as gravity, or those between particles). The result is that the relative probability of a 'microstate' of the system, specified by giving definite values of the positions and velocities of all N particles, is $\exp(-E/kT)$. Here E is the sum of the kinetic and potential energies of the particles, T is the absolute temperature and k is a constant now known as 'Boltzmann's constant'. The thermodynamic properties of a system in thermal equilibrium at a specified temperature and pressure or volume (defining a 'macrostate') can be computed by averaging over all microstates with probabilities given by this Maxwell–Boltzmann distribution law, also called simply the 'Boltzmann factor'.

Maxwell extended the kinetic theory to account for non-equilibrium 'transport phenomena' such as viscosity, heat conduction and diffusion. In particular, he deduced from his model of 'billiard ball' (elastic sphere) molecules that the viscosity coefficient, related to the mean free path, is independent of density or pressure and increases with temperature, contrary to the known behaviour of liquids. This prediction was later confirmed experimentally by Maxwell himself and by O. E. Meyer.

By finding a direct relation between the observable viscosity coefficient of a gas and the diameter of a molecule, through the mean-free-path formula which depends on Nd^2, Maxwell made it possible for the first time to obtain a reasonable estimate of molecular dimensions. This was actually done by the Austrian physicist and chemist Josef Loschmidt in 1865. Loschmidt pointed out that the volume occupied by molecules in the liquid state, if they just touch each other, is determined by Nd^3; by combining the value of the liquid volume with values of the viscosity one can find both N and d. Thomson then showed, in 1870, that several other methods give the same order of magnitude for d. Thus, he proclaimed, the concept of 'atom' – long dismissed as purely speculative because it was not subject to quantitative measurement – had now become a legitimate part of science.

In 1866, Maxwell proposed a more general theory of transport phenomena, going beyond the mean-free-path approximation, which was applicable to systems of particles interacting according to any force law. Boltzmann, in 1872, reformulated Maxwell's transport theory in terms of an integro-differential equation for the velocity distribution function f, regarded also as a function of space and time. But Maxwell and Boltzmann could not find exact solutions of their equations, except for the special case in which the interparticle force is inversely proportional to the fifth power of the distance. More general solutions (using infinite series) were not found until the early twentieth century, when David Hilbert analysed the Boltzmann equation by using his theory of integral equations (§3.10), and David Enskog and Sydney Chapman developed systematic approximations for the transport coefficients of systems governed by more general force laws (Brush *1966*; *1972*; *1976*, Chap. 12; Garber *et al. 1986*).

3 IRREVERSIBLE APPROACH TO EQUILIBRIUM, 1872–1913

In 1872, Boltzmann posed one of the fundamental questions of statistical mechanics: how do we know that a system in an arbitrary initial microstate will move toward thermal equilibrium? Only if it does can we use the Maxwell–Boltzmann distribution function to compute its properties (as long as there are no external influences that maintain inhomogeneities and fluxes of energy or matter). He answered the question by defining a functional $H = f \ln f$ (where ln denotes natural logarithm), depending on the general (non-equilibrium) distribution function f. He then used his integro-differential equation to show (subject to certain assumptions) that H must always decrease with time as a result of particle collisions, until it reaches the Maxwellian form; it then remains stationary. This result is known as Boltzmann's H-theorem.

Since the H-function for a Maxwellian f is equal to the negative of the thermodynamic entropy of the system (apart from a constant factor), it appears that the H-theorem is equivalent to the generalized form of the second law of thermodynamics announced by Clausius: the entropy of the universe increases to a maximum. Boltzmann's research throughout the last three decades of the nineteenth century showed that the generalized second law, or 'principle of dissipation of energy' as Thomson expressed it, can be derived from Newtonian mechanics simply by introducing statistical assumptions about molecular motions and collisions; those assumptions are not entirely consistent with Newton's laws. Entropy itself can be regarded as a measure of the disorder or randomness of a system, and the word 'entropy' has entered everyday language with this meaning. In Boltzmann's formula, as revised by the German physicist Max Planck, the entropy S of a macrostate is proportional to $k \ln W$, where W is the 'thermodynamic probability' (i.e. the number of microstates corresponding to that macrostate).

According to Boltzmann's formulation (sometimes called 'statistical thermodynamics'), the second law itself is only statistically valid; it is possible – though unlikely – that entropy can decrease as the result of a fluctuation. Indeed, if one waits long enough the chance of a reversal and a return to the initial state becomes almost a certainty. This is the 'recurrence paradox', discussed in the 1890s by the mathematicians Henri Poincaré and Ernst Zermelo (see §8.9 on the three-body problem). But in practice, any given microstate is almost certain to evolve in a short time period into a microstate belonging to the macrostate corresponding to thermal equilibrium, simply because that macrostate contains almost all microstates for the specified physical conditions (see Brush *1966* for translations of the papers by Poincaré, Boltzmann and Zermelo).

Although much of the early development of statistical mechanics was based on simplified models of point-mass particles, it was recognized early on by Clausius, Maxwell and Boltzmann that some properties of real substances would require a more complex model. In particular, it was known by 1860, from chemical evidence, that most gases are composed of diatomic molecules. The question then arose of whether the equipartition theorem applies to the motions of atoms *within* a molecule. If it does, then the ratio of the specific heat at constant pressure to that at constant volume, $\gamma = c_p/c_V$, a parameter easily measurable from the speed of sound in the gas, should be equal at all temperatures to

$$\gamma = (n + 2)/n, \tag{4}$$

where n is the number of mechanical degrees of freedom. Since each atom, regarded as a point mass, should have three degrees of freedom (corresponding to motion in the three spatial dimensions), the ratio for a diatomic molecule should be $\gamma = (6 + 2)/6 = 1.333\ldots$. However, experimental data showed that it is usually about $\gamma = 1.40$. One might suppose that a more 'realistic' model, in which for example the atoms are regarded as composite bodies rather than point masses, would give better agreement with experiment. But it is easily seen that any such model would have more degrees of freedom, and hence a value of γ closer to 1.000, making the discrepancy worse.

The 'specific-heats paradox', as it was later called, cast doubt on the validity of the equipartition theorem, since the simplest way to resolve the discrepancy was to assume that, for some reason, one of the six degrees of freedom of a diatomic molecule does not share the energy, for then $n = 5$ and $\gamma = 7/5 = 1.400$. This was actually proposed by Boltzmann, as a resolution of the paradox, but Maxwell rejected it since he did not see how it could be consistent with the basic principles of the theory.

The next problem was to establish under what conditions we can prove that a mechanical system in thermal equilibrium will satisfy the equipartition theorem. Maxwell and Boltzmann suggested, in the 1870s and 1880s respectively, that a sufficient condition would be that the system, in its evolution according to the laws of mechanics, passes through every microstate before returning to the initial microstate (Brush 1976: Chap. 10). This became known, following an extensive discussion by Paul and Tatiana Ehrenfest (1911), as the 'ergodic hypothesis'. Since a microstate of N point masses is specified by giving the values of $3N$ position and $3N$ momentum variables, and the microstates are assumed to have a fixed total energy, this means that a microstate is determined by a single point, called the 'phase point', in a $(6N - 1)$-dimensional space, usually called the 'phase space'. Taken literally (as Maxwell and Boltzmann probably did not), the ergodic hypothesis then requires that a 1-dimensional space, the time axis, can be put into 1–1 correspondence with a $(6N - 1)$-dimensional space.

After the Ehrenfests expressed scepticism about the mathematical plausibility of the ergodic hypothesis, two mathematicians – Artur Rosenthal and Michel Plancherel – proved in 1913 that ergodic mechanical systems cannot exist (English translations of their papers are in Brush 1971). This is not as obvious as it might seem to the non-mathematician, since Georg Cantor had shown that it was possible to map multi-dimensional sets onto 1-dimensional sets (§7.11). But specific properties of the mapping, such as continuity, that are required for dynamical systems, cannot be satisfied.

Attention later turned to a weaker version of the ergodic hypothesis, in which the phase point merely passes infinitesimally close to every point on the $(6N - 1)$-dimensional energy surface. This 'quasi-ergodic' hypothesis is adequate to justify the use of statistical mechanics for most physical systems; it was established in the 1930s by John von Neumann and Garrett Birkhoff. A stronger result was obtained for systems of elastic spheres by Ya. G. Sinai in the 1960s.

Although ergodic theory is still of considerable interest to mathematicians, it has lost most of its relevance to physics since the introduction of quantum theory. The specific-heats paradox and other apparent failures of the equipartition theorem could be explained more easily by showing that when the amount of energy needed to boost certain modes of intra-molecular motion to the first excited quantum state is as large as or larger than kT, the population of all the excited states will be small and essentially no energy will be transferred to those degrees of freedom in collisions.

4 QUANTUM STATISTICAL MECHANICS, 1900–1938

Quantum theory itself was born with the help of statistical mechanics. In 1900, after he had obtained a formula that accurately represented the frequency and temperature dependence of black-body radiation, Planck applied Boltzmann's theory to derive a statistical distribution of energy among atomic oscillators. Although Planck did not at that time explicitly introduce a physical discontinuity or quantization of radiation, his formula could easily be so interpreted after Albert Einstein proposed a quantum hypothesis for light in 1905. It was then found that statistical mechanics, especially in the generalized form developed by Gibbs (1902), was well suited to describe the distribution of energy quanta. Moreover, the Gibbs technique of 'ensembles', collections of systems in which quantities such as the energy and even the number of particles are no longer required to have fixed values, but are determined indirectly from other parameters, proved very useful in treating quantum systems.

In the 1920s, statistical mechanics was generalized to cover two new kinds of behaviour revealed by quantum theory (Brush *1983*: Chap. 4). These are most easily described in terms of the mathematical symmetry properties of the Schrödinger wave function ψ of a system of N identical particles (§9.15). Since the physical properties of the system depend on the absolute square $\psi * \bar{\psi}$, which must be invariant under any permutation of identical particles, ψ itself must change to either $+\psi$ or $-\psi$ when such a permutation is made. The first case is called 'Bose–Einstein statistics' (based on work by the Indian physicist Satyendra Bose, and Einstein), and particles that behave in this way are called 'bosons'. The second case is called 'Fermi–Dirac statistics' (developed by the Italian physicist Enrico Fermi and the British physicist Paul Dirac), and the corresponding particles are 'fermions'.

Quanta of electromagnetic radiation, also called 'photons', behave like bosons, and indeed this form of statistical mechanics was first introduced by Bose in order to provide a better derivation of Planck's formula for black-body radiation. Whereas photons have no rest mass and there is no constraint on how many of them can be present in a given system, Einstein showed that the same statistical formula can be adapted to systems of particles with finite rest mass, and that when their total number is fixed, a finite fraction of them will 'condense' into the lowest quantum state if the temperature drops below a certain value.

The German physicist Fritz London showed in 1938 that this 'Bose–Einstein condensation', deduced from statistical mechanics, can account qualitatively for some of the properties of the 'lambda transition' of helium into a superfluid state. In this case the helium-4 isotope, whose nucleus consists of two protons and two neutrons, behaves like a boson – unlike the helium-3 isotope (two protons and one neutron). Note, however, that the two isotopes are chemically indistinguishable. The fact that bulk helium-3 displays significantly different properties at very low temperatures shows that the statistical aspects peculiar to quantum theory have physical consequences for many-particle systems that go beyond the properties of individual particles.

Electrons, protons and neutrons behave like fermions, and under a wide range of physical conditions the ideal Fermi–Dirac gas model can be used, treating interactions as small perturbations. This means that no more than one identical particle may occupy a quantum state (defined by specifying the spin as well as other quantum numbers). As a result, even at zero temperature the excited states must be filled up to a certain level, known as the 'Fermi level', and the system has a finite pressure and energy. Fermi–Dirac statistics was used by the British physicist R. H. Fowler and the Indian astrophysicist Subrahmanyan Chandrasekhar to explain the properties of matter in stars, and by the German physicist Arnold Sommerfeld and the Swiss physicist Felix Bloch to explain the electronic properties of metals.

5 PHASE TRANSITIONS, 1925–1975

Quantum effects turned out to be essential in understanding the properties of individual atoms and important in interpreting the properties of macroscopic substances, especially at low temperatures and high pressures. But by 1940 it was clear that gases, liquids and non-metallic solids under normal conditions could be described without invoking quantum theory, except to explain the structure of molecules and the origin of intermolecular forces. For example, one could discuss the condensation of gases to liquids, and phase transitions involving atomic rearrangements, using only the theories of Maxwell, Boltzmann and Gibbs.

In the 1950s and 1960s, physical scientists came to appreciate the significance of the work of the Dutch physicist J. D. van der Waals, who in 1873 had proposed a simple kinetic-theory model for the gas–liquid transition. The same molecular model – elastic spheres with long-range attractive forces – could be used to describe both gases and liquids, depending on the temperature and pressure; previously it had been suspected that the molecule itself was different in the gaseous and the liquid states. Since the van der Waals equation, relating p, V and T, is a cubic in V, it has three real roots for some values of p and T but only one for other values. Van der Waals showed that this mathematical behaviour corresponds to a known property of a physical system: at low temperatures there are two possible volumes for a given pressure, corresponding to the gaseous or liquid state (the third root corresponds formally to a thermodynamically unstable state); at the 'critical temperature' the two roots merge into one, and above that temperature there is no longer a discontinuous transition between liquid and gas (van der Waals *1988*).

The theory of such phase transitions became a major area of research in statistical mechanics during the middle of the twentieth century (Brush *1983*: Chap 6). The single most important advance was made in the 1940s by the physical chemist Lars Onsager, in connection with a simple mathematical model of particles arranged on a lattice, interacting only with their nearest neighbours. The model had been proposed in the 1920s by Wilhelm Lenz and Ernst Ising for magnetism, and was subsequently adapted for alloys and gas–liquid phase transitions. Onsager used the abstract algebra of operators to obtain an exact solution of the Lenz–Ising model in two dimensions. (See Jones *1990* for some connections between this model and another area of modern mathematics.) This was the first successful attempt to determine precisely the effect of interatomic forces on the thermodynamic properties of a system undergoing a phase transition, and it showed that the results obtained by rigorous mathematical methods may be significantly different from those estimated by approximations.

In the early 1970s, Onsager's results were extended to systems of three or more dimensions, using an accurate approximation technique borrowed from quantum electrodynamics, the 'renormalization group'. Kenneth Wilson received a Nobel prize for his contribution to solving this problem in statistical mechanics (Anderson *1982*).

Onsager's spectacular achievement demonstrated that concepts such as 'fluid' and 'solid' do not have to be postulated as new properties emerging at the macroscopic level, but can be derived theoretically (with the aid of sufficiently powerful mathematics) from the properties of individual atoms which are not themselves either fluid or solid. Thus a major goal of statistical mechanics can be attained for a wide range of physical systems.

ACKNOWLEDGEMENTS

This article summarizes research sponsored by the History and Philosophy of Science Program of the US National Science Foundation. The author thanks M. E. Fisher for useful comments.

BIBLIOGRAPHY

Anderson, P. W. *1982*, 'The 1982 Nobel Prize in physics', *Science*, **218**, 763–4.

Brush, S. G. *1965*, *Kinetic Theory*, Vol. 1, *The Nature of Gases and of Heat*, Oxford: Pergamon.

—— *1966*, *Kinetic Theory*, Vol. 2, *Irreversible Processes*, Oxford: Pergamon.

—— *1971*, 'Proof of the impossibility of ergodic systems: The 1913 papers of Rosenthal and Plancherel', *Transport Theory and Statistical Physics*, **1**, 287–311.

—— *1972*, *Kinetic Theory*, Vol. 3, *The Chapman–Enskog Solution of the Transport Equation for Moderately Dense Gases*, Oxford: Pergamon.

—— *1976*, *The Kind of Motion We Call Heat: A History of the Kinetic Theory of Gases in the 19th Century*, 2 vols, Amsterdam: North-Holland. [Repr. 1986.]

—— *1983*, *Statistical Physics and the Atomic Theory of Matter, from Boyle and Newton to Landau and Onsager*, Princeton, NJ: Princeton University Press.

Cardwell, D. S. L. *1971*, *From Watt to Clausius: The Rise of Thermodynamics in the Early Industrial Age*, Ithaca, NY: Cornell University Press. [Repr. 1989, Ames, IA: Iowa State University Press.]

Ehrenfest, P. and Ehrenfest, T. *1911*, 'Begriffliche Grundlagen der statistischen Auffassung in der Mechanik', in *Encyklopädie der mathematischen Wissenschaften*, Vol. 4, Part 4, Appendix, 90 pp. (article IV 32). [English transl. by M. J. Moravcsik as *The Conceptual Foundations of the Statistical Approach in Mechanics*, 1959, Ithaca, NY: Cornell University Press.]

Garber, E. W. *1970*, 'Clausius and Maxwell's kinetic theory of gases', *Historical Studies in the Physical Sciences*, **2**, 299–319.

—— *1972*, 'Aspects of the introduction of probability into physics', *Centaurus*, **17**, 11–39.

Garber, E. W., Brush, S. G. and Everitt, C. W. F. (eds) *1986*, *Maxwell on Molecules and Gases*, Cambridge, MA: MIT Press.

Gibbs, J. W. *1902*, *Elementary Principles in Statistical Mechanics, Developed with Especial Reference to the Rational Foundations of Thermodynamics*, New Haven, CT: Yale University Press.

Gillispie, C. C. *1963*, 'Intellectual factors in the background of analysis by probabilities', in A. C. Crombie (ed.), *Scientific Change*, New York: Basic Books, 431–53.

Jones, V. F. R. *1990*, 'Knot theory and statistical mechanics', *Scientific American*, **263**, (5), 98–103.

Mendoza, E. *1982*, 'The kinetic theory of matter, 1845–1855', *Archives Internationales d'Histoire des Sciences*, **32**, 184–220.

Porter, T. M. *1986*, *The Rise of Statistical Thinking 1820–1900*, Princeton, NJ: Princeton University Press.

Sheynin, O. B. *1985*, 'On the history of the statistical method in physics', *Archive for History of Exact Sciences*, **33**, 351–82.

van der Waals, J. D. *1988*, *On the Continuity of the Gaseous and Liquid States* (ed. and with an introductory essay by J. S. Rowlinson), Amsterdam: North-Holland.

9.15

Quantum mechanics

L. M. BROWN

1 INTRODUCTION

Quantum theory, a product of the twentieth century, has revolutionized our conception of the physical world. While Albert Einstein's relativity theories were within the scientific traditions connected with the names of Isaac Newton and James Clerk Maxwell, quantum theory has given rise to a radical new epistemology. Although it is sometimes claimed to be a calculus of observables, the most widely used form of quantum mechanics introduces concepts such as Erwin Schrödinger's wave function (ψ-function) for n interacting particles, a complex-valued 'wave field' in (at least) $3n$ dimensions, which is the solution – under specified boundary conditions – of a partial differential equation, the Schrödinger wave equation. It is thus a completely determined spacetime description of the system in question. However, from this 'complete description' one can in general predict only the *probable* result of most physical measurements; only for certain 'compatible' measurements is a result predicted with certainty. This so-called indeterminacy of measurements is what distinguishes quantum theory from the classical theories of physics.

Since its characteristic scale is set by Planck's constant of action, $h = 6.626 \times 10^{-34}$ joule-seconds, quantum theory is important for determining the properties of microscopic systems such as elementary particles, atomic nuclei, atoms and molecules, and thus indirectly the properties of ordinary matter. But it also gives rise to unexpected macroscopic collected phenomena that are classically forbidden, such as superconductivity, the superfluidity of liquid helium, and the so-called degenerate matter that makes up white dwarf and neutron stars.

This article is concerned with that part of quantum theory, quantum mechanics, that deals with the non-relativistic interaction of stable elementary particles and systems of particles, both with each other and with given external fields. This subject is a generalization of classical point mechanics, and it yields the results of the latter whenever h is negligible and under other suitable limiting conditions, as described in Section 2. Other important branches of quantum theory, such as quantum statistical mechanics, the quantum theory of fields (including quantum electrodynamics) and relativistic theories of the subatomic particles, which involve particle creation and annihilation, are not considered here; also not treated are applications, as in theoretical chemistry or to the solid state of matter. Quantum mechanics usually deals with a fixed number of particles, interacting with each other via classical fields. General histories include Hund *1967*; a source book is van der Waerden *1967*.

Particles of small mass often have speeds approaching that of light; thus one might expect the special theory of relativity to be important in the microscopic domain. However, most atomic electrons and most of the protons and neutrons in nuclei have speeds less than one-tenth that of light, so that relativistic corrections are at most of the order of 1 per cent. Thus many interesting phenomena, including atomic structure and chemical binding, are not appreciably affected by the special theory of relativity. (The general theory of relativity (§9.13), a theory of gravitation, is not important for microscopic physics.)

2 PLANCK AND EINSTEIN, 1900–1913

The history of the quantum revolution is a tale full of irony. The quantum *h*, with its eventual implications of discontinuity, indeterminacy and at least partial acausality, was introduced in 1900 by Max Planck. A conservative middle-aged German physicist, Planck was trying to bring Maxwellian radiation into conformity with thermodynamics, on which he had published his doctoral dissertation in 1879, the year of Einstein's birth. It was Einstein who recognized that Planck's work implied (contrary to Planck's intention) that light was not a pure Maxwellian wave field, but that it had a kind of corpuscular character, much as Newton had believed. Following upon the triumph of wave optics in the nineteenth century (§9.1), this shocking blow to classical physics was not well received at the time. Neither Planck nor Einstein came to accept what has now become the standard theory of quantum mechanics, which incorporates the indeterminacy doctrine developed by Werner Heisenberg and Niels Bohr in the 1920s.

To study the role in radiation theory of the second law of thermodynamics, the law that the entropy of an isolated system never decreases, Planck set out in 1897 to deduce theoretically the frequency spectrum of the radiation in a closed isothermal cavity such as a furnace (also known as 'black-body radiation'), which had been shown earlier to depend only on the temperature, and not on the size or shape of the cavity or the material of its walls. In 1900 he conjectured, on the basis of empirical evidence, that the density of energy u at a frequency between ν and $\nu + d\nu$ was

$$u\,d\nu = \frac{8\pi h\nu^3/c^3}{\exp(h\nu/kT) - 1}.$$ (1)

Besides the absolute temperature T, this formula contains the universal constants h, c (the velocity of light in vacuum) and Boltzmann's constant k, which relates energy to temperature.

In obtaining the result (1), Planck assumed that the radiation in the cavity was in temperature equilibrium with the cavity wall, modelled as a collection of linear oscillators, and made a plausible guess about their total entropy. The formula proved to be an excellent fit to experiment; Planck then attempted a derivation using the methods of statistical mechanics that the physicist Ludwig Boltzmann had developed for gases. It is possible that Planck turned to the use of Boltzmann's combinatorial methods because Stirling's approximation for the factorial of large numbers provided the logarithmic form for the entropy that Planck was seeking. Planck's reasoning was controversial at the time, and remains so (Klein *1970*, Kuhn *1978*). What is clear, however, is that at one point in his derivation Planck assumed, as *a mathematical artifice*, following Boltzmann, that the otherwise classical oscillators (or resonators, as he called them), could only have energies proportional to $nh\nu$ (n being an integer), intending to let h tend to zero to obtain his final result. But the successful radiation formula (1) was reached only when h was assigned a non-zero value.

Einstein then proved that Planck's derivation was not valid unless radiation was itself quantized, in addition to Planck's model oscillators – that is, that radiation could be present in the cavity only in energy packages of size $h\nu$. Einstein showed further that if the same assumption were made for free as well as enclosed radiation, then other puzzling observations could be explained, including photoluminescence and the photoelectric effect.

In 1907, Einstein showed that Planck's quantum hypothesis also placed a general limitation on molecular motions, when he quantized the elastic vibrational modes of solid bodies in order to explain the non-classical behaviour of their specific heats. At the same time, he recognized (as Paul Ehrenfest re-emphasized in 1914) that Planck's statistical method for dealing with identical 'particles' – be they light quanta, elastic vibrations or molecules – was significantly different from Boltzmann's method, which had assumed that molecules were distinguishable from one another. As Martin Klein put it (*1970*: 257):

> The fact that Planck's distribution law is the correct one for such particles implies that they are not independent in any normal sense of the word, but

must show a kind of correlation ... these particles must also have lost that most basic of properties, their individuality.

3 THE BOHR ATOMIC MODEL AND THE OLD QUANTUM THEORY, 1913–1925

Planck's quantum of action made its next appearance in a model of the hydrogen atom proposed in 1913 by Bohr, who showed that it played an essential role in determining the structure of matter. While visiting Ernest Rutherford, Bohr (who had just completed his doctoral dissertation) became interested in the problem of the stability of the atom, which had been shown by Rutherford and his collaborators to be analogous to a planetary system. However, according to classical theory a system of electrons bound to a heavy, positive nucleus by the inverse-square Coulomb force, the analogue of the gravitational force, could be stable neither mechanically (against collisions) nor electrically (against radiation). To overcome these objections, Bohr proposed two new restrictions: in the nuclear atom, the electrons can occupy only a discrete subset of the planetary orbits allowed by classical mechanics; and, so long as they remain in those orbits, they do not radiate. Bohr's restrictions amounted to requiring that the angular momentum be an integer multiple of $h/2\pi$. With this model, Bohr could explain the line spectrum of hydrogen (and by implication, at least qualitatively, the line spectra of the other chemical elements) by assuming that the transition of an atom from one allowed energy state to another required either the emission or the absorption of a single quantum of radiation.

Bohr's original treatment was quantitatively successful only for hydrogen and other one-electron 'atoms' like ionized helium, but the next decade saw extensive application and generalization of Bohr's ideas, especially by Bohr and by the Munich school of Arnold Sommerfeld. There were some gratifying experimental confirmations, especially the dependence on atomic number of the characteristic X-rays observed by Henry Moseley in 1913 and the observation of new series of hydrogen spectral lines in the ultraviolet and infrared, which Bohr's theory had predicted. To explain the finer detail of the hydrogen spectrum and to treat more complex periodic systems, Sommerfeld introduced generalized quantum conditions and added elliptical orbits to Bohr's circular ones, characterizing states by additional integer quantities n_i (called quantum numbers). Each canonically conjugate coordinate–momentum pair q_i, p_i was assumed to satisfy the condition

$$\int p_i \, dq_i = n_i h, \quad n_i = 0, 1, 2, \ldots, \tag{2}$$

the integral being taken over one period of the motion. Sommerfeld's relativistic treatment of the hydrogen spectrum gave virtually complete and accurate agreement with the observed spectral frequencies. However, there remained serious discrepancies for atomic systems more complex than hydrogen.

In his first paper on the hydrogen spectrum, Bohr used what he later (in 1923) called the 'correspondence principle', which requires that in the limit of large quantum numbers the results of classical radiation theory should hold. To this, the practitioners of what we now call the old quantum theory added Ehrenfest's 'adiabatic principle', based on classical mechanics, according to which the ratio of the mean kinetic energy of a periodic system to its frequency is invariant under infinitely slow variations of the system's parameters. These two principles greatly extended the range of quantization procedures, allowing the treatment of systems other than those representable as harmonic oscillators.

Major experimental advances during this period were the proof of the quantum nature of X-rays (Arthur Compton, 1923) and the verification of 'space quantization' (Otto Stern and Walter Gerlach, 1922): in an external magnetic field the magnetic moment of an atom or a molecule can take up only a small number of alignments with respect to the field direction. A theoretical landmark was the 'exclusion principle': no two electrons in an atom can occupy the same state. This explains the nesting electron-shell structure of atoms, and of nuclei as well (Wolfgang Pauli, 1925). Also, the electron was found to have an intrinsic spin, its angular momentum being $h/4\pi$ (Samuel Goudsmit and George Uhlenbeck, 1925).

4 THE DEVELOPMENT OF QUANTUM MECHANICS, 1925–1927

By 1925, radical new measures were being considered in attempts to cure the problems with the old quantum theory: the difficulty in principle of reconciling the continuity of Maxwellian electrodynamics with the discontinuous quantum properties of matter and radiation (the 'wave–particle paradox'); the impossibility of predicting the intensities of spectral lines or of treating the related problem of the dispersion of light; and inaccurate calculation of the frequencies of spectral lines of atoms with more than one electron.

To deal with these issues, the 23-year-old Heisenberg invented, in 1925, the first form of quantum mechanics, which soon became known as 'matrix mechanics'. Heisenberg focused attention on an 'observable' set of quantized amplitudes, each depending on not one (as in the older theory) but two stationary states. These amplitudes resembled the coordinates of 'virtual' oscillators that had been introduced the previous year by Bohr, Hendrick Kramers and John Slater in an unsuccessful attempt to bring a statistical theory of atomic transitions to bear on the problem of wave–particle duality. Spectral-line intensities are proportional to the squares of the oscillator amplitudes, and Heisenberg was surprised to discover that his new amplitudes (bearing the quantum numbers of two states as indices) obeyed an algebra that was non-commutative under multiplication. Max Born and Pascual Jordan, in Göttingen, recognized that Heisenberg's algebra was that obeyed by matrices, and that Heisenberg's method was equivalent to replacing the coordinates and momenta of classical mechanics, the q's and p's with matrix $\mathbf{q}$'s and $\mathbf{p}$'s obeying the relation

$$\mathbf{pq} - \mathbf{qp} = h/2\pi i. \tag{3}$$

Born, Jordan, and Heisenberg then collaborated to show how Hamilton's classical equations of point mechanics could be fashioned into the new theory of matrix mechanics. By the end of 1925, Pauli had used Heisenberg's methods to obtain correctly the (non-relativistic) spectrum of hydrogen.

The second form of quantum mechanics, known as 'wave mechanics', was proposed in January 1926 by Schrödinger, then teaching in Zürich. His work was influenced by the doctoral thesis of a Parisian, Louis de Broglie, who in 1923 had suggested that particles are in general accompanied by waves, particle trajectories being related to these 'matter waves' as light rays are to light waves. The wavelength λ of the de Broglie wave for a particle of momentum p is given by the relation $\lambda = h/p$, which is the same relation as for a light quantum. De Broglie interpreted the stationary quantum states of the Bohr–Sommerfeld model as those in which the electron's 'guiding wave' formed a standing-wave pattern. (Matter waves with de Broglie's wavelengths were observed in 1927 by C. J. Davisson and L. H. Germer in the USA and by G. P. Thomson and A. Reid in Scotland.)

Schrödinger set out to merge the wave and particle aspects, emphasizing the wave aspects as fundamental, and thus extended the optical–mechanical analogy which had been developed by William Rowan Hamilton between 1828 and 1837, and which had also influenced de Broglie's thinking. Starting with the Hamilton–Jacobi equation for Hamilton's principal function from classical mechanics, Schrödinger used the variational calculus (§3.5) to infer the wave equation for a related function ψ:

$$H(\mathbf{p}_j, \mathbf{q}_j)\psi = i\hbar \frac{\partial \psi}{\partial t}, \tag{4}$$

where H is an operator obtained by replacing the momenta $\mathbf{p}_j$ in the classical Hamiltonian function by $-i\hbar$ grad$_j$. (It assumes that the coordinates $\mathbf{q}_j$ are Cartesian, and we have set $\hbar := h/2\pi$.) Most applications of quantum mechanics are based upon Schrödinger's equation (4). By imposing suitable regularity and either initial conditions or boundary conditions (for stationary states), equation (4) can be made to define an eigenvalue problem. Schrödinger learnt many of the requisite mathematical techniques from Schlesinger *1900* (Beller *1983*).

Shortly after learning of Heisenberg's introduction of matrix mechanics, the young Dirac, in Cambridge, saw that a 'royal road' to quantization is to replace each set of Poisson brackets occurring in classical mechanics (§8.9) with 'commutator brackets'. The commutator brackets of the operators $\mathbf{p}$ and $\mathbf{q}$, for example, are given by the left-hand side of equation (3). Dirac, who probably learnt of Poisson brackets from Whittaker *1917*, also showed in his 'transformation theory' that wave mechanics and matrix mechanics are both aspects (called 'pictures') of a more abstract mathematical structure. Dirac represented physical states by vectors in a multi-dimensional (often infinite-dimensional) complex function space (a Hilbert space; see §3.9); an observable physical quantity is represented by an operator in the space, which transforms one state into another. The eigenfunctions of the 'observable' in the Hilbert space represent the allowed stationary states of the system, and its eigenvalues are the possible values of the observable.

The mathematics needed to show the equivalence of wave mechanics and matrix mechanics had been developed twenty years earlier (1904–10) by David Hilbert in formulating his theory of linear integral equations, in which he related the latter to differential equations under specified boundary conditions and also to eigenvalue equations for (bounded) infinite matrices (§3.10). (Continental quantum theorists were familiar with Courant and Hilbert *1924*; Dirac learned of Hilbert spaces from Baker *1922*.) Hilbert's methods proved to be of great importance in quantum theory, and were later extended by John von Neumann and Norbert Wiener. Other branches of mathematics that received impetus from the problems of quantum theory were group theory, especially in the hands of Hermann Weyl and Eugene Wigner, the theory of distributions, by Laurent Schwartz, and the mathematics of quantum fields by, for example, Arthur Wightman.

The profound epistemological problems posed by quantum mechanics can only be hinted at here. Although some physicists tried to interpret Schrödinger's wave function as a classical wave, this is not possible for several reasons. For one thing, it is necessary for propagating systems (i.e. non-stationary states) that ψ be complex, and thus not a 'real' wave in either sense of the word. Furthermore, when n particles are involved, the 'wave' must propagate in mathematical $3n$-dimensional space. The standard interpretation of ψ is that of the Bohr school, the so-called 'Copenhagen interpretation', which attributes to the absolute square of ψ the meaning of probability density, an interpretation first proposed by Born in 1926.

Perhaps the most striking result of the theory is the indeterminacy relation of Heisenberg, which says that for non-commuting observables obeying a relation such as equation (3), it is impossible in principle to determine their exact values simultaneously, the error being limited from below by the value of Planck's constant, h. This restriction is usually insignificant at the macroscopic level, but it plays a major role in the interpretation of microscopic phenomena. A partial resolution of the vexing interpretational problems typified by the indeterminacy relations and the wave–particle duality (which, however, does not resolve the nature of 'reality') is Bohr's principle of complementarity. Briefly stated, it says that Nature presents herself in complementary pictures, which depend on observational methods; no one picture gives a complete description, which is given only by the totality of all possible pictures.

BIBLIOGRAPHY

Baker, H. F. *1922, Principles of Geometry*, Vol. 1, *Foundations*, Cambridge: Cambridge University Press.

Bell, J. S. *1987, Speakable and Unspeakable in Quantum Mechanics*, Cambridge: Cambridge University Press.

Beller, M. *1983*, 'Matrix theory before Schrödinger [...]', *Isis*, **74**, 469–91.

Courant, R. and Hilbert, D. *1924, Methoden der mathematischen Physik*, Berlin: Springer.

Dirac, P. A. M. *1930, The Principles of Quantum Mechanics*, London: Oxford University Press. [Three later editions.]

Feynman, R. P., Leighton, R. B. and Sands, M. *1963, The Feynman Lectures in Physics*, Vol. 3, Reading, MA: Addison-Wesley.

Fine, A. *1986, The Shaky Game*, Chicago, IL: University of Chicago Press.

Folse, H. J. *1985, The Philosophy of Niels Bohr*, Amsterdam: North-Holland.

Heisenberg, W. *1930, The Physical Principles of the Quantum Theory*, Chicago, IL: University of Chicago Press. [Repr. n.d., New York: Dover.]

Hund, F. *1967, Geschichte der Quantentheorie*, Mannheim: Bibliographisches Institut.

Klein, M. J. *1970, Paul Ehrenfest*, Amsterdam: North-Holland, Chap. 10.

Jammer, M. *1966, The Conceptual Development of Quantum Mechanics*, New York: McGraw Hill.

Kuhn, T. S. *1978, Black-Body Theory and the Quantum Discontinuity, 1894–1912*, Chicago, IL: University of Chicago Press.

Landau, L. D. and Lifshitz, E. M. *1958, Quantum Mechanics* (transl. J. B. Sykes and J. S. Bell), London: Pergamon.

Messiah, A. *1961, Quantum Mechanics*, 2 vols, Amsterdam: North-Holland.

Mehra, J. and Rechenberg, H. *1982–7, The Historical Development of Quantum Theory*, New York: Springer. [Especially Vols. 1–5.]

Schlesinger, L. *1900, Einführung in die Theorie der Differentialgleichungen mit einer unabhängichen Variabeln*, Leipzig: Göschensche.

Sommerfeld, A. *1919, Atombau und Spektrallinien*, Braunschweig: Vieweg. [Later editions.]

van der Waerden, B. L. (ed.) *1967, Sources of Quantum Mechanics*, Amsterdam: North-Holland.

Whittaker, E. T. *1917, A Treatise on the Analytical Dynamics of Particles and Rigid Bodies*, Cambridge: Cambridge University Press.

9.16

Mathematics in chemistry

I. S. DMITRIEV AND
T. B. ROMANOVSKAYA

1 INTRODUCTION

Although the intensive penetration of mathematical methods into chemistry can be traced back to the middle of the nineteenth century, the need to mathematicize chemical knowledge had been perceived much earlier. It was Isaac Newton who stressed the advantages of explaining all natural phenomena, including chemical ones, in terms of mathematical considerations based on mechanical foundations. The tendency to universalization inherent in mechanics ('mechanical philosophy') in the seventeenth and eighteenth centuries also affected chemistry, and thus promoted the emergence of the mathematico-mechanical ideal of chemical theory which retains its significance today. In addition, the Abbé Condillac's views on algebra (§6.9) influenced Antoine Lavoisier's introduction of the new notational system for chemicals which forms the basis of the modern notation.

But for this ideal to be realized, the conceptual basis of chemistry had to undergo a deep transformation. One of the landmarks in this process was John Dalton's one-to-one relation between the chemical elements and physical atoms (1803–10). He regarded atomic weight as a fundamental property of an atom, an atomic invariant whose value remained immutable in all possible physico-chemical transformations. In addition, the atomic weight was construed to be an additive property; the weight of a bulk sample was the sum of the weights of the smallest particles that entered into combination.

Dalton's atomism, together with the discovery of the main stoichiometric laws around 1800 (see e.g. J. B. Richter, *Reine Stöchyometrie*, 1792–4), opened the way for the application of quantitative methods to chemistry. However, until the first half of the nineteenth century the mathematicization of chemistry was limited to the most primitive calculations of the weight and volume relations of the reacting bodies and computations of their compositions, and to the most elementary apparatus of arithmetic and algebra (Freund *1904*). It was only as a result of thorough investigations of chemical reactions that chemistry was enriched with a new variable, time, making it possible to apply differential equations to some problems in chemistry.

That different chemical processes proceed at different rates had been known since Antiquity. However, only in the middle of the nineteenth century was it found that a chemical reaction was a complex process of many stages, and that each stage proceeded at its own rate; the time factor could thus yield important information.

In addition, in the second half of the nineteenth century some thermo-dynamical methods began to be used in chemistry; as a result the process of mathematicization became intensified. Some algebraic methods began to be used at the same time, both in structural organic chemistry and for the construction of the mathematical theory of chemical periodicity.

2 CHEMICAL KINETICS AND DIFFERENTIAL EQUATIONS

The first use of differential equations in chemistry is associated with L. F. Wilhelmy *1850*. From a study of experimental data on the variation of the angle of rotation of the plane of polarization of light passing through an acidulated dilute solution of cane sugar, he proposed the first mathematical expression for the velocity of the reaction of inversion of cane sugar:

$$- \frac{dZ}{dt} = MSZ, \tag{1}$$

where Z is the quantity of sugar, t is time, S the quantity of acid (presumed unchanging throughout the reaction), and M the average quantity of sugar being inverted in an infinitely small period of time. He also investigated the temperature dependence of the reaction, and found that it followed the same exponential law (1).

Wilhelmy's work was continued in research by Cato Guldberg and Peter Waage, Henri Sainte-Claire Deville, Marcelin Berthelot and L. Pean de Saint Gilles, V. Harcourt and W. Esson, and others. For example, Harcourt and Esson (Esson was a mathematician and professor of geometry, but in his youth had worked as a chemistry demonstrator under Harcourt at Oxford) showed in their *1866* that at constant temperature the rate of chemical change is directly proportional to the quantity of substance undergoing change:

$$y = a e^{-\alpha x}, \quad \alpha > 0, \tag{2}$$

where x is time, a the initial concentration, and α a constant.

The most important contribution to the mathematical foundations of phenomenological chemical kinetics was by Jacobus van't Hoff *1884*. He proposed a new classification of chemical reactions, and established a correspondence between each type of reaction and its related kinetic equation. Generalizing (and in some way modernizing) van't Hoff's approach, from the mathematical point of view the equations of the chemical kinetics can be said to form a system of ordinary differential equations with parameters

$$dx/dt = f(x, k), \quad x(0) = x_0 \quad \text{and} \quad 0 \leqslant t \leqslant t_k, \tag{3}$$

where x and x_0 are vectors representing the present and initial concentrations of reagents, $f(x, k)$ is a vector function of kinetic relations formed in accordance with the adopted mechanism of chemical transformation, and k is a parameter vector (e.g. the vector of the velocity constants). One should stress that the main mathematical expressions in phenomenological chemical kinetics are not deduced from basic principles, but are formal representations of experimental observations.

3 CHEMICAL THERMODYNAMICS: A NEW BRIDGE BETWEEN CHEMISTRY AND MATHEMATICS

Mathematical methods were also introduced into chemistry through the elaboration of chemical thermodynamics. General thermodynamics was originally called the 'mechanical theory of heat', and indeed had its origins in mechanics and in the theory of heat (§9.5). Mechanics gave rise first to general thermodynamics, and then (1870–80) to chemical thermodynamics with its elaborate formal apparatus, in particular the variational principle of equilibrium of Joseph Louis Lagrange (1788) (§8.1). As for the theory of heat, the most important landmark in its evolution was the formulation by Rudolf Clausius of the two principles of thermodynamics when, as a result of his researches, he came to the famous Clausius integral, $\oint dQ/T < 0$. It was in 1865 that he introduced the concept of entropy as the measure of energy transformed into work (§9.5).

The mathematical formulation generalizing the two principles of thermodynamics in the differential form

$$T dS = dU + dW \tag{4}$$

was obtained as the result of the works of Clausius *1867*, William Thomson (later Lord Kelvin) and others. Here S and U are the functions of the state of the entropy and internal energy of a system respectively, and T is its temperature and W its work.

In order to make a mathematical study of any kind of thermodynamic system including chemical ones (and such multi-component systems as solutions), both integral and differential forms of the second law were used. In the first case only reversible processes could be examined (i.e. ones in which an equilibrium is established); in the latter case the theorem of total differentials was applied to the functions of states (see §3.17 on potential theory).

Clausius was the first to apply thermodynamics to chemistry, having introduced the notion of chemical equilibrium. It was A. Horstmann in 1873 who succeeded in finding the general method of treating chemical equilibrium, defined by the mathematical condition $\delta S = 0$.

Having introduced in 1882 the notion of free energy ($F := U - TSH$), Hermann von Helmholtz *1883* made considerable progress in mathematical descriptions of thermodynamics. He succeeded in obtaining (with the help of the partial derivatives of F) the mathematical expressions for all the thermodynamic quantities:

$$S = -\frac{\partial F}{\partial T}, \qquad U = F + T\frac{\partial F}{\partial T}, \tag{5}$$

where F, S and U are F. Massieu's 'characteristic functions' (*1869, 1876*).

The mathematical apparatus of general and chemical thermodynamics was considerably improved by Josiah Willard Gibbs *1876–8*; he generalized the Clausius equation for systems of variable composition,

$$dU = T\,dS - p\,dV + \sum_i \mu_i\,dm_i, \tag{6}$$

where μ_i is the chemical potential of the ith component. His later scientific activity was largely devoted to the elaboration of mathematical analytical methods in chemical thermodynamics where properties of potentials, both as Massieu characteristic functions and as functions satisfying the stability conditions $(\delta^2 S)_U < 0$ and $(\delta^2 U)_S < 0$, were used. Gibbs introduced this condition in connection with the transformation of the nth component of a composition: $(\delta^2 S)_{U,n} < 0$. He also developed a geometrical approach by making state diagrams (§9.5), with thermodynamical functions used as coordinate axes.

It is through chemical thermodynamics (as well as the quantum chemistry), including the theory of irreversible processes initiated largely by Lars Onsager *1931*, that the mathematical apparatus of chemistry was enriched by the methods of mathematical physics and the theory of fields, as well as by the more sophisticated variational methods.

4 MATHEMATICAL MODELS IN
STRUCTURAL CHEMISTRY

Mathematical methods have been applied in structural chemistry since the second half of the nineteenth century. In the classical theory of chemical structure (due to Alexander Butlerov, August Kekulé and others), the chemical properties of compounds depended on how the interrelationship between their constituent atoms was perceived. A molecule was characterized by a 'special distribution of action of a chemical force' (affinity). What the scientists of the time implied by 'chemical structure' is what we now call the topological structure (or simply topology) of a molecule, usually represented in two dimensions as an undirected connected graph, expressing the order of the atomic bonds, the valence distribution over the chemically bound atoms (the edges of the graph representing chemical bonds, and its vertices the atoms).

So it was not surprising that methods of graph theory began to be used in structural chemistry (§7.13). Arthur Cayley *1874* was the first to use these methods to the problem of counting the possible isomers of an alkane, C_nH_{2n+2}. He introduced the notion of the enumerative polynominal for rooted trees – that is, the polynomial whose coefficients A_i specify the number of rooted trees with i vertices:

$$1 + A_1 x + A_2 x^2 + \cdots. \tag{7}$$

The abstract of Cayley's paper on counting the isomers of an alkane was published by the Deutsche Chemische Gesellschaft; the paper came to the attention of both chemists and mathematicians.

Later, beginning in the 1930s, analytical methods of finding the coefficients A_i were elaborated by J. H. Redfield (1927) and George Pólya (1937), while the American scientists H. R. Henze and C. M. Balir derived the formulas for determining the number of isomers in certain series of aliphatic compounds. The solution of the problem was greatly aided by Pólya's discovery of the so-called 'enumeration theorem' in 1937 (Harary and Palmer *1973*).

In 1878 the English mathematician J. J. Sylvester published in the USA an article on the application of algebraic methods in valence theory (Griffiths *1964*). It was devoted to the formal analogy which he and William Kingdon Clifford independently established between the theory of chemical structure and the theory of invariants (§6.8). His investigations of the 'chemico-algebraic analogy', taken up in 1901–04 by P. Gordan in

Germany and V. G. Alekseev in Russia, prepared the way for the spin–valence theory created a quarter of a century later. The authors of that theory (Walter Heitler, Fritz London, Max Born, Hermann Weyl and Yu. B. Rumer) were familiar with the work of Sylvester, Gordan and Alekseev, and made use of their results, adding to them quantum-chemical theory (Kuznetsov *1980*).

In particular, Weyl (*1949*: Appendix D) established a correspondence ('~') between each atom X, Y, ... and a two-place vector:

$$X \sim \mathbf{X} = (x_+, x_-), \quad Y \sim \mathbf{Y} = (y_+, y_-), \ldots . \tag{8}$$

One may then construct a kind of algebraic form $P(\mathbf{X}, \mathbf{Y})$ where the 'degree' V_x of each vector $\mathbf{X}$ is its number of places, given by the valence of atom $\mathbf{X}$. This form is chosen to be invariant in a unitary transformation of the components (x_+, x_-), (y_+, y_-), and so on. The simplest form of such an invariant for two vectors (atoms) $\mathbf{X}$ and $\mathbf{Y}$ would be

$$[x_+ y_- - x_- y_+] = [\mathbf{X}\,\mathbf{Y}], \tag{9}$$

which was formally linked to a single chemical bond and called the 'one-term invariant'. If there were several (k) chemical bonds between atoms X and Y, then the quantity $[\mathbf{X}\,\mathbf{Y}]^k$ corresponded mathematically to this case. A classical structural formula could thus be put into correspondence with the product of invariants. The significance of this analogy, and even its sense, became evident only after the emergence and development of quantum chemistry that took place in the later 1920s, after the creation of quantum mechanics (§9.15).

In the 1860s, Benjamin Brodie had also been guided by algebraic methods, especially Boolean algebra (§5.1). According to Brodie, who was antagonistic to the atomic theory, in chemistry a symbol could represent an operation in space which produced a weight. He was of the opinion that only weight should be considered, while other properties of matter should be neglected. For example, the decomposition of water vapour ($2H_2O = 2H_2 + O_2$ in modern terms) was represented by the symbols of the chemical unit of 'ponderable matter' of water, hydrogen and oxygen, respectively. However, Brodie's approach was not accepted by his contemporaries, who preferred an atomistic interpretation (Farrar *1964*, Brock *1967*).

5 QUANTUM CHEMISTRY

The birth of quantum chemistry, which led to the further introduction of the methods of mathematical physics in chemistry, can be dated to 1927. Then Heitler and London wrote a partial differential equation (the

Schrödinger equation for a hydrogen molecule) and solved it in the one-electron approximation, obtaining results in agreement with experimental data. The next crucial achievement in many-electron calculations was the development in 1928–1930 by Douglas Hartree and Vladimir Fock of the self-consistent field method with various modifications. The main idea of the method was the development of an exact solution of the wave equation (the Schrödinger equation) as a series of wave functions, each of which was the solution of an approximating non-linear differential equation. The apparatus of group theory (§6.4) was of great help in this development.

One of the methods of choosing the approximating solutions, the so-called 'valence-bond method', was interpreted within the framework of Weyl's chemical–algebraic analogy. A pair of one-electron spin functions $\alpha(\sigma)$ and $\beta(\sigma)$ were put into correspondence with the two-component vector $X(x_+, x_-)$, and the same type of correspondence was established between the algebraic invariants and an antisymmetrical product of spin functions.

Quantum chemistry is often labelled a science of approximative methods, for it examines many-electron problems admitting only approximate solutions. Something of a culmination of such an approach is to be found in work on *ab initio* calculations by E. Clementi *1980* and others.

In the 1970s, apart from the use of the computer in chemistry, there was a wide expansion of mathematical methods from functional analysis, topology, catastrophe theory and non-classical logic (Primas *1981*). There has been a hope to develop mathematical methods especially fitted to chemical problems, taking into consideration quantum arguments but not reducing the whole field of chemical knowledge to a set of homogeneous calculations or to purely physical considerations.

BIBLIOGRAPHY

Brock, W. H. *1967*, *The Atomic Debates*, Leicester: Leicester University Press.

Cayley, A. *1874*, 'On the mathematical theory of isomers', *Philosophical Magazine*, Series 4, **67**, 444–9. [Part in *Collected Mathematical Papers*, Vol. 9, 202–4.]

Clausius, R. *1867*, *Abhandlungen über die mechanische Wärmetheorie*, Braunschweig: Vieweg.

Clementi, E. *1980*, *Computational Aspects for Large Chemical Systems*, Berlin: Springer.

Farrar, W. *1964*, 'Sir B. C. Brodie and his calculus of chemical operations', *Chymia*, **9**, 169–79.

Freund, I. *1904*, *The Study of Chemical Composition: An Account of its Method and Historical Development*, Cambridge: Cambridge University Press. [Repr. 1968, New York: Dover.]

Gibbs, J. W. *1876–8*, 'On the equilibrium of heterogeneous substances', *Transactions of the Connecticut Academy of Arts and Sciences*, **3**, Part 1, 108–428; Part 2, 343–524. [Also in *Scientific Papers*, Vol. 1, 55–353.]

Griffiths, J. S. *1964*, 'Sylvester's chemico-algebraic theory', *Mathematical Gazette*, **48**, 57–65.

Harary, F. and Palmer, E. M. *1973*, *Graphical Enumeration*, New York and London: Academic Press.

Harcourt, V. and Esson, W. *1866*, 'On the laws of connection between the conditions of a chemical change and its amount', *Philosophical Transactions of the Royal Society of London*, **156**, 193–221.

Helmholtz, H. von *1883*, 'Die Thermodynamik chemischer Vorgänge', *Sitzungsberichte der Akademie der Wissenschaften zu Berlin*, 647–65. [Also in *Wissenschaftliche Abhandlungen*, Vol. 3, 92–114.]

Kuznetsov, V. I. *1980*, *Theory of Valency in Progress* [in Russian], Moscow: Mir. [See especially Chap. 5.]

Massieu, F. *1869*, 'Sur les fonctions caractéristiques des divers fluides', *Comptes Rendus de l'Académie des Sciences*, **69**, 858–62, 1057–61.

—— *1876*, 'Mémoire sur les fonctions caractéristiques des divers fluides et sur la théorie des vapeurs', *Memoires Présentés à l'Académie des Sciences par Savants Etrangers*, **12**, 1–92.

Onsager, L. *1931*, 'Reciprocal relations in irreversible processes', *Physics Review*, **38**, 405–26, 2265–79.

Primas, H. *1981*, *Chemistry, Quantum Mechanics and Reductionism: Perspectives in Theoretical Chemistry*, Berlin: Springer.

Sylvester, J. J. *1878*, 'On an application of the new atomic theory', *American Journal of Mathematics*, **1**, 64–104. [Also in *Collected Mathematical Papers*, Vol. 3, 148–206.]

van't Hoff, J. H. *1884*, *Etudes de dynamique chimique*, Amsterdam: Müller.

Weyl, H. *1949*, *Philosophy of Mathematics and Natural Sciences*, Princeton, NJ: Princeton University Press.

Wilhelmy, L. *1850*, 'Über das Gesetz, nach welchem die Einwirkung der Säuren auf den Rohrzucker stattfindet', *Annalen der Chimie*, **81**, 413–28, 499–526.

9.17

Crystallography

E. SCHOLZ

Crystallography, its roots in classical Antiquity, gained a profile of its own as a descriptive branch of natural history in early modern times, and was transformed during the nineteenth century into an empirical and strongly mathematicized science in the modern sense. Leaving aside the slow but important maturation of phenomenological knowledge of the nature of crystal forms from the sixteenth to the eighteenth century (Burke *1966*), a strong impetus for mathematicization was provided by the elaboration of symmetry concepts and their underlying mathematical theories during the late eighteenth and the nineteenth century.

1 INTRODUCTION OF SYMMETRY CONCEPTS IN CRYSTALLOGRAPHY

In 1815 the French scientist René Just Haüy introduced quite a refined concept of symmetry in his theory of crystal structure, which was to become the starting-point for most of the developments in the nineteenth century. Haüy adhered to an atomistic view of crystal structure, in the sense of a build-up of crystal matter by layers of densely packed, parallepipedal molecules which in turn were made up of more elementary polyhedral building blocks. He established a set of hypotheses about the laws of the building process: (a) explicit enumeration of a small number (up to 18) of primitive forms, which (b) defined the nuclei for a quantitatively well-determined process of building up more complicated shapes, which again was (c) governed by the internal symmetries of the primitive forms. The 'law of symmetry' (*loi de symétrie*) was one of Haüy's central principles; he had already used it implicitly years before he stated it explicitly, in 1815. He defined symmetry morphologically as equality of the appearance of a primitive form to the eye, when looked upon from different views, and used it as a regulating principle for equal parameters in the process of building the crystal form.

Thus Haüy introduced a far more complex symmetry concept into crystal structure theory than was being used in contemporary mathematics where symmetry was restricted to simple reflection (by Adrien Marie Legendre). Interpreted in group-theoretic terms, Haüy was able to characterize 8 point-symmetry types and (indirectly) all 14 space-lattice types underlying his threefold periodic brick-like construction scheme. (These are the so-called 'holoedric' crystal classes, O^*, D_{6h}, D_{3d}, D_{4h}, D_{2h}, C_{2h} and C_i, and the 'hemiedric' class T_d, in the notation later introduced by Arthur Schönflies.) This characterization was not direct, of course, and partially redundant. An internal difficulty of Haüy's approach lay in the hemiedric (or meriedric) crystal forms which, although they seemed to be derived from one of the primitive forms, do not respect the whole point symmetry of the latter, but only half of it, or even less (see §12.7, on symmetries in general).

The question of hemiedries was better understood in the vaguer but more abstract and open approach of the dynamistic crystallographers of the early nineteenth century, in most cases influenced by the philosophy of dialectical idealism then flowering in Germany (Scholz *1989a*: 48–73). The dynamists rejected the hypothesis of the atomic constitution of matter, and preferred to search for lawfully structured systems of attracting and repelling forces, which, according to their view, constituted what we consider to be matter. As one of the dynamists, C. S. Weiss pursued the hypothesis of three space-like main forces, from which all the other forces in a crystal are (linearly) derived by multiplication and superposition. The faces then ought to lie in orthogonal planes of such an internal force system of the crystal. Weiss's investigations led him to establish seven essentially different systems of possible crystal forms, characterized by a set of axes and its symmetries (roughly corresponding to the seven crystal families and their holoedries), and in addition several subsystems of reduced symmetry. This approach by axis systems gave a more direct expression of symmetry relations than Haüy's classification by primitive forms; and Weiss even started to develop a symbolic notation for the symmetry relations, using permutations of coordinates.

2 CONTRIBUTIONS FROM THE DYNAMISTS

Following Weiss, several theoreticians of the dynamistic school developed detailed and deep insights into the nature of point-symmetry systems in Euclidean 3-space, among them M. L. Frankenheim, J. F. C. Hessel, and J. G. Grassmann. None of them was as influential as they deserved; they knew of one another only vaguely; and they were almost completely ignored in France before Bravais. In 1826, Frankenheim analysed which subsystems of symmetry could arise from Weiss's seven crystal systems. He used a combined approach of geometric reasoning and a clever permutation notation for the normal coordinates of the directions of planes, and derived all the 32 point-symmetry systems which are relevant in classical three-dimensional crystallography, in later terminology the 32 'geometric crystal classes' (Burckhardt *1984*). The article in which he set out his analysis remained largely unread, however, so that the crystal classes had to be re-invented.

Justus G. Grassmann, the father of Hermann Grassmann (on whom see §6.2), developed at the end of the 1820s a nice little calculus for forces and directed straight-line segments. He defined composition and entire rational multiplication of these segments (in the later sense of three-dimensional vectorial operations), and introduced an appropriate multiplicative notation of the type $a^\alpha b^\beta c^\gamma$ for independent directed line segments a, b and c, and integers α, β and γ. He went on to represent the holoedric symmetry systems of Weiss, after an appropriate choice of the basis a, b, and c, by systems of permutations. This gave him a new geometric–algebraic system adapted for use in crystallography, which he called the 'calculus of geometrical combinations' (Scholz *1989b*).

J. F. C. Hessel, apparently not knowing of Frankenheim's result of the 32 crystal classes, started his own broad investigation of finite point-symmetry systems and polyhedral shapes in Euclidean 3-space. In 1830 he classified all possible point-symmetry systems, thus giving implicitly a complete classification of the finite orthogonal groups of dimension 3, and extended it by a discussion of simple polyhedral forms compatible with the respective symmetries as a 'doctrine of shapes' (*Gestaltenlehre*). Only at the end of his extended study did he single out which of these symmetry systems leave invariant a (geometrically) 3-dimensional rational vector subspace of Euclidean 3-space, and thus derived again the list of 32 crystal classes.

3 AUGUSTE BRAVAIS

The development of symmetry aspects remained more or less stagnant in France under the leading doctrine of Haüy's atomism. Actual developments in chemistry and the problem of hemiedries, the symmetry reduction of crystal forms, made it necessary to modernize the atomistic conception of crystal structure. This was done by Auguste Bravais in a series of memoirs between 1849 and 1851. Bravais was very receptive to German *Naturphilosophie*, and knew more of the dynamistic authors than anyone else in France. Nevertheless, he did not depend on their results, but built a solid foundation for a reorganization of atomistic crystal structure theory. In 1849 he started with a new, although not completely exhaustive, derivation of the finite point-symmetry systems in Euclidean 3-space which was clearer than Hessel's.

He continued with symmetry studies of point lattices in the plane and in space in 1850, and arrived at a classification of the space lattices into 7 lattice systems and 14 lattice types. In this work he spoke about infinite systems of symmetries and their representation by symmetric point lattices. And in 1851 he developed a crystal structure theory founded on three mathematical hypotheses about the allowed combinations of point (polyhedral) symmetries with lattice symmetries. From these principles he derived a detailed list of possible combinations of polyhedral and lattice symmetries, which implicitly contained a characterization of nearly all those crystallographic space-group types which arise as semidirect products of their translation lattices by a compatible orthogonal group (crystal class). In the terminology of Fedorov, these are the 'symmorphic space groups'.

4 FEDOROV AND SCHÖNFLIES

The explicit introduction of the group concept in geometry by Camille Jordan in 1869 (§6.4, Section 1) was motivated by Bravais's symmetry studies of crystal structures. But it took some time for crystallographers to start to adapt the group concept for their investigations. An early proponent of such investigations was the German physicist L. Sohncke, who had started to think along similar lines before he got to know about Jordan's memoir. He easily singled out those of the groups listed by Jordan which were potentially relevant for crystallography – that is, those which are discrete and contain a three-dimensional lattice of translations. Sohncke started to discuss the consequences for crystal physics of this point of view, but his approach was hampered by the fact that Jordan had restricted his study to proper Euclidean motions, leaving aside reflections, inversions and glide reflections. Felix Klein knew of this problem, and in the late 1880s invited Schönflies to attack this question when Schönflies was a young *Privatdozent* at Göttingen.

But Schönflies was not the only one to have thought in this direction. Very isolated, but very intensely, the Russian mineralogist Eugraf Fedorov had started, early in the 1880s, investigations of what he called 'regular systems of spatial figures', characterized by spatial symmetry systems like the later crystallographic groups. Thus Fedorov and Schönflies derived very similar results as far as symmetry systems are concerned, although they started from different points of view. In 1889, after their first separate publications in the early 1880s, they started an intense correspondence in which they worked out how their respective methods and notations translated into one another (Burckhardt *1967*). They discussed redundancies and gaps in their respective lists, Fedorov in most cases being a bit in advance of

Schönflies. In this manner they elaborated the complete list of 230 crystallographic space-group types, Fedorov late in 1890, Schönflies in early 1891. Fedorov's approach was technically difficult, and based on his use of regular figures, whereas Schönflies concentrated on the pure symmetry aspect and used the up-to-date language of group theory (Scholz *1989a*: 110–47). This made Schönflies's presentation on crystallographic space groups in a book published in 1892 much more readable for crystallographers and mathematicians, and it was soon accepted as a classic on the subject.

5 X-RAY SPECTROSCOPY AND TWENTIETH-CENTURY DEVELOPMENTS

Fedorov's crystal-structure theory and the theory of crystallographic space groups were considered very hypothetical among crystallographers of the late nineteenth century. This changed drastically from 1912 onwards, when W. Friedrich, P. Knipping and Max von Laue successfully put into practice Laue's plan to look for the diffraction of X-rays in crystal line matter, which would be analagous to optical diffraction by manufactured diffraction grids. In the following years the Laue effect was used to improve not only the understanding of the behaviour of X-rays, but also to gain detailed empirical knowledge of the internal structure of crystalline matter. The first instance of the determination of a crystal structure by X-rays was the investigation of diamond pursued by William Henry Bragg and Lawrence Bragg in 1913. This provided the starting-point for a broad development of mathematical and empirical methods and results in twentieth-century crystallography, the mathematics of space symmetry becoming something of a template for further studies. For the evaluation of diffraction patterns the analytical tools of Fourier analysis were introduced, and also more involved geometric–algebraic structures for pattern classification and identification. The more detailed the investigations became, the more questions were posed which started to extend or even undermine this framework – for example, investigations of higher-dimensional or colour symmetries (internal parameters), symmetry breaks, or the quasi-periodic behaviour of structures called 'quasi-crystals' which have non-periodic translational patterns and admit point symmetries which are not compatible with lattices.

BIBLIOGRAPHY

Burckhardt, J. J. *1967*, 'Der Briefwechsel von E. S. Fedorow and A. Schönflies, 1889–1908', *Archive for History of Exact Sciences*, 7, 91–141.

——*1984*, 'Die Entdeckung der 32 Kristallklassen durch M. L. Frankenheim im Jahre 1826', *Neues Jahrbuch für Mineralogie, Geologie und Paläontologie, Monatshefte*, **31**, 481–2.

Burke, J. G. *1966*, *Origins of the Science of Crystals*, Berkeley and Los Angeles, CA: University of California Press.

Lima-de-Faria, J. (ed.) *1990*, *Historical Atlas of Crystallography, International Union of Crystallography*, Dordrecht: Kluwer.

Scholz, E. *1989a*, *Symmetrie – Gruppe – Dualität. Zur Beziehung zwischen theoretischer Mathematik und Anwendungen in Kristallographie und Baustatik des 19. Jahrhunderts*, Basel: Birkhäuser.

——*1989b*, 'The rise of symmetry concepts in the atomistic and dynamistic schools of crystallography, 1815–1830', *Revue d'Histoire des Sciences*, **42**, 109–22.

9.18

Mathematical biology

GIORGIO ISRAEL

In the history of applied mathematics, there is an important turning-point in the mid-1920s. Before that time the contributions to mathematical biology (or 'biomathematics') can only be considered as isolated researches in the field of biology developed with the technical aid of mathematics. But from then on there is evidence of mathematics being regarded more as the conceptual core of a well-defined methodology of research, aimed at centring some classes of biological problems around a set of mathematical laws. From this point of view, the 1920s can be considered as the beginning of a reductionist approach inspired by mathematical physics in the field of biology; and it is in this sense that it is possible to speak of a scientific branch called 'mathematical biology'. These developments were undoubtedly helped by the emergence of the 'modelling' approach in applied mathematics. In this article, therefore, this phase is distinguished from the preceding one, and the prehistory of the discipline is examined only briefly.

1 PREHISTORY

The prehistory goes back to the seventeenth century, and the quantitative analysis of the physiology of motion developed by the so-called 'iatromathematical' school of Giovanni Borelli. In the nineteenth century the first attempts to study some aspects of biology from a quantitative viewpoint were the work on optics and physiological acoustics by Hermann von Helmholtz (§9.8), and on physiological thermodynamics by Paul Broca, as well as the classical research on the circulation of blood influenced by hydrodynamics (compare §8.5). However, in these cases mathematics had an almost marginal role, the central role being taken by the physical analogy.

Perhaps the first attempt to use mathematics directly was by the Italian astronomer Giovanni Schiaparelli, who in 1898 tried to give a geometrical representation of the variations occurring in the organic world, drawing a parallel between these variations and a special class of geometrical forms.

However, the most important work in which a systematic utilization of mathematics in biology is apparent was the mathematical contribution to the theory of evolution by Karl Pearson (for the most part published in the *Philosophical Transactions of the Royal Society of London* at the end of last century). Pearson also contributed (with W. R. Weldon and Francis Galton) in 1901 to the foundation and publication of the journal *Biometrika*, whose aim was to collect and spread knowledge of the statistical studies of biological problems (§10.10).

2 POPULATION DYNAMICS

The mathematical theory of population dynamics originated in attempts to formulate a law describing the growth of a human or animal population. The most elementary form is the exponential or 'Malthusian' law, asserting that the growth rate is a constant k and therefore that the derivative $\dot{x}$ of the population x with respect to time is given by the formula $\dot{x} = kx$. This very rough law was improved in 1837 by the Dutch mathematician and biologist Pierre Verhulst, with the introdution of the so-called 'logistic equation':

$$\dot{x} = kx - hx^2. \tag{1}$$

The term hx^2 takes account of the existence of an 'internal friction' due to overpopulation and, unlike the Malthusian law, implies the existence of a 'limiting population' k/h, such that the growth is negative when x exceeds this value.

In 1925 and 1926 two fundamental works giving rise to the systematic study of population dynamics were published. They tackled the problem of the interaction between different species, in terms of both predation and competition in the same ecological niche. These works were the *Elements of Physical Biology* (*1925*) by the American statistician A. J. Lotka, and a paper by the Italian mathematician Vito Volterra (*1926*), followed by other papers and books (*1931, 1935*). The starting-point of these contributions were the so-called 'Volterra–Lotka equations', formulated almost simultaneously by these two scientists and giving rise to a priority controversy (Israel *1982*). These differential equations describe the interaction between a population of predators y and a population of prey x which constitutes the total food supply of the predators. The growth rate of the prey and the decrease rate of the predators (in the absence of prey) are assumed to be exponential, while the mechanism of predation is represented by an analogy suggested by the kinetic theory of gases: the prey population decreases in proportion to the number of possible encounters between prey and predators (which is given by the product xy), and the increase in the predator population follows the same law. Therefore the equations are:

$$\dot{x} = Ax - Bxy \quad \text{and} \quad \dot{y} = -Cx + Dxy, \tag{2}$$

where A, B, C and D are constants.

Except for this common starting-point, the characteristics of the contributions of Volterra and Lotka are quite different (Kingsland *1985*, Israel *1988*). Lotka was more concerned with the energetic aspects of the question suggested by recent developments in physics, and drew upon analogies with thermodynamics (§9.5). By contrast, Volterra followed an approach of strict analogy with classical mechanics. In spite of the importance and the influence of the seminal work of Lotka, Volterra's contribution exerted a stronger influence on the subsequent development of mathematical modelling in biology.

Volterra's work can be divided into three areas: in the first place he tried to build a mechanics of biological associations in strict analogy with classical rational mechanics; it was necessary then to validate his mathematical results empirically; then to give a variational and Hamiltonian formulation of the results (Israel *1991*). His main mathematical tool was the theory of ordinary differential equations, which he studied from the qualitative viewpoint, though having recourse to quite elementary methods. Furthermore, he extended his theory to the case of systems with memory, by analogy with the 'hereditary' systems already studied by him in the field of elasticity theory. These are not strictly deterministic systems; their dynamic evolution is determined not only by their initial state but also by all of their past history. The mathematical analysis of such systems requires the use of integro-differential equations (a mathematical tool encouraged by Volterra himself in 1912; see §3.10).

This theory (called by Volterra the 'mathematical theory of the struggle for existence') is characterized also by the revaluation of the classical Darwinian theory of the struggle between different species whose ecological niches are overlapping, and particularly of the 'principle of competitive exclusion' (asserting that two competing species cannot coexist in the same ecological niche). This revival of Darwinism is much less evident in the work of Lotka.

3 EPIDEMIOLOGY

The modern mathematical theory of the spread of epidemics has a very important antecedent: the seventeenth-century attempt to discuss from the mathematical viewpoint the advantages of inoculation against smallpox. The practice of inoculating the active virus of smallpox became more and more widespread at this time, and it posed a difficult question: could the number of individuals becoming immune from smallpox after the inoculation be considered high enough to justify the deaths caused by the inoculation itself? Daniel Bernoulli developed a mathematical analysis of this question, following a probabilistic approach, in order to demonstrate the advantages of inoculation. Jean d'Alembert objected to Bernoulli's analysis, but the debate soon shifted on the problem of the legitimacy of the use of probability calculus in the analysis of empirical (and in particular non-physical) phenomena (Daston *1988*).

Modern developments began in the 1920s; they are characterized more by the intervention of the classical infinitesimal calculus than by the probabilistic approach. The main aim of the analysis is to find an answer to questions like the following. If an epidemic disease is spreading in a population, is it possible to calculate the number of infected individuals? Is it possible to determine the circumstances under which an epidemic will take place? The basic model of this theory was formulated by the American mathematicians W. O. Kermack and A. G. McKendrick in a work of 1927 which includes the so-called 'threshold theorem', which estimates the threshold beyond which an epidemic will spread. Although quite elementary, Kermack and McKendrick's model showed its effectiveness: for instance, its predictions were in good agreement with the observed numbers of individuals infected by the plague in Bombay in 1905 and 1906. This model was the starting-point for a long series of models of epidemics whose analysis was far more difficult, for example the exanthematic diseases, malaria and venereal diseases. (A very broad though no longer up-to-date review can be found in Bayley *1957*.) Nowadays the AIDS epidemic is also studied mathematically.

4 POPULATION GENETICS

The theoretical foundations of the modern theory of population genetics were laid down between 1918 and 1932 by Ronald Aylmer Fisher, J. B. S. Haldane and Sewall Wright. (For a careful review of these developments see Provine *1971*; for background, see Norton *1975*.) A parallel (although not strict) between the beginnings of this theory and the beginnings of population dynamics can be seen in the following two aspects. First, both theories refer to the Darwinian theory (which was seen as largely discredited at the beginning of the twentieth century); second, they both have recourse to a procedure of mechanical analogy, with reference to the model of the kinetic theory of gases (§9.14).

Undoubtedly the work of Fischer, Haldane and Wright was largely stimulated by the controversy over the continuity of evolution and the efficacy of Darwinian natural selection. There is, however, a difference with

Darwinism and Volterra's population dynamics: in fact, in the theory of population genetics, Darwinism is presented in a version completely renovated through a synthesis with Mendelism and the tradition of research in the field of biometry (on which see §10.7).

5 VAN DER POL'S MODEL

The mathematical model of heartbeat devised by the Dutch engineer Van der Pol and J. Van der Mark in *1928* deserves special mention, not only for its applicative importance but also because it introduced into mathematical modelling some mathematical tools whose use until then had been restricted to celestial mechanics. In a paper of 1926 Van der Pol had observed that some physical systems, like electrical circuits containing triodes, showed the presence of an internal source of energy which became negative when the amplitude of oscillation was greater than a fixed value, and positive when smaller than that value. The classical equations of the damped and forced oscillator were not suitable to describe such a phenomenon. The equation introduced in *1928*,

$$\dot{x} = y \quad \text{and} \quad \dot{y} = \varepsilon(1 - x^2)y - x, \quad \varepsilon > 0, \tag{3}$$

represented a system whose oscillations tended to a unique stable oscillation, therefore exhibiting an internal feedback mechanism (§8.4); for this reason the oscillation is called 'auto-oscillation'. A simpler example of such a system is the pendulum clock.

Some analogies between this phenomenon and the heartbeat suggested to Van der Pol that the behaviour of the electric potential generated by the heartbeat could be represented as a phenomenon of auto-oscillation. He devised a mathematical and physical model of the human heart to justify his hypothesis; it is the prototype of a long series of ever more sophisticated mathematical models of the heartbeat.

From the mathematical viewpoint, Van der Pol's equation shows the phenomenon of the 'limit cycle' (a periodic oscillation attracting all the other oscillations of the system) already studied by Henri Poincaré in celestial mechanics (§8.9). Furthermore, this cycle can disappear following the variation of the empirical parameter ε of the system, exhibiting in this way a mathematical phenomenon later studied by the German mathematician E. Hopf, author of a theorem of central importance in the qualitative theory of ordinary differential equations.

6 LATER WORK

The scale of the most recent research in biomathematics is so broad as to make impossible even a rough synthesis. Apart from the many books and journals devoted to the topic, an idea of the range of the field is given by the book series *Springer Lecture Notes on Biomathematics*. The mathematical tools used, besides ordinary differential equations, partial differential equations, and integral and integro-differential equations, now include reaction–diffusion equations and stochastic differential equations.

BIBLIOGRAPHY

Bailey, N. T. J. *1957*, *The Mathematical Theory of Epidemics*, New York: Hafner, and London: Griffin.

Daston, L. *1988*, *Classical Probability in the Enlightenment*, Princeton, NJ: Princeton University Press.

Israel, G. *1982*, 'Le equazioni di Volterra–Lotka: Una questione di priorità', in *Atti del Convegno su 'La storia delle Matematiche in Italia'* [. . .], Cagliari: University, 495–502.

——*1988*, 'On the contribution of Volterra and Lotka to the development of modern biomathematics', *History and Philosophy of the Life Sciences*, **10**, 37–49.

——*1991*, 'Volterra's analytical mechanics of biological associations', *Archives Internationales d'Histoire des Sciences*, **41**, 57–104, 307–52.

Kingsland, S. E. *1985*, *Modeling Nature: Episodes in the History of Population Ecology*, Chicago, IL: University of Chicago Press.

Lotka, A. J. *1925*, *Elements of Physical Biology*, Baltimore, MD: Williams & Wilkins. [Repr. as *Elements of Mathematical Biology*, 1956, New York: Dover.]

Norton, B. *1975*, 'Metaphysics and population genetics: Karl Pearson and the background to Fisher's multifactorial theory of inheritance', *Annals of Science*, **32**, 537–53.

Provine, W. B. *1971*, *The Origins of Theoretical Population Genetics*, Chicago, IL: University of Chicago Press.

Van der Pol, B. L. and Van der Mark, J. *1928*, 'The heartbeat considered as a relaxation oscillation, and an electrical model of the heart', *Philosophical Magazine*, Series 7, **6**, 763–75.

Volterra, V. *1926*, 'Variazioni e fluttuazioni del numero d'individui in specie animali conviventi', *Memorie della Reale Accademia dei Lincei*, Series 6, **2**, 31–113.

——*1931*, *Leçons sur la théorie mathématique de la lutte pour la vie* (ed. M. Brelot), Paris: Gauthier-Villars.

Volterra, V. and D'Ancona, U. *1935*, *Les Associations biologiques au point de vue mathématique*, Paris: Hermann.

Part 10
Probability and Statistics, and the Social Sciences

10.0

Introduction

The location of this Part well on in the encyclopedia reflects the historical fact that probability and statistics arrived relatively late as major branches of mathematics. While probability theory began to develop in the seventeenth century (with intuitive probabilistic *thinking* evident much earlier in some contexts), mathematical statistics did not really emerge until the late eighteenth and the nineteenth century, and many main theories and traditions belong to the twentieth century. Further, the nature and timing of their introductions varied with different disciplines (a curious phenomenon in itself, and not yet fully explained by historians); major progress in many sciences commenced only during the late nineteenth or even the twentieth century. This state of affairs is reflected, for example, in the rather small place granted to probability and statistics in the *Encyklopädie der mathematischen Wissenschaften*, which was highly praised in general in §0.

This Part begins in §10.1–10.2 with some of the early progress in probability theory, followed in §10.3 by one of its earliest applications, to actuarial practice. There follow two articles (§10.4–10.5) covering the emergence of mathematical statistics out of, and with, probability. (The principle of least squares appears here and in §10.2, treated each time from a different point of view.) Next, §10.6–10.7 look at two countries, Russia and England, which made distinctive contributions of their own.

The next seven articles (§10.8–10.14) cover the introduction of probability and statistics into various major disciplines. The order is governed more by related sciences than by chronology, which overall is extremely disjointed. (Note also that §9.13–9.14, on statistical and quantum mechanics, deal with areas of interaction with physics.) In §10.16 the sociological context is examined. Then, two articles survey the foundational and philosophical questions which have intrigued and beset probability from the outset. Finally, mathematical economics is treated; while this area is certainly concerned with the social sciences, the roles of probability and statistics were quite light.

The introduction of probability and statistics has been the largest single special input of mathematics into the social sciences, but it is not the only one. Thus sections of several articles cover the use of other branches or topics of mathematics, but not necessarily with a probabilistic or statistical import.

As a result of the lateness of the progress mentioned above, probability and statistics receive little or even no mention in most general histories of mathematics. Indeed, its own broad histories have begun to be written only in recent years (see the bibliography; Schneider *1988* is a source-book). In addition, the *Encyclopaedia of Statistical Sciences* has many historical articles, including biographical ones, in its nine volumes (1982–8).

BIBLIOGRAPHY

Dale, A. I. *1991*, *A History of Inverse Probability from Thomas Bayes to Karl Pearson*, New York: Springer.

Gigerenzer, G. *et al. 1989*, *The Empire of Chance. How Probability Changed Science and Everyday Life*, Cambridge: Cambridge University Press.

Hacking, I. *1975*, *The Emergence of Probability*, Cambridge: Cambridge University Press.

—— *1990*, *The Taming of Chance*, Cambridge: Cambridge University Press.

Hald, A. *1990*, *A History of Probability and Statistics and Their Applications Before 1750*, New York: Wiley.

Kendall, M. G. (ed.) *1970, 1977, Studies in the History of Statistics and Probability*, 2 vols (ed. with E. S. Pearson (first named) and R. L. Plackett, respectively), London: Griffin. [Reprints of articles, mostly from *Biometrika*.]

Krüger, L. *et al. 1987*, *The Probabilistic Revolution*, 2 vols, Cambridge, MA: MIT Press.

Lancaster, H. O. *1968*, *Bibliography of Statistical Bibliographies*, Edinburgh and London: Oliver & Boyd. [Includes obituaries and biographical notices. Continued by supplements in *Review of the International Statistical Institute*, 1969, 37 onwards.]

Pearson, K. *1978*, *The History of Statistics in the 17th and 18th Centuries* [...] (ed. E. S. Pearson), London: Griffin.

Schneider, I. (ed.) *1988*, *Die Entwicklung der Wahrscheinlichkeitstheorie von den Anfängen bis 1933*, Darmstadt: Wissenschaftliche Buchgesellschaft.

Sheynin, O. B. *1983*, 'Corrections and short notes on my papers', *Archive for History of Exact Sciences*, **28**, 171–95. [Appendix to, and index of, a long series of papers on the history of probability and statistics in this journal, which has continued since.]

Stigler, S. (ed.) *1980*, *American Contributions to Mathematical Statistics in the Nineteenth Century*, 2 vols, New York: Arno Press. [Reprints of original articles.]

—— *1986*, *The History of Statistics: The Measurement of Uncertainty Before 1900*, Cambridge, MA: Harvard University Press.

Todhunter, I. *1865*, *A History of the Mathematical Theory of Probability from the Time of Pascal to that of Laplace*, Cambridge and London: Macmillan. [Repr. 1949, New York: Chelsea.]

10.1

Combinatorial probability

EBERHARD KNOBLOCH

The doctrine of chances took a remarkably long time to develop, but once launched it proceeded very rapidly. When the calculus of probability got properly under way in the middle of the seventeenth century, a combinatorial algebra lay ready to hand.

1 EARLY COMBINATORICS

In the Far East, the development of combinatorial rules in India was far in advance of contemporary knowledge in the West (Biggs *1979*; see §1.12). As early as AD 850 the Jain mathematician Mahāvira gave the 'multiplicative' rule for the number of combinations $_nC_r$ of n things taken r at a time. It was repeated by the Hindu mathematician Bhāskara in his *Līlāvatī* (1150), where were given further combinatorial relations like $_nC_r = {_nC_{n-r}}$, or the number of possible arrangements of a things of one kind, b of another, c of another, and so on, there being n altogether, namely $n!/(a!\,b!\,c!\ldots)$. Other mathematicians of the tenth and eleventh centuries, like Halāyudha and Bhāttotpala, constructed the combinatorial or arithmetical triangle now known as Pascal's triangle, which was dealt with as the binomial triangle in China in about 1100.

In the Muslim world, combinatorial studies in the Maghreb (its Western part) were initiated in two areas outside mathematics – linguistics and astrology. At the end of the twelfth century or the beginning of the thirteenth, Ibn Mun'im placed in his *Fiqh al-Ḥisāb* some rules and results which had been obtained in linguistics, illustrating fundamental combinatorial operations such as permutations and combinations – special cases which could easily be generalized. Then, at the end of the thirteenth century, Ibn al-Bannā gave demonstrations of the rules expressing the number P_n of permutations of n objects, of combinations $_nC_r$ and arrangements $_nA_r$ of n things taken r at a time in at least two of his works, in the *Tanbīh al-Albāb* and in the *Raf' al-Hijāb*. The general formulation, argument and utilization show that the combinatorial operations and formulas were regarded as tools that could be used in different domains like astronomy, algebra, geometry or arithmetic. It is highly probable that in 1321 the Jewish mathematician Levi ben Gerson from southern France drew upon direct knowledge of these Arabic combinatorial studies when he included a chapter on combinatorics in his book *Ma'aseh Ḥoshev*. There had been Jewish writers of the twelfth century who had dealt with combinatorial problems, such as Abraham ibn Ezra and Moses Maimonides (Rabinovitch *1973*).

2 LULL AND LULLISM

Another combinatorial tradition originated with the Catalan missionary and poet Ramón Lull from Mallorca, who taught in his *Ars maior* (1273) and *Ars brevis* (1308) that knowledge can be increased by exploring pairwise combinations of certain categories. While he himself contributed little to the theory of combinatorics, the adherents of 'Lullism', which flourished in the seventeenth century, published numerous and voluminous books on the subject. These studies had little or no influence on the mainstream of mathematics: the authors were mostly not professional mathematicians; they gave rules without proofs; and the combinatorial studies were hidden away in philosophical, logical or music-theoretic treatises whose characteristic baroque peculiarity and Lullistic complacency deterred mathematicians from reading them.

These studies, however, show evidence of a deep combinatorial knowledge, which would be surpassed only by James Bernoulli at the end of the seventeenth century. The most important author among them was Marin Mersenne, who published six relevant works between 1623 and 1647. He strongly influenced the German Lullist Athanasius Kircher (*Musurgia universalis*, 1650; *Ars magna sciendi*, 1669), and Kircher's pupils Kaspar Schott (*Magia universalis*, Vol. 3, 1658) and Kaspar Knittel (*Via regia ad omnes scientias et artes*, 1682) (Knobloch *1979*). The Spanish Lullists Sebastian Izquierdo (*Pharus scientiarum*, 1659) and Juan Caramuel de Lobkowitz (*Mathesis biceps vetus et nova*, 1670) published their works in the same period.

3 THE ARITHMETICIANS OF THE RENAISSANCE AND THE SEVENTEENTH CENTURY

Combinatorial rules and problems are to be found in many textbooks by Renaissance arithmeticians, mostly again without any proof (Raymond *1975*). This applies, for example, to Italian, French, English and German authors like Luca Pacioli (*Summa de arithmetica geometria proportioni et proportionalita*, 1494), Girolamo Cardano (*De subtilitate*, 1550; *Opus novum de proportionibus*, 1570), Niccolò Tartaglia (*General trattato di numeri et misure*, 1556), Jean Buteo (Borrell) (*Logistica*, 1559) and William Buckley (*Arithmetica memorativa*, 1567).

In 1570 the German Jesuit Christoph Clavius published his commentary on Johannes de Sacrobosco's *De sphaera*. The few remarks on combinatorics it contained were implicitly or explicitly repeated not only by all the above-mentioned Lullist authors but also by many mathematicians, such as Pierre Hérigone (*Cursus mathematicus*, Vol. 2, 1634) and Gottfried Wilhelm Leibniz (*Dissertatio de arte combinatoria*, 1666; see Knobloch *1973*).

In the seventeenth century, combinatorical studies reached a completely new mathematical level. Mathematicians like Bernard Frénicle de Bessy (*Abrégé des combinaisons*, elaborated before 1675, published posthumously in 1693), Thomas Strode (*A Short Treatise of the Combinations, Elections, Permutations and Composition of Quantities*, 1678), and John Wallis ('A discourse of combinations, alternations and aliquot parts', an appendix to his *Treatise of Algebra*, 1685) discussed this subject intensively. It became a useful and valuable tool of the emerging calculus of probability. This is especially true of Blaise Pascal's *Traité du triangle arithmétique* (elaborated in 1654, published posthumously in 1665) and James Bernoulli's *Ars conjectandi* (elaborated *circa* 1685, also published posthumously in 1713). Unfortunately, Bernoulli knew of very few of his predecessors; he was unaware even that Pascal had proved the identity of the combinatorial numbers $_nC_r$, the figurate numbers and the binomial coefficients $\binom{n}{r}$.

4 PROBABILISTIC GLIMPSES

Games of chance (card games as well as games of dice) provide the first known examples of calculating possible outcomes of events in advance. Long before the calculus of probability was created by James Bernoulli between 1685 and 1705, it was known how to calculate in advance which of a certain number of events would be likely to happen more often than others – in other words, they could calculate relative frequencies. It was a question of enumeration, of finding combinatorial solutions.

The first known systematic studies of dice games date from the thirteenth century. A pseudo-Ovidian Latin poem, *De vetula*, written between 1222 and 1268, enumerates the 56 ways that three dice can fall, thus listing the 56 3-partitions of the numbers 3 to 18. The unknown author underlined the role of chance. The same list of 56 possible outcomes of a game of three dice is to be found in the *Libros de acedrez, dados e tablas*, compiled in 1283 by order of the Castilian king Alfonso el Sabio (see §12.2, on games).

Between 1307 and 1321 Dante wrote his *Divina commedia*. In the sixth canto of the 'Purgatorio' he mentions the 'game of hazard' (*giuoco della zara*) which was played with three dice. One player threw the dice, and the other had to guess the sum of the three numbers. The first commentator on Dante, Jacopo Giovanni della Lana, writing around 1324–8, not only tried to explain the game's rules but also looked for a mathematical analysis of the stochastic situation (Ineichen *1988*). The same is true of Lana's successors. The commentaries were only published from 1477 onwards, a hundred and fifty years later.

These first stochastic considerations based on combinatorial methods led to the insights that chance plays its part; that the number of possible ways of arriving at each of the different sums is crucial to a player seeking possible advantage; that there is a relation between this number of realizations and the frequency of outcomes; and that one can describe these realizations by partitions of the sums into three summands. In general, the early authors did not obtain equally possible cases, because they did not permute the summands. But the conceptualization of the perfect die and the equal frequency of each face coming up in a series of throws were explicit.

The problem of the three dice remained an important topic in the prehistory of the calculus of probability, and was later discussed by Cardano and Galileo Galilei. In about 1526, Cardano began collecting material for the book on games of chance he would write in about 1564 (but published only posthumously in 1663). Although it was entitled *Liber de ludo aleae* ('Book on Dice Games'), it discussed card games as well, dealing with the whole practice of games of chance. Cardano calculated the odds in dice and card games in order to discuss fair bets, and introduced certain probabilistic notions, like 'circuit' (the number of all equally possible outcomes) and 'equality' (half the circuit); however, these ideas had no influence on further development because it remained unknown. The same is true of Galileo's treatise *Sopra le scoperte dei dadi* ('On Outcomes in the Game of Dice'), written probably between 1613 and 1623.

5 PASCAL AND FERMAT: THE YEAR 1654

The famous correspondence between Pascal and Pierre de Fermat which initiated a new stage in the development of probability theory (David *1962*) began in 1654. They discussed a problem about a dice game, and the so-called gambler's problem of points, or division problem, which concerns the division of the stakes between two players when a game has to be left unfinished (§10.2). This problem was far from new at that time. A correct solution of a special case of it is to be found in a Florentine manuscript dating from 1400, but false solutions were given by many Italian arithmeticians of the Renaissance like Pacioli (1494), Cardano (1539), Tartaglia (1556), Francesco Peverone (1558) and Lorenzo Forestani (1603).

Pascal and Fermat advanced three correct methods for solving the division problem for any number of players. The first, the method of combinations, is to find the minimum number of single wins necessary to win the whole game. One has to list all possible sequences of won and lost games for players A and B, and determine by inspection who is the winner in each case. If A wants a points and B wants b, then the whole game must be over in at most $a + b - 1$ further single games, which may occur in 2^{a+b-1} equally probable ways (the number of arrangements, with repetition, of two things taken $a + b - 1$ at a time).

Pascal elaborated a second method, based on expectations. The stakes are to be divided according to the expectation of gain; that is, the value of a gamble is equal to its expectation. In the same year (1654), Pascal elaborated his *Traité du triangle arithmétique* in which he gave the binomial solution for two players. The combinations favourable to one player are enumerated by the partial sum or sums of the appropriate row of the arithmetical triangle of binomial coefficients. Thus the division ratio is, in modern terms:

$$\frac{\binom{a+b-1}{0} + \binom{a+b-1}{1} + \cdots + \binom{a+b-1}{b-1}}{\binom{a+b-1}{0} + \binom{a+b-1}{2} + \cdots + \binom{a+b-1}{a-1}}. \tag{1}$$

Pascal proved this result by mathematical induction in the *Traité*; it was given as a formula by Pierre de Montmort in the first edition of his *Essay d'analyse sur les jeux de hazard* (1708). This presupposes players of equal skill; Abraham De Moivre generalized the binomial solution for players of unequal skill in his *De mensura sortis* (1712) (see §4.1 on the binomial series).

In the meantime, Christiaan Huygens' essay 'De ratiociniis in ludo aleae' was published as an appendix to Frans van Schooten the Younger's *Exercitationes mathematicae* in 1657 (§10.2). It included the problem of points for any number of players of equal skill (basically a problem in combinatorics), and gave Pascal's expectation method as a means of solution.

6 JAMES BERNOULLI AND HIS IMMEDIATE SUCCESSORS

In 1713, James Bernoulli's nephew Nicholas Bernoulli published the treatise

Ars conjectandi of his uncle. The first chapter reprinted Huygens' booklet together with important editions and generalizations. Huygens' generalized 12th proposition led him, for example, to the binomial distribution. The second chapter presented the doctrine of permutations and combinations. James Bernoulli's stated aim was to develop the theory to the point where nothing remained unproved. He was aware that he was repeating at least partly known results, and this was true to a far larger extent than he realized. In the third chapter he nevertheless made a substantial step beyond the combinatorial studies written long before his time by using combinatorial rules as methods in various generalized games of chance, including dice games. The fourth and last chapter is doubtlessly the most important, and contains the weak theorem of large numbers which nowadays is called 'Bernoulli's theorem'. This started the controversy over inverse probability (§10.17). Bernoulli developed the notion of a measurable probability by means of the degree of certainty, and thus replaced the ratio of chances and the notion of expectation by the notion of probability. He corresponded about this with Leibniz, who had certain objections to Bernoulli's theorem; Leibniz was one of the leading scholars to take part in the philosophical debate on the notion of probability during the second half of the seventeenth century.

The mathematical community of that time knew that James Bernoulli was dealing with the calculation of chances, but it had to wait until 1713 for the publication of his work. In the meantime, Montmort published his *Essay*, whose second very enlarged edition also appeared in 1713, immediately after the publication of Bernoulli's book. In it he calculated the chances in various games of cards in much the same way as James Bernoulli had done.

The calculus of probability reached a new level in De Moivre's publications on the subject. In the foreword to his 1711 paper 'De mensura sortis', published in the *Philosophical Transactions of the Royal Society*, he stated that he had mainly applied the doctrine of combinations, which when well understood leads easily to the solution of many otherwise very difficult problems. The enlarged version of this paper appeared for the first time in 1718 as a book entitled 'Doctrine of chances', and again in 1738 and 1756. Among his major achievements were the limit theorem for binomial distributions, the application of generating functions for probabilities, and a concept of recurrent theories. Only in the introduction to the second edition did he state explicitly the measure of probability which his predecessors and he himself had been using for a long time: that it is 'the comparative magnitude of the number of chances to happen, in respect to the whole number of chances either to happen or to fail, which is the true measure of probability'. The definition remained valid until the end of the nineteenth

1291

century, and is even still nowadays erroneously attributed to Pierre Simon Laplace. De Moivre's introduction not only contained this definition of classical probability, but also consisted of a systematically ordered set of probabilistic rules and notions such as expectation, dependent and independent events, and the sum and product theorems of events. It was the beginning of a probability theory in the strict sense of the word, in so far as De Moivre began the discussion of the 90 problems of the second edition by saying that their solutions were deduced from the rules laid down in the introduction.

BIBLIOGRAPHY

Biggs, N. L. *1979*, 'The roots of combinatorics', *Historia mathematica*, **6**, 109–36.

David, F. N. *1962*, *Games, Gods, and Gambling. The Origin and History of Probability and Statistical Ideas from the Earliest Times to the Newtonian Era*, New York: Hafner.

Edwards, A. W. F. *1987*, *Pascal's Arithmetical Triangle*, London: Griffin, and New York: Oxford University Press.

Ineichen, R. *1988*, 'Dante-Kommentare und die Vorgeschichte der Stochastik', *Historia mathematica*, **15**, 264–9.

Knobloch, E. *1973*, *Die mathematischen Studien von G. W. Leibniz zur Kombinatorik*, Wiesbaden: Steiner.

—— *1979*, 'Musurgia universalis: Unknown combinatorial studies in the age of baroque absolutism', *History of Science*, **17**, 258–75.

Maistrov, L. E. *1974*, *Probability Theory. A Historical Sketch* (transl. and ed. S. Kotz), New York and London: Academic Press.

Pearson, E. S. and Kendall, M. G. (eds) *1970*, *Studies in the History of Statistics and Probability*, London: Griffin. [Several relevant articles.]

Rabinovitch, N. L. *1973*, *Probability and Statistical Inference in Ancient and Medieval Jewish Literature*, Toronto: University of Toronto Press.

Raymond, P. *1975*, *De la combinatoire aux probabilités. La Combinatoire de Cardan à Jacques Bernoulli*, Paris: Maspero.

Schneider, I. (ed.) *1988*, *Die Entwicklung der Wahrscheinlichkeitstheorie von den Anfängen bis 1933. Einführungen und Texte*, Darmstadt: Wissenschaftliche Buchgesellschaft. [Source-book.]

Sheynin, O. B. *1977*, 'Early history of the theory of probability', *Archive for History of Exact Sciences*, **17**, 201–59.

Todhunter, I. *1865*, *A History of the Mathematical Theory of Probability from the Time of Pascal to That of Laplace*, Cambridge and London: Macmillan. [Repr. 1949, New York; Chelsea.]

10.2

The early development of mathematical probability

GLENN SHAFER

This article is concerned with the development of the mathematical theory of probability, from its founding by Blaise Pascal and Pierre de Fermat in an exchange of letters in 1654 to its early-nineteenth-century apogee in the work of Pierre Simon Laplace. It traces how the meaning, mathematics and applications of the theory evolved over this period. The story is summarized in the first section.

1 SUMMARY

Pascal and Fermat are credited with founding mathematical probability because they solved the problem of points – the problem of equitably dividing the stakes when a fair game is halted before either player has enough points to win. This problem had been discussed for several centuries before 1654, but Pascal and Fermat were the first to give the solution we now consider correct. They also answered other questions about fair odds in games of chance. Some of their main ideas were popularized by Christiaan Huygens, in his *De ratiociniis in ludo aleae* (1657).

During the century that followed this work, other authors, including James and Nicholas Bernoulli, Pierre de Montmort and Abraham De Moivre, developed more powerful mathematical tools in order to calculate odds in more complicated games. De Moivre, Thomas Simpson and others also used the theory to calculate fair prices for annuities and insurance policies (§10.3).

James Bernoulli's *Ars conjectandi*, published in 1713, laid the philosophical foundations for broader applications. Bernoulli brought the philosophical idea of probability into the mathematical theory, formulated rules for combining the probabilities of arguments, and proved his famous theorem: that the probability of an event is morally certain to be approximated by the frequency with which it occurs. Bernoulli advanced this theorem (later called the 'law of large numbers' by Siméon-Denis Poisson) as a justification for using observed frequencies as probabilities, to be combined by his rules to settle practical questions. Bernoulli's ideas attracted philosophical and mathematical attention, but gambling – whether on games or on lives – remained the primary source of new ideas for probability theory during the first half of the eighteenth century.

In the second half of the eighteenth century, a new set of ideas came into play. Workers in astronomy and geodesy began to develop methods for reconciling observations, and students of probability theory began to seek probabilistic grounding for such methods. This work inspired Laplace's invention of the method of inverse probability, which benefited from the evolution of Bernoulli's law of large numbers into what we now call 'the central limit theorem'. It culminated in Adrien Marie Legendre's publication of the method of least squares in 1805 and in the probabilistic rationales for least squares developed (with different principles) by Laplace and by Carl Friedrich Gauss from 1809 to 1811 (§10.5). These ideas were brought together in Laplace's great treatise on probability, *Théorie analytique des probabilités* (1812).

2 GAMES OF CHANCE

Games of chance and the drawing of lots are discussed in several ancient texts, including the Talmud. A number of medieval authors, both mystics and mathematicians, enumerated the ways in which various games can come out. But most of these enumerations were not enumerations of equally likely cases, and they were not used to calculate odds (Kendall *1956*, David *1962*).

One genuine precursor of probability theory was Girolamo Cardano's *Liber de ludo aleae*. This book was probably written in the 1560s (Cardano died in 1576), but it was not published until after the work of Huygens. Cardano formulated the principle that the stakes in an equitable wager should be in proportion to the number of ways in which each player can win, and he applied this principle to find fair odds for wagering with dice. Another precursor was Galileo Galilei, who enumerated the possible outcomes for a throw of three dice in order to show that the faces add to ten more easily and hence more frequently than they add to nine. This work also remained unpublished until after the work of Huygens.

Galileo did not comment on the problem of points, but it was the most salient mathematical problem concerning games of chance during this period. It was discussed by Italian mathematicians as early as the fourteenth century, and Luca Pacioli, Niccolò Tartaglia, Francesco Peverone and Cardano all tried to solve it. Tartaglia concluded that it had no solution: 'the resolution of the question is judicial rather than mathematical, so that in whatever way the division is made there will be cause for litigation'. But Pascal and Fermat found a solution, and they gave several convincing arguments for it.

Fermat preferred to solve the problem by listing the ways in which the play might go. Suppose, for example, that Peter and Paul have staked equal money on being the first to win three points, and they want to stop the game and divide the stakes when Peter lacks two points and Paul lacks only one. Paul should get more than Peter because he is ahead, but how much more? If two more games were to be played, there would be four possibilities:

1 Paul wins the first and the second.
2 Paul wins the first, and Peter wins the second.
3 Peter wins the first, and Paul wins the second.
4 Peter wins the first and the second.

In the first three cases, Paul wins the game (in the first two, it is not even necessary to play for the second point), while in the fourth case Peter wins the game. By Cardano's principle, the stakes should be in the same proportion: three for Paul and one for Peter. If the stakes are divided now, Paul gets three-quarters.

Pascal preferred another method, the 'method of expectations'. This method relies on principles of equity instead of Cardano's principle, and it enabled Pascal to solve the problem for the case when the players lack so many points that listing all the possibilities is impractical. Since the game is fair, Pascal said, both players have an equal right to expect to win the next point. If Paul wins it, he wins all the stakes; this entitles him to half the stakes. If Peter wins, the two are tied; so both are entitled to half of what is left. Paul is therefore entitled to three-quarters altogether. Extending this backwards reasoning by mathematical induction, and using the recursive properties of what we now call Pascal's triangle (Figure 1), Pascal was able to solve the problem for any number of points the players might lack. He found, for example, that if Peter lacks four points and Paul lacks two, then their shares are found by adding the numbers in the base of the triangle: Paul's share is to Peter's as $1 + 5 + 10 + 10$ to $5 + 1$, or 13 to 3.

The combinatorial knowledge that underlay Pascal's reasoning was richer and less streamlined than the combinatorics now taught as a foundation for probability. Pascal organized much of this knowledge, together with his general solution of the problem of points, in his *Traité du triangle arithmétique*, which was published in 1665, three years after his death (Edwards *1987*).

```
1  1  1  1  1  1
1  2  3  4  5
1  3  6 10
1  4 10
1  5
1
```

Figure 1 Pascal's arithmetic triangle (as he laid it out)

Huygens heard about Pascal's and Fermat's ideas but had to work out the details for himself. His treatise *De ratiociniis in ludo aleae*, published in 1657, essentially followed Pascal's method of expectations. In one respect, he went further than Pascal: he justified the very idea that there was a fair price for a position in a game of chance by showing that, by paying that price, the player would end up with the same chances for the same net pay-offs that could be obtained in a fair game with side-payments.

At the end of his treatise, Huygens listed five problems about fair odds in games of chance, some of which had already been solved by Pascal and Fermat. These problems, together with similar questions inspired by other card and dice games popular at the time, set an agenda for research that continued for nearly a century. The most important landmarks of this work were James Bernoulli's *Ars conjectandi* (1713), Montmort's *Essay d'analyse sur les jeux de hazard* (editions in 1708 and 1711) and De Moivre's *Doctrine of Chances* (editions in 1718, 1738 and 1756). These authors investigated many of the problems still studied under the heading of discrete probability, including gambler's ruin, duration of play, handicaps, coincidences and runs. In order to solve these problems, they improved Pascal and Fermat's combinatorial reasoning, summed infinite series, developed the method of inclusion and exclusion, and developed methods for solving the linear difference equations that arise in using Pascal's method of expectations.

Perhaps the most important technical mathematical development in this work was the invention of generating functions. Although Laplace gave them their name much later, De Moivre invented them in 1733 in order to find the odds for different sums of the faces of a large number of dice. This was a first step in the development of what we now call Laplace and Fourier transforms (§4.8, §3.11).

From the time of Huygens onwards there was one important practical application of this work: the pricing of annuities and life-insurance policies. Huygens' ideas were first applied to pricing annuities by Jan De Witt in 1671 and Edmond Halley in 1694. De Moivre, in his *Annuities on Lives* (editions in 1725, 1743, 1750 and 1752), and Simpson, in his *Doctrine of Annuities and Reversions* (1742), extended this work to more complicated annuities, reversions, and insurance and assurance policies. This work made use of interest rates, but otherwise it remained conceptually close to the work on games of chance (§10.3).

3 PROBABILITY

Probability – the weighing of evidence and opinion – was an important topic in the seventeenth century, but it was not discussed in the letters between Pascal and Fermat or in Huygens' treatise. The word 'probability' did not appear in these works. For these authors, the number between zero and one that we now call probability was only the proportion of the stakes due to a player; they did not isolate it as a measure of belief (§10.17).

The desire to apply the new ideas to problems in the domain of probability arose almost immediately, however. Pascal himself used them in his famous argument for believing in God, and the Port Royal *Logic*, published by Pascal's religious colleagues in 1662, argued for using the theory of games of chance to weigh probabilities in everyday life. Soon the idea of probability as a number between zero and one was in the air. The English cleric George Hooper, writing in 1689, gave rules for combining such numbers to assess the reliability of concurring testimony and chains of testimony.

James Bernoulli formalized the connection between probability and Huygens' theory. In *Ars conjectandi*, published in 1713, eight years after his death, he contended that an argument is worth a portion of complete certainty, just as a position in a game of chance is worth a share of the total stakes. Take complete certainty to be a unit, he said, and probability is then a number between zero and one. By Huygens' rules, it will be the ratio of the number of favourable cases to the total number of cases.

Bernoulli tried to draw the traditional ideas of probability into the mathematical theory by formulating rules for combining the probabilities of arguments, rules that were similar to Hooper's rules for testimony but more general. Bernoulli hoped these rules would make probability a widely used practical tool. Probabilities would be found from observed frequencies, and then combined to make judicial, business and personal decisions.

Bernoulli buttressed this programme with his famous theorem, which says that it is morally certain (99.9% certain, say) that the frequency of an event in a large number of trials will approximate its probability. Although he proved this theorem rigorously, he gave only a disappointingly large upper bound on the number of trials needed in order to be morally certain that the frequency would be within a given distance of the probability. Nicholas Bernoulli improved the upper bound, and in 1733 De Moivre accurately estimated the number of trials needed, using a series expansion of the integral of what we now call the normal density. In retrospect, this can be seen as a demonstration that the binomial distribution can be approximated by the normal distribution, but De Moivre did not think of it in this way. He did not have our modern concept of a probability distribution; he was merely improving on Bernoulli in finding the number of trials needed in order to be sure that the observed frequency would approximate the probability.

This work by Bernoulli and De Moivre is also seen today as a precursor of the modern theory of confidence intervals: De Moivre's approximation allows us to state a degree of confidence, based on the observations, that the true probability lies between certain bounds. This is not the way Bernoulli and De Moivre saw the matter. They were aiming for moral certainty, not for some middling degree of confidence.

Bernoulli's theorem and his rules for combining probabilities did not achieve for him his goal of making probability a tool for everyday life and judicial affairs. His rules for combining probabilities were discussed in textbooks, but not used in practice; his theorem was discussed mainly in the context of speculation about the ratio of girl births to boy births. Yet his programme had great philosophical impact; it put probability, rather than equity, at the centre of the mathematical theory.

The increasing autonomy of probability from equity in the early eighteenth century can be seen in the contrasting attitudes of Nicholas Bernoulli and his younger cousin Daniel Bernoulli. In the 1730s, they discussed what we now call the St Petersburg problem. A person throws a die repeatedly, and wins a prize when he first gets a 6. The prize is doubled every time he fails to get a 6; he wins one crown if he gets a 6 on the first throw, two crowns if his first 6 is on the second throw, four crowns if it is on the third throw, and so on. How much should he pay to play this game? By Huygens' rules, he should pay an infinite amount, yet no one would be willing to pay more than a few crowns. Daniel explained this by formulating the idea of 'expected utility'. If a person's utility for money is only proportional to its logarithm, then the expected utility of the game is finite. Nicholas could not see the point of this explanation, because for him the theory of probability was based on equity. If we are talking about what a person wants to pay rather than what is fair, then there is no basis for calculation (Shafer *1988*).

4 THE COMBINATION OF OBSERVATIONS

Although the concept of probability had found a place by 1750 in what had been the mathematical theory of games of chance, the applications of this theory were still only to questions of equity. No one had learned how to use probability in data analysis. This was true even in the work on annuities and life insurance: De Moivre used theoretical mortality curves, and Simpson used mortality statistics, but no one used probabilistic methods to fit models in the way that modern demographers do.

It was work on combining observations in astronomy and geodesy that finally brought data analysis and probability together. In the eighteenth century, combining observations meant reconciling inconsistent equations. An observation typically yielded numbers that could serve as coefficients in a linear equation relating unknown quantities. A few such observations would give enough equations to find the unknowns. More observations would mean more equations than unknowns, and since the measurements contained errors, these equations would be inconsistent.

Galileo had struggled with this problem, and had vaguely formulated the principle that the most likely values of the unknowns would be those that made the errors in all the equations reasonably small; but the first formal method for estimating the unknown quantities was developed by Tobias Mayer in his study of the libration of the Moon in 1750. Mayer generalized the averaging of observations under identical circumstances to the averaging of groups of similar equations. Another early contributor was Roger Boscovich, who used the idea of minimizing the sum of absolute deviations in his work on the shape of the Earth, in 1755. Laplace, in an investigation of the motions of Saturn and Jupiter in 1787, generalized Mayer's method by using different linear combinations of a single group of equations. Then, in 1805, Legendre showed how to estimate the unknown quantities by minimizing the sum of squared deviations. This method of least squares was an immediate success, because of its conceptual and computational simplicity, its generality and its clear relationship to other methods that had been used in practice (§10.5). It was similar to Boscovich's principle of minimizing absolute deviations, but easier to implement. Like Laplace's generalization of Mayer's method, it entailed solving a set of linear combinations of the original equations, but it gave a generalizable rationale for the choice of these linear combinations.

Practical work on combining observations was not influenced by probability theory until after Legendre's publication of least squares, but attempts to base methods for combining observations on probability theory began very early. In 1755, Simpson made some arbitrary assumptions about the distribution of errors and deduced that the average of a set of observations of a single quantity is likely to be less in error than the individual observations. A number of authors, including Daniel Bernoulli, Joseph Louis Lagrange and Laplace, derived methods for combining observations from various assumptions about error distributions.

The most important fruit of this probabilistic work was the invention by Laplace of the method of inverse probability – what we now call the Bayesian method of statistical inference (§10.17). Laplace discovered inverse probability in the course of his work on the theory of errors in the 1770s. He realized that probabilities for errors, once the observations are fixed, translate into probabilities for the unknown quantities being estimated. He called them posterior probabilities, and he justified using them by adopting the principle that, after an observation, the probabilities of its possible causes are proportional to the probabilities given the observation by the causes. This was called 'the method of inverse probability' in the nineteenth century; we now call it 'Bayes's theorem', in honour of Thomas Bayes, who had enunciated a similar but more obscure principle in a paper published in 1764 (Shafer *1982*).

Inverse probability was recognized by Laplace's contemporaries as an important contribution to the theory of probability. During the 1770s and 1780s, moreover, Laplace made great strides in developing numerical methods for evaluating posterior probabilities. But this work did not immediately bring probability theory into contact with the practical problem of combining observations. When applied to the error distributions proposed by Laplace and his contemporaries, inverse probability produced methods for combining observations that were intractable or unattractive in comparison with the established methods.

This changed after Legendre's discovery of the least squares. In 1809, Gauss, who considered himself the inventor of the method, published his own account of it. Gauss provided a probabilistic justification for least squares: the least-squares estimate of a quantity is the value with the greatest posterior probability if the errors have what we now call a normal distribution – a distribution with probability density

$$f(\Delta) = (h/\sqrt{\pi}) \exp(-h^2\Delta^2) \tag{1}$$

for some constant h. Why should we expect the errors to have this distribution? Gauss gave a rather weak argument. It is the only error distribution, he said, that makes the arithmetic mean the most likely value when we have observations of a single unknown quantity. Thus the consensus in favour of the arithmetic mean must be a consensus in favour of this error distribution.

Laplace saw Gauss's work in 1810, just after he had completed a paper rounding out several decades of work on approximating the probability distributions of averages. Laplace had originally worked on this problem in the 1770s, in connection with the question of whether the average inclination of the planetary orbits to the ecliptic was too small to have happened by chance. In the 1780s, he found ways to use generating functions and the asymptotic analysis of integrals to approximate probabilities for averages of many observations, and by 1810 he had succeeded in showing that, no matter what the probability distribution of the original observations, probabilities for their average could be found by integrating functions of the form (1). This is a generalization of De Moivre's normal approximation for the binomial; it is the first statement of what we now call (after George Pólya) the central-limit theorem ('central' here meaning 'fundamental').

With this recent work in mind, Laplace saw immediately that he could improve on Gauss's argument for using (1) as an error distribution. Each individual error would have the distribution (1) if it was itself an average, the resultant of many independent additive influences. The following year, 1811, Laplace also pointed out that Legendre's least-square estimates, being weighted averages of the observations, would be approximately normal if there were many observations, no matter what the distributions of the individual errors. Moreover, Laplace showed, these estimates would have the least expected error of any estimates that were weighted averages of the observations. As Gauss pointed out in 1823, this last statement is true even if there are only a few observations.

5 LAPLACE'S SYNTHESIS

The work on the combination of observations brought to probability theory the main idea of modern mathematical statistics: data analysis by fitting models to observations. It also established the two main methods for such fitting: linear estimation and Bayesian analysis.

Laplace's great treatise on probability, *Théorie analytique des probabilités*, appeared in 1812, with later editions in 1814 and 1820. Its picture of probability theory was entirely different from the picture in 1750. On the philosophical side was Laplace's interpretation of probability as rational belief, with inverse probability as its underpinning. On the mathematical side was the method of generating functions, the central limit theorem, and Laplace's techniques for evaluating posterior probabilities. On the applied side, games of chance were still in evidence, but they were dominated by problems of data analysis and by Bayesian methods for combining probabilities of judgements, which replaced the earlier non-Bayesian methods of Hooper and Bernoulli (Shafer *1978*, Zabell *1988*).

Laplace's views dominated probability for a generation, but in time they gave way in turn to a different understanding of probability. The error distributions at the heart of Laplace's approach seemed to the empiricists of the nineteenth century to have a frequency interpretation incompatible with Laplace's philosophy, and Laplace's own rationale for least squares based on the central-limit theorem seemed to make his method of inverse probability unnecessary. In the empiricist view, which became dominant in the late nineteenth and early twentieth centuries, frequency is the real foundation of probability. On these philosophical questions, see §10.16–10.17.

Laplace's synthesis also came apart in a different way. His powerful mathematics of probability was aimed directly at applications. Mathematics and applications then constituted a single topic of study. Today probability has evolved into too vast a topic for such unity to be sustained – it is now a rich branch of pure mathematics, and its role as a foundation for mathematical and applied statistics is only one of its many roles in the sciences.

BIBLIOGRAPHY

Shapiro *1983* surveys non-mathematical probability in seventeenth-century England, while Hacking *1975* treats the broader intellectual antecedents of mathematical probability. Daston *1988* traces the interpretation of probability up to 1840. Hald *1990*, updating Todhunter *1865*, catalogues the mathematics and applications of probability before 1750. Stigler *1986* traces the development of error theory from 1750 onwards, as well as the further development of mathematical statistics in the nineteenth century.

Daston, L. *1988, Classical Probability in the Enlightenment*, Princeton, NJ: Princeton University Press.

David, F. N. *1962, Games, Gods, and Gambling* [...], London: Griffin.

Edwards, A. W. F. *1987, Pascal's Arithmetic Triangle*, London: Griffin, and New York: Oxford University Press.

Hacking, I. *1975, The Emergence of Probability*, Cambridge: Cambridge University Press.

Hald, A. *1990, A History of Probability and Statistics and Their Applications Before 1750*, New York: Wiley.

Kendall, M. G. *1956*, 'The beginnings of a probability calculus', *Biometrika*, **43**, 1–14. [Repr. in Pearson and Kendall *1970*: 19–34.]

Kendall, M. G. and Plackett, R. L. (eds) *1977, Studies in the History of Statistics and Probability*, Vol. 2, New York: Macmillan.

Pearson, E. S. and Kendall, M. G. (eds) *1970, Studies in the History of Statistics and Probability*, Vol. 1, Darien, CT; Hafner.

Shafer, G. *1978*, 'Non-additive probabilities in the work of Bernoulli and Lambert', *Archive for History of Exact Sciences*, **19**, 309–70.

——— *1982*, 'Bayes's two arguments for the rule of conditioning', *Annals of Statistics*, **10**, 1075–89.

——— *1988*, 'The St Petersburg paradox', in S. Kotz and N. L. Johnson (eds), *Encyclopedia of Statistical Sciences*, Vol. 8, New York: Wiley, 865–70.

Shapiro, B. J. *1983, Probability and Certainty in Seventeenth-Century England*, Princeton, NJ: Princeton University Press.

Stigler, S. M. *1986, The History of Statistics: The Measurement of Uncertainty Before 1900*, Cambridge, MA: Harvard University Press.

Todhunter, I. *1865, A History of the Mathematical Theory of Probability from the Time of Pascal to that of Laplace*, Cambridge and London: Macmillan. [Repr. 1949, New York: Chelsea.]

Zabell, S. L. *1988*, 'The probabilistic analysis of testimony', *Journal of Planning and Inference*, **20**, 327–54.

10.3

Actuarial mathematics

C. G. LEWIN

Actuaries predict future outcomes, based on assumptions (fixed in the light of past experience and anticipated future trends) about such matters as mortality rates, salaries, interest rates and dividends. Practical applications include the calculation of insurance premiums and pension-fund contributions. The most pertinent areas of mathematics are probability, statistics and compound interest.

1 EARLY TECHNIQUES

Some techniques for determining the capital values of life annuities were known in Roman times. For example, Ulpian's table of about AD 211 contained values which were probably based on assumed expectations of life with no allowance for interest. Money-lending was widely practised in the Middle Ages, and a typical rate of interest for pawnbroking transactions was twopence in the pound per week, equivalent to 43% per annum.

It was not until after AD 1500 that the mathematical techniques were developed. During the sixteenth century some of the Continental writers on arithmetic, such as Jean Trenchant and Simon Stevin, examined elementary problems in compound interest. In 1621 Gerard Malynes reported that the city of Amsterdam granted generous annuities, payable as long as at least one of several named lives survived.

There were no actuaries in those days: the nearest equivalent was the general mathematical practitioner. Richard Witt practised in London, and at the age of 44 wrote the first comprehensive book in English on compound interest, entitled *Arithmeticall Questions, touching The Buying or Exchange of Annuities* (1613); this was also one of the first books to use a form of decimal arithmetic. He understood compound interest thoroughly, giving numerous working examples and dealing not only with annual payments but also with payments at half-yearly and quarterly intervals (Lewin *1970, 1981*).

1303

Witt's book did not venture into life contingencies, and there is no evidence that he considered such problems. It was not until later in the seventeenth century that two additional necessary tools became available. One of these was the developing science of probability; the other was the concept of the life table, the first example of which was published by John Graunt of London in his *Natural and Political Observations* [...] (*1662*), which was based on a statistical analysis of the London Bills of Mortality for many years. The main purpose of these weekly Bills was to warn people when a plague epidemic was starting, so that they could flee to the countryside in good time (Pearson *1978*: Chap. 2). Had it not been for such epidemics, the history of actuarial mathematics might have taken a different course.

James Howell (*Letters* (1645), letter No. 7) had already used the Bills of Mortality for London and Amsterdam to infer that the latter had a much smaller population than the former; but Graunt, possibly employing an idea suggested to him by William Petty, was the first to demonstrate how statistics could be drawn up which would lead to a range of interesting conclusions. For example, he showed what proportions of deaths resulted from different causes. He estimated (probably correctly) that the population of London was about 384 000, rather than the millions which were commonly assumed. Above all, he produced the first ever life table, which showed, for every 100 children conceived, how many would reach successive ages. He then applied this table, in a manner foreshadowing the method of stationary populations, to estimate the numbers of men then alive at different ages.

Thus by 1670 the three main foundations of actuarial science were firmly in place: compound interest, probability and the life table. These tools were employed almost immediately by the Dutch Grand Pensionary (prime minister), Jan De Witt, to investigate the values of government life annuities. However, his treatise (although of great technical merit) remained unpublished for many years, and so did not influence the development of actuarial science. Johann Hudde also made a valuable contribution at about the same time.

It was not until 1693 that Edmond Halley had a crucial paper published by The Royal Society of London. It set out the method for determining the capital values of life annuities that is still employed today (Halley *1693*). Using the Bills of Mortality for the city of Breslau for 1687–91, in which the deaths were classified by age, he produced a life table showing, for every 1000 children in their first year of age, how many would survive to each subsequent year of life. His table (which was based on the implicit assumption that the population was stationary) was thus somewhat similar in principle to Graunt's, but based more firmly on actual data. It was also more detailed, in that he considered each single year of age, whereas Graunt had

had to work in broad age groups and make sweeping assumptions because of data limitations.

Halley then showed how to use the life table to calculate the present value of life annuities at 6% per annum interest. He calculated the present value of each annual payment as its discounted value at 6% multiplied by the probability (in the form of a ratio obtained from the life table) that it would be received, and then summed these present values. This led him to conclude that British Government life annuities, which were being sold on the basis of 7 years' purchase, were very cheap, as the true value of the annuity for a young life was over 13 years' purchase. He then described, with the aid of geometrical diagrams, a method for dealing with annuities depending on several lives. It is astonishing that such a huge step forward should have been made in a single short paper by someone who spent much of his time working in quite different fields.

2 DEVELOPMENTS IN THE EIGHTEENTH CENTURY

There were numerous writers on actuarial subjects in the eighteenth century. Abraham De Moivre's *Annuities on Lives* (1725) was based on the hypothesis that, for any given number of people alive at a certain age, the same number of them would die at each subsequent age. This was done in order to reduce the arithmetic required by Halley's method, and was sufficiently accurate for some practical purposes.

In the 1740s W. Kersseboom in Holland and M. Deparcieux in France calculated mortality tables based on the experience of long-established tontines (a type of financial investment where the annual payments increased as other participants died). These calculations had the merit that it was no longer necessary to make the doubtful assumption of a stationary population, since all the people exposed to risk of dying at each age were known.

In 1743 a fund (still existing today) was established to provide pensions for the widows and orphans of ministers of the Church of Scotland. A pamphlet of 1743 described its financial basis and gave projections of the size of the fund in each of the first 35 years, as well as a calculation of the benefits payable each year. These calculations implicitly assumed a stationary population and were based on statistical returns covering many years. Some further projections made in 1748 of the growth of the fund in future years proved extremely accurate: by 1765, for example, the accumulated fund amounted to £58,347, as compared to a forecast of £58,348!

In 1747 Corbyn Morris laid one of the foundation stones of modern insurance when he demonstrated, with the aid of elementary probability theory and the binomial theorem, that the overall risk was reduced if insurers spread their marine-insurance risks over a number of voyages at any one time, rather than just one or two.

In the 1750s James Dodson showed for the first time how to calculate level annual premiums for long-term life-assurance policies. This work formed the basis for the establishment of the Equitable Life Assurance Society in 1762 – the first organization to offer such policies on scientific principles.

One of the most popular books on actuarial science was Richard Price's *Observations on Reversionary Payments* (1771), which went through seven editions until 1812. Among other things, Price, a Nonconformist minister, advocated the establishment of properly financed schemes for providing old-age pensions, and showed how to calculate mortality tables correctly. He was the author of the Northampton Mortality Table, which was widely adopted by newly formed life-assurance offices, but was later found to be very inaccurate because the christenings recorded in Northampton fell far short of the actual number of births.

One of the mathematical devices adopted by the early actuaries to reduce the labour of computation was the technique known as commutation columns. Expressions combining mortality functions and compound interest functions were tabulated for each age, in such a way that it was possible to obtain the value of an insurance policy for any attained age and future duration by a single subtraction and division sum. This technique first appeared in William Dale's *Calculations Deduced from First Principles* (1772) (Seal *1950*).

3 MORE SOPHISTICATED METHODS

From time to time attempts have been made to find a mathematical 'law' of mortality, partly in order to simplify calculations but also to try to discover something about the nature of life itself. For example, Benjamin Gompertz suggested in 1825 that the annual force of mortality at each age x was proportional to c^x, where c was a constant (to be determined). Although such attempts have sometimes represented mortality closely for a time, no simple universal law has yet been discovered.

In 1848 the Institute of Actuaries was formed in England, and since then its *Journal* has published many ingenious mathematical theories covering various aspects of actuarial work. Around the beginning of the twentieth century, for example, H. W. Manly and later J. J. M'Lauchlan developed the theory of pension funds, including projections of model funds for many years. G. J. Lidstone put forward a useful formula for valuing life-assurance policies in groups, in order to simplify computation. The theory of the matching of the assets and liabilities in a financial institution was established by A. T. Haynes and R. J. Kirton in 1952, while in the same year F. M. Redington developed the mathematical theory of 'immunization', demonstrating that a fund could be made secure against a change in the rate of interest.

Many of the early advances in the subject were made in Britain, perhaps because conditions there may have been particularly favourable for the establishment of long-term financial institutions which needed such techniques if they were to be properly funded and achieve stability; but there are now actuaries in many countries and international congresses are held every few years. In the twentieth century European actuaries were among the pioneers in the development of stochastic processes (§10.2), and they later participated in the application of these processes to risk theory and general insurance.

In life-assurance and pension-fund work, the advent of computers has led to the increased use of techniques based on simulating large numbers of possible future outcomes on varying assumptions, looking at the size of the payment in each future year. Nevertheless, the earlier techniques (which expressed the values of streams of future payments in terms of present-day capital sums on a single set of assumptions about future experience) are still in widespread use, and are often a convenient way of presenting results.

BIBLIOGRAPHY

Cox, P. R. and Storr-Best, R. H. *1962, Surplus in British Life Assurance: Actuarial Control Over its Emergence and Distribution During 200 Years*, Cambridge: Cambridge University Press.

Farren, E. J. *1844, Historical Essay on the Doctrine of Life Contingencies in England*, London: Smith, Elder.

Graunt, J. *1662, Natural and Political Observations* [...] *Upon the Bills of Mortality*, London: Martin. [Repr. 1964, *Journal of the Institute of Actuaries*, **90**, 1–61.]

Hald, A. *1990, A History of Probability and Statistics and Their Applications Before 1750*, New York: Wiley.

Halley, E. *1693*, 'An estimation of the degrees of the mortality of mankind', *Philosophical Transactions of the Royal Society of London*, **17**, 596–610, 654–6. [Repr. in facsimile in G. Heywood, 1985, *Journal of the Institute of Actuaries*, **112**, 278–301.]

Hendriks, F. *1852, 1853*, 'Contributions to the history of insurance [...] with a restoration of the Grand Pensionary De Wit's treatise on life annuities', *Assurance Magazine*, **2**, 121–50, 222–58; **3**, 93–120.

Lewin, C. G. *1970*, 'An early book on compound interest – Richard Witt's *Arithmeticall Questions*', *Journal of the Institute of Actuaries*, **96**, 121–32.

—— *1981*, 'Compound interest in the seventeenth century', *Journal of the Institute of Actuaries*, **108**, 423–42.

—— 1986–, '1848 and all that', *Fiasco*. [1986 onwards. This was the magazine of the Staple Inn Actuarial Society until 1990, when it was succeeded by *The Actuary*, in which the series of articles continues under the same title. Other historical articles have appeared occasionally, notably those by T. A. Sibbett.]

Lewin, C. G., Evans, J. V., Goodare, K. J. and Packer, L. R. *1989*, 'Calculating devices and actuarial work', *Journal of the Institute of Actuaries*, **116**, 215–87.

Maseres, F. (ed.) *1804, Scriptores logarithmici*, Vol. 5, London: White. [Reprints several classical papers on interest.]

Ogborn, M. E. *1950*, 'The actuary in the eighteenth century', in *Proceedings of the Centenary Assembly of the Institute of Actuaries*, Vol. 3, Cambridge: Cambridge University Press, 357–86.

—— *1962, Equitable Assurances*, London: Allen & Unwin.

Pearson, E. S. (ed.) *1978, The History of Statistics in the 17th and 18th Centuries* [. . .] *Lectures by Karl Pearson* [. . .] *1921–1933*, London: Griffin.

Seal, H. L. *1950*, 'The columnar method – A historical note', in *Proceedings of the Centenary Assembly of the Institute of Actuaries*, Vol. 3, Cambridge: Cambridge University Press, 387–94.

10.4

Estimating and testing the standard linear statistical model

R. W. FAREBROTHER

1 INTRODUCTION

A fundamental problem of statistical data analysis which has concerned practical scientists since Antiquity is that of fitting a model of the form

$$y_i = f(\mathbf{x}_i, \boldsymbol{\beta}) + \varepsilon_i, \quad i = 1, 2, \ldots, n, \tag{1}$$

to n observations on $(\mathbf{x}_i, y_i)$, where f is a known linear or non-linear function, y_i is the ith observation on the dependent variable, $\mathbf{x}_i$ is an $h \times 1$ matrix of observations on the h explanatory variables, $\boldsymbol{\beta}$ is a $k \times 1$ matrix of parameters and ε_i is the ith disturbance term. (For conciseness, modern notations are used.) If it is assumed that f is linear and $h = k$, then model (1) becomes

$$y_i = \mathbf{x}_i^T \boldsymbol{\beta} + \varepsilon_i, \quad i = 1, 2, \ldots, n. \tag{2}$$

In particular, if $k = 1$ and $x_1 \equiv 1$, then

$$y_i = \beta_1 + \varepsilon_i, \quad i = 1, 2, \ldots, n. \tag{3}$$

The disturbance term was not an explicit feature of these models until recent times, so the model (2) should strictly be replaced by the n equations

$$y_i = \mathbf{x}_i^T \boldsymbol{\beta}, \quad i = 1, 2, \ldots, n, \tag{4}$$

and the model (3) by the n observations $y_1, y_2, \ldots, y_n$.

This problem had been studied in various guises since the fifth century BC (Plackett *1958, 1988*, Sheynin *1983*). For example, in his *History of the Peloponnesian War* (Book 3, Section 20), Thucydides records an early use of the mode: in 428 BC the Plataeans successfully estimated the height of an encircling siege wall by taking the modal value of several counts of the number of layers in the wall. But the story proper begins in 1750, when it was generally agreed that the model (3) should be estimated by the arithmetic mean of the observations (with the median and the mid-range as possible alternatives), and that the model (2) should be estimated by solving one (possibly two) sets of k typical equations (Hald *1990*).

2 THE METHODS OF MAYER AND BOSCOVICH

Tobias Mayer and Roger Boscovich both seem to have been worried by the discordance between these two rules and, after first suggesting simple compromises based on the arithmetic mean of n/k or $_nC_k$ sets of estimates, they proposed the practical procedures which are now named after them.

The method proposed by Mayer in 1750 used one of the explanatory variables to classify equations (4) into k groups of equal size; the equations in each group were then summed, and the k unknowns determined from the resulting k equations. This procedure was subsequently superseded by a variant suggested by Pierre Simon Laplace in which **b** is chosen to satisfy $\mathbf{Z}^T\mathbf{e} = \mathbf{0}$, where **Z** is an $n \times k$ matrix with entries which are either 1 or -1, defined by $z_{ij} = \text{sign}(x_{ij})$, and **e** is an $n \times 1$ matrix with its ith element $e_i = y_i - \mathbf{x}_i^T\mathbf{b}$. Mayer's procedure may also be written in this form, but with the rows of **Z** selected from the $k \times k$ identity matrix. Following Wald *1940*, this estimator has been associated with the errors in the variables model (2), in which $\mathbf{x}_i$ is observed with error η_i.

The method proposed by Boscovich (in his copious notes to Stay's Latin poem *Philosophiae recentioris*, 1760) estimates the parameters of the model

$$y_i = \beta_1 + \beta_2 x_i + \varepsilon_i, \quad i = 1, 2, \ldots, n \tag{5}$$

by choosing b_1 and b_2 so as to minimize $\Sigma |e_i|$ subject to $\Sigma e_i = 0$. His solution to this problem took a geometrical form, whereas the alternative developed by Laplace in 1793 and published in the second volume of his *Mécanique céleste* (1799) was analytic. Subsequently, in 1888 F. Y. Edgeworth supplied a procedure which is closely related to the simplex procedure of linear programming for the unconstrained minimization of $\Sigma |e_i|$ (Farebrother *1987*; and see §6.11, on linear optimization). It is sometimes suggested that Galileo Galilei had proposed a variant of this estimation procedure in 1632, but he was using a criterion function of the form $\Sigma |e_i - e_j|$ for testing purposes, and the summation did not extend over all pairs (i, j) (Hald *1990*).

3 THE METHOD OF LEAST SQUARES AND BAYESIAN PROCEDURES

The method of least squares was first proposed as an algebraic procedure by Adrien Marie Legendre in his *Nouvelles méthodes pour la détermination des orbites des comètes* (1805), and later justified as a statistical procedure by Carl Friedrich Gauss (*Theoria motus corporum coelestium*, 1809). Gauss noted that he could have used Legendre's argument to justify the minimization of $\Sigma \mid e_i \mid^p$ for $p = 2, 4, 6, \ldots$, or for $p = 1$. He also noted that he had been using Legendre's procedure since 1795, and thus provoked a bitter priority dispute (Plackett *1972*).

Gauss's 1809 derivation of the method of least squares is based on the Bayesian procedures developed some 35 years earlier by Laplace, then 25 years old. In this and a subsequent (1781) paper, he derived the posterior distribution of β_1 in model (3) based on a uniform prior, and suggested that β_1 could be estimated by its posterior mean, median or mode or by choosing b_1 so as to minimize the expected absolute error loss $E \mid \beta_1 - b_1 \mid$.

Gauss followed Laplace's suggestion and adopted the posterior mode as criterion function, but rejected his suggestion of the Laplace distribution $\phi(\varepsilon_i) = \frac{1}{2} m \exp(-m \mid \varepsilon_i \mid)$. Instead, he preferred a distribution which generated the arithmetic mean in model (3). He found that this requirement was uniquely satisfied by the normal distribution $\phi(\varepsilon_i) = (\tau^2/\pi)^{1/2} \exp(-\tau^2 \varepsilon_i^2)$. Applying this procedure in the context of model (2), Gauss found that the optimal estimator is given by

$$\mathbf{X}^T \mathbf{X} \hat{\beta} = \mathbf{X}^T \mathbf{y}, \tag{6}$$

and that the variance of $\hat{\beta}_k$ is given by the (k, k)th element of $\sigma^2 (\mathbf{X}^T \mathbf{X})^{-1}$, where $\mathbf{X}$ is an $n \times k$ matrix with $\mathbf{x}_i^T$ as its ith row, $\mathbf{y}$ is an $n \times 1$ matrix with ith element y_i, and $\sigma^2 = 2/\tau^2$.

The posterior mode corresponds to the maximum-likelihood criterion, and in this form is due to Johann Heinrich Lambert (1760) and Daniel Bernoulli (1778) (Sheynin *1983*, Kendall *1961*). For completeness, Gauss also describes a procedure (the Gauss–Newton procedure) for estimating the parameters of model (1).

4 ASYMPTOTIC DISTRIBUTION THEORY

In his statistical masterpiece *Théorie analytique des probabilités* (1812), Laplace offered an alternative non-Bayesian derivation of the method of least squares. Restricting himself to the case $k = 1$ and to estimators of the form

$$\tilde{\beta}_1 = (\mathbf{z}^T \mathbf{x})^{-1} \mathbf{z}^T \mathbf{y}, \tag{7}$$

where $\mathbf{x}$ and $\mathbf{z}$ are $n \times 1$ matrices, he showed to his own satisfaction that, for large n, $\xi_1 = \tilde{\beta}_1 - \beta_1$ satisfies

$$\Pr(0 < \xi_1 < d) = \frac{\tau_0}{\sqrt{\pi}} \int_0^d \exp(-\tau_0^2 \xi_1^2)\, d\xi_1, \tag{8}$$

where $\tau_0^2 = (\mathbf{z}^T\mathbf{x})^2/(2\sigma^2\mathbf{z}^T\mathbf{z})$ and $\sigma^2 = E(\varepsilon^2)$. Note that the definition of σ^2 used here is taken from Gauss (1816), and represents a considerable simplification of the expression actually used by Laplace (1812).

Laplace now suggested that the optimal estimator of β_1 may be obtained by choosing $\mathbf{z}$ so as to maximize (8) for a given value of d, or to minimize d for a given value of (8), or to minimize the mean positive error $E(\xi_1 \mid \xi_1 > 0) = \frac{1}{2}E(\mid \xi_1 \mid)$. In each case we have to choose $\mathbf{z}$ so as to maximize τ_0^2 and obtain the least-squares estimator

$$\hat{\beta}_1 = (\mathbf{x}^T\mathbf{x})^{-1}\mathbf{x}^T\mathbf{y}. \tag{9}$$

He asserted, without proof, that this derivation may readily be extended to larger values of k. However, he did establish that the least absolute-deviations estimator is also asymptotically normally distributed when $k = 1$, and showed that $\hat{\beta}_1$ is the better estimator when the disturbances are normally distributed.

5 FINITE SAMPLE THEORY

In his second derivation of the method of least squares in 1823 and 1828, Gauss adopted Laplace's class of estimators and a variant of his third optimality criterion to show, for all $k \times 1$ matrices $\mathbf{c}$, that $\mathbf{c}^T(\mathbf{X}^T\mathbf{X})^{-1}\mathbf{X}^T\mathbf{y}$ is the estimator of the form $\mathbf{c}^T(\mathbf{Z}^T\mathbf{X})^{-1}\mathbf{Z}^T\mathbf{y}$ with minimum mean-squared error $E(\mathbf{c}^T\tilde{\beta} - \mathbf{c}^T\beta)^2$. This is the so-called Gauss–Markov theorem. Note that there is no mention of unbiasedness in Gauss's derivation of $\hat{\beta}$, but only in his derivation of

$$\hat{\sigma}^2 = \sum_{i=1}^n \frac{e_i^2}{n-k}. \tag{10}$$

Other notable features of Gauss's contributions are his derivation of the stepwise updating formulas which have since been generalized as the Kalman filter, and his proof that the inverse of a symmetric matrix is symmetric, which came sixty years before the formalization of matrix algebra (§6.7).

6 SUBSEQUENT DEVELOPMENTS

The method of least squares rapidly came to be the accepted method of estimation in the years following 1805, with Laplace's asymptotic derivation as its principal statistical justification (Stigler *1986*). Gauss's second derivation seems to have been lost to the active literature, so that in 1938 F. N. David and J. Neyman attributed Gauss's theorem to Andrei Markov, and in 1897 T. N. Thiele offered an alternative, finite-sample derivation of the method based on the orthogonal–triangular decomposition of **X** (Hald *1981*). This decomposition, which is usually attributed to J. P. Gram and E. Schmidt, is in fact a result of Laplace's (Farebrother *1988*); a variant of it was used by Augustin Louis Cauchy in 1836 for computing his variant of the method of averages.

7 STATISTICAL TESTS OF SIGNIFICANCE

Before the advent of the English biometric school (§10.7), formal tests of statistical hypotheses seem to have been restricted to the deletion of outlying observations and tests of significance to the use of the format 'estimate plus probable error' in the statement of results, where the probable error is 0.6745 times the standard error (Stigler *1980*, *1986*). The familiar χ^2, t and F tests are due to Karl Pearson, W. S. Gosset, Ronald Aylmer Fisher and their successors, as are the likelihood ratio, and the score and Wald principles (Plackett *1983*). In particular, Galileo's procedure had no immediate successors, as it was couched in the form of an *ad hominem* argument and addressed a theologically sensitive hypothesis.

A simple test for outlying observations in model (3) due to W. Chauvenet is discussed by Mansfield Merriman in his standard textbook *The Method of Least Squares* (1913). At each stage of the procedure the observation corresponding to the most extreme residual is rejected if the absolute value of that residual divided by $0.6745\hat{\sigma}$ exceeds a critical value, which the absolute value of a standard normal variable divided by 0.6745 exceeds with probability $1/2n$. By contrast, it is interesting to note that the presence of large residuals prompted Laplace (1793, 1799) to reject the model hypothesis of an ellipsoidal figure for the Earth.

BIBLIOGRAPHY

Farebrother, R. W. *1985*, 'The statistical estimation of the standard linear model, 1756–1853', in T. Pukkila and S. Puntanen (eds), *Proceedings of the First International Tampere Conference on Linear Statistical Models and Their Applications*, Tampere: University of Tampere, 77–99.
—— *1987*, 'The historical development of the L_1 and L_∞ estimation procedures', in Y. Dodge (ed.), *Statistical Data Analysis Based on the L_1 Norm and Related*

Methods, Amsterdam: North-Holland, 37–63. [The L_1 estimator minimizes $\Sigma \mid e_i \mid$, and the L_∞ estimator minimizes max $\mid e_i \mid$.]

—— *1988, Linear Least Squares Computations*, New York: Marcel Dekker. [Historical details of computational procedures.]

Hald, A. *1981*, 'T. N. Thiele's contributions to statistics', *International Statistical Review*, **49**, 1–20.

—— *1990, A History of Probability and Statistics and Their Applications Before 1750*, New York: Wiley.

Harter, R. L. *1974–6*, 'Least squares and some alternatives', *International Statistical Review*, **42**, 147–74, 235–64, 273–8, 282; **43**, 1–44, 125–90, 269–72; **44**, 113–59. [A very detailed bibliography of the subject.]

Heyde, C. C. and Seneta, E. *1977, I. J. Bienaymé: Statistical Theory Anticipated*, New York: Springer. [For details of contributions by Bienaymé and Cauchy.]

Kendall, M. G. *1961*, 'The most probable choice between several discrepant observations and the formation therefrom of the most likely induction', *Biometrika*, **48**, 1–18. [Repr. in Pearson and Kendall *1970*: 155–172.]

Kendall, M. G. and Plackett, R. L. (eds) *1977, Studies in the History of Probability and Statistics*, Vol. 2, London: Griffin. [A second selection from *Biometrika* and other sources.]

Pearson, E. S. and Kendall, M. G. (eds) *1970, Studies in the History of Probability and Statistics*, Vol. 1, London: Griffin. [Reprint of several historical studies from *Biometrika*.]

Plackett, R. L. *1958*, 'The principle of the arithmetic mean', *Biometrika*, **45**, 130–35. [Repr. in Pearson and Kendall *1970*: 121–6.]

—— *1972*, 'The discovery of the method of least squares', *Biometrika*, **59**, 239–51. [Repr. in Kendall and Plackett *1977*: 279–91.]

—— *1983*, 'Karl Pearson and the chi-squared test', *International Statistical Review*, **51**, 59–72.

—— *1988*, 'Data analysis before 1750', *International Statistical Review*, **56**, 181–95.

Seal, H. L. *1967*, 'The historical development of the Gauss linear model', *Biometrika*, **54**, 1–24. [Repr. in Pearson and Kendall *1970*: 207–30.]

Sheynin, O. B., *1983*, 'Corrections and short notes on my papers', *Archive for History of Exact Sciences*, **28**, 171–95. [For details of his first 13 papers in this area; see also later issues.]

Stigler, S. M. (ed.) *1980, American Contributions to Mathematical Statistics in the Nineteenth Century*, 2 vols, New York: Arno Press. [Reprints of several interesting papers, including Robert Adrian's 1809 derivation of the method of least squares, Mansfield Merriman's bibliography and Benjamin Peirce's test for outliers.]

—— *1986, The History of Statistics: The Measurement of Uncertainty Before 1900*, Cambridge, MA: Harvard University Press.

Wald, A. *1940*, 'The fitting of straight lines if both variables are subject to error', *Annals of Mathematical Statistics*, **11**, 286–300. [Contains a valuable historical section.]

10.5

Theory of errors

O. B. SHEYNIN

1 TWO BRANCHES

The theory of errors is a discipline that aims at ensuring the plausibility of the results of experiments in the natural sciences, and their optimal (in the stochastic sense) mathematical treatment. In astronomy, the first goal was attained by (say) choosing such intervals of time during which observational errors resulted in the least inaccuracies in the quantities sought. The relevant branch of error theory, closely linked with experimental science in general (e.g. with the design of instruments), is determinate, and its development in essence started with the appearance of the differential calculus.

Beginning with Daniel Bernoulli (1780) and in particular, Carl Friedrich Gauss (1809), observational errors were divided into random (see Section 3.1) and systematic ones – either constant or changing according to some determinate law (such as the error produced by atmospheric refraction in the observed altitude of a star) (Sheynin *1972b*). The determinate theory of errors has to do with systematic errors, and its aims therefore partly coincide with the goals of preliminary data analysis. Again, this branch of the theory of errors is similar to the design of experiments; however, its concepts are much simpler, exactly because it is determinate.

The second aim of the theory of errors has led to the development of its stochastic branch. Suppose that, given the observations e_i, $i = 1, 2, \ldots, m$ (e.g. given the results of meridian arc measurements), it is required to determine, and to estimate the accuracy of the determination, of n ($n < m$; in particular, $n = 1$) constants (e.g. of two constants specifying the figure and size of the Earth) from the equations

$$a_i X + b_i Y + c_i Z + \cdots + e_i = 0 \tag{1}$$

with known coefficients. Systems (1) were linearly independent (and even well conditioned) and therefore inconsistent. Any set of quantities $(\bar{x}, \bar{y}, \bar{z}, \ldots)$ leading to admissible residuals (v_i, say) was assumed as their solution; actually, however, these systems were solved under some condition imposed on the v_i.

Adrien Marie Legendre (1805) was the first to formulate clearly the condition

$$v_1^2 + v_2^2 + \cdots + v_m^2 = \min \qquad (2)$$

(i.e. the principle of least squares), although without substantiating it by quantitative considerations or introducing any estimate of the precision of $(\bar{x}, \bar{y}, \bar{z}, \ldots)$. Gauss had used the principle (2) from 1794 or 1795 onwards, but his relevant publication appeared much later (see Section 3.4). In 1760 or 1761, in an unpublished manuscript, Thomas Simpson came close to the principle (2) while formulating a standard geodetic problem (Stigler *1984*). In 1778 Leonhard Euler had effectively come to the same principle (Sheynin *1972a*) and, finally, in 1808 or 1809 the condition (2) was introduced by the American mathematician R. Adrain, who very likely had read Legendre's memoir (Stigler *1986*: 374; Dutka *1990*).

Before 1809 the most widely used conditions were

$$v_1 + v_2 + \cdots + v_m = 0, \qquad |v_1| + |v_2| + \cdots + |v_m| = \min \qquad (3)$$

(established by Roger Boscovich in 1757). A special condition, that $|v_{\max}|$ is minimal among all permissible sets of $(\bar{x}, \bar{y}, \bar{z}, \ldots)$, does not belong to the theory of errors (§6.11). However, Euler (1749) and Pierre Simon Laplace (1792 and 1803) applied it in astronomy in order to ascertain whether observations agree with theory. Even Johannes Kepler's celebrated utterance that the Tycho Brahe's observations refute the Ptolemaic system should be interpreted as his attempt to use the same method.

The use of the condition (2) transformed equations (1) into a system of n normal equations which admitted one and only one solution. Before the advent of computers, such systems were successfully solved even for $n > 100$, usually by Gauss's algorithm of consecutively eliminating the unknowns. The general theory of such calculations, including for example the problem of determining the necessary number of decimal points to be retained, was developed in the twentieth century (§5.10).

The determination of $(\bar{x}, \bar{y}, \bar{z}, \ldots)$, is called the 'adjustment of indirect observations'; the determination of $\bar{x}$, for $n = 1$, is called the 'adjustment of direct observations'. We consider almost exclusively the second problem. The term 'theory of errors' (*Theorie der Fehler*) is due to Johann Heinrich Lambert (1765), who had defined it too formally (Sheynin *1971*); maybe for this reason neither Laplace nor Gauss ever used it. It gained currency in the second half of the nineteenth century. Lambert also defined the determinate branch of error theory, calling it the 'theory of consequences' (*Theorie der Folgen*). Lastly, while reasoning on the plausibility of observations, he indicated that their mathematical treatment aimed at determining the real values of the measured constants and at estimating the precision of observations (actually, he estimated the precision of these real values).

The concept of real values of $X, Y, Z, \ldots$ was characteristic of the theory of errors, whereas statistics used the idea of means, including fictitious ones (e.g. the mean height of a soldier). Again, the theory *determined* constants, while statistics *estimated* the parameters, X and σ^2 say, of the appropriate distribution. Thus, for the normal distribution, as it is now called,

$$\phi(x, X, \sigma) = \frac{1}{\sigma(2\pi)^{1/2}} \exp\left[-\frac{(x-X)^2}{2\sigma^2}\right]. \tag{4}$$

Both branches of the theory of errors were needed for the progress of experimental science. A most important condition for applying the theory was the availability of a large number of observations. Tycho Brahe, and even Muḥammad al-Bīrūnī (eleventh century) made regular observations, and James Bradley strongly advocated their use (Sheynin *1973*). Note, however, that natural scientists always considered the essence of their problems and often solved them in a non-standard manner, which obviously hindered the development of the theory of errors.

2 THE DETERMINATE BRANCH

The prehistory probably begins with al-Bīrūnī, who aimed at the greatest precision in astronomy, geodesy and meteorology. He listed the sources of errors in many kinds of observations, corrected the results of some of his measurements accordingly, and considered qualitatively the joint action of the errors of observation and calculation. However, even Ptolemy had recommended selecting the best observations; and it is possible that the century-old tradition, now rejected, of subjectively selecting the data, might be partly explained by the need to use doubtful observations only for a rough check of results obtained under optimal conditions and thus least corrupted by systematic and random influences.

Also part of the early story are Levi ben Gerson (fourteenth century), the practice of astronomical navigation (the end of the fifteenth century), and such scientists as Galileo Galilei (the free fall of bodies), Kepler, Christiaan Huygens (time-keeping and the free fall of bodies) and Isaac Newton (optics). The problems they set out to solve were not new, but their solution aided the development of experimental science.

In the context of his theory of consequences (Section 1), Lambert determined the optimal form of some geodetic figures. Thus he continued the work of Roger Cotes (1722), who calculated the errors of the unknown elements of rectilinear and spherical triangles, given the errors of their measured elements.

3 THE STOCHASTIC BRANCH

This has mainly to do with observations e_i corrupted only by random errors,

$$\xi_i := X - e_i, \quad E(\xi_i) = 0 \tag{5}$$

(where E is the expected value function), which are mutually independent $(E(\xi_i \xi_j) = E(\xi_i) E(\xi_j),\ i \neq j)$. In 1915 Jacobus Kapteyn, dissatisfied with the then new theory of correlation, introduced another, astronomical coefficient of correlation which enabled him to measure the dependence between two functions of partly coinciding arguments, $g(a_1, a_2, \ldots, a_k, b_1, b_2, \ldots)$ and $h(a_1, a_2, \ldots, a_k, c_1, c_2, \ldots)$ (Sheynin *1984*). But even he did not attempt to construct a theory of errors of dependent observations. (For a brief description of the adjustment of such measurements, see Linnik *1961*: Section 14.2.)

Condition $(5)_2$ means that systematic errors are excluded. Errors with zero expectation are called 'unbiased'.

3.1 Random errors

These are not completely indeterminate, chaotic variables, but rather errors obeying some law of distribution, or at least having certain stochastic properties. (For example, modulo small random errors of the 'usual' type tend to occur more often than large ones.) Note that although Siméon-Denis Poisson, in 1837, defined the concept of random variable, he called it by a provisional term, '*chose A*'. The modern definition of random variable appeared only in the twentieth century. Even al-Bīrūnī understood that random errors are inevitable and, while adjusting observations, obviously took into account their 'usual' properties (Sheynin *1973*: 116).

3.2 Sampling

Observations e_i (Section 1) belong to a universe of possible measurements, and may be considered as a sample from this universe. Andrei Markov, in the twentieth century, appears to have been the first to say so directly, but mathematicians and astronomers implicitly adhered to this principle even in the mid-eighteenth century. This is clear from their discussions about the choice of conditions imposed on the v_i or of the estimators (Section 3.3), and about the introduction of the laws of distribution (Sections 3.4 and 3.5). In each instance, scholars attempted to take into account the stochastic properties of observational errors, to treat a given set of measurements as a sample.

Sampling first entered science by way of population statistics. Jan De Witt suggested a certain stochastic law of mortality in 1671, and in 1694 Edmond Halley constructed the first mortality table (§10.3). Moreover, sample testing of new coins by weight and assay has been practised for many centuries (Stigler *1986*, 3). Again, the theory of sample surveys which is a part of theoretical statistics came into being in the twentieth century, independently of the theory of errors.

3.3 Estimators

Beginning in the seventeenth century, it became usual to choose the arithmetic mean of observations as the estimator of the constant X. (In his *Astronomia nova* (1609, Chap. 10), Kepler used as an estimate of X a quantity he called the *medium ex aequo et bono*. This expression was borrowed from the law, and meant 'in fairness and justice', with the added implication 'rather than according to the letter of the law'. Apparently, then, in 1609 the arithmetic mean already was the law, at least in astronomy.) The arithmetic mean corresponds to the principle of least squares since, for $n = 1$, the left-hand side of the condition (2) is then least. The median has also gained some currency. This is the observation e_{k+1} if $m = 2k + 1$ and any number in $[e_k, e_{k+1}]$ if $m = 2k$ ($e_1 \leqslant e_2 \leqslant \cdots \leqslant e_m$). The choice of the median entails some loss of information, but it makes the rejection of outliers (Section 3.6) almost unnecessary.

Finally, the mode is the point of maximal value of the density curve $\phi(x)$ of the errors, the most probable value of the constant sought. (If all the ξ_i obey the same law $\phi(x)$, then the corresponding e_i are called 'equally precise' in a wide sense; usually, however, equal precision of observations was understood only narrowly as the coincidence of the measure of their accuracy.) It is assumed that the mode exists, and that it is unique; for distributions encountered in the theory of errors this condition is usually fulfilled.

The only generally accepted measure of the plausibility of observations and their functions is the variance, introduced by Gauss. From the 1870s the variance also served as a measure of the scatter of random variables. For ξ_i, the variance is

$$E(\xi_i - E\xi_i)^2 = \sigma^2 = \int x^2 \phi(x)\,dx \tag{6}$$

(the integral is taken between the limits between which $\phi(x)$ exists). Laplace's main measure of precision was the absolute expectation

$$E\,|\,\xi_i\,| = \int |x|\,\phi(x)\,dx. \tag{7}$$

In practice, the unknown variance (6) is replaced by its unbiased sample estimator

$$S^2 = (v_1^2 + v_2^2 + \cdots + v_m^2)/(m-1) \tag{8}$$

(S is called the 'mean square error of unit weight'). The corresponding quantity for the arithmetic mean is

$$M^2 = (v_1^2 + v_2^2 + \cdots + v_m^2)/m(m-1). \tag{9}$$

Formulas (8) and (9) are due to Gauss (1823). According to him, and in keeping with modern ideas, the best choice for X is an unbiased estimator having least variance; for the normal distribution this estimator coincides with the arithmetic mean.

The principle of one of the main methods of estimating parameters, the method of maximum likelihood, goes back to Lambert (1760) (Sheynin *1971*). Daniel Bernoulli (1778) and Gauss (1809) used elements of this method along with the principle of inverse probability (and, broadly speaking, the Bayesian approach, on which see §10.17).

3.4 The method of least squares

This method (2) was adopted in various contexts:

1 Assuming that the arithmetic mean of observations is the best estimator of X, and that it coincides with the mode of the unknown differentiable distribution $\phi(x)$, Gauss (1809) determined this $\phi(x)$ (which was the normal law) and its mode, and arrived at the method of least squares.
2 In 1810, unrigorously proving the central-limit theorem, Laplace deduced the principle of least squares by indicating that the normal law existed on the strength of this theorem, but he had to assume that the number of observations was large.
3 In 1811, discarding the mode, Laplace determined the estimators sought by assuming that the measure (7) should be minimal, and by retaining the

normal law. Again, he arrived at the method of least squares. (Before 1810 he, like Daniel Bernoulli, restricted himself to adjusting a few observations since his method of treating them demanded complicated calculations. Even then, one of his approaches to determining the estimator for the unknown X was to minimize (7).)

4 In 1823, Gauss proved that the principle of least variance led to the method of least squares, irrespective of the form of $\phi(x)$ in (6). He discarded his original formulation since he believed that integral measures of precision, such as the variance, were better than differential estimators, and that the assumption of a single (normal) distribution for observational errors was too restrictive.

Nevertheless, the usual near-normality of distributions occurring in the theory of errors and the elegance of Gauss's original approach led to its popularity among astronomers until very recently. Be that as it may, in astronomy after 1809 the normal law reigned supreme, although in 1838 Bessel established that a certain instrumental error obeyed a U-shaped law.

3.5 Distribution

Other distributions were known in the theory of errors even before 1809. Apart from those introduced speculatively (Lambert, 1765; Laplace, 1774; Joseph Louis Lagrange, 1775 or 1776; Daniel Bernoulli, 1778), there were the uniform and triangular laws (both of them continuous) studied by Simpson (1756 and 1757), who aimed to prove that the arithmetic mean was better than a single observation. In 1781 Laplace introduced, again speculatively, the distribution

$$\phi(\alpha x) = q \quad (q > 0), \quad \alpha x = 0, \quad \text{and} \quad \phi(\alpha x) = 0, \quad \alpha x \neq 0, \quad \alpha \to 0, \quad (10)$$

with the assumption that $\int_0^\infty [\phi(\alpha x)]^m dx = A > 0$. He thus obtained Dirac's delta-function, and used it to prove that the number e_0 in the equality

$$\int_{-a}^{e_0} \phi[\alpha(x - e_1)] \dots \phi[\alpha(x - e_m)] \, dx$$
$$= \int_{e_0}^{a} \phi[\alpha(x - e_i)] \dots \phi[\alpha(x - e_m)] \, dx \quad (11)$$

(a was the greatest possible error) was the arithmetic mean. His reasoning might be repeated on the 'physical' level by taking

$$\phi(\alpha x) = \lim_{\lambda \to \infty} [\lambda \exp(-\tfrac{1}{2}\lambda^2)/\sqrt{\pi}], \quad (12)$$

but his equality, which was another condition for determining an estimator for X (compare context 3 in Section 3.4) loses meaning in the language of generalized functions.

In 1824, while studying the error of the arithmetic mean, Poisson discovered that its probability for the distribution

$$\phi(x) = 1/\pi(1 + x^2), \quad |x| < \infty \tag{13}$$

is independent of the number of observations. He thus effectively discovered the stability of this distribution, which is now named after Augustin Louis Cauchy.

In connection with the problem of rejecting outlying observations (Section 3.6), Ernst Abbe (1863) found that the squared sum of normally distributed errors obeys the chi-squared law, as it is now called. It is a most important distribution in present-day statistics.

So, the concept of random variable was used indirectly within the theory of errors long before Poisson formally introduced it in the theory of probability (Section 3.1). Still earlier, in the second half of the seventeenth century, mathematicians were studying such random variables as outcomes of games of chance (§10.1).

3.6 Rejection of outlying observations and posterior weights

Observations can be corrupted by blunders and, beginning with Benjamin Peirce in 1852, various tests were offered for their detection. In essence, they rejected an observation e_i if its deviation from the arithmetic mean ($\bar{e}$) had a sufficiently low probability. Thus in the three-sigma test (W. Jordan, 1877) e_i was rejected if $|\bar{e} - e_i| \geqslant 3\sigma$ where σ (or, in the notation of (8), S) was the sample variance of an observation (Harter *1977*: 63). For the normal distribution, the probability that this had happened to a sound observation was 0.003. The general concepts of hypothesis testing (tests of significance), the power of a test and the level of its significance were introduced in the twentieth century.

From 1763 (James Short), weights were sometimes assigned to observations in order to increase the influence of those results which were closer to the arithmetic mean. If $e_1 \leqslant e_2 \leqslant \cdots \leqslant e_m$, and $\bar{e}$ is this mean, then the posterior weight p_i of e_i depends on $|\bar{e} - e_i|$ and decreases with the increase of this difference. The estimator of X is then $\Sigma p_i e_i / \Sigma p_i$, the generalized mean. However, in 1778 Daniel Bernoulli recommended using, for a certain type of distributions, a generalized estimator of X with weights p_i increasing with $|\bar{e} - e_i|$ (Sheynin *1972b*). This unusual advice may now be justified.

3.7 Gauss onwards

The period from 1823 to the 1920s (the advent of mathematical statistics) was rather fruitless. The works of Johann Encke, Friedrich Wilhelm Bessel, J. W. L. Glaisher, Simon Newcomb, and F. R. Helmert did not essentially change the theory of errors. New features that did appear included the bivariate normal law and the related notion of the ellipse of errors; the mixtures of several normal distributions; tests for rejecting outliers; and practical formulas for adjusting typical geodetic figures. Nowadays, the stochastic theory of errors is a chapter of mathematical statistics.

As this article has shown, many important statistical and stochastic concepts grew out of the theory of errors, mostly after 1809. Statisticians and mathematicians, however, had to find (or modernize) the same concepts independently.

BIBLIOGRAPHY

Czuber, E. *1891, Theorie der Beobachtungsfehler*, Leipzig: Teubner.

Dutka, J. *1990*, 'Robert Adrain and the method of least squares', *Archive for History of Exact Sciences*, **41**, 171–84.

Harter, H. L. *1977, A Chronological Annotated Bibliography on Order Statistics*, Vol. 1, No place [several US Air Force establishments]. [Many items are highly relevant.]

Ku, H. H. (ed.) *1969, Precision Measurement and Calibration. Selected National Bureau of Standards Papers on Statistical Concepts and Procedures*, Washington, DC: National Bureau of Standards (NBS Special Publication No. 300, Vol. 1).

Linnik, Yu. V. *1961, Method of Least Squares and Principles of the Theory of Observations*, Oxford: Pergamon Press. [Original in Russian, 1958; 2nd Russian edn 1962.]

Petrov, V. V. *1954*, 'On the method of least squares and its extremal properties', *Uspekhi matematicheskikh nauk*, **9**, 41–62. [In Russian.]

Sheynin, O. B. *1971*, 'Lambert's work on probability', *Archive for History of Exact Sciences*, **7**, 244–56.

—— *1972a*, 'On the mathematical treatment of observations by Euler', *Archive for History of Exact Sciences*, **9**, 45–6.

—— *1972b*, 'D. Bernoulli's work on probability', *Rete*, **1**, 273–300. [Also in M. Kendall and R. L. Plackett (eds), 1977, *Studies in the History of Statistics and Probability*, Vol. 2, London: Griffin, 105–32.]

—— *1973*, 'Mathematical treatment of astronomical observations', *Archive for History of Exact Sciences*, **11**, 97–126.

—— *1984*, 'On the history of the statistical method in astronomy', *Archive for History of Exact Sciences*, **29**, 151–99.

Stigler, S. *1984*, 'Boscovich, Simpson, and a 1760 MS note on fitting a linear relation', *Biometrika*, **71**, 615–20.

—— 1986, *The History of Statistics: The Measurement of Uncertainty Before 1900*, Cambridge, MA: Harvard University Press.

Todhunter, I. *1865, A History of the Mathematical Theory of Probability from the Time of Pascal to that of Laplace*, Cambridge and London: Macmillan. [Repr. 1949, New York: Chelsea.]

Wolf, H. *1968, Ausgleichungsrechnung nach der Methode der kleinsten Quadrate*, Hannover: Dümmler.

10.6

Russian probability and statistics before Kolmogorov

E. SENETA

1 THE BEGINNINGS

The prime impetus for the initial development in the 1820s of probability theory in the Russian Empire (putting aside the eighteenth-century contributions of Leonhard Euler and Daniel Bernoulli) was the need for a proper basis for actuarial and demographic work, and for the statistical treatment of observations generally. Pierre Simon Laplace's classic work on probability (*Théorie analytique des probabilités*, 1812), which initiated the Paris school of probabilistic investigations, not only laid foundations for the subject, but also contained applications to real-world situations. Its ideology was brought to the Russian Empire, partly in response to the statistical needs mentioned above, by Viktor Yakovlevich Bunyakovsky (1804–89) and Mikhail Ostrogradsky, both of whom had studied in Paris.

Bunyakovsky's prime achievement was the first treatise on probability in the Russian language (Bunyakovsky *1846*). Its aim was the simplification and classification of existing theory; its lasting achievement was the creation of a Russian probabilistic terminology. His other notable achievement was the construction in about 1865 of mortality tables for the (Orthodox) population of Russia, beginning the Russian demographic tradition. Apart from work, important in its time, on pension funds in 1858, Ostrogradsky published several popular articles on insurance and games of chance, and theoretical articles on probability which corrected Laplace and anticipated Siméon-Denis Poisson's *Recherches sur la probabilité des jugements* [...] of 1837 (Gnedenko *1951*).

However, the first major results in Russian probability were to come from a younger St Petersburg contemporary, Pafnuty Chebyshev. (More detailed descriptions of the earlier era, including work in Kharkov and Moscow, are given by Maistrov *1980*: 170–78, and by Gnedenko and Sheynin *1978*: 213–16.) It is worth noting that probabilistic activity was only part of the mathematical *oeuvre* of Bunyakovsky, Ostrogradsky and Chebyshev, and established the Russian tradition of probability as a standard part of the study of mathematics.

2 PROBABILITY

2.1 Chebyshev's probabilistic work

Chebyshev's early mathematical education was at Moscow University, where he was enrolled from 1837. He came under the influence of N. D. Brashman (1796–1866), who had been active from the 1830s in stressing the importance of sound probabilistic foundations for statistical work in the Russian Empire. Under his influence, Chebyshev produced his first two contributions to probability theory in 1846: his master's thesis, and an analytical deduction of Poisson's weak law of large numbers (WLLN). This law states that if p_i, $i \geqslant 1$, is the probability of success in the ith of n independent trials, and the total number of successes is denoted by X (a random variable, since it will vary with repetitions of the n trials), then

$$\Pr\{|X/n - \bar{p}(n)| < \varepsilon\} \to 1 \quad \text{as} \quad n \to \infty, \tag{1}$$

for any $\varepsilon > 0$. Here $\bar{p}(n) = \Sigma_{i=1}^{n} p_i/n$. This publication remained unnoticed; consequently the law (in contrast to that of James Bernoulli, in which the probability is the same in each trial), because of the changing nature of $\bar{p}(n)$ with n, remained an object of controversy within the dominant French probabilistic activity. However, Chebyshev's methodology (the estimation for finite n of the deviation from its limit of a quantity which approaches a limit), would become a pivotal concept in the work of his illustrious disciples, Andrei Markov and Aleksandr Lyapunov.

In 1847 Chebyshev began to teach at St Petersburg University, and in 1860, on taking over the course in probability theory from Bunyakovsky, his interest in the subject revived. In an 1867 publication he obtained the inequality for the arithmetic mean $\bar{X} = \Sigma_{i=1}^{n} X_i/n$ of independently (but not necessarily identically distributed) random variables X_i, $i = 1, 2, \ldots, n$, in terms of their means $\mu_i = \mathrm{E} X_i$ and variances $\sigma_i^2 = \mathrm{Var}\, X_i$, namely

$$\Pr(|\bar{X} - \bar{\mu}| > \varepsilon) \leqslant \bar{\sigma}^2/(n\varepsilon^2), \quad \bar{\mu} = \frac{1}{n}\sum_{i=1}^{n} \mu_i, \quad \bar{\sigma}^2 = \frac{1}{n}\sum_{i=1}^{n} \sigma_i^2. \tag{2}$$

This inequality is important because, among other reasons, one can derive the WLLN, that $\bar{X} - \bar{\mu} \to 0$ ('in probability') as $n \to \infty$, provided $\bar{\sigma}^2/n \to 0$ as $n \to \infty$. This result includes the Bernoulli and Poisson WLLNs. It is properly termed the 'Bienaymé–Chebyshev inequality', for it was obtained earlier by the French statistician Irenée Jules Bienaymé in 1853 (Heyde and Seneta *1977*).

The essence of Bienaymé's proof, which later became known as the 'method of moments', is the determination of the value of an integral $\int_0^a f(x)\,dx$ by values of the integrals $\int_0^A x^r f(x)\,dx$, $r = 0, 1, 2, \ldots$, where $A > a$ and $f(x)$ is a positive function for $x \in (0, A)$. If f is the probability density of a positive random variable X, then the method is tantamount to the estimation of $\Pr(X \leqslant a)$ by using moments $E(X^r)$, $r \geqslant 1$. This technique was used by Chebyshev in 1887 to prove the first version of the central-limit theorem (CLT) for partial sums of independently (but not necessarily identically distributed) random variables X_i, $i \geqslant 1$, each described by a density (§10.2). Chebyshev's assertion is that if, for integers $r \geqslant 2$,

$$E X_i = 0, \quad i \geqslant 1, \quad \text{and} \quad |E\{X_i^r\}| \leqslant C, \quad i \geqslant 1, \tag{3}$$

then, as $n \to \infty$,

$$\Pr\{t < S_n/B_n < t'\} \to \frac{1}{(2\pi)^{1/2}} \int_t^{t'} \exp\left(-\frac{x^2}{2}\right) dx, \tag{4}$$

where

$$S_n := \sum_{i=1}^n X_i, \qquad B_n^2 := \sum_{i=1}^n \operatorname{Var} X_i. \tag{5}$$

2.2 The central-limit problem

Conditions (3) are inadequate to guarantee the conclusion (4), and the problem was taken up by other mathematicians in the Russian Empire, primarily Chebyshev's disciples Markov and Lyapunov from the strong group of mathematicians associated with Chebyshev which came to be known as the St Petersburg mathematical school (Seneta *1984*). The first rigorous proof of a restricted form of the CLT, together with a bound on the difference between the two sides of (4), however, was given in 1892 by Ivan Vladislavovich Sleshinsky (1854–1931). He was aware of the difficulties in Chebyshev's formulation, and instead followed the ideas of Cauchy, using characteristic-function methods.

Markov added several variants of a third condition to Chebyshev's two in 1898, while in 1901 Lyapunov used characteristic-function methods (essentially Fourier transforms) and what are now known as 'Lyapunov's inequalities' to prove the celebrated 'Lyapunov's theorem': that (4) holds if (3)$_2$ is replaced by

$$\left(\sum_{i=1}^n E\,|X_i|^{2+\delta}\right)^2 \bigg/ \left(\sum_{i=1}^n E(X_i^2)\right)^{2+\delta} \to 0 \tag{6}$$

as $n \to \infty$ for some $\delta > 0$. This led Markov, devoted to the work of Chebyshev and Bienaymé, to wonder whether the method of moments might not be suitably adapted to give the same result; he finally achieved this in the third edition of his *Ischislenie veroiatnostei* ('Calculus of Probabilities', 1913). The central new idea, now much used in probability theory, is the notion of truncation of random variables: put $X_k' = X_k$ if $|X_k| < N$, and $X_k' = 0$ if $|X_k| \geq N$. The random variables X_k', $k \geq 1$, being confined to a finite interval, are easily tractable, and the effect of truncation can be shown to disappear as $N \to \infty$, to give the conclusion (4) under Lyapunov's conditions.

Pavel Alekseevich Nekrasov (1853–1924), an eminent Moscow mathematician proficient in complex-variable theory, published in 1898 a paper dedicated to the memory of Chebyshev in which he attempted to use what we now call the saddlepoint and Laplacian-peaks methods, and the Lagrange inversion formula, to establish local and global-limit theorems of central-limit type for deviations of large value. This was much ahead of its time; it was never understood by Markov and Lyapunov, marking the beginning of a bitter controversy between the three (Seneta *1984*).

2.3 Markov chains

Nekrasov had noticed in 1902 that the Bienaymé–Chebyshev inequality continues to hold (and hence so does the WLLN, if $\bar{\sigma}^2/n \to 0$ as $n \to \infty$) if the X_i are merely assumed to be pairwise independent. He was under the delusion, however, that pairwise independence is necessary for the WLLN. This error led Markov to give thought to the WLLN for dependent random variables, and it is in a paper of 1906 on this topic that finite homogeneous 'Markov chains', which are sequences $\{X_n\}$ of dependent random variables with a simple dependence structure, first made their appearance in his writings. The probabilistic structure is as follows: X_n, the 'state' at time n, is one of a finite set of possible values $S = \{1, 2, \ldots, N\}$, and the conditional probability

$$\Pr(X_{n+1} = j \mid X_n = i, \ X_{n-1} = i_{n-1}, \ X_{n-2} = i_{n-2}, \ldots),$$

$$\text{for any} \quad j, i, i_{n-1}, i_{n-2}, \ldots \in S, \quad (7)$$

is independent of the sequence of past values $\{i_{n-1}, i_{n-2}, \ldots\}$, so that the probabilistic structure of the future is completely specified by a knowledge of the present state ($X_n = i$). The conditional probability is also taken independent of time n (time homogeneity), and thus can be written as p_{ij}. The array of numbers $\mathbf{P} = \{p_{ij}\}$, $i, j \in S$, is the transition matrix of the Markov chain, and together with the initial distribution (at time 0)

$\pi_i = \Pr(X_0 = i)$, $i \in S$, completely determines the joint distribution of the X_n.

Markov's interest in such chains of simply dependent random variables lay in investigating the WLLN and CLT for the partial sums $S_n = \sum_{i=1}^{n} X_i$. Nevertheless, his first paper on the topic also contains an elegant treatment of the averaging property of a finite row-stochastic matrix $\mathbf{P}$ (a matrix with non-negative entries whose row sums are unity), when applied to a vector $\mathbf{a}$ to produce a vector $\mathbf{b} = \mathbf{Pa}$. This property is used by him to establish ergodicity (the long-term independence of the initial state) of a homogeneous Markov chain, and is the essence of modern ergodicity theory for non-homogeneous chains (Seneta *1981*: 80–91).

Special instances of Markov chains were known before Markov; various urn models, in particular those due to Daniel Bernoulli in 1769 and Laplace in 1812, are much older; and the gambler's-ruin problem dates back to Pierre de Fermat and Blaise Pascal (§10.2). Nevertheless, Markov's outstanding contribution to probability theory was a general formulation (and basic general results): it is at the heart of much research in modern stochastic processes because of its mathematical tractability.

The analytical tenor of Markov's probability, very influential in Russia through his textbook, was brought to the USA, and thus to the English-speaking world, by an important book by his associate J. V. Uspensky (*1937*). His probabilistic work is described in Sheynin *1989*; Grodzensky *1987* presents a biographical study.

2.4 The beginning of the modern era

Sergei Bernstein received his mathematical education in Paris; he defended a doctoral dissertation at the Sorbonne in 1904 and taught at Kharkov University from 1908 to 1933. In his mathematical work he synthesized the traditions of the St Petersburg mathematical school of Chebyshev with modern Western European thinking. His probabilistic writings, including his textbook *Teoriia veroiatnostei* ('Theory of Probability'), which first appeared in 1927 and ran to a fourth edition in 1946, were largely responsible for determining probability's subsequent course of development in the USSR. In particular, in 1922 he was able to generalize Lyapunov's conditions for the validity of the CLT to ones which, when specialized to the same setting, are equivalent to those of J. W. Lindeberg, whose now-celebrated paper appeared in the same year. Further, in his best-known work Bernstein extended the CLT to 'weakly dependent' random variables

X_i, $i \geqslant 1$. The essence of this weak dependence is that the X_i form a martingale-difference sequence:

$$E(X_{k+1} \mid X_k, X_{k-1}, \ldots, X_1) = 0, \quad k \geqslant 1, \tag{8}$$

so that the sequence $S_n = \Sigma_{i=1}^n X_i$ is a martingale, an important concept in subsequent probability theory. His name is borne by the Bernstein polynomials; and Bernstein's inequality (a sharpening of Bienaymé–Chebyshev). He attempted an axiomatization of probability theory as early as 1917 from the standpoint of symbolic logic and propositional calculus; Andrei Kolmogorov's now standard set-theoretic axiomatization appeared in 1933 (§10.16).

Evgeny Evgenievich Slutsky (1880–1948), who was also an eminent mathematical economist and statistician, helped elucidate the notion of convergence in probability (of which the WLLN is an instance, and was Slutsky's motivation), and established a form of what is now known as Slutsky's theorem. The notion of stochastic limit (limit in probability) led him to study random functions more generally (as a result of which he developed the notion of stochastic continuity, differentiability and integrability). His approach, centred on moments of a random function, was in the tradition of Chebyshev, Markov and Slutsky's statistical colleague, A. A. Chuprov (see Section 3).

Vsevolod Ivanovich Romanovsky (1879–1954), like Chuprov better known as a mathematical statistician, developed the theory of finite Markov chains in a manner greatly influenced by Markov's ideas, but from the natural standpoint of the Perron–Frobenius theory of non-negative matrices (Seneta *1981*). His book on the topic was published in 1949.

3 STATISTICS

3.1 *Zemstvo* statistics and sample surveys

Early statistical investigations in the Russian Empire, demography and actuarial statistics apart, were performed on collections of official data pertaining to the rural economy. These date from the establishment of *zemstvos* (institutions of local government) in 1864, with which were associated the *zemstvo* statisticians, often, for political reasons, elements of the liberal and radical intelligentsia. The motivating force for those *zemstvo* investigations was provided by Aleksandr Ivanovich Chuprov (1842–1908), a professor of political economy and statistics at Moscow University. Although the data collected and conclusions drawn by *zemstvo* statisticians such as V. E. Postnikov (1844–1908), A. A. Rusov (1847–1915) and, especially, F. A. Shcherbina (1849–1936), had considerable influence as economic underpinning to the social revolutionary writings in the latter years of the Russian Empire (Kotz and Seneta *1990*), the analysis of the data extended little beyond grouping and the use of averages, and the data consisted merely of a complete enumeration. The writings of the father of the discipline of statistics in the Russian Empire, generally considered to be Yulii Eduardovich Ianson (or Yanson, or Jahnson) (1835–92), reflect this situation. His *Teoriia statistiki* ('The Theory of Statistics'), which appeared in five editions between 1885 and 1913, was studied diligently by Lenin.

A. I. Chuprov is generally credited with popularizing among the *zemstvo* statisticians, from 1894, the carrying out of sample surveys to complement complete enumeration. It appears that the first such sampling procedure (taking every tenth landholding from a list) was implemented in 1896 by A. V. Peshekhonov (1867–1933). The important step of achieving representative samples via random selection was exposited in 1910 by the son of A. I. Chuprov, Aleksandr Aleksandrovich Chuprov (1874–1926). The younger Chuprov recognized the contributions of sample size and variability to the size of probable error of estimates, and at this time already had the notions of cluster sampling and subsampling, and of stratified sampling. He investigated random sampling without replacement in detail; and in 1923 he anticipated several results of J. Neyman, especially the well-known formula for optimal allocation of a sample of fixed size n among t strata, stratum h $(h = 1, 2, \ldots, L)$ receiving sample size

$$n_h = n N_h S_h \bigg/ \sum_{r=1}^{t} N_r S_r, \tag{9}$$

N_h being the stratum size and S_h its population standard deviation. Although he was an *émigré* from May 1917, his work retained considerable influence in the USSR (Seneta *1985*) until the onset of Stalinism.

3.2 The dispersion coefficient

Possibly the first book in the Russian Empire on mathematical statistics was A. A. Chuprov's *Ocherki po teorii statistiki* ('A Précis of Statistical Theory', 1909). The initial probabilistic influence on him as a student at Moscow University (from whose physico-mathematical faculty he graduated in 1896 with a dissertation on probability theory as a basis for theoretical statistics) was Nekrasov. After graduation, Chuprov travelled to Germany to study political economy, where he met Ladislaus von Bortkiewicz (1868–1931), a student of Ianson's who had earlier done demographic work in Russia following Bunyakovsky; and through him he met Wilhelm Lexis (1837–1914), which led to a sharing of their primary interest – the stability theory of statistical series of trials (dispersion theory). On his life, see Karpenko *1957*.

In *Ocherki*, probabilities are regarded as objective (rather than subjective) and approximable by relative frequencies (proportions of successes) in accordance with the WLLN. Chuprov shows awareness of Chebyshev's work on the WLLN and Nekrasov's insight that only pairwise independence is required. An account of dispersion theory from a probabilistic standpoint is given; but it is clear that Chuprov was ignorant of current Russian probabilistic literature, in particular the work of Markov and Lyapunov. Markov learned by chance of the existence of *Ocherki*, and his correspondence with Chuprov (Ondar *1981*) began in November 1910 and lasted until February 1917. This led to Markov's grudging acceptance of mathematical statistics as a legitimate offspring of the marriage between statistics and probability (he had not been acquainted with the work of the English biometric school (§10.7)); and to Chuprov's development as a theoretical statistician with a strong interest in stochastic dependence and the mathematical expectation operation.

The point of contact was dispersion theory (Heyde and Seneta *1977*: Chap. 3). To illustrate the essence of the Markov–Chuprov interaction, notice that, in earlier notation, the quantity

$$\mathrm{Var}(X/n)/\{\bar{p}(n)(1-\bar{p}(n))/n\} \tag{10}$$

gives a theoretical measure of the comparison between n independent trials, in the ith of which the success probability is p_i $(i = 1, 2, \ldots, n)$; and n independent (Bernoulli) trials in each of which the success of probability is the same, specifically the number $\bar{p}(n)$, since the variance of the proportion of successes then is $\bar{p}(n)(1 - \bar{p}(n))/n$. If the block of n trials with varying probabilities is replicated (repeated independently) m times, and P_j is the proportion of successes in the jth replication, then the quantity

$$D' = \left\{ \sum_{j=1}^{m} (P_j - \bar{p}(n))^2/m \right\} \Big/ \{\bar{p}(n)(1 - \bar{p}(n))/n\} \tag{11}$$

has the value (10) as its expectation.

The expectation $E(D')$, given by (11), is in essence the dispersion coefficient of Lexis and Bortkiewicz; D' provides an empirical measure of (11), but it has the defect that it is not a statistic (since it cannot be calculated entirely from the data) because it involves the unknown number $\bar{p}(n)$. Markov insisted, correctly, that one must focus on genuine statistics (as we would now call them) as a measurement device; in particular, the quantity to replace D' should be

$$D = \left\{ \sum_{j=1}^{m} (P_j - P)^2/m \right\} \Big/ \{ P(1 - P)/n \}, \qquad (12)$$

where $P := \Sigma_{j=1}^{m} P_j/m$ is the proportion of successes in all mn trials and the expectation of D, its variability, and – indeed – distribution under the null hypothesis that all mn trials are indeed Bernoulli trials, should be investigated. This led Chuprov to give the first proof (in a letter of 1916) that under this null hypothesis

$$\mathrm{E} L = 1, \quad L = \{ (mn - 1)/n(m - 1) \} D, \qquad (13)$$

provided L is defined as unity when $P(1 - P) = 0$; and it led Markov (in two letters of 1917) to obtain the limiting distribution as $n \to \infty$ of $(m - 1)L$, and thus make a connection with the chi-squared distribution with $m - 1$ degrees of freedom.

3.3 The aftermath

Although the Markov–Chuprov correspondence marks the coming together of probability and statistics in mathematical statistics in the Russian Empire, growth was initially interrupted by the First World War and the ensuing Revolution, although important methodological advances were made in the USSR by workers such as Chuprov's student N. S. Chetverikov (1885–1973), Romanovsky and Slutsky, as well as the *émigré* Chuprov. The main Soviet vehicle for statistics, including mathematical statistics, was the first series (1919–29) of the journal *Vestnik statistiki* ('The Messenger of Statistics').

Generally speaking, however, the Stalinist ascendancy led to the almost total collapse of statistics by the late 1930s, several statisticians disappearing without trace; the situation in population genetics was similar. Probability, in the guise of pure mathematics and divorced from application, fared better. Demography, to which only fleeting attention has been given, also continued; one of the most eminent practitioners was Mikhailo Vasilovich Ptukha (1884–1961).

BIBLIOGRAPHY

Entries on the life and work of major figures mentioned in this article may be found in Kotz and Johnson *1982–8*. The titles of Gnedenko *1951*, Gnedenko and Sheynin *1978*, and Karpenko *1957* as given here are translations from the original Russian.

Bunyakovsky, V. Ya. *1846, Osnovania matematicheskoi teorii veroiatnostei*

['Foundations of the Mathematical Theory of Probabilities'], St Petersburg: Academy of Sciences.

Gnedenko, B. V. *1951*, 'On the writings of M. V. Ostrogradsky on the theory of probabilities', *Istoriko-matematicheskie issledovania*, **4**, 99–123.

Gnedenko, B. V. and Sheynin, O. B. *1978*, 'Probability theory', in A. N. Kolmogorov and A. P. Yushkevich (eds), *Matematika XIX veka* ['Mathematics of the 19th century'], Moscow: Nauka, 184–240.

Grodzensky, S. Ya. *1987*, *Andrei Andreevich Markov 1856–1922*, Moscow: Nauka.

Heyde, C. C. and Seneta, E. *1977*, *I. J. Bienaymé: Statistical Theory Anticipated*, New York: Springer.

Karpenko, B. I. *1957*, 'The life and work of A. A. Chuprov', *Uchenie zapiski po statistike*, **3**, 282–317. [Written with the cooperation of N. S. Chetverikov. Contains the most complete listing of Chuprov's writings.]

Kotz, S. and Johnson, N. L. (eds) *1982–8*, *Encyclopedia of Statistical Sciences*, 9 vols, New York: Wiley.

Kotz, S. and Seneta, E. *1990*, 'Lenin as a statistician: A non-Soviet view', *Journal of the Royal Statistical Society*, Series A, **153**, 73–94.

Maistrov, L. E. *1980*, *Razvitie poniatia veroiatnosti* ['Development of the Concept of Probability'], Moscow: Nauka. [There is substantial overlap in scope, content and level with the author's earlier *Probability Theory: A Historical Sketch* [in Russian] (1967) of which there is an English translation by S. Kotz, 1974, New York: Academic Press.]

Ondar, Kh. O. (ed.) *1981*, *The Correspondence between A. A. Markov and A. A. Chuprov on the Theory of Probability and Mathematical Statistics* (transl. C. Stein and M. Stein, from the Russian original of 1977), New York: Springer.

Seneta, E. *1981*, *Non-negative Matrices and Markov Chains*, 2nd edn, New York: Springer.

—— *1984*, 'The central limit problem and linear least squares in pre-revolutionary Russia: The background', *Mathematical Scientist*, **9**, 37–77.

—— *1985*, 'A sketch of the history of survey sampling in Russia', *Journal of the Royal Statistical Society*, Series A, **148**, 118–25.

Sheynin, O. B. *1989*, 'A. A. Markov's work on probability', *Archive for History of Exact Sciences*, **39**, 337–77.

Uspensky, J. V. *1937*, *Introduction to Mathematical Probability*, New York: McGraw-Hill.

10.7

The English biometric tradition

THEODORE M. PORTER

We owe to the English biometricians the creation of statistics as an area of mathematical enquiry, standing above every particular application. Yet biometry was itself a branch of natural science, not of mathematics, and some key statistical concepts of biometry originated as descriptions of the natural world rather than as mathematical formulations. This article is concerned both with how the biometricians used statistical methods to transform evolutionary biology, and with how biological principles were separated from their original subject-matter to become the basis of a new and more powerful framework for statistics.

1 GALTON AND THE STATISTICS OF EVOLUTION

The founding father of biometry was Francis Galton, Erasmus Darwin's second most famous grandson. Galton, while not an especially profound thinker, was a remarkably creative one. His creativity worked through a profusion of analogies. With them, he was able to apply ideas and methods worked out for data reduction in astronomy, or for identifying the processes and entities that underlay social statistics, to the study of heredity. His aim was to make the theory of evolution by natural selection explicitly quantitative and, more particularly, to establish a mathematical basis for human intervention in the evolutionary process. He looked forward to a brave new world in which mankind would take charge of its own evolutionary destiny through the agency of 'eugenics' – selective human breeding. Exact, statistical knowledge was in Galton's view the most promising basis for guiding the biological improvement of the human species. Eugenic commitments remained characteristic of the English biometric tradition long after Galton's death.

For Charles Darwin, evolution had meant changing frequencies of traits as the result of differential survival and reproduction, and it required no great leap of imagination to see his theory as in some sense statistical. But the quantitative aspect of Darwin's argument remained almost entirely implicit. In particular, the element of randomness is almost completely absent in Darwin's writings; indeed, he resisted the notion that evolution was anything but law-like. He almost never used the word 'chance' as a statistician would, referring to events whose causes (if any) were both irregular and outside the purview of his theory. Chance for Darwin, and for his defenders and critics, was simply the opposite of 'design' (Gigerenzer *et al. 1989*: Chap. 4). Thus Galton's achievement in founding a statistical biometry should not be minimized. He did so by merging evolutionary theory with 'statistics' in the nineteenth-century sense of the term – that is, social numbers.

Galton aimed in his earliest eugenic writings to demonstrate that exceptional achievement, 'hereditary genius', was biological in character, by showing that it tended to run in families. It was in this context that he first applied the 'error curve' or normal law of Pierre Simon Laplace, Carl Friedrich Gauss and Adolphe Quetelet (§10.2) to human variation. Subsequently he was inspired by Darwin's 'provisional hypothesis of Pangenesis' to identify an analogy between the transmission of traits from parents to offspring and certain social processes such as the conscription of soldiers or the formation of towns by migration from a nearby population. Through this analogy he came to see biological inheritance as a process of sampling, and thus probabilistic in nature (Galton *1869*; Porter *1986*: Chap. 9).

That insight was the key to the founding of the biometric school. It formed the basis of the first quantitative account of the evolutionary process, and of an empirical as well as a theoretical evolutionary research tradition. For Karl Pearson and others, it became something of a religion.

2 PEARSON AND THE GENERALIZATION OF STATISTICS

In Pearson's day, at least, the history of biometry was one of continuous controversy. Much of this clearly owed to Pearson's missionary ambitions, which linked statistics to biology and both to political order and social policy. His intolerance extended equally to deviants within the biometric camp, such as Udny Yule, who departed from the one true way on the matter of contingency analysis. Pearson's dispute with Ronald Aylmer Fisher, over somewhat deeper issues, endured for decades, and was one of the great controversies in the history of this remarkably contentious field. The disagreement between the biometricians and the Mendelians is another, and while its persistence doubtless owed much to the personalities of the protagonists, there was more at stake than a few touchy egos. Biometricians and Mendelians disagreed on deep and important issues which were simultaneously scientific, professional, methodological and ideological (Provine *1971*, MacKenzie *1981*; §10.10).

In fact, almost any controversy with Pearson was destined to become a wide-ranging one. His *Grammar of Science* (1892) was, after all, a philosophical tract designed to extend the reach of science into every domain. He idealized statistics, a field long nourished by similar ambitions, as the true form of all scientific knowledge. Statistics was descriptive and positivistic, not causal and metaphysical. Pearson's philosophy, which was positivistic in the sense of Ernst Mach, construed statistics as a fine model for all scientific knowledge. Statistical methodology implies the omnipresence of chance, or at least of uncertainty; Pearson held that uncertainty in the social and political realm is of the same character as in physics, though it is often greater in magnitude. In short, Pearson hoped to unify knowledge under a positivistic philosophy and statistical form. The metric of biometry was to embrace sociology and politics as well as the dimensions of pea-seeds and shrimp-skulls.

At issue in the disagreement between biometricians and Mendelians, then, was a method of science tied to a view of biology and politics. The genes of which Mendelians spoke seemed metaphysical to Pearson, and the identification of most traits as simple and Mendelian was hard to reconcile with the continuous distributions that Pearson and W. F. R. Weldon studied. Pearson subscribed to a resolutely Darwinian ethic and worldview, while the theories of discontinuous evolution preferred by the Mendelians William Bateson and Hugo de Vries were part of the revolt against Darwinism. Galton, incidentally, gave aid and support to the enemy, for while he never questioned the importance of natural selection, he became convinced that regression took place towards a point of stability, and hence that macro-evolutionary change required discontinuous shifts to new points of stability (Galton *1889*). This did not keep him from lending wholehearted support to a quantitative biology, however.

Pearson's opponents, in contrast, tended to see him as a mathematical interloper, largely untrained in biology, and the methods he advocated as a threat to their own specialized competence. The charge of biological amateurism would have been more difficult to sustain against Weldon, but Weldon died in the early phases of the controversy. Pearson was a confirmed scholarly buccaneer, unlikely to be daunted by the specialist knowledge of the biologists. He published on every topic from poetry and social history to rational mechanics, which he thought justifiable in part because he took science to be first of all a method, not a body of knowledge. Thus he placed little stock in intellectual boundaries. Probably, then, the controversy lasted so long because the issues it raised were so fundamental. Both sides had much at stake. Neither welcomed the first attempts at an 'evolutionary synthesis', first by Fisher and then by Sewall Wright, involving the application of statistics to complex Mendelian characters (§10.10).

3 THE BIOMETRIC LABORATORY

One should not suppose, however, that the biometricians gave up their work to engage in endless polemics. Instead they went their own way. Pearson was able to support a laboratory at University College, London, largely due to a bequest from the childless founder of eugenics himself, Galton. Students from a variety of specialities and nationalities visited there in order to learn the most modern methods of statistics. In 1901 Pearson, Weldon and Galton established a journal, *Biometrika*, where much of this work was published. This journal gives some idea of the range of subjects to which they applied statistical analysis: population genetics, eugenics, demography, social problems, astronomy, meteorology, agriculture, quality control in brewing, and on and on. It also became one of the most important places to publish work on the history of probability and statistics.

The importance of all this was twofold. First, Pearson's laboratory was to a large extent responsible for the spread of mathematical statistical methods in the early twentieth century. Second, Galton, Pearson and their collaborators formulated so much of the new statistical theory that we can reasonably credit the English biometricians with the creation of modern statistics.

Among their specific results, first place has to go to Galton's idea of correlation, subsequently formalized by Pearson and others. The history of correlation reveals that the connection between statistics and biology was no accident. Galton, eugenicist that he was, wanted to comprehend mathematically the way that unusual height, strength or intelligence was perpetuated from generation to generation. Using Darwin's explanation of inheritance in terms of independent genetic units (the otherwise ill-fated hypothesis of 'Pangenesis') and a social analogy that it suggested to him, Galton determined that the problem of heredity was statistical in nature. He formulated it in terms of sampling from a pool of 'gemmules' contributed by the parents and by more remote ancestors, and solved it in terms of a linear 'reversion' towards the ancestral traits, plus an error term to account for the sample (of gemmules) being finite. Later, he decided that this was not mainly reversion, but actually 'regression' towards a stable population mean. The important point here is that Galton understood the mathematics of regression strictly in terms of its biological meaning. Only at the end of 1888, when he realized that the same formalism applied to another, quite distinct, biological problem, the 'correlation' of parts, did he recognize his mathematics as a solution to the abstract problem of relating interconnected variables rather than as a biological law (Cowan *1972*, Hilts *1973*, Porter *1986*).

The statistical idea of correlation was of crucial importance in the emergence of mathematical statistics as the science whose business was the analysis of data from any discipline whatsoever. Pearson, who was attracted to statistics just a few years later, took for granted the importance of developing formal and standard techniques of data analysis. The other main statistical tools invented by the biometricians all emerged in the context of some specific problem. Pearson's chi-squared test of goodness of fit, for example, arose as part of his positivistic campaign to dethrone the normal curve. W. S. Gosset ('Student') developed the t-test, the first test that could be used with small samples, for quality control in brewing. Fisher's analysis of variance related again to biology and eugenics (§10.10); he wanted to be able to determine how much of human variation was determined by heredity and how much by environment. Thus the fruitful collaboration between *bios* and *metron*, between mathematical method and biological subject-matter, worked in important ways to transform both.

4 FISHER AND THE STATISTICS OF EXPERIMENT

The biometric school itself did not survive these changes. Fisher seems in some ways to belong to the same tradition, of evolutionary biology allied to statistics, but he had a very different idea from Pearson's about the direction in which this alliance should proceed. Even with personalities so tenaciously irascible as these, we would be wrong to suppose that personality alone was responsible for their lifelong antagonism. The fundamental difference was between observation and experiment, between correlation and causation. The positivist Pearson measured and counted, amassing huge samples from which he could calculate reliable correlations. These, after all, were according to his philosophy all that science could aspire to.

Fisher, in contrast, allied statistics to experiment in order to get beyond the correlations and find the causes. Like Gosset's earlier t-test, Fisher's statistical mathematics was designed to get a maximum of information from small experimental samples. In 1919, Fisher turned down the opportunity to become chief statistician under Karl Pearson at the Galton Laboratory, choosing instead to accept the newly created post of statistician at an agricultural experimental station in Rothamsted. There he developed methods of experimental design that in the ensuing four decades would revolutionize experimentation, not only in agriculture (§10.11) but also in most of the biological, behavioural and social sciences. With the important exception of non-experimental fields such as econometrics (itself explicitly modelled on biometrics), Fisherian methods have come almost to define what constitutes sound research in the quantitative human sciences. The Pearsonian and Fisherian variants of biometry, taken together, constitute the single most important source for the remarkable growth of the domain of calculation in the twentieth century (Gigerenzer *et al. 1989*).

BIBLIOGRAPHY

Cowan, R. S. *1972*, 'Francis Galton's statistical ideas: The influence of eugenics', *Isis*, **63**, 509–28.

Galton, F. *1869*, *Hereditary Genius*, London: Macmillan.

—— *1889*, *Natural Inheritance*, London: Macmillan.

Gigerenzer, G. *et al. 1989*, *The Empire of Chance. How Probability Changed Science and Everyday Life*, Cambridge: Cambridge University Press.

Hilts, V. *1973*, 'Statistics and social science', in R. N. Giere and R. S. Westfall (eds), *Foundations of Scientific Method: The Nineteenth Century*, Bloomington, IN: Indiana University Press, 206–33.

Kevles, D. J. *1985*, *In the Name of Eugenics: Genetics and the Uses of Human Heredity*, New York: Knopf.

MacKenzie, D. *1981*, *Statistics in Britain, 1865–1930: The Social Construction of Scientific Knowledge*, Edinburgh: Edinburgh University Press.

Norton, B. J. *1973*, 'The biometric defense of Darwinism', *Journal of the History of Biology*, **6**, 283–316.

Porter, T. M. *1986*, *The Rise of Statistical Thinking, 1820–1900*, Princeton, NJ: Princeton University Press.

Provine, W. B. *1971*, *The Origins of Theoretical Population Genetics*, Chicago, IL: University of Chicago Press.

Stigler, S. M. *1986*, *The History of Statistics: The Measurement of Uncertainty Before 1900*, Cambridge, MA: Harvard University Press.

10.8

Probability, statistics and the social sciences

THEODORE M. PORTER

1 PROBABILITY VERSUS STATISTICS

Probability has been, and continues to be, connected with the social sciences in two rather different ways: one highly individualistic and rationalistic, the other more nearly sociological. The individualistic variety was widely supported by students of probability in the Enlightenment, then was almost completely discredited early in the nineteenth century, and has re-emerged strongly only since the Second World War. It set up probability theory as a standard of rationality, and sought to understand individuals as maximizers of mathematical expectation, or of some other probabilistic expression related to utility. The other variety, clearly the more important from about 1820 through to the early twentieth century, is statistical, and aims to understand a whole society in terms of demographic, economic and 'moral' numbers. In this context, probability theory served not as a model or an ideal of rational individual behaviour, but as the basis for statistical inference concerning trends or causes.

To a large extent, these two forms of probabilistic social analysis have been rivals, and their history is all the more interesting and revealing on account of this disagreement. Moreover, a continuous line of social thinkers has rejected both. Such critics have called probability and statistics 'reductionistic', arguing that they are inadequate because they take society to be no more than the sum of the individuals who make it up. The shifting balance among these views in different countries, periods and disciplines tells a lot about the history of social science and the conditions that have encouraged or impeded its mathematicization.

2 PROBABILISTIC RATIONALISM

Pierre Simon Laplace proposed in 1814 that probability was merely good sense reduced to calculation. It is one of his most famous lines, but he wrote it just at the time when it was coming to seem implausible. Laplace's was in fact a thoroughly eighteenth-century or 'classical', conception of probability (§10.17). At least since James Bernoulli's *Ars conjectandi* (1713), probabilists had seen their calculus as a way of modelling rational judgement in cases where certainty was not available, which is to say in almost every situation in real life (§10.2). It was intended to be prescriptive. Calculation was to be a proxy for good sense, an invaluable aid to the unenlightened masses who lacked the wisdom or discernment to behave rationally. But probabilistic calculation was not thought to be constitutive of good sense; probabilists believed that truly enlightened individuals acted rationally as if by instinct, and that their inherent sensibility gave rationality its ultimate grounding. Probability, as a science of rational belief and action, should thus produce results consistent with the behaviour of these enlightened individuals.

This, as Daston *1988* argues, is why probabilists were so much troubled by such anomalous results as the St Petersburg paradox (§10.2). The St Petersburg problem arises from a coin game in which one player agrees to pay the other an amount that doubles with each successive toss of heads. The ordinary theory of probability assigns an infinite mathematical expectation to this game, even though the pay-off will in almost every case be rather small. Here the results of probability seemed clearly to be inconsistent with good sense, and generations of probabilists struggled to reconcile the calculus of probability with the dictates of rational behaviour. This, though, was more an anomaly than a serious threat to the classical programme, and on the whole the classical conception survived almost unscathed until some time after the French Revolution.

Classical probability was designed to serve a world of individuals aspiring to a universal standard of rationality, not to support the diverse customs, traditions and sentiments that bind together any particular human community. Enlightenment probabilists were concerned with gambling problems as a model for all decisions in cases of uncertainty. They wrote and argued about the rationality of smallpox inoculation, which, before Edward Jenner, carried a fair risk. Perhaps their favourite applied topic was the credibility of testimony, the degree of faith that a rational calculator would invest in truth claims as a function of their nature and the character of the people who made them. David Hume, in this tradition, argued that no amount of testimony could rationally justify belief in miracles; while George Hooper, recognizing that the credibility of a report must decay with successive retellings, calculated the time of the Second Coming on the premiss that Jesus Christ must return when faith had ceased to be probable on biblical evidence.

Towards the end of the eighteenth century, the Marquis de Condorcet argued for reform of the judicial system on probabilistic grounds, and the same issue continued to be treated after the French Revolution – though not with the same degree of moral conviction – by Laplace and by Siméon-Denis Poisson (Gillispie *1972*). Here, as elsewhere, social policy was at stake, but the grounds of argument had to do with rational belief by individuals.

3 MEASURING UNCERTAINTY

Enlightenment probabilists were somewhat less interested in statistics and 'political arithmetic' than in using the theory of games of chance as a model of rational behaviour under uncertainty, but they did apply their calculus to problems of demography and insurance. Here, too, their presuppositions were individualistic. James Bernoulli, Abraham De Moivre and Laplace worked out ways of extending the mathematics of chance so that it could be applied to very large numbers of events. This depended on treating each toss of the coin as independent, and all tosses as governed by the same probabilities of heads or tales. In the same way, Laplace calculated the accuracy of population estimates by assuming that every human birth was an independent event, and that a uniform probability governed the ratio of births to the total population. The eighteenth-century French government did not conduct a complete census, but it did collect records of births and deaths. The total of births in a year, multiplied by the ratio of births to total population from a well-chosen town, could be used to determine the population of France. Laplace used probability theory to compute a probable error for these estimates. The same formulas enabled Laplace, Condorcet and Poisson to calculate the probability that a jury of a certain size and requiring a certain majority to convict would reach an unjust verdict. Here again, they had to assume that each juror had the same fixed probability of voting in error and that every juror's vote was independent, so that a jury vote was like a series of tosses of an unbalanced coin.

By the time Poisson presented his work on judicial probabilities in the 1830s, it appeared to many as dangerous and implausible mathematical rationalism. The calculations, in fact, were possible only if one made assumptions of what would now be called a Bayesian sort about the mathematical structure of ignorance. To most people who wrote on the subject from about 1820 until the twentieth century, this seemed like mere speculation dressed up in mathematical garb. It would be better, argued John

Stuart Mill in 1842, to seek to remedy our ignorance than to base calculations on it; applications like these, he argued, had made probability the 'opprobrium of mathematics'. A host of distinguished thinkers agreed, among them Louis Poinsot, Auguste Comte, R. L. Ellis, J. D. Forbes and George Boole. These critics regarded classical probability the way Edmund Burke viewed the actions of the delegates of the Third Estate in the French Revolution, as rationalism unhinged from wisdom and experience. Some quite explicitly associated the rationalism of classical probability theory with the spirit of the Revolution. But even Mill, who praised the Revolution and admired Condorcet and the *Girondins*, wanted reason to be grounded in observation and experiment, principles far more sober and cautious than classical probability seemed to admit (Porter *1986*).

4 STATISTICS AND THE CONCEPT OF SOCIETY

It was indeed a sober form of social knowledge to which probability became attached in the nineteenth century. This was statistics, the science of the 'statist', and perhaps the strongest candidate to become the general science of society in early-nineteenth-century Britain and France. Few mathematicians and natural philosophers became seriously involved in the collection and analysis of social numbers. Numbers were appreciated by bureaucrats and reformers as a convenient medium for aggregating records and for the rhetorical advantages of their apparently unproblematical factuality. A revealingly extreme view of the philosophy of statistics was offered by the Statistical Society of London, partly in order to allay suspicions among natural philosophers that such a society would inevitably become a forum for political discord or radical enthusiasm. The society would, its council announced in the first volume of its journal (1838), aim to exclude all opinions, and direct its efforts at the establishment of solid reliable facts. 'Facts, facts, facts', parroted Thomas Gradgrind in Charles Dickens's *Hard Times* (1854). The complete exclusion of opinion was rather too lofty an ideal, but the goal of reducing speculation to a minimum was widely shared, even outside England, and placed constraints on the potential uses of probability.

In particular, nineteenth-century statisticians were almost unanimous in insisting that true statistical knowledge should be based on a complete census, and never on estimates from samples; sampling they regarded as too close to mere speculation. Hence Laplace's form of statistical demography essentially disappeared for about a century. There were other reasons too why probability had only a secondary role in social statistics. 'Statists' were on the whole practical men, men of business, medicine and the law, who were more interested in reform than scholarship and who commanded very little mathematics. Most nineteenth-century statistics had very little to do with any form of mathematics more exacting than elementary arithmetic. A few distinguished mathematicians, like Joseph Fourier, became seriously involved in statistical collection; but the *Recherches statistiques sur la Ville de Paris*, to which he devoted much of his attention for a decade around 1820, were notable for their exhaustive detail, not their use of probability – this despite the introductory essays contributed by Fourier himself, which aimed to show how the mathematics of error theory could be usefully applied.

5 THE ERROR LAW AND THE AVERAGE MAN

Still, probabilists had no trouble recognizing in statistics a potential field of application for their mathematics. This one was all the more needed because the most promising traditional application of probability, quantified rationality, now stood practically discredited. There was an obvious analogy, and indeed almost an identity, between statistics and life assurance, a standard use of probability dating to the seventeenth century (§10.3). Demography, even more interesting to Europeans after Thomas Malthus than before, involved probabilities of birth, death and marriage and the like. Such computations were not especially interesting from the standpoint of probability theory, but they were sufficiently difficult in other respects to require mathematical expertise. During the nineteenth century, demography and insurance provided the main link between probability and social science.

A more demanding role for probability was implied by a more tenuous analogy between statistics and the newest important application of the probability calculus, astronomical error theory. In 1809, Carl Friedrich Gauss published a probabilistic derivation of the method of least squares for fitting a curve to a series of astronomical observations (§10.5). Astronomers and geodesists quickly became convinced that the errors of individual observations were distributed according to what soon became known as the 'error law', and is now usually called the Gaussian or normal distribution. During the 1840s, a deep analogy between such measurements and human traits was proposed by the most vocal and influential advocate of a mathematical treatment of statistics in the nineteenth century, the Belgian astronomer and statistician Adolphe Quetelet (see e.g. Quetelet *1846*). He used statistical thinking to vindicate the central concept of his proposed 'social physics': *l'homme moyen*, the average man. The distribution of real humans around this fictitious average man, he suggested, is just like that of our imperfect measurements of a star's position to its true value. The average thus represents the true type of the human species, and deviations from it are simply a matter of error. Basic probability theory, he thought, provided essential grounding for a statistical science of society.

Less extreme in his positivism than the founders of the Statistical Society of London, Quetelet still wanted a social science that would follow directly from social facts rather than one presupposing some deep philosophical insight into the nature of society. The laws of society, according to his programme, were simply statistical. It was the business of the state to gather records of the physical, moral and demographic traits of the population. The numbers would permit calculation of the average man – characterized by a mean height, mean weight, mean propensity to marriage, crime, suicide, and so on. The error analogy justified him in fixing his gaze on the mean, and ignoring the teeming diversity of actual humans. In this way, he hoped, the social physicist could uncover the laws that govern the social system. Quetelet liked to compare society to the Solar System, and he declared repeatedly that similar laws would be discovered to obtain there. He found such analogies, in fact, in every kind of human activity, but the most fundamental involved the conservation laws. Since 'moral force' was conserved, one could calculate the future by extrapolating from the past. The most general law of history proved to be a progressive tendency for the merely physical to be effaced by the intellectual and scientific.

Statistical records, then, ought to show human improvement: slow, sometimes temporarily blocked by 'perturbing forces', but inevitable. Revolution Quetelet condemned as leading to a loss of the 'moral force' that brings progress; conditions of stability, he explained, are best for society. And it was a considerable consolation to him to learn that human actions of all kinds, even violent and irrational ones, show a wonderful stability when considered in the mass. That is, the number of births, deaths, crimes, marriages and even suicides is virtually constant in a given country from year to year. This he took to be a vindication of such statistical methods. There was no need to be troubled with explaining this or that suicide; it was much more profitable to attend only to what can be measured in large numbers.

The regularities of statistics also seemed to him a powerful testimony to the reality of society, and indeed its power over mere individuals. Every year, after all, the same number of people are driven to commit suicide, quite without respect to the desires and experiences of each individual. Quetelet's form of statistical determinism was anything but a counsel of despair; a wise legislator could reduce crime by changing the laws or by providing education, which would alter the customs and mores of the 'social body'. The existence of different crime and suicide rates in different countries proved that they were not fixed by nature, but rather were properties of each particular society. Comparative statistics would therefore be of the greatest interest to the scientist and reformer, for here, in the records of human aggregates, was the proper study of humankind. Mere individuals, in contrast, were quite without interest to the social scientist. Quetelet did not go so far as to deny that individuals have any freedom or autonomy. They have free will, but it produces nothing more than trifling accidental deviations, much like the atmospheric perturbations that cause telescopic measurements to vary irregularly, but only slightly, from the true position.

6 OPPOSITION TO STATISTICS

Quetelet's view, that statistical regularities provide powerful evidence of the reality of society, was a highly influential one. Emile Durkheim, for example, could think of no more compelling evidence of the existence of 'social facts' than the regularities of statistics (Turner *1986*). Quetelet's ideas about the nature of society, however, were widely criticized. To reduce society to an aggregation of essentially identical individuals seemed an antisociological move, and was denounced as such. Quetelet was regarded as a bit too much like the classical probabilists, who, like all Enlightenment *philosophes*, had since the triumph of European romanticism been charged with lacking any real conception of the social. Perhaps the sharpest criticism came from Comte, known to posterity as the founder of 'sociology'. Comte, in fact, only invented this term because Quetelet, in a book of 1835, had appropriated his preferred phrase, 'social physics', for 'mere statistics'. Comte was no friend of probability, nor indeed of social quantification, but he also gave a sociological criticism of statistical methods. When society is in transition from the theological and metaphysical stages to the positive one, as in his view it then was, an all-embracing average could only confuse matters by lumping together persons in very different stages of social development. Comte was scarcely the one to defend individualism, but he did subscribe to a more organic conception of society in which human variation was intrinsic to the social order, not an accidental deviation from it.

Such social organicism provided perhaps the most common reason for later sociologists to ignore or oppose statistics. More influential than the outright rejection of statistics, though, was the drive to reinterpret this science in a more fully sociological way. Especially in Germany, a powerful movement arose in the 1860s to rescue Quetelet from himself by reconciling statistics with a more holistic view (see the articles by Wise and Porter in Krüger *et al. 1987*). Part of the reason for doing so was to defend the proposition that individuals are responsible for their actions. At the same time, German statisticians took Quetelet's 'atomism' to imply a *laissez-faire* world in which individuals have no responsibility for one another. This was precisely the opposite of Quetelet's own intentions, but even the few who read him closely enough to recognize this believed that he had failed to gauge the implications of his own ideas. The historical economists and 'academic socialists' who dominated the German statistical community wanted to end the reign of the average man, and in place of broad averages to emphasize a minute analysis that would reveal the fine structure of human society. They would not, of course, take this analysis all the way to the level of individuals. What they wanted was still statistics; it was premissed on the value of 'mass observation' for the study of 'mass phenomena'. The purpose, however, was no longer to seek the broadest generalizations, but rather to dissect, to compare, to lay bare the relationships that accounted for the variation within every society. The results might not be laws. By about 1870, German statisticians had become convinced that 'natural law' was the opposite of history, and quite inapplicable to human societies. They were less troubled by the notion of social causation, and they hoped through careful analysis to gain knowledge that would permit reliable prediction and scientific reform.

7 MATHEMATICAL STATISTICS

German statisticians were emphatically social scientists, not mathematicians. The phrase 'mathematical statistics' came into use after about 1867, when it appeared as the title of a book by a German actuary, Theodor Wittstein. For several decades it was used to refer mainly to a more mathematical form of quantitative social science, usually demographic and insurance calculations. Probabilistic methods derived from error theory were also applied by Wilhelm Lexis and Ladislaus von Bortkiewicz to assess whether a model of free 'atomic' individuals was consistent with the degree of stability found in statistical tables of crime, suicide and marriage, and other moral acts. Lexis determined that the irregular fluctuations were too great to be reconciled with a probabilistic model of random independent acts, and concluded that probability provided rather a poor model of social behaviour.

The formal use of probability theory, while not central to the German statistical tradition, was gaining in prominence during the 1870s and 1880s. Similar tendencies toward increased reliance on probability are evident in the publications of the Société de Statistique de Paris, which was a relatively professionalized body if not an academic one. Even among the English social statisticians, some powerful spokesmen for a mathematicized statistics emerged late in the nineteenth century, most notable among them the economists William Stanley Jevons and Francis Edgeworth. Probability was also assuming an increasingly important role in a different variety of social science, anthropometry, many of whose practitioners looked to explain the character of different societies by discovering the racial characteristics of their members.

All of this did not yet amount to the integration of probability mathematics into the social sciences of the nineteenth century; it was rather that statistics had become crucial to a large part of social science. But the statistical approach came increasingly to be seen as raising problems that probability methods might help to solve. This interest in statistical mathematics was more than enough to make social scientists eager consumers of the new mathematical statistics developed by the English biometricians around 1900. The social statisticians were very interested in new techniques such as correlation, invented by Francis Galton and formalized by Karl Pearson (§10.7), and correlations involving crime, poverty, alcoholism and the like soon began to be offered in social debates. To a large extent, though, social statisticians nurtured their own mathematical tradition, one that drew from the biometricians but was not dominated by them. Perhaps their most distinctive contribution involved methods of sample surveys, first discussed systematically by A. T. Bowley in the 1900s, which quickly became one of the most important tools of American social science (Converse *1987*). The growing use of statistics in the social sciences, then, underwent no abrupt shifts under the influence of Karl Pearson and the biometric school. Fisherian analysis of variance and significance testing had perhaps a more rapid and sweeping effect, though the most radical development is the very recent re-emergence of probability mathematics to model rational behaviour under uncertainty.

The integration of statistical tools and concepts into social science is by no means a simple story of scientific progress through the development of increasingly powerful analytical methods. If natural science proceeds by laying aside old controversies, as an increasingly voluminous literature now suggests, controversy has been even more central to the development of the social sciences. And in social matters, controversies are rather more difficult to settle, in part because the implications for political ideology and the practice of governing are so pervasive.

Part of the appeal of quantification was that disagreements might be settled by calculation, but this was never easy. More often, perhaps, the effect was rather to sharpen disputes about the forms of quantification that were supposed to settle social controversy (Porter *1990*). As this article has shown, statistics was a social science long before it began to become a branch of mathematics, and its mathematics derives in part from social conceptions. Its bearing on society, moreover, has been very much a matter of practice – of the centralization of power, and the management of social problems – and not merely of theory. Statistical methods carried with them preconceptions about the relations of the individual to society, the importance of divisions within society according to race, social class, regional background or degree of enlightenment, the causes of social change, and a host of other factors that could not be separated from ideology. Of course, statistics is not simply reducible to ideology, but debates on these issues helped to determine where statistics would be found acceptable as a method of social investigation, and even the shape of statistical methodology itself.

BIBLIOGRAPHY

Baker, K. M. *1975, Condorcet: From Natural Philosophy to Social Mathematics*, Chicago, IL: University of Chicago Press.

Bannister, R. *1987, Sociology and Scientism*, Chapel Hill, NC: University of North Carolina Press.

Converse, J. M. *1987, Survey Research in the United States: Roots and Emergence*, Berkeley, CA: University of California Press.

Daston, L. J. *1988, Classical Probability in the Enlightenment*, Princeton, NJ: Princeton University Press.

Gigerenzer, G. *et al. 1989, The Empire of Chance. How Probability Changed Science and Everyday Life*, Cambridge: Cambridge University Press.

Gillispie, C. C. *1972*, 'Probability and politics: Laplace, Condorcet, and Turgot', *Proceedings of the American Philosophical Society*, **116**, 1–20.

Hacking, I. *1990, The Taming of Chance*, Cambridge: Cambridge University Press.

Krüger, L., Daston, L. and Heidelberger, M. (eds) *1987, The Probabilistic Revolution*, Vol. 1: *Ideas in History*, Cambridge, MA: MIT Press. [See especially Part 4.]

Lottin, J. *1912, Quetelet: Statisticien et sociologue*, Louvain: Institut Supérieur de Philosophie.

Oberschall, A. 1965, *Empirical Social Research in Germany, 1848–1914*, Paris: Mouton.

Porter, T. M. *1986, The Rise of Statistical Thinking, 1820–1900*, Princeton, NJ: Princeton University Press.

—— *1990*, 'Natural science and social theory', in R. Olby *et al.* (eds), *Companion to the History of Modern Science*, London: Routledge, 1024–43.

Pour une histoire de la statistique, Vol. 1: *Contributions 1987*, 2nd edn, Paris: Economica.

Quetelet, A. *1846, Lettres* [...] *sur la théorie des probabilités*, Brussels: Hayez. [English transl. by D. G. Downes, 1849, *Letters on Probability*, London: Layton.]

Ross, D. *1991, The Origins of American Social Science*, Cambridge: Cambridge University Press.

Turner, S. *1986, The Search for a Methodology of Social Science: Durkheim, Weber, and the Nineteenth-Century Problem of Cause, Probability, and Action*, Dordrecht: Reidel.

10.9

Psychology and probability: Two sides of the same coin

GERD GIGERENZER

1 ORIGINS

Probability and statistics helped to create psychology, rather than just having been applied to psychology at some point. Moreover, the support was not unidirectional: psychological and probabilistic ideas were at times two sides of the same coin, and the influence went in both directions. The associationist psychology of John Locke, David Hartley and David Hume created the psychological framework from which the classical interpretation of probability in the seventeenth and eighteenth centuries emerged: the assumption that subjective belief is apportioned to objective frequencies, and for some authors, to physical regularities. Only in the late eighteenth century, when associationist psychology shifted its emphasis from the rational mind to the illusions and distortions that passion introduces into mental calculations of probabilities, did the gap between subjective and objective probabilities become clear, forcing a choice between the two (Daston *1988*).

The alternative ultimately chosen is now known as the frequency interpretation of probability. Again, the shaping of the frequency view was interrelated with psychological ideas, through Gustav Theodor Fechner, the founder of psychophysics. Fechner's metaphysical indeterminism had its mathematical expression in his posthumous *Kollektivmasslehre* (1897). His theory of collectives became a standard topic for anyone working in probability and statistics in early-twentieth-century Germany, and stimulated Richard von Mises' development of the frequency view (Heidelberger *1987*).

2 PROBABILITY HELPS TO CREATE PSYCHOLOGY

Two major programmes of psychology emerged in the second half of the nineteenth century in Germany and England. Both were founded on the frequency view of probability, but they differed sharply in the interpretation of their respective frequency distributions. What is now known as experimental psychology originated in the German tradition of Fechner, Wilhelm Wundt and others. To support his psychophysical parallelism and refute Cartesianism, Fechner set out to demonstrate the mathematical relationship between the physical world (e.g. the weight of objects) and the subjective world (e.g. the perceived weight of objects). Frequency distributions (e.g. intra-individual repetitions of judgements of weight) provided both the definition and the measurement of the basic unit of perception, the 'just noticeable difference' or 'differential threshold': that difference in physical intensity that is correctly detected in a specified proportion (say 75%) of repetitions. Here, probability was not just a tool for the analysis of data, it provided the definition of the basic concepts. This established a long tradition of probabilistic models of perception, from L. L. Thurstone in the 1920s to R. D. Luce, J. C. Falmagne and others in the second half of the century (Luce *et al. 1963–5*).

In England, Francis Galton, Karl Pearson and others laid the foundations of what is called today 'differential' or 'correlational' psychology, for instance mental testing and the analysis of individual differences (§10.7). In his enthusiasm for the normal distribution, Galton outstripped even Adolphe Quetelet, using this distribution both to define natural ability (now called intelligence) and to make individual differences in intelligence measurable. His argument was merely an analogy: Quetelet had shown that physical stature is normally distributed, and, Galton reasoned, therefore so is brain size, and concluded that so too were talent and intelligence. Subsequently, the normal distribution became the most commonly used method of scaling psychological tests. Here again, probability was more than a tool: it helped to define the basic concepts such as intelligence. Reciprocally, the study of intelligence and heredity contributed directly to statistics: regression to the mean, correlation and, later, factor analysis were proposed first as concepts and theories of heredity and intelligence; only subsequently were they mathematically polished and abstracted from that specifically psychological context by Pearson and others (MacKenzie *1981*, Stigler *1986*).

Despite their common origin in a frequentist view of probability, the two research programmes in psychology differed sharply. First, in the experimental psychology that emerged from Fechner's psychophysics, statistics helped to create experimental designs and control experimental conditions. In the Galton–Pearson tradition, correlation and related techniques helped to provide a surrogate for experimental control and design. Second, variability was typically seen in experimental psychology, as in astronomy, as an *error* around a true value that characterizes a real state or process. In contrast, in the Galton–Pearson tradition variability was understood as *real*, as the motor of evolution. Third, Fechner and many of his German colleagues held that asymmetric frequency distributions are the normal case, whereas Galton believed in the symmetric 'law of frequency of error', whose very name, 'normal distribution', shows the wide acceptance of his conviction.

3 STATISTICS SHAPES EXPERIMENTAL PRACTICE

In the two decades following the publication of Ronald Aylmer Fisher's *Design of Experiments* in 1935, Fisher's link between experimental design and inferential statistics became institutionalized in American experimental psychology. Other kinds of experimentation that did not show the fingerprints of statistics (i.e. the elements of randomization, repetition and independence) were effectively ruled out: from Wundt's single-case experiment, and the Gestalt psychologist's demonstrational experiments, to Egon Brunswik's representative design. Even Fechner, who gave much thought to experimental design, had used not randomization but systematic variation of conditions. By 1970, for instance, about 70% of experimental articles in the major journals used Fisher's analysis of variance. But the institutionalized practice went beyond the adoption of analysis of variance and related techniques. In the 1920s and 1930s, American psychology (and, with a time lag of several decades, German psychology) started to replace the study of single persons (from Fechner, Wundt and Herman Ebbinghaus to Ivan Pavlov and B. F. Skinner) by studying mean differences between two (or more) groups of people. This was partly to serve the needs of American educational administrators, who wanted, for example, to justify that one curriculum was more effective than another (Danziger *1990*). This comparison of groups of individuals became the institutionalized practice for experimental design and statistical inference, and statistical inference thus became a major vehicle in a striking transition of experimental practice.

After the Second World War, Fisher's logic of significance testing became mishmashed with Neyman–Pearson theory (Jerzy Neyman and Egon Pearson, Karl Pearson's son) in curricula, textbooks and experimental practice, creating a hybrid logic of statistical inference. This hybrid logic was and still is presented as if it were a single monolithic logic of inference, resulting in the mechanical application of statistical inference and the foreclosing of statistical alternatives (Gigerenzer and Murray *1987*).

Abraham Wald's sequential testing and Bayesian inference, for instance, have almost never been used for statistical inference in psychology, whereas the hybrid logic became almost synonymous with scientific method in psychology and beyond (Gigerenzer *et al. 1989*).

4 THE MIND AS AN INTUITIVE STATISTICIAN

In the 1960s, the 'cognitive revolution' eroded the behaviourist hegemony in American psychology. The cognitive revolution was more than a return to the mentalist language ('insight', 'restructuring') of the earlier German-speaking psychology. The very meaning of the 'mental' changed through the use of two key metaphors, statistics and computers. Cognitive processes such as perception, memory and thinking became portrayed in terms of drawing random samples from nerve fibres, computing likelihoods, calculating analyses of variance, setting decision criteria and performing utility analyses. For instance, Fechner's thresholds were replaced by a form of Neyman–Pearson hypothesis testing: the theory of signal detectability assumes that the mind 'decides' whether two stimuli (e.g. weights) are different or the same, just as a statistician of the Neyman–Pearson school decides between two hypotheses. The very statistical tools that had become institutionalized in the practice of psychologists turned into theories of mind (Gigerenzer and Murray *1987*).

After the revival of Bayesian statistics around the middle of this century, Bayesians attempted to take over the institutionalized frequentist inferential statistics, but failed completely. However, Bayesian models of cognition were developed, and there exists an interesting body of research on whether and when intuition *is* Bayesian (see e.g. Kahneman *et al. 1982*), and controversial discussions on whether and when intuition *should* be Bayesian (see e.g. Cohen *1986*). Thus, the controversy within probability theory between frequentists and subjectivists has introduced a double standard: researchers themselves almost never use Bayesian statistics to make inferences about hypotheses, but often assume that their subjects should use Bayesian statistics to reason rationally.

Probability and statistics bridge psychological schools – the link between the cognitive revolution and the metaphor of the mind as an intuitive statistician is paralleled by many others, for example W. Estes's probabilistic version of behaviourist learning theory of the 1950s. They bridge not only competing schools, but also differences between continents. For instance, research on statistical intuitions started in the 1930s in Austria, Switzerland and other European countries, and in the 1960s in a Bayesian framework in the USA (with little interaction). But there are cultural differences, too. The enthusiasm with which American psychology promoted the hybrid logic of statistical inference to a methodological imperative and elevated the level of significance to a measure of the quality of research and regulator of publication has never been matched in Continental Europe.

5 VARIABILITY AND UNCERTAINTY: JUST ERROR OR REAL PROCESSES?

Despite the manifold interactions between probabilistic and psychological ideas, psychologists often hesitated to go beyond the interpretational framework of Laplacean determinism. The role of probability was to tame error, and error could in principle (although not in practice) be reduced to zero. That probability, uncertainty or unpredictability could mirror a real process in the subject under investigation as opposed to the experimenter's ignorance and lack of control was rarely considered – this despite role models such as the indeterminisms of Fechner, Charles Sanders Peirce and quantum physics, and the substantial role of variability in biology, from Darwin to Galton to random drift.

One historical counter-example is Egon Brunswik, who argued that even if God did not play dice – that is, even if Laplacian determinism did hold in the physical world – the human mind would none the less have to gamble. Our minds need to infer the physical world from perceptual cues that are in principle uncertain (Brunswik *1956*). According to Brunswik's probabilistic functionalism, the task of the mind is intrinsically and irreducibly probabilistic, and so is the task of psychologists: to find out how we come to terms with a world that presents itself only via uncertain cues.

BIBLIOGRAPHY

Acree, M. C. *1978*, 'Theories of statistical inference in psychological research: A historico-critical study', Ph.D. Dissertation, Clark University, Worcester, MA.

Brunswik, E. *1956*, *Perception and the Representative Design of Experiments*, Berkeley, CA: University of California Press.

Cohen, L. J. *1986*, *The Dialogue of Reason*, Oxford: Clarendon Press.

Danziger, K. *1990*, *Construing the Subject. Historical Origins of Psychological Research*, Cambridge: Cambridge University Press.

Daston, L. *1988*, *Classical Probability in the Enlightenment*, Princeton, NJ: Princeton University Press.

Gigerenzer, G. *1987*, 'Probabilistic thinking and the fight against subjectivity', in Krüger *et al. 1987b*: 11–33.

Gigerenzer, G. and Murray, D. J. *1987*, *Cognition as Intuitive Statistics*, Hillsdale, NJ: Erlbaum.

Gigerenzer, G. *et al. 1989, The Empire of Chance. How Probability Changed Science and Everyday Life*, Cambridge: Cambridge University Press.

Heidelberger, M. *1987*, 'Fechner's indeterminism: From freedom to laws of chance', in Krüger *et al. 1987a*: 117–56.

Kahneman, D., Slovic, P. and Tversky, A. (eds) *1982, Judgment Under Uncertainty: Heuristics and Biases*, Cambridge: Cambridge University Press.

Krüger, L., Daston, L. J. and Heidelberger, M. (eds) *1987a, The Probabilistic Revolution*, Vol. 1, *Ideas in History*, Cambridge, MA: MIT Press.

Krüger, L., Gigerenzer, G. and Morgan, M. S. (eds) *1987b, The Probabilistic Revolution*, Vol. 2, *Ideas in Sciences*, Cambridge, MA: MIT Press.

Luce, R. D., Bush, R. R. and Galanter, E. (eds) *1963–5, Handbook of Mathematical Psychology*, 3 vols, New York: Wiley.

MacKenzie, D. *1981, Statistics in Britain, 1865–1930: The Social Construction of Scientific Knowledge*, Edinburgh: Edinburgh University Press.

Piaget, J. and Inhelder, B. *1975, The Origin of the Idea of Chance in Children*, New York: Norton. [Originally published in France as *La Genèse de l'idée de hasard chez l'enfant*, 1951, Paris: Presses Universitaires de France.]

Porter, T. M. *1986, The Rise of Statistical Thinking 1820–1900*, Princeton, NJ: Princeton University Press.

Stigler, S. M. *1986, The History of Statistics: The Measurement of Uncertainty Before 1900*, Cambridge, MA: Harvard University Press.

10.10

Probability and statistics in genetics

A. W. F. EDWARDS

1 INTRODUCTION

The development of the calculus of probability in connection with the analysis of games of chance in the seventeenth century led, in the eighteenth, not only to a mathematical structure of increasing power and depth but also to a gradual appreciation that the application of the new calculus to problems of inference involved novel philosophical considerations. In the nineteenth century, however, the dominant driving force in the development of the new mathematical methods was science, with the physical and the biological sciences both playing major roles. In the early years of the twentieth century, with the basic calculus of probability firmly established, the physical, biological and inferential traditions were brought together, largely through the influence of two Englishmen: Karl Pearson and Ronald Aylmer Fisher, the first two holders of the Galton Professorship of Eugenics at University College, London. Both men were trained in mathematics and the physical sciences, both worked professionally in the biological sciences, and both thought deeply about the philosophical problems of inference.

The latter part of the twentieth century has not seen the resolution of the philosophical difficulties, about which debate still continues. This article traces the biological side of the development, with special reference to its major component, genetics. Far from probability and statistics having been imported into the biological sciences, much of the mathematical theory was actually developed within biology itself, only later being adopted by mathematicians as a branch of their own subject.

2 EARLY DEVELOPMENTS

The publication of John Graunt's *Natural and Political Observations Made upon the Bills of Mortality* in 1662 inaugurated a climate of statistical inquiry into biological phenomena (see §10.3 on actuarial mathematics). The question of the stability of the human sex ratio, raised by Graunt, attracted the attention of John Arbuthnott (1712) and Nicholas Bernoulli (1713). The former's discussion led to the idea of a formal test of significance based on the binomial distribution, while the latter's involved a sharpening of the approximation to the tail probability of the binomial which had been given earlier by his uncle, James Bernoulli. These early analyses of biological data were influential on the contemporary climate of enquiry which encouraged the theoretical advances of Pierre de Montmort and Abraham De Moivre (as well as those of Nicholas Bernoulli himself).

3 THE NINETEENTH CENTURY: FRANCIS GALTON

By the end of the eighteenth century the early advances had been consolidated into a recognizably modern theory of probability (§10.2), but little further progress on questions of statistical inference was made until the demands of the physical sciences led Carl Friedrich Gauss, Pierre Simon Laplace and their contemporaries to lay the foundations of the statistical theories of distribution and estimation. Not until the nineteenth century did this have an impact on biology, when Adolphe Quetelet applied statistical analysis to demographic and social data, making particular use of the normal distribution or 'error curve' which had originated as an approximation to the binomial distribution in the work of De Moivre and had rapidly become central to statistical theory. Quetelet, a Belgian whose professional career was as a mathematician and astronomer (§10.13), played a major role in the development of statistics through his part in the foundation of statistical societies in Europe, but it was his influence on the Englishman Francis Galton that affected the theoretical development the most (Keynes *1993*).

Galton applied the normal distribution to biological phenomena, not so much as a curve of 'error' but as a curve of variation; for he realized, in the wake of Charles Darwin's theory of evolution by natural selection, that biological variability – far from being of secondary importance compared with mean values – needed measuring, studying and ultimately partitioning and correlating. This fundamental shift in a point of view led to Galton's introduction of the notions of regression (when he was 55 years old) and correlation (12 years later), which represents the major contribution to statistical thought in the second half of the century.

Galton was to the inheritance of continuous variation what Gregor Mendel was to the inheritance of discrete variation. Each worked in ignorance of the contribution of the other (Mendel's paper, though read in 1865, did not come to general notice until 1900), and each worked with varieties of pea, Galton studying the size of the peas of the sweet pea, a continuous variable, and Mendel studying seven discrete characters in the edible pea, such as 'tall' versus 'dwarf'. Galton engaged the service of seven friends to grow sweet peas from seeds which he had separated into seven groups by size. From the results he discovered that the variability (or 'variance', as we would now say) in each group was sensibly the same, that the distributions were normal, and that the means of the groups were linearly related to the parental seed sizes, but with a slope of about one-third: the offspring means were not the same as the parental means (which would have meant a slope of unity), but were only one-third of the way from the overall mean towards the relevant parental mean. They regressed to that mean, and thus was born the statistical technique of 'regression'. The slope of the regression line of the relation is called the regression coefficient. In 1885 Galton achieved a full understanding of the true nature of the phenomenon of regression by developing the idea of a bivariate normal distribution for two associated variables.

Then, in 1889, came the fundamental notion of correlation. Galton realized that if the two associated variables were scaled so as to have the same standard deviation (as we would now say), then the two regression lines in a bivariate diagram (of x on y and y on x) would have the same slope, each with respect to its proper axis, which was therefore suitable as a dimensionless index of association, the correlation coefficient. Such a coefficient may be used to measure the intensity of the association between variables which are correlated though one does not actually cause the other, as with the heights of pairs of brothers, for example.

Galton also, with H. W. Watson, originated the mathematical theory of branching processes when he studied the probability of the extinction of families. This subject in the theory of probability discusses the properties of populations (of people or of things) whose members divide or give birth to new members, thus creating mathematical structures analogous to the branching of a tree.

Towards the end of his life Galton provided funds for the establishment of the Galton Professorship of Eugenics (now of Human Genetics) at University College, London. It was held in succession by Pearson and Fisher (§10.7).

4 THE TWENTIETH CENTURY:
THE MENDELIAN INHERITANCE

Mendel, like Pearson and Fisher after him, had received a training which included mathematics and physics, a fact which prepared his mind for the discovery of the genetical ratios which bear his name. Mendel's discovery led indirectly to many statistical developments because it provided a specific probability model for the central process of genetics, a process which was gradually to dominate all of biology from the beginning of the twentieth century onwards. Estimation theory, in particular, was developed largely in connection with genetic linkage, the first application of Fisher's method of maximum likelihood being to the estimation of a genetic recombination fraction.

In the design of experiments, Fisher's concept of factorial analysis (in which several treatments are tested simultaneously instead of one at a time, as had been the custom) was inspired by the idea of Mendel's factors. But the most influential developments arose out of the conflict between the 'biometricians', whose view of the inheritance of continuous characters derived from Galton's ideas and whose vigorous champion was Pearson, and the 'Mendelians', led by William Bateson of Cambridge, who sought to explain all inheritance in terms of Mendelian factors (Norton *1975*). The issue was resolved by Fisher's 1918 paper 'The correlation between relatives on the supposition of Mendelian inheritance', in which he demonstrated that a Mendelian genetic model with large numbers of genes could account for the familial correlations studied by the biometricians. In statistical theory this paper continued where Galton had left off by introducing the idea of (and the word) variance. Previously, variation had usually been described in terms of the probable error or standard deviation of a statistical distribution, which were linear measures; but by taking the square of the standard deviation and calling it the variance Fisher created a measure of variability which could be additively partitioned, thus enabling particular components of variability to be identified and measured through the 'analysis of variance'.

Pearson himself left a smaller mark on statistics and genetics than the bulk of his writings and his dominant position in Britain as the first Galton Professor might lead one to expect. His extensive work on non-normal statistical distributions, the method of moments for distribution-fitting, correlation and the mathematical theory of evolution was soon eclipsed by the analysis of variance, the theory of statistical estimation, small-sample distribution theory and the evolutionary studies of Fisher, Sewall Wright and J. B. S. Haldane. Pearson's most lasting contribution to statistical theory was his invention of the chi-squared goodness-of-fit test in 1900.

This significance test solved the problem of assessing whether a theoretical distribution adequately 'fitted' the data to which it was applied. The test made research workers familiar with the concept of a test of significance, and popularized the notion of a significance level. However, it was applicable only to large samples.

The foundation of small-sample distribution theory was the paper by W. S. Gossett in 1908 introducing the *t*-distribution. Gossett, who wrote under the pseudonym 'Student', was a biological scientist working at the Guinness Brewery in Dublin, and his work once again exemplifies the close involvement of biology in the development of statistical methods. The same is true of Wright's invention of path coefficients, which were introduced in connection with inbreeding in mammals but which have found subsequently many other uses in the analysis of multiple causes. Path coefficients are generalizations of regression coefficients which make allowance for the fact that, when a dependent variable is influenced by several independent variables at once, the regression on each needs to be measured while holding the remaining variables fixed.

In 1922 Fisher introduced a model for the variation in gene frequency from generation to generation in a finite population. It is regarded as the origin of stochastic diffusion theory, a branch of probability theory which represents probability changes by differential equations (§9.18).

5 STATISTICAL INFERENCE

Tests of significance, originating in the eighteenth century, acquired their modern dominance through the influence of Pearson and, particularly, Fisher, who with F. Yates published a famous set of statistical tables under the title *Statistical Tables for Biological, Agricultural, and Medical Research* (1938). Estimation theory, originally limited to means and the method of least squares, developed rapidly through Fisher's work in the 1920s especially, the inspiration being nearly always agricultural or genetical. Although the method of inverse probability associated principally with Thomas Bayes and Laplace had little initial connection with biology, its renaissance in the middle of the twentieth century owed something to its use in connection with genetic-linkage estimation.

The concept of likelihood, introduced by Fisher in 1921, and the associated likelihood principle, continue to play an important part in any modern discussion of statistical inference. The likelihood for the parameter of a probability model (such as the mean of a normal distribution, or a genetic recombination fraction) is defined as being proportional to the probability of the data. As a function of the parameter, it will usually possess a single maximum, giving the maximum-likelihood value of the parameter, while the whole function has a role to play both in Bayesian statistical theory and in the related likelihood theory of statistical inference.

6 CONCLUSION

From the time of Galton, biology, and especially genetics, took over from the physical sciences the role of the anvil on which statistical developments were forged. Although some of the associated mathematical advances were made by mathematicians whose advice was sought (such as Watson, whom Galton consulted, and J. D. Hamilton Dickson, to whom he turned for the mathematical form of the bivariate normal distribution), most of the important later mathematical theory was developed by the pioneers themselves, especially Fisher. Only with the work of Jerzy Neyman and Egon Pearson (Karl Pearson's son) in the 1920s and 1930s did statistics come to be seen by mathematicians as an important part of their own domain, leading to technical developments of great rigour and complexity.

The explanation for the fact that the development of probability and statistics in genetics between 1900 and 1930 took place almost wholly in England is that Karl Pearson's activities were at their peak when Mendel's work was rediscovered in 1900, plunging the English biometric scene into a vigorous controversy out of which Fisher's major advances developed with astonishing rapidity (Box *1978*). Several of these arose in the context of agriculture, as we shall now see.

BIBLIOGRAPHY

Box, J. F. *1978*, *R. A. Fisher: The Life of a Scientist*, New York: Wiley.

Gigerenzer, G. *et al.* *1989*, *The Empire of Chance. How Probability Changed Science and Everyday Life*, Cambridge: Cambridge University Press.

Hald, A. *1990*, *A History of Probability and Statistics and Their Applications Before 1750*, New York: Wiley.

Keynes, W. M. (ed.) *1993*, *Sir Francis Galton: The Legacy of His Ideas*, London: Macmillan. [See articles by A. W. F. Edwards and J. Edwards.]

MacKenzie, D. A. *1981*, *Statistics in Britain 1865–1930: The Social Construction of Scientific Knowledge*, Edinburgh: Edinburgh University Press.

Norton, B. J. *1975*, 'Metaphysics and population genetics: Karl Pearson and the background to Fisher's multi-factorial theory of inheritance', *Annals of Science*, **32**, 537–53.

Porter, T. M. *1986*, *The Rise of Statistical Thinking, 1820–1900*, Princeton, NJ: Princeton University Press.

Provine, W. B. *1971*, *The Origins of Theoretical Population Genetics*, Chicago, IL: University of Chicago Press.

Stigler, S. M. *1986*, *The History of Statistics: The Measurement of Uncertainty Before 1900*, Cambridge, MA: Harvard University Press.

10.11

Probability and statistics in agronomy

ZENO G. SWIJTINK

1 INTRODUCTION

Probability and statistical thinking have developed within biology in two areas. In genetics, the Mendelian laws were probabilistic laws and demanded a sophisticated theory of statistical inference for their elaboration and confirmation. In evolutionary theory, Charles Darwin, in his theory of evolution through natural selection, invoked chance as a cause of variation, while natural selection itself operated deterministically. Although Darwin's notion of chance was not influenced by probability theory or statistics, and although there were no probabilistic laws in his theory of evolution, further developments in the 1920s led to a stochastic theory of evolution, a synthesis of Mendelism and Darwinism. This theory not only specified a probabilistic mechanism for variation, based on Mendelian principles and without recourse to 'mutations', but also allowed for random drift and non-deterministic selection.

This article considers the relation between the growth of statistical and probabilistic thinking and agronomy, the scientific approach to agriculture. This relation has been remarkably close, but quite different from that with biology (§10.14). Probability was used not in an analysis of agricultural processes, but as a tool in the design and analysis of comparative experiments. Indeed, one of the founders of modern mathematical statistics, Ronald Aylmer Fisher, started his career as a statistician at an agricultural research institution in England. This has made agronomy the cradle of many of the experimental designs that are now adopted in so many other sciences, from psychology to medicine (§10.10, §10.13).

2 RATIONAL AGRICULTURE

Agriculture is one of the sciences allegorically represented on the frontispiece of the *Encyclopédie ou Dictionnaire raisonné des sciences, des arts et des métiers* (1751–1780) of Denis Diderot and Jean d'Alembert; she is depicted below Astronomy and Physics and alongside Optics and Chemistry. The association with chemistry became even more important with the rise of agricultural chemistry in the nineteenth century, when the discussion on plant nutrition and Justus von Liebig's mineral theory was put on a firm basis in the fields of Europe. Agronomists learned to value quantification and precision from their co-workers in chemistry.

One influential representative of the rationalization of agriculture was Albrecht Thaer (1752–1828), the author of *Grundsätze der rationellen Landwirtschaft* (1809–12). A gentleman by background and a physician by training, Thaer stood for the Enlightenment and rationalism within the German-speaking farming community. His book, written as a textbook, was based on his lectures at an agricultural institute he had founded in 1806 in Möglin near Berlin. This was the first such school, an idea that was later adopted elsewhere in Germany, Austria, Russia, France and Scandinavia.

Thaer saw farming as a 'business with the goal to maximize profit through the production of vegetable and animal substances'. He advocated experimentation to discover the determining factors of crop growth. These experiments had to be comparative, *ceteris paribus* trials, in which plots which were otherwise equal, were, for instance, dressed with different fertilizers. Ideally, agronomy would provide the farmer with a complete input–output analysis of the processes involved in, say, crop growth, and so help him solve his optimization problem. But since we cannot control, or even know, all the contributing factors, the best we can do is to improve upon existing practice by comparing various treatments or varieties side by side.

However, during the first half of the nineteenth century it became clear that it was practically impossible to perform perfect *ceteris paribus* experiments in which the other contributing factors are held constant from plot to plot. Plots with the same dressing would still show a variation in yield, no matter what effort was made to make them otherwise completely similar. Thus, the practice arose of determining the 'variation in natural productiveness of the land' by leaving two or three plots without fertilization, a replication of the no-treatment. The yield on a manured plot would only then indicate the beneficial effect of the fertilizer if it exceeded the range of natural productiveness. Sometimes, as in James F. W. Johnston's *Experimental Agriculture, Being the Results of Past, and Suggestions for Future Experiments in Scientific and Practical Agriculture* (1849), it was urged that each dressing be replicated, and that the means should determine which dressings had the more beneficial effects. But Johnston did not know how to use these mean values: 'As yet we do not possess any such system of mean results, though few things would at present do more to clear up our ideas as to the precise influence of this or that substance on the growth of plants.' This problem of Johnston, how to reason with means and variation, is the problem that agronomy confronted in the second half of the nineteenth century.

3 GERMAN DEVELOPMENTS

Under the influence of gentleman-farmers like Thaer and of the chemist von Liebig, who in 1840 published his immensely influential *Die organische Chemie und ihre Anwendung auf Agrikultur und Physiologie* ('Organic Chemistry and Its Applications to Agriculture and Physiology'), experimental farms with an in-house chemical laboratory were established in almost all countries in Europe. In Germany, by 1895 there were more than fifty of these stations, most of them state-funded. The German stations became the model for the American agricultural experimental stations established in the 1870s and 1880s. At the Versammlung der Deutschen Naturforscher (the German equivalent of the much younger British and American Associations for the Advancement of Science) in Wiesbaden in 1873, the chemist Adolf Mayer pleaded for national agricultural experiments to be conducted all over Germany. Their results should be assessed by averaging, an idea he later brought to The Netherlands when he took up the directorship of its national agronomical research station. In Saxony, this idea had already been put into practice by C. Maercker, a researcher with ties to the spirit industry, who in the early 1870s organized a number of experiments in which he compared the effects of different kinds of fertilizers on potatoes. With these large-scale experiments, Mayer hoped to understand better how far the results of local trials may be generalized. Although they praised Mayer's objectives, most agronomists at the time rejected his 'statistical method', the calculation of nationwide averages. Local circumstances differed too much for these averages to be meaningful, and these critics rejected an appeal to the 'power of large numbers', following the same line of argument as used by the French physiologist Claude Bernard earlier in the century, in his criticism of the so-called 'numerical method' employed by the Paris physician Pierre Louis (§10.12).

In his trials in the 1870s Maercker had replicated – as had become customary – the plots without any dressing, calculated the average, and determined the maximum deviation from the average. Only to the extent that a dressing gives yields beyond those limits would there be a positive effect. In the early 1880s, Paul Wagner, a critic of the statistical method, argued for a programme to make comparative, controlled experimentation more exact. First, each treatment had to be replicated; furthermore, the number of replications should be increased to six and be spread out evenly over the field. His attitude towards quantification and precision was explicitly derived from chemistry, in his strident exclamation that 'a method without known margins of error cannot be called exact', and in his goal to reduce the margin of error in agricultural experiments to 1%, the maximum error then permitted in chemical assays. Wagner came to his recommendation of six replications by studying so-called uniformity trials, in which a larger field is divided into many plots receiving the same treatment; only the average yield over six plots would differ by less than 1% from the grand average (Roemer *1930*).

In 1890, Georg Liebscher was asked to analyse the data obtained in the many trials the Deutsche Landwirtschaft Gesellschaft had made over previous years. He proposed a new way of calculating effects for such large-scale experiments consisting of many local trials spread over the country. The method of least squares (i.e. error theory) was inapplicable, according to Liebscher, because of the many missing data points. Instead of following Mayer in comparing average yields for each variety over all the trials, Liebscher tried to determine an error factor for the yield of each variety at each trial, by a sequence of averaging and subtracting. Lacking in all this was a coherent theoretical structure within which the different proposals could be compared and evaluated. To provide such a structure was possibly the most important contribution of error theory to the analysis of comparative experiments.

Liebscher's method was still in use when Hermann Rodewald, an agronomist at the University of Kiel, introduced in about 1905 the techniques of error theory into the analysis of comparative, controlled experiments. Rodewald was familiar with the calculations of error theory from his 1889 work on the determination of the purity of clover seed, an apparently new application that had obtained the stamp of approval of an astronomer, Paul Harzer, the director of the observatory in Kiel at the time. Similarly, in England, an agronomist (T. B. Wood) and an astronomer (F. J. M. Stratton) joined forces in 1910 to urge the use of error theory, which had been so useful in astronomy during the nineteenth century, in the interpretation of experimental results in agriculture. Studying uniformity trials, Wood and Stratton found sufficient agreement between observed plot-yields and the yields that were predicted by a fitted normal distribution. Thus, if chemistry gave agronomists the desire for precision, it was astronomy that set agronomists on the trail of error theory.

Not everyone, however, was convinced that the presuppositions of error theory (numerous independent observations from a Gaussian or normal distribution) were, even approximately, satisfied in agriculture. Some argued that error theory applied only to random, not systematic errors, the kind of errors to which agricultural trials were particularly prone. But what these concepts could mean in the context of variable material was never cleared up. The influential probabilist Emanuel Czuber, in a longish paper of 1918 in the *Zeitschrift für das landwirtschaftliche Versuchswesen in Österreich*, concluded that agriculture should make use instead of the methods of Wilhelm Lexis's *Kollektivmasslehre*, a non-parametric approach that made fewer distributional assumptions. One of his concerns was the small samples with which the agronomist had to work, which made the customary use of error theory less suitable. However, these objections were met by new developments in Ireland and Great Britain.

4 THE BRITISH BREAKTHROUGH

Agricultural experimentation was less advanced in Ireland and Great Britain. Still, it was here that the great leap forward in the design and analysis of comparative experiments in agriculture was made. How large an influence the Continental developments had on this advance is still a matter of speculation, but it is clear that the breakthrough was made possible by the higher level of statistical methodology in Great Britain, due to the contributions of Francis Galton, Karl Pearson and Francis Edgeworth (§10.13).

In the 1910s, one of the very few agricultural research stations in England was Rothamsted Experimental Station, established in 1837 as the research arm of a commercial fertilizer producer. In 1919 it hired Cambridge graduate Ronald Aylmer Fisher to head a new department of statistical analysis. At Cambridge, Fisher had been trained in mathematics and the physical sciences, and had studied the theory of errors under Stratton. Rothamsted was a large operation at the time, with departments of soil chemistry, soil physics, bacteriology, microbiology, entomology, insecticides, botany and plant pathology. During his tenure there, until 1934, Fisher worked on problems in almost all these fields, inventing new methods of statistical analysis as he went along. He was one of the first statistical consultants, and inspired many statisticians to perceive themselves as scientific generalists.

Fisher's pioneering contributions to probability and statistics ranged from conceptual papers on the nature of probability and statistical inference to work on blood-group systems. The two that concern us here, since they derived from his work on agricultural experimentation, dealt with exact sampling distributions and the theory of experimental design and analysis. The work on exact sampling distributions would answer an objection like Czuber's, mentioned above, while Johnston's plea for a 'method of means' is answered by Fisher's comprehensive theory of experimental design (Box *1980*).

Czuber's objection had been that the samples usually obtained in agricultural experiments were not large enough for the techniques of error theory to apply. Since the variance in an agricultural experiment would generally be unknown (each experiment having its own variance), it had to be estimated from the data themselves before the difference in means between various treatments could be tested on significance. But for a small sample this estimate would itself be subject to considerable variation, and the calculated significance level would be inexact. In 1908 this led W. S. Gosset, a chemist working at the Guinness Brewery in Dublin, to consider another approach by which this variation would be explicitly considered in the test of significance. Gosset, who had studied error theory under George Biddell Airy and taken classes with Karl Pearson at University College, London, discovered the distribution of the ratio of the sample mean and the sample standard deviation, the statistic of his famous *t*-test (§10.11). In 1912, while still at Cambridge, Fisher had given a mathematically rigorous derivation of Gosset's distribution.

In the years that followed, Fisher would publish the exact sampling distributions of a large number of other statistics important in significance testing. One of these distributions, of the difference of two means of normal distributions with possibly unequal variance, was independently discovered by Walter Behrens (1902–62), a German chemist who was head of the laboratory for agricultural chemistry at the company Kali-Chemie. Behrens worked under E. A. Mitscherlich, who did his dissertation under Rodewald and was one of the driving forces behind the use of error theory in German agriculture in the 1910s.

Experimental design and the analysis of experimental data are intimately connected, as Rodewald recognized in 1909, and Fisher made it a cornerstone of his theory of experimental design. If one knows what questions can be answered from what data, one can set up the experiment so that it is likely to produce the data that will answer one's questions, one way or another. Fisher described his theory in his *1926* paper 'The arrangement of field experiments'. The theory has three basic ideas, not all new from Fisher but developed by him into a coherent whole: replication, blocking and randomization.

Replication was already a well-known aspect of comparative experimentation in the fertilizer experiments of the nineteenth century, first only of the no-treatment, later of all treatments. It reduces the error while the variation among plots treated alike is the basis for an estimate of the error. In blocking, only yields from the same block are directly compared. This was the motivation behind Liebscher's averages and subtractions, and it allows measurement and removal of variation among blocks from experimental error. Randomization – the allocation of treatments over plots by use of a physical random device – is Fisher's original and still controversial contribution to experimental design, although the idea had been used before in experiments to test the telepathic powers of human subjects (Hacking *1988*). For Fisher, only randomization can guarantee that the estimate of error is valid and unbiased. A number of statisticians had objected that when the allocation is determined and has become known, testing should be conditional on the known allocation pattern, and how the pattern was obtained has become irrelevant. Some agronomists believed that systematic designs would be more precise since they can balance any known fertility trend on the field. But for Fisher, who stressed the importance of communication between research workers, without randomization there could be no valid estimate of error, a notion critically important in reporting experimental results. The ideal of precision in estimation, imbued by the chemists, clashed with this new scientific value, derived from the astronomical tradition of responsible discussion of observations.

The analysis of variance is the statistical technique developed by Fisher to test data from randomized-block experiments for significance. The method depends on the partition of both the degrees of freedom and the sum of squared deviations between a component called 'error' and another that may be called 'effect'. In the absence of the effect, on the null hypothesis, the expectation of the two sums will be in the ratio of their respective degrees of freedom. If there are real treatment effects, then the ratio of treatment mean square and of error mean square will be more than one, and tables will show the probability of so large or larger a value arising solely by chance.

Fisher's ideas on experimental design and the analysis of variance were given prominence in his book *The Design of Experiment* (1936), and were further elaborated upon in England and the USA. They have radically changed experimental practice in a number of sciences that work with variable material, especially medicine and psychology.

BIBLIOGRAPHY

Bennett, J. H. *1990, Statistical Inference and Analysis. Selected Correspondence of R. A. Fisher*, Oxford: Clarendon Press.

Box, J. F. *1980*, 'R. A. Fisher and the design of experiments, 1922–1926', *The American Statistician*, **34**, 1–7.

Fisher, R. A. *1926*, 'The arrangements of field experiments', *Journal of the Ministry of Agriculture of Great Britain*, **33**, 503–13.

Hacking, I. *1988*, 'Telepathy: Origins of randomization in experimental design', *Isis*, **79**, 427–51.

Roemer, T. *1930, Der Feldversuch. Eine kritische Studie auf naturwissenschaftlich- mathematischer Grundlage*, 3rd edn, Berlin: Deutsche Landwirtschafts- gesellschaft.

10.12

Probabilistic and statistical methods in medicine

J. ROSSER MATTHEWS

One of the most prominent contemporary uses for probabilistic and statistical methods in medicine is the clinical trial, for determining the effectiveness of prophylactic or therapeutic agents. The trial has certain standard features such as comparison between an experimental and a control group. Although the use of comparison between therapies to determine their relative efficacy had long been a part of medical practice, the analysis of these comparisons based on techniques derived from mathematical statistics had to await the professionalization of statistics in the twentieth century, as discussed in Section 3. Nevertheless, the mathematically trained had been offering applications of probabilistic methods to this problem almost from the inception of probability theory in the seventeenth century. This article describes these discussions of probability and statistics in medicine from the eighteenth through to the mid-twentieth century, and finishes with an account of the modern clinical trial (on which see Matthews *1992*).

1 EARLY USES OF COMPARATIVE STATISTICS IN MEDICINE

Since smallpox had taken the lives of over 10% of the populations of London and Paris during the seventeenth century, Lady Mary Wortley Montagu's introduction of the practice of inoculation from Constantinople (where, as wife of the British Ambassador, she had observed it) received considerable attention between 1718 and 1722, particularly among members of The Royal Society of London. Inoculation involved impregnating the skin with live smallpox pustules to immunize the individual against the disease; the problem this procedure presented was that some contracted the disease and died. Both the American clergyman Cotton Mather and the secretary of The Royal Society, James Jurin, argued in favour of the procedure on the basis of direct comparisons that showed a lower mortality among those who had been inoculated.

Following Voltaire's favourable observations on these experiments with inoculation, Continental mathematicians provided a more sophisticated analysis of the subject in the middle decades of the century. A notable exchange took place between the Swiss mathematician Daniel Bernoulli and the French *philosophe* Jean d'Alembert. In 1760, Bernoulli, then 60 years old, read an analysis of smallpox mortality to the Paris Academy of Sciences. He derived an expression for the number of people likely to succumb to smallpox in a given time period, and computed the average gain in life expectancy from inoculation for any given age. D'Alembert, Bernoulli's junior by 17 years, criticized this analysis in a paper subsequently published in his *Opuscules mathématiques* (1761); he argued that the psychological processes determining whether an individual decided to receive inoculation might lead that individual to arrive at a decision different from that required by Bernoulli's calculations (Daston *1988*: 82–9). The question was shown to be irrelevant by the English physician Edward Jenner, who in his *An Inquiry into the Causes and Effects of the Variolae Vaccineae* (1798) demonstrated that vaccination with cowpox offered protection from smallpox.

Both the mathematical theory of probability and clinical medicine made advances in late-eighteenth- and early-nineteenth-century France. Mathematicians such as the Marquis de Condorcet and Pierre Simon Laplace argued for the use of the calculus of probability as a tool in all areas of rational decision-making, including clinical medicine. In his popular *Essai philosophique sur les probabilités* (1814), the 65-year-old Laplace declared:

> It is sufficient to test each of them [methods of treatment] on an equal number of patients making all the conditions exactly similar; the superiority of the most advantageous treatment will manifest itself more and more in the measure that the number is increased.

Following the transformations of the French Revolution, medicine became centred in the study of pathological anatomy as revealed through physical examination, autopsy and the use of comparative statistics (Hacking *1990*: Chaps 8–12). The leading figure in this attempt to base clinical medicine on statistical comparison as recommended by Laplace was Pierre Louis, who advocated what he called the 'numerical method'. In work published in the 1820s on phthisis and typhoid fever, he studied each patient as thoroughly as possible at the bedside and at autopsy, collected his observations, grouped them in tabular form, and drew inferences about the value of a given therapy or the relations between clinical phenomena. He articulated his method most fully in *Recherches sur les effets de la saignée* (1835), in which he questioned the efficacy of the then common therapeuic procedure of bloodletting in pneumonia and other diseases. Such a statistical approach to medical therapy was criticized by the Montpellier-trained clinicians François Double and Risueño d'Amador. They emphasized a Hippocratic view of diagnosis and therapy according to which the physician learned from experience how to judge the idiosyncrasies of the individual patient rather than by forming a kind of statistical composite. In a paper read to the Paris Academy of Medicine and subsequently titled 'Mémoire sur le calcul des probabilités appliqué à la médecine' (1837), the 35-year-old d'Amador declared that 'good sense' was not calculable in the same manner as 'all that pertains to moral and intellectual life and affects human beings' (Murphy *1981*).

Louis's results were also criticized, but from a mathematical standpoint, by the 31-year-old Jules Gavarret, a former student at the Ecole Polytechnique who had changed careers to become a physician. In his *Principes généraux de statistique médicale* (1840), he utilized a formula derived by the mathematician Siméon-Denis Poisson which gave the margin of probable error for a statistical average *m* out of *u*:

$$\frac{m}{u} \pm 2\left(\frac{2mn}{u^3}\right)^{1/2}, \quad m+n=u, \tag{1}$$

assuming a confidence level of $p = 0.9953$. Gavarret noted that Louis had observed 140 cases of typhoid fever, of which 52 were deaths and 88 recoveries, and had merely taken the average to be 37% without considering that the results would vary between 26% and 49% for 140 cases observed.

2 THE RISE OF PHYSIOLOGY AND THE REJECTION OF STATISTICAL INDETERMINISM

The middle decades of the nineteenth century saw a shift in focus from observation in the clinic to experimental laboratory investigation of physiological processes as the key to providing medicine with a scientific foundation. This shift provided an alternative critique of statistical methods; they were rejected for offering probabilistic results as opposed to the deterministic certainty required by experimental science. One leader of this movement was the German clinician Carl August Wunderlich; in 1842, at the age of 27, he helped found the *Archiv für physiologische Heilkunde*, which he prefaced by declaring: 'The time has come when one must seek to establish... a positive science. This science will seek its roots not in the authorities but in the foundations and empirical evidence that the phenomena permit us to comprehend.' Statistical results were not science, but merely 'a material that must be examined' in order to arrive at scientific laws. The classic exposition of this view came from the French physiologist Claude Bernard in his *Introduction à l'étude de la médecine expérimentale* (1865), in which he declared that 'scientific law can be based only on certainty, on absolute determinism, not on probability'. In positivist fashion, medicine had to adopt the method of scientific experimentation if it was to advance beyond the empirical stage of 'a conjectural science based on statistics' (Schiller *1963*).

3 THE RISE OF MATHEMATICAL STATISTICS AND THE INSTITUTIONALIZATION OF THE MODERN CLINICAL TRIAL

The last third of the nineteenth century witnessed innovations in both medicine and mathematical statistics. Medically, the pioneering work of Robert Koch and Louis Pasteur helped to establish the germ theory of disease, ushering in bacteriology and the related field of immunology as important areas of research. Mathematically, Florence Nightingale's pioneering studies of the 1850s on mortality statistics, including her introduction of the polar-area diagram, led eventually to improvements in ward hygiene and hospital design in the British Empire. Furthermore, the biometrical tradition provided theoretical underpinnings for modern statistical methods (§10.7). These twin developments intersected in the first decade of the twentieth century, following studies conducted by the British bacteriologist Almroth Wright.

Wright advocated the continuing use of an anti-typhoid serum in the British Army on the basis of reduced mortality among immunized soldiers during the Boer War. However, in 1904, the biometrician Karl Pearson published an analysis in the *British Medical Journal* arguing that the procedure be discontinued because the degree of statistical correlation, as measured mathematically by a correlation coefficient, was lower than for other commonly adopted therapies like smallpox vaccination. Wright responded by professing ignorance of the techniques of 'coefficient correlation', and declared that Pearson required an unreasonable 'standard of perfection'. Such a lack of knowledge of statistical methods soon brought Wright into conflict with Pearson's student Major Greenwood over the effectiveness of the opsonic index for diagnostic purposes, a measure of a blood-serum substance which prepared bacteria to be ingested by white blood cells. In a 1908 article in the journal *The Practitioner*, Greenwood, then 28, determined the mean opsonic index and plotted the frequency distribution for a series of measurements, finding them to be markedly asymmetrical or 'skew'. These debates had important institutional consequences in that they led to Greenwood's appointment as medical statistician at the

Lister Institute for preventive medicine, the first such post in Great Britain. For further details on Wright's view of statistics, see Cope *1966*: 55–7. The modern institutionalization of the mathematically based clinical trial took place in Great Britain after the 1913 creation of the Medical Research Committee, which became the Medical Research Council in 1919. It included a medical statistics department headed initially by John Brownlee, a Pearson follower. In 1931, the Therapeutic Trials Committee was created which had access to expert opinion from the statistical committee, then supervised by Greenwood. On Greenwood's retirement in 1945, the statistical committee became the Statistical Research Unit, with Austin Bradford Hill as director. In 1946, Hill designed a trial to measure the effect of streptomycin in respiratory tuberculosis, one of the first trials to use the principle of the random allocation of patients to an experimental or control group as a technique in experimental design (Austoker and Bryder *1989*: 35–57).

BIBLIOGRAPHY

Armitage, P. *1983*, 'Trials and errors: The emergence of clinical statistics', *Journal of the Royal Statistical Society*, Series A, **146**, 321–34. [Discusses P. Pinel, Louis and Bernard.]

Austoker, J. and Bryder, L. *1989*, *Historical Perspectives on the Role of the MRC*, Oxford: Oxford University Press.

Cope, Z. *1966*, *Almroth Wright: Founder of Modern Vaccine-Therapy*, London: Nelson.

Daston, L. *1988*, *Classical Probability in the Enlightenment*, Princeton, NJ: Princeton University Press.

Greenwood, M. *1941–3*, 'Medical statistics from Graunt to Farr', *Biometrika*, **32**, 101–27, 203–25; **33**, 1–24. [Also in E. S. Pearson and M. G. Kendall (eds), 1970, *Studies in the History of Statistics and Probability*, Vol. 1, London: Griffin, 47–120.]

Hacking, I. *1990*, *The Taming of Chance*, Cambridge: Cambridge University Press.

Lilienfeld, A. *1982*, '*Ceteris paribus*: The evolution of the clinical trial', *Bulletin of the History of Medicine*, **56**, 1–18. [Discusses various features of clinical trials from the eighteenth century.]

Llewelyn, D. and Anderson, J. *1980*, 'The historical development of the concepts of diagnosis and prognosis and their relationship to probabilistic inference', *Medical Informatics*, **5**, 267–80. [Surveys developments from ancient Greece to set theory and computers.]

Matthews, J. R. *1992*, 'Mathematics and the quest for medical certainty: The emergence of the clinical trial, 1800–1950', Ph.D. Dissertation, Duke University, Durham, NC.

Murphy, T. *1981*, 'Medical knowledge and statistical methods in early nineteenth-century France', *Medical History*, **25**, 301–19.

Rosen, G. *1955*, 'Problems in the application of statistical analysis to questions of health: 1700–1880', *Bulletin of the History of Medicine*, **29**, 27–45. [Includes discussion of public health.]

Schiller, J. *1963*, 'Claude Bernard et la statistique', *Archives Internationales d'Histoire des Sciences*, **16**, 405–18. [Includes other nineteenth-century figures for context.]

Sheynin, O. *1982*, 'On the history of medical statistics', *Archive for History of Exact Sciences*, **26**, 241–86. [Survey from a mathematical standpoint.]

Shryock, R. *1961*, 'The history of quantification in medical science', *Isis*, **52**, 215–37.

—— *1979*, *The Development of Modern Medicine: An Interpretation of the Social and Scientific Factors Involved*, Madison, WI: University of Wisconsin Press. [Classic in medical history; see pp. 135–150.]

10.13

Probability and statistics in mechanics

ZENO G. SWIJTINK

Many concepts and techniques in probability theory and statistical inference have evolved in branches of mechanics, especially corporeal, celestial, planetary and engineering mechanics (including hydraulics), and in optics. Moreover, the use of probability theory and statistical techniques has led to important changes in the way observations are collected in these and related fields. This article focuses on the inferential use of probabilistic ideas, for the most part in error theory as a foundation for least squares, and on the broader influence on scientific practice. The use of probability theory in modelling mass phenomena, as in statistical mechanics, is covered in §9.14.

1 COMBINATION OF OBSERVATIONS BEFORE LEAST SQUARES

Repeated measurement of the same quantity usually gives a spread of values, and the data have to be combined in some way to get an estimate of the quantity measured. Before the 1750s, observations were combined in a variety of ways that left much choice to the calculator. A systematic theory of errors did not yet exist. By 1750 probability theory had been recognized as a mathematical discipline with its own problems and methods, as described by Abraham De Moivre in the *Doctrine of Chances* (1718) (§10.2). Thus, it was only in the 1750s that there can be said to have arisen a mathematical theory of errors and estimation in which the concept of mathematical probability provided a unifying element.

In contrast with this probabilistic approach to error theory is Galileo Galilei's discussion of a series of conflicting estimates of the location of the nova of 1572, early in the Third Day of his *Dialogue Concerning the Two Chief World Systems* (1632) (Hald *1990*). He held the view that the nova was located beyond the Moon, and thus showed the mutability of the heavens. Four years earlier, Scipione Chiaramonti, a professor of philosophy at the University of Pisa, had argued that the nova was sublunar. The immutability of the heavens was a serious problem, with philosophical and theological implications. Speaking through the mouth of Salviati, Galileo made some general statements about observational errors, such as that observers are 'equally ready to err in one direction than another', and that 'they are sooner to err little than much'. (Later, in the probabilistic treatment of observational error, these properties of errors would be encapsulated in the common assumption that an error distributed is unimodal and symmetric with mean zero.) Chiaramonti had based his case for a sublunary location on estimates from a dozen or so astronomers, and Galileo proposed a two-step approach to the combination of these estimates. First, he ranked the observations by stating that 'those ... must be called the more exact, or the less in error, which by the addition or subtraction of the fewest minutes restore the star to a possible position': an impossible position for a star would place it, for instance, at an infinite distance. The second step resembles a courtroom procedure in which credence is given to the opinion expressed by the largest group of concurring witnesses: 'among the possible places, the actual place must be believed to be that in which there concur the greatest number of distances, calculated on the most exact observations'.

Galileo used the term 'probability' only when he argued 'how much more clearly and with how much greater probability it is implied that the distance of the star placed it among the most remote heavens'. He made a probabilistic argument, not a demonstrative one: the conclusion was called 'probable', not the data.

In fact, the discussion about errors in the early part of the Third Day was mostly concerned with issues in determinate error theory, and intended to stress that 'at great distances a change of a very few minutes [of arc] will move a star through an immense distance', which was not true for sublunar altitudes. Determinate error theory is part of the theory of experimental design, and shows how to secure the most reliable experimental results through specially designed observational programmes, for example how to choose a set of triangles most advantageous for a derived arc measurement. Its mathematical tool was the calculus of difference equations, as in Roger Cotes's posthumous *Aestimatio errorum* (1722) (§4.3). Cotes, who had been Isaac Newton's assistant, used the term 'probable' in a sense similar to Galileo's when he proposed, as the most probable true position, a weighted average, with the weights determined by the errors to which the respective observations are due (Harter *1974*).

In two papers written in the 1750s the English mathematician Thomas Simpson made a conceptually pivotal step in a defence of the practice of taking the arithmetic mean of a series of observations. Using De Moivre's mathematics, he introduced the notion of 'a series for the chances of the happening of the different errors', or, as we would now say, the notion of the distribution of errors to which a single observation is due. By concentrating on the errors instead of the observations themselves, Simpson could assume that the error distribution was known, even if the quantity to be measured was not. This was mathematically important. Moreover, this proof assumed that the observations 'admit of being repeated under the same circumstances'; that is, that the observations from which a mean was calculated were subject to the same distribution of errors. Instead of the errors that may attach to the individual measurements, the procedure of observation and the circumstances under which the observations were collected became central. This conceptual shift would have radical consequences for scientific practice, as we shall see.

The Alsatian natural philosopher Johann Heinrich Lambert published, as part of his *Photometria* (1760), the first systematic discussion of the 'theory of errors', as he called it. Five years later he became a member of the Berlin Academy of Sciences, and in his *Beyträge zum Gebrauche der Mathematik* of that year he followed up the ideas of the Vienna court mathematician and military architect Giovanni Marinoni. He combined practical advice from Marinoni's *De re ichnographica* ('On Making Ground Plots') of 1751 with his interest in photometry to analyse the possible errors that arise from the limited resolution of the human eye and the construction of the instrument used. This led him to speculate about a law that describes the relative possibility of errors of different sizes. He was unable to solve the equation analytically, and thus proceeded empirically.

Lambert set off a line segment 20 times along a straight line, and repeated this 80 times, always starting at the same point. This led to a distribution of end-points, which he summarized in a frequency per interval of the same length. Thus he confirmed some qualitative conclusions about errors, for example that smaller errors occur more frequently, and are easier to make. He stressed that the error pattern of any instrument used in a measurement should be studied to find the instrument's reliability, a study which sixty years later would be brought to great perfection by the astronomer and geodesist Friedrich Wilhelm Bessel. Such an investigation, Lambert stated, will also make it possible to determine the reliability of a mean of several measurements of, say, the same angle. To do this, he proposed to compare the mean of all measurements with the mean of all but the two most extreme measurements (what is now called a 'trimmed mean'). Earlier, in his *Photometria*, he had outlined a graphical method of finding the estimate that gives the observations the greatest probability of happening. This is now called 'the maximum likelihood estimate'.

Joseph Louis Lagrange published a paper in 1774 on the usefulness of adopting the centre of a number of observations as an estimate. In this work we find the first known occurrence of the expression *loi de la facilité des erreurs*. The individual words of this expression (*Gesetz, leicht, Fehler*) can be found, in German, in the writings of Lambert discussed above; while 'facilities', as denoting tendencies in the world, had entered probability earlier in the writings of Gottfried Wilhelm Leibniz. Lagrange was then director of the mathematics section of the Berlin Academy, where he came to know Lambert. Lagrange continued the approach initiated by Simpson and applied it to several continuous-error laws, employing integral calculus. He showed for each case he considered that the mean error of a number of observations is less likely to exceed a given value than the error of a single observation is to exceed the same value.

In Paris, Pierre Simon Laplace, in several papers written in the period up to 1800, systematically explored the question of under what conditions the (arithmetic) mean, as opposed, say, to the median, is the best average. This is all just before the invention of an extension of taking the mean, the method of least squares, to which we now turn.

2 LEAST SQUARES

In combining observations according to the method of least squares, unknown quantities are estimated such that the sum of the squares of the errors is minimized, given a particular functional relation between unknowns and known observations (§10.5). Although Adrien Marie Legendre was the first to publish the method of least squares, in 1805, Carl Friedrich Gauss made the original invention, possibly as early as 1795. Gauss used the method to calculate the orbit of the newly discovered asteroid Ceres for the month of December 1801, using just a few patchy observations made by Giuseppe Piazzi earlier that year. Astronomers and geodesists immediately recognized the usefulness of the method of least squares, and by 1825 it was in general use in France, Italy, England, The Netherlands and Prussia (Stigler *1986*).

Legendre had not defended the method through a consideration of a law of the facility of errors, and had not attempted to measure the uncertainty to which a least-squares estimate is subject. Gauss did this in his *Theoria motus corporum coelestium* [...] (1809) on the motion of the planets around the Sun. In this pioneering work, he laid the foundations for the method of least squares, in which the normal of Gaussian distribution figured prominently, but he also developed new ways of manipulating the normal equations and rules to calculate weights and degrees of precision of the least-squares estimate. The expression 'probable error' was first defined and used in a 1815 paper by Bessel on the location of Polaris, probably on the instigation of his mentor Gauss, who a year later showed three ways of calculating the probable error from given observations. Bessel defined the probable error (of a single observation) as the 'limit separating a number of smaller errors from an equal number of larger errors, so that it is more probable to find an observation be mistaken but within any larger limit from the truth, than outside it'.

To justify the use of least squares, Bessel, in an 1818 analysis of the observations of the English astronomer James Bradley, compared an empirical error distribution with one given by normal theory, and noticed a relative preponderance of large errors in the empirical distribution. This, Bessel speculated, could have been due to unanticipated aberration of starlight or unnoticed movement of the instrument. Gauss and Bessel became advocates of the view that good data reduction should be based on a theory of the instrument with which the data were collected. Such a theory would provide a calculator with correction factors for constant errors, as collimation, and state an estimate of the probable accidental error. Bessel's error analysis of the seconds pendulum, published in 1826, is an exemplary study in this area that had a great impact outside mechanics (compare §8.13).

In 1810, Laplace had shown that the method of least squares can be asymptotically justified, practically independently of the law of the facility of errors to which the individual observations are subject. Using similar mathematics, the German hydrologist Gotthilf Hagen, in his *Grundzüge der Wahrscheinlichkeitsrechnung* (1837), proposed the 'hypothesis of elementary errors', in which assumptions about the causes of error are made explicit. According to Hagen's hypothesis, a measurement error is the algebraic sum of infinitely many elementary errors, of the same size and equally likely to be positive or negative. Hagen advocated the use of least squares and the calculation of the probable error of estimates in hydrology. His book stimulated research in the causes of errors and the resulting error laws, which led to the study of the causes and effects of the non-normality of statistical data in the so-called 'Scandinavian school', founded by the Swedish astronomer Carl Charlier (Särndal *1971*).

3 DISSEMINATION

Gauss's contribution to error theory and his stature as a mathematician were instrumental in the adoption of the method of least squares, and its strictures, not only in astronomy but also for the reduction of observational data in other physical sciences. In the German states, disciples of Gauss and their students applied the new reduction methods in their own work, corresponding with him on the finer details, or published expositions of the method. Johann Encke, a student of Gauss's in the early 1810s, became in 1825 director of the Berlin Observatory and professor at the Berlin Academy of Sciences. He taught a whole generation of astronomers, including foreigners such as the American B. A. Gould, who on his return to the USA in 1849 founded the *Astronomical Journal* and became head of the longitude department of the US Coast Survey.

Encke used his editorship of the *Berliner astronomisches Jahrbuch* as a platform to promote the method of least squares in astronomy and geodesy. In the early 1830s he published there a sequence of papers on least squares, based on his notes from Gauss's course in Göttingen. By recalculating the observations of J. F. Benzenberg's 1804 experiments on the law of fall in the mines near Schlebusch, Encke implicitly suggested that the method of least squares could be fruitfully applied outside its field of origin, planetary mechanics.

Bessel made Königsberg, where he was director of the observatory and Professor of Astronomy from 1810 until his death in 1846, another centre for least squares. Hagen was among his students there, but more important was the influence of his error analysis of the seconds pendulum on the adoption of strict reduction methods by the Königsberg physicist Franz Ernst Neumann (Olesko *1991*). Neumann took Bessel's instrument analysis and its use of least squares and transformed it into a new experimental and calculational approach to precision measurements in his study of the theories of heat, magnetism and optics. Gustav Kirchhoff wrote his dissertation under Neumann before he did his famous work on spectral analysis with Robert Wilhelm Bunsen and so helped to found the new discipline of astrophysics (§8.10).

In Göttingen, Gauss's joint work with the physicist Wilhelm Weber on geomagnetism led to another infusion of least-squares analysis into physics. Weber counted among his many successful students Ernst Abbe and Friedrich Kohlrausch. Abbe, who in his *Habilitationsschrift* in 1861 had derived the chi-squared distribution (§10.10), taught from 1882 to 1887 a course on least squares at the University of Jena. At the same time he worked on the development of the apochromatic lens system that brought the optical firm of Zeiss to international industrial prominence. Kohlrausch's *Leitfaden der praktischen Physik* (1870), an exercise handbook based on the classroom practices of experimental physics instruction in Göttingen, was immensely popular in Germany and provided a model for other laboratory courses.

Details of the emerging prominence of least squares in other fields of science and in other countries need more study, but it seems that the method of least squares was not so popular among French experimentalists; only later was it adopted in Great Britain, where the training in precision measurements began in William Thomson's Glasgow laboratory in the 1850s. George Biddell Airy, the English Astronomer Royal since 1835, spread the use of the method through a manual entitled *On the Algebraic and Numerical Theory of Errors of Observations and the Combinations of Observations* (three editions between 1861 and 1879). According to Mansfield Merriman *1877*, only Laplace was consulted in preparing Airy's book, and 'as a consequence it is unreadable except by those already thoroughly acquainted with the subject'. But it was still in use in Cambridge when Ronald Aylmer Fisher studied the theory of errors under F. J. M. Stratton in the early 1910s and embarked on his famous rethinking of the foundations of statistical inference (§10.10).

4 TABLES

To obtain the value of a probable error, a calculator would at some point have to calculate an integral. This was beyond the capacities of most practical workers, and even adept mathematicians would not want to spend the time necessary to calculate the required approximations. To make the probabilistic use of least squares practicable, a calculator would need to be able to consult a table of the normal distribution. The work that first granted it importance was C. Kramp's *Analyse des réfractions astronomiques et terrestres* (1798). He showed that for a calculation of refraction a table of the values of the function $\int_x^\infty \exp(-x^2)\,dx$ was necessary. Also, in the theory of heat conduction the function occupied a place of great importance (see §9.4, equation (4)). In the back of his book Kramp published three tables related to the error function.

The next table to appear was in Bessel's *Fundamenta astronomiae* (1818), in which he analysed Bradley's observations; he probably recomputed or interpolated it from Kramp's table. Encke derived the tables he published in 1834 in his *Jahrbuch* from Bessel's table, and was himself the source of the tables in *Elemente der Psychophysik* (1860) by the psychologist Gustav Theodor Fechner. The Belgian astronomer Adolphe Quetelet, who tried to develop a social physics in which the normal distribution and its mean were central conceptual tools (§10.8), took the tables in his *Lettres [. . .] sur la théorie des probabilités, appliquée aux sciences morales et politiques* (1846) directly from Kramp. Quetelet's work was immediately translated into English and appraised in 1850 by John Herschel for the *Edinburgh Review*, in a review that had an impact on James Clerk Maxwell's work in statistical mechanics (§9.14). This sequence of borrowings and adaptations of tables for the normal and binomial distributions is witness to the fact that the progress of probability and statistics in the nineteenth century was part of the development of many disciplines.

5 MORALITY

Wilhelm Jordan, the author of the standard German handbook on geodetics of the late nineteenth century, believed that the introduction of least squares in geodetic surveys had had a distinct moral advantage: only when people can follow a strict rule will they stop using the ambiguities in a situation to suit their own purposes. Jordan repeated a story about the French geodesist P. F. A. Méchain who had 'suppressed' some measurements made in surveying work done in Spain at the close of the eighteenth century and was rumoured to have died of the ensuing insanity. He may have been saved if he could have used the method of least squares!

The rise of probabilistic methods in dealing with observational and experimental data has indeed been a battle about proper morals and conduct. It was part of a movement toward increased quantification, more refined measurement and more rule-governed calculations. To become part of that movement required discipline – for some, the submission of one's individuality to a tightly organized experimental and calculational protocol. A new division of labour emerged, and the boundaries between those who had authority and those who submitted to it were redrawn: observers were often persons who saw without understanding, calculators or computers often innumerate beyond the operations they had been trained to perform. Those who wrote the protocol and designed the experiment possessed authority, but it was an authority limited by the broader requirements of least squares and other proper safeguards of objectively questioning nature. Simpson's 1755 proof of the advantage of taking the mean had assumed that observations 'admit of being repeated under the same circumstances'. The tightly woven disciplinary structure of experimental and observational investigations that emerged in the nineteenth century was meant to guarantee that the same circumstances are maintained and accidental errors are kept small.

BIBLIOGRAPHY

Hald, A. *1990, A History of Probability and Statistics and Their Applications Before 1750*, New York: Wiley.

Harter, H. L. *1974–6*, 'The method of least squares and some alternatives', *International Statistical Review*, **42**, 147–74, 235–64; **43**, 1–44, 125–90; **44**, 113–59.

Jordan, W. *1895, Handbuch der Vermessungskunde*, Stuttgart: Metzler.

Merriman, M. *1877*, 'A list of writings relating to the method of least squares, with historical and critical notes', *Transactions of the Connecticut Academy of Arts and Sciences*, **4**, 151–232.

Olesko, K. M. *1991*, *Physics as a Calling. Discipline and Practice in the Königsberg Seminar for Physics*, Ithaca, NY: Cornell University Press.

Särndal, C.-E. *1971*, 'The hypothesis of elementary errors and the Scandinavian school in statistical theory', *Biometrika*, **58**, 375–91. [Also in M. G. Kendall and R. L. Plackett (eds), 1977, *Studies in the History of Probability and Statistics*, Vol. 2, London: Griffin, 419–35.]

Stigler, S. M. *1986*, *The History of Statistics: The Measurement of Uncertainty Before 1900*, Cambridge, MA: Harvard University Press.

10.14

Statistical control of manufacture

DENIS BAYART AND PIERRE CRÉPEL

1 A GRADUAL DEVELOPMENT?

Although mechanical methods of control have been in widespread use in industry for a long time, probabilistic and statistical methods were introduced only in the twentieth century, both in manufacture, to control product quality, and in research and development. This vast domain covers reliability, quality control, signal processing, queuing theory, simulation, and so on, yet its history has been studied only in parts. This article deals with quality control.

In industrial batch manufacture, the aim is always to avoid undesirable variations in product characteristics. No two products from the same batch are ever strictly identical, but may be considered equivalent as long as the values of the characteristics defining their quality do not differ too much from the standard agreed upon between manufacturer and client. With this concern in mind, statistics furnishes tools to study and manage product variability, to reveal the causes of variability and, on the basis of sampling techniques, to reject defective products.

This, however, is a modern formulation of the problem, which emerged only in the 1920s. Earlier control practices were mostly based on intuition and empirical methods – far different from the stipulations that would have been made had probabilities been calculated. Remarkably, classical probability theory, set out by Pierre Simon Laplace in his *Théorie analytique des probabilités* (1812), and later innovations, were virtually ignored by the world of industry until 1920. There were a few exceptions: in France, the work of probabilists (Siméon-Denis Poisson *1837*, and P. J. P. Henry 1880, *1894* and other artillery officers on the accuracy of guns and the wear of gun barrels); and, at the Guinness Brewery in Dublin, the work of W. S. Gosset (alias 'Student') in the early twentieth century on fermentation processes, which introduced statistical methods for small samples.

After a few remarks on traditional practices, the article highlights the emergence of three concepts essential to the modern-day theory of the statistical control of manufacture: the representation of industrial production as random systems; the consumer's and manufacturer's notions of risk, and efficiency curves; and the optimization of sample size.

2 BRIEF OVERVIEW OF TRADITIONAL METHODS; THE FIRST APPROACH TO PROBABILISTIC THINKING

Random behaviour stems from several sources: from the manufacturing process itself, and also from product homogeneity (or lack of it), sampling, measurement errors, and so on. Sampling is necessary whenever products are numerous and testing is destructive, long or simply costly.

The first thing is to determine the size of samples, and decide what precautions need to be taken to ensure that they are representative. Before the explicit use of the calculus of probability, samples were usually small; except when obviously too inhomogeneous, the part was considered as representative of the whole. In general, in different areas (minting, military equipment, registrable substances, etc.), procedures were stipulated by official texts, increasing in sophistication over the nineteenth and early twentieth centuries, but fairly dissimilar from one country to another.

In minting coins, for example, sampling was used as a control procedure as far back as the Middle Ages, to check the composition and weight of coins made from precious metals (Stigler *1977*, Carcassonne *1988*). It was considered sufficient to take a small number of items from each batch, 'at random, not specifically chosen', with no consideration of their size, in order to subject them to tests carried out by different controllers. In 1750, a French regulation stipulated that the size of the sample would henceforth be proportional to the numbers delivered (for gold coins, one extra per batch of 200 coins manufactured).

Not until the 1920s did explicit probabilistic reasoning appear in Europe (Dumas *1925*, Vallery *1925*, Becker, Plaut and Runge *1927*). These first calculations (often based on elementary notions of the calculus of probability) show that, to protect the customer against risk of defects, samples had to be much bigger than previously thought, and not proportional to batch size.

3 THE RANDOM CHARACTER OF MANUFACTURING SYSTEMS

An essential step was taken at the Bell Laboratories in the USA between 1924 and 1929, when a viable theory of statistical control in manufacture was developed. Most articles published until then had included no more than 'remarks' on the matter, perhaps doubting established practices, but offering no means of developing a new system of control standards. At Bell, E. C. D. Molina (a mathematician) and W. A. Shewhart (a physicist), knowledgeable in probabilistic methods, applied these methods to the problems of organizing telephone services and controlling industrial variables, particularly in the manufacture of carbon microphones (each of which used 50 000 granules of material).

In a series of articles widely circulated in the *Bell System Technical Journal*, and in the book *1931*, Shewhart explained in detail hypotheses which legitimized a statistical study of product categories and defined conditions which rendered sampling methods valid. Producing examples to back up his statements, he showed first that, even with the most accurate manufacturing techniques, product variability was inevitable; and second that this uncertainty could be controlled using appropriate statistical techniques. From the point of view of the history of probabilistic thinking, Shewhart's theory was a direct continuation of the principles of Laplace: considering any set of machines to be a random system, just like some natural system, he attempted to maintain constant the 'system of chance causes' which generates randomness in product manufacture. If this can be done, the set of products will have stable statistical characteristics, so the manufacturer will know, in particular, the mean proportion of faulty items. Shewhart gave a definition of the concept of control directly applicable to the operations of industrial manufacture: 'a phenomenon will be said to be controlled when, through the use of past experience, we can predict, at least within limits, how the phenomenon may be expected to vary in the future'. On the operational level, this theory resulted in techniques which would show whether the production system had reached the status of a 'constant system of chance causes', or whether 'assignable causes of variation' still remained. If the latter, it is up to the engineers to find and eliminate the causes, provided the cost of this operation is not greater than the advantages which would result from it. Over the next few years, these techniques took on the almost standardized form of 'control charts'. In particular, they made it possible to detect any discrepancy in relation to the mean, and dispersal.

A measure of the upheaval Shewhart's approach caused in relation to the way most engineers thought at the time, is a comment by the eminent French metallurgist Henri Le Chatelier, who exerted a strong influence in industrial circles, having helped to disseminate F. W. Taylor's organizational methods. Believing only in determinism, both in the scientific domain and in the domain of work organization, Le Chatelier *1936* refused to believe in chance, and rejected any probabilistic approach: 'What we refer to as chance is only the result of our negligence which causes us to allow the essential conditions of our experiments to vary irregularly.'

4 THE THEORY OF TESTS

Outside the works of Shewhart, the first probabilistic thinking on the acceptance of equipment posed an essential question (albeit in an indirect and disordered manner): the efficiency of sampling methods. In modern practice, sampling plans are usually defined and represented by their operating characteristic curve, which indicates the probability of a batch being accepted in relation to the proportion of faulty items it contains. This curve contains all useful data about a plan, and in particular enables both the producer's risk (the probability of a good batch being rejected) and the consumer's risk (the probability of a poor batch being accepted) to be calculated. But awareness of the theoretical and practical importance of such curves developed only very slowly, as did a clear formulation of the problem of arbitration between consumer's and producer's risk. Attempting solely to determine the consumer's risk, Dumas *1925* worked out these curves, but without perceiving their full scope. By contrast, Becker, Plaut and Runge *1927* looked at matters from the manufacturer's viewpoint, without making the connection with the opposite viewpoint of the customer. 'Probability of acceptance curves' were in use at Bell, but their significance does not seem to have been fully grasped, since Dodge and Romig *1929* did not mention them in the article in which they presented their method; although they took consumer's risk into account in the acceptance indicated above, they defined the producer's risk only implicitly, through the financial penalties incurred in the event of a batch being rejected. The reasoning, certainly innovative (since it implied optimization of an economic function) did not contribute to clarifying the problems of acceptance control.

The theoretical work of Jerzy Neyman and Egon Pearson (Karl Pearson's son), starting in 1927, finally clarified the situation by explicitly tackling the question of comparison of tests. In their first article (1927) they defined the two types of risk (acceptance of a false hypothesis, and rejection of a true hypothesis), formally corresponding to consumer's and producer's risk, respectively, and showed how they fit together. Only after their subsequent series of works (1928–38) was the expression 'operating characteristic curve' coined in the USA, and its full importance recognized (Pearson *1935*).

5 OPTIMIZATION OF SAMPLE SIZE

Particularly with destructive testing or expensive control methods, it is extremely important to reduce the size of the sample, without however increasing the risk of an error in diagnosis. Rather than choose the sample size in advance (i.e. before verifying the products) on the basis of considerations of mathematical statistics, one can proceed in two stages. Double-sampling plans appeared fairly early on, in the 1920s. This procedure consists of checking a first sample: if it contains few faulty parts, the entire batch is accepted; if it contains too many, the batch is rejected; if it contains an intermediate number, a second sample is checked in the hope of gaining further information. A decision is then made in the light of both sets of data. It should be noted that this procedure is very different from the traditional 'expert versus counter' valuation, where the second check, serving for legal purposes, is used not as a complement but instead to overrule the first.

A considerable step forward was taken when, in 1943, at the request of army personnel and economists, Abraham Wald developed the theory of sequential analysis for the management of US military equipment (Wald *1947*). The procedure was inspired by the one described above: the parts are examined one after the other, and, at each stage, the decision is made whether to accept the batch, reject it or continue checking. The 'sequential probability ratio test' thus developed was a modification of the Neyman–Pearson test, and therefore introduced a random size for the sample. Wald showed that, for the same risks of error, the 'average sample number' (size of the inspected batch) was much less than for the ordinary Neyman–Pearson test. The general idea is fairly simple, but development of this theory and demonstrations required the use of the latest probabilistic techniques (such as stopping times and martingales). This is a remarkable instance of a practical question leading in just a few months to a new and difficult theory in advanced mathematics, which is then fed back as graphical methods or practices directly applicable by people with little mathematical skill. It should be remarked that G. A. Barnard in the UK arrived at similar though less extensive results at the same time.

The statistical techniques used in quality control have since diversified, particularly with the use of experimental plans. The work of the Bell engineers, and of Wald and the Statistical Research Group at Columbia University, have all contributed to an upheaval in the way that statistical knowledge is disseminated: thousands of people have been trained in the new techniques in intensive courses, in both factories and universities.

BIBLIOGRAPHY

Becker, R., Plaut, H. and Runge, I. *1927, Anwendungen der mathematischen Statistik auf Probleme des Massenfabrikation*, Berlin: Springer. [2nd edn 1930.]

Carcassonne, C. *1988*, 'Etalon, poids légal, pyx, boëtes, remèdes', in *L'À peu près*, Paris: Ecole des Hautes Etudes en Sciences Sociales, 47–58.

Dodge, H. F. and Romig, H. G. *1929*, 'A method of sampling inspection', *Bell System Technical Journal*, **8**, 613–31.

Dumas, M. *1925*, 'Sur une interprétation des conditions de recette', *Mémorial de l'Artillerie Française*, **4**, 395–420.

Fry, T. C. *1928, Probability and its Engineering Uses*, New York: Van Nostrand. [Various reprints.]

Grant, E. L. and Leavenworth, R. S. *1972, Statistical Quality Control*, 4th edn, New York: McGraw-Hill.

Hald, A. *1981, Statistical Theory of Sampling Inspection by Attributes*, London: Academic Press.

Henry, P. J. P. *1894, Cours de probabilité du tir à l'Ecole d'Application de l'Artillerie et du Génie*, Fontainebleau. [Repr. 1926, *Mémorial de l'Artillerie Française*, **5**, 294–447.]

Juran, J. M. *1951, Quality Control Handbook*, 1st edn, New York: McGraw-Hill.

Le Chatelier, H. *1936, De la méthode dans les sciences expérimentales*, Paris: Dunod.

Littauer, S. B. *1950*, 'The development of statistical quality control in the United States', *The American Statistician*, **4**, 14–20.

Pearson, E. S. *1935, The Application of Statistical Methods to Industrial Standardisation and Quality Control*, London: British Standards Institution.

Poisson, S. D. *1837*, 'Mémoire sur la probabilité du tir à la cible', *Mémorial de l'Artillerie*, **3**, 59–94.

Shewhart, W. A. *1931, Economic Control of Quality of Manufactured Products*, New York: Van Nostrand, and London: Macmillan.

Stigler, S. M. *1977*, 'Eight centuries of sampling inspection: The trial of the Pyx', *Journal of the American Statistical Association*, **72**, 493–500.

Vallery, L. *1925*, 'Le calcul des probabilités appliqué au prélèvements des échantillons en matière d'expertise', *Annales des Falsifications et des Fraudes*, **204**, 597–605.

Wald, A. *1947, Sequential Analysis*, New York: Wiley.

10.15

The social organization of probability and statistics

THEODORE M. PORTER

1 PROBABILITY AND IDEALS OF KNOWLEDGE

The history of probability and statistics in the last two centuries has been one of a massive expansion of influence, combined with far more resistance from institutions of learning than was typical even for the experimental sciences. There is no simple explanation for this slow acceptance, but plausible reasons can be given. For one, mathematics could be almost as important an ingredient of the conservatism of universities as Greek and Latin. Given how hesitant nineteenth-century Cambridge was even to grant algebra a place in the curriculum comparable to that of geometry, it is hardly surprising that probability and statistics were scarcely to be found there.

Perhaps the most important reason was the entanglement of probability and statistics in political and social discussion. Statistics as a numerical science was born early in the nineteenth century largely as a science of statecraft, pursued by unscholarly liberal reformers. Probability had a longer history, and had been widely accepted in the eighteenth century as the true science by which the mind apportioned belief when certainty was not possible. This conception, and the applications to such topics as elections and judicial decisions that went along with it, came to seem dangerously extravagant in the nineteenth century. The move towards identifying mathematics with rigour, reflecting in part a more conservative educational philosophy, did not encourage the study of probability. Neither probability nor statistics seemed well suited for the formation of young minds.

In fact, early in the nineteenth century probability became much less interesting to mathematicians than it had been in the eighteenth, from James Bernoulli to Pierre Simon Laplace. The efforts of several generations of Russian mathematicians, of whom the best known are perhaps Pafnuty Chebyshev, Andrei Markov and Andrei Kolmogorov, finally came to fruition in the 1920s and 1930s with the creation of a pure probability theory that could command the interest and respect of the international mathematical community (§10.6). But from Laplace to Kolmogorov, roughly 1810–1930, probability was pursued mainly by natural and social scientists, who aimed less to prove theorems of mathematical interest than to put probability to use in the study of natural and social distributions and the analysis of numerical data.

2 THE USES OF PROBABILITY

There were two primary sources of interest in probability. First was the association established by Carl Friedrich Gauss and Laplace between probability and the analysis of observational data in astronomy and geodesy (§10.13). The methods of error theory were readily applied also to other fields whose data could be seen as analogous to records of astronomical observations, such as surveying and meteorology. In the mid-nineteenth century, Gustav Theodor Fechner added psychophysics to this list (§10.9; and Stigler *1986*). No less important than error theory, and more widely visible, was the connection between probability and the burgeoning field of statistics – that is, social statistics. This was not quite new in the nineteenth century, for insurance calculations and related demographic problems had interested nearly all the most eminent classical probabilists. In the mid-eighteenth century, probability calculations even began to be used as the basis for setting rates by those life-insurance companies that, following the Equitable, offered financial security rather than a somewhat gruesome gambling opportunity to their clients (§10.3; and Daston *1988*). With the discrediting of the more ambitious attempts to mathematicize good judgement in the decades following the French Revolution, statistics became the standard application of probabilistic reasoning. The rise of a frequency interpretation of probability in the 1840s reflected this increasing prominence of statistics (Porter *1986*).

Because probability had so little standing of its own during the nineteenth century, the institutions that supported it were necessarily diverse (as was probability theory itself). For error theory, the key institutions were astronomical societies, observatories, and state-sponsored surveys such as geodetic ones. Although very few successful scientific careers in the nineteenth century were made through contributions to probability, error theory was a subject of explicit mathematical research, and the heightened concern with precise measurement inspired a steady stream of publications on probabilistic data analysis. The existence of textbooks shows that the subject was taught to some science students, but there were few if any teaching or research positions in error theory or probability.

3 THE ORGANIZATIONS OF STATISTICS

Statistics presents quite a different story. Statistics was conceived of in the nineteenth century as the numerical science of society, and only a few statisticians knew or cared much about probability theory. Still, the mathematical variety of statistics that emerged toward the end of the century was continuous in important ways with the work of these social statisticians, and many of the mathematicizers were associated with organizations dedicated to the promotion of social statistics.

These were of several sorts. Among the most important were government offices, for the collection of reliable statistics came to be seen as a normal and necessary function of government in the early nineteenth century. Statisticians were appointed in a host of agencies and bureaus in the governments of Europe and North America, and in the colonies. They could be found in offices concerned with trade, with public health, with crime and justice, with education, with military recruitment, and with taxation and revenue, to name only the most prominent. Statisticians also had their own agencies, most notably census bureaus, but also central statistical offices, such as the one in Belgium and its many imitators, that aimed to coordinate the activity of statisticians scattered around the rest of the government. State officials were among the most prominent and influential statisticians in every country, but perhaps they were most nearly dominant in France and Belgium.

Private statistical societies were the second main source of support for statistical activity. These were born as centres of civic virtue and public-spirited reform, but reflecting also a commitment to quantification nurtured by a sense that social problems had their roots in the increasing numbers of newly urbanized workers. The prototypes for such organizations were formed in the early 1830s in England, most notably in Manchester and London, with the USA not too far behind (Cullen *1975*). One might infer, and correctly, that the liberal spirit of *laissez-faire* infused these organizations, but that did not prevent ambitious civil servants such as William Farr and George Richardson Porter from playing a prominent role in them. The unity of voluntary activity with officialdom was even more evident in Paris, whose statistical society dated from 1865. Less successful statistical societies were formed elsewhere.

Universities were the third important source of support for social statistics. University statistics was almost unknown west of the Rhine, and was most important in Germany, Austria and Italy. A product of the utilitarian Enlightenment tradition of training civil servants, barely surviving a crisis in mid-century when the old descriptive *Statistik* was taken over by the quantifiers, statistics was from about 1860 to 1890 the most authoritative form of social science in Germany (Krüger *et al. 1987a*).

4 DEBATES ABOUT STATISTICAL METHOD

Even in the universities, and still more in state bureaucracies and private statistical societies, social statisticians were far more concerned with crime, sanitation and education than with mathematical theory. Nevertheless, strong ideological commitments made the interpretation of their numbers a matter of great import, and arguments about moral responsibility or the need for state intervention regularly spilled over into discussions of the best ways to present and analyse numerical data. The social statisticians did rather more than the error theorists to render commonplace the supposition that truly scientific generalizations, even laws, could be extracted from large numbers without worrying too much about what is happening at the level of individuals (Porter *1986*).

From the 1830s onwards, we can find writers who preferred to view statistics as a set of quantitative methods, applicable to any subject-matter. The more usual definition, though, was as a science of humankind in the mass – to which such methods were particularly appropriate. The social statistical tradition nurtured a few serious efforts to make statistics mathematical and base it on probability theory. The Belgian astronomer-turned-statistician Adolphe Quetelet and the French savant A. A. Cournot were the first, in the 1830s and early 1840s. In the 1870s the German professor Wilhelm Lexis initiated a mathematical approach that enjoyed far more institutional success, while in England the economist Francis Edgeworth wrote extensively on the uses of probability for analysing data. It was, however, mainly the English biometricians who fashioned a successful mathematical statistics out of the materials of error theory and social statistics, but guided by their own problems and ambitions.

5 BIOMETRIC AND STATISTICAL LABORATORIES

The intellectual influence of Francis Galton was as important as any, but Karl Pearson was the first to combine a viable conception of statistical method with the energy and insight to build the needed institutions. Galton's money was also helpful. With it, and other funds, Pearson established a biometric laboratory at University College, London, where students from all nations and disciplines could come to learn the methods of analysing data practised by Pearson and his associates (§10.7). A variety of fields were at least partially transformed by the work of the biometricians or other mathematical students of statistics. Prominent among them was social statistics itself, where the complete census had been *de rigueur* since 1800, but where sampling and even probabilistic estimates of accuracy began at last to be used in the early twentieth century. Two fields in which the methods of the biometricians played a crucial role were psychological testing and early forms of what in the 1930s would be labelled 'econometrics'. No less important than its role in the diffusion of new statistical conceptions and techniques, however, was the place of biometry as the first field to include in its business the investigation and refinement of the statistical method itself. The modern statistics department has its ancestor here.

Ronald Aylmer Fisher, like Pearson, encouraged students to examine the methods of statistics and experimental design, first at the experimental farm in Rothamsted, and then at University College, where he succeeded Pearson as Galton Professor of Eugenics. This personal influence, combined with his two enormously influential textbooks, *Statistical Methods for Research Workers* (1925) and *The Design of Experiments* (1935), promoted the incorporation of Fisherian methods into all areas of science, but especially the biological and human sciences. The statistical research laboratory of Pearson and Fisher also became a standard organizational model for statistics. Most crucially, it was naturalized in the USA by George Snedecor at Iowa State College, who also brought Fisher in to lecture at two summer sessions in the 1930s. Jerzy Neyman, champion of a rival conception to Fisher's understanding of statistical inference, and himself a former student at Pearson's laboratory, established a similar institution in the Department of Statistics at the University of California at Berkeley (Gigerenzer *et al. 1989*).

6 A MASTER DISCIPLINE?

The multidisciplinary statistical laboratory is no mere accident of history. It was adapted to the crucial role of statistics as a service discipline (or perhaps one should say master discipline) for agricultural researchers, biologists, psychologists, sociologists, anthropologists, medical researchers, and so on. The need for some autonomy also reflects the often weak position of statistics within departments of mathematics. But if statisticians placed too much value on application rather than rigour to satisfy the research ideals of militantly pure mathematicians, they were practically able to form the research ideals of certain of the human sciences that were most acutely afflicted with science envy. Fisher, for example, who had a gift (or perhaps a weakness) for the rhetorical power of over-simple formulations, wrote that the purpose of science is to reject null hypotheses; psychologists of the 1940s and 1950s took this laconic, subtly misleading formulation of

Fisher's approach much to heart, and came close to ruling out of the field all forms of experimentation but those that conformed to the structure invented by Fisher for agriculture (Gigerenzer *et al.* 1989; see also §10.9). A similar story could be told about other studies, including large parts of sociology and biology. It should be emphasized that the influence of statistics was felt most quickly and decisively in the USA and the other English-speaking countries. Even within Europe, differences in the conceptions and uses of statistics were extreme.

The relation between the statistics discipline and users in the scientific disciplines has, especially since Fisher's day, been one of at best imperfect comprehension. Statistics itself has been full of controversy; that between Pearson and Fisher was perhaps a prototype for the disputes still raging between the followers of Fisher and the followers of Neyman and Egon Pearson (Karl's son), and more recently the so-called Bayesians. Even within these factions, statisticians have been quite willing to recognize the subtleties of statistical method, the perpetual need for judgement, the impossibility of formula-based knowledge.

In contrast, statistical pioneers within the various disciplines (e.g. psychology, sociology and medical research) were not mathematicians, and did not care to be. Especially in the first flush of statistical enthusiasm, around 1940 to 1950 in many disciplines, what they wanted from the statisticians was precisely a mechanical recipe for designing experiments and obtaining knowledge. To get that they had to write their own textbooks, some of which attained almost scriptural authority and were, in a sense, more important than the actual works of Fisher and Neyman themselves. From them sprang not only a set of strategies for analysing data but also a style of research through which their fields were effectively redefined. The curious disciplinary standing of statistics, then, has mattered a great deal, not just or even mainly for the field of statistics itself, but especially for the multitude of disciplines for which certain statistical forms have become requisite for acceptable experimental work.

BIBLIOGRAPHY

Anderson, M. *1988, The American Census: A Social History*, New Haven, CT: Yale University Press.

Bourguet, M.-N. *1988, Déchiffrer la France: La Statistique départementale à l'époque Napoléonienne*, Paris: Editions des Archives Contemporains.

Cullen, M. *1975, The Statistical Movement in Early Victorian Britain*, Hassocks: Harvester.

Daston, L. *1988, Classical Probability in the Enlightenment*, Princeton, NJ: Princeton University Press.

Gigerenzer, G. *et al.* *1989*, *The Empire of Chance: How Probability Changed Science and Everyday Life*, Cambridge: Cambridge University Press.

Krüger, L., Daston, L. and Heidelberger, M. (eds) *1987a*, *The Probabilistic Revolution*, Vol. 1, *Ideas in History*, Cambridge, MA: MIT Press. [See especially papers by I. Schneider, L. Daston, T. Porter, I. Hacking and N. Wise.]

Krüger, L., Gigerenzer, G. and Morgan, M. (eds) *1987b*, *The Probabilistic Revolution*, Vol. 2, *Ideas in the Sciences*, Cambridge, MA: MIT Press. [See especially papers by G. Gigerenzer.]

Porter, T. M. *1986*, *The Rise of Statistical Thinking, 1820–1900*, Princeton, NJ: Princeton University Press.

Schneider, I. (ed.) *1988*, *Die Entwicklung der Wahrscheinlichkeitstheorie von den Anfängen bis 1933*, Darmstadt: Wissenschaftliche Buchgesellschaft. [A sourcebook.]

Stigler, S. *1986*, *The History of Statistics: The Measurement of Uncertainty Before 1900*, Cambridge, MA: Harvard University Press.

10.16

Foundations of probability

A. P. DAWID

Human activities involving intrinsically uncertain outcomes, such as gambling or the casting of lots, are of very ancient origin. However, their customary association with the workings-out of fate or divine will, as well as the apparent self-contradictions in the very idea of a 'theory of the uncertain', ensured that the concept of probability received little serious intellectual attention before the seventeenth century; and only in the present century has it been given a firm mathematical foundation. Hacking *1975* gives a good account of the early development of the concepts and laws of probability. For a thorough survey of the variety of mathematical structures and theories suggested by a range of philosophical approaches, see Fine *1973*.

1 CLASSICAL PROBABILITY

1.1 Formulation

The 'classical' approach to probability (§10.17) was well described by Pierre Simon Laplace:

> The theory of chance consists in reducing all the events of the same kind to a certain number of cases equally possible, that is to say, to such as we may be equally undecided about in regard to their existence, and in determining the number of cases favourable to the event whose probability is sought. The ratio of this number to that of all the cases possible is the measure of this probability. (*Essai philosophique sur les probabilités*, 1820 edn)

The theory of probability developed on this basis thus rests on counting cases, and can be regarded as a branch of combinatorial analysis (§10.1).

It has often been objected that the classical definition is circular, since the notion of 'equal possibility' it incorporates can only mean 'equal probability'. However, within its motivating context of gaming, devices such as well-balanced dice or well-shuffled decks of cards would seem to provide,

1399

in an intuitively satisfactory way, a set of 'cases equally possible', and this before any concept of probability is introduced. In rolling a fair die, for example, one would naturally take its six faces as 'equally possible' outcomes of a throw. The definition then assigns a probability of $\frac{1}{3}$ to the event that the throw results in an odd prime, since two of the six cases (3 and 5) are 'favourable' to that event. Unfortunately, the classical definition says little about how this probability should be interpreted. One can also object that the definition is so tied to games of chance that it does not naturally extend to other applications: it is not easy to understand the fall of a biased coin, or the safety of a nuclear reactor, in terms of an analysis into equally possible cases.

1.2 Paradoxes

If two coins are tossed, what is the probability of getting at least one head? The usual answer relies on the identification of four equally possible cases: the result may be HH, HT, TH or TT. Of these, the first three are favourable, yielding the answer $\frac{3}{4}$. This analysis depends on distinguishing the two coins somehow, so that TH differs from HT. Alternatively, we might treat the coins as indistinguishable, giving only three distinct cases – 0 heads, 1 head, or 2 heads – and leading to the answer $\frac{2}{3}$ (this approach is the basis of the Bose–Einstein statistics of particle physics: compare §9.15). Without further guidance as to what probability should mean, the classical definition gives no real reason to prefer either of these solutions over the other.

Still more paradoxes arise on trying to extend the definition to deal with an infinite set of cases. Suppose a quantity X is regarded as 'randomly drawn from the interval (0, 1)', in that any two possible values for X in this interval are considered 'equally possible'. One might take this statement to imply that the probability that $X \leqslant \frac{1}{4}$ is $\frac{1}{4}$. But if $Y = \sqrt{X}$, say, then any two values for Y may also be regarded as equally possible. Since $X \leqslant \frac{1}{4}$ is the same as $Y \leqslant \frac{1}{2}$, identical reasoning applied to Y yields the different answer $\frac{1}{2}$. Thornier still is Bertrand's paradox, concerning the probability that a 'random chord' of a circle will be shorter than its radius. At least six different interpretations of 'random' have been proposed in this context, all yielding different answers.

2 THE LAWS OF PROBABILITY

2.1 Formulation

Although unsatisfactory as a definition of probability, the classical theory is unproblematic when restricted to problems of calculation once the set of equally possible cases has been settled on. In order to build upon it, while avoiding arbitrariness and eschewing any attempt at interpretation, we might attempt to identify those consequences of the definition which hold good irrespective of the choice of the set of equally possible outcomes. Such properties may reasonably be called laws of probability.

A similar programme may be carried out from the standpoint of any other interpretation of probability, such as the frequency theory of Richard von Mises first presented in 1919 (published in *1936*), or the subjectivist theory of Frank Ramsey (1926) and Bruno de Finetti *1970* (§10.17). In the latter, for example, the assertion that the probability of A is p is an expression of the assertor's willingness, for any (positive or negative) amount k, to stake kp for the prospect of winning $k(1-p)$ should A occur. The only constraint is that one should never make a collection of probability assertions which allow the possibility of making a net loss no matter what occurs. Perhaps surprisingly, it turns out that this, and all other major interpretations, lead to exactly the same general laws of probability as does the classical approach. For simplicity, this article concentrates on the classical case.

2.2 Events

First to consider is the nature of the uncertain events to which the concept of probability is to be applied. What Laplace termed a 'case' is now renamed an 'elementary event' (ignoring henceforth the classical insistence on 'equal possibility'). The elementary events must be described in sufficient detail that when, eventually, it is observed which one obtains, the truth or falsity of any uncertain event A of interest will be determined, according to whether that elementary event is or is not 'favourable' to A. Thus A can be considered as a proposition about the eventual outcome, so that the propositional calculus, with its logical operations of conjunction, disjunction and negation, may be applied to events.

An alternative mathematical framework is in common use. The sample space S is defined to be the set of all elementary events, and with any event A is associated the subset of S consisting of all those elementary events favourable to A. By this mathematical device any event may be identified with a subset of S, and set-theoretic language may be used: conjunction becomes intersection, disjunction union, and negation complementation (with respect to S).

2.3 Laws

Laplace's definition easily implies the following general properties ($P(A)$ denoting the probability that the event A will happen):

$$\text{For any event } A, \ P(A) \geqslant 0. \tag{1}$$

$$\text{If } A \text{ is certain to happen, then } P(A) = 1. \tag{2}$$

If A and B cannot happen simultaneously, then

$$P(A \text{ or } B) = P(A) + P(B). \tag{3}$$

There are many other properties of classical probability which hold universally. For example, for arbitrary events A and B,

$$P(A \text{ or } B) = P(A) + P(B) - P(A \ \& \ B). \tag{4}$$

However, this and any other such properties may be deduced as logical consequence of properties (1)–(3).

3 CONDITIONAL PROBABILITY AND INDEPENDENCE

3.1 Definition

Conditional probability refers to the revision of uncertainty on obtaining partial information. Classically, if we learn that an event B has happened, then the original set of equally possible cases may be subdivided according to whether or not they were favourable to B. Those that were not are now ruled out, while those that were remain equally possible, and become the new basis for the calculation of possibilities. The revised probability for an event A is thus the number of cases initially favourable both to A and to B, divided by the number of cases initially favourable to B:

$$P(A \mid B) = P(A \ \& \ B)/P(B) \tag{5}$$

with $P(A \mid B)$ denoting the conditional probability of A, given B. Since (5) makes no mention of equally possible cases, it may be taken as a general definition of conditional probability in terms of unconditional (or 'marginal') probabilities (so long as $P(B) > 0$).

Many results follow easily from (1)–(3) and (5), including Bayes's theorem:

$$P(B \mid A) = P(A \mid B)P(B)/P(A). \tag{6}$$

Bayes's theorem underlies the method of inverse probability, prominent both in the classical and in the modern subjectivist ('Bayesian') and logical approaches to probability. This applies (6) to the case in which B is a tentative explanatory hypothesis (or law of nature), and A is observed evidence. The application is controversial, since some interpretations of probability reject the meaningfulness of 'the probability of a hypothesis'.

3.2 Independence

The event A is said to be independent of B when $P(A \mid B) = P(A)$. In this case, learning of the occurrence of B does not affect the probability of A. From (5), this is equivalent to

$$P(A \ \& \ B) = P(A) \times P(B), \tag{7}$$

and is thus a symmetrical relation between the two events A and B.

In the classical interpretation, on one throw of a die the events 'outcome is even' and 'outcome is 4 or less' are independent.

3.3 Repeated trials

Suppose a coin is tossed twice, and (taking the classical approach) the four possible sequences of results are treated as equally possible. Then 'H on toss 1' and 'H on toss 2' will be independent. In most frequency interpretations of probability, such 'repeated unrelated trials of the same phenomenon' are directly modelled as independent: then one need only specify probabilities for events relating to a single trial, extending to compound events using the product rule (7).

From other points of view, independence in repeated trials may not be appropriate. Even classically, while it may be reasonable to regard HH and TT as equally possible (each involves two identical outcomes), and similarly HT and TH, why should HH and HT be so regarded? Thomas Bayes, in a celebrated memoir of 1763 addressed to The Royal Society of London, analysed the alternative case where, in n trials of some phenomenon, each value $(0, 1, 2, \ldots, n)$ for the total number of 'successes' is assumed to be equally probable. From this follows Laplace's law of succession: a phenomenon, having resulted in success at every one of n trials, has a probability of success in the next of $(n + 1)/(n + 2)$. Laplace used this formula to calculate the probability that the Sun would rise the following day. Such frivolous applications, neglecting the appropriateness of its underlying assumption, brought the law of succession into disrepute. However, in any epistemic account of probability the assumption of independent trials is equally ridiculous, since it implies that we can learn nothing, from all our past experience, about the probability of tomorrow's sunrise. The modern subjectivist approach of de Finetti *1970* instead models 'unrelated trials' as 'exchangeable', meaning that any two sequences of outcomes containing the same total number of successes are assigned the same probability, but otherwise allowing freedom of probability specification; both independence and Laplace's assumption are included as special cases. It may be shown that, for a general exchangeable distribution, the conditional probability of success in the next trial, given r successes in the past n trials, will approximate to r/n for large n. Moreover, with probability 1, r/n will tend to some (random) limit, p say, conditional upon which the trials become independent with probability p of success. This purely subjectivist approach thus provides a rationale for modelling repeated trials as independent, and relates probabilities to observed relative frequencies.

4 UNCERTAIN QUANTITIES AND EXPECTATION

Much of the early interest in probability centred on determining the 'fair price' for a position in a game. In modern guise, this concept becomes the 'expectation' of an 'uncertain quantity' (i.e. the gain from the game). Here, 'uncertain quantity' (or, more commonly, 'random variable') refers to something that will take on a definite numerical value depending on which elementary event s is realized. It may thus be regarded as a real-valued function on the sample space.

Perhaps the nearest thing to an interpretation of classical probability is the implicit idea that the probability of an event A provides the 'fair price' for the gamble which pays 1 if A occurs, 0 if it does not. If it is further assumed that the fair price for $X + Y$ is the sum of those for X and Y, the fair price for an uncertain quantity X, having $P(X = x_i) = p_i$ ($i = 1, 2, \ldots, n$), is found to be $\sum_{i=1}^{n} p_i x_i$. This formula now defines the expectation of X (with extensions to continuously varying random quantities similarly defined in terms of integrals). While most modern treatments of probability theory start with uncertain events and their probabilities, some influential accounts (see e.g. de Finetti *1970*) emphasize the fundamental nature of uncertain quantities and their expectations, and define probability in terms of expectation.

5 AXIOMATIZATION

5.1 The axioms

Andrei Kolmogorov *1933* proposed an axiomatic approach to probability theory incorporating the general laws and concepts uncovered above. This is now widely accepted as providing the appropriate mathematical framework for probability. It involves an arbitrary set S, the sample space; a collection $\mathcal{A}$ of subsets of S representing the events of interest; and a function

$P: \mathscr{A} \to [0, 1]$ assigning probabilities to events. The only constraints are that

1 The collection $\mathscr{A}$ should be a σ-field: that is, it should include S, and be closed under complementation and finite or countable union; and
2 P should satisfy (a) $P(S) = 1$, and (b) $P(\cup_i A_i) = \Sigma_i P(A_i)$ whenever $\{A_i\}$ is a finite or countable collection of pairwise disjoint events in $\mathscr{A}$.

In particular, this theory does not favour any one probability assignment P over any other; it reduces mathematical probability theory to the study of normed measures on the measurable space $(S, \mathscr{A})$. A major success of Kolmogorov's approach is a mathematically rigorous account of the concept of the conditional expectation of X, given Y, when Y is a continuous random variable, so that, for all y, $P(Y = y) = 0$.

5.2 Some remarks on the axioms

The additivity property 2(b) follows from equation (3) (interpreted in terms of sets) so long as $\{A_i\}$ is a finite collection, but the extension to countably infinite collections does not. Is this extension appropriate? It prevents probabilities from being assigned in a countable sample space so as to give equal probability to every elementary event. It also makes it impossible to extend a general probability, defined for every interval on the real line, to every subset of the line. Such difficulties can be avoided by requiring only finite additivity, and some modern work (largely influenced by de Finetti) develops this approach. However, most workers happily accept countable additivity 'for convenience'.

One can similarly dispute the need for constraint 1. Von Mises argued that, in an infinite sequence of trials, the only events which can meaningfully be assigned probabilities are those approximable by events relating to only finitely many trials – a class not satisfying constraint 1. In any case, if countable additivity is rejected, the need for closure under countable union disappears.

We do not even need Kolmogorov's sample space to define events, but could model them on (say) an abstract Boolean algebra. This would allow straightforward extension of the class of events if desired; such extension is generally impossible within a fixed sample space, since the elementary events may not be described in sufficient detail for them to determine the new events. Since any Boolean algebra is isomorphic to the Boolean algebra of subsets of some set, the set-theoretic framework can be invoked if desired. However, this isomorphism does not extend to Boolean σ-algebras. Thus, if we do wish to follow Kolmogorov in requiring countable additivity, we might well regard his set-theoretic framework as insufficiently general.

BIBLIOGRAPHY

de Finetti, B. *1970, Teoria della probabilità*, Turin: Einaudi. [English transl. 1975, *The Theory of Probability*, New York: Wiley.]

Fine, T. L. *1973, Theories of Probability*, New York: Academic Press.

Hacking, I. *1975, The Emergence of Probability*, Cambridge: Cambridge University Press.

Kolmogorov, A. *1933, Grundbegriffe der Wahrscheinlichkeitsrechnung*, Berlin: Springer. [English transl. 1950, *Foundations of the Theory of Probability*, New York: Chelsea.]

von Mises, R. *1936, Wahrscheinlichkeit, Statistik und Wahrheit*, Vienna: Springer. [English transl. 1957, *Probability, Statistics and Truth*, London: Allen & Unwin.]

10.17

Philosophies of probability

DONALD A. GILLIES

1 THE CLASSICAL VIEW OF PROBABILITY

Probability theory has it in common with quantum mechanics that there is a generally agreed mathematical calculus regarding the interpretation of which there are profound, and seemingly irreconcilable, differences. This article gives a brief exposition of the various interpretations of probability, paying attention to the historical order in which they appeared.

The mathematical calculus of probability is normally, and with some justice, taken to date from a correspondence in the year 1654 between Blaise Pascal and Pierre de Fermat, concerning two problems about gambling odds posed to Pascal by the inveterate gambler the Chevalier de Méré (§10.2). As the introduction of mathematical probability is often attributed to Pascal, it should be pointed out that it was Fermat, not Pascal, who solved the problem. By 1812, when Pierre Simon Laplace first published his *Théorie analytique des probabilités*, there was a considerable body of mathematical knowledge about probability. Apart from Laplace himself, the main contributors were Christiaan Huygens, James Bernoulli, Abraham De Moivre, P. R. de Montmort and Thomas Bayes. On the early history, see §10.1–10.2, and Todhunter *1865*, David *1962* and Hacking *1975*.

In 1814 (and later editions) Laplace published his *Essai philosophique sur les probabilités* which presented an interpretation of probability which had developed along with the mathematical calculus. This is known as the classical view of probability. In fact some of the key ideas are nascent in part IV of James Bernoulli's *Ars conjectandi* (1713).

Surprisingly, the theory of probability first developed at a time when most mathematicians and scientists, because of the success of Newtonian mechanics, believed in universal determinism. Indeed, in his *Essai*, Laplace *1819* gave the classic exposition of universal determinism. He imagined a vast intelligence which knows the entire state of the universe at a particular instant. Such a being, Laplace thought, would be able to calculate the entire future course of the universe. This vast intelligence would not need probability theory which, however, is necessary for us humans because of our ignorance. As Laplace said:

> Probability is relative, in part to this ignorance, in part to our knowledge. We know that of three or a greater number of events, a single one ought to occur; but nothing induces us to believe that one of them will occur rather than the others. In this state of indecision, it is impossible for us to announce their occurrence with certainty. (*1819*: 6).

Suppose, however, that there are n possible events, of which 1 must occur. Suppose, further, that we can judge these n events to be equally possible, and that m of them favour a particular outcome, A say; then

$$\Pr(A) := m/n. \tag{1}$$

This is the famous classical definition of probability.

The definition is well suited to analysing games of chance. For a throw of a regular die, there are six equally possible outcomes: 1, 2, 3, 4, 5 and 6. Of these, three (1, 3, 5) are favourable to the outcome 'odd', whose probability is thus $3/6 = \frac{1}{2}$.

The appropriateness of the classical definition for games of chance is of course no accident since, in the period from Fermat and Pascal to Laplace, the main (though not the exclusive) application of mathematical probability was to such games. It may seem odd that such a powerful mathematical and scientific tool as probability theory should emerge from such a frivolous pursuit as gambling, but the activities of gambling-houses can be seen as a vast, if inadvertent, experimental investigation of the laws of chance. Yet once probability theory began to cut itself free from gambling, and to be applied elsewhere in science, the classical definition no longer appeared adequate. It is doubtful, moreover, whether it can be applied satisfactorily to the case of the biased die. Thus in the second half of the nineteenth century and in the twentieth century, new interpretations of probability were devised, but before describing these it will be useful to explain how they may be classified.

2 EPISTEMOLOGICAL VERSUS OBJECTIVE INTERPRETATIONS OF PROBABILITY

An epistemological interpretation of probability is one in which probability is taken as a measure of the degree of knowledge, or degree of rational belief, or degree of belief of a human subject or subjects. The classical view of probability clearly falls into this category. An objective interpretation of probability, on the other hand, is one in which probability is taken to be an objective feature of the material world. Just as a radioactive atom has a particular mass, so it has a particular probability of disintegrating in a given period of time. This probability is a human-independent feature of the physical world; such probabilities existed before there were human beings.

The difference between these two kinds of interpretation is nicely illustrated by an example given by Laplace (*1819*: 56). Suppose you are shown a coin and told by someone you can trust that it is definitely biased. Suppose, however, that you are *not* told the direction of the bias. You are then asked what is the probability of heads on the next toss. If you accept an epistemological interpretation of probability, you should answer (as Laplace does) that $Pr(heads) = \frac{1}{2}$. If you accept an objective interpretation of probability, however, you should answer that $Pr(heads) = p$, where all we know about the value of p is that $p \neq \frac{1}{2}$.

The two main epistemological interpretations of probability are described in Sections 3 and 4.

3 THE LOGICAL INTERPRETATION

The logical interpretation of probability was developed by the Cambridge school of W. E. Johnson, John Maynard Keynes and H. Jeffreys. Two important works expounding this viewpoint are Keynes *1921* and Jeffreys *1939*. The basic idea is that probability theory can be taken as an extension of deductive logic. A simple example will illustrate this.

Let L be the law that all ravens are black. Let e_0 be the statement that George is a raven, and c_0 the statement that George is black. Then c_0 follows logically from e_0 & L, or, in other words, e_0 & L entail c_0.

Now let e be the statement that all ravens so far observed have been black. As David Hume was the first to observe in formulating the problem of induction, L does not follow logically from e; none the less, it might be held that e partially entails L to some degree p. Here we suppose that there is a logical probability of L, given e, which is equal to p. This logical probability is further identified with the degree of rational belief in L given e. As Keynes says:

> The Theory of Probability is logical . . . because it is concerned with the degree of belief which it is *rational* to entertain in given conditions, and not merely with the actual beliefs of particular individuals, which may or may not be rational. (*1921*: 4)

But if there really are logical probabilities, how do we get to know about them? Keynes postulated the existence of a faculty of logical intuition which enables us, in some cases at least, to have something like a direct perception of logical probabilities. Thus he wrote (*1921*: 13): 'We pass from a knowledge of the proposition *a* to a knowledge about the proposition *b* by perceiving a logical relation between them. With this logical relation we have direct acquaintance', though he also added (*1921*: 18) that 'Some men — indeed it is obviously the case — may have a greater power of logical intuition than others'. The trouble here is that this alleged faculty of logical intuition seems distinctly mysterious, and not to provide a very sound justification for adopting the usual axioms of probability.

When it came to measuring probabilities, Keynes had to fall back on Laplace's device of equally possible cases. In his system cases were judged to be equi-possible by using what he called the 'principle of indifference', but this principle gave rise to a series of perplexing paradoxes. These difficulties in the logical interpretation of probability gave rise to the subjective view.

4 THE SUBJECTIVE INTERPRETATION

The subjective interpretation was introduced independently by Frank Ramsey in England and Bruno de Finetti in Italy; their classic accounts of the position are to be found in Ramsey *1931* and de Finetti *1937*. Ramsey was at Cambridge, and he introduced the new interpretation of probability in explicit reaction against the logical interpretation of the earlier Cambridge school. Indeed, he began his paper with a number of criticisms of Keynes, one of which was that

> there really do not seem to be any such things as the probability relations he describes. He supposes that, at any rate in certain cases, they can be perceived; but speaking for myself I feel confident that this is not true. I do not perceive them, ... moreover I shrewdly suspect that others do not perceive them either .... (*1931*: 65–6)

Had this criticism been uttered by a lesser mortal, Keynes might have replied that his critic was obviously one of those who have weak powers of logical intuition. But such a retort would hardly have been appropriate to Ramsey, one of the greatest logicians of his time (§5.2).

According to the logical view of probability, all rational people given the same evidence, *e*, will have the same degree of rational belief in a prediction, *a* say. The subjective theory drops this consensus assumption. Different subjects A, B, C, ..., although they are all rational, and all share the same evidence *e*, may none the less have different degrees of belief in *a*.

But how can we measure the degree of belief of Mr A in *a*, since, outside science fiction, there exists no electronic apparatus which we can attach to Mr A's head and which will register on a dial his exact degree of belief in a given proposition? Ramsey (*1931*: 73) suggested another, more realistic method: 'The old-established way of measuring a person's belief is to propose a bet, and see what are the lowest odds which he will accept. This method I regard as fundamentally sound.' If the bet situation is specified more precisely in a certain way, the rate at which Mr A bets is called his 'betting quotient'. On the subjective theory, then, the probability of a proposition a, $\Pr(a)$, is always relative to a particular person, Mr A. $\Pr(a)$ is then Mr A's degree of belief in a, and is measured by Mr A's betting quotient on a.

But do betting quotients satisfy the standard axioms of probability? To investigate this problem, Ramsey and de Finetti introduced the concept of coherence, which is defined as follows. If Mr A is betting against Ms B on events $E_1, \ldots, E_n$, his betting quotients $q_1, \ldots, q_n$ are said to be coherent if Ms B cannot choose stakes $S_1, \ldots, S_n$ such that she wins whatever happens. Now, Mr A, if he is at all reasonable, will surely wish to prevent his opponent Ms B winning money from him, whatever happens. He will therefore try to make his betting quotients coherent. But a remarkable result known as the Ramsey–de Finetti theorem shows that a set of betting quotients $q_1, \ldots, q_n$ is coherent if and only if it satisfies the standard axioms of probability.

This development establishes the subjective interpretation as one possible interpretation of the mathematical theory of probability. But is it the only interpretation? De Finetti thought it was, but such a monistic view is not forced on us – there may be more than one possible interpretation of the probability calculus. Take, for instance, the example mentioned in Section 2 of the probability of a particular radioactive atom disintegrating in a given period of time. Is it not more plausible to say that this probability is an objective feature of the physical world rather than the subjective betting quotient of a particular individual? To explore this conception further, it is necessary to examine the two main objective interpretations of probability.

5 THE FREQUENCY THEORY

The frequency theory was first developed by another Cambridge school, that of R. L. Ellis and John Venn, in the middle of the nineteenth century, and can be considered as a 'British empiricist' reaction against the 'Continental rationalism' of Laplace and his followers. Venn *1866* is the classic exposition of this early version of the theory. The leading advocate of the theory in the twentieth century was Richard von Mises, who made considerable advances beyond Venn from 1919 onwards. This section is based largely on von Mises *1939* (original text, 1928).

The fundamental idea of von Mises' theory (as he himself said) was to consider probability theory as a science of the same rank as theoretical mechanics. The science of mechanics deals with mass phenomena and repetitive events. An example of a mass phenomenon is provided by the molecules in a fixed volume of gas, while a series of tosses of a coin gives an example of repetitive events. In cases like these, a certain 'attribute' occurs at each of the 'events' which make up the mass phenomenon, but this attribute varies from one event to another. For a gas molecule, the attribute might be its velocity, while for a coin the two basic attributes are 'heads' and 'tails'.

Von Mises introduced the technical term 'collective' to describe mass phenomena or repetitive events of the above types. Suppose that the events of the collective can be ordered, and that in the first n events a particular attribute A occurs $m(A)$ times. Then von Mises took it to be an empirical law, established by experience, that the relative frequency of A, i.e. $m(A)/n$, tends to a fixed limit as n increases. In the mathematical treatment, a collective is represented by an infinite sequence, $\mathscr{L}$ say, and the first axiom is that, for each attribute A, $\lim_{n \to \infty} m(A)/n$ in $\mathscr{L}$ exists. The probability of A in $\mathscr{L}$ can then be defined as this limiting frequency.

This definition clearly does not apply to some cases where we speak of probabilities. Von Mises gave the example of the probability of Germany being at some time in the future involved in a war with Liberia (*1939*: 9). His answer to this difficulty was that such probabilities may occur in ordinary language, but that they have no place in the mathematical theory which applies only where there is a collective. Hence his famous maxim: 'first the collective – then the probability' (*1939*: 18).

Von Mises also stressed another feature of the collective, namely its lack of order, or its randomness. Again, he regarded the randomness of a collective as an empirical law, established by experience. In an ingenious fashion he related the disorder or randomness of a collective to the lack of success of attempted gambling systems, and so called the law of randomness the 'law of excluded gambling systems'. This approach led him to suggest a definition of randomness which was later refined and made mathematically rigorous using the theory of computable functions (§5.5) by Alonzo Church *1940*.

6 THE PROPENSITY THEORY

Some version of the frequency theory was accepted in the 1930s by most members of the Vienna circle, and even by Karl Popper. Later, however, Popper grew dissatisfied with the frequency theory and proposed a new objective, but non-frequency, theory of probability known as the propensity theory (see Popper *1959b*; for a fuller exposition see Popper *1983*: Part II).

Popper's starting-point was a consideration of the question of probabilities of singular events, or singular probabilities, within the frequency theory. Consider for example the probability of a German aged 40 in 1928 living to be 41. Such a probability is very easily defined within von Mises' frequency theory: we take as our collective the set of Germans aged 40 in 1928, and the frequency of those within the collective who live to be 41 gives us the required probability. But suppose a particular German, aged 40 in 1928 (Herr Schmidt, say), is interested in the probability of *his* living to be 41. Can we define such a singular probability within the frequency theory? Von Mises answered thus:

> We can say nothing about the probability of death of an individual even if we know his condition of life and health in detail. The phrase 'probability of death', when it refers to a single person, has no meaning at all for us. This is one of the most important consequences of our definition of probability.... (*1939*: 11)

Popper, however, wanted to introduce singular probabilities, particularly because of some problems in quantum mechanics. He therefore suggested that the singular probability of an event in a collective might be taken as the same as the probability in the collective as a whole. This suggestion, however, leads to difficulties – as Popper was one of the first to point out.

Consider (with Popper *1959b*) a collective which consists mainly of throws of a biased die for which $\Pr(5) = \frac{1}{4}$, but with a few throws of a normal die interspersed. According to Popper's original definition, the singular probability of 5 on a throw of the normal die will be $\frac{1}{4}$, whereas intuitively it ought to be $\frac{1}{6}$. To avoid this difficulty he suggested that we associate objective probabilities, not with collectives but with sets of repeatable conditions: 'we have to visualize the conditions as endowed with a tendency or disposition, or propensity, to produce sequences whose frequencies are equal to the probabilities; which is precisely what the propensity interpretation asserts' (*1959b*: 35).

Popper's suggestion of an objective, but non-frequency, interpretation of probability has aroused considerable interest, and there have been several attempts, along somewhat different lines, to develop such a view. In this connection see Hacking *1965*, Mellor *1971*, and Gillies *1973*.

BIBLIOGRAPHY

Church, A. *1940*, 'On the concept of a random sequence', *Bulletin of the American Mathematical Society*, **46**, 130–32.

David, F. N. *1962, Games, Gods, and Gambling*, New York: Hafner.

de Finetti, B. *1937*, 'La Prévision: Ses lois logiques, ses sources subjectives', *Annales de l'Institut Henri Poincaré*, 7, 1–68. [Transl. as 'Foresight: Its logical laws, its subjective sources', in H. E. Kyburg and H. E. Smokler (eds), 1964, *Studies in Subjective Probability*, New York: Wiley, 95–158.]

Gillies, D. A. *1973, An Objective Theory of Probability*, London: Methuen.

Hacking, I. *1965, The Logic of Statistical Inference*, Cambridge: Cambridge University Press.

—— *1975, The Emergence of Probability*, Cambridge: Cambridge University Press.

Jeffreys, H. *1939, Theory of Probability*, Oxford: Oxford University Press. [Repr. 1963.]

Keynes, J. M. *1921, A Treatise on Probability*, London: Macmillan. [Repr. 1963.]

Laplace, P. *1819, Essai philosophique sur les probabilités*, 4th edn, Paris: Courcier. [English transl. 1901, New York: Wiley; various reprints, cited here.]

Mellor, D. H. *1971, The Matter of Chance*, Cambridge: Cambridge University Press.

Popper, K. R. *1959a, The Logic of Scientific Discovery*, London: Hutchinson. [German original 1934.]

—— *1959b*, 'The propensity interpretation of probability', *British Journal for the Philosophy of Science*, **10**, 25–42.

—— *1983, Realism and the Aim of Science*, London: Hutchinson.

Ramsey, F. P. *1931*, 'Truth and probability', in his *Foundations of Mathematics and Other Logical Essays* (ed. R. B. Braithwaite), London: Routledge & Kegan Paul, 156–98. [Written in 1926; repr. in H. E. Kyburg and H. E. Smokler (eds), 1964, *Studies in Subjective Probability*, New York: Wiley, 63–92.]

Todhunter, I. *1865, A History of the Mathematical Theory of Probability from the Time of Pascal to that of Laplace*, Cambridge and London: Macmillan. [Repr. 1949, New York: Chelsea.]

Venn, H. *1866, The Logic of Chance*, 1st edn, London: Macmillan. [Repr. of 3rd edn of 1888, 1963, New York: Chelsea.]

von Mises, R. *1939, Probability, Statistics and Truth*, London: Hodge. [2nd rev. English edn, 1951, London: Allen & Unwin. German original 1928.]

10.18

Mathematical economics

GIORGIO ISRAEL

1 INTRODUCTION

The history of the applications of mathematics to economics shows some distinctive features which are also characteristic of the history of applications to biology (§9.18). In both cases it is possible to distinguish a phase in which mathematics was conceived of mainly as merely a technical aid to research, from a phase in which mathematics served as the conceptual core of a well-defined methodology of research. This last phase (which took on definite shape in the course of the nineteenth century) is characterized by a reductionistic approach in which the conceptual model of reference is mathematical physics, in particular mechanics. The two phases may also be characterized by the kind of mathematical methods used in each, the first being oriented towards statistical and probabilistic methods, while the second centred itself around a deterministic approach and made use of the infinitesimal calculus.

However, here end the analogies: the first phase cannot be correctly considered to be the 'prehistory' of the second. In fact, the technical level and the degree of sophistication of the mathematical methods used in the first phase (which we could call the period of research in the field of 'political arithmetic') were far higher than those at work in the first developments of the 'deterministic-mechanist' mathematical economics of the nineteenth century. When, at the end of the eighteenth century, the Marquis de Condorcet formulated a very general programme of a *mathématique sociale* based on the use of the calculus of probability, he was referring to a long tradition in research; the new trends of 'mathematical economics', however, lacked any previous tradition. This is one of those typical cases in the history of science in which a specific research trend prevails, not by virtue of a high level of technical achievement, but on the grounds of a programmatic choice (or 'paradigm'): in this case, the paradigm was certainly determined by the 'force of attraction' of the models of mechanics and classical mathematical physics (Mirowski *1989*).

2 POLITICAL ARITHMETICS AND SOCIAL MATHEMATICS

The first applications of mathematics in the social sciences took place in the context of what we now call population statistics, and were therefore strictly tied to the birth of statistics and the calculus of probability (Hacking *1975*, Daston *1988*). Among the first contributions were John Graunt's analysis of the London Bills of Mortality (1662) and the essays by William Petty on what he called 'political arithmetic' (see in particular his *Essay on Political Arithmetick*, 1682); that is, the mathematical study of the dynamics of population. The most significant problems in this context concerned annuities, in connection with insurances and mortality tables (§10.3). A very rigorous attempt to establish equitable prices for annuities was contained in the report presented in 1671 by Jan De Witt to the Estates-General of the United Provinces of Holland and West Friedland. As Daston *1988* observes, De Witt's contribution is original in that he tried to estimate the probability of death as a correlate of age, and to extend Christiaan Huygens' calculus of expectations.

The development of 'political arithmetic' (until its decline at the beginning of the nineteenth century) is connected with the development of the calculus of probability, but the concept of probability in this context was markedly subjectivist (on this, see §10.17). Some authors (see e.g. Dessì *1989*) have pointed out the apparently paradoxical fact that this subjectivist approach was the root cause of the general acceptance of the use of mathematics in the social sciences. In fact, the use of a mathematical tool having an inferior epistemological status (i.e. lacking any objective value) was acceptable even to the large number of those who believed that social and economic sciences could not be developed by following an objective approach having the same rank as the physico-mathematical sciences.

This state of affairs emerges very clearly from the debate which took place in the eighteenth century, in particular between Daniel Bernoulli and Jean d'Alembert, on the possibility of a rigorous mathematical analysis of the advantages of inoculation against smallpox (§10.12). The discussion concerned the legitimacy of using the calculus of probability in the scientific analysis of non-physical phenomena. This also explains why many outstanding scientists (such as Daniel Bernoulli, Joseph Louis Lagrange, d'Alembert himself and, later, Pierre Simon Laplace and Siméon-Denis Poisson) were interested in 'political arithmetic' and made some important contributions to this branch without discussing whether or not these results could be considered to be of the same epistemological level as in mathematical physics.

The research in the field of population statistics and 'political arithmetic' inspired a very important development, at least from the methodological viewpoint: Condorcet's programme of a *mathématique sociale* (Baker *1975*, Granger *1989*). His project, which was formulated with the important collaboration of E. E. Duvillard de Durand (Israel *1991*), was very ambitious and general: its aim was to provide the 'moral sciences' (i.e. the sciences whose purpose is human happiness) with a grounding, in the form of a mathematical rigour and an empirical methodology, no less solid than those of the physico-mathematical sciences. The concrete achievements of Condorcet's programme (in which the calculus of probability had a central role) were modest and restricted to the study of the probabilities that the decisions taken by majority vote at assemblies (such as a legislative assembly or a jury) will lead to correct judgements (§10.8; see also Crépel *1988*). In tackling this problem, Condorcet introduced the concept of *homo suffragans* (an abstraction, a kind of 'social atom' stripped of all qualities except the 'social' faculty of voting), suggested by the concept of mass-point in mechanics and from which derived the notion of *homo oeconomicus* which is at the centre of modern mathematical economics.

3 THE BEGINNINGS OF MATHEMATICAL ECONOMICS

The first developments of mathematical economics were in some ways indicative of an ambition more restricted than that of mathematicizing all moral, political, economic and social sciences, which was the distinctive feature of the programme of the *mathématique sociale*; However, they did demonstrate a clear methodological approach; and also a choice of the mathematical methods, through complete adherence to the model of mechanics, to the deterministic approach and to the use of the infinitesimal calculus. The concept of equilibrium was the main inspiration for these new developments (Ingrao and Israel *1990*). In fact, emphasis on the concept of equilibrium is a strong characteristic of the beginnings of mathematical economics, however sporadic and fragmentary they might be; this explains the central role that the theory of general economic equilibrium was to have throughout the history of mathematical economics. These trends were influenced mostly by the *physiocrates* (especially François Quesnay and A. R. J. Turgot) and by many members of the *idéologues* (another philosophical group), who developed their activities mostly in the context of the Classe de Sciences Morales et Politiques at the Institut de France near the end of the eighteenth century (Israel *1991*).

Among the first specific contributions was the first attempt to determine the economic equilibrium as the solution of a system of equations, by A. N. Isnard in his *Traité des richesses* (1781); and the first representation in geometric terms of a process of price adjustment, by the *physiocrate* Pierre

Samuel Du Pont de Nemours, in his *Des courbes politiques* (1774). But the more important contribution of the early developments are undoubtedly the *Principes d'économie politique* (1801) by the mathematician N. F. Canard, who was awarded a prize by the Institut de France. Canard's work, however naively based on elementary algebraic methods, contained the first complete example of a marginal analysis applied to the determination of prices deduced from an analysis of the competition mechanisms of the market, and a definition of market equilibrium suggested by both the equilibrium of a lever and by stationary equilibrium in hydrodynamics. The originality of Canard's contribution consists also in combining the static concept of equilibrium with a dynamic approach, and in making the first attempt at the analysis of monopoly, anticipating an analysis later developed by A. Cournot (Section 4).

Many other works contributed to the development of the applications of mathematics to economics, in particular in the context of engineering (Franksen and Grattan-Guinness *1989*; see also §6.11). The contribution of J. Dupuit (1844) to a modern definition of the measurability of utility is especially noteworthy.

4 COURNOT, WALRAS AND PARETO

Cournot's *Recherches sur les principes mathématiques de la théorie des richesses* (1838) is important both for its method and for its specific results. Concerning the first aspect, the originality of Cournot is the first clear formulation of the thesis that the purpose of mathematical economics is essentially theoretical: its aim was not to offer tools for numerical calculation and practical applications to the real economy, but to discover the general laws governing its evolution. He was the first to indicate how to apply mathematics (especially the theory of functions and the infinitesimal calculus) to the study of economic problems without having to specify more than the general properties of the functions involved. As for his specific results, there are essentially two: the first was a clear statement of the law regulating the relationships between the quantity of a good demanded and the monetary price of the good in a single market (*loi du débit*); the second was the first formal analysis of the two types of seller's monopoly and of perfect competition. Although ignored by his contemporaries, Cournot's contribution was to assume a fundamental role in the scientific training of Léon Walras.

Without ignoring the important contributions made by William Stanley Jevons (*Theory of Political Economy*, 1871) and Francis Edgeworth (*Mathematical Psychics*, 1881), Walras is undoubtedly of such central importance that he is generally considered to be the father of modern mathematical economics. His work, which started with the *Principes d'une théorie mathématique de l'échange* (1874), culminates in the *Eléments d'économie politique pure* (the fourth edition of 1900 is the most complete) and concludes with the short paper 'Economique et mécanique' (1909). Walras's scientific programme was based on the idea of building a political economy in mathematical form, following the example of Isaac Newton's mechanics, or in his phrase, an 'analytical economics' as a counterpart of analytical mechanics (see §8.1 on mechanics).

At the core of Walras's theory lay the concept of general economic equilibrium (formulated by analogy with the concept of static equilibrium) and the law of price formation (which for him played a role similar to that of the law of universal gravitation in mechanics). He was not a brilliant mathematician and did not solve any of the main technical problems of his theory; he did, however, make an outstanding contribution to the formulation of the modern structure of the theory of general economic equilibrium. In his breadth of vision and capacity to frame the problem of exchange in all its generality, he surpassed Jevons, whose priority in formulating the concept of 'marginal utility' (where the calculus was used) he acknowledged. In fact, one of the main specific achievements of Walras was the complete integration of utility theory into the general problem of the determination of the equilibrium prices and the formulation of the three major analytical problems of the theory: the existence, uniqueness and global stability of the equilibrium. The last problem is the equivalent in formal terms of Adam Smith's metaphor of the 'invisible hand' – the idea that the market is acted on by 'forces' tending to maintain it in a state of equilibrium which is consistent with the complete independence of the actions of the individual economic agents.

Walras encountered many difficulties in spreading his theories in France. He worked mostly in Lausanne (Switzerland), where he created the so-called 'Lausanne school'. His successor in the chair of political economics was Vilfredo Pareto, who developed his research programme while stressing its objective character, parallel to theories in classical mathematical physics. The most important of Pareto's contributions to mathematical economics are collected in his *Cours d'économie politique* (1896–7) and the *Manuale di economia politica* (1906). His most famous specific contribution is the notion of 'Pareto's optimum', a state of the economic system such that any shift from it worsens the position of at least one economic agent without improving that of others.

Pareto, more than Walras, considered as fundamental the empirical verification of the theory. This approach led him to deplore the absence of a good correspondence between the theory and the empirical facts, until attempting to resolve this difficulty by means of sociology. From this point of view, the last phase of Pareto's work was an important contribution to the analysis of the critical aspects of classical theory.

During this period a 'neo-classical' school developed in Britain, largely under the influence of Alfred Marshall. They focused on pricing, and introduced the concept of partial equilibrium (that is, dependent on time). Followers and disciples included John Maynard Keynes (Whitaker *1990*).

5 VON NEUMANN'S CONTRIBUTION

So, the main aspects of classical mathematical economics are, roughly, the idea of a strict analogy with the concepts and methods of classical mathematical physics, the centrality of the concept of equilibrium, the programme of demonstrating its existence, and the uniqueness and stability of general economic equilibrium. In spite of the fact that the concept of equilibrium still has the central position in mathematical economics, the total adherence to the model of physics (and in particular of classical mechanics) began to relax at the beginning of the twentieth century, as a more abstract and formal approach began to spread in all branches of science, particularly in the development of mathematical modelling. This abstract and formal approach inspired the research developed in Karl von Menger's colloquium in Vienna in the 1930s. In this context, the mathematician Abraham Wald achieved the demonstration of the first theorems of existence, uniqueness and global stability of equilibrium (1934–6), starting from the analysis of G. Cassel's and K. Schlesinger's version of Walras's equations. However, Wald's results were based on some strong hypotheses (in particular, all the commodities of economies were considered as 'substitutes') in order to make use of the techniques of calculus.

A turning-point was the contribution of John von Neumann, who in 1937 demonstrated a theorem of existence of equilibrium for a production economy which was far more general than Wald's theorems. Von Neumann's contribution to mathematical economics is fundamental, both methodologically and technically. From the point of view of economic theory, his approach had very little in common with Walras's: the analysis centred on the subjective behaviour of the economic agent as dictated by the purpose of maximizing utility and profit was replaced by a vision of the economy as a 'game' in which every 'player' follows some rules of rational behaviour, a 'strategy', designed to maximize winnings and minimize losses. He was therefore applying to economics the results of game theory which he had already developed mathematically with the demonstration of the minimax theorem, and which he considered the best framework in which to describe the interaction between individuals in the economic reality.

1420

The kind of mathematics used in this approach was markedly different from the classical approach, founded on the systematic use of the infinitesimal calculus: inequalities (rather than equations), functional analysis, convex analysis (see §6.11 on linear optimization) and topology. But the greatest modification to the course of mathematical economics, at least from the technical point of view, was the acknowledgement of the importance of the fixed-point theorems. In his 1937 paper, von Neumann made use of a lemma (today called 'Kakutani's theorem') which was a generalization, to the case of upper-semicontinuous multi-valued functions, of L. E. J. Brouwer's fixed-point theorem (1910) which was originally for continuous functions (§7.10).

The new direction indicated by von Neumann's game-theoretic approach to mathematical economics gave rise to a collaboration with Oskar Morgenstern, at the beginning of the 1940s, when both scientists were in the USA. The result was the seminal book *Theory of Games and Economic Behavior* (1944). This book also shed new light upon mathematical economics, in particular Walras's theory and his unsolved mathematical problems. Game theory had a strong development and is still very important today.

6 GENERAL EQUILIBRIUM IN THE USA

The emigration of many European scientists to the USA as a consequence of the spread of Nazism played a fundamental part also in the history of mathematical economics. A tradition in this field already existed in the USA. Particularly important were the work of I. Fisher (author of *Mathematical Investigations in the Theory of Values and Prices*, 1892), who was a pupil of the physicist Josiah Willard Gibbs; the researches of H. Hotelling in New York in the 1930s; and the activities of the Cowles Commission for Research in Economics, founded by A. Cowles in 1932. However, it is unquestionable that the emigration to the USA of scientists like Wald and, above all, Morgenstern and von Neumann greatly stimulated the development of research in this field. Following von Neumann's approach, and making use of the results obtained by J. F. Nash (1950), K. J. Arrow and G. Debreu demonstrated in 1954 the theorem of existence of equilibrium for the Walras model under very general hypotheses; this was perhaps the most important achievement of the theory. Two other lines of research were developed on the problem of the uniqueness of the equilibrium and on the problem of global stability (i.e. the convergence to equilibrium of the price-adjustment processes). The first line was developed mainly by Debreu, with the demonstration of a series of theorems showing the finiteness of the number of equilibria by means of techniques of differential topology. The stability problem was stated in a general form by P. Samuelson in 1941, and was studied by Arrow and L. Hurwicz at the end of the 1950s; the results were modest, until H. Scarf and H. Sonnenschein demonstrated that it was almost impossible to obtain satisfying results in this field.

A new trend in the field of general economic equilibrium was set by the systematic introduction of the methods of differential topology, but above all by the theory of differentiable dynamical systems, in particular by the mathematician S. Smale. It was a return of the 'mathematics of time' – that is, of the classical calculus, but reinterpreted from the point of view of the modern qualitative theory of ordinary differential equations – after the period of 'static mathematics', under the influence of von Neumann's conception. Of note also is the research in the 1970s by Scarf on the determination of equilibria by means of a numerical implementation of the techniques of the fixed-point theorems.

7 LATER WORK AND SOME OTHER DEVELOPMENTS

In spite of the central role of the microeconomic theory in mathematical economics (the 'tool box' of every good economist, following J. A. Schumpeter), many other lines of research have played their part. There have been linear programming (§6.11) and dynamic programming, and above all the research field which developed out of Keynes's macroeconomic theory. Macro- as opposed to microeconomics studies the economy not by analysing the behaviour of the single economic agent, but from a consideration of a few important aggregate variables.

Mathematically, theoretical macroeconomic models make use of techniques and trends quite analogous to those used in microeconomic theory; applied macroeconomics uses statistical, probabilistic and econometric methods. Also, so-called macroeconometric models have been developed for studying economies and for economic forecasting. The core of macroeconometric models is a system of macroeconomic equations, as many as are needed to achieve an accurate description. In order to analyse the trends of a real economy by means of econometric and numerical methods, the role of the computer is decisive. The interest in macroeconometric models is still very widespread, but their performance has been seriously questioned by the discovery of the so-called 'deterministic chaos', which has shown the difficulty of forecasting over anything other than the short term.

BIBLIOGRAPHY

Baker, K. M. *1975, Condorcet. From Natural Philosophy to Social Arithmetic*, Chicago, IL: Chicago University Press.

Baumol, W. J. and Goldfeld, S. M. (eds) *1968, Precursors in Mathematical Economics. An Anthology*, London: The London School of Economics and Political Science (Series of Reprints of Scarce Works on Political Economy, No. 19).

Crépel, P. *1988*, 'Condorcet, la théorie des probabilités et les calculs financiers', in *Sciences à l'Epoque de la Révolution Française* (ed. R. Rashed), Paris: Blanchard, 267–325.

Daston, L. *1988, Classical Probability in the Enlightenment*, Princeton, NJ: Princeton University Press.

Dessì, P. *1989, L'ordine e il caso. Discussioni epistemologiche e logiche sulla probabilità da Laplace a Peirce*, Bologna: Il Mulino.

Franksen, O. I. and Grattan-Guinness, I. *1989*, 'The earliest contribution to location theory? Spatio-economic equilibrium with Lamé and Clapeyron, 1829', *Mathematics and Computers in Simulation*, **31**, 195–220.

Granger, G.-G. *1989, La Mathématique sociale du Marquis de Condorcet*, Paris: Jacob.

Hacking, I. *1975, The Emergence of Probability*, Cambridge: Cambridge University Press.

Ingrao, B. and Israel, G. *1990, The Invisible Hand. Economic Equilibrium in the History of Science*, Cambridge, MA: MIT Press.

Israel, G. *1991*, 'El declive de la "mathématique sociale" y los inicios de la economia matematica en el contexto de los avatares del Institut de France', *Llull*, **14**, 59–116.

Mirowski, P. *1989, More Heat Than Light. Economics as Social Physics: Physics as Nature's Economics*, Cambridge: Cambridge University Press.

Schumpeter, J. A. *1954, A History of Economic Analysis*, New York: Oxford University Press.

Teocharis, R. D. *1983, Early Developments in Mathematical Economics*, London: Macmillan.

Weintraub, E. R. *1986, General Equilibrium Analysis: Studies in Appraisal*, Cambridge: Cambridge University Press.

Whitaker, J. K. (ed.) *1990, Centenary Essays on Alfred Marshall*, Cambridge: Cambridge University Press.

Part 11
Higher Education and Institutions

11.0

Introduction

This Part of the encyclopedia is devoted to surveys of mathematical institutions, societies and professionalization in the principal countries, usually since the sixteenth or seventeenth century (thereby overlapping a little with §2.11). The order in which countries are surveyed is very roughly the order of their overall mathematical importance combined with geographical proximity, but it is more arbitrary an ordering than those chosen for other Parts of the encyclopedia. The only specific strategy has been to take Europe first (with the influence of Spain and Portugal upon Ibero-America noted in §11.9) before considering the USA and Canada in §11.10; over recent decades the USA has become the leading country for mathematics, but its rise to importance started only in the late nineteenth century.

The principal concern is with the mathematics *in* a country, not of its natives; thus, for example, the work of the Italian Joseph Louis Lagrange in Paris is noted in §11.1 on France. It has not been possible to treat every country; for references to literature on many countries and some cities, see May *1973*: 647–79. Further noteworthy sources include Graf *1888–90* on the Bernese canton of Switzerland, and Plancherel *1960* for the later years of the country in general; Boucher *1973* for some details of developments in South Africa; Kuratowski *1980* on the remarkable rise of Poland in the 1920s; and Tee *1988* on Australia and New Zealand. On particular types of institution, Barnard *1862* contains many data, and other kinds of information, on the military schools of his time in many countries; usually it is very difficult to find material of this kind.

Finally, the journal *L'Enseignement mathématique* included articles on mathematical education and institutions from its inception in 1899 (see the references in Fehr *1955*). In particular, it included on a regular basis reports of the International Commission of Mathematical Education, which Felix Klein and others started in the 1900s; up to the early 1920s it produced a vast number of books and pamphlets on mathematical education in many countries and from numerous points of view (§12.13).

One feature of this multinational story is worth mentioning here: the considerable differences between the structures that were (or were not) developed in different countries, especially in the nineteenth century when institutional developments accelerated considerably. From these differences came a wide variation in the relative importance of countries and of the branches of mathematics in which a country might (or might not) specialize. For example, Italy and some German states rather dominated the early modern period; and the French were besotted with mathematical analysis, while the English liked algebras and some aspects of geometries in the nineteenth century. This point has emerged in some articles in earlier Parts; §10.6–10.7 were largely centred around Russian and English contributions to aspects of probability and statistics.

The type of history of mathematics treated in this Part is not well developed, and there are many large gaps in our knowledge. For example, partly for lack of information (but also due to the huge scale of the activities involved), very little is said here on education at school level; again, far too little is known about mathematics and science publishers for a companion piece on them to have been commissioned. However, this Part is completed by two other international themes (§11.11–11.12) – the place of women, and the general rise of mathematical journals.

This Part (and to some extent the next) is much concerned with the social history of mathematics; in Bos *et al. 1981* various aspects are treated for the nineteenth century. Mathematics turns up occasionally in the general literature of social history; in the books and journals devoted to the history of science or of education (and also the annual *History of Universities*); and those concerned with mathematical education. Many distinguished universities and academies in several countries have their own histories, in which mathematics is mentioned, usually only in brief.

BIBLIOGRAPHY

Barnard, H. *1862, Military Schools and Courses of Instruction in the Science and Art of War, in France, Prussia, Austria, Russia, Sweden, Switzerland, Sardinia, England and the United States*, 2 parts, Philadelphia, PA: Lippincott.

Bos, H. J. M., Mehrtens, H. and Schneider, I. (eds) *1981, Social History of Nineteenth Century Mathematics*, Basel: Birkhäuser.

Boucher, M. *1973, Spes in arduis. A History of the University of South Africa*, Pretoria: University of South Africa.

Fehr, H. *1955*, 'Table des matières recapitulative des tomes I à XL', *L'Enseignement Mathématique*, Series 1, **40**, 124–74.

Graf, J. H. *1888–90, Geschichte der Mathematik und der Naturwissenschaften in bernischen Länden* [. . .], 3 parts, Berne and Basel: Wyss.

Kuratowski, K. *1980, A Half Century of Polish Mathematics*, Oxford: Pergamon, and Warsaw: Polish Scientific Publishers.

May, K. O. *1973, Bibliography and Research Manual in the History of Mathematics*, Toronto: University of Toronto Press.

Plancherel, M. *1960*, 'Mathématiques et mathématiciens en Suisse (1850–1950)', *L'Enseignement Mathématique*, Series 2, **3**, 194–218.

Tee, G. J. *1988*, 'Mathematics in the Pacific basin', *British Journal for the History of Science*, **21**, 410–17.

11.1

France

I. GRATTAN-GUINNESS

1 A SURPRISING SYSTEM

The annual publication *The World of Learning* prepares its chapter on France in the usual way, giving most space to the universities and providing brief entries on other educational institutions. But this will not do for France, where, since the Revolution of 1789, most of the prestige has been assigned to the *grandes écoles* (as they came to be known).

The evolution of this structure is a product of the French Revolution, as shown in Sections 3–5, where the origins of these schools and of the modern universities are charted. Other aspects of institutional growth in the nineteenth century are noted in Sections 6 and 7, which bring us up to the twentieth century. To frame this material, Section 2 traces some main lines for the seventeenth and eighteenth centuries, while Section 8 considers the question of the alleged decline of French science in the light of the eminence of its mathematicians.

There are a large number of histories of French education, but as usual mathematics does not fare well: for example, mathematics in many of the *grandes écoles* has been poorly studied. Ponteil *1966* is quite a good example of the genre, which at least provides the general context; but otherwise the bibliography is confined to more pertinent writings.

The variations of French regime over the centuries has caused the adjectives *royal*, *impérial* and *national* to come and go in the names of institutions; for simplicity, they have usually been omitted.

2 THE SEVENTEENTH AND EIGHTEENTH CENTURIES

As in other countries, science did not enjoy a good professional base in France in the sixteenth and early seventeenth centuries: for example, until his death in 1648 Marin Mersenne was a major circulator of information via his correspondence. But important advances date from the 1660s, when the Académie des Sciences was founded and began to publish its journal the

Histoire (Hahn *1971*), and the *Journal des Scavans* was launched as a private but scholarly enterprise (Sergescu *1947*). Both journals were of major importance from the start, the latter especially for its reviews.

The Académie des Sciences was the main centre for science, with membership a cherished aim of the *savant* (Académie des Sciences *1979* is a valuable biographical record of members). Disputes and intrigues soon became the common fare: an important case in the 1730s and 1740s concerned the shape of the Earth (§8.14), when Isaac Newton's mechanics (but not his calculus!) were adopted and its consequences explored in detail (Greenberg *forthcoming*). By then the generation of Alexis Fontaine, Alexis Clairaut and Jean d'Alembert were installed, and the calculus, especially its extension to multivariate form, was a major preoccupation.

Of the publications, mention must be made of the famous *Encyclopédie* (1751–65), edited by Denis Diderot and d'Alembert. The latter was largely responsible for the mathematical articles, especially on the pure side. The work became a reference point for many subjects, although d'Alembert's view on particular aspects of mathematics was not always followed. It was succeeded from 1783 by the *Encyclopédie méthodique*, which treated all aspects of mathematics in various of its parts; it was not completed until 1828.

Science education in the eighteenth century is surveyed in Taton *1964*. For school-level mathematics teaching, three main types of institution were already in place in the seventeenth century: the Jesuit colleges, where a good level of instruction in algebra, geometry and trigonometry was given, with applications to optics; the *Oratoire*, founded in 1611 and by the eighteenth century comprising around thirty colleges, where elements of pure and applied mathematics were preached in impressive measure; and a few Benedictine colleges. At university level, Paris was of course the most important, but there was no faculty of science there (or at any of the other 21 universities): mathematics was taught in the constituent colleges, but not at a level much higher than at those institutions just mentioned. The highest level was probably practised at the Collège Royal, which was founded in 1530 to provide education for its own sake, without the pressure of career or examinations (Sédillot *1869*). Among the early mathematics professors were Petrus Ramus, a controversial figure there, who left money for a chair in mathematics. His successors included the physicist Joseph Sauveur (1653–1716), and the physics chair was held for a time by J. A. J. Cousin (1739–1800), who taught much applied mathematics and produced a good treatise on differential equations in 1796.

There was another tradition in place also: that of engineering schools. Among the military ones, that at Mézières, founded in 1748, became the most eminent, and had the Abbé Bossut and Gaspard Monge on its staff. Among the civil ones, the Ecole des Ponts et Chaussées, founded one year earlier, was the most important centre for mathematics. (The Ecole des Mines, which also dates from this time, came to mathematics only in the twentieth century.) Schools for 'hydrography', which also covered cartography, were set up. In all cases the staple fare was mechanics, and applications such as ballistics, hydraulics, structures and embankments; but pure mathematics was also taught. The *Cours de mathématiques* of Etienne Bézout, which became enormously popular and remained so for many years, was originally published (1764–9) to meet the needs of guardsmen and mariners.

3 REVOLUTIONARY REFORM: THE ÉCOLE POLYTECHNIQUE

After the Revolution all educational institutions were closed or suspended. This and the next three sections summarize the new system that was brought in and lasted for a hundred years. Its internal contrasts provide the theme of Grattan-Guinness *1988*; more details up to 1840 are provided in Grattan-Guinness *1990* (especially Chaps 2, 16 and 19).

This section is concerned with the engineering schools, of which the Ecole Polytechnique was the centre. A new institution founded in 1794, it took on definitive form around 1800, to which this summary is confined (on its rather chaotic start, see Langins *1987*). The Ecole offered a two-year preparatory course for civil and military engineers, thereby setting up a double aim which often was not resolved. Graduates of the school became known as *polytechniciens*.

Emperor Napoleon (as he was about to call himself) made a major change in 1804 when he militarized the school, made the governorship a permanent instead of a rotating post, and initiated a second-in-command post called 'Director of Studies' (which was often taken by a mathematician); but even after that the Ministère de l'Intérieur still exercised influence on the civil needs, and the balance of control changed from time to time between the ministry and the Ministère de la Guerre.

Mathematics was always the principal subject taught, but a tussle soon developed over curriculum policy. Tension was evident from the start, in that the founder-professors included the abstract(ed) Joseph Louis Lagrange and the engineer-mathematician Gaspard de Prony. Monge, very influential at the founding, argued for practical mathematics, especially his beloved descriptive geometry (§7.5); but gradually the opinion of Pierre Simon Laplace, that the more theoretical branches of mathematics should take prime place, prevailed. However, Monge's sentiments continued to attract adherents.

Laplace exercised his influence through the governing body, which he was instrumental in creating in 1800. He never taught there, but he served for a time as a graduation examiner, a post which, following a tradition of the eighteenth century, complemented that of the teachers. Mathematicians of great eminence filled one or sometimes both of these posts over the years; and because such a crop of brilliant mathematicians passed through the doors of the Ecole, especially during its first two decades, many started as students there (for a list to 1840, see Grattan-Guinness *1990*: 1363–7). Yet it always remained a preparatory establishment, and even though the mathematics syllabus was crammed full of calculus and mechanics, the amount covered was inevitably limited. There is an evident imbalance between its modest purpose and its high reputation.

This mismatch is well illustrated by its publications. The *Journal* began by fulfilling its original purpose of publishing lecture notes, but within a decade it was a research organ; so Monge's disciple J. N. P. Hachette founded a more elementary *Correspondance*, which ran until he was 'retired' by the new Bourbon regime in 1816 after the fall of Napoléon.

No major changes took place at that time; but *polytechnicien* Augustin Louis Cauchy, fanatic for the Bourbons, was made professor, and during the 1820s he set up there the basic elements of his mathematical analysis (§3.3). However, it was quite inappropriate for the students and was received poorly, by both students and the examiners de Prony and (*polytechnicien*) Siméon-Denis Poisson, who however were unable to effect any major alterations.

After the revolution of 1830, Cauchy fled into exile with the Bourbons, but the essentials of his doctrine remained largely in place. For the next thirty years the main posts of Professor, Examiner and Director of Studies were shared among Gabriel Lamé, Joseph Liouville, J. C. F. Sturm and J. M. C. Duhamel, all *polytechniciens* except the Swiss-born Sturm. Enrolment was maintained at around 130 per year, and the school entered a period of social glory (Shinn *1980*); yet it became somewhat stultified in its teaching, and ceased to produce major *savants* at the previous rate (although, of course, that was not its basic purpose). The battle over curriculum policy arose again in mid-century, when Urbain Le Verrier *1850* steered through a return to roughly Mongean lines; but some of the *savants* named above resigned in protest. When the school celebrated its centenary in 1894, the official history (in Pinet *1894–7*) described this reform as 'the second disorganization', the Emperor's changes of 1804 having apparently been the first!

4 ENGINEER EXPERTISE:
THE ÉCOLES D'APPLICATION

After his two years at the Ecole Polytechnique, the graduate usually went to a specialist school (called an *école d'application*) for three years of further study in some branch of engineering science, and then passed either into a civil corps or into the army or navy. Among the civil schools, Ponts et Chaussées and Mines carried on in Paris with changes only to accommodate the existence of the Ecole Polytechnique. The Ecole des Ponts et Chaussées is of importance here, as several major mathematicians attended it (Picon *1991*). De Prony ran it for forty years; but in contrast to his vigorous concurrent activity at the Ecole Polytechnique his control there was dilatory, and effective teaching of mechanics began only in 1819 when Claude Navier was appointed to teach mechanics. A distinguished tradition of *polytechniciens* was thereby launched, Navier being succeeded by Gaspard Coriolis and then Barré de Saint-Venant. Of the military schools, the most noteworthy is that at Metz in the Alsace, whither Mézières was moved in 1795 in order to be further away from the British. In particular, Jean Victor Poncelet (a *polytechnicien* who spent his career in the army) began teaching machine theory in 1824, and gathered around him a fine group of engineering *savants*.

One aspect of engineering which this system explicitly did not handle was the commercial and industrial side; but important improvements took place in the 1820s. The Ecole Centrale des Arts et Manufactures was launched in 1829, as a private-enterprise venture, to fill this gap, and after difficult early years it too became a *grande école* (Pothier *1887*). One of the founders was Théodore Olivier, one of Monge's most fervent followers, and descriptive geometry and mechanics were well represented in the syllabuses.

At a lower level, the Conservatoire des Arts et Métiers gained Charles Dupin as professor in a reform of 1819, to teach geometry and mechanics; he also gave great publicity to educational questions in general. In 1824 he proposed that *polytechniciens* should give free lectures on engineering subjects in their localities: this idea was institutionalized in 1830 for Paris in the Association Polytechnique, and technical education developed fairly steadily throughout the century (Day *1987*).

5 INSTITUTIONAL INNOVATION:
THE SO-CALLED UNIVERSITÉ

Among the casualties of the Revolution were all 22 universities, thus guillotining what seems to have been a good structure of mathematics teaching (Brockliss *1987*: 381–90). The first effort at replacement was an Ecole

Normale, founded in 1795; but in contrast to the subsequent success of the Ecole Polytechnique, this establishment was poorly conceived – crash-course production *à la absurde* of schoolteachers – and it folded within four months (Dupuy *1895*). However, it left an important publication: ten volumes of *Séances des Écoles Normales* (1800–01), in which both teaching and classroom discussion was published.

The failure of this school left a serious lacuna in the provision of teachers, and the school programme suffered, especially for science: a system of *écoles centrales* had to be replaced in the early 1800s by *lyçées* in which science was given a reduced place. Eventually, in a decree of 1808, the Emperor came up with the Université Impériale de France (Aulard *1911*).

The terminology of this institution was non-standard. In the first place, it was no university in the usual sense, but an Empire-wide system primarily concerned with schools (in the usual sense of that word). The Empire was divided into 26 regions, each containing an *académie* (another non-standard use of a term) run by a *recteur*. He was responsible to the supervisory *Commission* (after 1820, *Conseil*); he was likely to have been an *inspecteur* in an *académie*, and might be promoted to *inspecteur général*, covering several *académies*.

At the top of each *académie* tree were *facultés*, where higher education was carried out; but only the Académie de Paris had *facultés* in all subjects (including the new one of science). The professors were chosen from those already in posts at other institutions, such as the Ecole Polytechnique. But provincial *facultés* in science were usually mediocre, and student enrolments in science were small.

Outside the *académies*, the Ecole Normale was recreated in 1810 as the elite institution of the Université to give accelerated instruction to talented students. However, especially in science, it flowered only after 1830; but mathematics was the first scientific subject to prosper, and by 1840 a few noteworthy *normaliens* were appearing on the mathematical scene.

Overall, the Université was second class relative to the *grandes écoles*; but mathematics featured quite well in its structure. A mathematician was often a member of the *Conseil*, especially after its powers were strengthened in the 1820s: Poisson served for twenty years until his death in 1840, and was succeeded by Louis Poinsot. Various strong disagreements over general policy arose: after the revolution of 1848, for example, a policy of 'bifurcating' between literary and scientific (including mathematical) school education was adopted, though with strong objections which were later sustained (Hulin-Jung *1989*).

The final institution to be mentioned here is the Collège de France, which was the name given to the Collège Royal in 1795. Its basic function (described in Section 2) was not modified by the Revolution, though, and physics was taught there by Jean-Baptiste Biot (for 61 years) and for a decade from the mid-1820s by André Marie Ampère. However, mathematics was taught at a pretty elementary level until the 1840s, when figures such as Liouville became involved (Sédillot *1869*). The *grandes écoles* had yet to meet much of a challenge!

6 PROFESSIONAL INSTITUTIONS

The Revolution also affected professional bodies. In 1795 the Académie des Sciences was replaced by the scientific *classe* of a new umbrella organization called the Institut de France. Its function was not basically altered; but over the years the amount of work greatly increased, especially when *polytechniciens* became prolific *savants*. In 1803, two posts of *secrétaire perpétuel* were created, and one of them was often held by a mathematician (for example, Joseph Fourier in the 1820s). Académie des Sciences *1910–22* contains the record of the weekly meetings between 1795 and 1835, when the *Comptes rendus* was introduced to present this material in printed form. This journal supplemented the continuing *Mémoires* for members' work and *Savants étrangers* for the outsiders (for whom the winning of prize problems and the opportunity of publishing their work there was an attraction).

In addition to the corps and military bodies mentioned in Section 4, other organizations were set up. The Bureau des Longitudes was created to meet the needs of astronomy and navigation, and in Laplace's hands its journal *Connaissance des Temps*, which had been providing tables since 1702, now also became an important organ for celestial mechanics (Bigourdan *1928–33*). The Dépôt Général de la Guerre, a branch of the army, performed a similar service for cartography and topography, and its *Mémorial* was a publication of comparable importance (Berthaut *1902*). Louis Puissant, Laplace's successor at the Académie des Sciences, was very important in these areas. Another official (and overlooked) source is the newspaper *Moniteur universel*, which contains information frequently, and even the occasional short paper or book review.

Various 'private' societies were formed in which mathematics featured. The Société Philomatique was created in 1791, and profited from the closures taking place all around to become a major scientific society until the mid-1830s, by virtue of publishing (short) papers quickly in its *Bulletin* (Berthelot *1888*). In the mid-1800s, Laplace and the chemist Claude Berthollet gathered bright young *polytechniciens* to pursue research programmes in physics (the status of which they wished to improve) and chemistry (Crosland *1967*). Mathematical physics was Laplace's main interest there, and Biot and Poisson backed him up in the decade in which the group was in existence.

As for journals, the Baron Férrusac's *Bulletin Universel des Sciences et de l'Industrie*, the first comprehensive abstracting journal for science and technology, ran a remarkable series for 'mathematical sciences' during its short life between 1823 and 1831 (Taton *1947*). The first French journal devoted solely to mathematics was launched in 1810 by Joseph Gergonne as his *Annales des Mathématiques Pures et Appliquées*; he ran it until 1832, when Université commitments forced him to close it. Four years later, Liouville began his *Journal* of the same name, which is seen as a successor; but the level of work there is higher, and a closer correspondence is with the *Nouvelles Annales des Mathématiques*, which began in 1842 for 'candidates to the Ecole Polytechnique and the Ecole Normale'. Finally, over the border in Brussels Adolphe Quetelet launched his *Correspondance Mathématique et Physique* in 1826. By and large, Belgium followed course policies similar to France's, though the institutional structure differed (Dock *1911*).

The production of books was quite enormous during this period (see Dhombres *1985* for some information on this little-studied matter). There were three broad levels: for aspirants to the Ecole Polytechnique and the Ecole Normale; for use there; and as erudite treatises. The most eminent textbook writers were Sylvestre Lacroix, who covered all of pure mathematics at the first two levels and also produced a large treatise on the calculus (§3.2); and L. B. Francoeur, who concentrated on books in applied mathematics.

7 FROM NATIONAL TO WORLD WARS

Over the next four decades most of the institutions discussed above continued without major changes in structure. But the Franco-Prussian war of 1870–71 had profound consequences for French higher education. Among other things, it questioned the grandeur of those *écoles* when the defeat had been so decisive (and the military school at Metz was lost with its region). Partly as a consequence, further expansion occurred. Already in 1868 the Ecole Pratique des Hautes Etudes had been created to encourage research, and mathematics was one of its divisions (though not at first one of its strongest). In 1870 the *Bulletin des Sciences Mathématiques* was established, and five years later the Société Mathématique de France was founded, publishing a *Bulletin* (Gispert *1991*). Gaston Darboux was active in both these journals, and also in the production of editions of the works of predecessors (§12.13). Most of them were published by Gauthier-Villars, which became one of the world's principal houses for mathematics.

Educational institutions expanded, and mathematics teaching increased greatly (Gispert *1989*) especially in the Université, which gave way in the 1890s to universities (Shinn *1979*). The Ecole Normale was placed within the Université de Paris in 1904, and continued its rise to become a *grande école* (Ecole Normale *1895*), with a status equal to that of the Ecole Polytechnique (where professors, or *repetiteurs*, now included Camille Jordan, Henri Poincaré and Paul Painlevé).

During this period a brilliant generation gathered around Jacques Hadamard, including René Baire, Armand Denjoy, Maurice Fréchet, Emile Borel, Henri Lebesgue and the brothers Tannery (Lévy *1966*). Pure mathematics was favoured (although symbolic logics were taboo, and Louis Couturat was isolated in his enthusiasm for them). Borel edited a series of monographs published by Gauthier-Villars. Applied mathematics somewhat fell away; Poincaré and, before him, Emile Mathieu were rather isolated by their concern for it.

But the First World War was a setback, especially for France since the government did nothing to protect its most talented citizens, and some promising mathematicians became victims of ballistics. One consequence was that in the 1920s the new generation found themselves being taught by some *passé* figures. To avoid a repetition of this situation, some of them resolved to publish their version of mathematical theories, and so was born the Bourbaki group; it had to survive another war before its plans came to full fruition, but it started its work in the 1930s.

8 CONCLUDING REMARK: RISE AND FALL?

The period 1780–1830 is one of extraordinary glory for French mathematics: it dominated the world in a way which has never occurred since for any country. When other countries started to produce mathematicians in the 1820s, French texts were usually the main source; and foreign mathematicians came to Paris to study (J. P. G. Lejeune Dirichlet, Mikhail Ostrogradsky) or even stay for good (Sturm).

A well-known problem in French historiography is the decline of the country as a scientific centre after 1830. As far as mathematics is concerned (and unfortunately it rarely does concern historians), the tale is grossly exaggerated. After 1830 mathematics was much more international, but France was still among the top rank; indeed, engineering mathematics (which has always featured more prominently in science in France than in any other country) showed signs of improvement. Again, the heights recorded in Sections 3–5 were so exalted that downwards was an unavoidable direction. Furthermore, the decline is partly explicable as an illusion, caused by the rise in other countries. Historically, the decline thesis seems to have been stimulated especially by the Franco-Prussian War, when French scientific and educational leaders inflated genuine signs of decline as a ploy to gain more government support.

However, two related defects in the structuring of mathematical practice *are* worth mentioning; the failure to sponsor commercial and industrial engineering; and the excessive centralization of the bureaucracy (*central* is a technical term in French, referring either to Paris for the country or the capital of a *département*). In addition, the so-called *cumul* system, in which a *savant* would hold several positions at once, was very inefficient; for example, it led to excessive amounts of repetitive teaching and examining. Several mathematicians even gave time to politics, some rising to positions in government; for example Laplace, Arago and Dupin, and later Borel and Prime Minister Painlevé. The tendency for the French not to read the mathematics of foreigners also left them behind in some advancing areas.

Nevertheless, the record of the French in mathematics since the seventeenth century is extraordinary, worthy indeed of psycho-sociological study. And *plus ça change*; Paris today is still a major world centre for mathematical teaching and research.

BIBLIOGRAPHY

Académie des Sciences *1910–22, Procès-verbaux des séances de l'AS tenues depuis la fondation* [in 1795] *jusqu'au mois d'août, 1835*, 10 vols, Hendaye: Observatory.

—— *1979, Index biographique de l'Académie des Sciences* [. . .], Paris: Académie des Sciences.

Anonymous *1830*, [Survey of French educational system], *Bulletin Universel des Sciences et de l'Industrie, Sciences Géographiques, Canaux et Voyages*, **24**, 267–310. [English transl.: in *Quarterly Journal of Education*, 1831, **2**, 83–113. Remarkably good but little-known survey.]

Aulard, F. V. A. *1911, Napoléon et la monopole universitaire* [. . .], Paris: Colin.

Berthaut, H. M. A. *1902, Les Ingénieurs géographes militaires 1624–1831. Etude historique*, 2 vols, Paris: Service Géographique.

Berthelot, P. E. M. *1888*, 'Sur les publications de la Société Philomatique et sur ses origines', *Journal des Savants*, 477–83. [Also in *Mémoires publiés par la Société Philomatique à l'occasion du centenaire de sa fondation*, Paris: Gauthier-Villars, i–xvii.]

Bigourdan, C. G. *1928–33*, 'Le Bureau des Longitudes [. . .]', *Annuaire du Bureau des Longitudes*, (1928), A1–A72; (1929), C1–C92; (1930), A1–A110; (1931), A1–A145; (1932), A1–A117; (1933), A1–A91.

Brockliss, L. B. *1987, French Higher Education in the Seventeenth and Eighteenth Centuries* [. . .], Oxford: Clarendon Press.

Crosland, M. P. *1967, The Society of Arcueil* [. . .], London: Heinemann.

Day, C. R. *1987, Education for the Industrial World. The Ecoles d'Arts et Métiers and the Rise of French Industrial Engineering*, Cambridge, MA: MIT Press.

Dhombres, J. G. *1985*, 'French mathematical textbooks from Euler to Cauchy', *Historia scientiarum*, (28), 91–137.

Dock, M. *1911, Rapport sur l'enseignement des mathematiques dans les ecoles* [. . .] *belges*, Brussels: Goemaere. [Appendix by H. Ploumen on current trends.]

Dupuy, P. *1895*, 'L'Ecole Normale de l'an III', in Ecole Normale *1895*: 1–209.

Ecole Normale *1895, Le Centenaire de l'Ecole Normale 1795–1895*, Paris: Hachette.

Fourcy, A. L. *1828, Histoire de l'Ecole Polytechnique*, Paris: Ecole Polytechnique. [Repr. 1987, Paris: Belin, with introduction and notes by J. G. Dhombres. First major history of the school, and still indispensable.]

Gispert, H. *1989*, 'L'Enseignement scientifique supérieur et ses enseignants, 1860–1900: Les mathématiques', *Histoire de l'Education*, (41), 47–78.

—— *1991, La France mathématique. La Société Mathématique de France (1872–1914)*, Paris: Belin.

Grattan-Guinness, I. *1988*, '*Grandes écoles, petite Université*: Some puzzled remarks on higher education in mathematics in France, 1795–1840', *History of Universities*, **7**, 197–225.

—— *1990, Convolutions in French Mathematics, 1800–1840* [. . .], 3 vols, Basel: Birkhäuser, and Berlin: Deutscher Verlag der Wissenschaften.

Greenberg, J. *forthcoming, The Problem of the Shape of the Earth from Newton to Clairaut* [. . .], New York: Cambridge University Press.

Hahn, R. *1971, Anatomy of a Scientific Institution. The Paris Academy of Sciences, 1666–1803*, Berkeley and Los Angeles, CA: University of California Press. [Repr. 1986.]

Hulin-Jung, N. *1989, L'Organisation de l'enseignement des sciences: La Voie ouverte par le Second Empire*, Paris: Comité des Travaux Historiques et Scientifiques.

Langins, J. *1987, La République avait besoin des savants*, Paris: Belin.

Le Verrier, U. J. J. *1850, Rapport sur l'enseignement de l'Ecole Polytechnique* [. . .], Paris. [Rare; a remarkable study. Summary in *Moniteur universel*, 12 January 1851, suppl., i–xxxiv.]

Lévy, P. *1966*, 'Les mathématiques', in *Institut de France. Académie des Sciences. Troisième centenaire 1666–1966*, Vol. 1, Paris: Gauthier-Villars, 143–212. [Also skimpy articles here on branches of applied mathematics.]

Liard, L. *1888–94, L'Enseignement supérieur en France, 1789–1889*, 2 vols, Paris: Colin. [Includes documents.]

Picon, A. *1991*, 'L'invention de l'ingénieur moderne. L'Ecole des Ponts et Chaussées', 2 vols, Doctoral Dissertation, Ecole des Hautes Etudes en Sciences Sociales.

Pinet, G. (ed.) *1984–7, L'Ecole Polytechnique. Livre de centenaire 1794–1894*, 3 vols, Paris: Gauthier-Villars.

Ponteil, F. *1966, Histoire de l'enseignement en France. Les Grandes étapes 1789–1864*, Paris: Sirey.

Pothier, F. *1887, Histoire de l'Ecole Centrale des Arts et Manufactures* [...], Paris: Delamotte.

Sédillot, L. P. E. A. *1869, Les Professeurs de mathématiques et de physique générale au Collège de France*, Rome: Boncompagni. [More accessible in *Bullettino di bibliografia e storia delle scienze matematiche e fisiche*, **2** (1869), 343–68, 387–448, 461–510; **3** (1870), 107–70.]

Sergescu, P. *1947*, 'La littérature mathématique dans la première période (1665–1701) du *Journal des Savants*', *Archives Internationales d'Histoire des Sciences*, **1**, 60–99.

Shinn, T. *1979*, 'The French science faculty system, 1808–1914 [...]', *Historical Studies in the Physical Sciences*, **10**, 271–332.

—— *1980, Savoir scientifique et pouvoir social: L'Ecole Polytechnique et les polytechniciens, 1794–1914*, Paris: Presse Fondation Nationale des Sciences Politiques.

Taton, R. *1947*, 'Les mathématiques dans le «Bulletin de Férrusac»', *Archives Internationales d'Histoire des Sciences*, **26**, 100–25.

—— (ed.) *1964, Enseignement et diffusion des sciences en France au XVIIIe siècle*, Paris: Hermann. [Repr. 1986.]

11.2

Germany to 1933

GERT SCHUBRING

1 THE GERMAN SETTING

The period 1800–1933 constitutes the classical stage in the development of mathematics in Germany: during the first half of the nineteenth century, German mathematicians assumed the internationally leading role formerly held by the French. They attained this level mainly in pure mathematics. They were able to maintain this position until 1933, when Nazi political persecution forced a large number of mathematicians into emigration. The developmental process leading to this peak was, however, not homogeneous. In contrast to highly centralized France, the growth of mathematics as a discipline in the German states showed marked regional differences, in particular between the predominantly Catholic southern states and the northern states, where the majority of the population was Protestant. These regional differences corresponded to different realizations of the typically German institutional dichotomy between university and higher technical education, and resulted in diverging models of the discipline.

The German Empire persisted after the Thirty Years' War (1618–48) only as a collection of hundreds of sovereignties. Even after the Napoleonic period there was no national or cultural unity. Between 1815 and 1866 the 'Deutscher Bund' consisted of 39 sovereign states, the major ones being the kingdoms of Bavaria, Hanover, Prussia, Saxony and Württemberg, and the German territories of the Austro-Hungarian Empire. In 1866, Prussia annexed several states, among them Hanover. Even in the German *Reich*, constituted in 1871 under Prussian dominance, the individual states still maintained their sovereignty in cultural affairs.

It has to be emphasized that mathematics in Germany before 1800, as opposed to pre-Revolutionary France with its broad intellectual atmosphere (§11.1), was cultivated almost exclusively at its universities (35 in 1789), in particular the Protestant ones. Their teaching subject having been established since the humanist era, the mathematicians adhered to the traditional pattern of erudition. One of the better known is Erhard Weigel (1625–99) at Jena University. It is revealing that the most famous German mathematician before 1800, Gottfried Wilhelm Leibniz, not only worked outside the university system, as a rare universal genius, but also initiated the founding of the first academy in Germany for scientific research (it became in fact the only prominent German academy of the eighteenth century). This Royal Academy of Sciences in Berlin (1700) remained, however, rather atypical, and its leading scientists were foreigners. In fact, in 1809, Wilhelm von Humboldt characterized academies as alien to '*deutsche Wissenschaft*'; and one can understand the neohumanist reform of the Berlin University being seen as a usurpation of the academy's research task (Schubring *1991*).

2 THE TRANSITION PROCESS

Mathematics in Germany at the turn of the nineteenth century can be summed up as follows. In the secondary schools, mathematics was taught as a marginal subject at best, given the near-monopolistic status of the classical languages; in the universities, at least the Protestant ones, its position was more comfortable, being a discipline within the philosophical faculty. In the Protestant system, this faculty, having emerged from the former faculty of arts, provided general education on a level which was higher than that of the secondary schools, and differentiated from it by a broad range of courses required from all students before they could pass on to their professional studies. Of course, within this structure mathematics was far from autonomous: as they had no clientele of specialist students of their own, the mathematics professors were themselves generalists who held combined professorships, or 'ascended' to chairs in the higher faculties. Practitioners of mathematics in seventeenth- and eighteenth-century universities embraced several branches of knowledge, and they united theoretical and applied mathematics in their teaching and research. The character of the applications, however, altered immensely over the nineteenth century, so that what were formerly fields of application were now largely independent technical disciplines.

The situation where only elementary mathematical teaching was taught gradually began to change at some northern universities during the eighteenth century. Starting with the new reform-oriented universities of Halle (1694) and Göttingen (1737), professors increasingly assumed the additional role of researchers. Quite a number of mathematics professors, among them A. G. Kästner (Göttingen) and G. S. Klügel (Halle), who, while not 'extending the frontiers' of mathematics, made substantial contributions to research into foundational problems. From the 1760s there were students at Göttingen who matriculated specifically to study mathematics. One of them was J. G. Tralles, who obtained the chair of mathematics and physics at Bern (Switzerland) in 1785, after two years of specialized studies at Göttingen; he was called to the first chair of mathematics at the newly established Berlin University in 1810 (Schubring *forthcoming*).

These first steps towards specialization were strongly accelerated by the fundamental political and educational reforms in (predominantly Protestant) Prussia after 1806–10. One of the major effects was the secularization of teacher education. This function was removed from the theological faculty of a university and handed over to the philosophical faculty, which thereby won its independent status. Instead of being regarded as a lower faculty, it now had equal status with the other three since it was providing the graduates of its disciplines with careers as teachers at secondary schools. Their scientific authority, based on the new research orientation of the faculty's professors, ensured them social status as 'scholars' (*Gelehrte*), and thus helped to elevate teaching to one of the first modern professions (Schubring *1983*: 10–16).

Mathematics became one of the three principal subjects taught in the Prussian secondary schools, and in the universities mathematics attained its autonomy as a discipline by preparing students to become mathematics teachers in these schools. This close connection between mathematicians in the universities and in the secondary schools, supported by shared values regarding the structure and epistemology of their subject, developed into a 'disciplinary–professional complex'. Characterizing this unity was Karl Heinrich Schellbach (1804–92), a mathematics teacher at a Berlin *Gymnasium*, who stood on intimate terms with J. P. G. Lejeune Dirichlet and Carl Jacobi. For more than forty years (from 1843) he was the only examiner to graduate teachers in mathematics at Berlin. In 1855 he established a mathematical–pedagogical seminar which provided the newly graduated teachers with some teaching practice. In 1860, his influence became decisive in founding the mathematics seminar at Berlin University (Schubring *1988a*: 33–40).

Until around 1940, the teaching profession remained almost the sole basis for the disciplinary autonomy of mathematics at the universities.

3 THE RESEARCH ETHOS

Within this Prussian system, mathematics professors became first-rate researchers. Whereas among the German mathematicians of the 1800s only Carl Friedrich Gauss (at Göttingen) was internationally renowned (acting more as an academician than as a university professor, establishing no school and having no disciples), at the close of the 1820s the young Prussian professors Dirichlet (at Berlin) and Jacobi (at Königsberg) not only became leading mathematicians, but also trained numerous young scientists who later disseminated their research ethos. Mathematicians began to abandon not only the former domains of elementary mathematics but applied subjects as well, usually specializing mainly in pure mathematics, and then predominantly in the algebraic–analytical branch. The early proponents of disciplinary autonomy, Jacobi and Dirichlet, were also highly active in mechanics, which was however regarded as part of pure mathematics in being based on the notion of motion.

These developments entailed raising levels of rigour, which became famous as the arithmetization of analysis (§3.3). This was enhanced in two changes in the context of university mathematics. First, in the schools for which the universities were training the teachers, mathematics was taught within a cultural context which valued formal mental training. This stimulated teachers to strive for clarity in the basics, for a logical order and for purity in the use of methods. Second, university mathematics was an established part of the philosophical faculty, where each subject strove for independent status and aimed to sever all ties with other subjects, in particular to keep it from being an auxiliary to other disciplines.

It was in this vein that Jacobi rejected externally defined values like usefulness. He criticized the French mathematicians for putting too much stress on applied mathematics, and for confusing the true and the incidental reasons for progress in science (Koenigsberger *1904*: 131–4). Not only did mathematicians like Jacobi advocate the glory of pure science, being convinced that in the French way even the application of mathematics to physical problems suffered; during the first half of the nineteenth century even engineers placed pure mathematics at the top of the hierarchy of disciplines. A typical example is August Crelle: engineer, mathematics consultant to the Prussian Ministry of Instruction, and editor of mathematical and technical journals. In his view, pure mathematics provided the basis for meaningful and coherent applications (Eccarius *1977*). When Crelle was ordered by the Ministry to elaborate a concept for a polytechnic institute in 1828, this methodological conviction even led him to organize such an institute around a core of pure mathematics:

> it is ... important that pure mathematics should be explained in the first instance without regard for its applications and without it being interrupted by them. It should develop purely from within itself and for its own sake.... [Results by the applications of mathematics] will be extremely easy to find for the person who is trained in the science itself and who has adopted its spirit.
>
> (quoted in Schubring *1982*: 216)

It must be emphasized that the preference for pure mathematics within the universities came about through corporate processes that were a consequence of the institution's character; it was not part of the original plans of university reformers in the Prussian administration. For the two newly founded universities in Berlin (1810) and Bonn (1818), two chairs were provided for the discipline: one for pure mathematics, the other for a broad range of applications, including even military ones. The second assignment, however, did not prove feasible in university practice. In Berlin, the second chair was soon given to the expanding pure mathematics (on this eminent school, see Biermann *1988*), and in Bonn it went to astronomy.

4 RIVAL CONCEPTIONS AND INSTITUTIONAL SETTINGS

It is not generally appreciated that this functioning of the Prussian university system and the tendency of its disciplines towards autonomy was by no means typical of all of Germany; it was more a *Sonderweg* (a 'special way'), at least during the first half of the nineteenth century. (There is much discussion in political history about whether German national and cultural history can be explained by a *Sonderweg*, distinguishing it from the histories of its European neighbours: see Plessner *1982*.) In the other German states, there had been no comparable social and educational reforms. No profound institutional reforms took place in the universities and schools, and mathematics remained confined to a rather marginal or auxiliary status. Characteristic of the situation is the interaction between the grammar schools and the philosophical faculties: in almost all non-Prussian states, the Bavarian–Saxon–Holsteinian model of only partially reformed grammar schools was predominant. Mathematics figured only as an auxiliary subject, taught by teachers with little status. Correspondingly, the philosophical faculty and its disciplines enjoyed no independent functions within these states, with no 'research imperative' for its professors.

The example of Bavaria is telling: the disciplines taught in the 'lower' faculty were regarded as 'general sciences' which all students were required to study for two years before they could go on to the 'special' (i.e. professional) sciences offered by the higher faculties. Although the regulations for teacher examinations passed in 1833 made mathematical studies mandatory for future mathematics teachers, the number of such students was much too small to allow for any specialized development of mathematics within this institutional framework. Eventually, the obligatory preparatory courses in the philosophical faculty were abolished in Bavaria in 1848, but abolishing the traditional function left a vacuum, since no other ground had yet been prepared for the discipline to grow. It took several decades, and the 'importation' of Prussian professors, before a research orientation was finally established. In Württemberg, teacher education did not become secularized and was not transferred from the theological to the philosophical faculty before the early 1860s. Mathematics teaching in the non-Prussian universities during the first half of the nineteenth century thus remained largely elementary and encyclopedic, with a bias towards traditional geometrical methodology; research continued to follow the old-fashioned, non-expansive mode of erudition.

5 INSTITUTIONAL EXPANSION

The first step towards growth into an independent discipline had been achieved with the establishment of chairs (usually one or two per university) devoted entirely to mathematics, and separated from physics or other natural or technical sciences. However, several decades elapsed before the next step was taken. Seminars were established during the 1860s and 1870s, and in this regard the universities in the other German states quickly followed the Prussian institutional lead. Actually, in both its speed and manner of development as an independent discipline, mathematics for a long time followed the humanities more than it did the sciences. The dominant institutional form in the sciences became the *Institut*, emerging from the *Kabinett* (service facilities for teaching experimental sciences), often provided with a *Famulus* (amanuensis) and passing through the form of laboratory or *Praktikum* (allowing a considerable number of students to be introduced to modern research methodology; the professor as its director was assisted by a young scholar). The institutes were established from the 1850s onwards as 'incarnations' of the disciplinary autonomy of each of the natural sciences. Housed in new, imposing buildings, they were well endowed with the facilities to provide for practical exercises for all students, and with positions for junior faculty to guide the growing number of students and to assist the director in his research (Schubring *1989a*). Unlike the department system, institutes at German universities were hierarchically organized. If there were positions for professors besides the director, these were usually 'extraordinary' (or associate) and not full professorships.

On the other hand, the *Seminar* was the traditional form in the humanities, serving to complement the lectures by guided exercises in which students interpreted classical texts and gave prepared talks. In the framework of the Prussian university reforms, the *Seminar* system was extended to the three major *Gymnasium* disciplines: first, in the 1810s, to philology; then, from the 1830s, to mathematics; and last to history. (The process of further disciplinary differentiation and extension of the seminar system began in the 1870s; the seminar system allowed rather more flexibility with regard to hierarchy than the institutes.) After a stillborn attempt in Münster in 1831, the first mathematics seminar was founded in 1834 at Königsberg

University, directed by Jacobi. Extension to the Prussian universities was achieved in the 1860s. In the beginning, these seminars used to function as *Oberseminare*, for a restricted number of senior students. The major centre of communication was at first a reference library (including a reading-room), which provided more specialized literature than the university library. This service facility became the key to establishing an assistant's position. The transition from a merely remunerated student guardian of the reading-room to the position of salaried assistant for a young mathematician (with a doctoral degree) took several decades.

This process was initiated or accelerated where studies were intensified and the seminar exercises extended, in particular by introducing sections for younger students. The first positions for *Assistenten* as junior faculty members were established in 1872 and in 1881 – this time outside Prussia, at the universities of Würzburg and Leipzig. (Moreover, the positions seem to have been salaried during the first decades only as part-time functions.) In both cases, the initiating professors (F. Prym and Felix Klein) had previously taught at technical colleges, where the system of studies functioned much more intensively than at the universities. These two cases remained exceptions, however, in a period when it was more typical for mathematics professors to try to obtain bookcases for their seminars – while chemistry institutes were already endowed with, say, three assistant positions. It was not until the 1920s that almost all mathematics seminars were definitely granted assistant positions.

6 A STRUCTURAL AND CULTURAL CRISIS

In contrast to the other disciplines in the philosophical faculty, this slow but apparently continuous development suffered a serious crisis after the 1890s. A major cause was the rise of the technical colleges and their subsequent challenge to the universities, which had hitherto been the only institutions of higher education. These technical colleges emerged from polytechnical schools founded mainly in the 1820s and 1830s in non-Prussian states. Although they drew much on the fame of the Ecole Polytechnique in Paris, their resources and general level of instruction were incomparably poorer: until about the 1860s, most of them functioned merely as secondary schools. Nevertheless, in the non-Prussian states they made up for the marginal role of mathematics and the sciences in the classical schools, and even in the universities, by providing basic training in mathematics and the sciences. The 'polytechnical' character of mathematics as the basis of science and technology was exploited by the supporters of the polytechnical schools in order to promote these newcomers and to enhance their status. During the last third of the nineteenth century, these schools – soon renamed 'technical colleges' ('Technische Hochschule') – rose rapidly in status, number and enrolment, and became competitors to the university system. Some of them, particularly in Bavaria and Saxony, provided teacher education in mathematics for technical and 'realist' schools. By the end of the century, the technical and realist colleges had lost all vestiges of their former role as secondary schools. In 1899, after a long and fierce struggle, they achieved a status equal to that of the universities, and in 1901 they even won the right to confer doctoral degrees.

Mathematics had participated in the rise of status, and changed into a 'general science', in an analogous way to the earlier preparatory function it had fulfilled at the universities. In line with the transition to higher education, it was no longer teachers who the authorities entrusted with these fundamental courses but young university-trained mathematicians. In Germany, where traditions like the French *mathématico-physique* were lacking, this resulted in purist courses espousing the new standards of rigour in analysis. The mathematics professors were either unwilling or unprepared to adapt to the engineering needs of their students, and even reinforced an already growing 'anti-mathematical movement' (Hensel *1989*). As part of the struggle for a higher educational status for the polytechnical institutes, engineers and technologists increasingly challenged the privileged role of mathematics as the fundamental scientific basis for technology, and claimed autonomy for the technical disciplines. In consequence, they saw mathematics no longer as fundamental but as auxiliary, and demanded not only a reduction of the mathematics courses, but also that these be entrusted to engineers: the mathematical knowledge required would best be taught by the professors of the respective technical disciplines themselves. Since the technical colleges offered relatively more positions for mathematicians than the universities, and since the assistantships there were almost the only salaried positions for newly graduated mathematical scientists, giving in to these demands would have posed a major threat to the discipline.

University mathematics was also affected in another way by this crisis. Like the other disciplines of the philosophical faculty, mathematics suffered from a drastic fall in enrolment for teacher education in the mid-1880s: mathematicians were left with only a handful of students. Although the number of students began to increase after the early 1890s, a structural problem remained: during the 1890s, a rather stagnant period for the universities, enrolment at the technical colleges more than doubled – in Prussia, and throughout the entire German *Reich* (Schubring *1989b*: 182). Since the universities did not accept graduates of the similarly expanding 'realist' secondary schools, university mathematics might have been left with a diminishing audience, and without the better-prepared students from 'realist' schools (on this term, see Section 7).

What was required, therefore, was a professional policy for the discipline. Earlier attempts since the 1860s to found a mathematical society had failed, but now, during a time of crisis, success was achieved: in 1890, initiated by Georg Cantor and largely backed by Klein, the Deutsche Mathematiker-Vereinigung was constituted in Bremen and established itself as an effective body for developing the discipline (see Tobies *1986, 1991* for its founding, and Gericke *1980* for its further development).

The crisis in higher education, however, not only affected the institutional structures; it also concerned the conceptual structure of the discipline and the cultural role of mathematics in imperial Wilhelmian Germany. Pure mathematics, as established in Prussian universities and in its context of classical education, became increasingly isolated in Germany's economical and technological take-off and the subsequent stress on a scientific–technological view of the world and corresponding values. Educational systems in which mathematics had a relatively good position still coexisted with others allowing it only a marginal place.

Worse still, each subsystem was dominated by another practice according to the particular cultural context: arithmetization has usually been identified as the essence of Prussian university mathematics. As the same time, however, there was not only a predominance of geometry in school mathematics – due to the restriction to elementary subjects and to the bias towards 'purity of method', imposed by the fear of competition from philology teachers and university mathematicians (Schubring *1988b*); there was even a process of re-geometrization under way. A prominent example of this new process is Hermann Grassmann who, breaking with his father Justus's work on algebraization, propagated a geometrized view of mathematics (§6.2). While he could not affect the orientation of university mathematics (characteristic is the case of Berlin where, after the death of Jacob Steiner in 1863, geometry was no longer present in research or in teaching and Karl Weierstrass stressed mathematical analysis; §3.3), his work later merged, in the period of the *Reich*, with the more application-oriented traditions of geometry from the southern states: graphical statics, and synthetic and projective geometry. These new practices of geometry constituted a challenge both to traditional, static elementary school mathematics and to the distancing from applications of the dominant university mathematics. A landmark in this process was the creation of the first university chair consciously assigned to geometry in Leipzig (Saxony) in 1880, and the call to Klein. As a disciple of Julius Plücker, Klein represented the last vestiges of geometry at Prussian universities; he had also experienced, as professor at the Munich Polytechnic, a different cultural view of mathematics and understood the importance of applications for progress in mathematics. In fact, Klein was predestined for the new chair since he had set forth, in his famous *Erlanger Programm* of 1872, an innovative and unified conception of geometry (see Rowe *1983*; and §7.4).

7 FELIX KLEIN'S REFORM AGENDA

From the late 1880s, Klein developed at Göttingen an agenda which can be understood as an attempt to reconcile the different mathematical traditions in the various German states, and to establish a new balance within pure mathematics and with applications, integrating the reorganized mathematics into the overall culture of the *Reich*. The results of the combined efforts of professional associations, of mathematicians, teachers and industrialists, forged by Klein into a powerful movement, were considerable.

The major result was consolidation of the threatened position of mathematics in the technical colleges, in part due to their intention to achieve, after obtaining equal status with the universities around 1900, an integration into the culture of higher education. Mathematics and the sciences, and even the humanities, were defined as part of a 'general department' in these colleges: mathematics had a status there somewhat analogous to its former role in the philosophical faculty. Besides its traditional service function for the technical departments, it took on an independent status by training secondary-school teachers. This general trend is clearly reflected in the considerable number of mathematics students who enrolled from the 1910s onwards at the technical colleges. Moreover, from the early 1920s a second degree and course of studies in mathematics began to emerge which was independent of the teaching profession: the *Diplom-Mathematiker*, intended for careers in industry. Since higher degrees such as *Doktor* and *Habilitation* were also conferrable, it became possible to educate future mathematics professors in an applied context.

On the other hand, this consolidation was due to drastic changes in the mathematics curriculum: the more basic and elementary parts were consigned to the secondary schools, so that mathematics courses could begin at a more advanced level and be consistent with the subjects taught in the technical departments. This change was an incomparably more complex and important task, since it implied thawing out the frozen structure of school mathematics and introducing the elements of differential and integral calculus (banned since the 1820s). By introducing the pivotal notion of 'functional reasoning', Klein and his disciples were able from 1902 to initiate a movement 'from below'. Mathematics teachers all over Germany showed unprecedented energy and activity in curricular reform aimed at getting the function concept into all areas of school mathematics, thus replacing the former static view of mathematical quantities. Klein's ingenious strategical device to introduce the basic notions of calculus in all three types of secondary school as a natural consequence of the functional idea was likewise successful. The *de facto* teaching practice of the calculus already achieved by 1914 was officially acknowledged in 1925, even in the classical *Gymnasium* (Schubring *1989b*: 188–92).

The modernization of the mathematics curriculum and the integration of mathematics into the general culture of the Wilhelmian period were also made possible by structural decisions: the equivalence not only of the two institutional types of higher education, but also of the three types of secondary school; (humanistic) *Gymnasium* (Greek and Latin taught), *Real-Gymnasium* (only Latin), *Oberrealschule* (no classical language). Since their final examination allowed access to universities as well as to technical colleges, there was no institutional barrier restricting mathematics to the latter sector (and thus to only one of the cultural components), as had formerly been the case in Bavaria.

The most obstinate resistance to the reform movement came from the universities. When the introduction of a teaching licence in applied mathematics in 1898 permitted the creation of new positions, the response varied on a large scale: whereas mathematicians at some universities asked for additional professorships and even obtained new chairs for applied mathematics (Göttingen 1904, Halle 1905), some (like Bonn) asked only for an additional lecturer, and others refused to change their courses at all (Berlin).

8 MODERNIZATION AND ITS CONSEQUENCES

Gradually, however, the resistance to applied mathematics decreased. An expression of the change was the foundation in 1920 of the Institute for Applied Mathematics at Berlin University, with Richard von Mises as Director (Biermann *1988*: Chap. 8). An important innovation was a new degree and course of studies in university mathematics for insurance mathematicians, first established in 1896 at Göttingen (Tobies *1990*; see §10.3). This was a first step towards careers for mathematicians outside the educational system. A more decisive breakthrough in this sense was the introduction of the *Diplom* degree for university mathematics in 1942. The emphasis on applications of mathematics, achieved within German universities by 1914, could not be maintained after the First World War; some of the new positions were occupied by pure mathematicians. But even if this reduction is taken into account, one important but usually unnoticed outcome of the reform movement should be emphasized: the former dichotomy between an arithmetized pure mathematics and various geometrical programmes had been overcome by a more integrated practice of mathematics. An essential factor in this development was the redirection of journals as the most effective means of communication. By concerted effort, traditional regional peculiarities and adherences to particular schools were reduced in favour of more comprehensive, nationwide aims, and by giving the journals audiences differentiated into mathematical specialities (Tobies *1986–7*).

Göttingen became the rallying-point for the changes in mathematical research and teaching. After Klein's success in 1895 in calling David Hilbert to Göttingen, their cooperation led to an unprecedented mathematical creativity: Göttingen became the Mecca for mathematics students and scholars all over the world (Rowe *1989*). Young researchers prepared there in great numbers for their doctorate and *Habilitation*, and Berlin mathematics was entirely outrivalled; sometimes there were no doctoral students in Berlin. Numerous new positions for pure as well as applied mathematics were created by the Ministry, and institutes for technical disciplines were established, partly endowed by industry. The institutional expansion of mathematics at Göttingen induced a change from the *Seminar* form to the *Institut* model: from around 1910, Klein tried to obtain funds from industry to erect a separate building for a mathematical institute. The First World War halted these plans, but Klein's successor Richard Courant was luckier: in 1929, the impressive building for the Mathematics Department was inaugurated, funded by the Rockefeller Foundation in the USA.

Although Göttingen remained in this period the only university where mathematics expanded so much that it needed a building of its own, other universities followed suit in renaming the former seminars 'institutes' (Schubring *1990*). In fact, expansion was a great characteristic of the Weimar period: enrolment in mathematics increased considerably both in the universities and in the technical colleges, and helped create new positions for professors and assistantships which had by now become standard in order to ensure practical exercises, even for beginners. It was in particular the introduction of courses in applied mathematics which led to a growing need for practical exercises (e.g. in drawing) and a sufficient number of tutors and adequately equipped rooms (e.g. for models).

However, the modernization of mathematical research was not without its hazards. The virtual explosion of creativity also showed the limits of mathematics: the *Grundlagenkrise* (foundational crisis) shook the self-assuredness of mathematicians, especially their conviction that all problems are solvable within the discipline itself (§5.5, §7.4). But the outbreak of the crisis did not occur in the Wilhelmian period. Cantor and Hilbert knew quite early of the paradoxes of set theory, but without seeing dangers for the foundations of mathematics (§5.3). Mehrtens *1990* argues that it was the cultural context of the postwar period, the 'Weimar culture', that deepened the crisis. The loss of certainty in mathematics corresponded to the loss of what had appeared to be a stable social structure.

The costs of modernization led some of the mathematicians to try a return to older models and views of mathematics. In particular, Ludwig Bieberbach, Professor of Mathematics at Berlin University and influential in the mathematical association and in leading journals, propagated an 'anti-modern' return to *Anschauung* ('intuition') and to its alleged security. He not only paved the way for disciplinary support for Nazism, but also became the leading ideologist to integrate mathematics into the National Socialist system (Mehrtens *1987*). Bieberbach found a strong ally in institutional policy in Theodor Vahlen, a fervent adherent of National Socialism from the early 1920s, and active after 1933 in the Prussian Education Ministry and in the Berlin Academy of Sciences (Siegmund-Schultze *1984*). He developed the notion of a 'German' style of mathematics which was rooted in *Anschauung* and proceeded synthetically by intuition, as opposed to 'Jewish', abstract and analytical procedures. The journal *Deutsche Mathematik*, founded in order to develop this style, proved, however, to be a failure. The ideological hardliners were useful only during the first years to establish the regime in the universities and the scientific disciplines. In the preparations for war, the technical aspects of modern mathematics became welcome and most of the mathematicians who remained in Germany were integrated during the Second World War into *kriegswichtig* projects (i.e. projects purportedly essential to the war effort) of the various rival research institutions, ministries or *Wehrmacht* branches.

BIBLIOGRAPHY

Biermann, K.-R. *1988, Die Mathematik und ihre Dozenten an der Berliner Universität 1810–1933*, Berlin: Akademie-Verlag.

Eccarius, W. *1977*, 'August Leopold Crelle als Förderer bedeutender Mathematiker', *Jahresbericht der Deutschen Mathematiker-Vereinigung*, **79**, 137–74.

Gericke, H. *1980, Aus der Chronik der Deutschen Mathematiker-Vereinigung, Ergänzter Nachdruck*, Stuttgart: Teubner.

Hensel, S. *1989*, 'Die Auseinandersetzungen um die mathematische Ausbildung der Ingenieure an den Technischen Hochschulen in Deutschland Ende des 19. Jahrhunderts', in S. Hensel *et al.*, *Mathematik und Technik im 19. Jahrhundert in Deutschland*, Göttingen: Vandenhoeck & Ruprecht, 1–111 and appendix.

Koenigsberger, L. *1904, Carl Gustav Jacob Jacobi*, Leipzig: Teubner.

Lorey, W. *1916, Das Studium der Mathematik an den deutschen Universitäten im 19. Jahrhundert*, Leipzig: Teubner.

Manegold, K. H. *1970*, *Universität, Technische Hochschule und Industrie*, Berlin: Duncker & Humblot.

Mehrtens, H. *1987*, 'Ludwig Bieberbach und die "Deutsche Mathematik"', in E. R. Phillips (ed.), *Studies in the History of Mathematics*, Washington, DC: Mathematical Association of America, 195–241.

—— *1990*, *Moderne – Sprache – Mathematik. Eine Geschichte des Streits um die Grundlagen der Disziplin und des Subjekts formaler Systeme*, Frankfurt am Main: Suhrkamp.

Plessner, H. *1982*, *Die verspätete Nation*, Frankfurt am Main: Suhrkamp.

Rowe, D. *1983*, 'A forgotten chapter in the history of Felix Klein's "Erlanger Programm"', *Historia mathematica*, **10**, 448–57.

—— *1989*, 'Klein, Hilbert, and the Göttingen mathematical tradition', *Osiris*, Series 2, **5**, 186–213. [Whole volume on German science.]

Scharlau, W. (ed.) *1990*, *Mathematische Institute in Deutschland*, Braunschweig: Vieweg.

Schubring, G. *1981*, 'The conception of pure mathematics as an instrument in the professionalization of mathematics', in H. Mehrtens *et al.* (eds), *Social History of Nineteenth Century Mathematics*, Basel: Birkhäuser, 111–34.

—— *1982*, 'Pläne für ein Polytechnisches Institut in Berlin', in F. Rapp and H. W. Schütt (eds), *Philosophie und Wissenschaft in Preussen*, Berlin: Technische Hochschule, 201–24.

—— *1983*, *Die Entstehung des Mathematiklehrerberufs im 19. Jahrhundert, Studien und Materialien zum Prozess der Professionalisierung in Preussen (1810–1870)*, Weinheim: Beltz. [2nd edn 1991.]

—— *1988a*, 'Ein früher "Aufruf: Rettet die mathematisch–naturwissenschaftliche Bildung!" – Die Denkschrift Schellbachs von 1860', *Der Mathematikunterricht*, **34** (1), 30–72.

—— *1988b*, 'Differenzierung und Institutionalisierung von Wissen – Die Wirkung von Lehrplänen am Beispiel der Entstehung der Schulmathematik', in S. Hopmann (ed.), *Zugänge zur Geschichte staatlicher Lehrplanarbeit*, Kiel: Institut für die Pedagogik der Naturwissenschaften, 143–67.

—— *1989a*, 'The rise and decline of the Bonn Naturwissenschaften seminar: Conflicts between teacher education and disciplinary differentiation', *Osiris*, Series 2, **5**, 56–93.

—— *1989b*, 'Pure and applied mathematics in divergent institutional settings in Germany: The role and impact of Felix Klein', in D. Rowe and J. McCleary (eds), *The History of Modern Mathematics*, Vol. 2, Boston, MA; Academic Press, 171–220.

—— *1990*, 'Zur strukturellen Entwicklung der Mathematik an den deutschen Hochschulen 1800–1945', in Scharlau *1990*: 264–79.

—— *1991*, 'Spezialschulmodell versus Universitätsmodell. Die Institutionalisierung von Forschung', in G. Schubring (ed.), *'Einsamkeit und Freiheit' neu besichtigt*, Stuttgart: Steiner, 276–326.

—— *forthcoming*, 'Johann Georg Tralles: Der erste Ordinarius für Mathematik an der Universität Berlin – Eine Edition seiner Antrittsvorlesung', in H. Begehr (ed.), *Mathematik in Berlin*, Berlin, to appear.

Siegmund-Schultze, R. *1984*, 'Theodor Vahlen – zum Schuldanteil eines deutschen Mathematikers am faschistischen Missbrauch der Wissenschaft', *NTM – Schriftenreihe zur Geschichte der Naturwissenschaften, Technik und Medizin*, **21**, (1), 17–32.

Stäckel, P. *1915, Die mathematische Ausbildung der Architekten, Chemiker und Ingenieure an den deutschen Technischen Hochschulen*, Leipzig: Teubner.

Tobies, R. *1986*, 'Zur Geschichte deutscher mathematischer Gesellschaften', *Mitteilungen der Mathematischen Gesellschaft der DDR*, 112–34.

—— *1986–7*, 'Zu Veränderungen im deutschen mathematischen Zeitschriftenwesen um die Wende vom 19. zum 20. Jahrhundert', *NTM – Schriftenreihe zur Geschichte der Naturwissenschaften, Technik und Medizin*, **23**, (2), 19–33; **24**, (1), 31–49.

—— *1989*, 'Zur Stellung der angewandten Mathematik an der Wende vom 19. zum 20. Jahrhundert – allgemein und am Beispiel der Versicherungsmathematik', *Mitteilungsblatt des Fördervereins für mathematische Statistik und Versicherungsmathematik* (Göttingen), *Beilage zu Panem und Circenses*, Heft 2, 1–11.

—— *1991*, 'Warum wurde die Deutsche Mathematiker-Vereinigung innerhalb der Gesellschaft deutscher Naturforscher und Ärzte gegründet?', *Jahresbericht der Deutschen Mathematiker-Vereinigung*, **93**, 30–47.

11.3

Austria and Hungary

CHRISTA BINDER

It is not easy to define the region of 'Austria', or even of the Austro-Hungarian empire, for the latter emerged only in the second half of the nineteenth century. However, this article treats mathematics in today's Austria, the Czech and Slovak Republics, Hungary, and other parts of the former Austro-Hungarian Empire during its greatest extension, in so far as there was some connection with the German language and Vienna as capital. Vienna was, with some exceptions, the centre of interest until the start of the nationalist processes.

1 FROM 1000 TO 1770

The first records of mathematical activity in the region of Austria date back to the ninth century in Salzburg, where the books of Euclid and Boethius were studied in the Schola Sancti Petri, founded by St Rupert in the sixth century. In the following centuries mathematical knowledge was very scarce, but in every monastery there was at least one person who was able to calculate the date of Easter and who was familiar with the basic statics of buildings. From 1157 on, monks were not allowed to build churches any more; this led to the foundation of the so-called *Bauhütten*, a trade accumulating much knowledge in practical geometry, but keeping it secret – they even destroyed their plans after having completed a building (compare §2.8).

In 1348 the first German university, the University of Prague, was founded. It was followed by Cracow in 1364 and by Vienna in 1365. The University of Vienna soon gained worldwide reputation through the work of Johannes von Gmunden (1380–1442) (its first Professor of Mathematics and Astronomy) and Georg von Peuerbach (1423–61). Von Gmunden donated his private collection of books and astronomical instruments to the University of Vienna on the condition that it would be accessible to every interested student. This collection was the germ of the now famous library, since many colleagues followed his example (Peppenauer *1953*).

Johannes Müller (Regiomontanus) was only 13 years old, but already disappointed by the poor level of astronomy in Leipzig, when he chose Vienna for his studies and Peuerbach as his teacher. Together they studied the astronomy of Ptolemy, trigonometry (they calculated tables of the sine function with decimal fractions as values) and the calendar. After the death of Peuerbach, Regiomontanus went to Italy, where he participated in preparing the calendar reform by Pope Gregory XIII and wrote his best-known work, the *Dreieckslehre, de triangulis omnimodus libri quinque* (first published 1533).

After Regiomontanus had left, the level of mathematical sciences in Vienna declined. Nevertheless, tradition remained alive, and Austria still played a leading role in the age of humanism, during the development of book printing. Astronomy, cartography and elementary algebra were studied diligently. Heinrich Schreyber (Grammateus) (*circa* 1500) was the first university teacher to write successful German-language *Rechenbücher* (§2.3), and he was the last representative of the so-called 'second Vienna mathematical school'.

In the course of the political changes during the sixteenth century, notably the war against the Turks, universities in Austria lost their competence in mathematics. Moreover, schools (in the monasteries and in the cities) came under the influence of the Jesuits, and the teaching of mathematics – which was at a remarkably high level before the Thirty Years' War of the seventeenth century – lost its importance. Once again, as in the times before von Gmunden, university professors were not allowed to teach on their own initiative; they were only allowed to read verbatim from the text of approved books. Each semester they were allocated to courses by a drawing of lots. Thus, in most cases they could not even teach in their own field (Peppenauer *1953*, Obenrauch *1897*).

From this period, which lasted until the middle of the nineteenth century, only a few mathematicians deserve to be mentioned. Most of them made their contributions when working outside academic circles.

Johannes Kepler did all his famous work in Austria, first in Graz at the *Landschaftsschule*, then in Prague at the court of the Emperor Rudolph II, and lastly in Linz at the *Bürgerschule*. In Graz he established his fame as an astrologer; he had to increase his meagre income by calender-making and by astrological prophesies, as did every astronomer at the time. It was while in Linz that he wrote the *Stereometria doliorum vinorum*, a precursor of the integral calculus. In connection with his famous work in astronomy, he was also very much interested in the development of logarithmic tables. He was not satisfied with John Napier's tables, and could not wait for Jost Bürgi's; so he calculated his own, the *Centrobaryca*. He took base e, and used the functional concept of logarithm for the first time (§2.5).

Paul Guldin (1576–1643), a Jesuit and Professor of Mathematics at Graz University, calculated the volume of solids of revolution by a method using the centre of gravity (the 'theorem of Pappus or Guldin'). He was also interested in the development of logarithms. He owned a copy of the *Progress Tabulen* by Bürgi, printed in Prague in 1620, which is still in the library of Graz University (Gronau *1989*).

Georg Freiherr von Vega (1754–1802), a soldier born to a poor farmer's family, was to gain a knighthood and become a member of the most important scientific academies. He was an autodidact and had a great talent for calculations and for teaching. He wrote the best textbooks of his time on artillery, and is well known for his tables of logarithms and trigonometric functions. His tables of seven-figure logarithms were calculated with the help of trained soldiers, and they were famous for their accuracy – Vega promised one gold ducat to everyone who found a mistake. The first edition appeared in 1783, the last (the 90th!) in 1924.

2 FROM 1770 TO 1850

The end of this long period of stagnation came in 1773 when the influence of the Jesuits was eliminated, followed by an extensive reform of higher education. The beginning of the nineteenth century was marked by a renewed interest in the applications of mathematics, influenced by the needs of the applied sciences, the military and industry. Following the example set by the French Revolution, schools devoted mainly to applications were founded. The polytechnical institutes in Prague (1806), Graz (1811) and Vienna (1815) followed in quick succession. From these institutions, the technical high schools (nowadays universities) developed. They established themselves quickly, especially in mathematics, and from then on played a fundamental role in the development of mathematics in Austria. The level of instruction was high, especially in descriptive geometry, since the teachers, now themselves well educated in this field, could pass on their increasing knowledge to their pupils.

Let us now consider some figures whose careers developed during this period. Farkas Bólyai came from a family of impoverished landowners in Transylvania. He was to become a friend of Carl Friedrich Gauss, when they studied together in Göttingen, and maintained a lifelong correspondence with him. As professor of mathematics in the College of Maros-Vásárhely, his influence on mathematical education in Hungary was great. His book *Tentamen* (1832), a textbook for higher schools, gave a very clear and advanced insight into geometric problems, in which he had his main interest. An appendix of this book, written by his highly talented son, János Bolyai, contains the first publication of non-Euclidean geometry (§7.4).

Unfortunately János was not encouraged by Gauss to pursue his mathematical studies. He had taken up a military career; but having been pensioned off early on account of illness, he fell into a deep mental depression and never again published any results.

Bernard Bolzano, the famous logician, analyst, pedagogue and theologian, lived in Prague but he was isolated from the city's scientific life. His concepts of the continuity and differentiability of functions, and of limits, therefore remained almost unknown in his time, and the credit was given later to Augustin Louis Cauchy (§3.3). He was a professor of theology and never became professor of mathematics. He had many difficulties with the authorities, due to his revolutionary opinions on religion and education; he spent a decade away from Prague, starting in the early 1830s (Christian *1981*).

There was also Jakub Filip Kulik (1793–1863), an industrious compiler of extensive tables of prime divisors. His work still remains uninvestigated in the archives of the Academy of Sciences in Vienna; it found no echo, not even in Bohemia and Moravia (Folta *1977*).

3 FROM 1850 TO 1930 IN AUSTRIA

The old universities, starting with the University of Vienna, tried hard to reach the mathematical level of the universities in France and Germany. In a deliberate policy they succeeded in appointing some famous mathematicians as professors. Another, perhaps even more important stimulus was the fact that those young mathematicians who had already proved their abilities by publishing some mathematical papers received grants to study at the centres of the mathematical world: Berlin (later Göttingen) and Paris, sometimes also in Italy. Therefore Austrian mathematicians at the turn of the century had not only learned from the very best mathematicians of their time, but also had all the necessary personal connections. Another important factor in the improvement was the possibility of moving between universities. A typical career of a mathematician in Austria might have been as follows: studies in Vienna, first publications, a year in Göttingen followed by a year in Paris or Cremona, professor at Cernowitz, professor at Graz or Innsbruck, then Prague, and finally Vienna (Peppenauer *1953*, Einhorn *1985*).

Another important factor in the growth of mathematics in Austria was (and still is) the Austrian Academy of Sciences, founded – rather tardily – in 1848, following the example of the Academy in Berlin. There mathematicians from all universities could meet informally; they could publish in the *Sitzungsberichte* and discuss – sometimes with spirit – the merits of different contributions and future developments. Under the encouragement of Felix Klein, the Austrian Academy of Sciences joined forces with the academies in Göttingen and Munich to prepare, from 1894 on, the *Encyklopädie der mathematischen Wissenschaften* (§0). The leading Austrian contributors to this monumental work were Gustav von Escherich, and later Wilhelm Wirtinger and Philipp Furtwängler. In addition there was the journal *Monatshefte für Mathematik*, founded in 1890.

The tension between the expanding technical high schools (who wanted to be accepted as scientific institutions) and the classical universities (who did not wish to give up their exclusive privilege of awarding degrees, but wanted also to take part in the instruction of applied mathematicians) was settled by giving the technical high schools the right to award their own degrees (in 1901), and the universities the right to offer courses in the mathematics of insurance (in 1895) (Sequenz *1965*).

As indicated above, to become a professor at the University of Vienna was the final goal of most Austrian mathematicians. Below are mentioned a few of the most outstanding men (Peppenauer *1953*).

Josef Max Petzval (1807–91) developed new optical lenses which were a big step forward and allowed the construction of high-speed photographic cameras. In mathematics, he contributed to the Laplace transform (§4.8) (Gegenbauer *1903*).

Franz Mertens, a number theorist whose elementary proof of the Dirichlet theorem can still be found in the textbooks, came to Vienna in 1894 (after Cracow and Graz). Ludwig Boltzmann was Professor of Mathematics at the University of Vienna from 1873 to 1876; during this period he formulated the second law of thermodynamics (§9.5).

Of special reputation is the 'Czech (or Prague) or Austrian geometric school', with many members, who in the beginning were interested in problems in descriptive geometry, later in projective and in constructively synthetic geometry, and finally in the solution of geometric problems using analytical, algebraic and synthetic methods. Its main representatives were the brothers Emil and Eduard Weyr, who had studied at Vienna and in Italy and who, together with their colleagues and pupils, spread their knowledge to every part of the Empire. Emil Weyr dominated geometry in Vienna during the 1880s and 1890s.

In addition the algebraist Leopold Gegenbauer, a man of many mathematical, political and cultural interests, taught in Vienna as a professor from 1893 until his death in 1903. Finally, and reaching to the 1930s, Gustav Ritter von Escherich, a quiet man now almost forgotten, was of great influence during his 36 years as a professor at the University of Vienna. His use of Weierstrass's methods in functional analysis (§3.9) was masterly. He had many pupils, including Wirtinger, A. Tauber, Hans Hahn, and J. Radon.

One of the few exceptions to the rule of a career in Vienna was Otto Stolz who, after having studied at Innsbruck, Vienna, Berlin and Göttingen, became professor in his home town of Innsbruck, and preferred to stay there, rejecting all calls to Vienna. He made important contributions to many fields of mathematics. Most famous are his books on analysis, which were among the first textbooks on analysis in the sense of Weierstrass (§3.3). With only a few interested students, and without the funds for an extensive library, Stolz managed to build up a growing institute which attracted talented and already famous mathematicians like Gegenbauer and Wirtinger, both of whom went to Vienna after having worked at Innsbruck for some years (Huter *1971*).

The peak of mathematics in Austria was undoubtedly reached during the 1920s and 1930s. Then three mathematicians of worldwide fame were professors in Vienna (Einhorn *1985*): Wirtinger, still well known for his contributions to function theory; Furtwängler, whose talents extended from the measurement of earth gravity to algebraic number theory; and Hahn, the pioneer of set theory and functional analysis (the Hahn–Banach theorem, §3.9). Together with their students, co-workers and colleagues such as Wilhelm Blaschke, Hilda Geiringer, Kurt Gödel, Eduard Helly, Hans Hornich, Walter Mayer, Karl Menger, Richard von Mises, Kurt Reidemeister, Karl Strubecker, Olga Taussky, Heinrich Tietze and Leopold Vietoris, Vienna was to become one of the leading centres of mathematics of the time.

The 'Wiener Kreis', a loose union of philosophers, logicians, mathematicians and scientists, met regularly and attracted the deepest thinkers in this period. It gathered in a room at the Mathematical Institute of the University of Vienna, and kept in close touch with the development of mathematics in Austria (see §5.5 on metamathematics).

This prospering time for mathematics in Vienna came to a natural end in the 1930s with the death of Hahn, and the retirement of Furtwängler and Wirtinger. The intention of continuing the mathematical life of Vienna at the same level was frustrated by political developments, which forced some of the leading men (Gödel, Menger and others) to leave Austria.

The other universities in Austria included Innsbruck (already mentioned) and Graz (with a university and a technical high school). The universities at Prague, Bratislava and other towns developed similarly to Vienna, with further complications of nationalism; the German University in Prague had to close in the 1930s.

4 DEVELOPMENTS IN HUNGARY FROM 1850

Very similar developments took place in Hungary. It became independent in 1867, the year of the formation of the Austro-Hungarian dual monarchy. From this time on Hungary slowly became an industrialized country with better education, and there was substantial progress at school level. The universities of Budapest and Koloszvár (now Cluj, Romania) had raised the mathematical life of the country to the contemporary European level by the turn of the century (Mikolas *1975*). In this first generation of mathematicians were Julius König (still known for his contribution to the continuum problem; §5.4) at the Technical University of Budapest, and Ludwig Schlesinger at Cluj (and later Giessen), who worked on linear differential equations.

Hungarian mathematicians gained worldwide fame in the following generation with Lipót Fejér, Frigyes Riesz, Alfred Haar and their pupils. Fejér worked at Budapest; a very stimulating man, he achieved an exceptional scientific influence. His most important results dealt with the convergence of Fourier series (§3.10). Riesz's main fields were integral equations and subharmonic functions, and Haar worked mostly on the measure theory of locally compact groups.

After the First World War the University of Kolozsvár had to close down and move to a new home within the new Hungarian borders. It settled in Szeged, a town where there was no university before, and no mathematical library. Riesz and Haar, already famous, succeeded in a very short time in creating a mathematical centre, later to be compared with Göttingen. A very important role in this success was played by the international journal *Acta scientiarum mathematicarum* (founded in 1922). Many important journals exchanged their volumes, and many publishers gave free copies for reviewing. In this way the library became one of the best in the region. But from 1920 to 1949 many distinguished Hungarian mathematicians (for example, John von Neumann and Gabor Szegö) had to leave this small country, which was in a very poor social and economic condition; mostly they went to the USA. In this way the Hungarian mathematical school, with its emphasis on problem-solving, spread all over the world (Mikolas *1975*).

BIBLIOGRAPHY

Christian, C. (ed.) *1981, Bernard Bolzano, Leben and Wirkung*, Vienna: Verlag der Österreichischen Akademie der Wissenschaften.
Einhorn, R. *1985, Vertreter der Mathematik und Geometrie an den Wiener Hochschulen 1900 bis 1940*, Vienna: Verband der Wissenschaftlichen Gesellschaften Österreichs. [Dissertation, Technische Universität Wien 43/I und II.]

Folta, J. *1977*, 'Social conditions and the founding of scientific schools', *Acta historiae rerum naturalium necnon technicarum* (Prague), Special issue no. 10.

Gegenbauer, L. *1903*, 'Ein vergessener Österreicher', *Jahresbericht der Deutschen Mathematiker-Vereinigung*, **12**, 324–44. [On Petzval.]

Gronau, D. *1989*, 'The logarithms – From calculation to functional equations', in C. Binder (ed.), *II Österreichisches Symposium zur Geschichte der Mathematik*, Neuhofen an der Ybbs, 1–8.

Huter, F. (ed.) *1971*, *Die Fächer Mathematik, Physik und Chemie an der Philosophischen Fakultät zu Innsbruck bis 1945*, Innsbruck: Kommissionsverlag (Veröffentlichungen der Universität Innsbruck, Forschungen zur Innsbrucker Universitätsgeschichte, Vol. 10).

Mikolas, M. *1975*, 'Some historical aspects of the development of mathematical analysis in Hungary', *Historia mathematica*, **2**, 304–8.

Obenrauch, F. *1897*, *Geschichte der darstellenden und projektiven Geometrie mit besonderer Berücksichtigung ihrer Begründung in Frankreich und Deutschland und ihrer wissenschaftlichen Pflege in Österreich*, Brünn: Carl Winkler.

Peppenauer, H. *1953*, 'Geschichte des Studienfaches Mathematik an der Universität Wien von 1848 bis 1900', Dissertation, University of Vienna.

Sequenz, H. (ed.) *1965*, *150 Jahre Technische Hochschule Wien*, Vol. 1, *Geschichte und Ausstrahlungen*, Vienna: Technische Hochschule.

Szénássy, B. *1992*, *History of Mathematics in Hungary until the 20th Century*, Berlin: Springer. [Much bibliographical information.]

11.4

The Netherlands

J. A. VAN MAANEN

1 THE MIDDLE AGES

The oldest mathematical document from The Netherlands (in this article 'The Netherlands' refers to the present geographical situation) is a letter – about how the volume of the sphere depends on its diameter – from the Bishop of Utrecht, Adalboldus (970–1027) to Gerbert of Aurillac, by then Pope Sylvester II (*circa* 945–1003). In the early Middle Ages, however, institutional mathematical instruction did not yet exist. Schools, most of them connected to the Church, only treated reading and writing, followed for a minority of the pupils by the trivium (grammar, dialectic and rhetoric). Up to the end of the fifteenth century there are only a few traces of the quadrivium. Deventer, where Erasmus studied, and Zwolle (Hanseatic cities in the north-east) seem to have been 'proverbial' exceptions (for a further elaboration on this period, continued up to 1700, see Struik *1981*). From about 1500 onwards a new type of school emerged under influences from France and the southern Netherlands (the present Belgium, which separated from The Netherlands in 1830, and which is not considered in this article) where both commerce and education developed much earlier (Louvain had its university in 1425, a century and a half before the first university in the north). The new, so-called 'French' schools were set up in many towns next to the Latin school, the first one in 1482. They taught arithmetic and, as a business language, French (Post *1954*).

2 THE 'GOLDEN AGE' OF THE DUTCH REPUBLIC

In the second half of the sixteenth century further development went hand in hand with growing economic, political and spiritual independence. In a short period of time, commerce and trade increased considerably, and so did the need for practical mathematicians, such as surveyors and navigators, and their educators. At the political level the Dutch in 1568 started their armed struggle for independence from the Hapsburg monarchy. At

Leiden (1575) and in the northern province of Frisia at Franeker (1585), universities were founded. Other universities followed, at Groningen (1614), Utrecht (1636) and Harderwijk (1648), and several towns instituted an 'illustrious school' or 'athenaeum', which provided instruction slightly below the university level but did not have the right to award doctorates. Amsterdam had an athenaeum (1632), which gained university status in 1876. Higher education in the period 1575–1814 is amply documented, historically and sociologically, in Frijhoff *1981*.

Like all Dutch learning, mathematics in the beginning had its momentum at Leiden. There, in accordance with the classical university structure, it started as part of the arts faculty, which was subordinate to the three main faculties (theology, law and medicine), and took care of the preparatory instruction. The teaching language was Latin, the subjects taught were arithmetic and geometry.

This traditional structure did not respond to the growing need for practical mathematicians. At the basic level these were educated in the French schools and by the *rekenmeester* (literally, 'teacher of arithmetic', although most *rekenmeesters* also taught other subjects), who was a free entrepreneur. Before 1600 this basic instruction in practical mathematics had no academic counterpart. It was a major innovation when in 1600 the Leiden magistrate agreed to a request from Prince Maurits, the leader of the Dutch rebellion, and attached an engineering school to the university. (Van Maanen *1987* and Struik *1981* describe the process of tradition and innovation in seventeenth-century mathematics.) Surveyors, craftsmen and military engineers were trained in *Duytsche mathematicque*, which included basic constructions from Euclid's *Elements*, trigonometry, the 'practice of surveying', the solution of triangles, solid geometry and fortification. Field-work supplemented theoretical studies. '*Duytsche*' signifies not only that the curriculum, which Simon Stevin had outlined, was new, but also that the teaching was in the vernacular. The school started with two instructors (L. van Ceulen, famous for his approximation of π, and S. van Merwen), who were appointed 'professor extraordinarius', lower in rank than the full ('ordinary') professors who taught in Latin. Both men died in 1610, and were succeeded in 1615 by their former assistant Frans van Schooten the Elder, who further developed Stevin's plans.

Gradually, practical mathematics became a traditional element itself. But another important innovation took place when René Descartes, who worked in The Netherlands from 1629 to 1649, found an ardent propagator of his *Géométrie* (1637) in van Schooten's son, Frans van Schooten the Younger, who he knew personally (and on whom see Hofmann *1962*). Private students of Frans the Younger (Christiaan Huygens, Johann Hudde, Hendrick van Heuraet, Jan De Witt and others) were an active audience for

Cartesian mathematics. Sons of prosperous merchants and regents, they belonged to a new breed of student, who could afford to study subjects that were not of direct practical use. The structure of the group depended primarily on personal contact with van Schooten, who guided his chosen students into research and who included their results in his Latin editions (1649, 1659–61) of Descartes's *Géométrie*. But after his death, tradition prevailed again. True, Huygens remained active as an independent researcher, but from 1660 onwards he shifted his focus to Paris. Some interest in Cartesian mathematics survived (for example at Leiden, where Pieter van Schooten had in 1661 succeeded his half-brother Frans and in textbooks by G. Kinckhuysen and A. de Graaf), but it became a goal in itself, and was no longer a starting-point for new research.

The situation at the other universities was less favourable. Franeker at least had continuity in mathematics teaching (A. Metius, 1598–1635; B. Fullenius the Elder, 1636–57; A. de Grau, 1659–83; B. Fullenius the Younger, 1684–1706), but after Metius's time only practical mathematics was done, and only at an average level. At Groningen the chair of mathematics was regularly vacant (when John Bernoulli was appointed there in 1695 it had been vacant since 1669). Compared with Groningen, the position of mathematics at Utrecht was slightly better, but at Harderwijk it was even worse.

3 THE EIGHTEENTH CENTURY: DECLINE

Towards the end of the century the economy worsened and the numbers of students decreased. Yet, up to 1795, when the French took control of The Netherlands, little changed in higher education. After the death of Pieter van Schooten in 1679, the Leiden Engineering School was closed for financial reasons in 1681. In 1697, however, it was reopened, but a 'lector' (of lower rank than an 'extraordinary' professor) was appointed. New developments, the calculus for example, entered the institutions late. Some individual mathematicians mastered this new discipline, among them B. Nieuwentijt, who argued with Gottfried Wilhelm Leibniz in the late 1690s; John Bernoulli, who taught at Groningen (1695–1705), but without apparent result; and W. J. 's Gravesande, who worked at Leiden from 1717 until 1742, and who had a strong interest in Newtonian physics and mathematics. J. A. Fas (teaching at Leiden from 1763 until 1815, author of the first calculus text in Dutch, 1775) and J. F. Hennert (at Utrecht from 1764 until 1804) introduced calculus teaching in their private lectures.

A new mathematical institution (Baayen *1978* sketches its history) was the Amsterdam Wiskundig Genootschap ('mathematical society', the present Dutch 'MS'), which was founded in 1778 by the Amsterdam teacher of mathematics and astronomy A. B. Strabbe, with the motto 'Untiring labour overcomes all'. The first members of this society, a 'daughter' of the 1690 Hamburg Society, were found among the readers of Strabbe's book *Oeffenschool der mathematische weetenschappen*, which began to appear in two-monthly issues in 1700, and which organized its readers by publishing solutions to a problem section. The membership of the society increased from 101 (in 1783) to 133 (in 1813), 268 (in 1922) and 1285 (1990). No national institution to promote science existed until the foundation of the Royal Dutch Academy of Sciences (1851), which replaced the corresponding 'Institute' the French had created in 1808 (when in occupation), but several local societies partly satisfied the need. In later years, Huygens' *Oeuvres complètes* (1888–1950) were published by such a society: the Hollandsche Maatschappy at Haarlem.

4 THE FRENCH OCCUPATION AND ITS CONSEQUENCES

A major change in the educational system came during the French occupation (1795–1813). In December 1811 and September 1812, respectively, the universities of Franeker and Harderwijk had to close, and the University of Utrecht and several athenaea were downgraded. The two remaining universities were remodelled on French lines. In 1806 the French had already split the arts faculty into two parts: a science faculty, which had to provide preparatory instruction for medical students, but which also had students of its own; and a humaniora faculty, which students of theology and law had to pass first.

In 1813 Dutch sovereignty was restored and a royal monarchy under the House of Orange was instituted. The royal decree of November 1815, which set out the new regulations for higher education (the *Organiek Besluit*), maintained the arts faculty established under the French. It also restored the university status of Utrecht. Franeker and Harderwijk were given an athenaeum instead (1815 and 1816, respectively), but both schools eventually closed through shortage of students (1844 and 1818). The *Organiek Besluit* did not prescribe an explicit curriculum for the science faculty, which had only a few students of its own. Their number slowly increased from 1827, when science teaching at the Latin schools required a university degree. Yet in 1860 the total number of proper science students at the three universities was only 38. Two secondary-school reforms stimulated growth. In the new 'HBS' schools (Higher Burger Schools) of 1863, Latin and Greek were abandoned and more hours devoted to mathematics and natural sciences; and in 1878 the Latin school was transformed into the more science-oriented *Gymnasium*. The number of science students was 152 in

1870, rising to 285 in 1890. When in 1917 admission to the science faculty, which had been reserved for *Gymnasium* school-leavers, was also extended to those from the HBS, the numbers leapt to 800, and increased to 1800 in 1930. The number of ordinary professors increased from 17 in 1860 to 40 in 1876 and 60 in 1930.

The increase in 1876 is connected with the law of that year on higher education, which replaced the *Organiek Besluit* and remained in force until 1960. It gave university status to the Amsterdam Athenaeum, and it increased the number of science professors to ten, among whom were two mathematicians. New too in 1876 was the elimination of mathematics from the medical preparatory examination, and the subdivision of the science faculty into six autonomous disciplines (mathematics and astronomy, mathematics and physics, chemistry, geology, biology, and pharmacy). Other fields in which mathematics plays a role, like technology, were not yet considered as academic disciplines, but as part of secondary education. Higher education in technology had started at Delft in 1842 (the Royal Academy of Engineers) and was continued in Delft Polytechnic in 1864, which was recognized in 1905 as an academic institution with the right to award doctorates. Several other institutions existed or were founded where mathematics was represented: military schools (the Royal Military Academy in 1826, the Royal Institute for the Marine in 1829), two new denominational universities (in 1880 the Protestant Free University at Amsterdam, and in 1923 the Roman Catholic University of Nijmegen, where faculties of mathematics and physics were set up in 1933 and 1957, respectively), an agricultural college (Wageningen, 1917), two business schools (Rotterdam, 1913, and Tilburg, 1927; both received academic status in 1939) and two new polytechnics (Eindhoven, 1957, and Enschede, 1964).

Since no official curricula were prescribed for the science faculty in 1815 or 1876, we must look to other sources for the teaching programmes, such as the statement of the teaching task of newly appointed professors. In 1839, G. J. Verdam was appointed at Leiden to teach plane and solid geometry, trigonometry and calculus. When in 1854 his position was raised to that of ordinary professor, arithmetic, algebra and mechanics were added to the list. In 1892, J. C. Kluyver was appointed to teach higher mathematics, algebra, calculus, the theory of functions and probability theory, and the same statement was used in 1919 with the appointment of J. Droste. Other information about programmes is contained in the 1864 decree about the so-called MO examinations by which one could qualify as a teacher for the new HBS. The MO mathematics syllabus, which also influenced teaching at the universities, consisted of arithmetic; algebra (including higher-degree equations, continued fractions, the convergence and summation of infinite series, permutations and combinations, the binomial theorem, and the exponential, logarithmic and trigonometric functions); solid geometry; plane and spherical trigonometry; and the principles of descriptive and analytic geometry.

A survey of the mathematics lectures given in 1889 and 1910 (Groen *1988*: 37–41) shows that the 1864 MO programme, extended with the addition of the calculus, was still followed in the university syllabuses for the first two years. Special topics were offered for graduate students: the theory of functions, differential equations, the calculus of variations, elliptic functions, differential geometry, projective and non-Euclidean geometry, probability theory and mathematical physics.

5 THE TWENTIETH CENTURY: REVIVAL

In the two decades before the Second World War, the climate of academic mathematics changed. Whereas universities had served primarily as teacher-training institutes, now research became important too. L. E. J. Brouwer, who attempted to create in Amsterdam a research centre on the Göttingen model, provided the main impetus. He succeeded in doubling the size of the Amsterdam mathematics department, stimulated international contacts, and through outstanding work on the foundations of mathematics and on topology raised mathematics in The Netherlands to an international level (§5.6, §7.11) (research during this period, by Brouwer and others, is amply documented in Bertin *et al. 1978*). Academic mathematicians began to find employment outside education as mathematics slowly became an independent academic discipline.

BIBLIOGRAPHY

Baayen, P. C. *1978*, ' "Wiskundig genootschap" 1778–1978: Some facts and figures concerning two centuries of the Dutch Mathematical Society "Een onvermoeide arbeid komt alles te boven" ', *Nieuw Archief voor Wiskunde*, Series 3, **26**, 177–205.

Bertin, E. M. J., Bos, H. J. M. and Grootendorst, A. W. *1978*, *Two Decades of Mathematics in the Netherlands 1920–1940*, Amsterdam: Mathematical Centre.

Bierens de Haan, D. *1883*, *Bibliographie néerlandaise historique–scientifique* [. . .] *sur les sciences mathématiques et physiques* [. . .], Rome: Boncompagni. [Repr. 1960, 1965, Nieuwkoop: De Graaf. Lists most of the books published in the Netherlands up to 1850; originally published in Boncompagni's *Bullettino*, **14** (1881), **15** (1882), **16** (1883).]

Frijhoff, W. Th. M. *1981*, *La Société néerlandaise et ses gradués, 1575–1814*, Amsterdam: Academic Publishers Associated.

Groen, M. *1987, 1988, 1989, Het wetenschappelijk onderwijs in Nederland van 1815 tot 1980*, 3 vols, Eindhoven: the author.

Hofmann, J. E. *1962, Frans van Schooten der Jüngere*, Wiesbaden: Steiner.

Maanen, J. A. van *1987*, 'Facets of seventeenth-century mathematics in the Netherlands', Doctoral thesis, Utrecht. [Contains a general survey of the period; chapters about John Pell and Hendrick van Heuraet, and a chapter about the seventeenth-century mathematical manuscripts preserved in Leiden University Library.]

Post, R. R. *1954, Scholen en onderwijs in Nederland gedurende de Middeleeuwen*, Utrecht and Antwerp: Spectrum.

Struik, D. J. *1981, The Land of Stevin and Huygens*, Dordrecht: Reidel. [Includes a rich bibliography of publications in English on the general history of The Netherlands and on the history of science related to The Netherlands.]

11.5

Scandinavia

I. GRATTAN-GUINNESS

1 GEOGRAPHIES

The countries considered here are Sweden, Denmark, Norway and Finland; and the relationship between them has to be noted first, for at various times they were in conflict with one another and with Germany and Russia. Denmark was the dominant power until the mid-seventeenth century, when Sweden gradually assumed the leading role. Norway was under Danish rule from 1534 to 1814, when it was taken over by Sweden, who ruled it until 1905. Similarly, Finland was always a buffer state between Sweden and Russia; it fell under Swedish rule from the fourteenth century until 1808, although Russia occupied it for periods in the eighteenth century. It sided with the White Russians against the Revolution of 1917, thereby escaping annexation to the Soviet Union. It became a republic in 1918, and adopted the Finnish version 'Helsinki' of the name of its capital, Helsingfors.

A major factor in educational developments was the religious conversion of Scandinavia under the reformation from the 1570s: existing educational institutions converted from Catholicism to Lutheranism. In addition, sons of the nobility, accompanied by tutors, often studied in Germany, especially at the Lutheran stronghold of Wittenberg University.

2 PROGRESS UP TO THE MID-NINTEENTH CENTURY

Interest in mathematics in the seventeenth century was confined to elementary arithmetic and (some) basic algebra (see Dahlin *1875* and Vanäs *1955* on Sweden, and Brun *1962* on Norway). Catholic universities were founded at Uppsala in Sweden in 1477 (Lindroth *1976*), and at Copenhagen two years later, although little emphasis was placed on research; for example, Tycho Brahe studied in Copenhagen in the 1560s, but continued his education at Leipzig, Wittenberg and Rostock. His career as an astronomer in Denmark (§2.7) depended not only on his fortunate financial circumstances but also on national support furnished on a scale never surpassed afterwards in any country; for example, in the 1570s the King of Denmark gave him the island of Hveen near Copenhagen upon which he built an observatory. Among his successors was Ole Römer, who edited his manuscripts for publication and also standardized Danish measures. During the 1670s Römer worked in Paris, designing clocks and other instruments, and also demonstrating that the speed of light was finite (§9.2).

During the seventeenth century, new universities were founded by the Swedes in 1640 at Åbo (now called 'Turku', and then the capital of the part of Sweden known as Finland) and in Sweden at Lund (1668). Students could take a doctorate at a university, but it was common (in all subjects) for the supervisor to write the thesis (in Latin), for payment, and then to dispute it with his professorial colleagues in the 'examination'. The candidate was just a spectator; sometimes he entered into the spirit by, say, dedicating the fruits of his labour to his parents in an irrelevant language such as English (Kohler *1985*); such antics also occurred in other European countries. The mathematical 'doctorates' usually treated some fairly elementary questions in algebra or geometry; during the eighteenth century calculus or mechanics was also rehearsed.

Some scientific academies were formed (for Sweden in Stockholm in 1739, for Denmark in Copenhagen three years later, and at Trondheim in Norway in 1767), and scientists of major stature emerged from Sweden (the chemist Jöns Jakob Berzelius, the naturalist Carolus Linnaeus, and the physicist Anders Celsius). But the mathematicians were still of modest calibre (for Denmark, see Christensen *1895* and Nielsen *1912*, and the survey in Andersen *1980*), although a good standard of calculus teaching was achieved at Copenhagen University. The highest level is exemplified by the substantial lectures on mechanics given by the Norwegian–Danish professor at the 'academy for the nobility' at Sorø, J. Kraft; his *Forelaesninger over Mekanik* (1763–4) covered most areas of the subject (including machine theory) in its two volumes, and was translated into Latin and German. Another notable figure is Anders Lexell, who taught at the university of his home town Åbo but moved around 1770 to St Petersburg, where he assisted the (then blind) Leonhard Euler. Again, the Danish astronomer T. Bugge produced some textbooks, and made more of a stir with his excellent study of French science written in connection with the international conference held in Paris in 1799 to determine the new weights and measures (Crosland *1969*). However, the professional situation is still well exemplified by the surveyor Caspar Wessel, who was motivated by his concerns to produce in 1799 his geometrical interpretation of complex numbers (§6.2) – but made no impact.

As the sciences progressed rapidly in the early nineteenth century and became much more international, Scandinavia responded with some new institutions: for example, institutes of technology were founded in Sweden and in Denmark in the 1820s. The capital city of Finland was moved from Åbo to Helsingfors in 1821, and after being destroyed by fire the university followed six years later. But even then, mathematics still lagged behind. Norway produced a major figure in the 1820s in Niels Abel, but after studying at the University of Christiania (as Oslo was then called) – a university recently founded under the introduction of Swedish rule mentioned in Section 1 – he had to travel to France and Germany (on a governmental scholarship) for his genius to develop (Ore *1957*). No other Scandinavian mathematician of that calibre emerged in this period (for Denmark, see Nielsen *1910*), although the Dane P. A. Hansen was to become a respectable figure in mathematical astronomy, and the Swede C. J. Malmsten gained comparable status in analysis. Book authorship was usually confined to textbooks; Abel's teacher B. Holmboe (Piene *1937*) wrote a textbook, and also prepared in 1839 the first edition of his protégé's works.

3 THE IMPACT OF THE WEIERSTRASSIAN TRADITION

In the later nineteenth century, Scandinavia rose substantially in international mathematical standing. The principal single stimulus was the teaching of Karl Weierstrass at the University of Berlin from the 1860s onwards (§11.2); as with those from many other nations, Scandinavian mathematicians came to learn at his feet, so they could then go home and do likewise. Now they wrote their own theses – and often in their national language.

The most important mathematician of the first wave was the Swede G. Mittag-Leffler, founding Professor of Mathematics at the new University of Stockholm (formed in 1878) after teaching at the University of Helsinki. He applied Weierstrassian technologies to complex-variable analysis (§3.12), and in 1882 he launched the journal *Acta mathematica*, a major serial from the start with fundamental papers by Henri Poincaré, Georg Cantor and many others, including himself (Riesz *1913*, Nörlund *1958*). A man of great wealth (his wife's), he also built up a major mathematical library and collection of manuscripts (Grattan-Guinness *1971*). In addition, he secured a post at Stockholm for the female Russian mathematician Sofya Kovalevskaya in 1883, which she held until her early death in 1891.

Other Scandinavian contributors to the *Acta* included Mittag-Leffler's compatriots Ivar Bendixson, Helga von Koch (§3.8), Ivar Fredholm (a major figure in integral equations: §3.10), A. V. Bäcklund and E. Phragmén

(Gårding *1987*); the Danish analyst J. P. Gram and the graph-theorist (and influential textbook writer) J. Petersen; and the Finnish analysists E. L. Lindelöf and R. H. Mellin (Elving *1981*). Both in the *Acta* and elsewhere, these and other mathematicians (such as E. Holmgren) gave Scandinavia a prominent name in analysis.

More in tune with Felix Klein's approach to analysis was the Norwegian Sophus Lie, who studied and then taught at the University of Christiania before moving to Leipzig in 1889. One of his concerns was the second (1881) edition of the works of Abel, which he prepared with the algebraist L. Sylow, who himself gained a Christiania chair in 1898.

In applied mathematics, Norwegian father and son C. A. and W. B. Bjerknes specialized in hydrodynamics. Celestial mechanics benefited from the attention of the Finn H. Gyldén, and meteorology from methods developed around 1900 by the Norwegian V. Bjerknes (§9.7). In addition, a prize for mathematics proposed by King Oscar II of Sweden and Norway was won in 1889 by Poincaré with a major paper on the three-body problem (§8.9); after revisions, it appeared in the *Acta*.

Interest in the history of mathematics also developed at this time. For example, Mittag-Leffler's regard for the historical development of analysis was strong; he was also involved in the work of G. Eneström, whose important journal (§12.13) *Bibliotheca mathematica* (1884–1915) started off as a supplement to the *Acta*. Two other historians from Denmark were J. L. Heiberg and H. G. Zeuthen, who produced important studies of Greek mathematics (§1.3); Zeuthen also studied the medieval period, and made important contributions to algebraic geometry. Later mathematicians with historical interests included the Norwegian number theorist V. Brun and the Dane N. Nielsen; one can mention also the Danish historian and philosopher of logic J. Jørgensen.

In 1911, Mittag-Leffler brought to Stockholm the Hungarian analyst M. Riesz, whose students were to include the (American-born) analyst E. Hille and the statistician H. Cramer. Other figures of this time include the Danish analyst Harald Bohr (brother of the physicist Niels); and from Norway Thoralf Skolem, who worked in number theory and the (rather new) area of mathematical logic. Enough activity was taking place for congresses of Scandinavian mathematics to be mounted from time to time from 1909. A Mathematical Society was formed in Norway in 1918, launching a journal the following year (Thalberg *1943*). From that time on the progress of mathematics in Scandinavia has been steady, and *Acta mathematica* is still a major journal.

ACKNOWLEDGEMENTS

For advice I am indebted to K. Andersen, Y. Domar and O. I. Franksen.

BIBLIOGRAPHY

Anderson, K. *1980*, 'An impression of mathematics in Denmark in the period 1600–1800', *Centaurus*, **24**, 316–34.

Brun, V. *1962*, *Regnekunsten i det Gamle Norge. Fra Arilds tid til Abel*, Oslo and Bergen: Universitetsforlaget.

Christensen, S. A. *1895*, *Matematikens Udvikling i Danmark og Norge i det XVIII. Aarhundrede* [. . .], Odense: Hempel.

Crosland, M. P. (ed. and transl.) *1969*, *Science in France in the Revolutionary Era Described by Thomas Bugge* [. . .], Cambridge, MA: MIT Press. [Danish original: 1800, Copenhagen.]

Dahlin, E. M. *1875*, *Bidrag till de Matematiska Vetenskapernas Historia i Sverige före 1679*, Uppsala: Edqvist (Universitets Årsskrift).

Elving, G. *1981*, *The History of Mathematics in Finland 1828–1918*, Helsinki: Societas Scientiarum Fennicae.

Gårding, L. *1987*, 'Svenska matematiker', *Elementa* (Sweden), **4**, 182–90.

Grattan-Guinness, I. *1971*, 'Materials for the history of mathematics in the Institut Mittag-Leffler', *Isis*, **62**, 363–74.

Kohler, C. C. *1985*, *The History of Mathematics. A Collection of 208 Mathematical Dissertations Published in Sweden During the 17th, 18th and 19th Centuries*, Dorking, Surrey: Kohler. [Bookseller's catalogue; introduction by I. Grattan-Guinness. Second list of 107 titles, 1986.]

Lindroth, S. *1976*, *A History of Uppsala University*, Stockholm: Almqvist & Wiksell.

Nielsen, N. *1912*, *1910*, *Matematiken i Danmark*, 2 vols, Copenhagen: Gyldendal. [Published in this order; cover the period 1528–1800 and 1801–1908; include bibliographical information.]

Nörlund, N. E. *1958*, *Acta mathematica. Table des tomes 1–100*, Uppsala and Stockholm: Almqvist & Wiksell.

Ore, O. *1957*, *Niels Henrik Abel: Mathematician Extraordinary*, Minneapolis, MN: University of Minnesota Press. [Repr. 1974, New York: Chelsea.]

Piene, K. *1937*, 'Matematikkens stilling i den høiere skole i Norge efter 1800', *Norsk matematisk tidsskrift*, **19**, 52–68.

Riesz, M. (ed.) *1913*, *Acta mathematica 1882–1912. Table générale des tomes 1–35*, Uppsala and Stockholm: Almqvist & Wiksell. [Contains a full catalogue of portraits of contributors.]

Thalberg, O. M. *1943*, 'Norsk Matematisk Forening gjennom 25 år', *Norsk matematisk tidsskrift*, **25**, 65–75.

Vanäs, E. *1955*, 'Divisionens historia i Sverige', *Lychnos* (1954–1955), 141–64.

11.6

Russia and the Soviet Union

I. GRATTAN-GUINNESS AND
ROGER COOKE

1 GEOGRAPHY AND LITERATURE

The principal territory and period covered in this article is Tsarist Russia. Over the decades the region has changed, with the Baltic States and Finland sometimes belonging to Russia or to Sweden (§11.5), and Poland enjoying periods of autonomy. The article also takes in the early years of the Soviet Union up to 1930.

Since 1917 much writing on the history of mathematics in Russia (and everywhere) has been done in the Soviet Union, especially in the journal *Istoriko-matematischeskiie issledovaniya* (1948–) and in various books (see May *1973*: 664–79 *passim* for references). Of these the principal one on Russia is Yushkevich *1968*, and Gnedenko *1946* may also be consulted; Shtokalo *1964* deals with the Ukraine. There are also a number of biographies and/or editions of works of major Russian or Soviet mathematicians. Vucinich *1963* and Mikulinsky and Yushkevich *1977* are valuable surveys of Russian science in general to 1860 and to 1900 respectively.

2 THE MIDDLE AGES

In the early Middle Ages the major parts of Russia formed a kind of fiefdom for the Rurik dynasty, with Kiev as the capital. A mathematical culture was evident from the tenth century onwards, when the Orthodox Church was established. An interesting document is a sixteenth-century copy of a twelfth-century manuscript on the calendar by the scholar Kirik of Novgorod; this was a respectable topic (in all countries) for its utility in determining Easter.

However, following the Mongol and other invasions from the thirteenth century, Russia lost her autonomy, and intellectual activity fell away. The principal centre for 'research' was in Samarkand in Uzbekistan, where

Ulugh Beg gathered some scholars around him in the early fifteenth century.

Some revival took place in the early sixteenth century when the needs of commerce led to the production of textbooks in geometry and arithmetic; but general standards remained low. One obstacle was the now hostile attitude of the Russian Orthodox Church to education and science (and other subjects such as logic; see Anellis *1992*).

3 FROM PETER THE GREAT TO CATHERINE THE GREAT

The main effort to raise the standards of science was due to Peter the Great (1682–1725), especially after his tour of Europe in 1697–8. He founded a school for 'mathematics and navigational arts', appointing as its director Jacob Bruce, a Russian born to a Scottish father; he had been educated in Britain and became a major figure in Russian geodesy. Peter also brought back some Britons, in particular Henry Farquharson, to teach elementary mathematics. In 1703 L. F. Magnitsky published an influential textbook on arithmetic which tried to raise the quality of instruction; he also produced tables of logarithmic and trigonometric functions, as well as geographical and astronomical tables for use in navigation.

Peter was in contact with Gottfried Wilhelm Leibniz and later with Christian Wolff, and in his new city of St Petersburg he formed an Academy of Sciences, following Leibniz's advice on the need to import scholars and scientific artefacts and instruments from abroad. Thus when it opened its doors in 1725, shortly after his death, founder or early mathematicians included Daniel and Nicholas Bernoulli, Jacob Hermann and Euler from Switzerland.

An Academy journal was founded, which ran under various titles: initially a *Commentarii* (first published 1728), then *Novi commentarii* (1750), *Acta* (1778) and *Nova acta* (1788). Its aim was to equal in calibre the publications of The Royal Society of London and the Paris Académie des Sciences; and, thanks especially to the continuous flow of papers from Euler, it achieved its aim, for not only members but also foreign scientists contributed to its pages. However, the indigenous growth of Russian mathematics (and science in general) was not so clearly served by the Academy. For example, the language of publication was always French or Latin.

The first corresponding secretary was Christian Goldbach, who moved to Moscow in 1728 to become a tutor to the Royal family; but he kept up a correspondence with his mathematical friends, and it is from this that we have his famous 'conjecture' about even integers always being expressible as the sum of two primes (§6.1, Section 6); he dropped this in Euler's lap in 1742.

Nicholas Bernoulli died soon after arrival; Daniel and Hermann did not stay at the Academy for many years, and political difficulties led Euler to accept a post at the Berlin Academy in 1741. However, in 1766 he returned to St Petersburg, to stay for the rest of his life, and formed an active circle to discuss scientific developments (Home *1979*). His blindness did not reduce his production, and he surrounded himself with a circle of helpers, such as his son Jean, the Swiss N. Fuss (1755–1826) and the Swede Anders Lexell (1741–84). However, after Euler's death in 1783 the quality of work done at the Academy declined considerably, his own posthumous papers (appearing until 1830, no less) often being the most interesting items. Under his influence, the calculus and mechanics (including applications to engineering) remained the chief topics of study; new figures included the German F. I. Shubert (1758–1825) and S. F. Gur'ev (1764–1813) (Ozhigova *1980*).

Following Wolff's suggestion, an 'Academic University' was set up within the Academy in 1726; but the quality and quantity of students was very poor, and around 1800 it was closed after little success. In some compensation a university was opened in 1755 in Moscow (with A. A. Barsov as first professor of mathematics), but it was largely oriented to teaching and the popularization of science (differential equations entered the syllabus only in 1835). Staff were often poorly paid, and kept in arrears; the students fared better, as they often came from the gentry or even the aristocracy.

During her reign (1762–96), Catherine the Great, strongly influenced by the French Enlightenment, tried to increase the availability of education for her people; in particular, in 1782 she launched a programme to raise the general level of education, including founding a pedagogical seminary in St Petersburg. But she concentrated more on primary and secondary schools than on higher education, and was less effective than Peter had been in implementing decisions. Some of these were not happy; for example, in 1767 she decreed that the teaching at Moscow University had to be done in Russian. The return of Euler to St Petersburg, mentioned above, was one of her main successes for Russian science.

4 ALEXANDER I AND A PERIOD OF LIBERALISM

New initiatives were taken in 1801 when the liberal Alexander I ascended to the throne. In the light of some of the Marquis de Condorcet's post-Revolutionary views on the organization of education in France, the student enrolment was somewhat widened. Alexander soon founded new universities at Dorpat (1802, in succession to an institution that had operated under previous Swedish rule), Vilna (1802), Kazan' (1804, where S. Y. Rumovsky (1732–1812) was a key figure), Kharkov (1804), and eventually St Petersburg (1819). At Moscow University a separate physico-mathematical faculty was established in 1804 (previously science had been taught within the philosophy faculty); in addition, a Society of Mathematicians ran for some years from 1811 to translate major foreign works into Russian.

In 1809 the St Petersburg Academy changed the name of its journal again, to *Mémoires*, and published solely in French. Contacts with France were initiated; when the Ecole Polytechnique opened in Paris in 1794 (§11.1), some Russian officers and princes were sent there to study. The links with engineering increased in 1810 when an Institute of Ways and Communications was created in St Petersburg, modelled on the Paris Ecole des Ponts et Chaussées; for Napoleon accepted Alexander's request that graduates of the Ecole Polytechnique be sent over to teach and conduct research. The most notable immigrants were Gabriel Lamé and Emile Clapeyron, who came in 1820 when the programme was restarted after the fall of Napoleon; during their decade of residence they published a mass of excellent collaborative work in the *Journal* of the Institute, which was launched in 1826 (Bradley *1981*).

In the opposite direction, Mikhail Ostrogradsky and Viktor Bunyakovsky spent part of the 1820s in Paris, studying current research. Bunyakovsky worked on analysis and mechanics, and later on calculating machines; Ostrogradsky made notable contributions to complex-variable analysis (§3.12) and potential theory (§3.17), much of which he published upon his return to St Petersburg in the Academy's *Mémoires* (and also in its recently founded *Bulletin*). He continued his researches in these areas; and around 1840 he published on ballistics, which excited some interest in this topic (§8.11) in Russia. However, by birth a Ukrainian, his Russian was not strong, and he always published in French. He also did not respond warmly to the proposal of non-Euclidean geometry made by Nikolai Lobachevsky far away at Kazan' University (§7.4). Lobachevsky's awkward personality did not aid the reception of his ideas, and much of his contemporary influence lay within the affairs of his university (Aleksandrov and Laptev *1976*); his most notable student was the mathematical physicist A. F. Popov (1815–78).

Despite the fine achievements of these mathematicians, and the contributions of lesser figures such as I. M. Simonov (1794–1855) and O. I. Somov (1815–76), Russia did not fully participate in the rapid internationalization of mathematics from the late 1820s onwards; for after Alexander's death in 1825 the regime returned to its more usual authoritarian and even repressive style, with the annexation of Poland (so that 'officially' the geometer and geodeter E. F. A. Minding (1806–85) at Dorpat was Russian). Thus, for example, the French links were largely broken (and Polish ones increased, with more Poles studying at the Ecole Polytechnique).

5 FROM 1850 TO 1900

At Moscow University the Moravian N. D. Brashman (1796–1866) and the Russian K. M. Petersen (1828–81) developed a speciality in differential geometry (§3.4), while N. V. Bugaiev (1837–1903) worked in number theory, and N. E. Zhukovsky (1847–1921) ran the Department of Mechanics and from the 1900s became a major figure in aerodynamics (§8.12). In addition, Brashman led the founding of the Moscow Mathematical Society in 1864 and the launch of its journal soon afterwards (Nekrasov *1904*). This was one of the first national or city-based societies for the subject (compare §11.12 on journals), and Zhukovsky was a most effective President from 1905 to his death.

However, the main mathematical centre remained St Petersburg, where a strong school arose with Pafnuty Chebyshev (a Moscow graduate) as a central figure; colleagues and students included Andrei Markov, Aleksandr Lyapunov and A. V. Vasiliev (1853–1929). The Russian contributions to probability theory and statistics which he led are described in §10.6; Bunyakovsky had led the way, and under Chebyshev major contributions were made to the central-limit theorem and the conception of Markov chains and processes. Chebyshev also worked in approximation theory and number theory, and several areas of applied mathematics fell within their concerns (see §8.4 on dynamical stability).

Elsewhere, universities were set up at Odessa (1865) and Tomsk (1888), though not always with science prominent. A Mathematical Society was founded at Kharkov in 1892, where the analyst and mathematical physicist V. Steklov (1862–1926) was active, as Lyapunov had been earlier. Kazan's principal mathematical claims to fame were Vasiliev, who was promoted there from St Petersburg and worked in elasticity theory and function theory; and the astronomer and logician P. S. Poretsky (1846–1907). Dorpat could now boast A. Kneser (1862–1930, the first of a dynasty) for geometry and I. Schur for algebra. This field was also encouraged at Kiev, by D. A. Grave (1863–1939) and his student O. Y. Schmidt (1891–1956).

Excluded from these developments by her sex was Sofya Kovalevskaya, a descendent of Shubert. She studied in Germany and became one of Karl Weierstrass's best followers. Through G. Mittag-Leffler, she obtained a post at Stockholm University in 1883, which she held until her death eight years later (Koblitz *1983*). Her main mathematical contacts with Russia were maintained via correspondence with Chebyshev and Vasiliev.

Several of these mathematicians were elected to the St Petersburg Academy as foreign members of academies and societies elsewhere; many published abroad as well as in Russian journals. Russia was beginning to enter the international scene.

6 THE EARLY TWENTIETH CENTURY

Another university was founded, at Saratov in 1909, and Russia participated in international surveys of mathematical education at various levels (Bobynin *1903*, Possé *1910*). A major figure at the St Petersburg School of Mines was A. Krilov (1863–1945), who specialized in several areas of engineering mathematics. Kharkov was graced by the presence of the probabilist S. N. Bernshtein (1880–1968), who had an interesting student of Lebesgue-type set topology called Jerzy Neyman (Reid *1982*).

At Moscow University, some unusual developments took place. In the late nineteenth century idealistic philosophy took such a hold that even mystical notions such as the identification of discontinuity with free will were seriously debated. But in the mid-1900s, a more orthodox change occurred: the general acceptance of set theory (§3.6, §5.4) by mathematicians was well evident there. The chief figures were I. I. Jegalkin (1869–1947), and especially N. N. Luzin (1883–1950). After studying at Paris and Göttingen, Luzin not only made significant contributions to descriptive set topology but also inspired a strong circle of mathematicians (Phillips *1978*, Demidov *1985*): M. Y. Suslin (1894–1919) in the same area, P. S. Aleksandrov and P. S. Urysohn in topology (§7.11), D. E. Menshov in measure theory and N. K. Bary (1901–61) in trigonometric series, and Andrei Kolmogorov in probability theory (§10.16) and various other fields.

After the Russian Revolution of 1917, followed by civil wars and the end of the First World War, higher education became much more widely available (15 new universities had been set up by 1925) but also politicized. At academy level, the St Petersburg and other institutions were merged in 1919 into one Union-wide organization, which officially became the principal centre for scientific research in the mid-1920s (Vucinich *1984*). The changes initially dislocated mathematics as much as anything else. For example, the complications in the Ukraine caused the subject largely to disappear from the university, and the Mathematical Society to cease operating until well into the 1920s.

In Moscow the mathematicians were largely apolitical, and survived relatively peacefully. But the inevitable 'attention' occurred in 1930, when a strong attack was launched on the current President of the Moscow

Mathematical Society, the analyst D. Egorov; as a result he was arrested and died the following year in exile in Kazan'. From then on mathematics developed under conditions of which the history has yet to be written.

BIBLIOGRAPHY

Aleksandrov, P. S. and Laptev, V. L. (eds) *1976, N. I. Lobachevsky. Nauchno-pedagogicheskoe nasledie*, Moscow: Nauka.

Anellis, I. *1992*, 'Theology against logic: The origins of logic in Old Russia', *History and Philosophy of Logic*, **14**, 15–42.

Bobynin, V. V. *1903*, 'L'Enseignement mathématique en Russie', *L'Enseignement Mathématique*, Series 1, **5**, 237–61, 397–414. [Also many articles by him in Russian; he edited a journal on the history of the 'fisico-mathematical sciences', 1885–1904.]

Bradley, M. *1981*, 'Franco-Russian engineering links: The careers of Lamé and Clapeyron, 1820–1830', *Annals of Science*, **38**, 291–312.

Demidov, S. S. *1985*, 'N. V. Bougaiev et la création de l'école de Moscou de la théoreie des fonctions d'une variable réelle', in M. Folkerts and U. Lindgren (eds), *Mathemata. Festschrift für Helmuth Gericke*, Munich: Steiner, 651–73.

Gnedenko, B. V. *1946, Ocherki po istorii matematiki v Rossii*, Moscow: Technical Theoretical Literature.

Home, R. *1979*, 'Introduction', in *Aepinus's Essay on the Theory of Electricity and Magnetism* (ed. R. Home, transl. P. J. Connor), Princeton, NJ: Princeton University Press, 1–224.

Koblitz, A. H. *1983, A Convergence of Lives. Sofia Kovalevskaia* [...], Boston, MA: Birkhaüser.

May, K. O. *1973, Bibliography and Research Manual in the History of Mathematics*, Toronto: University of Toronto Press.

Mikulinsky, S. R. and Yushkevich, A. P. *1977, Razvitie estestvoznaniya b Rossii*, Moscow: Nauka.

Nekrasov, P. A. *1904, Moskovskaya filosofsko-matematicheskaya shkola i eya osnovateli*, Moscow: Moscow University Press.

Ozhigova, E. P. *1980, Matematika v Peterburgskoi Akademii nauk v kontse XVIII – pervoi polovine XIX veka*, Leningrad: Nauka.

Phillips, E. *1978*, 'Nicolai Nicolaevich Luzin and the Moscow school of the theory of functions', *Historia mathematica*, **5**, 275–305.

Possé, C. *1910, Rapport sur l'enseignement mathématique dans les universités, les écoles techniques supérieures et quelques-unes des écoles militaires en Russie*, St Petersburg: Trenké & Fusnot.

Reid, C. *1982, Neyman – From Life*, New York: Springer.

Shtokalo, I. Z. (ed.) *1964, Z istorii vitchiznyanogo prirodo znavstva*, Kiev: Naukogo Dumka.

Vucinich, A. *1963, Science in Russian Culture. A History to 1860*, London: Owen.

—— *1984, Empire of Knowledge. The Academy of Sciences of the USSR (1917–1970)*, Berkeley, CA: University of California Press.

Yushkevich, A. P. *1968, Istoriya matematiki v Rossii do 1917 goda*, Moscow: Nauka.

11.7

The British Isles

I. GRATTAN-GUINNESS

1 FOUR DIFFERENT COUNTRIES

The differences between nations, which is a major feature of this Part of the encyclopedia, is especially marked in Britain; for while it is natural to group together England, Scotland, Ireland and Wales, substantial variations in their respective histories are evident, sufficient to form a major theme of the article and the subject of its last section. Developments up to about 1800 in England and Scotland are surveyed in the next three sections; then progress to the mid-nineteenth century, and thereafter, are treated in Sections 5 and 6 respectively.

The discussion naturally concentrates on universities, but professional societies and other organizations are noted, and a few remarks on more elementary instruction are contained in Section 5. There are numerous histories of all these institutions, but as usual the mathematical aspects have been given a minor or misunderstood place, even in contexts where the subject was quite important. The bibliography is confined to the most substantial and pertinent items.

2 ENGLAND: THE CAMBRIDGE DOMINANCE

It is imperative to begin with this city, for it has always exercised an extraordinary degree of power over British mathematics. The main account of its development to the mid-nineteenth century is Ball *1889*.

Mathematics began to gain some status at Cambridge around the mid-sixteenth century, with figures such as Robert Recorde, John Dee and Henry Billingsley (the first English translator of Euclid); within a few decades the navigator Edward Wright was there, and also his friend Henry Briggs, specialist in the related topic of tables. But many of these men also worked at other institutions. On the period, see Taylor *1954*; on the importance of navigation, see §8.18.

The scene during the seventeenth century is of course dominated by the presence of Isaac Newton in its second half; but his predecessor as (the first) Lucasian Professor, Isaac Barrow, and John Pell are also of note. Even then their significance lay more in research than in teaching (where Newton is said to have addressed the walls on occasion). Of the various colleges, Trinity College was the most important.

The first generation of Newtonians formed a considerable group: for example, Brook Taylor (of series fame; §4.3) who produced results in several areas of Newton's concerns; and Roger Cotes, the first Plumian Professor of 'astronomy and experimental philosophy', who prepared the second edition of Newton's *Principia* in 1713. But in general the effect of Newton's achievements in mathematics, and especially the Newton-inspired priority dispute of the 1710s over the invention of the calculus, were rather unfortunate for British mathematics in general: the fluxional calculus (§3.2) and the elementary parts of his mechanics (§8.1) became monotonous fare, and the massive achievements of the Continental mathematicians did not penetrate this little island (Guicciardini *1989*: Chaps 1–6). The two figures of greatest note are the blind mathematician Nicholas Saunderson in the first half of the century, and Edward Waring in the second half. But, by and large, research was mediocre, and teaching worse. This tale is resumed in Section 5.

The examination system acquired some peculiar jargon. The examination became known as the 'tripos', after the practice of a certain fifteenth-century University representative of sitting on a three-legged stool and interrogating candidates. Graduation was granted at three levels: 'wranglers', 'junior optimaes' and 'poll-men'. But even the highest level did not reflect great mathematical knowledge: basic fluxional calculus and astronomy were the hardest topics. In some compensation, a prize for proficiency in mathematics was instituted around 1770 in the will of Plumian Professor Robert Smith.

3 OTHER ENGLISH CENTRES

Some of the Cambridge men mentioned earlier also taught at Oxford, especially at Merton College; in particular, the city was graced for the second half of the seventeenth century by the presence of Wallis, who held the chair in 'geometry' founded by Henry Savile in 1621. He was succeeded by Edmond Halley. But mediocrity befell the university in general for a long time: even as late as the polite survey by De Morgan *1830* the story is not exciting.

The Royal Society of London was created in 1666. It has never carried out research or teaching in science, and until the 1830s it was not short of aristocratic nonentities among its membership; but from the start it was the main national focus for science, and the chief coveted election for British scientists. Mathematics was prominent, especially during Newton's regime (1703–1727) as President, and has always been well represented there. Royal Society *1940* is a valuable source of facts on its organization, and records all the Fellows up to the time of its writings.

It is clear that the degree of professionalization of mathematics was extremely modest (Taylor *1966*): high-level training was available only at Cambridge, and few posts were available anywhere. One source of employment was with (or even as) Astronomer Royal at Greenwich Observatory, a post created in 1675; but, for example, Thomas Harriot, the best British mathematician of the early seventeenth century, was in aristocratic service. Gresham College was founded in London in 1597 as a centre of higher teaching, and the most significant chairs were those of astronomy and geometry (of which Henry Briggs was the first holder); but its significance, never great, declined from the mid-seventeenth century onwards. A Spitalfields Mathematical Society ran for a time from 1717 (Cassels *1979*), but did not have a major impact.

Around this time there emerged a type of mathematics teacher and practitioner known as a 'philomath', teaching in London coffee-houses and/or serving as itinerant lecturer. Their achievements were modest (actuarial mathematics (§10.3) was a speciality for some), though the amount of publication is remarkably large (Wallis and Wallis *1986*). The most distinguished of their number was Thomas Simpson, of the well-known interpolative rule (§4.13). A number of interesting journals for, or including, mathematics were published, some continuing into the early nineteenth century (Archibald *1929*).

4 SCOTLAND TO 1800

Scottish universities have as long a history as the oldest English ones, with St Andrews and Aberdeen founded respectively in the early and late fifteenth century, and Edinburgh and Glasgow a hundred years after that. But Scotland was rather slower than England in developing significant mathematicians (Gibson *1927*, Wilson *1935*). Even John Napier did not influence developments, despite the importance of his tables when published in the late sixteenth century (§2.5).

Perhaps the first notable figures were the Gregorys, James I and his nephews James II and David. All followers of Newton, they held chairs at Edinburgh between 1664 and 1725, James I serving at St Andrews first and James II afterwards, and David moving on to Oxford. For half a century from 1711 Glasgow had Robert Simson, who produced a highly regarded edition of Euclid and restored lost Greek works. Aberdeen and then

Edinburgh were graced by the effective teaching of Colin Maclaurin from 1717 until his death in 1746. However, while trigonometry and navigation were quite well favoured in teaching from early on, the level of instruction was normally no higher than in England: for example, algebra was taught only from the eighteenth century.

5 REFORMS AROUND 1800, AND EXPANSION

Around 1800, British mathematics at last began to take serious notice of Continental work, especially in the calculus. Guicciardini (*1989*: Chaps 7–9) gives the details, which are much richer and more varied than has hitherto been realized. The countries are treated here in turn.

The principal single stimulus to change seems to have been the publication in 1799 of the first two volumes of Pierre Simon Laplace's *Traité de mécanique céleste*. At all events, around that time various British mathematicians began to learn and use either the differential form of the calculus or the algebraized version propounded by Joseph Louis Lagrange (§3.2). In England, at the two military schools men such as Peter Barlow, Olinthus Gregory and Charles Hutton took note, while at Cambridge Robert Woodhouse tried to be more Lagrangian than Lagrange.

Then Charles Babbage, John Herschel and George Peacock, undergraduates at Cambridge, formed in 1813 the 'Analytical Society' (Enros *1983*). They preferred Lagrange's methods (hence their title), but differential techniques were also brought in – and the fluxional calculus met a rapid death. The first two men soon carried out research in Lagrangian topics, while Peacock pushed the changes at the University (more oriented around the Leibniz/Euler form of the calculus). Eminent mathematicians of the future soon arrived as undergraduates: Augustus De Morgan, William Whewell and George Biddell Airy were among the first crop, and Duncan Gregory (of the dynasty), Arthur Cayley, J. J. Sylvester, George Green (as a mature student, and rather an outsider), William Thomson (later Lord Kelvin) and George Stokes in the next. Cambridge strengthened its habitual place as the premier British centre (Grattan-Guinness *1985*); operator methods for solving differential equations became an industry (§4.7).

New textbooks in almost all areas of pure and applied mathematics were written over the years; especially notable were Whewell's on mechanics, which brought new philosophical views to Newton's formulation (§8.1). As Master of Trinity College, he came to have a great influence within the University (Becher *1980*). As for papers, the Cambridge Philosophical Society, founded in 1819, became an important new venue, while a wider audience could be addressed in the *Reports* of the British Association for the Advancement of Science, founded in 1831.

Another general publishing activity of that time was multi-volume encyclopedias, in which substantial articles on branches of mathematics were included. Most notable were the *Encyclopaedia Metropolitana* (1817–45), the supplementary volumes (1815–24) to the *Encyclopaedia Britannica* (later editions also contained important mathematical articles), and the *Edinburgh Encyclopaedia* (1808–30).

These last two encyclopedias were produced in Scotland. There, in the 1800s, John Playfair lamented the inability of the British to read Laplace, and tried to generate some interest. In 1805 he was succeeded in his Edinburgh chair by John Leslie (Morrell *1975*). In later decades Cambridge graduates such as Philip Kelland and, especially, Thomson took chairs, and so did home-grown products such as W. J. M. Rankine. The Royal Society of Edinburgh, founded in 1783, was a principal home for the publication of research papers.

To Trinity College Dublin (founded in 1592, and then still the only university-level institution in Ireland), John Brinkley came from Cambridge in 1790 to be Professor of Astronomy, and Bartholomew Lloyd started to make changes in 1813 upon his appointment as Professor of Mathematics (Grattan-Guinness *1988*). Soon a group of mathematicians of remarkable quality was in place, most notably William Rowan Hamilton, but including also Lloyd's son Humphrey and James MacCullagh (McConnell *1945*). A favoured home for papers was the *Transactions* of the Royal Irish Academy, an institution founded in 1786.

In the late 1840s, three Queen's Colleges were set up, in Cork, Belfast and Galway, to supplement the protestant Trinity College by providing educational opportunities for Catholics. They opened just after the potato famine, during which half of Ireland's population of eight million either died or went abroad (like most of the food which Ireland produced); so it is not surprising that their growth was slow (and, incidentally, with a dominating Protestant representation, the church and the convent still being the preferred aspiration for Catholic parents). George Boole (a self-taught English mathematician) was the founding professor at Cork, but he did not make much impact in Ireland itself.

Meanwhile, London had at last entered higher education with the creation in 1828 of its University, soon to become known as University College, London when King's College was created. A theological rivalry was involved in that King's, like Oxford and Cambridge, required adherence to the articles of the Church of England, which University College expressly avoided. The founder professor there was De Morgan. In 1866, towards the end of his life, he became the first president of the London Mathematical Society, the first major organization for professional mathematicians in the British Isles (Collingwood *1966*). The creation of this society is an example of the worldwide expansion of sciences, to the extent that societies for individual disciplines were being created. Prior to that, the Statistical Society of London was formed in 1834 (by Babbage, among others); it was more concerned with the need to collect data than to analyse them mathematically (Hilts *1978*).

The standards of examination rose: for example, at Cambridge the theorem in potential theory known now after Stokes (§3.17) was first published as a Smith's Prize question in 1854 (Cross *1985*). Written forms of examination came to replace entirely the method of oral testing; and a remarkably rapid innovation took place in the mid-1850s when written examinations at more elementary levels were institutionalized, both at technical-school level and for school-leavers. Mathematics was a prominent subject here, and a leading figure was the Irish-born mathematician James Booth (Foden *1989*). In the later 1860s The Association for the Improvement of Geometrical Teaching was created for school-teachers, and started the important journal *The Mathematical Gazette*; in 1897 the Association changed its name to The Mathematical Association (Price *1983*). The interesting original choice of name reflected the preoccupation with geometry in English education (§7.8); another example was the advocacy of squared paper (Brock and Price *1980*).

6 EXPANSION DURING THE LATE NINETEENTH CENTURY

Some of the events just described take us into the last third of the nineteenth century; this period and later is the concern here. Changes or additions took place at the established centres. Oxford woke up at last, especially with H. J. S. Smith there from the 1870s, and succeeded in 1883 by Sylvester. University College, London gained William Kingdon Clifford, who in the 1870s distinguished himself in both applied mathematics and philosophy before his early death. It also became the major British centre for mathematical statistics, thanks to Francis Galton (§10.7): in addition to his own work, he funded in his will a chair of eugenics, which was taken in 1911 by Karl Pearson. He had already been in the field for twenty years, having founded the journal *Biometrika* in 1901. In 1933 Pearson was succeeded by Ronald Aylmer Fisher, his own son Egon taking a new chair in statistics there.

From the 1870s onwards Cambridge began to function more like a proper university instead of as a group of rather disparate colleges, and professors were given more influence and power. Nevertheless, the system of coaches (or 'pupil-mongers'), who trained up tripos candidates for the race with great skill and at good profit, continued to be very influential: William

Hopkins and E. J. Routh (who wrote several standard textbooks in mechanics) were distinguished examples of this genre. Changes in the tripos scheme were brought in at various times, especially in 1907 when a new Part II was introduced which was more realistic for the educational needs of undergraduates (Ball *1918*: Chap. 15). Among journals, J. W. L. Glaisher is notable for editing both the *Messenger of Mathematics* and the *Quarterly Journal of Pure and Applied Mathematics*.

Research work in applied mathematics was of high quality, the contributions of James Clerk Maxwell (professor for eight years until his death in 1879) inspiring his successor the third Lord Rayleigh (who also specialized in acoustics), J. J. Thomson, Joseph Larmor and others; meanwhile George Darwin was occupied with geophysics. Later figures included G. I. Taylor, a major figure in aerodynamics from the 1920s; and Harold Jeffreys, who worked in mathematical physics and also probability theory.

By contrast, under the influence of Arthur Cayley the pure side had become bogged down in the dreariest ends of invariant theory (§6.8). Aspiring mathematicians either went abroad, like Grace Chisholm, who took a doctorate at Göttingen in 1895 under a Prussian programme for female education (Grattan-Guinness *1972*); or else, like Bertrand Russell, they despised the system and turned to other subjects, in his case the philosophy of mathematics with his former tutor Alfred North Whitehead (Griffin and Lewis *1990*; §5.2).

In a repeat of the fluxions show, one especially notable absentee from Cambridge mathematics was Continental real- and complex-variable analysis; the nearest approximation was the writing of solid but stodgy textbooks on related topics by men such as Isaac Todhunter (who did better work in the history of mathematics: see §12.13), E. W. Hobson and A. R. Forsyth. News from abroad was brought to Cambridge awareness in the early 1900s by them, and especially by the young G. H. Hardy, soon joined by J. E. Littlewood – who then set up such a regime of their own in England that it in its turn was to lose sight of foreign advances in major areas such as abstract algebras and topology.

The creation of university-level colleges elsewhere in England increased the diffusion of mathematics, although sometimes they functioned as preparatory schools for Cambridge. Owens College, Manchester (later Manchester University) was particularly notable, and had Horace Lamb (a Cambridge graduate) as professor from 1885 to 1919. He was succeeded by Sydney Chapman, who moved to Imperial College London five years later, to succeed Whitehead. A fellow-professor there was the fourth Lord Rayleigh, who specialized in optics. Ireland itself continued to produce distinguished figures (Purson *1903*), especially in applied mathematics;

G. F. Fitzgerald is the best-remembered. However, Cambridge still maintained most of its dominance: for example, graduates from the newer universities (especially in England) sometimes took a further first degree there, as 'affiliated students'.

In Scotland the Edinburgh Mathematical Society was formed in 1882, and notable figures continued to appear both there and in Ireland. A particularly interesting case was the Cambridge graduate E. T. Whittaker, who was appointed Astronomer Royal of Ireland in 1906, and then moved to a chair at Edinburgh six years later. During more than thirty years there he was a pioneer for Britain in developing numerical analysis, and a rarity in writing important works in the history of applied mathematics and physics.

Wales at last had its University Colleges, formed in the early 1880s, but mathematics did not flourish well. G. H. Bryan and G. B. Mathews at Bangor were the first notable figures (respectively, in thermodynamics and number theory), followed by the itinerant W. H. Young at Aberystwyth after the First World War.

The 'career' of Young is a good example of the limited opportunities in British mathematics. A leading Cambridge pupil-monger, he married Chisholm soon after her return from Göttingen, and they returned to the Continent to discover what mathematics was. They soon formed a formidable partnership – the first such in mathematics, in fact – but in his home country he obtained only appointments at Aberystwyth and Liverpool. Further, thirty years after Chisholm's German doctorate their daughter could only take the *title* of a doctorate in mathematics at Cambridge...

Finally, a major bibliographical project must be noted. In 1857 the Royal Society commenced a 'Catalogue of Scientific Papers' for the nineteenth century; eventually 19 volumes were published between 1867 and 1925 (§11.12). They were much enhanced by four index volumes covering mathematics, mechanics and physics (the first two compiled with help from Mathews), which came out between 1908 and 1914 – priceless sources for the historian, and far too little known (Royal Society *1940*: 180–82). At that time the catalogue was continued with other organizations on an international basis: 14 (smaller) volumes for mathematics, covering the period 1901–14 appeared between 1906 and 1921 (§13.1, items 3.1 and 3.6).

7 NATIONAL DIFFERENCES

I have often thought that an interesting essay might be written on the influence of race in the selection of mathematical methods. (Ball *1889*: 123).

It is clear from even this brief survey that, while the usual kind of international contacts were taking place, there were still significant differences between the four constituent countries of the British Isles. Rouse Ball continued the above quotation by claiming, among other things, that 'English mathematics might be characterized as analytical.' He had in mind the emphasis on algebra, from the time of Wallis through the post-Lagrangian interests of Babbage and Herschel, and further. This feature is apparent especially in the nineteenth century, when all the major algebras received attention from English mathematicians: in addition to the one just mentioned, there were the algebraic logics of Boole and De Morgan (§5.1), the foundations of common algebras (§6.9), the contributions to invariant theory and linear algebra (§6.7–6.8) and probability theory.

Now the English mathematicians were also greatly concerned with geometry, though more for foundational reasons such as axiom systems in Euclidean and non-Euclidean geometry (Richards 1988). Scottish mathematics, by contrast, was much more intrinsically geometrical in character; that is, it not only showed an interest in the subject of geometry but especially displayed a preference for its methods and ways of thinking. The reasons are not happenstance: from the late eighteenth century much influence was enjoyed by a movement called 'common-sense philosophy', in which, together with a nativist psychology, empirical factors were favoured and analogies between different contexts stressed. The view was not just a whim of philosophers: it was extensively taught, and influenced local science. Olson 1975 gives an excellent account of its influence upon British physics in general, and a similar story could be told for Scottish mathematicians: for example, it is quite explicit in Kelvin and Maxwell, and the place of geometry was already marked in Maclaurin.

Meanwhile, the Irish had their own answer, often a mixture of algebra and geometry, of which Hamilton's quaternions (§6.2) are only the best-known example. Finally, the Welsh bred few mathematicians, although they produced many teachers.

BIBLIOGRAPHY

Archibald, R. C. 1929, 'Notes on some minor English mathematical serials', The Mathematical Gazette, 14, 179–200.

Ball, W. W. Rouse 1889, A History of the Study of Mathematics at Cambridge, Cambridge: Cambridge University Press.

—— 1918, Cambridge papers [. . .], Cambridge: Cambridge University Press.

Becher, H. 1980, 'William Whewell and Cambridge mathematics', Historical Studies in the Physical Sciences, 11, 1–48.

Brock, W. H. and Price, M. 1980, 'Squared paper in the nineteenth century [. . .]', Educational Studies in Mathematics, 11, 365–81.

Cassels, J. W. S. 1979, 'The Spitalfields Mathematical Society', Bulletin of the London Mathematical Society, 2, 241–58. [Addendum in Ibid., 1980, 3, 343.]

Collingwood, E. F. *1966*, 'A century of the London Mathematical Society', *Journal of the London Mathematical Society*, **41**, 577–94.

Cross, J. J. *1985*, 'Integral theorems in Cambridge mathematical physics, 1830–55', in Harman *1985*: 112–48.

De Morgan, A. *1830*, 'State of the mathematical and physical sciences in the University of Oxford', *Quarterly Journal of Education*, **4**, 191–208. [Authorship attributed.]

Enros, P. J. *1983*, 'The Analytical Society (1812–1813): Precursor of the renewal of Cambridge mathematics', *Historia mathematica*, **10**, 24–47.

Foden, F. *1989*, *The Examiner. James Booth and the Origins of Common Examinations*, Leeds: University.

Gibson, G. A. *1927*, 'Sketch of the history of mathematics in Scotland to the end of the 18th century', *Proceedings of the Edinburgh Mathematical Society*, Series 2, **1**, 1–18, 71–93.

Grattan-Guinness, I. *1972*, 'A mathematical union: William Henry and Grace Chisholm Young', *Annals of Science*, **29**, 105–86.

—— *1985*, 'Mathematics and mathematical physics at Cambridge, 1815–40: A survey of the achievements and of the French influences', in Harman *1985*: 84–111.

—— *1988*, 'Mathematics instruction and research in Ireland, 1782–1842', in J. Nudds, N. McMillan, D. Weaire and S. McKenna Lawlor (eds), *Science and Engineering in Ireland 1800–1930; Tradition and Reform*, Dublin: Trinity College, 11–30. [Some other articles here are also pertinent.]

Griffin, N. and Lewis, A. C. *1990*, 'Bertrand Russell's mathematical education', *Notes and Records of the Royal Society*, **44**, 51–71.

Guicciardini, N. *1989*, *The Development of Newtonian Calculus in Britain 1700–1800*, Cambridge: Cambridge University Press. [Complete bibliography of fluxional writings. Appendices C and D list all chairs in British universities and military academies in the eighteenth century.]

Gunther, R. T. *1937*, *Early Science at Oxford*, Vol. 11, *Oxford Colleges and their Men of Science*, Oxford: Clarendon Press. [Useful *passim*.]

Harman, P. M. (ed.) *1985*, *Wranglers and Physicists* [. . .], Manchester: Manchester University Press.

Hilts, V. L. *1978*, '*Aliis exterendum*, or the origins of the Statistical Society of London', *Isis*, **78**, 21–44.

Howson, A. G. *1982*, *A History of Mathematics Education in England*, Cambridge: Cambridge University Press. [More a history of nine mathematics educators, including Recorde and De Morgan.]

McConnell, A. J. *1945*, 'The Dublin mathematical school in the first half of the nineteenth century', *Proceedings of the Royal Irish Academy*, Series A, **50**, 75–88.

Macfarlane, A. *1916*, *Lectures on Ten British Mathematicians of the Nineteenth Century*, London: Chapman & Hall.

Morrell, J. B. *1975*, 'The Leslie affair: Career, kirk and politics in Edinburgh in 1805', *Scottish Historical Review*, **54**, 63–82.

Olson, R. *1975*, *Scottish Philosophy and British Physics, 1750–1880* [. . .], Princeton, NJ: Princeton University Press.

Panteki, M. *1987*, 'William Wallace and the introduction of Continental calculus to Britain [...]', *Historia mathematica*, **14**, 119–32. [Text of an important letter.]

Price, M. H. *1983*, 'Mathematics in English education 1860–1914: Some questions and explanations in curriculum history', *History of Education*, **12**, 271–84.

Purson, J. *1903*, [Address on nineteenth-century Irish mathematicians], *Report of the British Association for the Advancement of Science*, (1902), 499–511.

Richards, J. *1988*, *Mathematical Visions: The Pursuit of Geometry in Victorian England*, New York: Academic Press.

Royal Society *1940*, *Record of the Royal Society of London* [...], 4th edn, London: The Royal Society.

Taylor, E. G. R. *1954*, *The Mathematical Practitioners of Tudor and Stuart England, 1485–1714*, Cambridge: Cambridge University Press.

—— *1966*, *The Mathematical Practitioners of Hanoverian England, 1714–1840*, Cambridge: Cambridge University Press. [This and the previous volume to be used with some caution.]

Wallis, P. and Wallis, J. *1986*, *Biobibliography of British Mathematics and Its Applications, Part II, 1701–1760*, Letchworth: Epsilon Press.

Wilson, D. K. *1935*, *The History of Mathematical Teaching in Scotland to the End of the 18th Century*, London: University of London Press.

11.8

The Italian states

U. BOTTAZZINI

1 INTRODUCTION

As Dirk Struik wrote in the preface to the Italian edition *1981* of his *Concise History of Mathematics*, a mathematician thinking of Italy will certainly remember the names of Enrico Betti and Eugenio Beltrami, for instance, and also those of Federigo Enriques, Giuseppe Peano, Vito Volterra, Tullio Levi-Città, Gregorio Ricci-Curbastro and their colleagues, all leading figures in modern mathematics. An interest in history will probably help the reader recall the rise of modern algebra and the names of Niccolò Tartaglia and Girolamo Cardano. But the tradition of mathematical research in Italy is much richer than that. According to Struik, no country in the world, apart perhaps from China, has a longer mathematical tradition than Italy, much of it of fundamental importance.

One could start by going back to Boethius (*circa* AD 500), or to Gerbert (*circa* 1000) who, as Pope Sylvester II, created in Rome one of the first schools of Western mathematics. A lasting interest in mathematics, however, developed only during the Middle Ages under the influence of Arabic science. This tradition began with Leonardo of Pisa (Fibonacci) in the early thirteenth century (§2.4). From that time to the Renaissance, mathematics was cultivated in commercial schools rather than in universities. In towns like Genoa, Pisa, Florence, Milan and Venice, the abbacists taught the art of counting with the positional system and Arabic numerals to the sons of the merchants. Even though the abbacists were mainly interested in solving equations arising from practical problems, they also played an important role in mathematics by opening the way to the further development of algebra (Franci and Toti Rigatelli *1985*).

2 UNIVERSITIES AND COURTS IN THE RENAISSANCE

It can hardly be disputed that modern mathematics began with the Renaissance, when the Italian algebraists discovered the solution of the general cubic equation (§6.1), a task which had challenged mathematicians since Antiquity. Scipione Del Ferro, who is credited with finding the 'Tartaglia–Cardano' formula for the roots of the cubic, was a 'lecturer' (as the post was then called) at the University of Bologna. Among his colleagues there were Luca Pacioli, who gained universal fame for his *Summa de arithmetica* (1494), and the astronomer Antonio Maria Novara, the teacher of Nicolaus Copernicus.

The University of Bologna, which is among the oldest in the world, was at the beginning of the sixteenth century one of the most famous in Europe, attracting students from all Western countries. In addition to Copernicus, Albrecht Dürer and Ludovico Ferrari studied there. After a period Ferrari spent in Milan as a pupil of Cardano, both men eventually obtained positions at Bologna, where they taught respectively mathematics and medicine for some years. Even though he was never listed among the 'lecturers' of mathematics at the University, Rafael Bombelli too has to be numbered among the mathematicians belonging to the Bolognese school. His *Algebra* (1572), in which he introduced 'imaginary numbers', was for decades one of the more influential mathematics books. At the turn of the century the mathematical tradition of the University of Bologna was still living in the works of Antonio Cataldi.

Rome too had acquired an increasing importance in the mathematical world, since Christoph Clavius had taught there for some thirty years at the *collegio* of the Jesuits. This became the centre for the mathematical education of many Jesuits, who were to play a primary role in seventeenth-century science. One of Clavius's pupils there was Luca Valerio, who later became 'lecturer' at the University of Rome. A deep interest in mathematics and the natural sciences led Prince Federico Cesi to found in Rome in 1603 the Accademia dei Lincei, which is still the most important Italian academy.

During the Renaissance, mathematics was also cultivated at the courts of the princes and dukes of the Italian states. The court of Duke Montefeltro at Urbino was for decades one of the most important centres for mathematics and the liberal arts; it was here that Francesco Commandino spent his life working at editing classic texts of Greek science (including the works of Archimedes and Euclid's *Elements*; §2.1). Among his pupils were Guidobaldo Del Monte and Bernardino Baldi, whose *Cronica de' mathematici* (published only in 1707) was one of the first sources for the history of mathematics. The long cultural tradition of the Montefeltro court was also fed by the presence of celebrated painters, from Piero della Francesca, who had quite an interest in geometry (and perspective in particular; §12.6) and arithmetic, to Raphael.

3 THE GALILEAN SCHOOL

Apart from Bologna, the most important universities in seventeenth-century Italy were at Pisa and Padua. Galileo Galilei taught first at Pisa, then at Padua for some twenty years, before being called to the court of the Medicis in Florence as 'Mathematician of the Grand Duke'. He dominated Italian science, not only in physics but also in mathematics; among his pupils were Benedetto Castelli (the teacher of Bonaventura Cavalieri), Evangelista Torricelli and Vincenzo Viviani, who liked to call himself 'Galileo's last pupil'.

As a 'lecturer' at the University of Bologna, Cavalieri continued the high mathematical tradition. One of his most brilliant students, Stefano degli Angeli, was called to the University of Padua, where Isaac Barrow and James Gregory attended his lectures. Torricelli succeeded Galileo as 'Mathematician of the Grand Duke' in Florence. This position was eventually also filled by Guido Grandi, former 'lecturer' of mathematics in Pisa.

The influence of Galileo's work and methods in mathematics was still dominant at the end of the seventeenth century. The primary role he attributed to the theory of proportions, together with the idea that Euclidean geometry was the key to understanding the 'language of the world', contributed to keep his later followers away from the methods of Cartesian geometry and Leibnizian calculus. Thus, the leading position of the Galilean school could have been one of the reasons for the delay in accepting the new mathematics in Italy (Pepe *1982*).

4 UNIVERSITIES AND SCIENTIFIC SOCIETIES IN THE EIGHTEENTH CENTURY

After the publication of Leibniz's 'Nova methodus' in 1684, it was some thirty years before the new calculus was taught by Jacob Hermann at the University of Padua (§3.2). From the beginning of the eighteenth century, however, the Leibnizian calculus was studied by individuals and groups, noblemen like Jacopo Riccati or Giulio Fagnano, young gifted students like Gabriele Verzaglia and the brothers Gabriele and Eustachio Manfredi in Bologna, and others. The *Giornale dei letterati in Italia*, a journal founded in Venice by the literary man Apostolo Zeno, often published their research papers, and thus served for some thirty years as a vehicle for spreading the new methods of Leibniz and Newton among educated Italians. In addition to this *Giornale*, the journals of academies included papers of mathematical content. Academies and societies were indeed much more active centres of mathematical research than the universities. With a few exceptions, until the middle of the century the level of mathematical teaching at the Italian universities was rather poor; a better mathematical education was to be had at the military schools. Thus it is not surprising that the young Joseph Louis Lagrange should have been appointed to a chair for teaching higher calculus at the Royal School of Artillery in Turin. Together with some friends, he launched in 1757 a private society which published a scientific journal; this was the origin of the Academy of Sciences of Turin, still active today.

Major changes began only in the last decades of the eighteenth century, when Gregorio Fontana and Lorenzo Mascheroni began teaching at Pavia, Pietro Paoli at Pisa, Paolo Ruffini at Modena, and so on. During this period, as a result of its political history, Italy was divided into several independent states. For example, Piedmont and Sardinia constituted a separate kingdom with Turin as a capital, Lombardy and Venetia belonged to the Austro-Hungarian Empire, Rome was the capital of the Papal States, which included a great part of central Italy, while Naples was the capital of the Kingdom of South Italy and Sicily. In addition, there were a number of grand duchies, such as Tuscany and Modena. This state of affairs was reflected in the scientific development of Italy. In order to overcome the difficulties caused by the political division, in 1782 was founded the Società Italiana di Scienze, intended as the first national scientific society. In fact, the Società played an essential role by promoting personal contacts and exchange of ideas among Italian scientists. In addition to the journals of the academies, the *Memorie* of the Società became one the favourite publishing places for Italian mathematicians (Grattan-Guinness *1986*). This society, soon to be called the Società dei XL from the number of its effective members, was to play an important role throughout the nineteenth century, and the collection of its *Memorie* represents one of the best sources for the history of nineteenth-century Italian mathematics.

As a consequence of the French Revolution, the political situation changed when in 1797 the French army headed by Napoleon launched its campaign in Italy. Napoleon liked to consider himself a patron of scientists, and in the newly founded Cisalpine Republic he created a National Institute modelled on the French one (§11.1). The Institute had its seat in Bologna, and its short life ended with the Restoration in 1816.

5 THE FIRST HALF OF NINETEENTH CENTURY

After the dramatic changes of the Napoleonic era, the 'Restoration' re-established in Italy the previous political division. Removing the political presence of the French did not mean, however, banning the influence of French science on Italian scientists, all the more so because one of the leading figures of French mathematics, Lagrange, was a native Italian, having been born at Turin. In addition, some of the most gifted Italian students had studied at the Ecole Polytechnique in Paris, among them Gaetano Giorgini and Giovanni Plana. Giorgini was later appointed Superintendent of Education in the Grand Duchy of Tuscany; while Plana, the originator of a theory of the Moon awarded a prize by the London Astronomical Society, became professor at the University and Director of the Observatory at Turin.

As at Turin, so at Milan was the Observatory a centre of mathematical research, directed for some thirty years by Francesco Carlini. At that time Milan had no university (one would be founded there only in 1926); 'its' university was in the nearby town of Pavia. Throughout the eighteenth century, prominent mathematicians like Girolamo Saccheri, Roger Boscovich, Gregorio Fontana and Lorenzo Mascheroni taught at Pavia. At the turn of the century this tradition had been continued by Vincenzo Brunacci, former professor at Pisa. As a convinced follower of Lagrange, Brunacci strongly contributed to the diffusion in Italy of the Lagrangian methods in calculus and mechanics. Among his students there were Gabrio Piola, Antonio Bordoni and Fabrizio Ottaviano Mossotti.

Piola was of noble descent, and cultivated mathematics for its own sake. Even though he never held a teaching position, together with his friend Bordoni he played an important role in the development of the Italian mathematical community. In the early 1830s he edited the *Opuscoli matematici e fisici*, the journal where there first appeared in print (in Italian translation) the memoir on celestial mechanics and the *calcul des limites* that Augustin Louis Cauchy presented to the Turin Academy during his self-exile in Italy (§3.12). Piola's translation was enriched by explanatory footnotes and accompanied by papers in which he surveyed Cauchy's most important contributions to analysis.

From 1816, for over thirty years, Bordoni taught calculus, mechanics and geodesy at the University of Pavia, creating a school where some of the leading Italian mathematicians received their education. Gaspare Mainardi and Delfino Codazzi were students of his before becoming colleagues, as later did Francesco Brioschi, Felice Casorati, Luigi Cremona and Beltrami.

In 1818 Mossotti was forced into exile for his opposition to Austrian domination in Italy. After twenty years in London and Argentina, he was recalled in 1843 by the Grand Duke of Tuscany to teach at the newly reopened Scuola Normale Superiore in Pisa, originally founded by Napoleon in 1808 on the model of the Parisian Ecole Normale (§11.1). Under the leadership of Mossotti and his pupil Betti, the Scuola Normale rapidly grew in prestige. Directed by Betti for some thirty years, the 'Normale' became in the second half of the nineteenth century the leading centre in Italy for mathematical education and research.

The political broad-mindedness of the Grand Duke (as well as the role played by Giorgini) had been witnessed at the first congress of Italian scientists, held at Pisa in 1839. From then on, annual meetings took place in different states (Turin, Milan, Padua, Naples, and so on) until 1848, when nationalist uprisings broke this tradition. The congresses had perhaps provided an opportunity for political meetings, as the secret police of the various states suspected, and as the Pope feared to such a degree that he prevented scientists from the Papal States from participating (or even exchanging letters with the participants). It would be misleading, however, to consider this the principal aspect of the congresses. Their scientific significance was that they provided the opportunity to establish personal contacts and a forum for the exchange of ideas and results. Italian scientists also had the opportunity to meet foreign colleagues attending the congresses; Charles Babbage, for example, at Turin in 1840, Carl Jacobi at Lucca in 1843 (he stayed a year in Italy for the sake of his health, with his friends Jacob Steiner, J. P. G. Lejeune Dirichlet and Ernst Kummer).

The published proceedings of those congresses provide an excellent picture of the state of mathematical research in Italy towards the middle of the nineteenth century (Bottazzini *1983*). The diffusion of the mathematical research in Italy was aided by the publication of the *Annali di scienze matematiche e fisiche*, edited from 1850 by Barnaba Tortolini, Professor of Calculus at the University of Rome. This was the first Italian journal entirely devoted to mathematics and physics. In addition to papers by prominent Italian mathematicians, the *Annali* published papers by foreign mathematicians like C. Gudermann, Arthur Cayley and J. J. Sylvester.

While in the rest of Italy mathematicians were directing their attention to the new developments in France or Germany, in Naples Nicola Fergola and his pupil Vincenzo Flauti developed a 'school' exclusively interested in classical and synthetic geometry. According to Michel Chasles, the Neapolitan geometers succeeded in re-establishing 'in its primary purity' the geometrical analysis of the ancients. At the same time however, Flauti's strong opposition to everything new led to an increasing isolation, and it prevented young mathematicians from bringing their mathematical education up to date. The decline of the synthetic school and the emergence of an 'analytic' one took place in the years of the Risorgimento, which led to the political unification of Italy.

6 FROM UNIFICATION TO THE FIRST WORLD WAR

In the autumn of 1858, Betti, Francesco Brioschi and Felice Casorati travelled to Göttingen, Berlin and Paris to visit the leading European mathematicians. This journey has often been taken to symbolize the entry of Italian mathematics on the European scene. On the initiative of Brioschi, Betti and Angelo Genocchi, in the same year Barnaba Tortolini's *Annali* became the *Annali di matematica pura e applicata*, a new name modelled on the German *Journal* of Crelle and the French *Journal* of Liouville (see §11.12 on journals).

Most of the young Italian mathematicians participated enthusiastically in the Risorgimento, and subsequently in the political life of the new unified State. As members of parliament, some were involved in the reform of national education, including high schools and universities. New chairs were created. In 1860 Cremona was called to teach higher geometry at Bologna, while Giuseppe Battaglini was called to Naples. Three years later Battaglini, who had long been prevented from taking up a teaching position by Flauti's school, founded the *Giornale di matematiche*, which was to become the chief vehicle for the diffusion of non-Euclidean geometry in Italy.

With both the French Ecole Polytechnique and the German Technische Hochschulen (§11.1–11.2) as models, Brioschi founded in Milan the Instituto Tecnico Superiore (today's 'polytechnic') for the training of engineers in 1863. A succeeding teacher there was Cremona, who moved to Rome in 1873 as the director of the school for engineers.

It was to Cremona's credit that in Italy there arose a school of geometry of internationally recognized excellence. Among his pupils there were geometers like R. De Paolis, E. Bertini, G. Veronese and G. B. Guccia. In 1884 Guccia founded the celebrated Circolo Matematico di Palermo, a private society which came to attract as members mathematicians from all over the world. The *Rendicondi* of the Circolo became at the turn of the century one of the most authoritative journals of mathematics, where leading mathematicians like Henri Poincaré were pleased to publish their papers (Brigaglia and Masotto *1981*).

Many mathematicians of the Risorgimento generation, like Brioschi, Betti and Beltrami, were involved in the political life of the new State. For a short time Cremona was even Minister of Public Education, before ending his political life as a Vice-President of the Senate. This tradition of political involvement of mathematicians was followed by Ulisse Dini, Volterra and others until the early decades of the twentieth century.

Dini completed his studies in 1864 at the Scuola Normale Superiore in Pisa under supervision of Beltrami, who began his teaching career there before moving to Bologna, Pavia and, eventually, Rome. After a postgraduate scholarship in Paris, Dini obtained a chair at Pisa, and began there a teaching career which lasted uninterrupted for more than thirty years. In those days the 'Normale' in Pisa was (and still is) a major centre of mathematics. The impressive list of graduates under the supervision of Betti and Dini included Volterra, Salvatore Pincherle, Luigi Bianchi, Enriques, Giulio Ascoli, Cesare Arzelà, Ricci and Alberto Tonelli. They became in turn teachers for the next generation of mathematicians.

In 1883 at the age of 23, Volterra was appointed to a chair of mechanics at Pisa, before moving to Turin and eventually to Rome where he was called to succeed Beltrami. Ricci was called to the University of Padua, where he had as colleagues Veronese and his former pupil Levi-Cività. In 1880 both Pincherle and Arzelà obtained chairs in Bologna, where they spent their entire academic life. They were joined by Enriques in 1894. After a postgraduate scholarship in Munich under Felix Klein, Bianchi was called to Pisa, where he set about creating a prominent school for differential geometry and number theory.

Thus, in addition to Pisa, many centres were acquiring an increasing importance, for instance Rome, Bologna, Padua, Palermo and Turin. From the 1880s in Turin, Corrado Segre had founded the celebrated Italian school of algebraic geometry, which was to be developed by Gino Fano, Guido Castelnuovo, Enriques and Francesco Severi, and their pupils (§7.9). Contemporary with the 'geometrical' school of Segre, in Turin there gathered around Peano and his *Rivista di matematica* a school of mathematicians mostly interested in logic and foundations (§5.2), including Cesare Burali-Forti, Giovanni Vailati, Alessandro Padoa and Mario Pieri. They were also involved in the *Formulario mathematico*, edited by Peano. The *Formulario* was planned as an encyclopedia of the whole of mathematics, written in Peanian symbolism and, eventually, in *Latino sine flexione*, the artificial international language Peano himself invented.

At the end of the nineteenth century, Italian mathematicians had achieved a leading position in many fields, as was to be confirmed in a symbolic way by the Fourth International Congress of Mathematicians which, following the ones in Paris and Heidelberg, took place in Rome in 1908.

7 THE INTERWAR PERIOD

Among the victims of the First World War was the optimistic ideal inherited from positivism, that science did unify nations beyond their borders. Strong nationalist feelings had an impact on the history of mathematics institutions; in particular, ten years had to pass before mathematicians could again meet in an International Congress. This was thanks to Pincherle's diplomatic abilities, for the first postwar Congress took place in Bologna in 1928. Similar difficulties were encountered by other international institutions; the irreversible decline of an international society like the Circolo Matematico di Palermo did not depend only on the death of Guccia in 1914.

This led to the creation in Italy of national structures which still today support mathematical research (on the interwar period, see Guerraggio *1987*). With a considerable delay in comparison to other countries, Pincherle in 1922 established in Bologna the Unione Matematica Italiana. As a national society for mathematicians, the Unione joined the Mathesis, a society founded some 25 years before, whose members however were mostly teachers of mathematics in secondary schools. At the same time Volterra created in Rome the Consiglio Nazionale delle Ricerche (CNR), which had existed in embryonic form since the years of the war. In the first decades of the twentieth century, Volterra played a tremendous role in organizing the Italian scientific community. In 1906 he had set up the Società Italiana per il Progresso delle Scienze, which brought new life to the old tradition of the congresses of Italian scientists. During the 1920s, besides the presidency of the CNR, he was also appointed first to the vice-presidency, then the presidency of the Accademia dei Lincei.

On the institutional level, during the 1930s an effort was made to set up in Rome what would be the leading centre for mathematical research. In addition to Castelnuovo, Severi and Volterra, prominent personalities like Enriques, Levi-Città and Mauro Picone were called there. Deeply interested in the numerical applications of analysis, Picone created in Rome an Istituto per le Applicazioni del Calcolo, sponsored by the CNR. Directed by Picone himself for some thirty years, the Istituto played a pioneering role in the development of numerical analysis and applied mathematics.

Picone's institute was joined by the Istituto Italiano degli Attuari, founded by Francesco Paolo Cantelli for the applications of mathematics (including probability) to economics. Its *Giornale*, edited by Cantelli from 1930 to 1958, became one of the most authoritative international journals in its field.

It would be hard to say whether Fascism had any impact on the development of mathematics in Italy. Political nationalism certainly spread into science. It did lead to an emphasis on the national character of various mathematical schools, for instance Severi's school of algebraic geometry. But rather than directly influencing the policy of the development of mathematics, Fascism seems to have had an intermediate political impact. In 1931, university professors were required to swear fidelity to the Fascist regime. Among the 13 who refused, the only mathematician was Volterra; at once he had to give up all his institutional positions, including the chair at the University of Rome. Seven years later, following Nazi Germany, the Fascist regime adopted racial laws; Jewish professors, including Enriques, Beppo Levi and Beniamino Segre, were forced to leave their university chairs. But by then the Second World War was about to begin.

BIBLIOGRAPHY

Bottazzini, U. *1981*, 'Il diciannovesimo secolo in Italia', in Struik *1981*: 249–312.

—— *1982*, 'Enrico Betti e la formazione della scuola matematica pisana', in *La storia delle matematiche in Italia, Atti del convegno*, Cagliari, 229–76.

—— *1983*, 'La matematica e le sue "utili applicazioni" nei congressi degli scienziati italiani, 1839–1847', in G. Pancaldi (ed.), *I congressi degli scienziati italiani nell'età del Positivismo*, Bologna: CLUEB, 11–68.

Brigaglia, A. and Masotto, G. *1981*, *Il circolo matematico di Palermo*, Bari: Dedalo.

Franci, R. and Toti Rigatelli, L. *1985*, 'Towards a history of algebra from Leonardo of Pisa to Luca Pacioli', *Janus*, **72**, 17–82.

Grattan-Guinness, I. *1986*, 'The Società Italiana 1782–1815 [...]', *Symposia mathematica*, **27**, 147–68. [Whole volume devoted to mathematics in Italy.]

Guerraggio, A. (ed.) *1987*, *La matematica italiana tra le due guerre mondiali*, Bologna: Pitagora.

Pepe, L. *1982*, 'Note sulla diffusione della *Géométrie* di Descartes in Italia nel secolo XVII', *Bollettino di storia delle scienze matematiche*, **2**, 249–88.

Struik, D. J. *1981*, *Matematica: Un profilo storico*, Bologna: Il Mulino.

11.9

Spain, Portugal and Ibero-America, 1780–1930

E. L. ORTIZ

1 MATHEMATICS AND THE ENLIGHTENMENT, 1780–1800

Around 1780, Spain and Portugal started to incorporate foreign science on a large scale, and the institutional structure improved considerably. The geographical sciences and the areas of science and technology related to them, all close to the interests of the navy, were those in which Spain, and to a lesser extent Portugal, made their most significant scientific contributions in the late eighteenth century and the early years of the nineteenth (Delambre *1810*). They included mathematics-related topics such as the construction of tables, maps (§8.15) and instruments for astronomical navigation (§8.18), and the design of special machines and mechanisms.

The pioneers of applied mathematics and astronomy in Spain were Jorge Juan and Antonio de Ulloa, and their junior contemporary Benito Bails. He was the author of a mathematical treatise, based on Bézout; it dominated mathematics teaching for several decades from the 1770s in special institutions where mathematics was required for military and naval engineering training.

In Portugal the university played a more definite role in the development of mathematics than in Spain. At Coimbra University, two Portuguese mathematicians, José Monteiro da Rocha and Anastacio da Cunha, were responsible for the revitalization of the exact sciences (Guimarães *1909*, Ferraz *et al. 1990*). The first of these was a Jesuit priest, and the second a follower of the *philosophes*. Their divergent views reflected the tensions of contemporary Portuguese society.

2 THE IMPACT OF THE FRENCH REVOLUTION, 1800–1830

In the last decade of the eighteenth century, the Spanish navy sponsored a group of young Spanish scientists to study abroad. Among them were the astronomer-mathematicians José de Mendoza Ríos and Felipe Bauzá, and the mechanics experts Agustín de Betancourt and José María Lanz (the latter born in Mexico). This was the first highly qualified generation of Spanish applied mathematicians, and also the most brilliant (García Diego and Ortiz *1988*).

The textbooks of the Frenchman Sylvestre Lacroix were adopted in the Iberian world in the first decade of the 1800s; translations into Spanish appeared from 1807. Gaspard Monge's descriptive geometry, L. B. Francoeur's mechanics and several other French mathematics books relevant to engineering were translated and printed before 1810. In Portugal, pure mathematics, and in particular mathematical analysis, developed more strongly than in Spain, where interests focused more generally on applications. Lagrange's textbook on the theory of functions (§3.2) was translated into Portuguese in 1798, just a year after it was published in Paris.

The main characteristics of early-nineteenth-century mathematics teaching in the Peninsula are the development of descriptive geometry following Monge's approach and a large expansion in the teaching of infinitesimal calculus. Such expansion caused an increase in the size of the mathematical community there.

José Mariano Vallejo, younger than Betancourt and Lanz, followed Cauchy's ideas (§3.3). He published a mathematical treatise, reprinted several times, in which he took a more modern approach than that of Bails.

Except for brief periods of development, the end of the Napoleonic Wars signalled the start of a reversal of these progressive trends. Spain lost several of the distinguished mathematicians mentioned above: Mendoza Ríos to England, where he published a collection of nautical tables used by the navies of many countries; Lanz to France; and Betancourt to Russia. A second wave of scientists left Spain after the collapse of a liberal government in 1823, Bauzá and Vallejo among them. Several of these mathematicians had achieved recognition from The Royal Society of London or the Paris Academy of Sciences. The fragile scientific system which had been developing since the last decades of the eighteenth century was severely damaged by this second substantial loss in a period of some ten years. A similar process took place in Portugal.

For Portuguese mathematicians, life turned out to be very different from that of their Spanish contemporaries. In 1807, as Napoleon's troops invaded Portugal, the court decided to move to Brazil and to take with it part of its valuable royal library and specialized personnel, including a good number of the best teachers at the naval school. With these teachers, a military school was created at Río de Janeiro in 1810. Mathematics was taught there in the first year of study. A printing press was also taken to Brazil; a large number of translations were published in this period, among them the elementary parts of the books by Lacroix, and Adrien Marie Legendre's textbook on geometry.

Mathematical analyst Francisco de Borja Garção Stockler, a contemporary of Mendoza Ríos, belongs to the generation of Portuguese mathematicians trained in da Cunha and da Rocha's time. He also moved to Brazil, where he continued with his analytic research and finished an interesting brief history of mathematics in Portugal up to the end of the eighteenth century. This book is said to be the first national history of mathematics; it was published in Portuguese in Paris (!) as Stockler *1819*. For Brazil, the end of the Napoleonic Wars and the return to Europe of the Portuguese court resulted in a loss of momentum in the rapid progress made since 1807.

The progress made in mathematics was not sustained in either the Peninsula or their newly independent former colonies in the Americas. The period from the 1830s to the late 1840s is generally marked by consolidation and re-elaboration of ideas, when not by actual regression.

3 MATHEMATICS AT THE ENGINEERING SCHOOLS, 1840–1870

In the 1850s there slowly opened an era of relative prosperity and stability. More specialized engineering training than that offered by military schools began to be required to satisfy the desire in Spain, Portugal and some of the Ibero-American countries to become closer to the international economic markets. Such demand was met through new civil engineering schools, created around 1840.

The creation of these schools had a very positive effect on the quality of advanced mathematics teaching and, to a certain extent, on the contents of its syllabuses, which now included regular courses on topics of advanced calculus and new courses on applied geometry. The main emphasis of mathematical work produced in this period lay in the dissemination of recent results in pure and applied mathematics published abroad. José Echegaray made interesting efforts in this direction (for his theatre plays he received a share of the 1904 Nobel Prize for Literature). Several journals which published mathematics began to appear from around 1850.

The combined effect of rigorous entrance examinations to professional schools and an attempt to relax the involvement of the state in education within the new free-market ideas, in vogue in the second part of the nineteenth century, created a new and thriving market for private teaching in mathematics. At least from the last third of the century it became an important source of professionalization for mathematicians in Spain and Portugal, and to some extent also in Ibero-America.

4 MATHEMATICS UNDER POSITIVISM, 1870–1900

Advanced mathematics teaching made great strides at the University of Madrid from the 1870s. Eduardo Torroja y Caballé, the leading mathematician there, imported into Spain Carl von Staudt's projective geometry. Even though the topic he chose was not at the forefront of mathematics research in the late 1870s, Torroja must be credited for advocating a more modern approach, characterized by *doing* mathematics rather than just being well informed or even enthusiastic about it. More geometry and its applications entered into the schools of engineering through graphical statics and other branches; then, and for some decades to follow, these were the main computational tools available to engineers. Descriptive geometry continued to be a principal language in engineering, and its teaching continued to develop (Ortiz *1988b*).

Secondary schools also became an important source of professionalization for mathematicians all over the Iberian world from the 1870s. The most interesting example of a research mathematician working at a secondary school in Spain is that of Ventura Reyes Prósper, perhaps the leading Spanish pure mathematician of the nineteenth century; he was interested in the foundations of geometry.

Pure mathematics began to develop as a research area also in Portugal from the 1870s through Francisco Gomes Teixeira, a formidable force of change in the Peninsula. In him the Portuguese classical analytical tradition reached its highest point. His comprehensive treatise on curves was translated and reprinted in several countries. He created the first exclusively mathematical and astronomical journal in Portugal, *Jornal de Ciências Matemàticas y Astronòmicas*, in 1877.

At Buenos Aires University, the demands on mathematics from engineering had increased considerably. Valentín Balbín, who had trained for several years in England, introduced graphical statics, quaternions and other advanced geometric and analytic methods in science and engineering courses.

In the last quarter of the century, statistical offices began to be set up in a number of countries in the Iberian world. The use of inappropriate mortality tables had caused financial failures, so from the 1870s mathematics began also to be deployed in the financial analyses using locally collected statistical data to test their viability. A number of journals catering mainly for the training of aspirants to the new professional schools, which were also a vehicle for the publication of other, more advanced, mathematical works, began to be published by the end of the century. Balbín started *Revista de Matemáticas Elementales* in Argentina in 1889, and in 1891 Zoel García de Galdeano launched in Spain *El Progreso Matemático*, an appropriate name in the time of positivism.

5 MATHEMATICS IN THE NEW CENTURY

Leonardo Torres Quevedo, a distinguished applied mathematician and inventor at the turn of the century, trained at the engineering school in Madrid, which was still a powerful force in mathematics. He is best known for his chess-playing automata and his innovative mechanical and electromechanical calculating machines (Ortiz *1993b*). His work attracted the attention of the state, which had begun to show a deeper commitment to sponsorship of science. A *Junta* (or science research council) was created in Spain in 1907. In the same decade a number of more specific scientific societies were founded, including the Sociedad Matemática Española and its journal.

From 1915, Julio Rey Pastor became a dominant figure in mathematics in Spain, and later in Argentina. In that year the *Junta* created for him in Madrid the first mathematics research institute in Spain which was outside a university. There he started a research seminar, and contributed to the advance and modernization of mathematics research and teaching. He was also the author of a series of advanced mathematics textbooks which had a very considerable influence in the Iberian world.

From 1921 Rey Pastor settled permanently in Argentina, whose economy and industrial base were expanding rapidly and making heavy demands on science and engineering. His influence in Argentina was as important as it had been in Spain. Again, he was instrumental in the modernization and development of research and of mathematics teaching. In the mid-1930s he directed a substantial group of research students in Buenos Aires. His continued interest in the mathematics life of Spain created two parallel schools, on both sides of the Atlantic, working on similar research topics. Several Spanish and Argentinian students of his school have acquired an international reputation, among them Ricardo San Juan and Sixto Ríos in Spain, and Mischa Cotlar, Luis A. Santaló, Eduardo Zarantonello and Alberto González Domínguez in Argentina; the last of these was the teacher of Alberto P. Calderón. The Uruguayan mathematician José Luis Massera belongs to the same school (Ríos *et al. 1979*, Ortiz *1993a*).

After the formative period of 1900–15 and the consolidation of the

1920s, the mid-1930s was a crucial period for the development of mathematics in the Iberian world. Although young gifted mathematicians were emerging in Spain, such development was largely disrupted by the Civil War. Not only did it cause immense human and material damage, but also in consequence a number of scientists left their country.

In Portugal, modern pure mathematics advanced rapidly after the return in 1936 of Antonio A. Monteiro from Paris, where he had worked under Maurice Fréchet. In rapid succession, there were organized a research seminar on modern abstract analysis and topology, the journal *Portugaliae Mathematica* (the first exclusively mathematical journal to be published in Portugal) and the Sociedade Portuguesa de Matemática. Several members of this group have acquired an international reputation, among them Ruy Luis Gomes, Alfredo Pereira Gomes, Hugo Baptista Ribero and Jose Sebastião e Silva. However, these promising developments were substantially based on the personal efforts of a small group of dedicated mathematicians. Official support was extremely thin, and often obstructive on grounds of political security. In 1945, Monteiro – then regarded as the most distinguished research mathematician in the Iberian world – was forced to leave Portugal for Brazil. There, and later in Argentina, he had an extraordinary impact, comparable only to that of Rey Pastor. Among his Brazilian students was Leopoldo Nachbin, who later acquired an international reputation (Gomes *et al. 1980*).

In the late 1930s and 1940s several leading US mathematicians (G. Birkhoff, M. Stone, S. Lefschetz and A. Zygmund) visited Buenos Aires and Mexico. They established a close contact which had profound impact on the development of pure mathematics in these countries: it moved in the direction of the research that was then being done in the USA, particularly in modern analysis and topology.

BIBLIOGRAPHY

Babini, J. *1949, Historia de la ciencia Argentina*, Mexico: Fondo de Cultura.

Barajas, A. *1960*, 'La investigación físico-matemática', in *México: Cincuenta años de revolución (IV. La cultura)*, Mexico: Fondo de Cultura, 171–6.

Delambre, J. B. J. *1810, Rapport historique sur le progrès des sciences mathématiques depuis 1789*, Paris: Imprimerie Impériale.

Ferraz, M. L., Rodriguez, J. F. and Saraiva, L. (eds) *1990, Anastacio da Cunha (1744/1787)*, Lisbon: Impresa Nacional and Casa da Moeda.

García Camarero, E. and García Camarero, E. (eds) *1970, La polémica de la ciencia Española*, Madrid: Alianza. [A selection of texts by Spanish scientists, including mathematicians.]

García Diego, J. A. and Ortiz, E. L. 1988, 'On a mechanical problem of Lanz', *History and Technology*, **5**, 301–13.

Gomes Teixeira, F. *1934*, *Historia das Matemáticas em Portugal*, Lisbon: Academia das Ciências.

Gomez, R. L. *et al.* *1980*, 'A. Monteiro', *Portugaliae mathematica*, **29**, 1–38.

Gortari, E. de *1956*, *La ciencia en la historia de México*, Mexico: Fondo de Cultura.

Guimarães, R. *1909*, *Les Mathématiques en Portugal*, Coimbra: Imprimerie de l'Université. [Contains an extensive list of mathematical publications by Portuguese mathematicians up to 1909.]

Loria, G. *1919*, 'La matematiche in Ispagna, ieri ed oggi' and 'Le matematiche in Portogallo, ciò che fuorono, ciò che sono', *Scientia*, **25**, 353–59, 441–9; **26**, 1–9. [A presentation deeply influenced by Echegaray and Rey Pastor's views, in which the Mendoza Ríos generation is disregarded.]

Oliveira Castro, F. M. de *1954* [?], 'A matematica no Brasil' in F. de Acevedo (ed.), *As Sciencias no Brasil*, Vol. 1, São Paulo and Rio de Janeiro: Ediçoes Melhoramentos, 41–77.

Ortiz, E. L. *1988a*, 'El krauso-positivismo, la Junta y la nueva ciencia en España', in *El krausismo y su influencia en America Latina*, Madrid: Fundación Friedrich Ebert, 137–67. [On the local philosophical background to the creation of some key scientific institutions.]

—— *1988b*, 'Introduction', in *The Works of Julio Rey Pastor*, Vol. 1, London: The Humboldt Library, 1–22. [The impact of Rey Pastor on the development of mathematics in Argentina and Spain.]

—— *1993a*, 'Leonardo Torres Quevedo', in A. Ralston and E. D. Reilly, Jr (eds), *Encyclopedia of Computer Science*, London: Chapman & Hall, to appear.

—— *1993b*, 'La formación científica de Julio Rey Pastor en Alemania: 1911–1914', *Memorias de la Academia Nacional de Ciencias Exactas, Físicas y Naturales*, to appear. [Contains an extensive bibliography on the development of mathematics in Argentina and Spain in the early years of this century.]

Peset, J. L., Garma, S. and Pérez Garzón, J. S. *1978*, *Ciencias y enseñanza en la revolución burguesa*, Madrid: Siglo XXI. [Political and economical background in Spain in the second part of the nineteenth century.]

Rey Pastor, J. and Babini, J. *1951*, *Historia de la matemática*, Buenos Aires: Espasa-Calpe. [This and other important works by Rey Pastor on the history of mathematics, with special reference to Spain, are included in E. L. Ortiz (ed.), 1988, *The Works of Julio Rey Pastor*, 8 vols, London: The Humboldt Library.]

Rios, S., Santaló, L. and Balanzat, M. *1979*, *Julio Rey Pastor, matemático*, Madrid: Instituto de España.

Stockler, F. de Borja Garção *1819*, *Ensaio historico sobre a origem e progressos das mathematicas em Portugal*, Paris: Rougeron.

Vernet Ginés, J. *1975*, *Historia de la ciencia Española*, Madrid: Instituto de España.

11.10

The United States of America, and Canada

KAREN HUNGER PARSHALL AND

DAVID E. ROWE

1 AMERICAN MATHEMATICS PRIOR TO 1876

During the colonial period, mathematics played very little part in American intellectual life. In the earliest years of America's oldest institution of higher education, Harvard College (founded in 1636), the student encountered mathematics only in his final year and then only in the form of arithmetic, geometry and astronomy. Out of roughly thirty hours of studies weekly, philosophy claimed the most with ten hours, while mathematics received the least attention with only two hours (Cajori *1890*: 19). The clearly defined emphases of this curriculum, however, were perfectly consistent with the primary aim of the early Harvard College – namely, the training of young men for the ministry.

Although the typical course of study was extended to four years around the middle of the seventeenth century, no significant change occurred in Harvard's mathematical instruction until 1726. In that year, Thomas Hollis endowed a 'Professorship of the Mathematicks, and of Natural and Experimental Phylosophy', which took at least some of the mathematics teaching out of the hands of the tutors and put it into the presumably more qualified hands of the professor. The chair's first incumbent, Harvard-trained Isaac Greenwood, introduced elementary algebra into the mathematical curriculum; and his successor, John Winthrop, taught the theory of fluxions (or Newtonian calculus) as early as 1756 (Smith and Ginsburg *1934*: 52). While mathematics teaching at Yale (founded in 1701) followed roughly this same developmental pattern, the mathematical levels of the other colonial colleges tended to be comparable (as at Princeton, founded in 1746) or lower (as at the College of William and Mary, America's second-oldest college, founded in 1693). By the eve of the War of Independence, higher mathematics education in America mimicked its British model, consisting principally of algebra, trigonometry and the geometry of Euclid. Calculus entered the curriculum only in the presence of a professor or tutor competent and willing to present it (Smith and Ginsburg *1934*).

The British influence on American mathematics flagged little, if at all, as a result of the war; American educators continued to rely primarily on British texts for their courses until well into the nineteenth century. By 1820, however, Americans had begun to turn increasingly to France for their mathematical inspiration. Indicative of this shifting emphasis, Sylvanus Thayer overhauled the United States Military Academy at West Point (founded in 1802), after becoming its superintendent in 1817, and refitted it to resemble more closely France's Ecole Polytechnique. This brought with it the introduction of French mathematical texts in translation, which presented such new topics as the descriptive geometry of Gaspard Monge and such classical topics as the differential and integral calculus, but from the Leibnizian as opposed to the Newtonian point of view. As this French approach became more firmly established alongside the continuing British tradition, Americans began to produce not only their own hybridized texts, such as those by West Point's Charles Davies, but also remarkably original texts, like those by Harvard's Benjamin Peirce (Pycior *1989*). While this indigenous activity may have had an impact on American mathematics teaching, it had little or no impact on research-level mathematics as defined by the European – and particularly the Continental – model.

Nevertheless, some Americans did contribute to mathematical research during the first three-quarters of the nineteenth century (Smith and Ginsburg *1934*). Robert Adrain, a transplanted Irishman, edited the short-lived journal, *The Analyst, or Mathematical Museum* (1808–9), from his home base in New York City; in it he published (independently of Carl Friedrich Gauss) his proof of the exponential law of errors, as well as his deduction from it of the law of least squares. Nathaniel Bowditch, a self-taught mathematician from Salem, Massachusetts, prepared an English translation with extensive commentary (1828–39) of Pierre Simon Laplace's *Mécanique céleste*. In 1870 Peirce privately published the paper 'Linear associative algebra', in which he classified all algebras of dimensions 1–6 over the field of complex numbers using his notions of idempotent and nilpotent elements (§6.4).

Although significant, mathematical research such as this was the exception and by no means the rule throughout most of the nineteenth century. In mathematics, as in all the sciences, the first three-quarters of the century marked not so much a period of great results, but rather one of professionalization and the reorganization of higher education along Continental lines. With governmental support of higher education through the

Morrill Act of 1862, with the era of large private fortunes and educational philanthropy, and with the adoption of a research ethic by the scientific community, original research and the training of graduate students had become at least part of the goal of the best institutions of higher education in the USA when the twentieth century arrived. For mathematics, the Johns Hopkins University (opened in 1876) set the early standard.

2 THE PERIOD OF UNIVERSITY BUILDING, 1876–1900

Under its first professor of mathematics, the English algebraist J. J. Sylvester, Johns Hopkins supported an active and productive graduate programme, which trained several dozen students in many of the latest advances in mathematical research (Parshall *1988*). Among them, Fabian Franklin, Christine Ladd Franklin, Thomas Craig, George Bruce Halsted, and W. Irving Stringham particularly stood out. In addition, Sylvester and his associate William E. Story founded in 1878 the *American Journal of Mathematics*, America's oldest continuous research-level mathematical journal, and used it not only as a publication outlet for themselves, their students and other American mathematicians, but also for a significant number of Europeans.

With Sylvester's return to England in 1883, however, the Hopkins programme lost its momentum and fell from pre-eminence as a training ground for American mathematicians (Parshall and Rowe *forthcoming*). Nevertheless, its educational ideals would come to exert considerable long-term influence on the older universities like Harvard and Yale, on prominent state universities like the Universities of Michigan and Wisconsin, and particularly on two soon-to-be-formed institutions, Clark University (opened in 1889) and the University of Chicago (opened in 1892). Until these ideals took hold more widely, however, many would-be American mathematicians travelled to Europe, especially to Germany, to pursue their advanced studies.

In the 1880s and 1890s, a few Americans journeyed to Paris to work under Camille Jordan; some went to Berlin to study under Karl Weierstrass, Leopold Kronecker and Lazarus Fuchs; and a number sought out Sophus Lie in Leipzig. During this period, however, Felix Klein attracted most of the mathematical itinerants, first to Leipzig and then, after 1885, to Göttingen. Under Klein, Americans such as Frank Nelson Cole, William Fogg Osgood, Maxime Bôcher, Henry Burchard Fine and Mary Francis Winston (later Newson), gained exposure to the full range of late-nineteenth-century mathematical research. By bringing this learning as well as a commitment to mathematical research back home with them, these foreign-trained Americans helped to redirect mathematics at institutions like the University of Michigan, Columbia, Harvard, the Massachusetts Institute of Technology and Princeton (Parshall and Rowe *forthcoming*). At the new research-oriented schools, however, programmes were set up according to these standards from the start.

Clark University opened in Worcester, Massachusetts, as a purely graduate-level institution. Under the leadership of the former Hopkins professor Story, its mathematics department also included one of Klein's German students, Oskar Bolza (Cooke and Rickey *1989*). Together these men and their colleagues crafted a programme which rivalled that of Hopkins in its heyday and which seemed destined to shape American mathematics. Unfortunately, internal political tensions caused the university's near collapse in 1892, and much of the faculty, including Bolza in mathematics, left for the newly forming University of Chicago. Thus, in mathematics as in other areas, the promise of Clark became the reality of Chicago. When Chicago opened, in fact, two of its three mathematicians – Bolza and Heinrich Maschke – were German students of Klein. The third member and leader of the department, however, was an American: Eliakim Hastings Moore.

3 E. H. MOORE AND THE UNIVERSITY OF CHICAGO

Under Moore's guidance, the Chicago department dominated American mathematics by virtue of its programme, its research output, and its impact on the broader mathematical community (Parshall *1984*). The first modern research-oriented mathematics department in the country, the Chicago group covered virtually all areas of mathematics at both undergraduate and graduate levels. Moore worked and gave instruction in abstract algebra, foundational studies and, later, general function theory; Bolza concentrated primarily on the calculus of variations; and Maschke focused on differential geometry, while also teaching various topics in applied mathematics. In addition to teaching and pursuing their own research work in these areas, the Chicago mathematicians, particularly Moore, worked vigorously to organize American mathematics.

In 1893 the Chicago faculty, together with Henry White of Northwestern University in nearby Evanston, arranged a mathematical congress in connection with the World's Columbian Exposition in Chicago. A modest but none the less important event, the Congress drew over forty participants, including Klein and Eduard Study, and highlighted the mathematical research endeavours of Americans alongside those of Klein and the German community (Parshall and Rowe *forthcoming*). In 1894, Moore approached the New York Mathematical Society, an organization formed at Columbia College in 1888, to seek its support in publishing the Congress proceedings.

This lobbying effort spurred the Society to change its name to the American Mathematical Society (AMS) as a more proper reflection of its national goals (Archibald *1938a*: 7). By 1899, Moore and others further realized these goals by setting up a new research-level publication, the *Transactions of the American Mathematical Society*. Thus, by 1900 the USA supported enough active mathematicians to sustain four serious research-level periodicals: the *American Journal*, the *Annals of Mathematics* (founded in 1884), the *Bulletin of the American Mathematical Society* (begun in 1891), and the *Transactions*. Among the contributors to these journals were the members of the first generation of Chicago students.

In the years from 1896 to 1908, Moore, Bolza and Maschke guided the research of some two dozen Ph.D.s. The most notable of these – L. E. Dickson (Ph.D. 1896), G. A. Bliss (1900), Oswald Veblen (1903), R. L. Moore (1905) and G. D. Birkhoff (1907) – went on to shape decisively the course of higher mathematics in twentieth-century America from their various institutional vantage points (Parshall and Rowe *forthcoming*).

Initially, Chicago maintained its leading position. In 1900 its faculty expanded to include its first Ph.D. in mathematics, Dickson. On the basis of his seminal book, *Linear Groups with an Exposition of the Galois Field Theory* (1901), as well as his prolific contributions to algebra and number theory, Dickson attracted many students (among them A. Adrian Albert) to Chicago until his retirement in 1937. In 1908, however, the University suffered major losses with Maschke's death and Bolza's subsequent return to Germany. Although Ernst Wilczynski, a graduate of Berlin University, was called from the University of California at Berkeley to maintain the geometrical tradition established by Maschke, and although Bliss, who had studied the calculus of variations under Bolza, assumed his mentor's mantle, neither of these men was as strong as his predecessor.

Meanwhile, Moore's interests had shifted to what he called 'general analysis', an area which utilized an axiomatic approach based on classes of functions defined on a 'general range' to study the properties of integral equations. Considering the burgeoning interest in functional analysis generated by the work of David Hilbert, Maurice Fréchet, Henri Lebesgue and others, Moore's new-found field of research seemed timely and promising. Ultimately, though, it offered few significant insights (Moore–Smith convergence being a notable exception), and increasingly left Moore out of touch with mainstream developments. Although the University of Chicago continued to produce impressive numbers of mathematicians (over 115 Ph.D.s from 1910 to 1930), including a few like Albert and E. J. McShane who rose to international prominence, its proportion of truly first-rate graduates declined. In the words of Saunders Mac Lane, Chicago became something of a 'Ph.D. mill' in the interwar years (Mac Lane *1989*: 138).

4 THE REAWAKENING OF HARVARD UNIVERSITY

If Chicago opened the century as America's foremost centre of mathematics training and research, Harvard followed a somewhat distant second. Under the direction of Osgood and Bôcher, its mathematics department excelled in analysis but fell short in most other areas. Perfectly complementing one another, Osgood pursued the function-theoretic side of Klein's teaching legacy, while Bôcher concentrated on generalized series expansions in potential theory. Although both men were prolific and gifted writers, Bôcher proved far more inspiring as a teacher, and consequently attracted many more disciples (Parshall and Rowe *forthcoming*). Most notable of his 17 doctoral students, Griffith Evans (Ph.D. 1910) succeeded in bringing the Mathematics Department at the University of California, Berkeley, to international prominence. Bôcher also exerted a strong influence on G. D. Birkhoff during his undergraduate years, as evidenced by the latter's earliest work on asymptotic expansions, boundary-value problems and Sturm–Liouville theory, topics at the very heart of Bôcher's research programme.

In 1912, Harvard greatly strengthened its programme and its reputation when it appointed its newly acclaimed alumnus to its faculty. That year, G. D. Birkhoff established his international reputation by proving Poincaré's conjecture that any 1–1 area-preserving transformation of an annulus which moves the two boundary circles in opposite directions must have two fixed points. Birkhoff came to dominate the Harvard department in the 1920s and 1930s, and supervised the dissertations of several prominent students, including Marston Morse, Marshall Stone, J. L. Walsh, C. B. Morrey and Hassler Whitney. Of these, all but Morrey subsequently joined the Harvard staff: Morse was appointed in 1926 with his seminal work on the calculus of variations in the large (published as an AMS Colloquium volume in 1934) well under way and stayed until he accepted a professorship at the Institute for Advanced Study in 1935; except for a two-year sojourn at Yale, Stone pursued his work on Fourier analysis and Hilbert space from 1927 to 1946, when he left to revivify the University of Chicago; and Walsh carried on the tradition in analysis established by Osgood and Bôcher beginning in 1921. Harvard further assured its dominance in analysis in 1920 with the acquisition of Hilbert's student, Oliver Kellogg, an expert in mechanics and potential theory (G. Birkhoff *1989*).

5 PRINCETON AND THE RISE OF TOPOLOGY

The 1900s also witnessed the emergence of a third leading centre of American mathematics, at Princeton. As a member of the Mathematics

Department (beginning in 1885) and subsequently as Dean first of the Faculty (1903–12) and then of the School of Science (from 1911 until his death in 1928), Henry Fine masterminded numerous key appointments and promotions. Fresh from his Hopkins Ph.D., Luther P. Eisenhart joined the faculty as an instructor in 1900; Veblen arrived in 1905 and stayed until moving to the Institute for Advanced Study in 1932; the Scottish algebraist Joseph H. M. Wedderburn served from 1909 until his death in 1947; and Bliss and Birkhoff took three-year preceptorships in 1905 and 1909, respectively (Aspray *1988*).

Of this new blood, the two geometers, Eisenhart and Veblen, had emerged as the powers in the revitalized department by 1920. In that same year, the department added the services of one of Veblen's star students, the topologist James W. Alexander. While Veblen himself had made important contributions to this new field (§7.10) in 1905 when he gave the first rigorous proof of the Jordan-curve theorem, after the First World War his interests had shifted, and he now pursued a geometric model aimed at unifying Einstein's new relativistic theory of gravitation with electromagnetic theory. Thus, it fell to Alexander to inaugurate the rich tradition in topology that has since become a byword for Princeton mathematics. For example, in 1915 he essentially solved the invariance problem (i.e. that the homology of a simplicial complex is a topological invariant); by 1922 he had established Alexander duality, a generalization of the Jordan-curve theorem; and in 1928 he discovered the so-called Alexander polynomial for knots.

Augmenting Alexander's research productivity along these lines was Solomon Lefschetz, a Russian-born Jew trained as an engineer at the *grandes écoles* in Paris. After losing both hands in an industrial accident, he took a Ph.D. under Story at Clark, and eventually came to Princeton in 1924 on Alexander's recommendation. A brilliantly intuitive thinker with little patience (or gift) for details, Lefschetz's novel use of topological methods in connection with algebraic varieties not only profoundly influenced the research of algebraic geometers like the Englishman W. V. D. Hodge, but also helped craft modern algebraic topology through such key notions as relative homology groups. This work attracted an impressive array of students, including Paul A. Smith and Norman Steenrod, who settled at Columbia and Princeton, respectively; and Albert W. Tucker, who moved out of topology to become a pioneering figure in non-linear programming and game theory at Princeton (§6.11). Finally, as editor of the *Annals of Mathematics* between 1928 and 1958, Lefschetz was largely responsible for transforming it into one of the world's élite mathematics journals (Aspray *1988*).

Whereas algebraic topology flourished at Princeton, general and point-set topology became the nearly exclusive province of an expanding school that developed around the fifth and final of the first crop of influential Chicago students, R. L. Moore. A teaching legend first at the University of Pennsylvania but especially at the University of Texas for several decades beginning in 1920, Moore developed the so-called 'Moore method', a pedagogical approach which required students to 'discover' proofs of theorems independently. The doctoral students who thrived under Moore's unique guidance – J. R. Kline, R. L. Wilder, G. T. Whyburn and R. H. Bing – significantly advanced his brand of topological research (Wilder *1982*).

6 AMERICANS ABROAD AFTER 1900

Somewhat paradoxically, Göttingen – which had awarded some 34 doctorates to American mathematicians between 1862 and 1930 – defined a fourth major centre of American mathematics after 1900. As it had during the crucial post-Sylvester decade from 1884 to 1894, Göttingen continued to attract American students by virtue of the strength of its faculty, which at the turn of the century included David Hilbert. Between 1899 and 1907, Hilbert directed 11 Americans – among them Earle Raymond Hedrick, Kellogg and Max Mason – who wrote dissertations concerned primarily with aspects of the theory of integral equations (§3.10). A particularly active and energetic figure in the American mathematical community, Hedrick held, among other posts, the first presidency of the student-and-teacher-oriented Mathematical Association of America (founded in 1915).

Although Americans continued to venture abroad for training until August 1914, they only rarely followed such a course thereafter. The professed ideals of German *Wissenschaft* (§11.2) that had inspired two generations of Americans rang hollow to those who lived through the years of the First World War. Although the German defeat brought an end to the fighting, embittered enemies continued to confront one another as the issue of nationalist loyalties seriously undermined international co-operation. By the 1920s, for example, the once vibrant 'American colony' in Göttingen had all but vanished from the scene (Parshall and Rowe *forthcoming*).

7 THE INTERWAR YEARS AND IMMIGRATION

The war years also marked an interlude during which many American mathematicians took a sabbatical from their theoretical endeavours. Max Mason, an applied mathematician at the University of Wisconsin, made key contributions to research on submarine detection as part of a secret project sponsored by the recently established National Research Council. Also, the army employed some thirty mathematicians to analyse ordnance tests conducted at the Aberdeen Proving Grounds. Of these, Veblen, Evans, Morse and Warren Weaver drew from their experience to play significant roles in mobilizing the country's mathematical expertise during the Second World War. However, none of this 'hands-on' research appears to have persuaded American mathematicians to pursue direct applications after their return to civilian life (Price *1988*).

The 1920s and early 1930s marked a period of consolidation within American mathematics in which many established programmes around the country expanded into solid departments. In particular, the Massachusetts Institute of Technology (MIT) emerged as a major force in American mathematics. Although the department at MIT supported a number of geometers – C. L. E. Moore, F. S. Woods, P. Franklin, Dirk Struik and F. L. Hitchcock – Norbert Wiener, a polymath whose work on generalized harmonic analysis and Tauberian theorems won the Bôcher Prize in 1933, stood out as its leading figure.

Outside the more élite institutions, however, the typical mathematician carried a weekly teaching load of 12 to 18 hours each semester, with a 9-hour load generally reserved as an inducement for research. During this interwar period, slightly less than one-third of those who taught mathematics at college level held doctoral degrees, and only about one-fifth of these had conducted serious research. According to long-time Secretary of the AMS, R. G. D. Richardson, half the production of this era's American mathematcal community issued from only some 60 individuals (Richardson *1936*).

The complexion of American mathematics underwent a profound change after 1933. Hitler's rise to power in Germany triggered a European 'brain drain' of truly staggering proportions which primarily benefited the USA (Reingold *1981*). Two of the most famous émigrés, Albert Einstein and Hermann Weyl, accepted positions on the original faculty of the Institute for Advanced Study (opened in 1932), where they were later joined by Carl Ludwig Siegel. Unfortunately, owing partly to nationalism and some anti-semitism within the American academic community prior to the Second World War, the reception afforded to less prominent displaced scholars was not always a warm one. Through the efforts of a few energetic and well-connected individuals, most notably Veblen, many of the immigrants encountered only the usual hardships of adjusting to life in a foreign country.

8 THE RISE OF APPLIED MATHEMATICS

Although this influx of European mathematicians affected the American mathematical climate in many ways, it quite noticeably altered the

American commitment to applied mathematics. In the last half of the nineteenth century, the United States had boasted such virtually self-taught applied mathematicians as Benjamin Peirce, Simon Newcomb and George William Hill in mathematical astronomy, and Josiah Willard Gibbs in physics and physical chemistry, but after 1900 fewer Americans engaged in mathematical research of an applied nature. In fact, this deficiency had become so pronounced that in 1938 G. D. Birkhoff could name only six Americans with serious interests in traditional applied mathematics, four of whom had received their training in Britain (G. D. Birkhoff *1938*: 313). This situation changed dramatically with the wave of immigration from Europe.

Richard Courant, who had headed the Mathematical Institute at Göttingen before leaving first for England in 1933 and then for the USA in 1934, guided a major centre of war-related research at New York University (NYU) (Reid *1976*). Through the appointment of such scholars as Kurt Friedrichs and James Stoker, Courant established applied mathematics in the Göttingen tradition at NYU. After the war, Courant continued to build up his faculty by adding Lipman Bers, Wilhelm Magnus and Harold Grad, among others. In 1954, the group working under Grad on fusion energy and plasma physics became the Courant Institute of Mathematical Sciences, a centre specializing in research on partial differential equations, fluid mechanics and applied mathematics generally.

A much earlier émigré (although from Canada and not from Europe), R. G. D. Richardson, joined the faculty at Rhode Island's Brown University in 1907. During the Second World War and under the sponsorship of the Navy's Office of Scientific Research and Development, he directed a centre for applied research at Brown which employed such distinguished foreign-born mathematicians as Bers, J. D. Tamarkin, George Pólya and Willy Feller. Unlike Courant at NYU, however, Richardson was unable to maintain his department's wartime momentum (Rees *1980*).

As at NYU and Brown, the war spurred on other institutions around the country to engage in applied mathematics. At the Institute for Advanced Study, one of the founders of the theory of operator algebras, John von Neumann, also made seminal contributions to the development and utilization of high-speed electronic computers (§5.12). At the University of Michigan, Stephen Timoshenko elevated engineering mathematics to a new level of sophistication, while Theodor von Kármán did the same for aerodynamics at the California Institute of Technology. Richard von Mises, formerly the first occupant of the chair of applied mathematics in Berlin, took a similar position at Harvard. Yugoslavian-born and Göttingen-educated Feller arrived at Brown in 1939, served as the first executive editor of the *Mathematical Reviews* (founded in 1939), and went on to spread the modern theory of probability to America via positions first at Cornell and then at Princeton. After fleeing Poland in 1934, Jerzy Neyman did much the same thing for statistics once he reached the University of California at Berkeley, in 1938 (Greenberg and Goldstein *1983*).

9 DEVELOPMENTS AFTER THE SECOND WORLD WAR

Emerging from the tumult of the war, American mathematics bore little resemblance to the image it projected during the tranquil interwar years. The universities no longer served as the unique sources of institutional and financial support for research mathematics. Specialized institutes like the Institute for Advanced Study in Princeton (founded in 1930), and much later the Mathematics Institutes in Minneapolis and Berkeley (both opened in 1981), encouraged mathematical research outside the university context. Industry also established its own 'think tanks' for mathematical (and other) research, like the Bell Laboratories in Murray Hill, New Jersey (already an important centre for the Laplace transform (§4.8) and statistics (§10.14)). Relative to funding, American mathematicians continued to supplement their university support with federal grants from such organizations as the Office of Naval Research during the immediate postwar years. This source expanded markedly after the establishment of the National Science Foundation in 1950 and with the later institutionalization of offices such as the National Security Agency.

Furthermore, the cast of leading players in American mathematics as well as the character of American mathematical research changed dramatically after the war. New waves of immigrants from China, the Soviet Union and elsewhere further diversified the American mathematical community. Mathematics became a highly international endeavour, with Americans working and excelling in virtually all branches and subspecialties of the discipline.

Indicative of this, the American mathematical community made important breakthroughs in key outstanding problems: Hilbert's first problem of 1900 on the continuum hypothesis was solved by the combined work of Kurt Gödel and Paul Cohen; Andrew Gleason, Deane Montgomery and Leo Zippin resolved the fifth problem, dealing with locally Euclidean groups; the tenth problem on Diophantine equations yielded to the efforts of Martin Davis, Hilary Putnam and Julia Robinson, together with the Soviet mathematician Yuri Matijasevic; at the University of Illinois at Urbana-Champaign, Kenneth Appell and Wolfgang Haken gave a positive solution to the four-colour problem; and Purdue University's Louis de Branges solved the Bieberbach conjecture in complex analysis, to name but a few. For more information, see Ewing *et al. 1977*.

10 MATHEMATICS IN CANADA

As in the USA, the years since the close of the Second World War have witnessed great mathematical growth in Canada. Whereas prior to the war Canadian mathematicians had to utilize the meetings and publications of the neighbouring American Mathematical Society to maintain their mathematical contacts, in 1945 the Canadian Mathematical Congress met for the first time in Montréal. Conceived by Lloyd Williams and Gordon Pall of McGill University, the Congress served as a vital forum for the country's far-flung mathematicians. Furthermore, by 1949 the Congress had brought out the first issue of the *Canadian Journal of Mathematics*, Canada's first specialized journal for mathematical research, and had supplemented this periodical with its *Bulletin* by 1958 (Keeping *1968*).

Although Canada developed these manifestations of a professional mathematical community more than fifty years after their appearance in the United States, they grew out of similar changes in higher education. During the last quarter of the nineteenth century, specialized programmes of study had evolved within both the French- and English-speaking universities of Canada, subsequently resulting in departments of varying strengths and influence. By 1920, the Mathematics Department at the University of Toronto had emerged as the nation's leader through the energies of such individuals as J. C. Fields. Noted for his researches in algebraic function theory, Fields succeeded in revivifying the International Congress of Mathematicians, when the Americans, who had been scheduled to host the 1924 meetings in New York City, backed out. Thanks to his diplomatic efforts, he and his colleagues hosted the Seventh Congress in Toronto. Fields also lobbied to set up an international award in mathematics comparable in prestige to the Nobel prizes. His efforts came to fruition shortly after his death in 1932, when the Fields medal was named in his honour. Thanks to the foundation laid by Fields, Williams, Pall and others, Canada has developed and maintained strong departments of mathematics from coast to coast since the 1950s, which have nurtured such prominent mathematicians as Irving Kaplansky, I. N. Herstein, Robert Langlands and Robert Moody (Charbonneau *1988*).

BIBLIOGRAPHY

Archibald, R. C. *1938a*, *A Semicentennial History of the American Mathematical Society, 1888–1938*, New York: American Mathematical Society.
—— (ed.) *1938b*, *Semicentennial Addresses of the American Mathematical Society*, New York: American Mathematical Society.
Aspray, W. *1988*, 'The emergence of Princeton as a world center for mathematical research, 1896–1939', in W. Aspray and P. Kitcher (eds), *History and*

Philosophy of Modern Mathematics, Minneapolis, MN: University of Minnesota Press, 346–66. [Also in Duren *1988–9*: Part 2, 195–215.]

Birkhoff, G. *1989*, 'Mathematics at Harvard, 1836–1944', in Duren *1988–9*: Part 2, 3–58.

Birkhoff, G. D. *1938*, 'Fifty years of American mathematics', in Archibald *1938b*: 270–315.

Cajori, F. *1890*, *The Teaching and History of Mathematics in the United States*, Washington, DC: Government Printing Office.

Charbonneau, L. *1988*, 'Mathematics: History in Canada', in J. H. Marsh (ed.), *The Canadian Encyclopedia*, Edmonton: Hurtig, 1098–9.

Cooke, R. and Rickey, V. F. *1989*, 'W. E. Story of Hopkins and Clark', in Duren *1988–9*: Part 3, 29–76.

Duren, P. (ed.) *1988–9*, *A Century of Mathematics in America*, 3 parts (Part 1, 1988; Parts 2 and 3, 1989), Providence, RI: American Mathematical Society. [This work brings together over a hundred new and reprinted articles on American mathematics by historians of science as well as mathematicians.]

Ewing, J. H. *et al.* *1977*, 'American mathematics from 1940 to the day before yesterday', in D. Tarwater (ed.), *The Bicentennial Tribute to American Mathematics, 1776–1976*, (no place given), The Mathematical Association of America, 79–99.

Greenberg, J. and Goldstein, J. R. *1983*, 'Theodore von Kármán and applied mathematics in America', *Science*, **222**, 1300–04. [Also in Duren *1988–9*: Part 2, 467–77.]

Keeping, E. S. *1968*, *Twenty-one Years of the Canadian Mathematical Congress 1945–1966*, Montreal: Canadian Mathematical Congress.

Mac Lane, S. *1989*, 'Mathematics at the University of Chicago: A brief history', in Duren *1988–9*: Part 2, 127–54.

Parshall, K. H. *1984*, 'Eliakim Hastings Moore and the founding of a mathematical community in America, 1892–1902', *Annals of Science*, **41**, 313–33. [Also in Duren *1988–9*: Part 2, 155–75.]

—— *1988*, 'America's first school of mathematical research: James Joseph Sylvester at The Johns Hopkins University 1876–1883', *Archive for History of Exact Sciences*, **38**, 153–96.

Parshall, K. H. and Rowe, D. E. *forthcoming*, *The Emergence of the American Mathematical Research Community: J. J. Sylvester, Felix Klein, and E. H. Moore*, Providence, RI: American Mathematical Society, and London: London Mathematical Society.

Price, G. B. *1988*, 'American mathematics in World War I', in Duren *1988–9*: Part 1, 267–8.

Pycior, H. *1989*, 'British synthetic vs. French analytic styles of algebra in the early American Republic', in D. E. Rowe and J. McCleary (eds), *The History of Modern Mathematics*, Vol. 1, Boston, MA: Academic Press, 125–54.

Rees, M. *1980*, 'The mathematical sciences and World War II', *The American Mathematical Monthly*, **87**, 607–21. [Also in Duren *1988–9*: Part 1, 275–89.]

Reid, C. *1976*, *Courant in Göttingen and New York: The Story of an Improbable Mathematician*, New York: Springer.

Reingold, N. *1981*, 'Refugee mathematicians in the United States of America, 1933–1941: Reception and reaction', *Annals of Science*, **38**, 313–38. [Also in Duren *1988–9*: Part 1, 175–200.]

Richardson, R. *1936*, 'The Ph.D. degree and mathematical research', *American Mathematical Monthly*, **43**, 199–215. [Also in Duren *1988–9*: Part 2, 361–78.]

Smith, D. E. and Ginsburg, J. *1934*, *A History of Mathematics in America Before 1900*, Chicago, IL: Mathematical Association of America.

Wilder, R. L. *1982*, 'The mathematical work of R. L. Moore: Its background, nature and influence', *Archive for History of Exact Sciences*, **26**, 73–97. [Also in Duren *1988–9*: Part 3, 265–92.]

11.11

Women and mathematics

JOHN FAUVEL

The participation of women in mathematical activity at any particular time is closely related to their position and role in society. If we consider the cutting-edge of mathematical research activity, then, in broad outline, women made few mathematical contributions before the nineteenth century, but provide several examples of notable mathematics thereafter. During the present century, women have increasingly benefited from wider educational opportunities and contributed accordingly, although there remains a considerable imbalance. To account for this pattern, it is helpful to be aware of social and ideological factors influencing the participation of women in intellectual activity, and mathematical and scientific pursuits in particular.

1 WOMEN AND MATHEMATICS BEFORE THE NINETEENTH CENTURY

The only woman mathematician of significance whose name has come down to us from Antiquity is Hypatia of Alexandria, who was murdered by a Christian mob in AD 415. (Evidence for the tradition that the wife of Pythagoras, nine centuries earlier, was a mathematician is too slender to be relied upon; and compare §1.5.) Although none of Hypatia's work survives, she is reported to have written commentaries on Apollonius's *Conics*, Ptolemy's *Almagest* and Diophantus's *Arithmetic*; and Knorr (*1989*: 753–804) has recently argued that her editorial hand may be detected in the extant text of Archimedes's *Dimension of the Circle*.

Hypatia was clearly exceptional in her opportunities: her father, Theon of Alexandria, was a distinguished mathematician and astronomer. At this period, the influence of Classical Greek ideas permeated European beliefs. The apparently more equitable views of Plato were offset by the clear Aristotelian message that women were intellectually as well as politically inferior. In his *Politics*, for example, Aristotle wrote: 'The slave has no deliberative faculty at all; the woman has, but it is without authority, and the child has, but it is immature.' None the less, from the Middle Ages onwards some women (especially in higher social classes) were educated as well as men were. After the scientific revolution, in seventeenth- and eighteenth-century Italy in particular, learned women rose to intellectual prominence (Alic *1986*). Mathematically the most distinguished was Maria Agnesi, whose *Analytical Institutions* (1748) was the first general textbook of the new mathematics of the period (Truesdell *1989*).

If the scope of the term 'mathematics' is broadened to include astronomy and natural philosophy, as would have been understood at the time, several other women enter the historical record. In particular, there were many women astronomers, several associated with male astronomers as wife, daughter or sister: in the seventeenth and eighteenth centuries these include Sophia Brahe, Maria Cunitz, Caroline Herschel, Elizabeth Hevelius, Maria Kirch, Christine Kirch, Maria Lalande and Nicole Lepaute (Ogilvie *1986*). It is not uncommon even today for mathematical male relatives to provide opportunity, help and encouragement to successful women mathematicians, although it is true that some women in the past have had to overcome considerable obstacles to their interest in mathematics from within their families.

Women's participation in mathematical activity extended beyond the celebrated few. In eighteenth-century Britain, for example, quite a number of women took a constructive interest in mathematics – attending lectures, subscribing to books, and posing and answering problems in magazines (Perl *1979*, Wallis and Wallis *1980*). The editor of *The Ladies' Diary* wrote in 1718 that 'foreigners would be amaz'd when I show them no less than 4 or 500 several letters from so many several women, with solutions geometrical, arithmetical, algebraical, astronomical, and philosophical'.

2 THE NINETEENTH CENTURY

One of the most penetrating judgements about women's participation in nineteenth-century mathematics was made by Carl Friedrich Gauss, in a letter to Sophie Germain of 1807. After commenting that the taste for the abstract sciences was in any case very rare, Gauss continued:

> But when a woman, because of her sex, our customs and prejudices, encounters infinitely more obstacles than men in familiarizing herself with their knotty problems, yet overcomes these fetters and penetrates that which is most hidden, she doubtless has the most noble courage, extraordinary talent, and genius. (Bucciarelli and Dworski *1980*: 25)

The first part of the century continued the earlier pattern, of a few women whose talents and opportunities enabled them to make a contribution unusual for their gender. Later in the century, various institutions were created to advance women's education, and women were able to gain access to higher education somewhat more freely in European countries and North America. The question of women's education became more widely discussed, and in the process many assumptions hindering their progress came out into the open. A remark by Augustus De Morgan to the mother of Ada Lovelace is typical of many:

> All women who have published mathematics hitherto have shown knowledge, and the power of getting it, but no one, except perhaps (I speak doubtfully) Maria Agnesi, has wrestled with difficulties and shown a man's strength in getting over them. The reason is obvious: the very great tension of mind which they require is beyond the strength of a woman's power of application. (Stein 1985: 82–3)

This rejection of the possibility of women being mathematically creative was internalized by some. For example, Mary Somerville, translator of Pierre Simon Laplace's *Traité de mécanique céleste*, wrote in her old age, 'I have perseverance and intelligence but no genius, that spark from heaven is not granted to the sex'. Women's lacking confidence in their creative ability has long been a barrier to their full participation in mathematics, and may be a factor in explaining why women became known as commentators, textbook writers and translators before they were recognized as creative mathematicians.

The most distinguished woman mathematician of the nineteenth century was the Russian Sofya Kovalevskaya (also known as Sonja Kovalevsky). She gained a doctorate in mathematics in 1874 at Göttingen University for her work on partial differential equations, Abelian integrals and the form of Saturn's rings. With her appointment as Professor of Mathematics at Stockholm, and as an editor of the journal *Acta mathematica*, she was the first female professional mathematician, and was to many people a symbol of what a woman could achieve under the right circumstances (Koblitz 1983).

As the century drew to a close, more women began to achieve mathematics research degrees. A good example is Grace Chisholm, who was adjudged to have achieved the level of a first-class degree at Cambridge in 1892 (although as a woman she could not formally be awarded a degree). She wrote her Ph.D. dissertation in Göttingen in 1895 under the direction of Felix Klein, who was active in a government programme to encourage higher education for women. Soon after her return to England she married the Cambridge coach W. H. Young and drew him into research; they became the first significant married partnership in mathematics (§11.7), with distinguished work to their credit in measure theory (§3.7) and Fourier analysis (Grattan-Guinness *1972*).

3 THE TWENTIETH CENTURY

The present century has seen some important mathematics created by women (Kramer *1957*, Green and LaDuke *1990*), and there are now more women with mathematics Ph.D.s and in leading academic positions than ever before. The story has not been one of constant upward progress, however; for instance, only in the 1970s did American women begin to regain the position they held, in terms of relative Ph.D. rate, in the early part of the century (Kenschaft *1982*).

The most influential woman mathematician of the early twentieth century, Emmy Noether, was an algebraist who — like Kovalevskaya — died early, two years after her dismissal from Göttingen University by the Nazis. Her career illustrates several themes not uncommon to women mathematicians: her father was a mathematician; she found great difficulty in obtaining a paid position as a mathematician because of traditional academic prejudice; her work was encouraged by supportive male mathematicians, but even friendly male colleagues seem to have mocked her appearance, which did not coincide with traditional norms of femininity; she found a haven in her last 18 months in a female academic community, Bryn Mawr College in Pennsylvania, USA.

The mathematical standards of Bryn Mawr had been set by its first head of mathematics (from 1885 to 1925), the Englishwoman Charlotte Scott. Scott, the second woman anywhere (after Kovalevskaya) to obtain a doctorate in mathematics (a D.Sc. at London University in 1885) was also a key figure in the development of what became the American Mathematical Society, and had a considerable influence on the growing American mathematical community, and the position of women within it. In the early decades of this century, American women came to form a substantial community of mathematicians – 229 women gained Ph.D.s before 1940 (Green and LaDuke *1987*), but factors connected with the Second World War depressed numbers, both actually and relatively, for over thirty years.

Factors militating against women's participation in mathematics apply all the more so to black women. Only in 1949 did two black women, Marjorie Lee Browne and Evelyn Boyd Granville, become the first to receive Ph.D.s in mathematics in the USA (the first black man was Elbert Cox, in 1925), and by 1980, 21 had done so (Kenschaft *1981*); for comparison, in the period 1946–70, 663 women (and 9787 men) gained mathematics Ph.D.s from American universities.

Although it may be the best documented, the situation of women mathematicians in the USA is not necessarily typical of their situation in other countries, where the role and position of women is different (see e.g. Fenaroli *et al. 1990*). Statistics for astronomy are quite revealing: the proportion of professional astronomers who are women ranges from 2% in Japan to 25% in Argentina and Mexico; the US figure is 8%. In most countries, the proportion of women declines as more senior positions are reached.

Most women mathematicians in history have had to be conscious of their role as female mathematicians, in a way that men are not self-conscious about being male mathematicians. Various organizations, such as the American Association for Women in Mathematics and European Women in Mathematics, have been formed to promote the interests and continued development of mathematical women.

4 HISTORIOGRAPHY

As with the related study of women in science, the historical study of women mathematicians has passed through several phases. A few pioneer works, of which Mozans *1913* is a fine example, set a trend for rediscovering women who contributed importantly to mathematics. Several works (see e.g. Osen *1974*, Perl *1978*, Tee *1983*) have in this tradition retold the stories of essentially the same half-dozen pre-twentieth-century great women mathematicians. The development of feminist approaches to the history of science (Merchant *1982*, Shiebinger *1987*) has led to deeper exploration of factors affecting the participation of women in mathematics (see e.g. Koblitz *1987*), a desire to escape from mirroring the masculine tradition of 'great man' history, and indeed a growing disposition towards reinterpreting history 'from the perspective of women as a sociological group challenging cultural norms that militate against their participation in science' (Merchant *1982*: 406).

Some researchers have attempted to attribute the imbalance of male and female contributions to mathematics, as well as differences in school mathematical performance between boys and girls, to intrinsic genetic factors such as may show up in psychological tests. It seems best to regard such studies and arguments as continuations of the earlier tradition of spurious pseudo-scientific rationalizations for the inferiority of women. It is increasingly acknowledged (Ernest *1976*, Badger *1981*, Fennema *1985*) that social attitudes and expectations, expressed in a variety of complex ways, are the fundamental factors determining relative mathematical performance.

BIBLIOGRAPHY

Alic, M. *1986, Hypatia's Heritage: A History of Women in Science from Antiquity to the Late Nineteenth Century*, London: The Women's Press.

Badger, M. E. *1981*, 'Why aren't girls better at maths? A review of research', *Educational Research*, **24**, (1), 11–23.

Bucciarelli, L. L. and Dworski, N. *1980, Sophie Germain: An Essay in the History of the Theory of Elasticity*, Dordrecht: Reidel.

Ernest, J. *1976*, 'Mathematics and sex', *American Mathematical Monthly*, **83**, 595–614.

Fenaroli, G., Furinghetti, F., Garibaldi, A. C. and Somaglia, A. M. *1990*, 'Women and mathematical research in Italy during the period 1887–1946', in L. Burton (ed.), *Gender and Mathematics: An International Perspective*, London: Cassell, 144–55.

Fennema, E. (ed.) *1985*, 'Explaining sex-related differences in mathematics: Theoretical models', *Educational Studies in Mathematics*, **16**, 303–20.

Grattan-Guinness, I. *1972*, 'A mathematical union: William Henry and Grace Chisholm Young', *Annals of Science*, **29**, 105–86.

Green, J. and LaDuke, J. *1987*, 'Women in the American mathematical community: The pre-1940 PhDs', *The Mathematical Intelligencer*, **9**, 11–23.

—— *1990*, 'Contributors to American mathematics: An overview and selection', in G. Kass-Simon and P. Farnes (eds), *Women of Science: Righting the Record*, Bloomington, IN: Indiana University Press, 117–46.

Grinstein, L. S. and Campbell, P. J. (eds) *1987, Women of Mathematics: A Bio-bibliographic Sourcebook*, New York: Greenwood Press.

Kenschaft, P. C. *1981*, 'Black women in mathematics in the United States', *American Mathematical Monthly*, **88**, 592–604.

—— *1982*, 'Women in mathematics around 1900', *Signs: Journal of Women in Culture and Society*, **7**, 906–9.

Koblitz, A. H. *1983, A Convergence of Lives. Sofia Kovalevskaia: Scientist, Writer, Revolutionary*, Boston, MA: Birkhäuser.

—— *1987*, 'Career and home life in the 1880s: The choices of mathematician Sofia Kovalevskaia', in P. G. Abir-Am and D. Outram (eds), *Uneasy Careers and Intimate Lives: Women in Science, 1789–1979*, New Brunswick, NJ: Rutgers University Press, 172–90. [See also the article by Abir-Am on D. Wrinch in this volume.]

Knorr, W. R. *1989, Textual Studies in Ancient and Medieval Geometry*, Boston, MA; Birkhäuser.

Kramer, E. E. *1957*, 'Six more female mathematicians', *Scripta mathematica*, **23**, 83–95.

Merchant, C. *1982*, 'Isis' consciousness raised', *Isis*, **73**, 398–409.

Mozans, H. J. *1913, Woman in Science: With an Introductory Chapter on Women's Long Struggle for Things of the Mind*, Cambridge, MA: MIT Press. [Repr. 1974.]

Ogilvie, M. B. *1986, Women in Science, from Antiquity Through the Nineteenth Century: A Biographical Dictionary with Annotated Bibliography*, Cambridge, MA: MIT Press.

Osen, L. M. *1974*, *Women in Mathematics*, Cambridge, MA: MIT Press.

Perl, T. *1978*, *Math Equals: Biographies of Women Mathematicians + Related Activities*, Menlo Park, CA: Addison-Wesley.

—— *1979*, 'The Ladies' Diary or Woman's Almanack, 1704–1841', *Historia mathematica*, **6**, 36–53.

Phillips, P. *1990*, *The Scientific Lady: A Social History of Woman's Scientific Interests 1520–1918*, London: Weidenfeld & Nicholson.

Schiebinger, L. *1987*, 'The history and philosophy of women in science: A review essay', in S. Harding and J. F. O'Barr (eds), *Sex and Scientific Inquiry*, Chicago, IL: University of Chicago Press, 7–34.

Stein, D. *1985*, *Ada: A Life and a Legacy*, Cambridge, MA: MIT Press.

Tee, G. *1983*, 'The pioneering women mathematicians', *The Mathematical Intelligencer*, **5**, 27–36.

Truesdell, C. *1989*, 'Maria Gaetana Agnesi', *Archive for History of Exact Sciences*, **40**, 113–42.

Wallis, R. and Wallis, P. *1980*, 'Female philomaths', *Historia mathematica*, **7**, 57–64.

11.12

Mathematical journals

E. NEUENSCHWANDER

The importance of professional mathematical journals for scientific inter-change in the present-day mathematical community is unquestioned. Nevertheless, there has been no comprehensive study of the historical development of mathematical journals. The special studies thus far avail-able on this theme are practically unknown, and are mainly confined to a single journal. This article can, therefore, provide only a preliminary overview of the subject, and compile the literature for further research.

1 THE GROWTH OF SCIENTIFIC JOURNALS

The first scientific journals appeared in the seventeenth century, and were published predominantly by academies and other learned societies. In them, mathematical articles were mixed with articles on other disciplines. Among the oldest publications of this sort are the *Journal des Savants* (1665–), the *Philosophical Transactions* of the Royal Society of London (1665–), the *Miscellanea curiosa medico-physica* of the Academia Naturae Curiosorum (1670–1706), and the *Acta eruditorum* (1682–1731), in which Gottfried Wilhelm Leibniz published a number of papers. These were followed in the eighteenth century by numerous other academy journals, as well as several journals with a more popular appeal, such as *The Ladies' Diary* (1704–1840), aimed at the mathematically interested laity.

The increasing professionalization and specialization of science towards the end of the eighteenth century led to the rise of subject-area journals. In the earliest of these, pure mathematics was frequently to be found along-side its fields of application, such as physics, astronomy and geography. Most of the early journals of this kind appeared for short periods only and were quite often of a relatively low standard. For this reason, *Annales de Mathématiques Pures et Appliquées* (1810–31), edited by Joseph Diaz Gergonne, may be regarded as the first significant mathematical journal which was not associated with any academic institution. Gergonne's journal soon inspired a number of analogous publications in various countries of

Europe, each seeking to offer its own countrymen a similar outlet for their mathematical writings; in most cases, both the conception and the title were closely modelled on Gergonne's own. Some examples of these are the *Journal für die reine und angewandte Mathematik* (1826–), founded by August Leopold Crelle in Germany; Joseph Liouville's *Journal de Mathématiques Pures et Appliquées* (1836–) in France; *The Cambridge (and Dublin) Mathematical Journal* (1839–54) in Great Britain, which was the forerunner of J. J. Sylvester's *The Quarterly Journal of Pure and Applied Mathematics* (1855–1927); and the *Annali di Scienze Matematiche e Fisiche* (1850–57) in Italy, the forerunner of *Annali di Matematica Pura ed Applicata* (1858–). Other early mathematical organs included the *Journal de l'Ecole Polytechnique* (1795–); the *Correspondance Mathématique et Physique* (1825–39), edited by J. G. Garnier and Adolphe Quetelet; the *Archiv der Mathematik und Physik* (1841–1920); and the *Nouvelles Annales de Mathématiques* (1842–1927) (Neuenschwander *1986*).

The second half of the nineteenth century saw a further upturn in the sciences, together with the founding of numerous new mathematical societies. Such societies had already existed in Hamburg since 1690 and in Amsterdam since 1778, to mention just two well-known examples. Many of them issued journals to publish members' contributions and the proceedings of the society, and this led to the appearance of several new mathematical journals. Among these were the *Proceedings of the London Mathematical Society* (1865–); the *Matematicheskii sbornik* (1866–) of the Moscow Mathematical Society; the *Bulletin de la Société Mathématique de France* (1872–); the journals of the mathematical and mathematical–physical societies in Prague (1872–), Kharkov (1879–), Edinburgh (1883–), Palermo (1884–), Tokyo (1884–) and elsewhere; as well as the official publications of the Deutsche Mathematiker-Vereinigung (1890) and of the American Mathematical Society (1894). For a comprehensive overview of mathematical journals up to 1900, see Gascoigne *1985*: 31–5, 139–41; Müller *1903*; and Müller *1909*: 21–40. (The first and final publication dates given for journals may vary by one or two years in different bibliographical sources, since they may refer to the date of issue of the complete volume, or to the last or first number included in it, or to the date of the proceedings published in that volume.)

2 SPECIALIST JOURNALS IN MATHEMATICS

While there were only some fifteen journals which contained mathematical articles up to the year 1700, in the eighteenth century there were more than two hundred, and towards the end of the nineteenth century already more than six hundred (Müller *1903*: 439; *1904*: 106). After the First World War, the further growth of scientific activity was accompanied by the founding of topic-oriented journals in pure mathematics, among the first of which were *Fundamenta mathematicae* (1920–), *Acta arithmetica* (1936–) and the *Journal of Symbolic Logic* (1936–).

In a broader sense, it is, of course, possible to find much earlier specialized journals; this demonstrates the multi-faceted variety of the history of mathematical journals – something that cannot be adequately described in this short article. Thus, the eighteenth and nineteenth centuries already provide examples of journals devoted to elementary mathematics and the teaching of mathematics, which were aimed at lay mathematicians, mathematics students or mathematics teachers. Examples are (including some which have already been mentioned): *The Ladies' Diary: or Woman's Almanack* (1704–1804), *The Gentleman's Diary, or the Mathematical Repository* (1741–1840), *Archiv der Mathematik und Physik* (1841–1920), *Nouvelles Annales de Mathématiques* (1842–1927), *The Mathematical Gazette* (1894–) and *L'Enseignement Mathématique* (1899–). Furthermore, in the nineteenth and early twentieth centuries, there already existed several journals devoted exclusively to special subjects such as mathematical statistics, biomathematics and the history of mathematics. These included the *Journal of the Statistical Society of London* (1839–), the *Journal of the American Statistical Association* (1888–), *Biometrika* (1901–), the *Bullettino di Bibliografia e di Storia delle Scienze Matematiche e Fisiche* (1868–87), the *Abhandlungen zur Geschichte der mathematischen Wissenschaften mit Einschluss ihrer Anwendungen* (1877–1913) and *Bibliotheca mathematica* (1884–1914).

The real breakthrough in the growth of such specialist journals, however, took place only after the Second World War, when the number of publications displayed an exponential rate of increase. Today, there are probably some three thousand journals which predominantly or occasionally publish mathematical papers. For more details of the variety of mathematical journals, see, for example, the Library of Congress's *Union List of Serials* and the *British Union-Catalogue of Periodicals*, as well as the lists of periodicals in the index volumes of the mathematical review journals.

3 REVIEW JOURNALS IN MATHEMATICS

The continual increase in the number of mathematical publications has given rise, since the eighteenth and especially in the nineteenth and twentieth centuries, to subject-oriented review journals. Baron de Férussac's *Bulletin des Sciences Mathématiques, Astronomiques, Physiques et Chimiques* (1824–31) was one of the more important and often-imitated forerunners of this genre. This journal was revived for mathematics in 1870 by

Gaston Darboux under the title *Bulletin des Sciences Mathématiques et Astronomiques* (1870–84), and later continued as the *Bulletin des Sciences Mathématiques* (1885–). At the same time, Carl Ohrtmann and Felix Müller in Germany founded the *Jahrbuch über die Fortschritte der Mathematik* (1868–1942), which sought to review all the mathematical publications of a given year, though the reviews appeared only two or three years after the publication dates of the papers themselves. Meanwhile, there was also the *Revue Semestrielle des Publications Mathématiques* (1893–1934) in Amsterdam. In the twentieth century, however, the mathematical world grew dissatisfied with the considerable time-lag between the appearance of publications and their reviews in the *Jahrbuch*, and for this reason Otto Neugebauer founded the monthly *Zentralblatt für Mathematik und ihre Grenzgebiete* in 1931. After he emigrated to the USA in 1940, he also founded *Mathematical Reviews*. Other important contemporary review journals are the *Referativnyi Zhurnal* (1953–) and the *Bulletin Signalétique* (1956–). These journals keep the mathematical world informed about an ever-increasing flood of literature, and in fact, it is only with the aid of the electronic databases associated with them since the 1970s that it is possible nowadays to keep oneself reasonably abreast of the latest developments.

BIBLIOGRAPHY

Archibald, R. C. *1929*, 'Notes on some minor English mathematical serials', *The Mathematical Gazette*, **14**, 379–400.

Baayen, P. C. *1978*, '"Wiskundig Genootschap" 1778–1978: Some facts and figures concerning two centuries of the Dutch Mathematical Society', *Nieuw archief voor wiskunde*, Series 3, **26**, 177–205.

Behnke, H. *1973*, 'Rückblick auf die Geschichte der Mathematischen Annalen', *The Mathematische Annalen*, **200**, i–vii.

Black, M. P. and Howson, A. G. *1979*, 'A source of much rational entertainment', *Mathematical Gazette*, **63**, 90–8.

Bolton, H. C. *1897*, *A Catalogue of Scientific and Technical Periodicals, 1665–1895*. 2nd edn, Washington, DC: Smithsonian Institution. [Repr. 1965, New York: Johnson.]

Brigaglia, A. and Masotto, G. *1982*, *Il Circolo Matematico di Palermo*, Bari: Dedalo.

Broadbent, T. A. A. *1946*, '"The Mathematical Gazette": Our history and aims', *The Mathematical Gazette*, **30**, 186–94.

Bubendey, J. F. *1889*, 'Beiträge zur Geschichte der Mathematischen Gesellschaft in Hamburg. 1690–1790', *Mitteilungen der Mathematischen Gesellschaft in Hamburg*, **1**, 8–16.

Collingwood, E. F. *1966*, 'A century of the London Mathematical Society', *The Journal of the London Mathematical Society*, **41**, 577–94.

Domar, Y. *1982*, 'On the foundation of *Acta mathematica*', *Acta mathematica*, **148**, 3–8.

Eccarius, W. *1976*, 'August Leopold Crelle als Herausgeber wissenschaftlicher Fachzeitschriften', *Annals of Science*, **33**, 229–61.

—— *1976–7*, 'August Leopold Crelle als Herausgeber des Crelleschen Journals', *Journal für die reine und angewandte Mathematik*, **286–287**, 5–25; **289**, 214.

Elkhadem, H. *1978*, 'Histoire de la "Correspondance mathématique et physique" d'après les lettres de Jean-Guillaume Garnier et Adolphe Quetelet', *Académie Royale de Belgique, Bulletin de la Classe des Lettres et des Sciences Morales et Politiques*, Series 5, **64**, 316–66.

Fiske, T. S. *1988*, 'The beginnings of the American Mathematical Society', in P. Duren (ed.), *A Century of Mathematics in America*, Part 1, Providence, RI: American Mathematical Society, 13–17.

Folta, J. and Rozsíval, M. *1972*, 'Sto let od zaluženi časopisu Jednoty čs. matematiků a fyziků [Centenary of the founding of the journal of the Union of Czechoslovak Mathematicians and Physicists]', *Pokroky Matematiky Fyziky a Astronomie*, **17**, 61–7.

Gani, J. *1982*, 'An adventure in publishing: Eighteen years of the Applied Probability Trust', *Australian Journal of Statistics*, **24**, 1–17.

—— *1988*, 'Adventures in applied probability', *Journal of Applied Probability*, **25A**, 3–23. [Special volume, entitled 'A celebration of applied probability'.]

Gascoigne, R. M. *1985*, *A Historical Calalogue of Scientific Periodicals, 1665–1900 with a Survey of Their Development*, New York and London: Garland.

Gispert, H. *1985*, 'Sur la production mathématique française en 1870 (Étude du tome premier du "Bulletin des sciences mathématiques")', *Archives Internationales d'Histoire des Sciences*, **35**, 380–99.

Goldsmith, N. A. *1953*, 'The Englishman's mathematics as seen in general periodicals in the eighteenth century', *The Mathematics Teacher*, **46**, 253–59.

Grattan-Guinness, I. *1986*, 'The "Società Italiana", 1782–1815: A survey of its mathematics and mechanics', *Symposia mathematica*, **27**, 147–68.

—— *1992*, 'A note on "The Educational Times" and "Mathematical Questions"', *Historia mathematica*, **19**, 76–8.

Guggenbuhl, L. *1959*, 'Gergonne, founder of the Annales de Mathématiques', *The Mathematics Teacher*, **52**, 621–9.

Hermann, D. B. *1972*, *Die Entstehung der astronomischen Fachzeitschriften in Deutschland (1798–1821)*, Berlin: Veröffentlichungen der Archenhold-Sternwarte Berlin-Treptow, No. 5.

Hogan, E. R. *1976*, 'George Baron and the "Mathematical Correspondent"', *Historia mathematica*, **3**, 403–15.

—— *1985*, 'The Mathematical Miscellany (1836–1839)', *Historia mathematica*, **12**, 245–57.

Houghton, B. *1975*, *Scientific Periodicals: Their Historical Development, Characteristics and Control*, London: Bingley.

James, I. M. *1986*, 'Topology: Past, present and future', in *Algebraic Topology. Proceedings of an International Conference held in Arcata, California, July 27–August 2, 1986*, New York: Springer (Lecture Notes in Mathematics, Vol. 1370), 1–9.

Johnstone, P. T. *1986*, '100 not out', *Mathematical Proceedings of the Cambridge Philosophical Society*, **100**, 1–4.

Kronick, D. A. *1976*, *A History of Scientific & Technical Periodicals. The Origins and Development of the Scientific and Technical Press, 1665–1790*, 2nd edn, Metuchen, NJ: Scarecrow.

Lampe, E. *1904*, 'Das Jahrbuch über die Fortschritte der Mathematik. Rückblick und Ausblick', in *Atti del Congresso Internazionale di Scienze storiche (Roma, 1–9 Aprile 1903)*, Vol. 12, Rome: Loescher, 97–104.

Lorey, W. *1916*, *Das Studium der Mathematik an den deutschen Universitäten seit Anfang des 19. Jahrhunderts*, Leipzig and Berlin: Teubner (Abhandlungen über den mathematischen Unterricht in Deutschland, Vol. 3, No. 9).

Meadows, A. J. (ed.) *1980*, *Development of Science Publishing in Europe*, Amsterdam: Elsevier.

Müller, F. *1903*, 'Abgekürzte Titel von Zeitschriften mathematischen Inhalts', *Jahresbericht der Deutschen Mathematiker-Vereinigung*, **12**, 427–44. [Includes explanations and historical notes.]

—— *1904*, 'Über mathematische Zeitschriften', in *Atti del Congresso Internazionale di Scienze storiche (Roma, 1–9 Aprile 1903)*, Vol. 12, Rome: Loescher, 105–13.

—— *1909*, 'Führer durch die mathematische Literatur mit besonderer Berücksichtigung der historisch wichtigen Schriften', *Abhandlungen zur Geschichte der mathematischen Wissenschaften mit Einschluss ihrer Anwendungen*, **27**.

Neuenschwander, E. *1984*, 'Die Edition mathematischer Zeitschriften im 19. Jahrhundert und ihr Beitrag zum wissenschaftlichen Austausch zwischen Frankreich und Deutschland'. Preprint, Mathematisches Institut der Universität Göttingen.

—— *1986*, 'Der Aufschwung der italienischen Mathematik zur Zeit der politischen Einigung Italiens und seine Auswirkungen auf Deutschland', *Symposia mathematica*, **27**, 213–37.

Perl, T. *1979*, 'The Ladies' Diary or Woman's Almanack, 1704–1841', *Historia mathematica*, **6**, 36–53.

Rankin, R. A. *1983*, 'The first hundred years (1883–1983)', *Proceedings of the Edinburgh Mathematical Society*, Series 2, **26**, 135–50.

Remmert, R. *1991*, 'Inventiones mathematicae: Die ersten Jahre', in P. Hilton, F. Hirzebruch and R. Remmert (eds), *Miscellanea mathematica*, Berlin: Springer, 269–75.

Rohrbach, H. *1982*, 'Helmut Hasse und Crelles Journal', *Mitteilungen der Mathematischen Gesellschaft in Hamburg*, **11**, (1), 155–66.

Royal Society *1867–1925*, *Catalogue of Scientific Papers. 1800–1900*, 19 vols, London: HMSO, and Cambridge: Cambridge University Press; *Subject index*, 3 vols, Cambridge: Cambridge University Press. [Repr. 1965, New York: Johnson; and 1968, Metuchen, NJ: Scarecrow.]

Schimank, H. *1952*, 'Mittel und Wege wissenschaftlicher, insbesondere naturwissenschaftlicher Überlieferung bis zum Aufkommen der ersten wissenschaftlichen Zeitschriften', *Sudhoffs Archiv für Geschichte der Medizin und der Naturwissenschaften*, **36**, (3/4), 159–82.

Schmidt, F. K. *1967*, 'Reorganization of mathematical reviewing', *Notices of the American Mathematical Society*, **14**, 606–9.

Scudder, S. H. *1879*, *Catalogue of Scientific Serials* [...], *1633–1876*, Cambridge, MA: Library of Harvard University. [Repr. 1965, New York: Kraus.]

Sergescu, P. *1947–8*, 'La Littérature mathématique dans la première période (1665–1701) du "Journal des Savants"', *Archives Internationales d'Histoire des Sciences*, **1**, 60–99.

Siegmund-Schultze, R. *1984*, 'Das Ende des Jahrbuchs über die Fortschritte der Mathematik und die Brechung des deutschen Referatemonopols', *Mathematische Gesellschaft der DDR. Mitteilungen*, No. 1, 91–101.

Stäckel, P. *1906*, 'Das Archiv der Mathematik und Physik, ein Geleitwort zu den ersten zehn Bänden der dritten Folge', *Jahresbericht der Deutschen Mathematiker-Vereinigung*, **15**, 323–9.

Taton, R. *1947–8*, 'Les Mathématiques dans le "Bulletin de Férussac"', *Archives Internationales d'Histoire des Sciences*, **1**, 100–25.

Tobies, R. *1986–7*, 'Zu Veränderungen im deutschen mathematischen Zeitschriftenwesen um die Wende vom 19. zum 20. Jahrhundert', *NTM – Schriftenreihe für Geschichte der Naturwissenschaften, Technik und Medizin*, **23**, (2), 19–33; **24**, (1), 31–49.

Topsøe, F. *1986*, 'EUROMATH – The integrated database and communications system for European mathematicians', *Nieuw archief voor wiskunde*, Series 4, **4**, 155–60.

Wilkinson, T. T. *1848–53*, 'Mathematical periodicals', *The Mechanics' Magazine*, **48–59**. [ca. fifty articles on thirty periodicals.]

Wölffing, E. *1903*, 'Mathematischer Bücherschatz. Systematisches Verzeichnis der wichtigsten deutschen und ausländischen Lehrbücher und Monographien des 19. Jahrhunderts auf dem Gebiete der mathematischen Wissenschaften', *Abhandlungen zur Geschichte der mathematischen Wissenschaften mit Einschluss ihrer Anwendungen*, **16**, (1).

Part 12
Mathematics and culture

12.0

Introduction

This, the final main Part of the encyclopedia, is devoted to a miscellany of topics in which mathematics has come to bear upon general cultural and artistic questions; some of them are of ancient origin. The first three articles cover the broadest areas, with the first one relating mostly to non-European cultures. The next two articles, §12.4–12.5, treat topics which have even had some mystical or occult (in the original meaning of the word, 'hidden') significance; the second of these leads in a natural way to the use of mathematics in art and architecture in §12.6. Note also the place of cultural questions in several articles in Part 1, and in §2.10 on early music theory.

Geometry is a major branch of mathematics in these contexts, and §12.7–12.8 take two aspects, symmetry and tilings, in which they continue to feature even in modern mathematics. The next two articles consider mathematics and mathematicians as they have appeared in literature, followed by two manifestations of the bestowal of honour. To a limited extent this Part returns to the ambit of the first, in that non-Western cultures are considered occasionally here.

The final article, §12.13, is a self-reflective piece on the history of mathematics itself. Its own history is considered, and the concerns of §0 are revisited at the end, to make the self-reflection into a self-reference.

12.1

Ethnomathematics

M. ASCHER AND R. ASCHER

1 INTRODUCTION

As a field of enquiry, ethnomathematics began in the 1960s when mathematicians and mathematics educators, both Western and Western-trained from emergent and recently industrialized nations, became interested in the mathematical ideas of indigenous people and in the mathematics practised outside Western-style educational systems. As distinct from the usual areas over which discussions about mathematics range, ethnomathematics draws heavily upon the theories, knowledge and methods of culture history, cognitive studies and, above all, linguistics and anthropology.

Ethnomathematics holds that mathematical ideas are pan-human and are developed within cultures. Mathematical ideas are taken to be those that involve number, logic, spatial configuration and, most importantly, the combination or organization of these into systems or structures. The imposition of order on space, for example, is universal, but its particular expression varies with and within culture, and may change over time.

During the past three hundred years, using the criterion of mutually exclusive speech communities, there have existed between five and six thousand different cultures. Well over 95% of these had no writing as we generally understand the term. Today, indigenous traditions are blended with or are yielding to a few dominant cultures. And, in the midst of any complex, industrialized nation-state, large or small, there are part-, sub- and composite cultures which develop vernacular ways of doing things that are particular to them even if they derive from the Western mode. For earlier traditions, artefacts, and the conversations and observations reported by contributors to the ethnographic literature provide the starting-points for analysis from a mathematical perspective. For contemporary groups, fieldwork is also possible.

The category 'mathematics' is Western and is not found as such elsewhere. That is not to say that mathematical ideas do not exist; it is rather that others do not distinguish them in the same way. From culture to culture, and within any culture, mathematical ideas appear in various contexts which are neither clear-cut nor mutually exclusive. For example, art, religion and food preparation may be separable from an outsider's viewpoint, but not for those living within the culture. Further, the contexts of an idea in one culture need not be its contexts in another, and what is important in one may be unimportant or non-existent in another. If context and context cross-overs are borne in mind, mathematical ideas may be recognized in cultures in different and sometimes unexpected places, from record-keeping and trade to myth, decoration or drumming. Whatever the context, mathematical ideas and the ways in which they are expressed are part of the intricate web out of which a culture is woven. How and where the mathematical ideas fit in must be understood if they are to be properly appreciated.

The search for mathematical ideas in other cultures is necessarily constrained by one's own cultural and mathematical frameworks. Ideas that are different may escape recognition, while those that are in some way similar to one's own are more likely to be recognized. Descriptions must eventually be cast in one's own terms, but clear distinction must be maintained between the concepts used to describe or explain and those attributed to people in other cultures.

For some, the primary goal of ethnomathematics is to broaden the history of mathematics by gaining a global, multicultural perspective; for others (Bishop *1985*, D'Ambrosio *1988*, Gerdes *1985*) the pedagogical implications and possibilities are paramount. Some examples are presented below in order to convey the sense and perspective of ethnomathematics and to introduce cultural diversity into the realm of discussion about mathematical ideas.

2 NUMBER

2.1 Words

In histories of mathematics, traditional peoples, if they are acknowledged at all, are usually to be found in a preliminary chapter devoted to number. The focus on number-words in such chapters requires re-examination (see §1.16).

Excessive importance has been placed on number-words because it was held that for all cultures there was a single linear path *beginning with number* along which mathematical ideas progressed. A culture with few number-words could have no further mathematical ideas. It is now known that all cultures have their own history, and that how high number-words go is simply a reflection of the importance a culture places on counting things. Number-words are a language phenomenon, and contemporary linguists view the capacity to count as part of the *universal* language capacity. The set of number-words in a language generally follows a pattern and contains implicit arithmetic relationships. The important feature of all the diverse number-word construction patterns is that the numbers formed are an arithmetically interrelated set which is open-ended and extendable.

For many of the world's languages, numeral classifiers are intimately connected with number-words. Numeral classifiers are words or prefixes or suffixes that must be included when a number word is *spoken* with a noun. It was the inclusion of these classifiers, mistaken by Europeans for different number-words, that led to the notion that counting was concrete for some while abstract for others. Classifiers are now known to exist in many languages including Chinese, Japanese, almost all the languages in Southeast Asia, numerous languages in Oceania, some African languages, and in a few European languages such as Gaelic and Hungarian.

Some languages have as few as two classifiers, others as many as two hundred. As a simple example, in the language of the Maori, the indigenous people of New Zealand, a classifier must be used when a numerical statement is made about human beings. Saying 'three *humans* mathematicians' rather than 'three mathematicians' makes the concept of three no more or less abstract; it is merely augmented by the classifier concept 'human'. For other languages, particularly those with a greater number of classifiers, the nouns that take a particular classifier can be understood only from a complete lexicon of them. Something many cultures have in common, however, is that classificatory characteristics often distinguish the animate from the inanimate, or involve geometric properties. On the whole, classifiers convey information that is qualitative rather than quantitative, but they are nevertheless a *necessary* part of making a quantitative statement.

2.2 Number records

Quite distinct from number-words are symbolic representations of numerals. The familiar numerals (e.g. 1, 2, 25, 73) and the Roman numerals (e.g. I, IV, XI, MCM) are usually associated with writing and with inscriptions on flat surfaces. That writing is not necessary for symbolic representation is shown by the system used by the fifteenth-century Inca of South America (Ascher and Ascher *1981*). The Inca had no writing as we generally use the term; nevertheless, they were intensive data-users and kept extensive records.

For the Inca, textiles were of great importance; they accompanied life passages from birth to death and were essential in gift-giving, trade, and religion. Knowing this, the choice of cotton for the medium of symbolic

Figure 1 A quipu in the collection of the Smithsonian National Museum, Washington, DC

representation seems natural. The cotton was woven into cords; the cords were dyed different colours, and then arranged, tied and knotted. A sophisticated logical–numerical recording system was expressed through these spatial arrays of knotted, coloured cords, called *quipus* (Figure 1).

It is rare to be able to associate a given *quipu* with a particular usage. All extant specimens – some five hundred in museum collections – come from graves. In general, *quipus* served the needs of the Inca state, which, in the century preceding the conquest, had taken control of hundreds of cultures spread over the entire western flank of South America. These needs included, at the very least, population counts, architectural plans, astronomical and calendric recording, and records of food and goods moved with regularity in and out of vast storage areas maintained by the state bureaucracy.

Characteristic of Inca culture were an interest in portability, a great concern with spatial arrangement and symmetry, and a methodical and conservative outlook. The *quipu* embodies this ethos, and the ethos can be appreciated through an understanding of the logical–numerical system of the *quipu*.

A *quipu* can consist of as few as three cords, or as many as two thousand. Each *quipu* has a main cord; spread out along it and suspended from it are other cords. Relative to the main cord, some suspended cords fall downward, others fall upward, and some dangle from its end. The relative spacing along the main cord forms some suspended cords into groups and even into groups of groups. Yet other cords attached to the suspended cords, and others in turn suspended from them, form levels of cords. The colours of the cords further associate some cords and distinguish them from others. Cord placement, relative spacing and colour are the elements of a symbolic system. The arrangement of any *quipu* is a particular selection from this vocabulary, which provides the logical structure for the numerical data.

Numbers are represented by clusters of knots spaced along the cords. Basically the knot clusters represent digits and, where the clusters on a cord form one number, it is an integer in the base 10 positional system. The units position of a number is always distinguished by the type of knot used. Because of this distinction, a cord can contain multiple numbers, each units position indicating a new number.

From cord to cord, knot clusters are carefully aligned. Hence, if cords within a group carry multiple numbers, the level of the number on a cord is comparable from cord to cord. More importantly, though, the overall alignment of digits makes it clear when there is a digit position with no knots within a number; it is the absence of a knot that represents 'zero'. Just as the absence within a number is unambiguous, a particular use of colour patterning clarifies when a cord without any knots should be interpreted as a zero.

The numbers on the *quipus* are more than just quantitative data. Numbers can be magnitudes but numbers can also be labels, as they are in the modern world in telephone numbers and the numbers on footballers' shirts. *Quipus* contain structured arrangements of numerical data made up of both magnitudes and labels.

Examples of the data arrangements found on *quipus* are multidimensional charts (such as we represented by a_{ijk}, where $i = 1, \ldots, m$; $j = 1, \ldots, n$; $k = 1, \ldots, p$) and data trees (such as we represent with tree diagrams: see §7.13 on combinatorics), and combinations into charts of data trees. There are numerous instances of data on a *quipu* interrelated by summation (as is found in ledgers and subledgers) and other more intricate logical and numerical relationships, sometimes between sets of *quipus*.

2.3 Numerical practices

The arithmetic practices of counting, measuring and calculating, and the significance attached to the resulting quantities, are all a part of the domain of number. The importance of numbers varies considerably from culture to culture, and from group to group within a culture, and changes with changes in the culture. That arithmetic techniques also develop and vary with culture and context has been shown for indigenous groups in Africa and Papua New Guinea; tailors, cloth merchants, and farmers on the Ivory Coast; workers in a commercial dairy and shoppers in supermarkets in the USA; and sugarcane farmers, carpenters, and street-vendor children in Brazil.

3 SPACE

3.1 Concepts of space

People of all cultures define the space around them by both the physical and the mental imposition of order, and these orderings play a large part in the individual's perception and interpretation of experience. In Western culture, until the late nineteenth century it was believed that Euclidean geometry describes truths about the physical world. Basic to Euclidean geometry are points, lines, surfaces and solids. It assumes that space has three dimensions, is continuous and uniform, and has zero curvature everywhere. This view influenced and was reinforced by Western art, architecture, the design of objects and interiors, measuring and mapping schemes, and ways of seeing and describing. Although Western scholars came to understand that Euclidean geometry is but one of many possible mental constructs (§7.4), the Euclidean model is still the prevalent world-view in Western culture.

By contrast, for the Navajo, a native American culture, space and time are so inextricably interwoven that one cannot be discussed without the other (Pinxten *et al.* 1983). They believe in a dynamic universe made up of processes rather than of objects and situations. Central to the Western mode of thought is the idea that things are separable entities which can be subdivided into smaller, discrete units. Things that change through time do so by going from one specific state to another. While Westerners believe time to be continuous, it is often broken into discrete units or frozen, such as when one speaks about an instant or a point in time. Quite often, time is just ignored: for example, when a line is said to be divided by a point, the description is of a static situation in which time plays no role at all. Among the Navajo, where the focus is on process, change is ever-present;

interrelationship and motion are of primary importance, for they incorporate and subsume space and time.

Consider the idea of subdividing things into parts. The human body, for example, is seen by Westerners as a unit but it can also be described as having distinct parts such as arms, legs, or heart. Although blood pressure is recognized as important, it is not a *part* of the body in the same sense – it has no specific location or specific boundaries. To the Navajo, blood pressure *is* a part of the body. The body is regarded as a dynamic whole, a system of interrelated parts, and to be a part of the body is to be involved in making the body work.

For the Navajo, spatial boundaries and borders do exist, but they too have dynamic components. There are boundaries that require one to take action in order to go around or over them, but whatever one was doing on one side can be continued on the other side; other boundaries require that actions be modified or reoriented. A mountain ridge within Navajo territory is a boundary of the first type, while one separating Navajo and non-Navajo territory is of the second type. As for the actual position of the mountain ridge, to Westerners position is defined as where something *is*; the Navajo view it as the resultant of the withdrawal of motion – the mountain ridge is in the process of being in a particular place. The ridge itself is an interrelated system of moving and changing parts, and indeed the entire Earth, of which the mountain ridge is an integral part, is also in process. The Earth and sky are undergoing a continual expansion and contraction. Starting from a centre, Earth and sky are being stretched outward in a clockwise spiral and will eventually shrink back to the centre along the same path. The motion is continuous, systematic and gradual.

To Westerners, what is significant about the overlapping of two surfaces is that they have a region in common; to the Navajo, the fact that two things overlap is just one aspect of an active process. Of primary importance is whether or how the elements are in contact during the happening. The overlap of a snake and a rock, for example, is of one type if the snake is sleeping on the rock but of another if it is slithering over the rock.

For the Navajo, space is continuous, has three dimensions and is finite in that the universe will eventually retract to the centre. But above all it is the dynamic character of space and everything within it that is of primary significance. In general, descriptions of events and objects, even statements of concepts, emphasize movement and contain detailed elaboration of the kind and direction of the movement.

3.2 Planar graphs

The sand-tracing tradition of the Malekula of Oceania provides evidence of ideas akin to what Westerners categorize as graph theory, and associated with them are other geometric and topological ideas (Ascher *1991*).

Graph theory, described geometrically, is concerned with points (called vertices) interconnected by lines (called edges) (§7.13). In a connected planar graph, every vertex is joined to every other via some set of edges, and the graph lies entirely in the plane. The degree of a vertex is the number of edges emanating from it, and a vertex is odd or even depending on its degree. A classical question in graph theory is whether, for a connected planar graph, a continuous path can be found that covers every edge once and only once (such a path is called an Eulerian path), and, if such a path exists, whether it can end at the same vertex as it started from. The answer, first of all, is that Eulerian paths do not exist for all connected planar graphs. Such a path does exist if a graph has one and only one pair of odd vertices, provided the path begins at one of them and ends at the other. Also, a Eulerian path exists if all the graph's vertices are even and, in this case, the path can start from any vertex and end where it began.

The concern for continuous path tracing is tightly enmeshed in the Malekula ethos. According to the Malekula, in order to get to the Land of the Dead one must pass a ghost or spider-like ogre who poses a challenge to all who try to enter – to trace in the sand a figure learned during life. The stipulation is that the figure must be traced without lifting the finger, covering each line once and only once, if possible beginning and ending at the same point. Other Malekula myths also place emphasis on knowing one's figures and being able to trace them according to these rules.

From the ethnographic literature, about a hundred figures and the exact paths for tracing them are known. Viewed as graphs, the figures range from simple closed curves to those with more than a hundred vertices, some with degrees of 10 or 12. The tracing courses used corroborate the Malekula's concern for what Westerners call Euclidean paths, and for ending at the starting-point when that is possible.

Within individual figures, the Malekula tracing procedures are quite systematic, but, even more importantly, the systems extend to groups of figures. There are three or four extended systems; the basic idea of one is briefly described here. For each figure there is an initial procedure – some ordered sequence of motions (call it A). This is followed by the procedure modified by formal transformations. (Call a transformed procedure A_T.) For the group of figures, only a specific set of transformations is used: rotation through $90°$, $180°$ or $270°$, vertical reflection (V) and horizontal reflection (H), each alone or in combination with inversion; inversion ($\bar{A}$) reverses the order of the tracing procedure. In Figure 2, a figure and its initial procedure, A, are shown. In terms of A, the figure can be succinctly described as $AA_{90}A_{180}A_{270}$. The figure in Figure 3, in terms of its initial

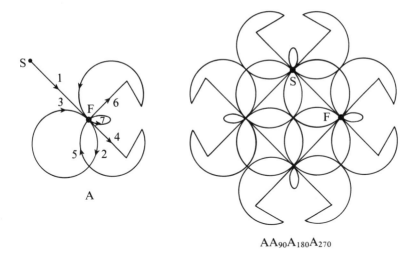

$$AA_{90}A_{180}A_{270}$$

Figure 2 A Malekula sand-tracing

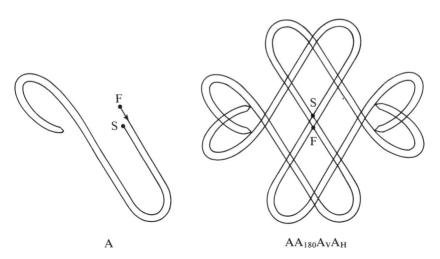

A $\qquad$ $AA_{180}A_\vee A_H$

Figure 3 Another Malekula sand-tracing

procedure, is $AA_{180}A_\vee A_H$. The two figures exemplify the Malekula interest in symmetry of effect combined with the self-imposed constraint of a cyclic Eulerian tracing path and the use of systematic tracing procedures that are particular expressions of larger general systems.

BIBLIOGRAPHY

Ascher, M. *1991, Ethnomathematics: A Multicultural View of Mathematical Ideas,* Pacific Grove, CA: Brooks & Cole.

Ascher, M. and Ascher, R. *1981, Code of the Quipu: A Study in Media, Mathematics and Culture,* Ann Arbor, MI: University of Michigan Press.

—— *1986,* 'Ethnomathematics', *History of Science,* **24,** 125–44.

Bishop, A. *1985,* 'Culture and mathematics education', in M. J. Johnson-Høines and S. Mellin-Olsen (eds), *Mathematics and Culture,* Radal: Caspar, Norway, 59–67.

Carraher, T. N., Carraher, D. W. and Schliemann, A. D. *1985,* 'Mathematics in the streets and in schools', *British Journal of Developmental Psychology,* **3,** 21–9.

Closs, M. (ed.) *1986, Native American Mathematics,* Austin, TX: University of Texas Press.

Cole, M., Gay, J., Glick, J. A. and Sharp, D. W. *1971, The Cultural Context of Learning and Thinking,* New York: Basic Books.

D'Ambrosio, U. *1988,* 'Ethnomathematics: A research program in the history of ideas and in cognition', *International Study Group on Ethnomathematics Newsletter,* **4,** 5–8.

Gerdes, P. *1985,* 'Conditions and strategies for emancipatory mathematics education in undeveloped countries', *For the Learning of Mathematics,* **5,** 15–20.

Hurford, J. R. *1987, Language and Number: The Emergence of a Cognitive System,* London: Blackwell.

Pinxten, R., van Dooren, I. and Harvey, F. *1983, The Anthropology of Space,* Philadelphia, PA: University of Pennsylvania Press.

Washburn, D. K. and Crowe, D. W. *1988, Symmetries of Culture,* Seattle, WA: University of Washington Press.

Zaslavsky, C. *1973, Africa Counts: Number and Pattern in African Culture,* Boston, MA: Prindle, Weber & Schmidt.

12.2

Mathematical games

RÜDIGER THIELE

1 WHAT ARE MATHEMATICAL GAMES?

This article looks at the variety of games which have some mathematical element. Typical examples are briefly described, with their origin, development, mode of play, and links with mathematics: strategy, optimum number of moves, achievability of a set aim, and so on.

Play and games are found in various spheres of human activity. What is the essence of a mathematical game? A game proceeds according to a set of rules, but tempered with the whim of chance – in the guise of 'good luck' and 'hard luck' – lacing the game with a spirit of novelty. Between these two poles, regimen and chance, the game steers its course. Card games are an excellent example. A game may be regarded as 'mathematical' when the players can perceive and/or influence the course of the game on the basis of mathematical considerations.

Play provides an escape from reality into an ideal world, a world of order, yet not utterly predictable (a good example is chess). A game's links with mathematics will depend on the regulating quality of its set of rules, and will be stronger where mental skill takes precedence over physical skill. However, as well as rules-dominated games like chess, solitaire and nim, with their strictly logico-mathematical structure, games of chance, too, have a close connection with mathematics. Games of chance, with their simple mathematical structure and psychological motivation (the pursuit of profit), have figured in the elaboration of the theory of probability, as witness the work of Girolamo Cardano, Luca Pacioli, Niccolò Tartaglia, Blaise Pascal, Pierre de Fermat, Christiaan Huygens and James Bernoulli (see Section 5, and §10.2). Among the problems these men dealt with was the fair apportionment of stakes in a game of chance that for some reason is not played out.

In Newtonian mechanics, the result of a random event such as tossing a coin or rolling a die is completely predetermined by the initial conditions, the uncertainty or chance result we experience being merely a consequence of the incompleteness of our information. T. Bass, J. Farmer and N. Packard empirically determined the initial conditions in the game of roulette and thus broke the bank, thereby robbing the game of its interest as a game (Bass *1980*).

In many games, it is the incompleteness of information that produces the element of chance, where otherwise the outcome would be quite certain. Chess is a case in point. Played according to the optimum strategy (known to exist but as yet unknown to man or machine), the outcome would never be in doubt. Lack of complete information can constitute one of the chief appeals of a game, as in bridge.

The mathematical theory of games analyses games in which the outcome depends not only on the behaviour of the players, but also on nature (theory of decision-making); it seeks the optimum strategy for the game (von Neumann and Morgenstern *1944*; see §6.11). M. Eigen and R. Winkler *1975* regard evolution as a game determined by the effect of chance events on the laws of nature.

There are also a number of games for one person which are regularly counted among the mathematical games, but are better defined as mathematical problems in disguise. The level of these games, as of all others, may have reached considerable heights even in the early days of the mathematical sciences. African string figures, which may come from the advanced culture of Egypt, are based on visual problems but their solution calls for a command of advanced topological relations (§1.8; see also §12.1). In addition, the philosopher George Berkeley invented a game, called 'De ludo algebraico', in order to drill students in solving algebraic equations.

In recent times, the computer has entered the realm of games. There are good programs for a number of board games, in particular for chess and backgammon. Other games, calling cases to be distinguished, such as puzzles with plane and solid pieces, make use of the computer. The possibility of testing numerous alternatives in a very short interval of time has opened up new fields. John Conway has even made use of the idea of play to construct real numbers, generalizing the Dedekind cut.

2 BOARD GAMES

Board games probably had their origin in the use of a dicing board, over which counters were moved according to the fall of the dice (possibly for cultic or oracular purposes). As time progressed, moving the pieces on the board became more and more divorced from the element of dicing, until rules developed to the extent that from the original simple games there arose games of mental agility. Chess is an example of a game whose earlier forms included the employment of dice; backgammon is an intermediate form, in which chance dicing is combined with strategic thinking.

The earliest-known form of board is one of six by three squares, measuring 18 cm by 7 cm, discovered together with its pieces near the Egyptian town of Abydus, and dating from the Old Kingdom. Egyptian temples built in about 2500 BC contain representations of board games being played. At Ur, Leonard Woolley excavated board games with pieces and dice (truncated tetrahedrons) dating from 2560 BC. On the roof of the temple at Ķûrnah, in Upper Egypt, as at other temples, game boards were scratched into the stonework, and holes made to take mancala pieces, an indication of how the workmen spent their leisure moments.

Murray 1952 distinguishes five kinds of board games, with examples: games of alignment and configuration (go bang, halma); war games (chess, go, leap-frog); hunt games (fox and geese); race games (backgammon, snakes and ladders); and mancala (see Section 8). As a rule, the board is rectangular, triangular or circular, or has a plan deriving from one of these, as in solitaire and riman-riman. These plans are frequently reduplicated, as in alquerque and rithmomachia. The modes of taking opposing pieces are manifold: by occupation of a square (chess), jumping over to a vacant square (draughts), interception (tablut), completion of a figure (nine-men's-morris), arithmetical relations (rithmomachia) or even huffing – capture by default (draughts). Murray lists no fewer than ten possibilities.

Attempts to apply logical analysis to games go back a long time. Islamic *mansubs* (roughly translated as 'end-games') could be regarded as precursors of chess problems. The *Codex Alfonso* (Alfonso 1283) gives optimum strategies for a variety of games: in three-men's-morris (*alquerque de tres*), in English better known in the form of noughts and crosses, the central square must be occupied at the outset, or an experienced opponent will at the very least force a draw. In a game of twelve-men's-morris, two equally matched players, using the optimum strategy, will always arrive at a draw.

In addition to the old-established board games with their numerous variants, new ideas have also been successfully introduced: reversi (Mollett and Waterman, *circa* 1870), Halma (Monks, *circa* 1880), salta (Büttgenbach, 1899), laska (Lasker, before 1932) and the L game (de Bono, 1969).

3 RITHMOMACHIA

The earliest-known report on rithmomachia, or the *pugna numerorum* ('battle of the numbers'), describes a game which unites strategic thinking and a sense of harmony. It comes from eleventh-century southern Germany (manuscript of Asilo von Würzburg, *circa* 1030), where it originated in a monastery. The name 'rithmomachia' has been in common use since the thirteenth century (Evans *1976*); and as the scholastics were particularly addicted to the game, it has also acquired the name 'the philosopher's game'.

The earliest rules are quite incomplete, and in part contradictory, Asilo not even describing the board on which the game is played. This is a possible indication that the game is of still older origin, gaps being indicative of common knowledge, inconsistencies resulting from lack of understanding. From the fifteenth century the rules developed almost beyond control, mostly because of the admission of new numerical proportions (see below). Although Asilo expressly referred to the *Institutio arithmetica* of Boethius, dating from around AD 500, the players' knowledge of the arithmetic of the ancients was anything but uniform.

The modes of play rest largely on a thorough knowledge of the Pythagorean law of ratios: the numbered shapes used as pieces are moved with the aim of achieving a numerical proportion (arithmetic, geometric,

Figure 1 The game of rithmomachia (from Smith *1923*: 199)

harmonic means) in the opponent's field of play (Figure 1). The vertical, horizontal and diagonal moves are vaguely reminiscent of moves in chess, but completely at variance with chess is the mode of taking opposing pieces. The piece making the capture remains immobile, calculating the victim by means of arithmetical operations, and thus capturing it. For example, a white 12 beats a black 36 at a distance of three squares (Borst *1986*).

The game enjoyed a certain popularity among educated people from medieval times up to the beginning of the eighteenth century, passing through monastic, feudal, courtly and bourgeois phases. The theory has been advanced that it was devised originally as an introduction to arithmetic; there was without doubt a close link with music and philosophy, via the Pythagorean theory of harmony. When, in the eighteenth century, fractions began to supplant proportions in arithmetic, interest in this game diminished rapidly.

4 SOLITAIRE

Apart from a pictorial representation of the game of solitaire dating from 1697 (Bereys), the first definite reference to the game is in Leibniz *1710*. Although its origin is obscure and has given rise to various conjectures, it seems certain that solitaire was well known at the court of the Sun King, Louis XIV. Three forms of board are common, each having the form of a cross, and with 33, 37 and 41 squares (nodes). The object of the game is, beginning with all the squares filled except one, to remove all but one of the pieces by over-jumping in horizontal and vertical moves to a vacant square. Leibniz loved the game, and was in the habit of playing it the other way round, starting with one piece on the board and filling the over-jumped squares. This constructive variant allows a simpler notation. Other varieties of the game require the player to execute the moves so as to produce a particular final pattern, such as a cross.

Reiss *1871* elaborated a theory of solitaire, which was later taken further by Hermary (1879). The chief conclusion is that, starting from any square, the final square must fulfil certain requirements if the problem is to have any solution at all. For a board of 33 squares there are 16 possible solutions (not counting purely symmetrical variants). A further question is to find the least number of moves in which the object may be achieved. For a 33-square board, J. Beasley showed in 1962 that, starting from the central square, no solution is possible in less than 18 moves, 'move' being taken to mean either a single jump or a multiple jump in which a further piece or pieces may be removed at the same time (Beasley *1985*). From the basic cross form, the game has been adapted to boards of other shapes, for example square and triangular.

Some games of the Tafl family are played on the solitaire board, such as the English fox and geese, the French *jeu d'assaut* and the German *Belagerungsspiel*. In this group of games, in which a small but mobile minority faces superior numbers of slower-moving opponents (David and Goliath), there is usually a strategy by which the minority may win. Such games have enjoyed great popularity since the nineteenth century; their roots lead back to Old Icelandic sagas (*circa* AD 1000). Whether all these games share a common origin is unknown. Fox and geese can also be played on other boards, including the chessboard, which version appears in the *Codex Alfonso* under the name of *el inego de cercor la liebre*.

5 DICE

Dice have been used by all peoples at all periods, and have served not only for gaming but also for other purposes (oracular, in drawing lots, *I ching*). Religious structures such as the altar at Delos (the Delian problem) and the Kaaba in Mecca are illustrations of the cult significance of the cube.

Dice have not always been cubic, with the dots on opposite sides adding to seven (a practice obligatory in the manufacture of dice since medieval times). Anything that can roll and yet be capable of arriving at a stable, identifiable position of rest can be employed, such as cowrie shells, bones (astralagus) or beans provided with markings. Dicing has not always been a matter of pure chance, since at one time, given the necessary skill, it was possible to intervene in the 'result' of a throw, should an unfavourable result appear to be on its way, or should it be a matter of being the first to count the numerical value of a throw.

Dice games, even with several dice, did not offer very much scope. In medieval France, before the rise of playing cards, they made do with such simple games of chance as *le drinquet*, in which the dice had to land on a predetermined colour of a checkerboard, and *les plus poinz*, which had as its aim simply to roll a higher number of spots than the other players. Playing cards, in Europe from about 1380 and originating in Italy, considerably augmented the number of chance initial positions possible.

Between 1750 and 1800, some pieces of music were published (C. P. E. Bach 1754; Mozart 1793) in which minuets and waltzes were in the strictest sense of the word 'composed' with the aid of dice from a given set of bars (Kirnberger *1757*).

From dice to cubes is a small step. In 1921, P. MacMahon introduced the idea of matching coloured cubes. This three-dimensional idea has spread since then to plane figures (e.g. regular polygons), making possible problems based on the domino idea (MacMahon *1921*; collection of problems in Hardy *1983*). In 1936, P. Hein invented the Soma cube, whose seven pieces are made from three or four cubes joined along their faces. The idea of constructing a solid body out of components which are themselves built up from the basic structure has been generalized. In place of cubes, other bodies – spheres, right parallelepipeds, Platonic solids, regular octagonal prisms – have been employed. The puzzler is asked to reconstitute given solids, or to prove the impossibility of reconstituting them.

The concept was 'reduced' to one plane by S. W. Golomb in 1954 in polyominoes, and in 1961 by T. H. O'Beirne in polyaminoes (Golomb *1965*). Polyominoes are contiguous arrangements of squares. The best-known set of polyominoes is pentominoes, in which all twelve parts are built up of five squares. The checkerboard puzzle of H. Luers (1880) could be regarded as the forerunner of pentominoes, since it contains all of them. Polyiamonds uses triangles instead of squares. Here too there are forerunners, in T. Scrutchin's puzzle (1908) for example, built on equilateral triangles. The most interesting set of polyiamonds is the twelve hexiamonds of six equilateral triangles. The number of problems possible with these sets is legion.

6 DOMINOES

Sources (of uncertain validity, as is usual with games) suggest that dominoes originated in China around 300 BC; the game can certainly be traced back to fifteenth-century Italy. Dominoes are small rectangular flat pieces with the upper face divided into two halves, each half marked with spots. The usual set of dominoes contains 28 pieces, marked from zero ('double blank') to six ('double six') spots, but there are sets which run up to nine spots, on 55 pieces (another source describes a set running to twelve spots, on 91 pieces). The game is played by matching the spots on one half of a domino against those on one half of another. Other rules may be added; in matador, the ends of dominoes matched in play must total seven.

Dominoes have proved a fruitful ground for plane puzzles. How many chains may be laid out according to the principal rule was asked in the *Nouvelles Annales de Mathématiques* in 1849. The answer was provided by Reiss *1871*, and in 1886 by Tarry: playing with a set of 28 dominoes, there are $28 \times 37 \times 64\,988\,160$ possibilities. Besides simple chains, other configurations such as Latin squares and magic squares can be formed. E. Lucas invented quadrilles – squares consisting of the halves of four dominoes on these half pieces have equal number of spots (Lucas *1882–94*: Vol. 2). H. Delannoy demonstrated that there are 34 such possibilities. O. S. Adler thought out a reconstitution game (1901): after a set of dominoes is laid out in a square, the number and arrangement of the spots is noted, but the disposition of the pieces themselves is ignored. The problem is then to reconstitute the boundaries of the dominoes. Dominoes may be regarded also as polyominoes in which the number of spots stands for a colour.

7 PUZZLES IN DISSECTION AND COMBINATION

Superficially, plane puzzles in dissection and combination resemble jigsaw puzzles. However, in the jigsaw puzzle, invented in 1760 by John Salisbury as an aid in acquiring familiarity with maps, and having perhaps as many as 15 000 pieces, the mechanical testing of myriad combinations is the essence of the exercise, whereas mathematical dissection and combination problems are not primarily concerned with sheer numbers of pieces.

Planimetric puzzles based on geometric figures have been known (and to the layman) for many centuries, as demonstrated by the medieval manuscript by Abu-l-Wafa dealing with the knowledge required of an artisan. In it is treated the problem of dividing three squares of equal size into components such that a new square can be formed from them. Geometrical proofs of Pythagoras's theorem follow this scheme, and have attracted interest outside the world of mathematics (e.g. US President James Garfield). In the third century BC there existed in Greece a complicated puzzle, the *stomachion*, attributed to Archimedes. The pedagogic utility of these puzzles has not been neglected (Williams *1817*, Kunze *1842*).

A prototype of the dissection/combination puzzle is recognizable in tangrams. Tangrams consist of a set of seven pieces cut out of a square (five triangles, a square and a rhomboid) which can be laid out to form prescribed designs. The game originated in China, but we have no trustworthy information on its origin. From Japan, a woodcut by Kitagawa Utamaro from the year 1780 shows tangram players; the earliest known reference in literature is Chinese and dates from 1813; but the game is evidently older. Its popularity can be gauged from the wide variety of Europeans and Americans who have owned sets of tangrams – they include Henry James, Thomas Alva Edison, Lewis Carroll, Ernest Dudeney and the German Kaiser Friedrich III. Towards the end of the nineteenth century, the German firm Anker produced tangram-like puzzles (36 as late as 1950), but here too there were predecessors of Chinese origin (Read *1965*).

In 1903, Sam Loyd published a book with around 700 tangrams, accompanied, in his usual way, by a story incorporating a 'history' of tangrams. Since then there have been collections of up to 1600 patterns, setting mathematical problems (Elffers *1976*). Besides the reconstitution of given patterns, there are also problems requiring proof of impossibility, as for example when all possible convex tangrams are sought (solution 13; Wang and Hsiung *1942*), and classification problems (convex hull).

1562

8 NIM

A general description of nim would be 'any game in which the players take turns in drawing from piles of counters according to certain rules of play'. According to the rules, the player who draws the last counter may be either the winner or the loser. In a simple form of the game, slightly different in its make-up (Bachet *1612*), the players took it in turns to add from one to ten counters to a pile in a race to see who could add the hundredth. Then Bouton *1901* introduced the name 'nim' for the following game: two players take it in turns to draw any number of counters from three piles, but from one pile only in any particular move. He developed a theory on the initial circumstances under which the opening player or the following player can be sure of winning. The winning positions are easily memorized if the numbers of counters in each pile are noted in pairs.

This game has been variously generalized. W. A. Wythoff *1907* presented a variant for two piles: each player draws any number of counters, either from one or from both piles, in the second case drawing the same number from each. In 1931, world chess champion Emanuel Lasker enriched nim with the rule that in one move a pile should be divided in two. P. M. Grundy played nim with the rule that the existing piles should be thus divided for as long as this is feasible, and Claude Shannon required of nim that the number of counters drawn in any move should be a prime number. Conway *1976* contributed an original version called 'whim' in which the object, to 'win' or to 'lose', is undecided at the start of the game, and in the course of play the players must by one move determine in which way the game is to be played out.

The arithmetic nim games were clothed in geometric guise, for the first time, it seems, in 1907 in a game in which counters in a number of squares had to reach a goal. In this version, the number of squares distant from the goal represents the number of counters in a pile (Ahrens *1910–18*: Vol. 2). Turning an arrangement of coins to produce a particular head/tail distribution relates to problems in nim (Berlekamp *et al. 1982*). 'Kayles' is a game for which P. M. Grundy, R. K. Guy and C. A. B. Smith first delivered a complete analysis in the 1950s (Grundy and Smith *1956*), Dudeney, however, having examined a particular case in *The Canterbury Puzzles* in 1907. From a row of 'kayles', one or two neighbouring pieces must be removed. The winner is the person who 'bowls over' the last pin.

Numerous derivatives of nim are to be found in the book *Winning Ways* (Berlekamp *et al. 1982*) in which complete theories with new arithmetical operations (nim addition and nim sums, nim multiplication and nim products) are presented (Conway *1976*). A computer ('Nimatron') that will play a perfect game of nim was invented by E. U. Condon in about 1940.

Finally, mancalas are predecessors of a similar game in which counters are distributed under varying rules (see §1.8 on African mathematics).

9 DISPLACEMENT PUZZLES

In about 1880, a mathematical game spread around the world like wildfire, a sliding puzzle known as 'the fifteen puzzle', *jeu de taquin* in French, *Boss puzzle* in German (the 'cubomania' of the 1980s arising from Rubik's Cube bears comparison). The game is played on a square, edged so as to accept fifteen small square pieces numbered from 1 to 15, and thus leaving one position vacant. These pieces are laid out randomly on the board. The basic game is to arrange them in numerical order, starting with 1 at top left, and leaving the bottom right-hand corner vacant, by sliding pieces into the position at any one time left vacant. These are in all 16! different permutational possibilities (more than 2×10^{12}), only half of which can be arrived at by sliding the pieces according to the rule (Johnson *1879*, Story *1879*).

The inventor of the puzzle (probably invented in the USA in 1878) is unknown. It came onto the American market in 1879, and soon reached Europe. The (unsolvable) problem set by Loyd (in 1880 or 1881), calling for the numbers 14 and 15 to change places, particularly captured public attention.

The game was treated in papers by American and European mathematicians (Ahrens *1910–18*: Vol. 2), and by 1879 the Americans W. W. Johnson and W. E. Story had already produced an exhaustive theory of the permutation puzzle. The mathematical facts governing the theory rest on a theorem set by Etienne Bézout according to which changing the places of two elements of a permutation means changing the sum of the inversions by an odd number (Johnson *1879*, Story *1879*). Generalization of the game to other networks of polygons was also the subject of attention (Kowalewski *1930*, Wilson *1974*).

The enormous success of the fifteen puzzle spawned many other sliding puzzles, and the trend has continued to the present day. Hordern *1986* lists more than 150 patents granted in the USA during the period 1869–1978. The form of the pieces has undergone changes: from uniformly square pieces have developed rectangles (Dad's puzzle, 1909; Washington's career, 1920/30; *l'Âne rouge*, 1932), and L shapes (Ma's puzzle, 1927; Traffic jam puzzle, 1928; Climb pro 24 (474 moves!), 1985). Following Rubik's Cube, three-dimensional displacement puzzles appeared, in part old games in new guise (Varikon box, 1905 and 1974; Babylon tower, around 1980; Pionir cube, 1984).

The two questions that arise in all mathematical puzzles are first, whether a particular position can be arrived at from a given initial position; and second, how many moves are necessary. In many games, these questions remain unanswered, even in the age of computers.

Geometric dissections (i.e. the cutting of geometric figures into pieces that can be rearranged to form other figures) were introduced to recreational mathematics by Montucla *1754*, who called attention to these problems. Lindgren *1972* describes some of the most interesting dissections.

10 SOURCES

Play and games go back to the start of civilization, and therein lies the first difficulty: the lack of continuity in the written record, together with the frequent denigration of play and games as belonging to the realm of childish things, means that so much contemporary testimony has simply vanished among the refuse of culture. We are left with few sources, and many problems. It is only from the Middle Ages that there have survived details of rules of play, and here too the written documentation tends to conceal rather than reveal the basic ideas of half-forgotten games; the chroniclers, like ourselves, being too ill-informed about the rules.

The first extensive account is that of Alfonso the Tenth of Castile, in the *Codex Alfonso* (Alfonso *1283*). In it, he categorizes board games into games of chance, games of mental agility and games which combine both elements, as we would today. Accounts of mathematical elements in games can be found in earlier works on jurisprudence, strategy and ethics, as in the ancient Indian manual on political science by Kautilya, the *Arthaśāstra*, and in medieval Islamic manuscripts. There has always been an evident link between strategic thinking and planning and the group of war games, from the forerunners of chess (chaturanga) to the Colonel Blotto of the modern theory of games.

In the medieval monasteries, collections of problems were compiled which would correspond to the recreational mathematics of modern times. One such is the *Propositiones ad acuendos iuvenes* ('Whetstones for the Budding Intellect'), attributed to Alcuin of York (735–804). The later collection of Bachet de Méziriac (*1612*) naturally takes things a stage further, and presents nearly all games known to that time.

That mathematicians were well aware of the link between games and mathematics can be demonstrated by Leibniz's classification *1710* of games, and the interest in chess shown by Leonhard Euler and Carl Friedrich Gauss. In more modern times, there have been various classic compilations of mathematical games (Lucas *1882–94*, Ball *1892*, Ahrens *1910–18*); more recent are the publications by Martin Gardner, and Berlekamp, Conway and Guy's *Winning Ways* (Berlekamp *et al. 1982*). See also the extensive bibliography on recreational mathematics in the next article.

BIBLIOGRAPHY

Ahrens *1910–18*, Berlekamp *et al. 1982* and Elffers *1976* each contain a substantial bibliography.

Ahrens, W. *1910–18, Mathematische Unterhaltungen und Spiele*, 2 vols, Leipzig: Teubner. [Also later editions.]

Alfonso el Sabio *1283, Libros de acedrex dados e tablas. Codex Alfonso, Escorial Madrid*. [Published 1941, A. Steiger (ed.), Geneva and Zurich: Droz, with German transl.].

Bachet de Méziriac *1612, Problesmes plaisans et délectables qui se font par les nombres*, Paris. [2nd edn 1624, Lyon; also later edns.]

Ball, W. W. Rouse *1892, Mathematical Recreations and Problems of Past and Present Times*, London: Macmillan. [12th edn 1974, Toronto: University of Toronto Press. On the various editions, see the entry in the bibliography of §12.3.]

Bass, T. A. *1990, The Newtonian Casino*, London: Longman.

Beasley, J. D. *1985, The Ins and Outs of Peg Solitaire*, Oxford: Oxford University Press.

Berlekamp, E. R., Conway, J. H. and Guy, R. K. *1982, Winning Ways for Your Mathematical Plays*, London: Academic Press. [German transl. 1985–6 as *Gewinnen*, 4 vols, Braunschweig: Vieweg.]

Borst, A. *1986, Das mittelalterliche Zahlenkampfspiel*, Heidelberg: Winter.

Bouton, C. L. *1901*, 'Nim, a game with a complete mathematical theory', *Annals of Mathematics*, **3**, 35–9.

Conway, J. H. *1976, On Number and Games*, London: Academic Press. [German transl. 1983 as *Über Zahlen und Spiele*, Braunschweig: Vieweg.]

Eigen, M. and Winkler, R. *1975, Das Spiel, Naturgesetze steuern den Zufall*, Munich: Piper.

Elffers, J. *1976, Tangram*, Cologne: du Mont. [In Dutch and German.]

Evans, G. R. *1976*, 'The rithmomachia: A mediaeval mathematical teaching aid?', *Janus*, **63**, 257–73.

Golomb, S. W. *1965, Polyominoes*, New York: Scribner's.

Grundy, P. M. and Smith, C. A. B. *1956*, 'Disjunctive games with the last player losing', *Proceedings of the Cambridge Philosophical Society*, **52**, 527–33.

Hardy, R. *1983, Gry w figury*, Warsaw: Krajowa Agencja Wydawnicza.

Hordern, L. E. *1986, Sliding Piece Puzzles*, Oxford: Oxford University Press.

Johnson, W. W. *1879*, 'Notes on the "15" puzzle 1', *American Journal of Mathematics*, **2**, 397–9.

Kirnberger, J. P. *1757, Der allezeit fertige Polonoisen- und Menuettencomponist*, Berlin: Winter.

Kowalewski, G. *1930, Alte und neue mathematische Spiele*, Leipzig: Teubner.

Kraitchik, M. *1943, Mathematical Recreations*, London: Allen & Unwin.

Kunze, C. L. A. *1842, Das geometrische Figurenspiel*, Weimar: Böhlau.

Leibniz, G. W. *1710*, 'Annotatio de quibusdam ludis [...]', *Miscellanea Berolinensia*, 22–6. [Also in *Opera omnia Leibnitii*, Vol. 5 (ed. Dutens), 1768, Geneva, 203–5.]

Lindgren, H. *1972, Recreational Problems in Geometric Dissections and How to Solve Them*, New York: Dover.

Loyd, S. *1903, The Eighth Book of Tan*, New York: Loyd. [Repr. 1968, New York: Dover.]

Lucas, E. *1882–94, Récréations mathématiques*, 4 vols, Paris: Gauthier-Villars. [Repr. 1960, Paris: Blanchard.]

MacMahon, P. A. *1921, New Mathematical Pastimes*, Cambridge: Cambridge University Press.

Montucla, J. E. *1754, Histoire des recherches sur la quadrature du cercle*, Paris: Agasse.

Murray, H. J. R. *1952, A History of Board-games Other than Chess*, Oxford: Clarendon Press.

Ozanam, J. *1694, Récréations mathématiques*, Paris: Jombert [Enlarged edn by E. Montucla, 1750, Berlin: Rollin. English transl. 1814 as *Recreations in Mathematics* [...], London: Murray.]

Read, R. C. *1965, Tangrams*, New York: Dover. [German transl. 1985 as *Tangram*, Munich: Hugendubel.]

Reiss, M. *1871*, 'Evaluation du nombre de combinaisons [...]', *Annali di matematica*, Series 2, **5**, 63–120.

Silvermann, D. L. *1971, Your Move*, New York: McGraw-Hill. [German transl. 1972 as *Spielend denken – denkend spielen*, Munich: Moderne Verlag.]

Slocum, J. and Botermans, J. *1986, Puzzles Old and New, How to Make and Solve Them*, Seattle, WA: University of Washington Press. [German transl. 1987 as *Geduldspiele der Welt*, Munich: Hugendubel.]

Smith, D. E. *1923, History of Mathematics*, Vol. 1, Boston: Ginn. [Repr. 1958, New York: Dover.]

Story, W. E. *1879*, 'Notes on the "15" puzzle 2', *American Journal of Mathematics*, **2**, 399–404.

Thiele, R. *1984, Die gefesselte Zeit*, Leipzig: Urania. [Also published as *Das grosse Spielvergnügen*, 1984, Munich: Hugendubel.]

von Neumann, J. and Morgenstern, O. *1944, Theory of Games and Economic Behavior*, 1st edn, Princeton, NJ: Princeton University Press.

Wythoff, W. A. *1907*, 'A modification of the game of Nim', *Nieuw Archief voor Wiskunde*, Series 2, **7**, 199–202.

Wang, F. T. and Hsiung, C.-S. *1942*, 'A theorem on the tangram', *American Mathematical Monthly*, **49**, 596–9.

Williams, W. *1817, New Mathematical Demonstrations of Euclid* [...] *with* [...] *Commonly Called the Chinese Puzzle*, London: Maynard.

Wilson, R. *1974*, 'Graph puzzles, homotopy and the alternating group', *Journal of Combinatorial Theory*, series B, **16**, 86–96.

12.3

Recreational mathematics

DAVID SINGMASTER

1 THE NATURE OF RECREATIONAL MATHEMATICS

To begin with, it is worth considering what is meant by recreational mathematics. An obvious definition is that it is mathematics that is fun; but many mathematicians claim they do mathematics because it is fun, so this definition would encompass almost all mathematics and hence is too general. There are two, somewhat overlapping definitions that cover most of what is meant by recreational mathematics.

First, recreational mathematics is mathematics that is fun and popular – that is, the problems should be understandable to the interested layman, though the solutions may be harder. However, if the solution is too hard, this may shift the topic from the recreational toward the serious – e.g. Fermat's last theorem (§6.10), the four-colour problem (§7.10) or the Mandelbrot set (§3.8).

Second, recreational mathematics is mathematics that is fun and used either as a diversion from serious mathematics or as a way of making serious mathematics understandable or palatable. These are the pedagogic uses of recreational mathematics. They are already present in the oldest-known mathematics and continue to the present day.

Problem 79 of the Rhind Papyrus (§1.2) leads to adding 7 + 49 + 343 + 2401 in a context similar to the 'As I was going to St Ives' children's rhyme, which is known in several European countries. Although there is some question as to whether this problem is really a fanciful exercise in summing a geometric progression, it has no connection with other problems in the papyrus and seems to have been inserted as a diversion or recreation.

Old Babylonian tablets give fanciful problems leading to quadratic equations. The one known as AO 8862 describes a field where the length plus the width is known, and the area plus the difference of the length and the width is known. This can hardly be considered a practical problem; rather it is a

1568

way of presenting two equations in two unknowns which should make the problem more interesting for the student.

These two aspects of recreational mathematics – the popular and the pedagogic – overlap considerably, and there is no clear boundary between them and 'serious' mathematics. In addition, there are two other independent fields which contain much recreational mathematics: games and mechanical puzzles.

Games of chance and games of strategy seem to be about as old as human civilization (§12.2). The mathematics of games of chance began in the Middle Ages, and its development by Pierre de Fermat and Blaise Pascal rapidly led to probability theory (§10.2). The mathematics of games of strategy started only around the beginning of the twentieth century, but soon developed into game theory (§6.11).

Mechanical puzzles range widely in mathematical content. Some require only a certain amount of dexterity; others require ingenuity and logical thought; others require systematic application of mathematical ideas or patterns, such as Rubik's Cube or the Chinese rings. The earliest surviving mechanical puzzles seem to be Phoenician puzzle-jugs from about 1500 BC found in Cyprus; such jugs have been around ever since. The Loculus of Archimedes is a set of 14 pieces which can be assembled into various shapes – an elephant, a boat, and so on, like the more recent seven-piece tangrams. The Loculus was known to Archimedes, and is mentioned in Classical literature until the sixth century AD. Legend ascribes the Chinese rings to about AD 200, but the earliest records seem to be from the Sung Dynasty (960–1279).

This outlines the conventional scope of recreational mathematics; there is some variation according to personal taste.

2 THE UTILITY OF RECREATIONAL MATHEMATICS

Why is recreational mathematics of interest to the historian of mathematics? First, recreational problems are often the basis of serious mathematics. The most obvious fields are probability and graph theory, where popular problems have been a major (or the dominant) stimulus to the creation and evolution of the subject. Number theory, topology, geometry and algebra have also been strongly stimulated by recreational problems. Although geometry has its origins in practical surveying, the Greeks treated it as an intellectual game and much of their work must be considered as recreational in nature, although they viewed it more seriously as reflecting the nature of the world (§1.3). From the time of the Babylonians, algebraists tried to solve cubic equations, though they had no practical problems which led to cubics. There are even recreational aspects of calculus, for example the great variety of curves studied since the sixteenth century. Consequently, the study of recreational topics is necessary to understanding the history of many, perhaps most, topics in mathematics.

Second, recreational mathematics has frequently turned up ideas of genuine but obscure utility. The Königsberg bridge problem (§7.13) and Hamilton's Icosian Game have led to basic techniques of combinatorial optimization, and the Icosian Game led also to the first presentation of a group in terms of generators and relations. Knots have recently become of interest in molecular biology because DNA can form closed loops which may be knotted. The Chinese-rings puzzle utilizes the binary coding known as the Gray code, patented by Bell Laboratories in 1953. The Möbius strip has been patented several times. Penrose's pieces (§12.8) have led to the discovery of a new kind of solid – the 'quasicrystals'. Such unusual developments, and the more straightforward developments of the previous paragraph, demonstrate the historical principle of 'the unreasonable utility of recreational mathematics'. This and similar ideas are the historical and social justification of mathematical research.

Third, because recreational problems are often of great age and can usually be clearly recognized, they serve as useful historical markers, tracing the development and transmission of mathematics (and culture in general) in place and time. The Chinese remainder theorem, magic squares (§7.13), the cistern problem and the 'hundred fowls' problem are excellent examples of this process. (The original 'hundred fowls' problem, from fifth-century China, has a man buying 100 fowls for 100 *cash*, cocks cost 5 *cash*, hens 3 *cash*, and chicks are 3 for a *cash* – how many of each did he buy?) The number of topics which have their origins in China or India is surprising, and emphasizes our increasing realization that modern algebra and arithmetic may derive more from Babylonia, China, India and the Arabs than from Greece. Such easily understood examples are extremely useful in conveying the antiquity and the cultural diversity of the origins of mathematics, particularly to school students, school teachers and the general public.

Finally, a more pedagogic reason – so obvious that it might be overlooked – is that recreational mathematics provides a resource of problems which have demonstrably engaged the interest of students for many years and will continue to do so.

3 COLLECTING INFORMATION

Recreational mathematics presents many difficulties to the historian. The range of time and cultures to be surveyed means that few (possibly no one) can read all the relevant languages. Much material is in manuscript form, in libraries throughout Europe, Asia and America, so it is fortunate that many editions – facsimiles, transcriptions and/or translations – have appeared and are available, though much material remains difficult to access.

A different and perhaps more serious difficulty is that much of the relevant history is not be be found in the usual kinds of source. Relevant material appears in the many books on puzzles and games, but also in books on chess, magic, entertaining, things to do, scientific recreations, woodworking, and so on, often aimed at children. Many of these books had limited publication. Few academic libraries collect such books, and many in the British Museum's holdings were destroyed in the Second World War. The University of Calgary in Canada is the first library to start collecting such books seriously, beginning with the acquisition of the Strens Collection of some 2200 books on recreational mathematics in 1983, at the instigation of Richard Guy, who also assisted in financing the purchase. (A catalogue of this collection is available from the University.) The collections of several mathematicians, notably C. W. Trigg, have since been left to Calgary, and Martin Gardner intends to donate his collections and files. There are presently a handful of puzzle collectors who are collecting historical material of interest.

Some older and not so obvious works of relevance are the *I Ching*, the *Confucian Analects*, *The Greek Anthology* and the *Annales Stadenses* (from thirteenth-century Germany). Because some recreational topics had religious or mystical significance, particularly magic squares and Latin squares, they turn up in academic journals such as *History of Religions* or *Der Islam*, and also in journals on anthropology and folklore. Important material appears in general magazines such as the *Ladies' Diary*, *Boy's Own Paper*, *American Agriculturist*, *Tit-Bits*, *Strand Magazine*, *The Woodworker*, *Industrial Arts*, *Hugard's Magic Monthly*, *Sphinx* and *Our Puzzle Magazine* (this is so little known that it is not certain how many issues appeared), as well as newspapers such as the *Brooklyn Daily Eagle* and the *Weekly Dispatch*. In two cases, early evidence is found in works of art; the earliest evidence for Solitaire includes manuscripts by Gottfried Wilhelm Leibniz from the late seventeenth century (de Mora-Charles *1992*); one of the earliest depictions of tangrams in the Orient is a 1780 woodcut by Kitagawa Utamaro.

An important source is the patent literature, which is very accessible, but very voluminous. Patents must be used with care as the issuing of a patent is no guarantee of its novelty or validity: they have been issued for ideas which had been published in the same form several times beforehand. As an extreme example, the Gray code was patented in 1953, though the pattern was known a millennium earlier and its application to communications had already been made by J. E. Baudot in 1873.

An even greater difficulty for the historian is that much important material is ephemeral. Puzzles were made and sold, and the instructions or packaging – if they survive – are often the earliest printed material about them. Often, such material does not even indicate the manufacturer's name, much less the place or date; dating must often be done from the typography. Sometimes the earliest sources are advertisements or catalogue entries for the item. The 1801–7 catalogue of the Nuremberg toy manufacturer G. H. Bestelmeier is the earliest source for several puzzles – sadly, no other such catalogues seem to have survived from before the 1880s. Other puzzles were made into advertising cards or premiums. Some broadsheets of puzzles, called 'Nuts to Crack', were produced in London somewhere between 1840 and 1860. At least thirty of these appeared, and they contain much material which was new or relatively new at the time. Other material may turn up in due course.

However, the greatest difficulty is that relevant material often does not exist. The first known appearance of a problem usually shows it already fully developed, and sometimes even described as an old problem! This may be because the author is the actual creator of the problem, but most times it is clear that he is not the creator, sometimes stating this explicitly. It is rare, but illuminating, to find a problem which is half-formed and even rarer (though very satisfying) to find enough early material to see how the problem evolved from previous ideas. Perhaps the earlier versions never made it into print.

In recent times, a problem-creator may describe his invention of a problem, but faulty memories and large egos can make such reminiscences extremely unreliable. Sometimes it works the other way in that the creator wishes to preserve the mystery of the source (E. Lucas never admitted in writing that he was the inventor of the tower of Hanoi), or to avoid being associated with such trivia. Or perhaps the creator did not really appreciate the originality of the creation, which may have rapidly developed to the point where he did not recognize his own offspring. H. Davenport seems to have invented the birthday problem (how big does a group of people have to be for the probability of any two of them having the same birthday to exceed 50%?), but denied it, possibly because his version was slightly different or because he thought it was so natural that someone else must have posed it before. Evidence is generally sparse in such cases. Recreational mathematics even has its own priority battle between the supporters of Sam Loyd and of Henry Dudeney over which of them invented certain puzzles. The dispute is hampered by lack of evidence, but books are being prepared on both Loyd and Dudeney which should help clarify the situation.

1572

Even for items produced within living memory, information can be hard to find. With the twelve coins problem, for which the earliest printed reference seems to be definitely known (January 1945), its author, E. D. Schell, revealed that he had not invented the problem and that the journal in which is was published, the *American Mathematical Monthly*, had considerably modified his proposal. Despite extensive discussion of the problem in the literature in 1945–1950 and later, no one has claimed the invention of the problem, possibly because it developed rapidly through several stages such that no one contributor felt like the real inventor.

The phenomennon of a problem appearing fully developed is common for older problems, but with them there is often also uncertainty as to how the problem was subsequently transmitted. The 'St Ives' type problem in the Rhind Papyrus turned up next with Fibonacci – a gap of about three thousand years! Tangrams are often claimed to be several thousand years old, but they first appear, essentially simultaneously, in China and in Europe in about 1800, a slightly different version having appeared in Japan in the mid-eighteenth century. The more complex Loculus of Archimedes is the same basic idea, but is only known from the Mediterranean countries up to the sixth century AD (though there are two seventeenth-century Arabic manuscripts of Archimedes on it). A monkey-and-coconuts problem appears in ninth-century India, but its next appearance is in eighteenth-century France. The Josephus problem, usually in the form of counting out every ninth person in a circle until 15 of the original 30 remain, is well known from about the ninth century in Europe, and appears in very similar forms in Japan, certainly in 1627 and traditionally in about 1159.

As mentioned in Section 2, the history of recreational mathematics clearly shows more transmission from China to India and the West than has been generally recognized. There are also examples of apparent transmission between the West and Japan and/or China, in both directions, which has left no trace in India or the Arab world. Perhaps the transmission of popular ideas happened along the Silk Routes. Investigation of Central Asian folklore may demonstrate this, but there may well be no surviving evidence.

4 PRINCIPAL LITERATURE

Because recreational mathematics encompasses so many diverse topics, no general history has been attempted. Smith *1923* and Sanford *1927, 1930* have traced the history of some topics. The recreational works of Lucas *1882–94*, Ball *1892* and Gardner *1956–* contain much incidental historical information, but the most extensive attempt to date has been by Ahrens *1910–18*. The recent appearance of numerous editions of old texts has greatly clarified the history of many topics, though actual origins tend to remain elusive. The most notable work has been that of Joseph Needham and his colleagues in opening up the history of Chinese mathematics, that of Kurt Vogel and his colleagues in publishing early texts and correlating their problems, and that of Gino Arrighi in publishing medieval Italian texts. Vogel, Reich and Gericke's rewriting of Tropfke's *Geschichte der Elementarmathematik*, of which one volume, *Arithmetik und Algebra* (Tropfke *1980*), has so far appeared, contains many sections on recreational problems, with extensive outlines of their history. Since 1955, W. L. Schaaf has been compiling bibliographies of recreational mathematics, concentrating on covering all current articles; four volumes appeared (Schaaf *1970–78*).

In 1982 I embarked on 'Sources in recreational mathematics – An annotated bibliography'; five preliminary editions have been circulated since 1986, and the sixth is now in preparation (Singmaster *1986–*). This is more historical than Schaaf's work and attempts to trace the history of each topic by selecting and briefly describing the important material. Detailed references for all the sources mentioned in this article (and many more) can be found in it.

BIBLIOGRAPHY

Ahrens, W. *1910–18*, *Mathematische Unterhaltungen und Spiele*, 2nd edn, 2 vols, Leipzig: Teubner. [1st edn 1901. The 3rd edn of Vol. 1 (1921) is a reprint of the 2nd edn with 2 pp of extra notes.]

Ball, W. W. Rouse *1892–*, *Mathematical Recreations and Essays*, 1st–14th edns [1st–3rd edns entitled *Mathematical Recreations and Problems*; 11th–13th edns revised by H. S. M. Coxeter], London: Macmillan for 1st–11th edns; Toronto: University of Toronto Press for 12th and 13th edns; New York: Dover for 14th edn. [There were also two French edns, the 2nd in 3 vols, 1907–09, with much additional material.]

Gardner, M. *1956–* 'Mathematical games', column in *Scientific American*. [Although Gardner stopped doing these regularly in December 1980, he has contributed occasional columns since. He has also collected and extended these columns in numerous books, starting with *The Scientific American Book of Mathematical Puzzles and Diversions*, 1959, New York: Simon & Schuster; UK edn *Mathematical Puzzles and Diversions from Scientific American*, 1961, London: Bell. Several of the earlier collections have recently been revised and some have been retitled.]

Lucas, E. *1882–94*, *Récréations mathématiques*, 4 vols, Paris: Gauthier-Villars. [2nd edns of Vol. 1 (1891), Vol. 2 (1893). Repr. several times by Blanchard (Paris).]

de Mora-Charles, S. *1992*, 'Quelques jeux de hazard selon Leibniz [. . .]', *Historia mathematica*, **19**, 125–57.

Sanford, V. *1927, The History and Significance of Certain Standard Problems in Algebra*, New York: Teachers College, Columbia University (Contributions to Education, No. 251). [Repr. 1972, New York: AMS Press.]

—— *1930, A Short History of Mathematics*, Boston, MA: Houghton Mifflin.

Schaaf, W. L. *1970–78, Bibliography of Recreational Mathematics*, 4 vols, Washington, DC: National Council of Teachers of Mathematic. [Vol. 1 has four editions: 1955, 1958, 1963, 1970.]

Singmaster, D. *1986–*, 'Sources in recreational mathematics – An annotated bibliography'. [Available from the author, South Bank University, London, SE1 0AA, UK.]

Smith, D. E. *1923, History of Mathematics*, 2 vols, New York: Ginn. [Repr. 1958, New York: Dover.]

Tropfke, J. *1980, Geschichte der Elementarmathematik*, Vol. 1, *Arithmetik und Algebra*, 4th edn (revised by K. Vogel, K. Reich and H. Gericke), Berlin: De Gruyter.

12.4

The golden number, and division in extreme and mean ratio

ROGER HERZ-FISCHLER

Two distinct expressions for the same mathematical concept are used in this article in order to distinguish clearly between two aspects of the subject. The expression 'division in extreme and mean ratio' (DEMR) is used when talking about the strictly mathematical history of the subject, whereas the expression 'golden number' (GN) is used when talking about statements which assert that the concept is somehow related to a natural phenomenon or a man-made object. The distinction leads to a better understanding of the historical and, in the second case, of the philosophical and sociological aspects of the subject.

1 DIVISION IN EXTREME AND MEAN RATIO

1.1 Definition, origin, use

The basic concept of DEMR, and the Greek terminology of which 'division in extreme and mean ratio' is a translation, is introduced in Book VI (Def. 3) of Euclid's *Elements* (*circa* 300 BC). A line AB is to be divided at a point C in such a way that $AB:AC = AC:CB$; that is, so that the ratio of the whole line to the larger segment equals the ratio of the larger segment to the smaller segment (Figure 1). Although the association of a number with this ratio is foreign to the *Elements*, note that the common ratio is the positive solution $G = (1 + \sqrt{5})/2$ of the equation $x^2 - x - 1 = 0$.

A   C   B

Figure 1 Division in extreme and mean ratio:
the basic concept – AB:AC = AC:CB

1576

Following this definition, we are shown in VI: 30 how to go about dividing a line in EMR. It is at this point that the question of the historical origins of the concept becomes interesting, for it turns out that there is another construction involving the division of a line in EMR in Book II. Whereas the construction in Book VI uses concepts from the theory of proportions developed in Book V, the earlier construction is stated in the 'geometry-of-areas' language of Book II. Thus it is now stated that we are to find a division point C such that the area of the square with side AC (the larger segment) has the same area as the rectangle with sides AB (the whole line) and CB (the smaller segment) (Figure 2). Since the two constructions are different, one must enquire not only about the dates of and the reasons for their introduction, but also why there were two constructions.

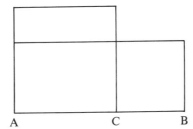

Figure 2 Division in extreme and mean ratio: the geometrical interpretation – the rectangle on AB equals in area the square on AC

Some authors connect the origins of DEMR with the quadratic equation $x^2 + ax = a^2$. If we think of a as the length of the line, then the positive solution of this equation is $x = a/G$. The relationship of DEMR to equations is made either via II: 5, 6 (the latter being used in II: 11), which have been interpreted as being algebraic identities in geometrical dress, or via VI: 28, 29 (the latter being used in VI: 30), which are theorems dealing with what is referred to as 'applications of areas'. Other authors suggest that the concept of DEMR arose in connection with the concept of incommensurability or in connection with operations associated with the Euclidean algorithm.

To delve further into the historical development of DEMR, one must find where the concept is used elsewhere in the *Elements*. The first place is in IV: 10, 11, where the regular pentagon is constructed using the area formulation of II: 11. After Book IV, DEMR does not appear again – except for VI: def. 3, 30 – until Book XIII, whose main goal is the construction of the five regular polyhedra. DEMR, and results related to it, are used in the construction of the icosahedron (XIII: 16) and dodecahedron (XIII: 17), both of which involve the pentagon. In particular, in XIII: 8 we learn that the diagonals of a pentagon cut each other in EMR, the larger segment being equal to the side of the pentagon.

Some authors associate the development of DEMR with the discovery of this last result. In support of this conclusion they cite ancient sources which state either that the pentagram was a symbol of the Pythagoreans or that members of the sect were associated with the construction of the dodecahedron.

For reasons that can be found in Herz-Fischler *1987*, I favour the view that what we find in Euclid represents the historical process: that the origins of the concept of DEMR date from attempts by early mathematicians to rigorously construct the pentagon. The earliest form would be that of II: 11, the definition and construction of Book VI having been added in the context of the theory of proportions. I also suggest that the discovery is due to Theaetetus and his school, independent of Plato's Academy, in about 386 BC.

1.2 The post-Euclidean Greek period (300 to 350 BC)

The Greek world after Euclid produced several mathematical developments involving DEMR. The most impressive appear in the 'Supplement' to the *Elements*. This text, often called Book XIV, is associated with Hypsicles, but appears to contain work by Aristaeus and Apollonius; the sources and chronology are very tentative (§1.3). The main result is theorem 2, which states that if an icosahedron and a dodecahedron are inscribed in the same sphere, then the circle circumscribing a triangular face of the icosahedron has the same radius as the circle which circumscribes the pentagonal face of the dodecahedron. An early Greek manuscript of a work related to Book XIV seems to have contained the result, similar in nature to XIII: 9, that when the radius of a circle is divided in EMR then the larger segment is the side of the decagon (Herz-Fischler *1988*).

In the works of Hero and Ptolemy we begin to find numerical values associated with DEMR. Hero's work is interesting because of the approximations – often unexplained – which he uses to find the areas of the pentagon and decagon. Hero also gave the method shown in Figure 3 of dividing a line in EMR ($DB = \frac{1}{2}AB$). In the *Collection* of Pappus we find new constructions of the icosahedron and dodecahedron. Contained in the latter construction is an absolutely wonderful proof of theorem XIV: 2 cited above. Pappus also compares the volumes of the icosahedron and dodecahedron and in the process presents a series of lemmas, some of which may be much older, involving properties of DEMR.

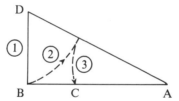

Figure 3 Hero's method of dividing a line in EMR

1.3 The Arabic world and India

In the *Algebra* of al-Khwārizmī and that of Abū Kāmil are equations whose solution involve DEMR, but since the concept itself is not introduced we cannot be sure that this is not accidental. Abū Kāmil's *On the Pentagon and Decagon* contains 20 problems involving calculation of the areas, sides and diameters, though surprisingly sometimes without the use of properties of DEMR. Various trigonometric calculations involving the pentagon and decagon appear in Arabic and Indian texts.

1.4 Fibonacci and the 'Fibonacci sequence'

Evidence of a great mastery of the properties of DEMR is found in the works of Leonardo of Pisa (Fibonacci). There are planar computations involving the pentagon and decagon, and volume calculations for the icosahedron and dodecahedron, as well as a reworking of Abū Kāmil's problems.

Fibonacci is probably best known for his 'rabbit problem'; a translation is given in Struik (*1969*: 2). We start with one pair of rabbits who give birth to a new pair from the first month on, and every succeeding pair gives birth to new pair in the second month after their birth. Fibonacci shows that this leads to the sequence 1, 2, 3, 5, 8, 13, 21, 34, 55, 89, 144, 233, 377, ..., 'and in this way you can do it for the case of infinite numbers of months'. The rule of formation of the 'Fibonacci numbers' is $f_{n+2} = f_n + f_{n+1}$, and because of this it turns out that $\lim_{n \to \infty} (f_{n+1}/f_n) = G$. There is, however, absolutely no indication that Fibonacci was aware of this property. The first definite indication of a knowledge of this fact is an annotation in an early sixteenth-century copy of the *Elements* (Curchin and Herz-Fischler *1985*). This result was also known to Johannes Kepler (1608), Albert Girard (1634) and Thomas Simpson (1753). The explicit expression

$$f_n = \frac{1}{\sqrt{5}} \left[G^{n+1} - \left(-\frac{1}{G} \right)^{n+1} \right], \quad n \geqslant 1 \tag{1}$$

for f_n in terms of G was apparently first published by J. P. M. Binet in 1843.

Other aspects of the history of results involving the Fibonacci numbers are given in Dickson (*1919*: Chap. 17).

1.5 Europe

After Fibonacci there is a hiatus until the fifteenth century, with the works of Piero della Francesca and Rafael Bombelli. They both contain various calculations involving DEMR, including very interesting volume and surface computations for the icosahedron and dodecahedron. The 1509 work by Luca Pacioli, *Divina proportione* (an expression referring to DEMR), is eloquent in its exaltation of the *mathematical* properties of DEMR ('inestimable', 'ineffable' . . .), but there is only a low-level discussion of the mathematics itself. The tradition of praising the mathematical properties of DEMR goes back to the thirteenth century when Campanus of Novara, in his edition of the *Elements*, wrote that it was worthy of the philosophers' admiration. We find similar remarks in later texts, including those of Kepler.

2 GOLDEN NUMBERISM

2.1 Examples and early history

The expression 'golden numberism' refers to the ensemble of doctrines and claims relating the GN to natural phenomena or the use of the GN in human constructs. Examples of these claims can be found in Zeising *1854* and Röber *1855*, the cofounders of golden numberism. Zeising saw in the GN a primary morphological quantity to which various dimensions (e.g. the relative position of the navel) of the human body were related. Röber claimed that the GN was used to design almost all the pyramids of Egypt (but not the Great Pyramid). In Figure 4 Röber is claiming, via a geometrical construction, that the theoretical angle of inclination of the side of the pyramid is given by $\sec \alpha = G$.

The earliest examples that can be said to be a form of golden numberism are found in the works of Kepler. These are his use (1596) of the icosahedron and the dodecahedron, and thus the GN, in his polyhedral model of the Solar System, and perhaps (interpretation is difficult because of other remarks) his statement (1608) that the ratio of the orbital periods for Earth and Venus 'comes very close to the divine proportion', and some remarks (1608, 1619) relating the 'laws of generation' to the GN.

After Kepler is the claim made in the second edition (1799) of the Montucla–Lalande *Histoire des mathématiques* that Pacioli had advocated the use of the GN in determining the proportions of works of art and

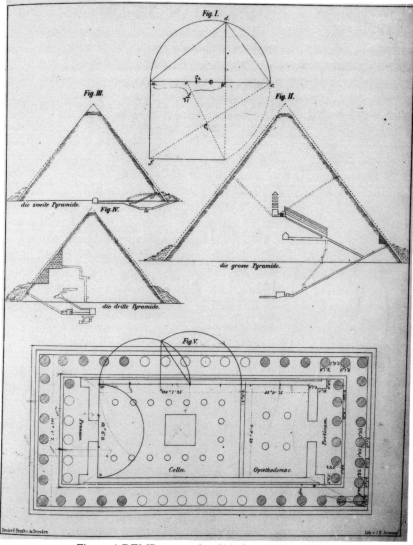

Figure 4 DEMR meets the GN (from Röber *1855*)

architecture. Not only is this oft-repeated claim false, but, to the contrary, Pacioli explicitly states that simple ratios are best.

Thus, historically the situation is this: there does not seem to be the slightest piece of documentary evidence that anyone before Zeising and Röber, apart from Kepler and Montucla/Lalande, wrote or did anything which could be interpreted as golden numberism. As far as the name 'golden number' and related forms ('section', 'cut', and so on) are concerned, the earliest attested use – albeit in a strictly mathematical context – is by Martin Ohm in 1835.

2.2 Propagation and later history

The 'success' of golden numberism seems to be due to several factors. One is perhaps the connotations of the adjective 'golden', which is also used in many other contexts. The geometrical construction, which is simple yet not mundane, and the defining quadratic equation, with its irrational and closed-form solution, were accessible to the general public. The infinite-reproduction property of XIII: 5 (if we divide a line in EMR and add the larger segment, we again have a line divided in EMR) and the related connection with the Fibonacci numbers also increase its mathematical and exotic appeal. A scientific basis for golden numberism has been claimed by means of references to aesthetic experiments, usually misinterpreted in the GN literature, performed in the 1870s by the pyscho-physicist Gustav Fechner. Research has indicated that there is simply no GN basis for the very complicated processes involved in aesthetic preference (Zusne *1970*: 399).

Golden numberism seems to have been limited to Germany until the early part of this century, but after about 1910 it spread rapidly to other countries as 'a sudden and devastating disease which has shown no signs of stopping' (Cook, 1922). An impetus was given in the USA by J. Hambridge ('dynamic symmetry', *circa* 1920) and in France by M. Ghyka (from 1927). Under the latter's influence the architect Le Corbusier, previously anti-GN, published drawings indicating that an already constructured building – for which he had used another system – had been designed using the GN! His book *Le Modulor* (1948), which described an architectural design system based on the GN, was in turn influential in the spread of golden numberism (§12.5).

At present, golden numberism is flourishing, and even in some otherwise austere and serious mathematics books we find, often in connection with the Fibonacci numbers or Fibonacci search, the completely false assertion that the ancient Greeks used the GN in designing the Parthenon and other temples (see Figure 4) or the statement that the GN has special aesthetic properties.

2.3 Philosophical basis

Röber did not claim that the GN had been used to design the Great Pyramid, having, in this case, accepted a competing theory. Indeed, there are at least nine theories concerning the shape of the Great Pyramid, four of which agree with the observed value to the first decimal place. Two of the nine theories involve the GN, one based on a non-existent passage from Herodotus that gives rise to the GN in a most surprising way. This multitude of theories requires an examination of the philosophical difficulties connected with golden numberism.

The GN literature abounds with claims wherein the author does not realize that approximate measurements or drawings do not suffice to prove the claim. A proof requires some sort of documentary evidence that the designer of the object in question had the GN in mind as a theoretical basis. Sometimes, as with the painting *Le Cirque* by Georges Seurat (1890–91), it is known that the simple ratio 8/5 was used, yet a claim is made that the GN is involved. Similarly, the cubist painter Juan Gris (*circa* 1920) explicitly stated that he had not used the GN; yet claims to the contrary, based on measurements, have been made. In the same vein authors who discuss natural phenomena rely on *a posteriori* measurements instead of providing an acceptable, testable primary hypothesis which in turn would imply the presence of the GN. Incorrect treatment of data has also sometimes led to an apparent clustering of observed values around the GN.

BIBLIOGRAPHY

A complete review of the literature, both ancient and modern, related to DEMR may be found in Herz-Fischler *1987*. I am currently in the process of writing a book (Herz-Fischler *forthcoming*) which considers the history of the GN as well as the philosophical and sociological aspects. Partial studies, which discuss some of the topics mentioned, are listed below.

Curchin, L. and Herz-Fischler, R. *1985*, 'De quand date le premier rapprochement entre la suite de Fibonacci et la division en extrême et moyenne raison?', *Centaurus*, **28**, 129–38.

Dickson, L. *1919*, *History of the Theory of Numbers*, Vol. 1, Washington, DC: Carnegie Institution. [Repr. 1952, New York: Chelsea.]

Fischler, E. and Fischler, R. *1980*, 'Juan Gris, son milieu et le "nombre d'or"', *Canadian Art Review*, **7**, 33–6.

Fischler, R. *1979*, 'What did Herodotus really say? Or how to build (a theory of) the Great Pyramid', *Environment and Planning* B, **6**, 89–93.

——*1981a*, 'On applications of the golden ratio in the visual arts', *Leonardo*, **14**, 31–2.

——*1981b*, 'How to find the "golden number" without really trying', *Fibonacci Quarterly*, **19**, 406–10.

Herz-Fischler, R. *1983*, 'An examination of claims concerning Seurat and "the golden number"', *Gazette des Beaux-Arts*, **125**, 109–12.

—— *1984*, 'Le Corbusier's "regulating lines" for the villa at Garches (1927) and other early works', *Journal of the Society of Architectural Historians*, **43**, 53–9.

—— *1987*, *A Mathematical History of Division in Extreme and Mean Ratio*, Waterloo, Ontario: Wilfrid Laurier University Press.

—— *1988*, 'Theorem XIV** of the first "supplement" to the *Elements*', *Archives Internationales d'Histoire des Sciences*, **38**, 3–66.

—— *forthcoming*, *The Golden Number: A Philosophical, Historical, Sociological and Analytical Study*, Waterloo, Ontario: Wilfrid Laurier University Press.

Röber, F. (the Younger) *1855*, *Die äegyptischen Pyramiden in ihren ursprünglichen Bildungen, nebst einer Darstellung der proportionalen Verhältnisse am Parthenon zu Athen*, Dresden: Franke.

Struik, D. *1969*, *A Source Book in Mathematics 1200–1800*, Cambridge, MA: Harvard University Press.

Zeising, A. *1854*, *Neue Lehre von den Proportionen des menschlichen Körpers, aus ein bisher unerkannt [. . .]*, Leipzig: Weigel.

Zusne, L. *1970*, *Visual Perception of Form*, New York: Academic Press.

12.5

Numerology and gematria

I. GRATTAN-GUINNESS

1 HIDDEN KNOWLEDGE

The word 'numerology' refers to an ensemble of doctrines in which features of the world are associated with numbers, and sometimes also with their powers, sums and products. (The word 'number' is used here to refer only to positive integers.) Often these associations are given a mystical or religious character, and the resulting doctrine is regarded by its adherents to be secret (or occult, in the original meaning of the word). Some traditions draw on the 'reduced number' of a number, obtained by adding the constituent numbers of, say, a date as follows:

$$\text{for } 23/6/1941: \quad 2 + 3 + 6 + 1 + 9 + 4 + 1 = 26$$
$$\text{and } 2 + 6 = 8 \text{ (the reduced number).} \tag{1}$$

Gematria is a special version of numerology, in which numbers are assigned to letters of the alphabet of a language (usually Greek, Latin or Hebrew); numbers are then found for particular words and phrases by adding together the numbers of the constituent letters. An example is given in Section 3. One of the original inspirations may have been cryptographical, in that secret codes and messages could be buried in the numbers obtained.

Even histories of number theory have little to tell us about the development of numerology; the very secrecy has made it hard to detect. But evidence of it can be found in ancient times, although the influence of one culture upon a later one is extremely hard to delineate. For example, gematria may have Greek roots, and the word comes from their language; but Hebrew examples appear to go back to the seventh century BC (Scholem *1971*), and its later prosecution seems to owe much to medieval Hebrew sources. Hopper *1938* provides an excellent survey of 'medieval number symbolism' in Christianity up to the influential case of Dante (early fourteenth century); the general adoption at that time of the Hindu–Arabic number system (§1.16) must have aided this development.

There is some rapport between numerology and Fibonacci numbers (§12.4), but only when the appropriate numbers are in place: Ghyka *1927* and *1931* are well-known and influential sources, although their reliability is in doubt. In certain cultures and traditions particular geometrical figures were also interpreted in occult ways (Pennick *1980*), and some forms of (so-called!) recreational mathematics (§12.3) are also related. However, numerological theories do not have much mathematical interest in themselves; no deep theorems or general properties are involved. They are noted in this encyclopedia as unusual examples of applications and influences of mathematics.

This article gives examples from ancient sources, architecture and music. They all belong to the Western tradition; some pertinent points on other parts of the world are made in various articles in Part 1. In addition, Birkhoff *1933* is recommended for its study of vases and tilings from various cultures, Albarn *et al.* *1974* as a (modern) study of pattern inspired by numerology in Islamic decoration, and Cavendish *1975* on Tarot cards as a clear-cut case of numerology acting together with occultism in that the designs often relate to the assigned number.

2 SOME MYSTICAL NUMBERS

For many cultures, one source of numerology would have been the manifestation of small numbers in the universe and in human and animal forms: for example, 2 (male/female, Sun/Moon), 3 (male genitalia, past/present/future) or 4 (seasons, elements). The number 7 gained high status, perhaps originally as the number of perceived planets, but also (for example) in Christianity with the number of deadly sins, of Jesus's last words on the cross, and so on.

In some cultures properties were also assigned to relatively large numbers. For example, 42 played various roles for the Egyptians (among other things, it was the number of gods who assessed the dead); and in Babylonian mathematics 30, 60 and 180 were linked with (mis)perceived lengths of the month and year and the division of time in units of measurement (§1.1).

An interesting example is the quartet 3, 37, 40 and 111 (the Trinity number), which are related by

$$40 = 5 \times 2^3 = 37 + 3 \quad \text{and} \quad 37 \times 3 = 111;$$
$$\text{in addition,} \quad 111 \times 6 = 666 \ (= 1 + 2 + \cdots + 36), \qquad (2)$$

the 'number of the Beast'. The origin of this unpleasant appellation of 666 is typically obscure; it has some Hebrew roots (in particular, it is the number of 'Neron Caesar' in Hebrew), but the number is also so named in

1586

the Book of Revelations 13:18, and much gematriac effort has been expended since the sixteenth century to associate it with the names of disliked personalities.

Among the Greeks, the Pythagorean world-view based upon numbers and their interpretation as musical proportions maintained its mystical elements later (§2.10), despite the mishap over $\sqrt{2}$ not being a rational number (§1.3). Plato upheld in the *Timaeus* a tradition shown in Figure 1(a) (not from his text), the 'Pythagorean *tetraktys*' of 10 dots in a combination of $1 + 2 + 3 + 4$, with various symmetries evident in the arrangement. In addition, the 'sloping sides' formed what became known as the 'Platonic lambda', 7 locations to which sides could be assigned the powers of generation: 2, 4, 8 of 2 and 3, 9, 27 of 3. Figure 1(b) shows a three-dimensional version, where the 10 dots now produce a pyramid; it may have been this case which inspired the name 'triangular numbers' to be given to the numbers of Figure 1(a) (that is, numbers which can be expressed in the additive form $1 + 2 + 3 + \cdots + n$).

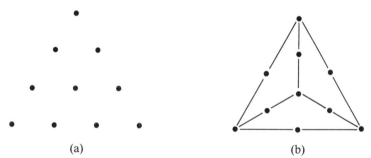

(a) (b)

Figure 1 (a) The Pythagorean *tetraktys* (two dimensions); (b) arrangement of ten dots in a pyramid (three dimensions)

A good example of a numerological account of small numbers is the text on 'the theology of arithmetic' by Iamblichus (flourished fourth century AD); *1988* is a valuable translation into English, with commentary. He writes on the 'monad', 'dyad', and so on, up to the 'decad', ascribing properties to these numbers and their factors. For example, the octad 'is the first actual cube'; there are 8 spheres for the heavenly bodies, the teeth in the human head divide into two 8's top and bottom, and so on.

3 SACRED GEOMETRY AND RELIGIOUS ARCHITECTURE

Neither Greek architecture nor its geometry seems to display much numerology. This was made more prominent in Vitruvius's *De architectura* (first century AD). He proposed a 'modulus', a basic unit of measure of which multiples should be manifest in aesthetically satisfying buildings and works of art. Among the uses of proportion, triangles were prominent, both in the large (the top of a spire and the ends of the base when viewing the cathedral from the side) and in the small (as in an architrave over a door: Rosevaal *1944*).

According to Bannister *1968*, an important example of both proportion and gematria was the Basilica of St Peter, which was built for Constantine the Great in the fourth century and stood until the present cathedral was begun in the sixteenth century. Its basic unit was the 'Egyptian' (later, the 'Ptolemiac') foot, and in this measure (which is about 350 mm) various key numbers are evident in certain measurements: 153 and its multiples, and Ptolemy's approximation $377 \div 120$ to π. Further, Greek gematria may underlie the lengths of the interior perimeters of the nave, the nave with the aisles, and the transept; for they yield respectively 666, 888 ('Jesus') and 755 ('Peter'). In addition, the perfect numbers 6, 28 and 496 seem to have been involved.

These traditions were continued and enriched in the Middle Ages, especially in cathedral design (Lund *1921*: Chaps 10–13). An important example was Chartres, at its rebuilding around 1300. The motto '*assumptio virginis beate mariae*' was adopted, and its Latin gematria determined (from the so-called 'Lesser canon' system, in which a = 1, b = 2, and so on, for the 22 letters of the Latin alphabet). The total, 306, was used as the number of Roman feet of a major length in the design of the cathedral (James *1979*: esp. 150–65). Gematriac numbers were often of Biblical significance; here 306 is twice the number of fish (itself an important symbol, of course) caught by Simon Peter (John 21: 11). In addition, 153 is the triangular number to 17.

Much of the technical and religious knowledge involved was held by the masons, and secretly so. One of their tools was the 'masons' square', a triangle in the proportions 3 : 4 : 5, which was used not only to lay out right angles but also to construct rectangles with sides in the proportions 2 : 1 and 3 : 1 (the *ad quadratum* and *ad triangulatum*, respectively). Figure 2 contains a representation of these relationships.

In the Renaissance, Vitruvius's views became influential (§2.8), and the use of proportions was sometimes taken to extremes. His ideas on city planning were also adopted, with proposed cities divided into 8, and even up to 32, equal sectors of the circle.

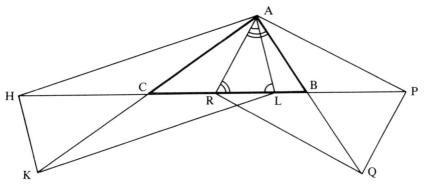

Figure 2 The mason's square, a 3–4–5 triangle, used to construct rectangles with sides in proportion 2:1 (APQR) and 3:1 (AHKL).
AB:AC:BC = 3:4:5, AP:PQ = 2:1, AH:HK = 3:1

Another role for numbers in the Renaissance concerned the aesthetic appreciation of buildings and squares (Zucker *1959*). The 'angle of fusion' *a*, specifying the measure of elevation (or declination) over which one could 'accurately' appraise a building, was given by

$$\tan a := \frac{\text{height of building}}{\text{distance from viewpoint}} = \frac{1}{3}, \tag{3}$$

giving a value for *a* of about 18°. The angle of vertical 'perception' *b*, and also that of horizontal 'reception' which one could readily encompass to left or to right, were both held to be given by

$$\tan b = \tfrac{1}{2}; \quad \text{thus } b = 27° \text{ (more or less)}. \tag{4}$$

Threads of this kind of thinking in architecture and beyond linger on in our times with a few architects. In particular, seemingly under the influence of Ghyka, Le Corbusier *1954* proposed his 'modulor' as a measuring unit based upon human proportions and applicable to the design of buildings. For the purpose he devised two series of Fibonacci-like numbers, the 'red' and the 'blue', working out respectively from (70, 113) and (140, 226):

$$5, 11, 16, 27, 43, 70, 113, 186, \ldots \quad \text{and} \quad 13, 20, 33, 53, 86, 140, 226, 336, \ldots. \tag{5}$$

4 MUSIC, ESPECIALLY MOZART'S

Music has been prone to numerology ever since the discoveries of the Pythagoreans. Medieval music contains intricate procedures about the numbers of notes, rhythms, and so on (see Crevel *1959* for the case of

Obrecht); they make some pieces exceedingly hard to perform! The 'settled' version of Western harmony gave us an octave divided into 12 principal notes, 8 of them giving the prime intervals, in which the octave is divided into 5 tones with 2 diatonic semitones. Some modern composers, such as Partch *1974*, have advocated that the octave be divided into other numbers of intervals.

One great composer who unites both orthodox musical considerations with the secret knowledge of the cathedral builders is Wolfgang Amadeus Mozart. By the late eighteenth century their lore had been transformed into the rituals and wisdom of the Freemasons. Mozart joined them in 1784, and over the remaining seven years of his life he was an ardent Mason, writing for them several works and showing Masonic influences in many others.

The outstanding case is the opera *The Magic Flute*, which he wrote shortly before his death in 1791 to a libretto by fellow Mason Emanuel Schikaneder. The Freemasonry movement was in crisis at that time in Austria due to the hostility of the Emperor; the opera was written as a defence of its noble values. Chailley *1972* has shown that its apparently silly story is in fact a most sophisticated allegory on Masonic norms and practices. Furthermore, the numerology upheld by Austrian Masons is built deeply into the construction of the work, the design of the music and libretto, and even the circumstances of the first performance (Grattan-Guinness *1992*). Briefly, 3 is the dominating number, exhibiting a masculine character shared also by 6; 5 and 8 are feminine; 18, 7 and 30 have religious connotations; and 10 is also prominent. There are thus 10 numbers; and in a nice self-reference they group into $6 + 3 + 1$. This also gives the pyramidal arrangement of Figure 1(b), appropriate for an opera set in Egypt.

The main key of the music is E♭ major; that is, with 3 flats. The numerology is manifest principally in the numbers of notes which make up the melodies in a musical item, but other aspects are sometimes involved (such as the number of bars in an item, or of staves of music). The most conspicuous example occurs in the middle of the Overture (and in a few places in the opera proper). Here the 15 wind instruments of the orchestra play across 6 bars 3 statements of the '3-fold accord', a suite of 3 chords in a particular rhythm which imitates a Masonic knock used to indicate the elevation of a Mason to one of the 3 lowest levels of his craft.

In the arias, Papageno's bird-catcher song runs on 3s and 5s, including his 5-note pipe and his 3 verses of text. The Queen of the Night's aria of rage has 8s all over it, most conspicuously in the famous *coloratura* passage. Sarastro, the High Priest, works on 18s; he has 18 singing entries in the whole opera and sings in all in 180 bars. His two arias, 'O Isis und Osiris' and 'Heiligen Hallen', each use melodies of 18 notes (plus upbeats).

The first is sung with his chorus of 18 priests, who also have their own item, also called 'O Isis und Osiris', in which their melody lasts 18 bars.

The first performance took place at the end of the 3rd 4ter of the calendrial year, on 30 9ber (as Masons would have thought of it) at 7 o'clock in the evening (itself a normal time); but the day changed for them at 6 o'clock (i.e. 18.00 hours), so that the date was actually 1 8ber. The opera lasted about 3 hours, or 180 minutes. It was Mozart's last opera, but not his last completed work; that was to be the Masonic cantata entitled 'Laut verkünde unsre Freunde', written out on 18 folios of manuscript and performed on 18 9ber to open the new temple of his lodge. In a strange stroke of fate, he died 18 days later.

Numerology is not well recognized among musicologists. Much of the information above about *The Magic Flute* escaped even Chailley; the case of J. S. Bach has only recently been clarified (Tatlow *1991*). But later composers have used it in some form or other (and not necessarily linked with Freemasonry). In our century, Béla Bartók is an excellent example, for he seems to have deployed Fibonacci numbers in many works (Lendvai *1971* argues strongly for this interpretation). A fine example is his *Music for Strings, Percussion and Celesta* (1936), which begins with a theme in 5 + 8 notes with the stresses falling respectively on the 3rd and 5th notes, and then unwinds a neo-fugue on intervals of the 5th before reaching climax at the 55th of its 88 bars (to which, according to Lendvai, one follows a tradition in adding on a bar's rest).

5 THE TRAVESTY OF NUMEROLOGY

These doctrines are still alive; the engineer Ambrose Fleming *1928* elaborated them into a unifier of science and religion. But in general the religious background has declined so much in importance that a proper basis no longer exists for them. Thus we find only moronic versions serving pastiche versions of astrology suitable for tabloid newspapers, or as satires of ancient lore barely understood by gurus of alternative life-styles. Even the apparently more sensible texts are trivial in their content (e.g. Gibson *1927* or Bell *1933*). 'Rational' science and philosophy despises it all, without being burdened with knowledge of its content or context.

BIBLIOGRAPHY

Albarn, K., Miall Smith, J., Steel, S. and Walker, D. *1974, The Language of Pattern*, London: Thames & Hudson.

Bannister, T. C. *1968*, 'The Constantinian Basilica of Saint Peter at Rome', *Journal of the Society of Architectural Historians*, **27**, 3–32. [The whole issue is devoted to proportions in building.]

Bell, E. T. *1933*, *Numerology*, Baltimore, MD: Williams & Wilkins.

Birkhoff, G. D. *1933*, *Aesthetic Measure*, Cambridge, MA: Harvard University Press.

Cavendish, R. *1975*, *The Tarot*, London: Michael Joseph.

Chailley, J. *1972*, *The Magic Flute. Masonic Opera* [. . .] (transl. H. Weinstock), London: Gollancz.

Crevel, M. van *1959*, 'Secret structure', in J. Obrecht, *Opera omnia*, Vol. 1, Part 6, Amsterdam: Alsbach, xvii–xxv.

Fleming, A. *1928*, 'Number in nature and in the Biblical literature indicating a common origin in a supreme intelligence', *Transactions of the Victoria Institute*, **60**, 11–44. [Includes discussion.]

Ghyka, M. *1927*, *Estétique des proportions dans la nature et dans les arts*, Paris: Gallimard.

—— *1931*, *Le Nombre d'or. Rites et rhythmes Pythagoriciens dans le développement de la civilisation occidentale*, 2 vols, Paris: Gallimard.

Gibson, W. B. *1927*, *The Science of Numerology: What Numbers Mean to You*, London: Rider.

Grattan-Guinness, I. *1992*, 'Counting the notes: Numerology in the works of Mozart, especially *Die Zauberflöte*', *Annals of Science*, **49**, 201–32.

Hopper, V. F. *1938*, *Medieval Number Symbolism* [. . .], New York: Cooper Square. [Repr. 1969.]

Iamblichus *1988*, *The Theology of Arithmetic* (transl. R. Waterfield), Grand Rapids, MI: Phanes.

James, J. *1979*, *The Contractors of Chartres*, Vol. 1, Dooralong, Australia: Mandorla, and London: Croom Helm.

Le Corbusier *1954*, *The Modulor*, Cambridge, MA: Harvard University Press. [French original 1951, Paris; various translations and related books.]

Lendvai, E. *1971*, *Béla Bartók. An Analysis of His Music*, London: Kahn & Everill.

Lund, J. L. M. *1921*, *Ad quadratum. A Study of the Geometric Bases of Classic and Medieval Religious Architecture*, Vol. 1, London: Batsford.

Partch, H. *1974*, *Genesis of a Music*, New York: Da Capo.

Pennick, N. *1980*, *Sacred Geometry. Symbolism and Purpose in Religious Structures*, Wellingborough: Turnstone.

Rosevaal, J. *1944*, 'Ad triangulum, ad quadratum', *Gazette des Beaux-Arts*, **26**, 149–62. [Expresses reservations over Lund *1921*.]

Scholem, G. *1971*, 'Gematria', in *Encyclopaedia Judaica*, Vol. 4, Jerusalem: Keter, cols. 369–74.

Tatlow, R. *1991*, *Bach and the Riddle of the Number Alphabet*, Cambridge: Cambridge University Press. [Contains valuable information on cabbalistic numerology.]

Zucker, P. *1959*, *Town and Square* [. . .], New York: Columbia University Press.

12.6

Art and architecture

PETER SCHREIBER

1 INTRODUCTION

K. Menninger *1959* wrote in his little book *Mathematik und Kunst* that there are three possible points of view about the relation between mathematics and the arts:

1 They have nothing in common; moreover, there is some hostility between them.
2 Arts and architecture use some applications of mathematics (e.g. perspective); mathematics therefore is a servant which may be called upon by the arts – or not. Modern art does not often call upon it.
3 Both subjects are concerned with discovery and presentation of truth and beauty, albeit from different points of view, and therefore they are intimately connected.

If we accept the second and third of these, we must add an important remark. Wherever mathematics touches human activity, an exchange takes place: the mathematics is applied, but it also receives inspiration to create new notions, theories and methods. The relationship between mathematics and the arts is no exception.

Up to the beginning of the nineteenth century, architecture in encyclopedias and academic lectures was considered a regular part of applied mathematics. Many mathematicians worked as practical architects, either as experts or as leading state officials (e.g. Christopher Wren in London, or Johann Heinrich Lambert in Berlin). Through the ages, some architects and painters have been creative mathematicians, or at least have given important hints to mathematics.

This article can only give examples of the long history of cross-fertilization between mathematics and the arts. These are: the development of mathematical perspectives by Renaissance painters; the many other contributions to mathematics by Albrecht Dürer and Leonardo da Vinci; the discovery of projective geometry by Girard Desargues (§7.6); and, recently

1593

the two-way flow of ideas between the Dutch artist Maurits Escher and mathematics.

2 BUILDING AND GEOMETRICAL CONSTRUCTION AS ALLIED RITUALS

In ancient cultures like the Egyptian or the Babylonian, the surveying of new temples was governed by strict geometrical and astronomical procedures (§1.1–1.2) and was the privilege of the Pharaoh and the highest priests (Figure 1). Therefore regular geometrical shapes were sacred and reserved for ritual and official buildings, whereas profane buildings were often intentionally made skew and irregular. Geometrical instruments and basic figures, such as the Jewish Star of David, were regarded as sacred symbols and were used as amulets (Figure 2) (Kadeřávek *1991*).

Figure 1 Ramses II (1388–1322 BC) and Seschat, the goddess of surveying, pegging out the site of a new temple. It is remarkable that the Egyptians had a special goddess for surveying; of course invisible to the people, she helped him in his sacred work (from Ricken *1977*: figure 1)

In the European Middle Ages the building of churches and monasteries was again subject to fixed rules. Equilateral triangles and quadrilaterals, and also regular hexagons and octagons, were dominant in the design of such buildings, and again the profane parts of monasteries were built askew. A painting of Giotto di Bondone and his pupils in the lower church of St Francis in Assisi shows the personified virtues 'Obedientia', 'Humilitas' and 'Prudentia' with haloes in the shape of regular quadrilaterals and

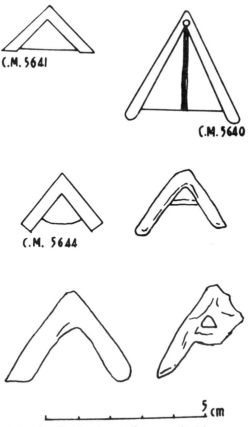

Figure 2 Some amulets in the shape of geometrical instruments from an
Egyptian tomb (from Ricken *1977*: figure 2)

regular hexagons. There is much symbolism and mysticism in the geometrical codes of gothic art; the proportions of the cathedral at Chartres,
for example, are said to correspond to the first Gregorian scale (compare
James *1979–81*). But there is also a tendency in the modern literature on
art theory to overstate such interpretations.

Outside the sacred context, geometrical regularity has been cultivated as
an expression of beauty and harmony since ancient Greece. The term
'symmetry', today associated with an essentially group-theoretic interpretation in mathematics and science (§12.7), originally referred to the
repetition of shapes and ratios from the smallest parts of a building, sculpture or painting to the whole work. (On the overstated role of the golden
number in this connection, see §12.4.) Such a view of the role of mathematics in art and architecture was propagated in the famous *Ten*

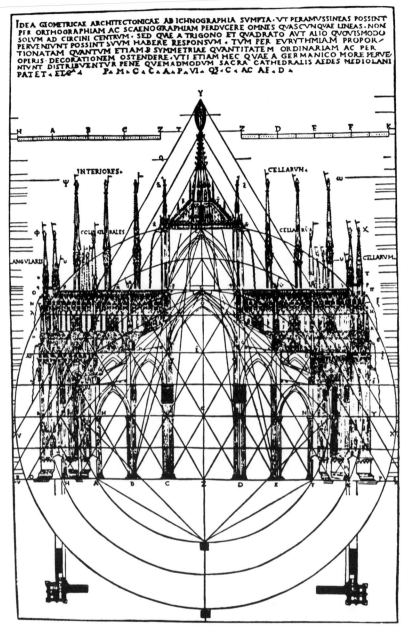

Figure 3 Elevation of the cathedral of Milan, with geometrical constructions revealing the hidden principles of the design, from Vitruvius's *Architectura* (here, by Cesariano, 1540). Because the book's original illustrations were lost in early Roman times, all later printed editions of the work had contemporary illustrations (from Kadeřávek *1935*)

Books on Architecture written by the Roman Vitruvius Pollio in the first century AD. This book had a major influence on the theory and practice of building up to the end of the eighteenth century (Figure 3). Another famous architect, Pietro da Padova, known as 'Palladio', wrote in 1570 that

> The pure proportions of the tones are harmonies for the ear, the corresponding ones of the spatial ratios are harmonies for the eye. Such harmonies give us the feeling of happiness, but nobody knows why, except he who is searching for the causes of the matters. (Leonhardt *1984*)

3 CENTRAL PERSPECTIVE

Democritus is said to have been the first to study the laws of perspective, in about 400 BC (they may have been applied in the production of theatre scenery, but there is no written evidence). The great age of central perspective was the European Renaissance, when painters tried to make their pictures realistic (this is neither a general nor an obvious intention of artists in all times and all cultures). It is quite characteristic of the difficulty in finding a mathematical basis for the production of realistic pictures that Leonardo da Vinci, in his *Treatise on Painting* (*circa* 1510), lists three sorts of perspective which seem to him to be of equal importance: the diminution of the size of more distant objects (with which mathematical perspective is concerned), the fading of colours, and the loss of sharpness of shape. (This shows Leonardo as a scientist more than a mathematician; compare §2.8.)

The architect Filippo Brunelleschi, famous for the cupola of the cathedral of Florence, is said to have been the first to have an exact method for constructing perspective pictures from the plan and elevation of a scene. This method consisted of adding the eye position and the plane of the picture to the plan and the elevation, and pointwise constructing the intersection of the cone of vision and the picture plane (Figure 4). But the only evidence for the priority of Brunelleschi is some remarks in his biography, written in 1550 by Giorgi Vasari (*Lives of the Most Famous Painters, Sculptors and Architects of the Renaissance*).

The first written textbook on perspective was by Leon Battista Alberti (in Latin 1435, in Italian 1536), a man of universal interests and talents. He presented Brunelleschi's method, but without mentioning him, and some other methods based on the correct construction of the so-called *pavimento*, a chessboard-patterned ground which Renaissance painters liked to use for approximating the diminution of people and objects. In the years that followed, many artists laboured at various constructions, mostly

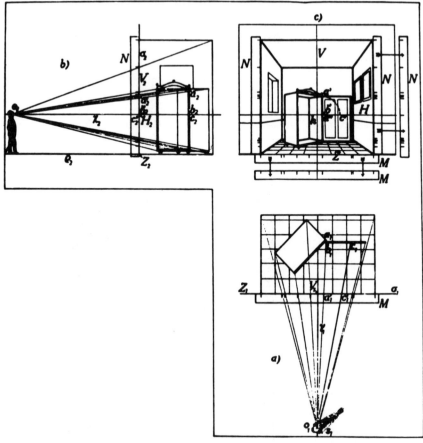

Figure 4 Brunelleschi's method of constructing an image:
(a) plan of the scene, with the picture plane and the eye position;
(b) elevation of the scene also with picture plane and eye position;
(c) the pointwise constructed image
(from Kadeřávek *1935*)

without any proof of correctness and mixed with false instructions: among others, Piero della Francesca, Antonio Filarete, E. Danti, and Guidobaldo del Monte in Italy. At the beginning of the sixteenth century, the idea of central perspective began to spread into other countries, first to France (where Jean Pelerin published in 1505 the first mostly correct textbook), then to Germany (Albrecht Dürer, H. Rodler and others), and later to The Netherlands (Simon Stevin and Frans van Schooten the Younger) and England (Brook Taylor).

A qualitative leap was made by Lambert in his book *Freye Perspective* (1759, second enlarged edition 1774) and some papers on perspective. (The 1774 edition, reproduced in *1943*, also contains a short history of the earlier contributions to perspective.) He tried to construct perspective pictures by using an implicit form of projective geometry (§7.6), without a given plan and elevation of the scence. In some sense his *Freye Perspective* contains the most fruitful use of the plane as a model of space; that is, to work on the plane with projections of the spatial objects instead of the objects themselves. This idea was later developed by Gaspard Monge (§7.5).

Figure 5 The ceiling of the town church at Badia (Arezzo), by Andrea Pozzo (seventeenth century), an outstanding master of perspective painting. It gives the perfect illusion of a cupola in three dimensions
(from Schomann *1990*: figure 29)

The practical use of central perspective in the seventeenth and eighteenth centuries reached a high level, especially in paintings and other decorations inside churches and palaces (Figure 5), in art (e.g. Canaletto), and in theatre scenery. The use of impressive perspective effects was emphasized by the Jesuits, some of whom wrote excellent books on perspective. But most artists of the time also favoured mechanical and optical methods, some of which were recommended in the oldest textbooks (e.g. by Dürer; see Figure 6). Canaletto produced all his admirable realistic pictures with the help of a camera obscura.

Figure 6 The Designer of the Lute: perspective drawing by mechanical means (from Dürer's *Underweysung*, 1525)

Lambert lamented in 1768 that most artists tried to avoid subjects which required perspective, that they produced errors in their works, and that they were unwilling to make a proper study of the laws of mathematical perspective (Lambert *1943*: 398–401). Only once descriptive geometry had begun to be taught regularly at the academies of the fine arts and the higher engineering schools from the beginning of the nineteenth century did the central perspective (including special topics such as the construction of shadows, or the relief perspective) become a serious topic of geometry. The invention of photography and the trends of modern art seem to have reduced the application of central perspective, but recently some of its aspects have been revived by the needs of computer image processing.

Figure 7 Dürer's copper engraving *Melencolia I* (1514)

4 ALBRECHT DÜRER AS A MATHEMATICIAN

Unlike many other Renaissance artists, Dürer's relation to mathematics was not bound to linear perspective; unlike Leonardo da Vinci, who was interested not only in several aspects of mathematics, but also in technology, and was an inductive thinker. Dürer had an affection for deep theoretical, systematic and in some sense deductive thought. His contributions to mathematics concern the construction of (partly new) curves, surfaces and solids, very good approximative solutions of constructional problems, geometrical transformations and some notions of combinatorial thinking (e.g. combinations of faces from given sets of different eyes or noses) – but also some mistakes. These contributions are contained in his books *Underweysung der messung mit dem zirckel un richtscheyt* (1525), *Etliche underricht zu befestigung der Stett Schloss und flecken* (1527) and *Vier Bücher von menschlicher Proportion* (1528). There are many discussions of Dürer as a mathematician (see especially Matvievskaya *1987*), but often only the first of these works is cited.

His deep affection for mathematics is strongly reflected in his copper engraving *Melencolia I* (1514; see Figure 7). Beyond such external features as consistent applications of central perspective, the magic square on the wall, the complicated polyhedron, and the shadowed sphere, it shows symbolically the melancholy produced by extended and eventually fruitless thinking about difficult mathematical problems. This was the best self-portrait he ever painted.

5 RELIEF PERSPECTIVE

Given an eye position E and a trace plane t, a relief perspective effects a $1-1$ mapping of the half-space behind t onto that part of the space between t and a given vanishing plane v parallel to t (Figure 8). Let any line g intersect t at S, and take the line g_E parallel to g and passing through E. The vanishing point V_g of g is then the point where g_E meets v, and any point P of g maps onto the point P_E of intersection of EP with SV_g.

Since the beginning of the Renaissance in Italy reliefs have played an important role in sculpture. A famous example is the so-called 'Paradise door' at the baptistry in Florence designed by Lorenzo Ghiberti in about 1450; but in spite of its admirable realism it seems impossible that it could have been produced with the help of exact constructions. The situation altered in 1482, when Donato Bramante used a large relief in St Mary's chapel at St Satiro in Milan to produce the illusion of an extended apse, as there was not room enough to actually build one. This masterpiece of relief art seems to be the first constructed relief perspective. Later on, relief

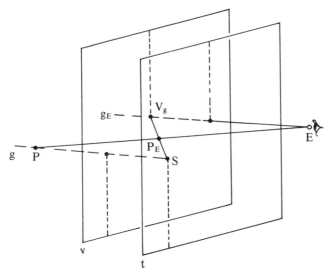

Figure 8 The principle of relief perspective

perspective was frequently used in designing the scenery for baroque operas, and it also gave good occasion to artists and architects to think more deeply about perspective and related elements of projective geometry (§7.5). But only at the beginning of the nineteenth century did relief perspective become an accepted discipline of geometry (Burmester *1883*).

6 POHLKE'S THEOREM

Pohlke's theorem is an impressive example of art inspiring mathematics. It states that to any trihedral (three concurrent lines) in a plane a spatial Cartesian trihedral and a direction of a parallel projection can be found such that the given plane trihedral is the image of the spatial one. This statement gives the theoretical justification for the usual practice of drawing the image of, say, a cube without considering from which point of view the cube may be seen. Therefore it is surprising that such a fundamental theorem of descriptive geometry was formulated only in 1853 (published in 1860), and not by a professional mathematician but by the German painter and late professor of descriptive geometry at the Berlin Academy of Fine Arts, Karl Wilhelm Pohlke. He was not able to prove his theorem; the first proof was published by the well-known mathematician K. H. A. Schwarz in 1864. It was complicated, and many other proofs were subsequently found (Wendling *1912*, Salenius *1978*). But this is a result which textbooks in descriptive geometry usually omit.

1603

7 INCONSISTENT IMAGES OR IMPOSSIBLE FIGURES

Mathematicians today know and (mostly) love the works of the Dutch artist Escher. Among them, the *Belvedere* (Figure 9) and the so-called 'Escher cube' are especially popular. Escher was not the first to create such images. Similar but simpler constructions were produced as early as 1934 by the Swedish artist Oscar Reutersvärd, whose *Opus 1* (Figure 10), so he says, was made accidentally when he was a pupil. Many painters of today produce images of this sort (Ernst *1989*); we know too of similar effects causing inconsistencies in medieval paintings, apparently caused partly by the inability of the artists to master perspective, and partly by them wishing not to obscure essential parts of the design with foreground objects.

It is possible to construct objects and photograph them from such a direction that the photograph seems to show an impossible object

Figure 9 Maurits Escher: *Belvedere* (1958) (from Ernst *1986*)

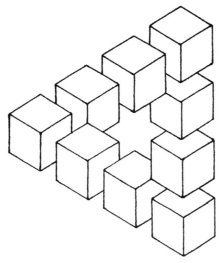

Figure 10 Oscar Reutersvärd: *Opus 1* (1934) (from Ernst *1989*)

Figure 11 Mathieu Hamaekers: the same object from different points of view; one view seems to show an impossible object (from Ernst *1989*)

(Figure 11). This demonstrates that a perspective picture can never be a 1–1 image of a spatial construct. Nevertheless, in everyday life photographs and other two-dimensional perspective pictures serve as indispensible means of information, and communication in general depends on a complicated process of interaction between the visual impression and some previous knowledge of objects or situations. (The Escher cube teaches us that this knowledge may be false.) This cooperation is hitherto insufficiently clarified by theory. It seems to be important in the field of pattern recognition (as a branch of artificial intelligence). Is this another chance for art to inspire mathematics? Geometry in cooperation with the art of Reutersvärd and Escher can teach us that things may not be as they seem, and at school level could help to develop a critical sense also in other matters.

8 ORNAMENTS, FRIEZE GROUPS, WALLPAPER GROUPS AND TILINGS

Menninger *1959* claimed that there is some unconscious sort of mathematics, such as counting and comparison, identification and classification of shapes, and combinatorial enumeration of possibilities, which must be developed, heuristically and mentally, before objective, systematic and even deductive mathematics comes through. One of the oldest ways in which such 'unconscious mathematics' has been expressed is in the decoration of ordinary objects with ornamental patterns. This we find in all periods and all cultures. A first quasi-mathematical treatment of this subject was apparently made by the Pythagoreans, who examined the question of which tilings of the plane by regular polygons are possible. This question is related to regular and semiregular polyhedra and to space-filling polyhedra and crystallography. Johannes Kepler in his *Harmonice mundi* (1619) and Dürer in his *Underweysung* later discovered many non-trivial tilings of the plane (Figure 12).

Ancient Egyptian and Chinese ornamentation provide an important early example of the (unconscious) enumeration of frieze and wallpaper groups. A frieze group is a discrete subgroup of the group of all plane motions which contain a translation, but not two linearly independent translations. A wallpaper group is a discrete subgroup of the group of all plane motions which contain two linearly independent translations. A frieze is a potentially unbounded ornament which is invariant under a frieze group, and a wallpaper ornament is a potentially unbounded ornament which is invariant under a wallpaper group. Today it is well known that there are exactly 7 frieze groups and 17 wallpaper groups. After the early enumerations work in the framework of crystallography in the nineteenth century (§9.17), the

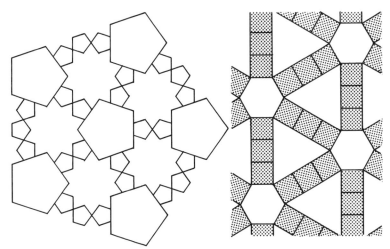

Figure 12 Two of Kepler's tilings which go beyond the concept of semiregularity (from Flachsmeyer *et al. 1990*)

general theory was first explained in terms of group theory by George Pólya and G. Niggli in 1924, and in the well-known book Speiser *1927*.

How problematic a naive treatment of this question may be is apparent from the incomplete independent attempt by Wilhelm Ostwald (an outstanding scientist!) in *1922–5* to offer a broad set of ornaments for users (Flachsmeyer *et al. 1990*). The highest power in empirical discovering of possibilities we find in the Islamic ornamental art and especially in the famous Alhambra in Granada (*circa* 1400) which influenced Escher when he visited it in 1922 and 1936. Coxeter and Moser *1957* claimed that all 17 possible wallpaper groups are represented in the ornamentation of the Alhambra; in 1981 Martin Gardner showed that only 11 are represented there.

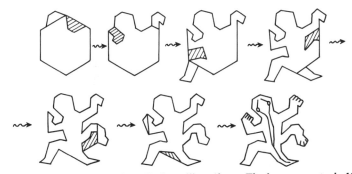

Figure 13 Construction of an Escher tiling (from Flachsmeyer *et al. 1990*)

1607

The German mathematician H. Heesch found in 1933 the first example of a tile which tiles the entire plane, giving a tiling for which there is no motion from one tile onto another that leaves it invariant. An analogous result for space was found by K. Reinhardt as early as 1928 (solving one of David Hilbert's problems from 1900). From Heesch's theory of irregular tilings, he and others have made many applications (Bigalke *1988*); rather different are the fantasies of Escher that are based on tilings (Figure 13).

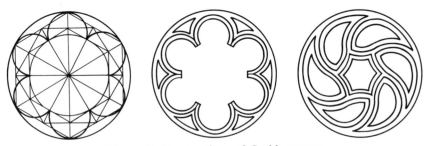

Figure 14 Constructions of Gothic tracery

Figure 15 Pisa, St Caterina (from Schomann *1990*: figure 203)

Recently, interest in tilings has been stimulated by tiles discovered by Roger Penrose (§12.8).

Another sort of ornamentation is tracery, the patterns of intersecting ribs found in Gothic architecture. These patterns, which were constructed by compass alone, have few symmetries overall, but parts of them are invariant under the so-called 'dihedral' groups; that is, discrete subgroups of plane motions which have a common fixed point (Figure 14). Typical ornaments, on the whole invariant under some dihedral group, are the rosette windows of Gothic churches (Figure 15).

9 FURTHER USES OF GEOMETRICAL SHAPES IN ART AND ARCHITECTURE

The goldsmith and designer Wentzel Jamnitzer *1568* found regular, semi-regular and star polyhedra attractive subjects for use in design. His book contains 120 beautiful drawings of such bodies (Figure 16). The eighteenth-century architect Etienne-Louis Boullée tried to use regular shapes in architecture, but it was not until 1928 that a building in the shape of a sphere was erected, in Dresden. (It was devastated in the Second World War, after which Buckminster Fuller constructed 'geodesic domes'.) Other new shapes in buildings were first used in bridges, but after the Second

Figure 16 Star polyhedra used as ornamentation (from Jamnitzer *1568*)

World War also in roofs and wall-facings. Lack of knowledge of geometrical shapes has produced a monotonous, cubical architecture; however, new shapes such as fractals (§3.8) may stimulate designers to find new styles.

BIBLIOGRAPHY

Bigalke, H.-G. *1988, Heinrich Heesch*, Basel: Birkhäuser.

Burmester, L. *1883, Grundzüge der Reliefperspektive nebst Anwendung*, Leipzig: Teubner.

Coxeter, H. S. M. and Moser, W. O. J. *1957, Generators and Relations for Discrete Groups*, Berlin: Springer.

Deken, J. *1983, Computer Images. State of the Art*, London: Tabori & Chang.

Ernst, B. *1986, Der Zauberspiegel des M. C. Escher*, Berlin: Taco.

—— *1989, Das verzauberte Auge*, Berlin: Taco.

Flachsmeyer, J., Feiste, U. and Manteuffel, K. *1990, Mathematik und ornamentale Kunstformen*, Leipzig: Teubner.

James, J. *1979–81, The Contractors of Chartres*, 2 vols, London: Croom Helm. [Elaborate uses of perspective.]

Jamnitzer, W. *1568, Perspectiva corporum regularium*. [Facsimile repr. 1973, Graz: Akademische Druck- und Verlagsanstalt.]

Kaděřávek, F. *1935, Geometry and Art in Old Times*. [In Czech; very rare book. German edn 1992, Leipzig and Stuttgart: Teubner.]

Lambert, J. H. *1943, Schriften zur Perspektive* (ed. M. Steck), Berlin: Lüttke.

Leonhardt, F. *1984*, 'Zu den Grundfragen der Ästhetik bei Bauwerken', *Sitzungsberichte der Heidelberger Akademie der Wissenschaften, Mathematischnaturwissenschaftliche Klasse*, No. 2.

Lietzmann, W. *1931, Mathematik und bildende Kunst*, Breslau: Hirt.

Locher, J. L. *1971, De werelden van M. C. Escher*, Amsterdam: Neulenhoff. [English transl. 1982 as *Escher: The Complete Graphic Works*, London: Thames & Hudson.]

Matvievskaya, G. P. *1987, Albrecht Dürer – Uchenyi* [learned man], Moscow: Nauka. [In Russian. The most recent and deepest study of the mathematical work of Dürer.]

Menninger, K. *1959, Mathematik und Kunst*, Göttingen: Vandenhoeck & Ruprecht.

Ostwald, W. *1922, Die Harmonie der Formen*, Leipzig: Unesma.

—— *1922–5, Die Welt der Formen. Entwicklung und Ordnung der gesetzlichschönen Gebilde*, Leipzig: Unesma.

Panofsky, E. *1956*, 'Dürer as a mathematician', in *The World of Mathematics* (ed. J. R. Newman), Vol. 1, New York: Simon & Schuster, 603–21.

Ricken, H. *1977, Der Architekt – Geschichte eines Berufs*, East Berlin: Henschelverlag.

Salenius, T. *1978*, 'Elementärt Bevis för Pohlkes Sats', *Nordisk Matematisk Tidskrift*, **25–26**, 150–52. [The first really elementary proof of Pohlke's theorem.]

Schomann, H. *1990, Kunstdenkmäler in der Toskana*, Darmstadt: Wissenschaftliche Buchgesellschaft.

Schröder, E. *1980, Dürer – Kunst und Geometrie*, Basel: Birkhäuser, and Berlin: Akademie-Verlag.

Speiser, A. *1927, Die Theorie der Gruppen von endlicher Ordnung*, 2nd edn, Berlin: Springer.

Wendling, E. *1912, Der Fundamentalsatz der Axonometrie*, Zurich: Speidel.

12.7

Symmetries in mathematics

KLAUS MAINZER

1 HISTORY AND DEFINITION

Historically, ideas of symmetry have been very influential in the development of human thought. In technology, the practical advantage of balance soon became apparent. In early cosmology, the planets were at first assumed to move uniformly on the spheres of a geocentric model. Irregularities in the motion of planets which were observed later on were interpreted as a kind of symmetry-breaking, which had to be reduced in order to fit modified symmetric models. Philosophers, mathematicians and astronomers believed in the simplicity and symmetry of nature; therefore the retrograde movements of planets were geometrically explained by epicycles, the centres of which moved on deferents round the Earth (§2.7). In Platonic physics the variety of observed phenomena were reduced to the regular bodies of Euclidean geometry.

Even in the natural sciences of modern times, symmetry models have often been used to illustrate and visualize natural regularities, from Johannes Kepler's heliocentric model of the Solar System to Ernest Rutherford and Niels Bohr's model of atoms and their electron orbitals. Since the nineteenth century, symmetries have been not only features of geometric models, but also essential components of natural laws and theories. In this sense symmetry means the invariance of a theory with respect to a transformation of its coordinates by a mathematical transformation group (§7.4).

Mathematically, the unification of natural science can be described by structures of symmetry. Atoms and elementary particles in physics are characterized by particular symmetry groups. Symmetry, dissymmetry and chirality are well-known topics in chemistry. Even living organisms and populations on the macroscopic level have functional properties of symmetry. The whole of physical, chemical and biological evolution seems to be regulated by the emergence of new symmetries and the breaking of old ones.

The fascination of symmetry comes not only from science, but also from art and religion. 'Symmetry' is derived from the Greek συμμετρία, and originally (around the fifth or sixth century BC) meant a common measure or harmony of proportion, whether in the universe or in architecture and art. Later its meaning was enlarged to cover reflection, rotation, periodicity, cyclicity, parity (e.g. crystallographic symmetry), conservation, invariance (e.g. symmetry laws in physics) and even balance and equilibrium in biology (Mainzer *1988*). All these meanings are satisfied by the mathematical axioms of an automorphism group. Thus symmetries in general are defined by automorphisms that represent self-mappings of figures, spaces, and so on which leave the structure unchanged.

In geometry the mapping of similarity is an example of an automorphism which leaves the form of a figure invariant. The relation of similarity $F \sim F'$ (i.e. figure F is similar to figure F') satisfies the three conditions of an equivalence relation: (a) $F \sim F'$ (reflexivity); (b) if $F \sim F'$, then $F' \sim F$ (symmetry); and (c) if $F \sim F'$ and $F' \sim F''$, then $F \sim F''$ (transitivity).

In general, the composition of automorphisms satisfies the axioms of a mathematical group A: (a) the identity I which maps a figure onto itself is an element of A ($I \in A$); (b) for every mapping $T \in A$ there is an inverse $T^{-1} \in A$ with $T \cdot T^{-1} = T^{-1} \cdot T = I$; and (c) if S and T are automorphisms, then the composition $S \cdot T$ is itself an automorphism (Weyl *1952*: 47). Examples of discrete groups are the finite rotation groups of polygons (Figure 1). An example of a continuous group is the rotation $R(\theta)$ with the continuous parameter θ, which satisfies the axioms of a group with $R(\theta_1 + \theta_2) = R(\theta_1) \cdot R(\theta_2)$, $R(0) = I$, $R(\theta)^{-1} = R(2\pi - \theta)$ and $R(\theta) \cdot R(\theta)^{-1} = R(\theta)^{-1} \cdot R(\theta) = I$ (Figure 2).

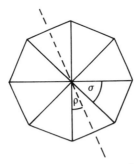

Figure 1 The finite rotation group of a polygon: for n sides, $\sigma = 360°/n$ and $\rho = 360°/2n$ (cyclical group C_8, dieder group D_8 with rotation and reflection symmetries)

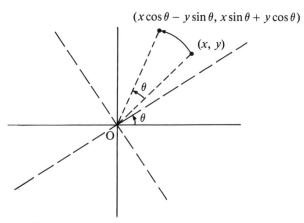

Figure 2 A continuous group: the rotation $R(\theta)$

2 SYMMETRIES OF ORNAMENTS AND CRYSTALS

Discrete groups of general interest are the so-called ornamental groups, discernible in the picturesque ornaments of Indian and Arabic culture. Ornaments of stripes can be classified by the seven 'frieze groups' which are systematically produced by periodic translations in one direction and reflections transverse to the longitudinal axis of translation (Figure 3). In reliefs

Figure 3 The seven 'frieze groups', with an example of each

1614

there may be lines that overlap; we must then consider both sides of a plane, and the number of groups then increases to 31 (Shubnikov and Koptsik *1974*: 79; Speiser *1956*: §28).

In two dimensions there are 17 'wallpaper groups' produced by translations in two directions, reflections, inversions and rotations. Some examples were already well-known in the history of art. In Book II of Kepler's *Harmonice mundi* (1618) he sought regular polygons that would fill the plane. After partial classifications by Camille Jordan (1869) and L. Sohncke (1874), it was the Russian crystallographer Eugraf Fedorov who proved in 1891 that there are exactly 17 types of ornament in the plane (§9.17); George Pólya (1924) delivered the group-theoretical classification. If we use the plane as mirror and consider both its sides, as with the frieze patterns, Pólya's classification can be enlarged to 80 symmetry groups.

The classification of 3-dimensional crystals was an important problem of chemists in the nineteenth century. The French chemist Auguste Bravais suggested there were 14 types of 3-dimensional grid built up from regular cells of parallelepipeds.

In order to classify the discrete space groups, one starts with the regular Platonic solids of Euclidean geometry. Whereas, in the plane, there is a regular n-sided polygon for each $n \geqslant 2$, there are only five regular polyhedra in 3-dimensional Euclidean space. Furthermore, if one considers the proper rotation groups of a point in 3-dimensional space, one finds that there are only three new groups which leave invariant the regular tetrahedron, the cube or octahedron, and the dodecahedron or icosahedron. An analysis of the corners, edges and faces of the Platonic solids bears out this result. Each rotation which leaves the cube invariant also leaves the octahedron invariant, and vice versa (Figure 4). The same applies for the dodecahedron and the icosahedron; the tetrahedron corresponds to itself.

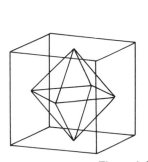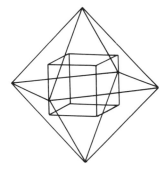

Figure 4 Cube and octahedron

As with the plane, one may ask which of the finite point groups of movements leave spatial grids invariant. In two dimensions there are 10 groups; in three dimensions there are the 32 groups that are of such importance in crystallography. If one considers translations too, one finds the previously mentioned 17 frieze groups with two independent translations in the plane, and the 230 space groups with three independent translations first classified by Fedorov (1890) and Arthur Schönflies (1891) (Burckhardt *1967*). A famous application was Max von Laue's analysis of crystals by X-ray diffraction. He calculated the symmetry groups of crystals from the corresponding diffraction patterns recorded on a photographic plate.

3 SYMMETRIES OF GALOIS THEORY

In geometry, symmetric properties of figures and bodies indicate invariance with respect to automorphisms like rotations, translations and reflections. It was the ingenious idea of the young Evariste Galois to characterize the solutions of equations by their symmetric properties ('permutation groups') (§6.1; see Birkhoff *1937*).

Consider, for example, the polynomial equation of fifth degree $x^5 - 1 = 0$ with the five solutions ('roots'):

$$\left. \begin{aligned}
a_1 &= \tfrac{1}{4}[-1 + \sqrt{5} + (-10 - 2\sqrt{5})^{1/2}], \\
a_2 &= \tfrac{1}{4}[-1 - \sqrt{5} + (-10 - 2\sqrt{5})^{1/2}], \\
a_3 &= \tfrac{1}{4}[-1 - \sqrt{5} - (-10 + 2\sqrt{5})^{1/2}], \\
a_4 &= \tfrac{1}{4}[-1 + \sqrt{5} - (-10 - 2\sqrt{5})^{1/2}], \\
a_5 &= 1.
\end{aligned} \right\} \tag{1}$$

As the coefficients of the polynomial are integers, $x^5 - 1 = 0$ is a polynomial on the field Q of rational numbers. But only the solution a_5 is well defined in Q; the roots a_1, a_2, a_3 and a_4 cannot be distinguished in Q. Thus all relations in Q remain invariant if these four roots are permuted. Besides the identity 1, there are the following permutations:

$$\left. \begin{aligned}
1 &= \begin{pmatrix} a_1 & a_2 & a_3 & a_4 & a_5 \\ a_1 & a_2 & a_3 & a_4 & a_5 \end{pmatrix}, & \delta_1 &= \begin{pmatrix} a_1 & a_2 & a_3 & a_4 & a_5 \\ a_2 & a_4 & a_1 & a_3 & a_5 \end{pmatrix}, \\
\delta_2 &= \begin{pmatrix} a_1 & a_2 & a_3 & a_4 & a_5 \\ a_3 & a_1 & a_4 & a_2 & a_5 \end{pmatrix}, & \delta_3 &= \begin{pmatrix} a_1 & a_2 & a_3 & a_4 & a_5 \\ a_4 & a_3 & a_2 & a_1 & a_5 \end{pmatrix}.
\end{aligned} \right\} \tag{2}$$

Only 4 out of $5! = 120$ possible permutations are automorphisms with the required properties of invariance, and they satisfy the axioms of a group, which is called the Galois group of the polynomial $x^5 - 1 = 0$ on the chosen field. Thus if the field Q is enlarged step by step by one of these roots, the corresponding permutation group is reduced to those roots, which remain indistinguishable.

In general, the increasing chain of (normal) enlargement $F \subset F_1 \subset F_2 \subset \cdots \subset F_n = K(\alpha_1, \ldots, \alpha_n)$ of a field F by roots $\alpha_1, \ldots, \alpha_n$ corresponds to a decreasing chain of permutation groups $G = G_0 \supset G_1 \supset G_2 \supset \cdots \supset G_n = \{1\}$. It can be proved that a polynomial f is resolvable by roots if and only if the Galois group G of f is solvable in the manner described.

For a general equation of nth degree

$$f(x) = x^n + a_1 x^{n-1} + \cdots + a_n = 0 \qquad (3)$$

on the field F, the Galois group is the whole symmetric group S_n with all $n!$ possible permutations of the n solutions. It can be proved that for $n \geqslant 5$ the symmetric group is not solvable, which is why for $n \geqslant 5$ the general equation of nth degree is not solvable by roots (Abel's theorem).

Thus a problem famous in algebra since the Renaissance was resolved by Galois theory. In his manuscripts (published by J. Tannery in 1908), Galois called his theory 'an intellectual simplification: ... the mind grasps a large number of operations promptly and at one go'. Indeed, many famous problems in the history of mathematics were decided by his theory 'at one go': among others, the old problems of constructions by circle and ruler in Greek geometry (Artin *1965*). So it is no wonder that Galois's 'top-down' view of symmetry became very influential in modern algebra.

4 SYMMETRIES OF GEOMETRY AND SPACETIME

In the nineteenth century several theories offering alternatives to Euclidean geometry became established. Felix Klein proposed to order these theories under the general viewpoint of 'geometrical invariants' which remain unchanged by metric, affine, projective, topological and other transformations (the *Erlanger Programm* of 1872; §7.6). As an example the concept of a regular triangle is an invariant of Euclidean, but not of projective geometry: it remains invariant with respect to metric transformations, while a projective transformation distorts the triangle's sides.

In this sense the different concepts of spacetime which were proposed in natural philosophy from Isaac Newton and Gottfried Wilhelm Leibniz to the nineteenth century are mathematically more or less complex structures of symmetry. Galilean invariance means that the form of equations of motion ('natural laws') in classical mechanics is preserved ('invariant') with respect to transformations of the Galilean group (i.e. Galilean transformations of the coordinates in uniformly moving inertial systems) (Figure 5). Intuitively, it means that a natural law is true independently of the

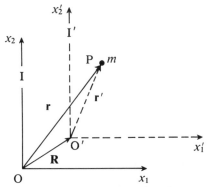

Figure 5 Galilean transformation: $m\,\mathrm{d}^2\mathbf{r}/\mathrm{d}t^2 = m\,\mathrm{d}^2\mathbf{r}'/\mathrm{d}t^2 = \mathbf{R}$

particular reference system of an observer. In this sense Albert Einstein's special relativistic spacetime is an extension to a richer structure of symmetry, the Lorentz-invariant Minkowskian geometry in which the speed of light is constant (§9.13).

Classical mechanics and special relativity are examples of global symmetry; that is, the equations are invariant if all the coordinates are transformed simultaneously. Similarly, the form of a sphere is invariant with respect to a rotation if the coordinates of all its points are changed through the same angle (Figure 6).

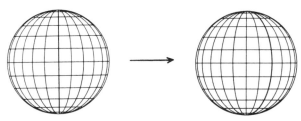

Figure 6 Invariance of the sphere

In general relativity the inertial systems are accelerated and observers seem to be influenced by gravitational forces. Geometrically, we may say that the local deviations of the global symmetry (caused by acceleration) are compensated by fields of force which preserve the symmetry ('form-invariance') of the gravitational law ('local symmetry'). Similarly, there are distortions on the surface of a sphere by local changes of the coordinates. The form of the sphere is preserved by the introduction of forces (Figure

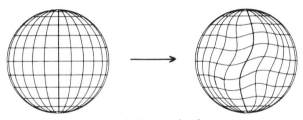

Figure 7 Sphere under forces

7); in short, gravitational forces are a consequence of the transition from global to local symmetry.

In the standard model of relativistic cosmology, the evolution of the universe is determined by the cosmological principle (Weinberg *1972*: 409). It postulates that, at each moment of time, the spatial state of the universe is homogeneous and isotropic to an observer. This principle of symmetry can be made precise by a continuous group of isometries under which physical states like gravitational potential and the energy–momentum tensor are form-invariant. The construction of homogeneous manifolds by groups of isometries is a generalization of the differential geometry of Bernhard Riemann, Hermann von Helmholtz and Sophus Lie which was mathematically introduced by Elie Cartan in 1932 in his theory of symmetric spaces (§6.5), and physically applied by H. P. Robertson and A. G. Walker to cosmology.

5 SYMMETRIES OF ELEMENTARY PARTICLE PHYSICS

As well as the gravitational force, the other fundamental forces of physics (Table 1) can be described in terms of a transition from global to local symmetry (Doncel *et al. 1987*; Mainzer *1988*: Chap. 4). Forces are interpreted as gauge fields which compensate local deviations of a global symmetry.

Table 1 The fundamental forces of physics

	Gravitation	Electro-magnetism	Strong force	Weak force
Range	∞	∞	10^{-13}–10^{-14} cm	10^{-14} cm
Example	Forces between astronomical objects	Forces between charges (in an atom)	Connection of nuclei	β-decay of atomic nuclei
Numerical value	G_{Newton} $= 5.9 \times 10^{-39}$	$e^2 = 1/137$	$g^2 \approx 1$	G_{Fermi} $= 1.02 \times 10^{-5}$
Particles	All	Charged	Hadrons	Hadrons

Each one may be characterized by a dimensionless number; for example, the force of gravitation is 39 orders of magnitude weaker than the strong nuclear force. In James Clerk Maxwell's electrodynamics, a magnetic field compensates a local change of an electric field (i.e. the movement of a charged body) and preserves the invariance of the electromagnetic field equations. In quantum electrodynamics, an electromagnetic field compensates a local change of a material field (i.e. the phase deviation of an electronic field) and preserves the invariance of the corresponding field equations. Mathematically, the phase deviations are described by transformations $\psi \rightarrow e^{i\alpha}\psi$ with a 1×1 unitary matrix as the phase factor. So the electromagnetic force is specified by what is called U(1) symmetry.

The complex variety of particles like hadrons which interact with strong forces can be reduced to the so-called 'quarks' with new degrees of freedom ('colour states'). The colour state of a hadron preserves invariance with respect to a global transformation of the colour states of all quarks in the hadron. But a local transformation of a colour state (i.e. a colour change of only some quarks) needs a gauge field (a 'gluon') in order to compensate the local change and to preserve the invariance of whole hadrons. With three colour states, R (red), B (blue) and G (green), there is a local SU(3) symmetry (Figure 8).

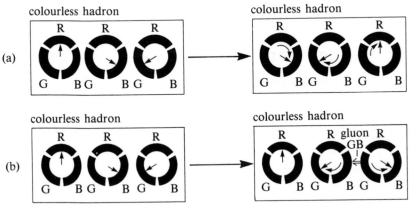

Figure 8 (a) Global and (b) local colour transformations in particle physics

In 1918 and 1919, Herman Weyl had the mathematical idea of unifying physical forces by gauge groups, but he failed for gravitation and electrodynamics because of the lack of quantum-mechanical knowledge in those days. Today the electromagnetic and weak nuclear forces can be unified at very high energies in accelerators – that is to say, at a state of very high energy the interacting particles of both forces cannot be distinguished. Thus mathematically they are described by the same symmetry group, $U(1) \times SU(2)$. At a particular critical value of energy the symmetry breaks down into two partial symmetries, $U(1)$ and $SU(2)$, which correspond to the electromagnetic and weak nuclear forces.

The physical research programme of so-called 'grand unification' (the electromagnetic, and the weak and strong nuclear forces), and ultimately the 'superunification' of all forces is mathematically founded in an extension to richer structures of symmetry ('gauge groups'). Even the very general symmetry group $E_8 \times E_8$ of superstring theory with 10 dimensions is a mathematically well-known continuous group, introduced by Cartan (Helgason *1962*, de Rújula *1986*). It describes a fully symmetric situation of very high energy in which no particles can be distinguished, but they all can be transformed into one another. Following the initial inflationary phase in the expansion of the universe it began to cool, passing through a series of critical temperatures at which symmetries broke down and new particles and forces emerged. 'It is the dissymmetry which creates the phenomenon', said Pierre Curie in 1894.

6 SYMMETRY IN THE MATHEMATICAL SCIENCES

The analysis of molecular symmetry is a basic research area of modern chemistry. In biochemistry, macromolecules (for instance L-amino acids or D-sugars) possess a characteristic 'homochirality' (dissymmetry) which is assumed to be caused by parity violations of weak atomic forces (Figure 9). In the nineteenth century, Louis Pasteur already presumed that living systems are characterized by typical dissymmetries of their molecular building blocks.

The emergence of a pattern or structure may be described by symmetry-breaking not only in chemistry, but also in biology (Engström and Strandberg *1969*). Since the pioneering work by Alan Turing on the

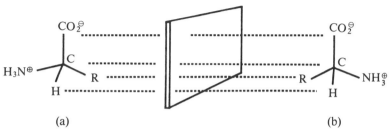

(a)   (b)

Figure 9 Dissymmetry of (a) L- and (b) D-α-amino acids

chemical basis of morphogenesis in biology (1952), there is an increasing interest in this topic. Even the growth of macroscopic populations (i.e. of animals) can be described by symmetry-breaking. A population is regarded as a dynamical system whose growth remains in ecological balance with its environment until this symmetry is broken by some irreversible disturbance. The symmetric pattern is seen in the rhythmical curves of the corresponding non-linear differential equations of populations (Figure 10; see also §9.18). The non-linearity leads to the emergence of global chaotic states caused by tiny fluctuations of the system. Nowadays these chaotic states can be visualized by computer graphics, using coordinates calculated by digital recursion equations. These computer calculations show that mathematical chaos is basically characterized by a fundamental symmetry. As a repeated calculation of coordinates enlarges a computer image, the typical 'self-similarity' pattern of a chaotic figure (for example the Mandelbrot set, shown as Figure 5 of §3.8) emerges again and again. Mathematical chaos is determined by the automorphism of self-similarity (Mandelbrot *1982*). In this sense the famous idea of Heraclitus, that even behind the chaos of our world there is hidden harmony and (mathematical) order, seems to be verified.

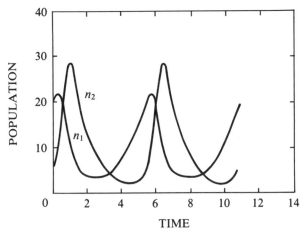

Figure 10 Symmetry in populations: plots of the Lotka–Volterra equations for co-existing populations of prey fish n_1 and predator fish n_2

BIBLIOGRAPHY

Artin, E. *1965, Galoissche Theorie*, Leipzig: Teubner.

Birkhoff, G. *1937*, 'Galois and group theory', *Osiris*, **3**, 260–68.

Burckhardt, J. J. *1967*, 'Zur Geschichte der Entstehung der 230 Raumgruppen', *Archive for History of Exact Sciences*, **4**, 235–46.

de Rújula, A. *1986*, 'Superstrings and supersymmetry', *Nature*, **320**, 678.

Doncel, M. G., Hermann, A., Michel, L. and Pais, A. (eds) *1987, Symmetries in Physics (1600–1980)*, Singapore: World Scientific.

Dyson, F. J. *1966, Symmetry Groups in Nuclear and Particle Physics*, New York: Benjamin.

Engström, A. and Strandberg, B. (eds) *1969, Symmetry and Function of Biological Systems at the Macromolecular Level*, New York: Wiley.

Flurry, R. L. *1980, Symmetry Groups. Theory and Chemical Application*, Englewood Cliffs, NJ: Prentice Hall.

Helgason, S. *1962, Differential Geometry and Symmetric Spaces*, New York: Academic Press.

Mainzer, K. *1988, Symmetrien der Natur. Ein Handbuch zur Natur- und Wissenschaftsphilosophie*, Berlin and New York: De Gruyter.

Mandelbrot, B. *1982, The Fractal Geometry of Nature*, San Francisco, CA: Freeman.

Shubnikov, A. V. and Koptsik, V. A. *1974, Symmetry in Science and Art*, New York and London: Plenum Press.

Speiser, A. *1956, Die Theorie der Gruppen von endlicher Ordnung*, 4th edn, Basel and Stuttgart: Birkhäuser.

Weinberg, S. *1972, Gravitation and Cosmology. Principles and Applications of the General Theory of Relativity*, New York: Wiley.

Weyl, H. *1952, Symmetrie*, Basel and Stuttgart: Birkhäuser.

12.8

Tilings

JOSEPH MALKEVITCH

1 SOME BASIC QUESTIONS

Can a given collection of shapes be used to fill (cover) a geometric region without holes or overlaps? This is the basic question that arises in investigations of tilings or, synonymously, tessellations (Figure 1). The richness of the theory, from the mathematical point of view, arises from the many choices available for the type of shape used (e.g. convex, triangular, regular, starlike, not all in one piece), and for the type of region (e.g. plane, rectangle, cylinder, sphere) to be filled.

To give a feeling for the variety of possibilities, these are some of the questions one could ask. Which regular convex polygons will tile the Euclidean plane? How does the issue of whether the edges of the polygonal tiles abut (i.e. the tiling need not be edge-to-edge) affect what tilings are

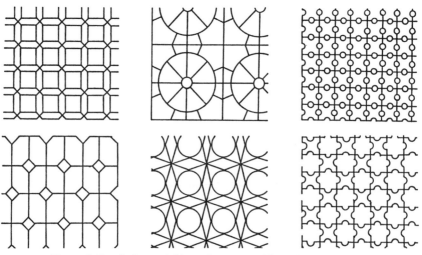

Figure 1 Renderings of fifteenth-century tilings from Portugal
(from Grünbaum and Shephard *1987*: figure 10)

possible? When will congruent copies of one convex polygon tile the plane? Can a square be tiled with squares, all of whose side lengths are different? Which shapes of tetrahedra will tile 3-dimensional Euclidean space? Which convex polyhedra with regular polygons as faces will tile 3-dimensional space?

There are many sets of rules that can be laid down about what constitutes a tile and how the tiles can be juxtaposed. For example, with rectangles as tiles, one can tile the plane in a variety of different ways, as shown in Figure 2. One needs to decide whether one will allow a tile to consist of a region such as those shown in Figure 3. Many strange phenomena can happen when the tiles are permitted to belong to a collection with areas that get smaller and smaller (e.g. different types of long thin tiles). For ease of exposition here, it is assumed that tiles are bounded by finitely many straight-line segments, and that only a finite number of tiles are permitted to meet at a vertex. Intuitively, we are interested in the kinds of tile that might be manufactured for a flooring.

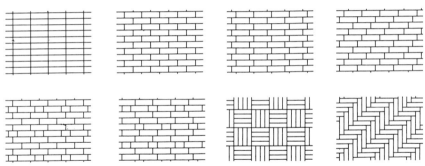

Figure 2 A variety of tilings using rectangles
(from Grünbaum and Shephard 1987: figure 11)

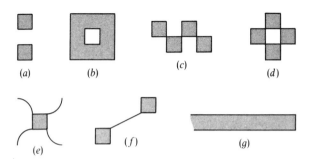

Figure 3 Non-standard tiles, including some consisting of more than one
piece (from Grünbaum and Shephard 1987: figure 1.1.1)

Among the many issues that arise are these. (a) When are two 'different-looking' tilings to be considered equivalent (Figure 4)? (b) How are all possible 'inequivalent' tilings to be enumerated, subject to a particular set of rules? (c) How many different ways are there of colouring the tilings? (d) How does one classify how symmetric a particular tiling may be?

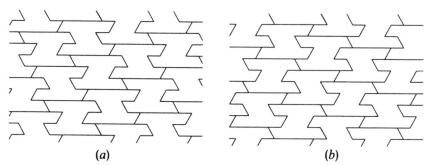

(a)   *(b)*

Figure 4 Tilings which result from using a tile and its mirror image
(from Grünbaum and Shephard *1987*: figure 1.5.2)

2 EXAMPLES FROM ART AND ARCHITECTURE

Historical analysis of tiling is complex because contributions to the 'theory' of tilings have been carried out not only by mathematicians (and other scientists) but also by artists, artisans and craftsmen responsible for creating designs and for tiling the floors and walls of houses, churches and public buildings. A good recent example of the interaction of artists and mathematics is the influence of the Dutch graphic artist Maurits Escher (1898–1972) on tiling theory (Coxeter *et al. 1986*; see also §12.6).

Consider the case of the Arab-influenced Alhambra, in Granada, Spain (Müller *1940*). This building is decorated with a wide variety of symmetric coloured tilings of the plane (i.e. on floors; Figure 5 shows such tilings, in black and white) and of cylinders (i.e. on pillars). If one were to locate (neglecting colours) representatives of the 17 different so-called wallpaper groups, would this mean that the creators of these decorations were in some way aware of the conceptual framework in which we now view these tilings, and that the presence of the tilings in this building was a demonstration of this awareness? It is almost certain that in many cases examples of individual tilings that were constructed for the Alhambra, and also for a house in Pompeii, were not constructed with current theoretical mathematical viewpoints in mind. Unfortunately, it has been common to allow current perspectives to colour our analysis of what was or might have been accomplished in earlier times. It is unlikely that mathematicians prior to the

Figure 5 Details of several tilings from the Alhambra
(from Grünbaum and Shephard *1987*: figure 3)

nineteenth century thought about tilings as we do today; yet some sketchy information is known to us about early specific tilings and mathematical work on tilings.

3 MATHEMATICAL DEVELOPMENTS

What are the origins of a mathematical theory of tilings? Tilings with regular polygons are clearly of mathematical interest. One problem is to find what patterns of regular polygons will fit together around a point. Only a single number of sides for the polygon (i.e. all regular p-gons) might be allowed; the Greek mathematician Proclus (fifth century BC) attributed to the Pythagoreans the discovery that 6 triangles, 4 squares or 3 hexagons are the only regular polygons to fill space around a point (Pythagoras and/or his followers were often cited as the source of 'facts' to which no particular discoverer could be ascribed.) In the context of trying to explain why bees' honeycombs are hexagonal in shape, Pappus (late third century AD), in the preface to Book V of his *Collection*, also mentions that only regular 3-gons, 4-gons and 6-gons can fill space around a point. His discussion certainly suggests that he understood the concept of a regular tiling with one type of polygon, and that there were precisely three of them (Pappus *1933*). It seems, however, to be impossible to attribute the origins of these basic ideas and facts to a particular individual or time. The theory of polyhedra, which from a modern perspective has close links to the theory of tilings, has a sizeable literature starting from Greek times. Activity in the study of polyhedra picked up during the Renaissance, increased in momentum with Johannes Kepler's contributions, but exploded with results only after Leonhard Euler's work around 1750 on the polyhedral formula (vertices − edges + faces = 2; §7.10). Yet by comparison, tilings are barely mentioned in Greek works, seem not to have attracted attention as a mathematical question during the Renaissance, and were only picked up for study again by Kepler.

From a current viewpoint, Kepler's treatment of tilings in *Harmonice mundi* (1619) looks very modern. Just as he built on his knowledge of Pappus's work to 'rediscover' and extend Archimedes' work on polyhedra with regular polygons as faces and with the same pattern of polygons at each vertex, he investigated what some today call the Archimedean tilings, those with regular polygons and the same pattern of polygons about each vertex. He found 11 such tilings which, in more recent terminology, would be called 'uniform', as the vertices all look the same. He gave a 'proof' that his enumeration was complete, but to achieve this he would have had to have a list of all patterns of regular polygons (perhaps of different numbers of sides) which would fit at a point. Here, his analysis does not seem to be totally accurate, but what he knew and what he said he knew might be different. Furthermore, the diagrams that Kepler drew and what he wrote indicate his interest in allowing star polygons and self-intersecting polygons as tiles, in allowing non-edge-to-edge tilings and in other concerns similar to those of today's systematic investigations (Figure 6). Kepler's work, despite its remarkable range of concerns, appears to have exerted virtually no influence on the development of a theory of tilings (just as his work on polyhedra had surprisingly little long-term effect; Malkevitch *1988*).

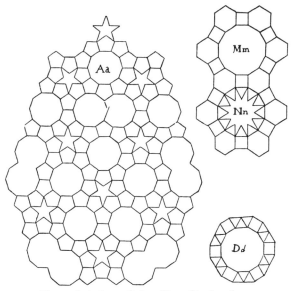

Figure 6 A few of the tilings Kepler drew.
The one marked Aa may be extended to cover the plane
(from Grünbaum and Shephard *1987*: frontispiece)

Although in retrospect one can see antecedents of modern ideas on tilings in pre-twentieth-century mathematics, there was in fact relatively little work on the subject. What work there was is not discussed here because it was almost always faulty in ways that many other parts of nineteenth-century mathematics were not. Specifically, many of the attempts to enumerate different types of tilings were incomplete, or the rules being used to define what was being enumerated were so carelessly set forth, that it is impossible to tell whether these early workers truly accomplished what they claimed (Grünbaum and Shephard *1987*).

4 DEVELOPMENTS IN THE TWENTIETH CENTURY

Progress in the early twentieth century on tilings was heavily influenced by late-nineteenth-century developments in using group theory as a tool for the study of 'symmetry', and by crystallography (§9.17). For example, Georg Pólya (in 1924) and P. Niggli (in 1924) independently rediscovered Eugraf Fedorov's enumeration of the wallpaper groups, and Niggli (in 1926) was apparently the first to enumerate the seven so-called frieze groups. What emerged from this research was a basis for a theory of periodic tilings – those with translational symmetry in independent directions. The group-theoretic point of view encouraged the study of the enumeration of tilings on the basis of their symmetry group being transitive on the tiles, edges or vertices of the tilings. It also encouraged the study of, say, edge-to-edge tilings with regular polygons and with two transitivity classes of vertices. These interesting questions are very much the sort that are under intensive study today.

Independently, as the 18th of his mathematical problems posed in 1900, David Hilbert asked whether or not there existed a single type of polyhedron which tiled 3-dimensional space, yet did not tile space such that the symmetry group of the tiling acted in a transitive manner on the tiles. The fact that he posed a tiling problem which was not about the plane suggests that he may have thought that plane-tiling problems would have straightforward answers. However, his doctoral student Kurt Reinhardt *1918* showed the incredible richness of problems involving tilings of the plane – in particular, that for 'non-pathological' tilings of the plane with convex polygons, there world have to be a tile with at most six sides. Some years previously, Duncan Sommerville *1905* proved that there were 11 Archimedean tilings; he seems to have been unaware of earlier work of a related kind.

Despite the impressive start given to tiling theory early in the twentieth century, and unlike what was to happen for polyhedra, tilings were studied relatively little. However, since about 1975 the remarkable collaboration of Branko Grünbaum and G. C. Shephard has revitalized interest in tilings and related questions. Intriguingly, their work grew out of their earlier (usually) separate investigations into the theory of polyhedra. In a monumental book (Grünbaum and Shephard *1987*) they have energized the study of the theory of tilings. First, they have produced a bibliography referencing nearly all relevant work of both a mathematical and pictorial (artisan-generated) nature. Second, they have examined the proofs of earlier work, correcting and extending it as necessary. Third, they have codified and strengthened methods of earlier workers, and developed new methods and ideas of their own. The result has been to put the foundations of the theory of tilings on a firm basis, to summarize what is currently known, and to lay the foundation for new discoveries by explaining their powerful methods and describing enormous numbers of unsolved questions. There is no doubt that this book has revolutionized the field. In another recent development, Grünbaum *1990* has indicated a framework within which artisans' tilings may be subjected to a mathematical approach based on various schemes involving local repetition of pattern rather than group theory (see §7.3 on regular polyhedra).

One development, reported at length by Grünbaum and Shephard *1987*, deserves independent mention. It may be possible to tile the plane with a given collection of tiles in a manner which is periodic (i.e. has translational symmetries in two directions) or by placing them in a different manner which is not periodic. In 1974, Roger Penrose discovered a surprisingly simple set of two tiles that can tile the plane, but never in a periodic manner! (Figure 7 shows a small piece of one such tiling.) Such a set of tiles is called aperiodic. His discovery accompanies the surprising discovery of larger sets of such tiles by H. Wang, G. Aumann, A. Mackay and R. Robinson. Many new discoveries about the properties of Penrose tiles have been found

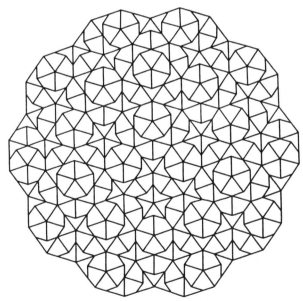

Figure 7 A portion of an aperiodic tiling with two tiles discovered by Roger Penrose (from Grünbaum and Shephard *1987*: figure 10.5.2)

1631

by John Conway. Interest in aperiodic plane tilings has been paralleled by an interest in related phenomena in 3-dimensional space and the study of quasi-crystals. As an example of the simplicity of stating unsolved problems in the area of tilings, it is still not known whether there exists a single tile which constitutes an aperiodic set for tiling the plane!

The study of tilings began in the work of ancient craftsmen and mathematicians and, due to the influence of Grünbaum and Shephard, can be expected to be vigorously pursued into the next century. Tilings generate both beautiful mathematics and beautiful images.

BIBLIOGRAPHY

Coxeter, H. S. M. *et al. 1986*, *M. C. Escher: Art and Science*, New York: North-Holland.

Grünbaum, B. *1990*, 'Periodic ornamentation of the fabric plane: Lessons from Peruvian fabrics', *Symmetry*, **1**, 45–68.

Grünbaum, B. and Shephard, G. C. *1987*, *Tilings and Patterns*, New York: Freeman.

Malkevitch, J. *1988*, 'Milestones in the history of polyhedra', in M. Senechal and G. Fleck (eds), *Shaping Space*, Boston: Birkhäuser, 80–92.

Müller, E. 1940, *Gruppentheoretische und strukturanalytische Untersuchungen der maurischen Ornamente aus der Alhambra in Granada*, Rüschlikon: Baublatt. [Originally Ph.D. Thesis, University of Zurich.]

Pappus *1933*, *La Collection mathématique* (ed. and transl. P. Van Eecke), Paris: Blanchard.

Reinhardt, K. *1918*, *Über die Zerlegung der Ebene in Polygone*, Borna and Liepzig: Noska. [Inaugural Dissertation, University of Frankfurt am Main.]

Schattschneider, D. *1978*, 'Tiling the plane with congruent pentagons', *Mathematics Magazine*, **51**, 29–44.

Senechal, M. *1981*, 'Which tetrahedra tile space?', *Mathematics Magazine*, **54**, 227–43.

—— *1989*, 'A brief introduction to tilings', in M. V. Jaric (ed.), *Introduction to the Mathematics of Quasicrystals*, San Diego, CA: Academic Press, 1–51.

Sommerville, D. M. Y. *1905*, 'Semi-regular networks of the plane in absolute geometry', *Transactions of the Royal Society of Edinburgh*, **41**, 725–47 and plates.

12.9

Mathematics and prose literature

HELENA M. PYCIOR

Although popularly viewed as antithetical, mathematics and literature have shared themes, methods and prose styles. Indeed, authors from Aristophanes to William Boyd have freely borrowed the problems, concepts and symbols of mathematics for their literary works. Thus in his *Birds* (fifth century BC), Aristophanes crafts a play on words around the problem of squaring the circle. In *A La Recherche du temps perdu*, written early in the twentieth century, Marcel Proust regularly uses mathematical metaphors and similes, including the comparison of a moving flock of sheep to a triangle, and the mysterious Madame de Guermantes to a theoretical treatise on geometry (Graham *1966*: 73, 145–6). However, despite the persistent literary fascination with mathematics, there has been little scholarly study of 'mathematics and literature', none addressing all the aspects of the subject sketched below.

1 MATHEMATICAL THEMES IN PROSE LITERATURE

Some major authors have turned to mathematics for secondary or even primary themes. In *Hard Times* (1854), Charles Dickens uses Sissy Jupe's study of proportion to protest contemporary enthusiasm for arithmetical and statistical analysis of the economic and social conditions of Industrial England (Holloway *1962*: 160). In *Alice in Wonderland* and *Through the Looking-Glass*, Charles Dodgson – a Victorian mathematical lecturer at Christ Church, Oxford, writing under the pen-name Lewis Carroll – played with mathematical and logical problems, to the point where mathematics and logic were stripped of their traditional meaning and chaos prevailed. In *Through the Looking-Glass*, for example, the Red Queen not only asks Alice to subtract nine from eight (Alice responds, 'I can't, you know') but also proposes a series of preposterous pseudo-mathematical problems such as dividing a loaf by a knife and subtracting a bone from a dog (Carroll *1960*: 320–21). The strange mathematics and logic of the *Alice* books seem to speak to a theme of what Donald Rackin has described as 'man's absurd condition in an apparently meaningless world' (Rackin *1966*: 314; see Pycior *1984*).

Another Victorian, Edwin Abbott Abbott, constructed his highly successful fantasy *Flatland (1963)* around the notion of dimensionality. The narrator of *Flatland* is a square who lives in a two-dimensional world. There is a caste system, going from women, who are mere straight lines; to soldiers and workmen, who are isosceles triangles; to middle-class men, who are equilateral triangles; and so on to priests, who are regular polygons with so many sides that they are practically indistinguishable from circles. Through the apparition of a sphere the narrator is made aware of three-dimensional space, but his attempts to convince fellow Flatlanders of the third dimension result only in his imprisonment. In short, a mathematical theme permitted Abbott to satirize the class and gender systems of Victorian society even as he made a general plea for open-mindedness, including receptivity to the idea of the fourth and higher dimensions.

If Dodgson spoke to meaninglessness and Abbott to narrow-mindedness, the *Alice* books and *Flatland* were still fantasies. In the late nineteenth and early twentieth centuries, however, mathematics became associated on the Continent and in the Russian Empire with more serious (sometimes naturalistic) prose. Fyodor Dostoevsky, Robert Musil and Yevgeny Zamyatin – all of whom had studied engineering – seized on non-Euclidean geometry and imaginary numbers as useful metaphors for their fictional accounts of the coming-to-terms with the modern human condition. Taken together, the writings of these three implied that mathematics was at least an accessory to the breakdown of traditional beliefs in God, the loss of certainty in human knowledge, and at the same time an extreme and false rationality. Both Ivan and his devil in Dostoevsky's *Brothers Karamazov* (1880), for example, explore the relationship between changing ideas about geometry and changing ideas about God. Paralleling the common 'if, then' arguments of mathematicians, Ivan declares to Alyosha that if there is a God and he created the world, he created it according to Euclidean geometry. Dostoevsky, however, was acquainted with non-Euclidean geometry, and so Ivan soon thereafter refers to 'geometricians and philosophers, and even some of the most distinguished, who doubt whether the whole universe ... was only created in Euclid's geometry; ... [and who] dare to dream that two parallel lines ... may meet somewhere in infinity' (Dostoevsky *1976*: 216). Accepting the non-Euclidean geometries even as he temporarily ignores their human creators, Ivan suggests furthermore that minds that can entertain only a three-dimensional Euclidean Earth are incapable of deciding if God exists or not. 'What Ivan is really aiming at', according to D. E. O. Thompson (*1987*: 77), 'is the idea that since non-Euclidean geometry is not of this world, God is not of this world.'

In the novels of Musil and Zamyatin, imaginary numbers are symbols of a positive irrationality that can counter such dehumanizing forces as mechanization and totalitarianism. Both Musil and Zamyatin initially posit orderly worlds, in which traditional mathematics is recognized as a prime underpinning of certainty and stability, and then trace the breakdown of these worlds as even mathematics loses its claim to absolute certainty and rationality. For example, in Musil's *The Young Törless* (1906), an encounter with the imaginary numbers turns Törless away from the simple knowledge of his youth and towards the complexities and ambiguities of life. Initially viewing mathematics as a 'preparation for life', Törless is troubled by an imaginary number, which (he notes) cannot be a real number; and at the same time he is fascinated by the real results that come from operations on imaginary numbers. Such operations, he confesses, make him feel somewhat 'giddy' and seem to lead 'part of the way God knows where' – at which point Beineberg, a schoolmate, tells Törless, who has argued earlier that there is no God, that he is starting to talk like a chaplain. The association between mathematics and religion continues as Beineberg later reflects that, even when using imaginary numbers, humans have a 'feeling of certainty', and, once aware of this feeling, they cannot deny the existence of the soul. For Törless, who however is unwilling to jump from the imaginary numbers to a soul, the real and imaginary numbers come to symbolize the two sides of human existence, the rational and the irrational. When in a final confrontation with his schoolmasters he tries to explain his behaviour in the case of Basini (a fellow-student who has been degraded by his classmates), Törless explains that existence involves more than thought alone. He says that he will probably always see things two ways – with reason's eyes and with 'those other eyes' (Musil *1986*: 84–6, 141, 168, 171; see White *1966*).

Zamyatin's motives were more political than Musil's, but he (like Musil's Beineberg) also linked the imaginaries to a human soul. Zamyatin's anti-utopian *We*, published in English in 1924, four years after completion, posits a totalitarian state that is regulated by mathematics, in which, for example, the ruler or 'Well-Doer' is 'that Number of all Numbers' and poems are written on the happiness of the multiplication table. The basic story centres around a state mathematician, D-503, whose rationality is challenged as the result of an affair with a female revolutionary, I-330. Even before meeting I-330, D-503 finds the imaginary numbers 'strange, foreign, terrible', and after their meeting the imaginary number becomes a symbol of D-503's emotional and intellectual chaos and of his defiance of the state's rules. D-503 is diagnosed as having a soul and, alternately, fancy, which is eventually excised by a newly invented operation that restores him to the mathematical order of the state (Zamyatin *1952*: 65, 63, 37–8, 96–7; see White *1966*). In this strange historic twist, then, the imaginary numbers – which were problematic for earlier mathematicians – were wholeheartedly embraced by Musil and Zamyatin as entities that defied the apparent rationality of mathematics. In so doing they were addressing an essential human irrationality – of belief, of passions, and (sometimes) of a soul.

Of course, in depicting the modern scene authors have turned to mathematical themes drawn not only from newer mathematical concepts, like non-Euclidean geometry and the imaginary numbers, but from classical concepts as well. Thus the pattern underlying *Petersburg*, which was published in book form in 1916 by Andrei Bely (whose father was the mathematician N. V. Bugaev), is a sphere. Here lines, polygons and cubes symbolize regularity and stability, whereas widening circles and expanding spheres symbolize 'disintegration and death' (Maguire and Malmstad in Bely *1978*: xxi–xxii). Thus Apollon Apollonovich, a government official, finds great comfort in rows of cube-like houses and, in general, loves 'the rectilineal prospect'. However, as Apollon Apollonovich's son, Nikolai, struggles with a hastily given promise to murder his father, father and son are besieged by a series of unsettling objects and situations that embody a menacing sphericity, including Apollon Apollonovich's diseased heart and the instrument of patricide given to Nikolai by his radical associates – a time bomb that is contained in a sphere-like sardine can and meant to explode in a 'sphericality of ... expansion' (Bely *1978*: 10, 14, 157).

Unlike *Flatland*, *The Young Törless*, *We* and *Petersburg*, Thomas Pynchon's *Gravity's Rainbow*, a postmodern novel of 1973, is dominated by no single mathematical theme. Described variously as metafiction or Menippean satire, the novel offers 'songs and stories, mysteries, clues, signs and counter-signs [that] imply some kind of a plot, but because they point so many different ways they generate almost as much confusion as coherence' (Ozier *1975*: 193). Among the clues and signs, which critics are still struggling to interpret, are the scientific and the mathematical, the latter drawn from geometry, probability theory, cybernetics and, especially, analysis. Although an explosive style with multiple mathematical metaphors sets Pynchon's work apart from the earlier mathematical novels, some of his metaphors (especially Δt, the double integral and the singular point) seem to stand for 'the idea of transformation from one world order into another or from one state of being into another' (Ozier *1975*: 203) – a theme somewhat resonant of *The Young Törless*, *We* and *Petersburg*.

Less frenetic, but filled with mathematical snippets, is William Boyd's *Brazzaville Beach* (*1990*). Of the two major stories that are spliced together to form this novel, one concerns the relationship between Hope and John Clearwater. John is a brilliant, if overly stereotypical mathematician; his research into mathematics (depicted as a discipline for the initiates) seems to drive a wedge between him and Hope; still, despite frenzies of research John is unable to achieve a major mathematical breakthrough. Although claiming little mathematical understanding, Hope as narrator provides interesting prose snippets on mathematics, which will probably generate a variety of interpretations from future critics. The snippets stress that traditional mathematics is associated with a regularity that is unattainable in the physical world; it is this abstract mathematics that helps distance John from real life. John, whose research centres on the new area of chaos (here Boyd borrows from the actual history of chaos), finally tires of studying change and turns to 'concepts of permanence'. Hope, on the other hand, finds everyday relevance in catastrophe theory and Fermat's last theorem (her favourite). Fermat's last theorem is probably true but mathematicians cannot prove it, she emphasizes. How like 'statements about the world and our lives', she reflects near the novel's end (Boyd *1990*: 291, 282). Perhaps Fermat's last theorem is for Hope what the imaginary numbers are for Törless and D-503: mathematical entities that transcend a strict rationality and thus symbolize an irrational side of human experience.

2 MATHEMATICAL METHODS IN PROSE LITERATURE

Literature has adapted not only the concepts of mathematics but also its methods. In the seventeenth and eighteenth centuries, Euclidean geometry – with its clear definitions, self-evident axioms, deductive method and hence necessary conclusions – was thought to be the paradigm of human knowledge, and moralists and philosophers (including Henry More and Spinoza) as well as political theorists hastened to craft their writings in the deductive style. By the nineteenth century, however, analysis had blossomed in mathematics and belief in the self-evidence of geometrical axioms had waned (especially with the invention of non-Euclidean geometries). Curiously, as deduction then lost much of its special appeal to scholars outside mathematics, the methods of mathematics found a literary home in detective fiction. The most famous of the nineteenth-century detectives, Edgar Allan Poe's Auguste Dupin and Arthur Conan Doyle's Sherlock Holmes, were known to use methods similar to those of the mathematician but not exclusively deductive.

Thus Poe's *Murders in the Rue Morgue*, first published in 1841 and later in the *Prose Romances* of 1843, opens with a discussion of the power of analysis, or 'disentangling', in which Dupin excels. According to the narrator, 'the faculty in question [analysis] is possibly much invigorated by

mathematical study, and especially by that highest branch of it, . . . analysis'
(Poe *1843*: 9–10). As is made clear in 'The Mystery of Marie Rogêt' (of the
same period), the analysis in question is the calculus of probabilities. Thus
Dupin explains, 'We make chance a matter of absolute calculation. We subject the unlooked for and unimagined, to the mathematical *formulae* of the
schools' (Poe *1965*: 2, 39).

Holmes, for his part, appears no less eager to associate his detection with
the methods of mathematics. In Chapter 2 ('The Science of Deduction') of
A Study in Scarlet, Dr Watson happens upon an essay that Holmes has
published in a magazine. In the essay, Holmes has endorsed the 'Science of
Deduction and Analysis', and claimed that 'his conclusions were as infal-
lible as so many propositions of Euclid' (Doyle *1930*: 23). Similarly, *The
Sign of Four* opens with a chapter on 'The Science of Deduction', in which
Holmes opines that 'Detection is, or ought to be, an exact science', and that
the ideal detective needs the powers of observation and deduction. Further
discussion with Watson, however, suggests that, like Dupin, Holmes relies
on probability as well as deduction. Thus, when pressed by Watson to
explain how he has been able to identify and describe the deceased owner
of a watch, Holmes states, 'that is good luck. I could only say what was
the balance of probability. I did not at all expect to be so accurate.' Continuing to probe Holmes's methods, Watson next asks, 'But it was not mere
guesswork?' Holmes responds, 'No, no: I never guess' (Doyle *1930*: 90–93).

Although Dupin and Holmes themselves stress deduction and probability, episodes such as the dialogue on guesswork from *The Sign of Four*
have suggested to modern semioticians a new interpretation of the methods
of the premier detectives of literature. Umberto Eco, Thomas Sebeok, Jean
Umiker-Sebeok and Nancy Harrowitz have recently argued that Dupin's
and Holmes's basic method was neither deduction nor induction but rather
abduction; that is, a third method of reasoning described by Charles
Sanders Peirce, the early-twentieth-century mathematician, logician and
semiotician, and also an avid reader of Poe (Eco and Sebeok *1983*).
Explained crudely, Peirce's abduction is good guesswork, and it is now
claimed that, despite Holmes's protestations, he and Dupin guess
frequently and well.

3 INFLUENCE OF MATHEMATICS ON PROSE STYLE AND LITERARY AESTHETICS

Not only have mathematical concepts and methods been absorbed into literature, but mathematics has influenced prose style and possibly literary
aesthetics as well. Indeed, some seventeenth-century English men of letters,
scientists and mathematicians argued against ornate prose and for a plain,

straightforward and mathematical style of writing. Francis Bacon called for concise and unadorned accounts of scientific observations. Subsequently, Thomas Sprat and Robert Hooke endorsed plain and terse prose, citing mathematics as an exemplar. In his *History of the Royal-Society* Sprat claimed that the society 'exacted from all their members a close, naked, natural way of speaking, . . . bringing all things as near the Mathematical plainness as they can' (Sprat *1667*: 113). With the new symbolic style of algebra in mind, Hooke urged that the registering of experiments be like 'Geometrical Algebra, the expressing of many and very perplex Quantities by a few obvious and plain Symbols' (Hooke *1705*: 64). John Wilkins and other reformers called for a universal language, but this proved infeasible, and it is now doubtful that the style of post-Restoration literature matched the ideals of the reformers. Whereas Richard Foster Jones maintained that in the seventeenth century there was a broad reform of English prose, which covered even sermons (Jones *1951*), Brian Vickers has argued more recently that little actual reform occurred, but Sprat's *History* 'can still be taken as a prophetic document – that is, what Sprat falsely announced as a *fait accompli* then is what scientific prose has become' (Vickers *1987*: 16, 21).

Even more fundamentally, it has been suggested (without a careful tracing of causal links) that modern mathematics has contributed to major changes in the theory of meaning and in literary aesthetics. Just as, from the nineteenth century on, mathematicians emphasized relations and structures, so in the twentieth have other modern thinkers moved away from the view that meaning is found in things, and towards structuralism. Literary critics, and indeed authors like James Joyce, came to see works of literature less as fictionalized accounts of life told in chronological sequence and in 'the words people actually use' and more as 'patterns gotten by selecting elements from a closed set and then arranging them inside a closed field' (Kenner *1962*: 598–9; see Cassedy *1988*).

4 MATHEMATICAL PEDAGOGY AND PROSE-FORMS

Although mathematicians and scientists have remained largely faithful to the tight prose style developed in the early modern period, mathematical educators have occasionally experimented with variant styles, more commonly associated with fiction. Thus the dialogue and the play, in particular, have been lauded as effective alternate prose-forms to awaken and sustain interest in mathematics as well as to explain its fundamental concepts. A paradigm of its kind, Plato's *Meno* is a dialogue on the philosophy of mathematics. The sixteenth-century British mathematician Robert Recorde also saw pedagogical advantages in the dialogue form, and so adopted it in three of his mathematical textbooks (Howson *1982*: 19–20), as did

Augustus De Morgan in his *Connexion of Number and Magnitude: An Attempt to Explain the Fifth Book of Euclid* (1836). Nearly a half-century later Dodgson, using here his given name, published *Euclid and His Modern Rivals*, an elementary textbook in the form of a play. Moreover, recent accounts of the success of plays in mathematical classrooms attest to continued pedagogical interest in variant styles (Fraser and Koop *1981*).

5 MATHEMATICIANS WITH LITERARY TALENT, AND LITERARY AUTHORS WITH MATHEMATICAL TALENT

As some mathematicians have experimented with variant prose-forms for pedagogy, so have a few first-rate mathematicians composed fictional prose of literary merit. Using the pen-name Dr Paul Mongré, Felix Hausdorff, who laid the basis for general topology, published poetry, aphorisms and a farce, *Der Artz seiner Ehre*, which was successfully staged in 1912 (Katětov *1972*). Sofya Kovalevskaya, a major mathematical analyst, found in literature a primary outlet for her political radicalism. More distinguished for their content than style, her plays and novellas explored such topics as rebellion against tradition, feminism, socialism and communism. She also wrote popular essays that appeared in Russian and European journals and newspapers, as well as *Memories of Childhood*, a fictionalized account of her life that was left incomplete at her premature death in 1891 but nevertheless proved of enduring literary merit (Koblitz *1983*: 257–67).

On the other side, quite a few literary figures evidenced strong mathematical interests, if not talents as well. 'Writers ... of the first rank who received respectable or extensive training in mathematics', Martin Dyck has noted, 'include Novalis, Dostoevsky, Valéry, Musil, Broch, Robbe-Grillet, Arno Schmidt, Solzhenitsyn' (Dyck *1980*: 509 (quote); *1977*: 98–9). Still other authors, like Abraham Gotthelf Kästner, Thomas Carlyle and even Dodgson, made notable contributions to mathematical education. Kästner was both a popular mathematics professor and an epigrammatist. As a professor of mathematics at Leipzig, he taught Gotthold Ephraim Lessing, who himself displayed dual talents and whose 'literary writings ... are informed by a certain mathematicity in their form of organization and argumentation'. Later, as a professor at Göttingen, Kästner taught Carl Friedrich Gauss, who as an undergraduate seriously considered philology as well as mathematics for his major area of study (Dyck *1977*: 96, 103–5, 107 (quote); *1980*: 510). Carlyle concentrated on mathematics from 1809 to the early 1820s, prepared a translation of Legendre's *Geometry* (for which he composed an essay on proportion), and also published *Sartor Resartus* of 1833–4, literary essays, and historical books, on which his fame rests. Carlyle seems, moreover, to have extended the concept of proportion from mathematics to *Sartor*, which has been described as 'a working-out of the proportionality of things material and spiritual' (Moore *1976*: 80).

6 EXPLANATIONS OF MATHEMATICAL–LITERARY DUALITY

There are various partial and indirect explanations of such mathematical–literary duality. Thus Novalis, Leopold Kronecker and others have stressed affinities between the mathematical and literary creative processes, and compared the mathematician to a poet (see e.g. Dyck *1960*: 86–7). Arguing for 'an analogy between literature and mathematics' in his *Anatomy of Criticism* of 1957, Northrop Frye observed that both mathematical and literary works are 'concerned more and more with . . . inner integrity, and less and less with . . . reference to external criteria' (see Section 3). On a specific level, Frye compared the irrational numbers, which 'depend for their meaning solely on the interrelations of the subject itself', to prepositions in verbal language, and mathematical equations to literary metaphors. From Frye's perspective, mathematics – like literature – is an art, falling between architecture and music (Frye *1957*: 350–52, 364). Quite differently, in an essay of 1977 Martin Dyck spoke of mathematics and literature as 'supplementary pursuits', mathematics representing convergent creativity, and literature divergent creativity. Thus, somewhat like Törless, Musil himself perhaps wrestled with two sides, for, according to Dyck, Musil and many other major authors studied mathematics early – 'all seeking convergent patterns, then digressing toward antimathematical states so profuse as to approach confusion' (Dyck *1977*: 98).

BIBLIOGRAPHY

Abbott, E. A. *1963*, *Flatland: A Romance of Many Dimensions*, 5th edn, New York: Barnes & Noble. [1st edn 1884.]

Bely, A. *1978*, *Petersburg* (transl. and with introduction by R. A. Maguire and J. E. Malmstad), Bloomington, IN: Indiana University Press.

Boyd, W. *1990*, *Brazzaville Beach: A Novel*, New York: William Morrow.

Carroll, L. *1960*, *The Annotated Alice* (ed. M. Gardner), New York: Clarkson N. Potter.

Cassedy, S. *1988*, 'Mathematics, relationalism, and the rise of modern literary aesthetics', *Journal of the History of Ideas*, **49**, 109–32.

Dostoevsky, F. *1976*, *The Brothers Karamazov* (transl. C. Garnett, ed. R. E. Matlaw), New York: Norton.

Doyle, A. C. *1930*, *The Complete Sherlock Holmes*, Garden City, NY: Doubleday.

Dyck, M. *1960*, *Novalis and Mathematics*, Chapel Hill, NC: University of North Carolina Press.

—— *1977*, 'Lessing and mathematics', *Lessing Yearbook*, **9**, 96–117.

—— *1980*. 'Mathematics and literature in the German enlightenment: Abraham Gotthelf Kästner (1719–1800)', *Studies on Voltaire and the Eighteenth Century*, **190**, 508–12.

Eco, U. and Sebeok, T. (eds), *1983*, *The Sign of Three: Dupin, Holmes, Peirce*, Bloomington, IN: Indiana University Press.

Fraser, B. J. and Koop, A. J. *1981*, 'Changes in affective and cognitive outcomes among students using a mathematical play', *School Science and Mathematics*, **81**, 55–60.

Frye, N. *1957*, *Anatomy of Criticism: Four Essays*, Princeton, NJ: Princeton University Press.

Graham, V. E. *1966*, *The Imagery of Proust*, New York: Barnes & Noble.

Holloway, J. *1962*, '*Hard Times*: A history and a criticism', in J. Gross and G. Pearson (eds), *Dickens and the Twentieth Century*, Toronto: University of Toronto Press, 159–74.

Hooke, R. *1705*, *The Posthumous Works of Robert Hooke*, London. [Repr. 1969, New York and London: Johnson.]

Howson, G. *1982*, *A History of Mathematics Education in England*, Cambridge: Cambridge University Press.

Jones, R. F. *1951*, *The Seventeenth Century: Studies in the History of English Thought and Literature from Bacon to Pope*, Stanford, CA: Stanford University Press.

Katětov, M. *1972*, 'Felix Hausdorff', in *Dictionary of Scientific Biography*, Vol. 6, New York: Scribner's, 176–7.

Kenner, H. *1962*, 'Art in a closed field', *The Virginia Quarterly Review*, **38**, 597–613.

Koblitz, A. H. *1983*, *A Convergence of Lives, Sofia Kovalevskaia: Scientist, Writer, Revolutionary*, Boston, MA: Birkhäuser.

Koehler, D. O. *1982*, 'Mathematics and literature', *Mathematics Magazine*, **55**, 81–95.

Moore, C. *1976*, 'Carlyle, mathematics and "mathesis" ', in K. J. Fielding and R. L. Tarr (eds), *Carlyle Past and Present: A Collection of New Essays*, London: Vision Press, 61–95.

Moss, S. P. *1970*, 'Poe as probabilist in Forgues' critique of the *Tales*', in R. P. Benton (ed.), *New Approaches to Poe: A Symposium*, Hartford, CT: Transcendental Books, 4–13.

Musil, R. *1986*, *Selected Writings* (ed. B. Pike), New York: Continuum.

Ozier, L. W. *1975*, 'The calculus of transformation: More mathematical imagery in *Gravity's Rainbow*', *Twentieth Century Literature*, **21**, 193–210.

Poe, E. A. *1843*, *The Prose Romances of Edgar A. Poe*, Philadephia, PA: William H. Graham. [Facsimile edn 1968, New York: Saint John's University Press.]

—— *1965*, *The Complete Works of Edgar Allan Poe* (ed. J. A. Harrison), Vol. 5, New York: AMS Press.

Pycior, H. M. *1984*, 'At the intersection of mathematics and humor: Lewis Carroll's *Alices* and symbolical algebra', *Victorian Studies*, **28**, 149–70.

Rackin, D. *1966*, 'Alice's journey to the end of night', *Proceedings of the Modern Language Association*, **81**, 313–26.

Rucker, R. *1984, The Fourth Dimension: Toward a Geometry of Higher Reality*, Boston, MA: Houghton Mifflin.

Sprat, T. *1667, The History of the Royal-Society of London, for the Improving of Natural Knowledge*, London: Martyn. [Facsimile edn 1958 (ed. J. I. Cope and H. W. Jones), St Louis, WA: Washington University Press.]

Stark. J. O. *1980, Pynchon's Fictions: Thomas Pynchon and the Literature of Information*, Athens, OH: Ohio University Press.

Thompson, D. E. O. *1987*, 'Poetic transformations of scientific facts in *Brat'ja Karamazovy*', *Dostoevsky Studies*, **8**, 73–91.

Vickers, B. (ed.) *1987, English Science, Bacon to Newton*, Cambridge: Cambridge University Press.

White, J. J. *1966*, 'Mathematical imagery in Musil's *Young Törless* and Zamyatin's *We*', *Comparative Literature*, **18**, 71–8.

Zamyatin, E. *1952, We* (transl. G. Zilboorg), New York: Dutton.

12.10

Mathematics and poetry

JOHN FAUVEL

There are numerous ways in which two apparently disparate forms of cultural expression, mathematics and poetry, have interacted. Although differences between the practices of poetry and mathematics have been widely recognized, there are similarities too, of style, of activity and of psychology, which have been the subject of comment and analysis (Buchanan *1962*, Tahta *1981*, Priestley *1990*). This article surveys the different modes of interaction between poetry and mathematics.

1 MATHEMATICS THROUGH VERSE

Among the earliest mathematics to have come down to us is that contained in early Indian texts written in condensed poetic style; for example the *Śulbasūtras* of 800–600 BC, which discuss the geometry of ritual altar construction. Storing mathematical knowledge in verse enabled easy memorization, but proved as fragile as any other oral tradition if the chain of transmission was impaired. Later Hindu astronomers wrote in verse; for example, *Śuryasiddhānta* (*circa* AD 500) contains a sine table in verse form.

Among the notable mathematical results preserved in verse is the solution of the cubic equation which Niccolò Tartaglia entrusted to Girolamo Cardano in 1539 (§6.1), explaining that 'to enable me to remember the result in any unforeseen circumstance, I have arranged it as a verse in rhyme, because if I had not taken this precaution, I would frequently have forgotten it'. For several subsequent centuries there was a tradition of displaying mathematical problems in verse. This was especially prevalent when such problems formed a major part of the fare of lay magazines such as the *Ladies' Diary* (1704–1840). Some writers of textbooks have, indeed, chosen to communicate much of their text in verse, trusting no doubt to the greater memorability of their poetry. Thomas Hylles's *The Art of Vulgar*

Arithmetic, both in Integers and Fractions (1600) is an early example:

Addition of fractions and likewise subtraction
Requireth that first they all have like bases
Which by reduction is brought to perfection
And being once done as ought in like cases,
Then add or subtract their tops and no more
Subscribing the base made common before.

2 POETRY ABOUT MATHEMATICS

As well as the display of mathematical problems or techniques in not especially inspired verse, there has been a tradition of writing poetry about
mathematicians. From the sixteenth century to the nineteenth, poems were
written to eulogize the writers of mathematical textbooks and other
worthies. A notable example is the writing of poems in praise of, or inspired
by, Isaac Newton, a popular pursuit during the eighteenth century and after
(Bush *1950*, Nicolson *1966*). And accounts of episodes in the history of
mathematics have been the subject of poems, of which a not untypical
example is the exposition of Hindu number-words in Edwin Arnold's *Light
of Asia* (1879):

And Viswamitra said, 'It is enough,
Let us to numbers.
   After me repeat
Your numeration till we reach the lakh,
One, two, three, four, to ten, and then by tens
To hundreds, thousands.' After him the child
Named digits, decads, centuries; nor paused,
The round lakh reached, but softly murmured on,
'Then comes the koti, nahut, ninnahut,
Khamba, viskhama, abab, attata, . . .

Poems are of interest to the historian in reflecting how mathematics
has been perceived within strands of culture. In particular, following the
eighteenth-century success of the Newtonian world-view, there was a reaction
in which mathematics was taken to epitomize narrow blinkeredness and
cruel mechanism in the face of humanity. This view is seen in the work of
Romantics such as Blake, Wordsworth and Keats, and in a late exemplar of

this tradition, Walt Whitman (1855):

> When I heard the learn'd astronomer,
> When the proofs, the figures, were ranged in columns before me,
> When I was shown the charts and diagrams, to add, divide, and measure
> them,
> When I sitting heard the astronomer where he lectured with much applause
> in the lecture-room,
> How soon unaccountable I became tired and sick,
> Till rising and gliding out I wander'd off by myself,
> In the mystical moist night-air, and from time to time,
> Look'd up in perfect silence at the stars.

Another kind of reaction to mathematics is shown in the sonnet by the American poet Edna St V. Millay, 'Euclid Alone Has Looked on Beauty Bare' (1923). Poems may also present evidence of mathematics education, similar to that found in prose autobiographies. Victor Hugo, for example, wrote in 1846 (translated by M. Ashby) of his childhood experiences of mathematics:

> I was a living sacrifice to numbers, black executioners;
> I was force-fed with algebra,
> They tied me to a rack of Bois-bertrand
> They tortured me from the wings to the beak
> On the terrible rack of X and Y: . . .

3 POETRY BY MATHEMATICIANS

Mathematicians reflect the age in which they live, not least during periods when poetry is a common mode of expression. During the nineteenth century, poetic composition was a frequent pastime, and poetry was written by several mathematicians: notably, in the UK, by George Boole, William Rowan Hamilton, James Clerk Maxwell, Thomas Penyngton Kirkman and J. J. Sylvester, and elsewhere in Europe, by Augustin Louis Cauchy and Karl Weierstrass. Little of this poetry is especially memorable – 'Umar al-Khayyāmī is the only person to be remembered equally as a poet and a mathematician, although some poets have been amateur mathematicians (Henry Wordsworth Longfellow is an example). There is little of mathematical import in poems written by mathematicians of the nineteenth century. The twentieth century, however, has seen the growth of much light and serious verse directly on mathematics, published in anthologies (see e.g. Robson and Wimp *1980*) as well as in occasional corners of journals addressed to the mathematical community. An example is T. Apostol's song, of a loosely historical character, whose first stanza runs:

Where are the zeros of zeta of s?
G. F. B. Riemann has made a good guess,
They're all on the critical line, said he,
And their density's one over $2\pi \log t$.

4 MATHEMATICAL ANALYSIS OF POETRY

Mathematicians interested in poetry have explored the field in another way, the mathematical analysis of poetic style. A seminal work was Sylvester *1870*, exploring such concepts as 'phonetic syzygy' in order to elucidate and characterize the poetic quality in poetry, while Birkhoff *1933* provided a quantitative measure of phonetic syzygy in an 'aesthetic formula'

$$M = (aa + 2r + 2m - 2ae - 2ce)/c, \tag{1}$$

according to which Samuel Taylor Coleridge's 'Kubla Khan' has the value $M = 0.83$, while for 'Onward, Christian Soldiers' $M = 0.51$; this produces an order relation, at least, acceptable to most readers of these two poems. Studies of poetic style have also been made during this century by Russians, notably by Andrei Markov and Andrei Kolmogorov (Kendall *1990*: 40).

5 MATHEMATICS IN POETRY

The deepest use of mathematics in poetry has been where mathematical concepts or imagery have been integral to the poetic expression. Examples of this occur throughout the history of poetry. The work of two of the great medieval poets, Dante and Geoffrey Chaucer, is permeated with their mathematical–scientific learning. The seventeenth-century metaphysical poets John Donne and Andrew Marvell saw aspects of love through mathematical analogies such as the compass, and the parallel lines of Euclidean geometry. Referring to the early nineteenth century, Alfred North Whitehead (*1926*: 105) has drawn attention to a stanza in Shelley's *Prometheus Unbound*, which he claims 'could only have been written by someone with a definite geometrical diagram before his inward eye – a diagram which it has often been my business to demonstrate to mathematical classes':

I spin beneath my pyramid of night,
Which points into the heavens, – dreaming delight,
Murmuring victorious joy in my enchanted sleep;
As a youth lulled in love-dreams faintly sighing,
Under the shadow of his beauty lying,
Which round his rest a watch of light and warmth doth keep.

In the twentieth century, the work of poets such as William Empson and Wallace Stevens is notable for mathematical resonances. Empson, for example, uses concepts of projective geometry in a self-reflexive conceit:

> Duality too has its Principal.
> These lines you grant me may invert to points;
> Or paired, poor grazing misses, at your joints,
> Cross you on painless arrows to the wall.

Besides its use in the writing of poems, mathematics has been significantly called upon in the work of modernist literary theoreticians from the late nineteenth century onwards, such as the poets Stéphane Mallarmé, Andrei Bely and Paul Valéry (Cassedy *1988*).

6 MATHEMATICS AS POETRY

In Western culture, 'poetry' can be used almost metaphorically, as a term of high aesthetic appreciation and to say something particular about the special qualities of imagination and beauty in a creation. Several examples of mathematics itself have been seen as poetry, from the view of W. M. Priestley (*1990*: 15) that 'Leibniz's way of writing the calculus approaches the poetic', to the judgement of G. Mittag-Leffler that Niels Abel's last works were 'truly lyric poems of a subtle beauty whose perfection of form reveal profundity of thought' (Tahta *1981*: 43). And Sofya Kovalevskaya, who herself cultivated both mathematics and literature, concurred with the view that mathematicians and poets share vital attributes. 'It seems to me', she wrote, 'that the poet has only to perceive that which others do not perceive, to look deeper than others look. And the mathematician must do the same thing'. It is arguable to what extent such claims are merely metaphorical, observing that doing mathematics is like doing poetry, and to what extent they capture a deep epistemological truth about poetry and mathematics.

BIBLIOGRAPHY

Beer, G. *1990*, 'Translation or transformation? The relations of literature and science', *Notes and Records of the Royal Society of London*, **44**, 81–99.

Birkhoff, G. D. *1933*, *Aesthetic Measure*, Cambridge, MA: Harvard University Press.

Buchanan, S. *1962*, *Poetry and Mathematics*, Philadelphia, PA: Lippincott. [1st edn 1929.]

Bush, D. *1950*, *Science and English Poetry: A Historical Sketch 1590–1950*, New York: Oxford University Press.

Byard, M. M. *1977*, 'Poetic responses to the Copernican revolution', *Scientific American*, **236**, (June), 120–29.

Cassedy, S. *1988*, 'Mathematics, relationalism, and the rise of modern literary aesthetics', *Journal of the History of Ideas*, **49**, 109–32.

Kendall, D. G. *1990*, 'Kolmogorov: The man and his work', *Bulletin of the London Mathematical Society*, **22**, 31–47.

Koehler, D. O. *1982*, 'Mathematics and literature', *Mathematics Magazine*, **55**, 81–95.

Nicolson, M. H. *1966*, *Newton Demands the Muse: Newton's Opticks and the Eighteenth Century Poets*, Princeton, NJ: Princeton University Press.

Priestley, W. M. *1990*, 'Mathematics and poetry: How wide the gap?', *The Mathematical Intelligencer*, **12**, 14–19.

Robson, E. and Wimp, J. (eds), *1980*, *Against Infinity*, Parker Ford, PA: Primary Press.

Sylvester, J. J. *1870*, *The Laws of Verse, or, Principles of Versification Exemplified in Metrical Translations*, London: Longmans, Green.

Tahta, D. *1981*, 'On poetry and mathematics', *For the Learning of Mathematics*, **1**, 43–7.

Whitehead, A. N. *1926*, *Science and the Modern World*, Cambridge: Cambridge University Press.

12.11

Stamping mathematics

H. WUSSING

1 A RANGE OF ILLUSTRATIONS

Stamp collecting is a worldwide pursuit. As the number of stamps issued has made it impossible for the average enthusiast to build up a complete national collection going right back to the beginning, thematic philately has become increasingly popular in the decades since the Second World War. Subjects such as sport and the Olympic Games, railways, flowers and space currently occupy leading positions in thematically orientated philately.

The theme of mathematics as reflected by philately is also very rewarding for collectors, and experts in the subject can confirm and demonstrate that the number and variety of related motifs is surprisingly high. As would be expected, there are many stamps bearing portraits of prominent mathematicians, along with related mathematical achievements and instruments. Other aspects that have been depicted on stamps are mathematics in education, its link with the natural sciences and technology, and also particular events in the history of mathematics. It has became difficult to find a firm, meaningful place to draw the line. Even philosophers, princes of the church and politicians have been concerned with mathematics. Mathematics is closely bound up with the history of human culture in every part of the world; here too the observant collector educated in the history of mathematics will be able to discover many significant and interesting motifs. Last but not least should be mentioned related postmarks, first-day covers, postcards and suchlike.

Postal authorities in every country of the world have resorted to the mathematics theme, albeit to varying extents and on widely different grounds. Some countries have illustrated the contribution of individuals to mathematics on a world scale, irrespective of their nationality, while others have been moved to concentrate almost exclusively on their national traditions. Particularly noteworthy is the effort of numerous African and Ibero-American states to highlight scientific and cultural achievements in stamp issues, including those in the field of mathematics and natural sciences, even though they have been unable to make any significant contribution themselves.

In short, the highly interesting philatelic theme of mathematics is practically inexhaustible. In this article it is possible to do little more than give examples of the aspects indicated above, and of these to describe only a few. For each stamp mentioned or illustrated the country and year of issue are given so that it can be identified in any good stamp catalogue. Book-length surveys of stamps are listed in the bibliography.

The stamps illustrated are in eight groups on four pages at the end of the article. Each group has its own caption; the numbering of the individual stamps is across the rows, from top to bottom. Reference is made to some of them in the text, thus: '(2.4)'.

2 PERSONS

Many mathematicians appear in portrait or figurative representation, some on stamps from several countries. To name just some: Pythagoras, Zu Chongzhi, al-Ṭūsī, al-Fārabī, al-Khwārizmī, Regiomontanus, Ries, Nuñez, Kepler, Pascal, Descartes, Euler, Leibniz, Newton, Diderot, d'Alembert, Lagrange, Laplace, Monge, F. J. Gerstner, Gauss, J. Bólyai, Lobachevsky, Poincaré, Bolzano, Cauchy, Galois, Abel, Ostrogradsky, Chebyshev, Lyapunov, Quetelet, Kovalevskaya, Dedekind, Russell and Banach. There are some remarkable omissions, including Euclid, Georg Cantor and Hilbert.

Particularly effective are those stamps showing the portrait together with the scientific achievement. This group includes, for example, the stamp for Blaise Pascal (France 1962) showing him framed between mathematics (conic section) and passionate religiosity (4.3); and the one for Euler (Switzerland 1957) with his portrait and the symbols e and i he introduced in the famous formula $e^{i\pi} + 1 = 0$ (2.3). There are curiosities too: the French stamp commemorating the 300th anniversary of the appearance of Descartes's *Discours de la méthode* (1637) originally appeared with the erroneous title *Discours sur la méthode*.

Personalities whom mathematicians would automatically include in the story of mathematics have often, in the opinion of the postal authorities, been deemed worthy of honour in quite a different connection. For example, there is Leibniz the philosopher (German Reich 1926) (2.5), Diderot and d'Alembert as editors (France 1958–9), and O. Y. Schmidt not as a group theorist but as a polar explorer (Soviet Union 1953), while Gerbert von Aurillac, who learnt of the Hindu–Arabic numerals in eleventh-century Spain, was celebrated as the later Pope Sylvester II (Hungary 1938, France 1964).

Among the subjects which are somewhat tricky to classify are portrait designs from the meeting-ground of mathematics with natural sciences and technology. There exist commemorative stamps for Democritus, Aristotle, Aristarchus, Archimedes, Ptolemy, Zhang Heng, Avicenna, Thomas Aquinas, Lull, Leonardo da Vinci, Dürer, Brunelleschi, von Gmunden, Mercator, Brahe, Kepler, Galileo, Torricelli, Huygens, Vauban, L. M. N. Carnot, Ampère, Bessel, Foucault, Le Verrier, Helmholtz, Maxwell, Hertz, Steinmetz, Boltzmann, Planck, Nernst, Stefan, Lorentz, von Laue, Schrödinger, Pauli and Bohr, and many others. Here too, even from the viewpoint of the history of mathematics, many stamps are very expressively designed.

Some personalities, including those closely connected with mathematics, have enjoyed an arguably inflated worldwide tribute in their memorial years, for instance Copernicus in 1973 (500th anniversary of his birth) and Einstein in 1979 (100th anniversary of his birth) (6.3, 8.3). For a catalogue of mathematical subjects, particularly portrait stamps, see V. Grossmann's admirable research in Schreiber *1980*: 80–98.

3 FACTS

Geometric figures, formulas and mathematical symbols form another group within the general subject of mathematics. A selection of examples will give an idea of its extent. First, there is the set of ten from Nicaragua (1971) bearing 'the ten mathematical formulas that have governed the face of the Earth', including elementary calculation, Pythagoras's theorem, the law of gravity (2.1), Einstein's formula $E = mc^2$ (6.3), and logarithmic tables, sometimes accompanied by symbolic images of their application. Further examples include the Gaussian theory of numbers (Federal Republic of Germany (hereafter FRG) 1977) (3.6), the Möbius strip (Brazil 1967), the geometric representation of Pythagoras's theorem (among others Greece 1955 (5.4), Surinam 1972), Hamilton's theory of quaternions (Ireland 1983) (3.3), a set of four to commemorate the appearance of Newton's *Principia* (Great Britain 1987), Roman numerals (Czechoslovakia 1968), computer graphics (among others, Finland 1978 to mark the International Mathematical Congress in Helsinki in 1978 (6.4), Netherlands 1970), integral signs (Soviet Union 1966, to mark the International Mathematical Congress in Moscow), the 'age pyramid' in population statistics (Austria 1979) and fundamental arithmetical operations (Colombia 1968).

4 INSTRUMENTS AND APPARATUS

Images of mathematical instruments and equipment are often associated with astronomy, geography, measuring techniques, navigation and other fields of natural science or practical application. Here too there is enormous variety. Suffice it to name, next to the oft-appearing circle (e.g. Archimedes, Greece 1983; and even in the first-day cover postmark for Leon Battista Alberti, Italy 1972), the triquetrum, an astronomical measuring instrument (Mongolia 1973), the astrolabe (e.g. Syria 1978, 1980; Iran 1956 (1.1); Portugal 1975 (1.4); Poland 1982), the armillary sphere (e.g. China 1953; Austria, German Democratic Republic (hereafter GDR) 1972 (1.3)), the nautical cross-staff (e.g. Netherlands 1986 (1.5)), the *quipu* (Inca device for recording information, see §12.1) (Peru 1972 (1.2)), the quadrant (e.g. Cuba 1982), the theodolite (e.g. West Berlin 1981) and the sextant (e.g. Nicaragua 1971); also the illustrations of calculating devices from the abacus (e.g. Australia 1972) to the hand-reckoner by W. Schickard (FRG 1973) and modern computers and data-processors (e.g. Bulgaria 1980). In addition, famous clocks can be included among mathematical and astronomical instruments in the wider sense of the term, such as the global clock by Bürgi (e.g. Switzerland 1983) and the artistic clocks modelled on the astrolabe, for example the one in the Old Town Hall in Prague (Czechoslovakia 1978) and the table-clock (GDR 1975) in the Dresden *Zwinger* (a museum).

5 NATURAL SCIENCE AND TECHNOLOGY

The connections between mathematics and the natural sciences and technology are variously reflected in postage-stamp themes; a few examples may at least give some idea of the extent: electromagnetic field lines (Czechoslovakia 1959, FRG 1983), light beams in the eye by al-Haytham (Pakistan 1965), meridian measurement in Lapland by Maupertuis (France/Finland 1986), deciphering German military radio messages in the Second World War by exiled Polish mathematicians (Poland 1983), the Royal Greenwich Observatory (Great Britain 1984), Kirchhoff and the law of radiation (West Berlin 1974), the Hertzsprung–Russell diagram for correlating data concerning stars (Mexico 1942), the Inca observatory (Peru 1960), Halley's Comet (e.g. Nicaragua 1985), Petzval's photographic object lens (Austria 1973), postal land maps and nautical sea charts (e.g. Portugal 1960, Spain 1974; see also 4.4), early maps (e.g. Bermuda 1979, Brazil 1972), metrical systems of measurement (e.g. France 1954, 1975), education in mathematics (Japan 1984, Haiti 1967), the Brahe observatory (Ascension 1971), the sundial (e.g. GDR 1983) and the thermometer (Sweden 1982).

Decoding the scientific and mathematical content of the stamp design often requires a sharp sense of perception and fairly advanced specialized scientific knowledge on the part of the collector.

6 CULTURAL HISTORY

Mathematics has played a considerable part in humankind's cultural history. It is therefore obvious that a thematic collection of mathematics' many aspects will cover an abundance of related subjects. To mention but a few: geometric decoration on pre-Columbian Indian earthenware (among others, Brazil 1981), a pre-Columbian codex of astrological–astronomical content (Mexico 1982), the dome of St Peter's (Vatican 1972), personalities of the French Revolution 1789, with Condorcet (France 1989), the Abel memorial in Oslo (Norway 1983), Alfonso the Wise of Castille (Spain 1965), Moses Maimonides (Israel 1953), Bishop Berkeley (Ireland 1985), chess (among others, Paraguay 1982) and the world chess champion (among others the mathematician Lasker, 1976), impossible geometric figures by the Swedish artist O. Reutersvärd (Sweden 1982) (6.7–6.9; see also §12.6, Figure 10) and in the style of M. Escher (6.5), and signs of the zodiac (e.g. San Marino 1969).

BIBLIOGRAPHY

Robin Wilson contributes a regular 'Stamp corner' in *The Mathematical Intelligencer*.

Schaff, W. L. *1978, Mathematics and Science. An Adventure in Postage Stamps*, Reston, VA: National Council of Teachers of Mathematics.
Schreiber, P. *1980, Die Mathematik und ihre Geschichte im Spiegel der Philatelie*, Leipzig: Teubner.
Wussing, H. and Remane, H. *1989, Wissenschaftsgeschichte en miniature*, Berlin: Deutscher Verlag der Wissenschaften.

1.1   Astrolabe (Iran 1956)
1.2   In the Inca kingdom information was recorded on *quipus* (§12.1);
differently coloured knotted cords indicated the different concerns
(news, trade, numerical data) (Peru 1972)
1.3   Armillary sphere, seventeenth century (GDR 1972)
1.4   With astronomical knowledge, the astrolabe can be used for location-
finding, which classifies it as a practical navigational aid (Portugal 1975)
1.5   Nautical cross-staff (Netherlands 1986)

2.1   Newton's law of gravitation, with allusion to the legend of the
falling apple (Nicaragua 1971)
2.2   Newton and the second reflecting telescope (Ascension 1971)
2.3   Euler (Switzerland 1957)
2.4   Youthful portrait of Huygens (Netherlands 1928)
2.5   Leibniz (German Reich 1926)

3.1 Dedekind: prime ideal division (erroneous formula!) (GDR 1981)
3.2 Gauss's construction of the regular 17-sided polygon (GDR 1977)
3.3 Hamilton's quaternions (Ireland 1983)
3.4 Kovalevskaya (Soviet Union 1951)
3.5 Chebyshev (Soviet Union 1957)
3.6 Gaussian numerical plane (FRG 1977)

4.1 Galileo and some of his main scientific achievements:
the ballistic curve, isochronal pendulum, hydrostatics and
experiments with falling objects (Panama 1965)
4.2 Kepler's laws of planetary orbits (FRG 1971)
4.3 Pascal, split between mathematics (conic section) and
religious passion (France 1962)
4.4 The Portuguese universal scholar Nuñez:
algebra, the loxodrome (Portugal 1978)

5.1 Archimedes as geometrician and physicist (Greece 1983)
5.2 The Chinese mathematician Zu Chongzhi who, among other things,
calculated π accurately to six decimal places (China 1956)
5.3 The legendary Imhotep, supposed to have been the constructor of the
first great pyramid, the stepped pyramid of Saqqara, Old Kingdom
(Egypt 1928)
5.4 Pythagoras's theorem (Greece 1955)
5.5 The stepped pyramid of Saqqara (Egypt 1973)
5.6 An abacus and the electronic circuit of a modern calculating machine
(Australia 1972)

6.1 Russell (India 1972)
6.2 Schrödinger (Austria 1987)
6.3 Einstein (China 1979)
6.4 The International Mathematical Congress in Helsinki, 1978
(Finland 1978)
6.5 Impossible geometric figure, on stamp for the 10th International
Austrian Mathematical Congress, Innsbruck 1981 (Austria 1981)
6.6 Banach (Poland 1982)
6.7–6.9 Impossible geometric figures. Related drawings and paintings
created by the Swedish artist O. Reutersvärd (Sweden 1982)

7.1   Monge (France 1953)
7.2   Lagrange (France 1958)
7.3   Bolzano (Czechoslovakia 1981)
7.4   Abel (Norway 1929)
7.5   Galois (France 1984)
7.6   Cauchy (France 1989)

8.1   Maxwell's formula for electromagnetic fields (Nicaragua 1971)
8.2   János Bólyai (not an authentic portrait) (Hungary 1960)
8.3   Einstein's formula $E = mc^2$ (Vietnam 1979)
8.4   Mach (Austria 1988)
8.5   Lobachevsky (Soviet Union 1951)
8.6   Lorentz (Netherlands 1928)
8.7   Hertz and Maxwell (Mexico 1967)

12.12

Monuments to mathematics and mathematicians

DAVID SINGMASTER

1 TYPES OF MONUMENT

The true monument to any mathematician is his work. Nonetheless, few of us are immune to the peculiar fascination of the more solid variety of monuments to our predecessors. The fact that 'The Mathematical Tourist' is one of the longest-running features in *The Mathematical Intelligencer* demonstrates the popularity of this fascination, but I know of only one attempt to gather such information in a single article (Alexanderson *1982*). After reading it, I began collecting and collating further material in a computer file. The current version runs to 112 pages, which includes considerable material of the form 'A taught at B', recording associations rather than actual monuments (Singmaster *1992*). This compilation is most complete for the British Isles. I would be delighted to hear from anyone with information about monuments, particularly outside the British Isles.

Mathematical monuments may be classified into five types:

1 Monuments to mathematics itself.
2 Monuments with some incidental mathematical content.
3 Sites where actual discoveries were made (generally encrusted by legend).
4 Artefacts accumulated in local, regional and national museums and learned societies.
5 Monuments associated with a particular mathematician. Most fall into this category: birthplace, house(s) lived in, institution(s) worked in, place of death, grave and other memorials. The mathematician may be honoured by a museum or a section of a museum near any of these sites.

2 EXAMPLES OF MONUMENTS

Here are described some of the rarer types 1–3, and some examples of the most interesting or characteristic of types 4 and 5.

2.1 Mathematics itself

The only monument to mathematics *per se* that I know of is the 'Monumento à Matemática' at Itaocara, Brazil (Figure 1). There is also a sculpture garden of mathematicians and physicists at the Otto von Guericke Technische Hochschule in Magdeburg.

Figure 1 The 'Monumento à Matemática' at Itaocara, Estado do Rio, Brazil, erected in 1946 by Dr Carlos Moacir de Faria Souto (photograph provided by Jaime Poniachik). It is not too hard to read the names Hamilton, Galois, Hermite, Riemann, Dedekind, Cantor and Poincaré on the front slope. The symbols d_x, C_m^n, $f(x)$ and $e = 2.718281$ can be seen on the upper right face, and $\frac{0}{0}$, lim, $(x + a)^m$, $\sin^2 x + \cos^2 x = 1$ on the upper left face

2.2 Incidental mathematical content

Salisbury Cathedral in England has two monuments with some mathematical interest. There is the grave of Thomas Lambert, 'who was borne May y^e 13 An. Do. 1683 & dyed Feb 19 the same year'. The tomb of the Gorges family, *circa* 1610, has the skeletons of icosahedra, dodecahedra and cuboctahedra on the tops of its columns. These seem to be modelled on Leonardo's drawings.

The tombstone of the educator Friedrich Froebel at Schweina, near Jena, Germany, shows a sphere on a cylinder on a cube, thus commemorating his advocacy of blocks for children to play with at school.

The Houses of Parliament in London have an odd mathematical interest. The previous building burned down in 1834 in a fire caused by burning obsolete tally-sticks; also destroyed were the Imperial standard weights and measures, which had to be reconstructed. Consequently, secondary standards were set up at various other sites such as the north side of Trafalgar Square and at the Royal Greenwich Observatory, where they can still be seen.

Many modern artists have used mathematical concepts such as prime numbers, graphs, the Möbius strip and the Golden Number in their work, but such art is outside the scope of this article. However, the Brazilian artist Antonio Peticov has plans for a monument to Fibonacci using the Fibonacci numbers and the Golden Number (12.4), while the Shobana Jeyasingh Dance Company recently created a 'dance opera', *Correspondences*, inspired by the life and work of Srinivasa Ramanujan.

2.3 Sites of discovery

'Galileo's lamp' in the Cathedral of Pisa was made some years after Galileo's discovery of the pendulum (it has the date 1587 on it), but perhaps he was watching the previous lamp. The Leaning Tower still leans, and opinion is still divided as to whether Galileo ever dropped weights off it.

The celebrated meeting of Henry Briggs with John Napier, leading to the adoption of base 10 logarithms, took place in Merchiston Tower, Napier's birthplace, now handsomely restored as part of Napier Polytechnic, Edinburgh.

'Newton's apple tree' is long dead, but cuttings were propagated and the tree under 'Newton's windows', outside Trinity College, Cambridge, is a descendent. The Royal Astronomical Society in London has a piece of the original tree.

William Rowan Hamilton invented quaternions on 16 October 1843 at Brougham Bridge, Dublin. There is a plaque facing the footpath under the bridge, but the formulas he scratched into the stonework with his knife have long since vanished.

There is a plaque on the Moore School of Electrical Engineering, at the University of Pennsylvania, Philadelphia, recording the invention and construction of ENIAC and EDVAC (§5.12).

There are instances where a mathematician has recorded the date of a discovery, but the actual site is unknown. Niccolò Tartaglia solved the cubic on the night of 12–13 February 1535, presumably at his house in Venice. Richard Dedekind records that he invented Dedekind cuts (§3.3) on 24 November 1858, while he was teaching at the Polytechnic School in Zurich.

2.4 Mathematical artefacts

Although anonymous, the Rhind Mathematical Papyrus in the Third Egyptian Room of the British Museum and the Old Babylonian tablet Plimpton 322 in the Rare Book Room of the Butler Library at Columbia University must be mentioned as the greatest monuments of ancient mathematics.

Charles Babbage's prototype difference engine and several fragments of analytical engines (§5.11) are in the Science Museum in London. His Difference Engine No. 2 was constructed in 1991 to celebrate the bicentenary of his birth.

2.5 Memorials

Graves produce various emotions, especially if one knows the life of the individual. The graves of Niels Abel, Bernhard Riemann and Georg Cantor are especially moving. Evariste Galois's grave is unknown; Ramanujan and Alan Turing were cremated and their ashes scattered. But a grave or monument is more fascinating when it displays some mathematics.

Archimedes' tombstone in Syracuse showing the sphere, cone and cylinder is the classic example. Cicero describes his uncovering of the monument in 75 BC in his *Tusculan Disputations* (V: xxiii). Recent excavations may have uncovered the tomb, but the tombstone has not been found. A modern version has been erected at San Jose State University in California, with the Golden Section worked into the design.

Van Ceulen's moment in St Peter's Church, Leyden, with his 35 digits of π, is lost. James I Bernoulli's 1705 monument in the Peterskirche, Basel, with the logarithmic spiral, survives, though the spiral looks more Archimedean than logarithmic (Figure 2).

Figure 2 James I (Jacobus) Bernoulli's tombstone in the Peterskirche, Basel, with a spiral (from the dustjacket of his *Werke*, Vol. 2 (1989); photograph kindly supplied by Dr E. A. Fellmann)

The Gauss statue at Braunschweig has a star 17-gon on the side of the base. At Göttingen, there is a statue of Carl Friedrich Gauss and Wilhelm Weber apparently discussing magnetism and their invention of the telegraph, but it has a greater mathematical significance in that David Hilbert prepared his famous study on the foundations of geometry (§7.4) for its unveiling in 1899.

The base of the bust of Carl Lindemann at the Mathematical Institute in Munich shows a squared circle. Josip Plemelj's monument in Bled, near Ljubljana, Slovenia, has his extensions of the Cauchy integral formula carved upon it (Figure 3).

The only mathematician to be commemorated in a feature film is Sofya Kovalevskaya, in the Swedish film *A Hill on the Dark Side of the*

Figure 3 Josip Plemelj's tombstone at Bled, Slovenia, showing his formulas (photograph by David Kirby). The third line of the inscription is partially obscured by a vein in the stone.

Moon (1983). The names of numerous mathematicians have been attached to asteroids and features on the Moon and planets, but it is difficult to visit these at present.

Collective memorials at colleges, universities and 'pantheons' tend to be less notable, but no discussion of mathematical monuments can omit mention of Trinity College, Cambridge, where the Antechapel has the statue of Isaac Newton with his prism (as celebrated by Wordsworth), along with statues of Isaac Barrow, Francis Bacon and William Whewell. Roger Cotes is buried in the Chapel, and there is a wall plaque. On the walls, brass plaques record some fifteen Trinity mathematicians and many others of interest – physicists, philosophers and historians of science. This is without doubt the most extensive collective monument to mathematicians.

BIBLIOGRAPHY

Alexanderson, G. L. *1982*, 'On mathematical monuments, or the mathematician's Baedeker', *California Mathematics*, 7, (1), 3–11 and cover.

Singmaster, D. *1992*. [Computer file on monuments, available from the author, South Bank University, London SE1 0AA, UK.]

12.13

Talepiece:
The history of mathematics
and its own history

I. GRATTAN-GUINNESS

The reader who has followed the order in which the articles are presented in this encyclopedia is at the end of a long trail. The next item, §13.1, is a bibliography of the most pertinent writings in, or for, the field of the history of mathematics; this article is a bridge, in that the field itself is the subject-matter.

Naturally only a selective history can be furnished, and potential boredom forbids the mere production of long lists of names, so it is mainly the general histories and principal journals that are mentioned. However, Sections 4 and 5 take some note of the views of historians down the ages on the nature of their task, especially the coverage of mathematics they attempted and the sorts of audience they addressed. One aim is to show that the field has its own history, including its creative and philosophical aspects.

Before that, Section 1 picks out some major texts written up in 1870. Section 2 continues the tale up to the First World War, and Section 3 records certain developments up to our time. Some note is also taken of the history of the philosophy of mathematics.

Since §13.1 is full of information, the bibliography here exceptionally excludes the substantial number of histories which are mentioned in the text but which can easily be found there. Few works like this article have been written; this is a pity, for, as it shows, the history is a curious, up-and-down story.

1 FROM EUDEMUS TO BONCOMPAGNI

The history of mathematics itself has a long history; for the first historian of whom we have any indication is Eudemus, who flourished in the fourth century BC. His work is lost, but it was used by Proclus seven centuries later. Historical work played a role in medieval researches in mathematics, for example in the Arabic translations of, and commentaries on, Greek texts. In particular, several Arabs not only translated and edited Greek texts and developed the ideas in them, but also wrote histories of mathematics (al-Birūnī was especially noteworthy in this regard).

Thus our field was born in the East; however, its later developments have been dominated by Western traditions, from the seventeenth century, although non-Western mathematics has never been forgotten. But the emergence of history of mathematics as an activity distinct from mathematics as such, and from other kinds of history, starts in the seventeenth century, and in the Western tradition (Struik *1980*). For example, John Wallis's *Treatise of Algebra* (1685) included an extensive (although not unbiased) history of the subject. Among the (few) works of the eighteenth century, J. C. Heilbronner's *Historia matheseos universae* [...] (1742) gave an odd (though informative) general survey, for he presented a melange of historical accounts of certain branches and also listed publications and manuscripts.

Soon afterwards there appeared a work which is rightly seen as a milestone: Etienne Montucla's *Histoire des mathématiques* (1758) (Sarton *1936b*). His first volume concentrated on Greek, Roman and Eastern traditions, while the second covered geometry, mechanics and optics up to the seventeenth century. Forty years later the second edition began to appear, although its author then died and the task of completion fell to Jerome Lalande. With help from various colleagues, he completed the third volume and wrote the fourth within four years. The first two volumes broadly treated the same material as before, while the other two handled all aspects of the eighteenth century; the whole encompassed around 3000 pages, including indexes (themselves an excellent feature).

One curious feature of the work, especially in its chapters on applied mathematics, is the lack of mathematical symbolism in the text: Montucla and Lalande were often content with a verbal account of the work without entering into symbolic renderings. Of course many basic aspects of the history of mathematics were thus left out. In some ways a closer contact with history is found in the *Traité du calcul différentiel et du calcul intégral* (1797–1800) by his contemporary Sylvestre Lacroix. Lacroix was the textbook writer *par excellence* of his day, and this book was in fact a treatise; but he possessed vast historical knowledge of mathematics (he had helped Lalande with parts of Montucla), which he presented in the extensive tables of contents, and he discussed the technical details in a thorough way.

The treatment of non-Western sources by historians of that time is also worth noting. Montucla, and even more so the astronomer and historian of astronomer Jean-Baptiste Delambre, show rather clearly a certain arrogance concerning language: when treating very ancient traditions such as the Egyptian, they frankly admitted their ignorance, but for (medieval) Arabic sources they drew on Latin editions without demur. One might say that for the ancients these historians knew that they did not know, whereas for the Arabs they did not even know that! Nevertheless, the French maintained a good record in Eastern scholarship, especially at the Collège de France. Among other non-Western cultures, study of Sanskrit mathematics was pioneered by figures such as the Englishman H. T. Colebrooke.

Other nations produced historians; for example, the German Abraham Kästner issued a characteristically encyclopedic *Geschichte der Mathematik* in four volumes (1796–1800), and Johann Poppe followed suit in 1828 (but with a mere 660 pages). The Italians have always shown a strong interest in the history of mathematics (Barbieri and Cattelani Degani *1989*), and Guglielmo Libri produced a major national *Histoire des sciences mathématiques en Italie* in four volumes (1838–41), covering the time from the Romans to Galileo. But the importance of France was evident even in the language in which Libri wrote; and the first national history, *Ensaio Historico Sobre a Origem e Progressos das Matematicas em Portugal* (1819) by F. Garçao-Stockler, was published in Paris.

The French took an initiative of another kind in 1855 when Olry Terquem added an historical *Bulletin* to the monthly numbers of his journal *Nouvelles Annales des Mathématiques*. But the first major journal devoted to the history of mathematics was the *Bullettino di Bibliografia e di Storia delle Scienze Mathematiche*, launched in 1868 by Baldassarre Boncompagni. A rich prince able to indulge his eccentricities, he even changed text while passing proofs (so that apparently not all copies are the same), and insisted on authors indicating line-breaks when rendering titles of works! But his twenty annual volumes contain a mass of invaluable researches, especially for the Renaissance period, his own major interest. He even indexed each volume, a source of information to be enhanced by the cumulative index recently prepared by S. A. Jayawardene and currently in the press.

2 1870–1914: A GOLDEN AGE

There now began a period of over four decades when work in the history of mathematics reached a peak. German-language scholarship led the way. The field-marshal was Moritz Cantor, who produced three hefty volumes of *Vorlesungen über Geschichte der Mathematik* between 1880 and 1898, when he began the second edition. By then in his seventies, he edited a mammoth fourth volume written by colleagues, which covered the period 1759–99; it even included a chapter on the history of mathematics up to that time (Günther *1908*). He also edited a series of monographs under the title *Abhandlungen zur Geschichte der mathematischen Wissenschaften* (1877–1913), which is especially useful for the medieval period and non-Western mathematics, and also for important bibliographies. Another significant work was the *Geschichte der Elementar-Mathematik* by Johannes Tropfke, which came out in augmented editions between 1902 and 1940, and is being updated now.

Among mathematicians with an historical interest, Felix Klein played the role corresponding to Cantor's with the historians. He was a major initiator of the *Encyklopädie der mathematischen Wissenschaften*, the vast German project published between 1898 and 1935, singled out in §0 as an irreplaceable source of information about its time and to some extent about its prehistory. He was also a stimulus to the International Commission for Mathematical Education, which produced many fine studies (mostly in German) from the late 1900s onwards. Several volumes were partly historical, and some of its topics have hardly been treated since: for example, H. E. Timerding, *Die Mathematik in den physikalishen Lehrbüchern* (1910), M. Gebhardt, *Die Geschichte der Mathematik in mathematischen Unterrichte* [...] (1912) and J. Schröder, *Die* [...] *mathematischen Unterrichts an den höheren Mädchenschulen Deutschlands* [...] (1913).

Among journals, the *Jahrbuch über die Fortschritte der Mathematik* began in 1868 as the first comprehensive abstracting journal for mathematics, and from the start it contained divisions for the history and philosophy of mathematics. Then the Swede Gøsta Eneström launched another specialist journal for the field, *Bibliotheca mathematica*. It began in 1884 as an appendix to the (new) mathematical journal *Acta mathematica* edited by Gøsta Mittag-Leffler (who published there his own historical articles on his favourite mathematicians); but three years later it continued as an independent publication, with many important pieces including much new material on the eighteenth century. Eneström also introduced a series of interesting queries about details of the past. Unfortunately, over the last decade to 1913 he became obsessed with reporting quibbles over errors in Cantor's *Vorlesungen*; the points made are of very uneven quality.

Meanwhile, the Italians continued the tradition established by Boncompagni. G. Battaglini ran a *Bollettino di Storia e Bibliografia di Matematica* from 1892 to 1897, when Gino Loria launched a *Bollettino* which exactly followed the rest of Boncompagni's title and general purpose. Loria's journal also lasted for twenty years, although it did not reach the same level of importance. However, he published an interesting 'guide' to our field, the first of its kind, containing not only various classified bibliographies but also reflections on its prosecution and on its own history (Loria *1916*).

This period also saw a rise in the preparation of editions, selected or complete, of the works of mathematicians. There is an extraordinarily large number of such editions, far more than for other sciences. A few date back to the sixteenth and seventeenth centuries; Greeks so honoured were Ptolemy (in Latin, 1541) and Archimedes (1544), while Western mathematicians included Christoph Clavius (1611–12), François Viète (1646) and Girolamo Cardano (1663). Some editions appeared in the eighteenth century, for example for James and John Bernoulli (1742 and 1744, respectively) and Isaac Newton (1779–85). The nineteenth century saw the first Niels Abel edition (1839), the books of Pierre Simon Laplace (1843–7) and the mathematics of Gottfried Wilhelm Leibniz (1849–63); but the main flow began around 1860. The first edition of Johannes Kepler (1858–71) and the editions of Augustin Jean Fresnel (1866–8) and Carl Friedrich Gauss (begun 1866) appeared, and far too many more to list here (Sarton *1936a* gives a good listing for more modern mathematicians). The quality was very variable, especially in France: the Fresnel is excellent, and those for René Descartes (1897–1913) and Pierre de Fermat (1891–1912) respectable if controversial on some policy matters; but Joseph Louis Lagrange, Laplace (in the second version, including papers also), Augustin Louis Cauchy and Joseph Fourier were badly served. German, Italian and British editions are usually more reliable. As already with John Bernoulli, a few mathematicians prepared, or at least started, their own editions (for example, Arthur Cayley, Lords Rayleigh and Kelvin, and Karl Weierstrass).

A different type of edition was launched in Germany in 1889: *Ostwald's Klassiker der exakten Wissenschaften*, in which major scientific (often mathematical) texts were reprinted (or, where necessary, translated into German) and furnished with notes. Generally the standard of editing was impressive, and the series has been durable: from the Second World War to 1989, each Germany ran its own continuation. In addition, editions/ translations of some ancient Greek mathematics were produced by several figures – for example into German by J. L. Heiberg, English by T. L. Heath and French by Paul Tannery.

3 DECLINE AND RENAISSANCE

After the First World War and until the 1960s our field was at a relatively low ebb, although the war itself was only partly responsible. Some projects continued, especially the *Encyklopädie der mathematischen Wissenschaften*; but, for example, the death in 1914 of Jules Molk saw the demise of the French edition of that encyclopedia, which under his dedicated direction had even surpassed in quality those parts of the original that had been reworked (Eneström *1915*). Among other projects, the edition of the works of Leonhard Euler had begun to appear in 1911, and has been a centre of activity down to today; among other editions, the works of Christiaan Huygens reached completion in 1950.

However, the old impetus had gone: in particular, even in Germany the historians became isolated figures. Heinrich Wieleitner produced in the 1920s a *Geschichte der Mathematik* of his own and a *Mathematischen Quellenbücher* of original sources, and Otto Neugebauer launched in 1929 a series of monograph articles on *Quellen und Studien* in the history of mathematics, physics and astronomy, and was a leader of an excellent group of historians of ancient and Arabic mathematics. But work on most other periods declined quite considerably. One reason may have been the rise of positivistic world-views in philosophy, which convinced mathematicians, like other scientists, that 'history was bunk'.

At this stage the USA grew in importance. Figures such as Florian Cajori, D. E. Smith and R. C. Archibald had been active there for some years in producing both research articles and introductory books: Cajori had even published *The Teaching and History of Mathematics in the United States* in 1890. Smith now put out his own *History of Mathematics* in 1923, concentrating on elementary topics (and including a section on 'the historians of mathematics'). Six years later his *A Source Book in Mathematics* came out; unfortunately it excluded works before 1500, following a policy of the series to which it belonged. E. T. Bell published various books which are still well known (much more so than is warranted by the notable element of fiction in them). In addition, the journal *Scripta mathematica* was launched in 1932 by Jekuthiel Ginsburg; as a consequence of the recent progress in the philosophy of mathematics (Part 5), it covered philosophical as well as historical topics. Other US historians include L. C. Karpinski and J. L. Coolidge.

Then George Sarton, one of the few historians of science who took the history of mathematics seriously, produced a valuable handbook *1936a*. His journal *Isis*, devoted to the history of science, played a major part in keeping our field alive. Two new abstracting journals for mathematics started at this time, *Zentralblatt der Mathematik* (Germany, 1931) and *Mathematical Reviews* (USA, 1940): the presence on both projects of Neugebauer, who was one of their principal organizers, ensured that history was included.

After the Second World War the field gradually picked up, especially in the USA, with figures such as Carl Boyer, Morris Kline, Neugebauer (who even founded a department for the history of mathematics at Brown University) and Dirk Struik. Other countries in which there was activity included the Federal Republic of Germany with J. E. Hofmann (who inaugurated there a regular conference on the field which still gathers) and H. Gericke; France with René Taton; and the Soviet Union with A. P. Yushkevich (who launched a journal for the field, *Istoriko-mathematicheskogo issledovaniya*, in 1948).

From 1960 the growth of work in the history of science in general was quite rapid; it was led by the USA, where the journal *Archive for History of Exact Sciences* was founded that year by C. A. Truesdell III. However, the history of mathematics was rather left out of this change, for unfortunately it has always been viewed as marginal by historians of science (Grattan-Guinness *1990*). The main advances, at least in the West, have occurred only since the mid-1970s. Since then, though, progress has been rapid: the increase has been especially marked in France, Italy and the (former) German Democratic Republic. Several countries have come newly to the field, and their interests include the history of their own regions: for example, Australasia, Ibero-America and Africa. Quite a number of women have come into the field – for the first time.

A major initiative led by K. O. May was the founding in 1972 of a Commissioner for the History of Mathematics, under the auspices of the International Union of the History and Philosophy of Science. Its most important achievement was to launch in 1974 the journal *Historia mathematica*, which not only publishes articles in the field but also reviews current activity – publications (including a department of abstracts), meetings and projects. Several other journals have commenced in certain countries, some of them also concerned with the (history of the) philosophy of mathematics. Books are also fairly numerous; there are even several book series for, or involving, the field. Editions are another prominent form of work: Newton's mathematical manuscripts have been handsomely done, the Euler edition is now on to his manuscripts, and the Bernoulli dynasty is at last making a collective bow, with both previously published and unpublished writings.

4 REFLECTIONS

The main thing is to understand that in a world dominated by scientific methods the history of science should be the keystone of higher education.
(Sarton *1936a*, preface to the 1957 reprint)

Here are four groups of comment on facets of the history just outlined.

4.1 Mathematics becomes pure mathematics

As mathematics became more professionally organized during the nineteenth century, especially the second half, its pure sides gained in prestige over its applications, and among its applications the more theoretical ones became detached from those prevailing in engineering and technology. During the same period the same tendency became apparent in the history of mathematics, and it still is. The history of applied mathematics has always been fairly buoyant, especially for mechanics and astronomy, but it still does not attract the attention appropriate to its place in history. For example, the most neglected part of the field is mathematics connected with engineering and technology.

Like everything else, this situation has a history. Signs of the favouring of pure mathematics are already evident in Montucla, who began the preface of the second (1799) edition of his history with the news that he had intended to write the history of 'geometry and pure mathematics' (note this phrase) but that friends had prevailed upon him to write a 'general' history. In the end the four volumes give the majority of space to applications, as befitted the mathematics of his time. But there was to be a change of balance during the nineteenth century. Contrast, for example, two *Geschichten der Mathematik*: in 1828 Poppe devoted the last 300 of 550 pages of main text to applications (his own main interest), whereas H. Suter's two volumes (1873–5) were larged devoted to the pure side. Most of the later general histories (and source-books) manifest this imbalance.

The great mistake made here has been to overlook the fact that, until the growth of professionalization just noted, applications – or at least applicability – dominated in mathematics, and they still remained very prominent afterwards. This imbalance must be reflected in general (even if summary) histories. The extent to which distortion occurs can be appraised from the *Abrégé d'histoire des mathématiques 1700–1900* (1978) produced under the editorship of J. Dieudonné: a modern view was adopted, so that about three-quarters of all mathematics of the period mentioned in the title was omitted, and moreover in silence!

We have here examples of anachronism: the view of the historian's own time imposed upon his period of study. A similar situation holds for probability and statistics, which, as detailed in Part 10, arrived late on the general mathematical scene. Study of the subject's history has been comparably tardy, despite pioneers such as I. Todhunter's *A History of the Mathematical Theory of Probability* (1865), for it 'arrived' only in the 1980s. It is still usually, but indefensibly, placed on the margin in general histories of mathematics.

4.2 Where is the nineteenth century?

Partly connected with the above considerations is the history of the nineteenth century. Histories produced early in the twentieth century, such as Cantor's and Rouse Ball's (even his fourth edition of 1908), scampered through the nineteenth with lists of names and a few results. This neglect is rather strange, since by then the early decades were already far behind. Gradually the situation has improved, thanks to innovations such as Klein's lectures on the century (1926–7) and the third volume (1933) of Loria's general *Storia della matematica*; and Sarton's handbook *1936a* is noteworthy for its stress on the nineteenth and twentieth centuries. Even then, the period was still pretty deserted in the mid-1960s.

By contrast, the nineteenth century and the early twentieth, at least as far as the interwar period, is now a major focus of attention, with a substantial survey of many aspects provided in Rowe and McCleary *1989*. Some general histories reflect this change: for example, Kline's *Mathematical Thought from Ancient to Modern Times* (1972) is only half way through when 1800 is reached. (However, the criticisms made above about imbalance between pure and applied mathematics apply to this book.) This concentration of interest enabled me – thankfully – to let my conception of this encyclopedia run that far forward.

4.3 Philosophy of history and history of philosophy

In the jargon of the history of science, traditional history of mathematics was 'internalist', recounting results and proofs as such but little else. Such narrow-mindedness has largely given way to serious attention being granted to the social, educational and institutional (or 'external') aspects: for example, these appear not only in Part 11 of this encyclopedia. Biography had also gained in popularity, and welcomely: the melange of internal and external issues is well exemplified there, and it also has educational utility (Pycior *1987*). A few mathematicians have even written autobiographies (Cardano, Charles Babbage, George Biddell Airy, Norbert Wiener), though none has produced a major work of this kind.

In tandem with these concerns, general questions in the history of science have been considered, such as: is mathematics cumulative? (No, after some reflection, but do 'revolutions' occur in it?) In addition, greater care is often taken in stressing the differences between reconstructing a historical situation as it seems to have been, and reformulating it in some modern version (a perfectly legitimate activity, but one which mathematicians usually misconstrue as the 'real' historical task). William Whewell's *History* and

Philosophy of the Inductive Sciences (1837, 1840) deserve mention in these contexts.

In addition, the philosophies of mathematics have gained much interest and also vigour in recent years, after the rather specialized obsessions recorded in Part 5. Its own history has been studied, not only as a record but also as a source for new positions. In addition, the varied scope of philosophical questions in mathematics has been better appreciated, following the work on heuristics in mathematics initiated by George Pólya and continued by followers such as Imre Lakatos.

4.4 History and education

With historians like Heilbronner and Montucla, the history of mathematics was a branch of scholarship, and much work has always been of this character. During the last hundred years, however, educational stimuli have become more noticeable. Cantor's big work was based on lectures, as his title states, and American authors have been particularly motivated by education (e.g. Cajori in 1893, and Smith in 1923). This concern has continued to grow: most historians of my acquaintance have come to the field (and also to philosophy, in some cases) from an educational experience, as was mentioned in §0. Their reactions to education have often been negative, but various positive threads are also evident: the history and philosophy of mathematics education, and its two-way interaction with research.

5 SELF-REFLECTION

How better to close this encyclopedia than with some remarks about this encyclopedia? Certain specific points have been made above: suffice it to suggest here that, for all its gaps and imperfections, the most striking feature is that it exists at all; it could not have been set in train even ten years ago. However, the reports of rapid recent progress in Section 4 should not distort the general context. The field is still marginal to the disciplines where it deserves a major place, and fails as ever to gain academic posts; it has even lost some recently in former socialist countries.

Thus we are truly back with the concerns stated at the head of §0. Mathematics occupies a peculiar position in cultural life today. 'Everybody knows' that it is one of most basic, and also ancient, types of knowledge; yet it is not part of normal cultural discourse, and few people know much about its historical development, or even that it *has* a history...

BIBLIOGRAPHY

Barbieri, F. and Cattelani Degani, C. (eds) *1989, Pietro Riccardi (1828–1898) e la storia delle matematiche in Italia* [...], Modena: Department of Mathematics, University.

Demidov, S. S. and Folkerts, M. (eds) *1992*, 'Historiography and the history of mathematics', *Archives Internationales d'Histoire des Sciences*, **42**, 1–144. [Special issue on history in various countries and periods.]

Eneström, G. *1915*, 'Jules Molk als Förderer der exakten mathematisch-historischen Forschung', *Bibliotheca mathematica*, Series 3, **14**, 336–40.

Grattan-Guinness, I. *1990*, 'Does History of Science treat of the history of science? The case of mathematics', *History of Science*, **28**, 149–73.

Günther, S. *1908*, 'Geschichte der Mathematik', in M. Cantor (ed.), *Vorlesungen über Geschichte der Mathematik*, Vol. 4, Leipzig: Teubner, 1–36.

Loria, G. *1916, Guida allo studio della storia delle matematiche*, 1st edn, Milan: Hoepli. [2nd edn 1946. Many errors in rendering of names and titles.]

Pycior, H. *1987*, 'Biography in the mathematics classroom', in I. Grattan-Guinness (ed.), *History in mathematics education* [...], Paris: Belin, 170–86.

Rowe, D. and McCleary, J. (eds) *1989, The History of Modern Mathematics*, 2 vols, Boston, MA; Academic Press.

Sarton, G. A. L. *1936a, The Study of the History of Mathematics*, Cambridge, MA. Harvard University Press. [Includes references on historians of mathematics. Repr. 1957, New York: Dover, bound with his *The Study of the History of Science*.]

—— *1936b*, 'Montucla (1725–1799). His life and works', *Osiris*, **1**, 519–67.

Struik, D. J. *1980*, 'The historiography of mathematics from Proklos [sic] to Cantor', *NTM – Schriftenreihe zur Geschichte der Naturwissenschaften, Technik and Medizin*, **17**, (2), 1–22.

Part 13
Reference and information

13.0

Introduction

The encyclopedia is completed by articles to scan rather than to read. §13.1 lists important general books and periodicals (not only current ones) in the history of mathematics. There follows in §13.2 a chronology of some principal births, deaths, publications and (more rarely) other events; and in §13.3 a concise record of basic facts about the major mathematicians. The authors are listed in a similar way in §13.4; finally, §13.5 is a general index of persons and subjects.

13.1

Select bibliography of general sources

ALBERT C. LEWIS

This article provides an introduction both to the literature in the history of mathematics and to bibliographical sources for history and for mathematics itself. Items are ordered alphabetically by author or title, as appropriate. Limitations of space have forced selection to be severe, and omissions should not be taken as value judgements. For a higher degree of detail see Dauben (1.1) and May (1.5), both of which also list some major manuscript collections. See also the sources department of *Historia mathematica* (2.9). Many mathematical items are among the thousands of original printed books and periodicals in microform in D. Roller (ed.) 1968– , *Landmarks of Science* and *Landmarks II*, New Canaan, CT: Readex.

1 GUIDES AND BIBLIOGRAPHIES FOR HISTORY OF MATHEMATICS

1.1 Dauben, J. W. (ed.) *1985, The History of Mathematics from Antiquity to the Present: A Selective Bibliography*, New York: Garland. [Forty-nine historians of mathematics contributed to this indexed, subject-arranged and heavily annotated bibliography.]

1.2 Grattan-Guinness, I. *1977*, 'History of mathematics', in A. R. Dorling (ed.), *Use of Mathematical Literature*, London and Boston, MA: Butterworth, 60–77. [Other articles are also useful.]

1.3 Jayawardene, S. A. *1983*, 'Mathematical sciences', in P. Corsi and P. Weindling (eds), *Information Sources in the History of Science and Medicine*, London: Butterworth, 259–84.

1.4 Loria, G. *1946, Guida allo studio della storia delle matematiche; generalità, didattica, bibliografia*, 2nd edn, Milan: Hoepli. [Particularly useful bibliographies.]

1.5 May, K. O. *1973, Bibliography and Research Manual of the History of Mathematics*, Toronto: University of Toronto Press. [The most comprehensive single bibliography available, covering publications from 1868 to 1965;

however, it lacks indexes and has other shortcomings. The abstracts department of *Historia mathematica* (2.9) seeks to continue the same comprehensive coverage.]

1.6 Müller, F. *1909, Führer durch die mathematische Literatur, mit besonderer Berücksichtigung der historisch wichtigen Schriften*, Leipzig: Teubner (*Abhandlungen zur Geschichte der mathematischen Wissenschaften mit Einschluss ihren Anwendungen*, Vol. 27).

1.7 Russo, F. *1969, Éléments de bibliographie de l'histoire des sciences et des techniques*, 2nd edn, Paris: Hermann. [Similar to Sarton (1.9), but not so wide-ranging.]

1.8 Sarton, G. *1936, The Study of the History of Mathematics*, Cambridge, MA: Harvard University Press. [The classic historiographical treatise by one of the founders of the modern discipline of the history of science. Repr. 1954, New York: Dover, and bound with his *The Study of the History of Science*.]

1.9 Sarton, G. *1952, Horus. A Guide to the History of Science; a First Guide for the Study of the History of Science, with Introductory Essays on Science and Tradition*, Waltham, MA: Chronica Botanica, and New York: Ronald.

1.10 Whitrow, M. (ed.) *1971, Isis Cumulative Bibliography of the History of Science Formed from Isis Cumulative Bibliographies 1–90, 1913–65*, London: Mansell. [Similar volumes covering 1965–84 were published as J. Neu (ed.) 1980, London: Mansell, and 1990, Boston, MA: G. K. Hall.]

1.11 Wölffing, E. *1903, Mathematischer Bücherschatz, Teil 1. Reine Mathematik*, Leipzig: Teubner (*Abhandlungen zur Geschichte der mathematischen Wissenschaften mit Einschluss ihrer Anwendungen*, Vol. 16).

2 JOURNALS IN OR FOR THE HISTORY OF MATHEMATICS

Various journals in the history of science, and some in mathematics and philosophy, also publish pertinent articles from time to time. The abstracting journals listed in Section 3 are the best general sources to consult first.

2.1 *Abhandlungen zur Geschichte der Mathematik*, ed. M. Cantor, 30 vols, Leipzig: Teubner, 1877–1913. [Vols 1–10 were published irregularly as a supplement to the *Zeitschrift für Mathematik und Physik* and are usually bound with it.]

2.2 *Archiv für die Geschichte der Naturwissenschaften und der Technik*, 13 vols, Leipzig: Teubner, 1908–31. [Changed title to *Archiv für Geschichte der Mathematik, der Naturwissenschaften und der Technik*; continued by *Quellen und Studien* (2.13).]

2.3 *Archive for History of Exact Sciences*, ed. C. A. Truesdell III, Berlin: Springer, 1960– .

2.4 *Bibliotheca mathematica*, ed. G. Eneström, 30 vols, Stockholm and Leipzig: Teubner, 1884–1915. [Published quarterly, the first series was a supplement to *Acta mathematica* and eventually took upon itself the role of correcting mistakes in M. Cantor's *Vorlesungen* (4.7).]

2.5 *Bolletino di storia delle scienze matematiche*, ed. E. Giusti Florence: Unione Matematica Italiana, 1981– .

2.6 *Bollettino di bibliografia e storia delle scienze matematiche*, ed. G. Loria, 27 vols, Turin: Clausen, 1898–1919. [The first series of 6 vols was published under the title *Bolletino di storia e bibliografia matematica* as an appendix to the *Giornale di matematiche*. It continued as an appendix in the *Bolletino di matematica*.]

2.7 *Bulletin signalétique, 522. Histoire des sciences et des techniques*, Paris: Centre National de la Recherche Scientifique, 1941– .

2.8 *Bullettino di bibliografia e di storia delle scienze matematiche e fisiche*, ed. B. Boncompagni, 20 vols, Rome: Tipografia delle Scienze Mathematiche e Fisiche, 1868–87. [Not all copies are identical since the editor changed the text after proof sheets were passed. Sometimes spelt '*Bollettino*'. Repr. 1964, New York: Johnson.]

2.9 *Historia mathematica*, founded by K. O. May, San Diego, CA and New York: now Academic Press, 1974– . [Readers' contributions to the abstracts department cover all literature in the field.]

2.10 *Istoriko-matematicheskie issledovaniya*, Moscow: Nauka, 1948– . [The major source of Russian-language articles in the history of mathematics.]

2.11 *Mathesis*, ed. A. Garciadiego, Mexico: National University, 1986– . [Mostly translations into Spanish of articles, often on the foundations of mathematics and the history of set theory.]

2.12 *Philosophia mathematica*, ed. R. Thomas, Toronto: University of Toronto Press, 1964– . [Many articles have a historical aspect.]

2.13 *Quellen und Studien zur Geschichte der Mathematik*, in two parts, *Quellen* and *Studien*, 4 vols each, ed. O. Neugebauer and others, Berlin: Springer, 1929–38. [Continued from *Archiv für die Geschichte* (2.2).]

2.14 *Scripta mathematica: A Quarterly Journal Devoted to the Philosophy, History, and Expository Treatment of Mathematics*, 29 vols, New York: Yeshiva University, 1932–73. [R. C. Archibald's 'Bibliographia de mathematicis' in Vols 1–4 is a useful biobibliography of early-twentieth-century mathematicians.]

3 BIBLIOGRAPHIES FOR MATHEMATICS (INCLUDING ITS HISTORY)

3.1 *International Catalogue of Scientific Literature 1901–1914, Section A, Mathematics*, 14 vols, London: The Royal Society, 1902–21. [Repr. in 1 vol, 1968, New York: Johnson. Successor to the Society's *Catalogue* (3.6).]

3.2 *Jahrbuch über die Fortschritte der Mathematik*, 68 vols, Berlin: de Gruyter, 1871–1944. [Essential for the years covered, 1868–1942.]

3.3 *Mathematical Reviews*, founded by O. Neugebauer, Providence, RI: American Mathematical Society, 1940– . [Includes history but, along with its ancestor, the *Zentralblatt* (3.8), though founded by a historian of mathematics the coverage of that subject has not been as thorough as for mathematics.]

3.4 Reuss, J. D. *1801–21, Repertorium commentationum a societatibus litterarius editarum secundum disciplinarum ordinem*, 16 vols, Göttingen: Dietrich. [Coverage of all mathematical topics from society journals. Repr. 1962, New York: Franklin.]

3.5 *Revue semestrielle des publications mathématiques*, 39 vols, Amsterdam: Versluys, and Groningen: Noordhoff, 1893–1935. [Published under the auspices of the Mathematics Society of Amsterdam.]

3.6 *Royal Society of London Catalogue of Scientific Papers 1800–1900*, 19 vols plus 4 index vols, Cambridge and London: Cambridge University Press and others, 1867–1925. [Repr. 1965, New York: Johnson & Kraus; and 1968, Metuchen, NJ: Scarecrow. Continued by the *International Catalogue* (3.1).]

3.7 *ZDM. Zentralblatt für Didaktik der Mathematik Analysen*, Köln: Saarbach, 1969– . [Abstracts works in mathematics education and its history.]

3.8 *Zentralblatt für Mathematik und ihre Grenzgebiete*, founded by O. Neugebauer, Berlin: Springer, 1931– .

4 GENERAL HISTORIES OF MATHEMATICS

4.1 Becker, O. *1983, Grundlagen der Mathematik in geschichtlicher Entwicklung*, 2nd edn, Frankfurt-am-Main: Suhrkamp. [1st edn, 1954.]

4.2 Bell, E. T. *1945, The Development of Mathematics*, 2nd edn, New York and London: McGraw-Hill. [Perhaps the earliest and most influential mathematician who wrote what are still models of popular mathematical exposition. This work is freer of historical myths than some others by the author. 1st edn, 1940.]

4.3 Bottazzini, U. *1990, Il flauto di Hilbert. Storia della matematica moderna e contemporanea*, Turin: UTET. [Covers many topics in the period 1750–1940. Based on the author's chapters in P. Rossi (ed.), 1988, *Storia della scienza moderna e contemporanea*, 5 vols, Turin: UTET.]

4.4 Bourbaki, N. *1974, Eléments d'histoire des mathématiques*, 3rd edn, Paris: Hermann. [Essentially the historical notes from the Bourbaki *Eléments de mathématique*. Useful mostly for the twentieth century. 1st edn 1960.]

4.5 Boyer, C. B. *1989, A History of Mathematics*, 2nd edn, revised by U. C. Merzbach, New York: Wiley. [Updates the bibliographies and augments the twentieth-century section of the 1st edn of 1968. A reliable introduction.]

4.6 Cajori, F. *1928–9, A History of Mathematical Notations*, 2 vols, Chicago, IL: Open Court.

4.7 Cantor, M. *1880–1908, Vorlesungen über Geschichte der Mathematik*, 4 vols, Leipzig: Teubner. [Still a valuable reference, covering developments up to 1799. Repr. 1965, New York: Johnson.]

4.8 *Encyclopaedia Britannica, 1910–11*, 11th edn (ed. H. Chisholm), 29 vols, Cambridge: Cambridge University Press. [Several very substantial articles. Earlier editions are also useful.]

4.9 *Encyklopädie der mathematischen Wissenschaften mit Einschluss ihrer Anwendungen, 1898–1935*, 6 vols in 23 parts, Leipzig: Teubner. [A valuable survey of post-1800 literature with some historical interpretation. On it was based the *Encyclopédie des sciences mathématiques pures et appliquées*

(1904–14), started by the French, but with no volume complete. Early *Berichte* of the *Jahresberichte der Deutschen Mathematiker-Vereinigung* contained further substantial articles. 2nd edn was begun in the 1930s.]

4.10 *Encyclopaedia metropolitana, 1817–45*, 26 vols, London: Fellowes. [Several substantial articles on branches of mathematics.]

4.11 *Encyclopédie, ou dictionnaire raisonné des sciences, des arts et des métiers, 1751–65* (ed. J. le R. d'Alembert and D. Diderot), 28 vols, Paris and elsewhere. [Various reprints. Important sources for Enlightenment views of science.]

4.12 Hofmann, J. E. *1953–7, Geschichte der Mathematik*, 3 vols, Berlin: de Gruyter. [Useful biobibliographical indexes which were omitted from the English translation of 1967, *The History of Mathematics to 1800*, 2 vols, Totowa, NJ: Littlefield, Adams.]

4.13 Kaestner, A. G. *1796–1800, Geschichte der Mathematik seit der Wiederherstellung der Wissenschaften bis an das Ende des achtzehnten Jahrhunderts*, 4 vols, Göttingen: Rosenbusch. [Includes mechanics, optics and astronomy. Repr. with a foreword by J. E. Hofmann, 1970, Hildesheim: Olms.]

4.14 Klein, F. *1925–8, Elementarmathematik vom höheren Standpunkte aus*, 3rd edn, 3 vols, Berlin: Springer. [Vols 1 and 2 translated into English by E. R. Hedrick and C. A. Noble as *Elementary Mathematics from an Advanced Standpoint*, 1932, London: Macmillan; repr. 1939, New York: Dover. Vol. 3 (on numerical mathematics) is strangely overlooked. Not a history as such, but contains good historical insights.]

4.15 Klein, F. *1926–7, Vorlesungen über die Entwicklung der Mathematik im 19. Jahrhundert*, 2 vols, Berlin: Springer. [The foremost overview of the nineteenth century. Repr. 1967 in 1 vol, New York: Chelsea. Transl. into English by M. Ackerman as *Development of Mathematics in the Nineteenth Century*, 1979, Brookline, MA: Math Sci Press.]

4.16 Kline, M. *1972, Mathematical Thought from Ancient to Modern Times*, New York and Oxford: Oxford University Press. [Somewhat broader in scope than Boyer (4.5), and with more references to applied topics.]

4.17 Montucla, J. F. *1799–1802, Histoire des mathématiques*, 2nd edn, 4 vols, Paris: Agasse. [Presents the well-rounded eighteenth-century view of mathematics.]

4.18 Struik, D. J. *1987, A Concise History of Mathematics*, 4th edn, New York: Dover. [Translations into various languages have been made of this most reliable of concise histories. 1st edn, 1948.]

5 BIOGRAPHICAL SOURCES

The existence of individual biographies of mathematicians is indicated in the biographical notes (§13.3).

5.1 Gillispie, C. C. (ed.) *1970–90, Dictionary of Scientific Biography*, 18 vols, New York: Scribner's. [Historical biographies and some subject essays. Includes supplement volumes.]

5.2 Gottwald, S., Ilgauds, H. J. and Schlote, K.-H. (eds) *1990, Lexikon bedeutender Mathematiker*, Leipzig: Bibliographisches Institut. [About 1800 short but useful entries. Weak on engineering mathematicians.]

5.3 Grinstein, L. S. and Campbell, P. J. (eds) *1987, Women of Mathematics. A Biobibliographic Sourcebook*, New York: Greenwood.

5.4 Poggendorff, J. C. (ed.) *1863– , Bibliographische-Literarisches Handwörterbuch der exakten Wissenschaften*, Berlin: Akademie. [Arranged as a biographical dictionary; an indispensable source, especially for the nineteenth century. Various editors continued the series with changes in title and publisher. First 6 vols (1863–1940) repr. 1965, Amsterdam: Israel.]

5.5 Wussing, H. and Arnold, W. (eds) *1983, Biographien bedeutender Mathematiker. Eine Sammlung von Biographien*, 3rd edn, Berlin: Volk und Wissen. [Chronologically arranged from Antiquity to David Hilbert and Emmy Noether.]

6 GENERAL SOURCE-BOOKS AND ANTHOLOGIES

All the following contain historical introductions and commentaries by the editors.

6.1 Bottazzini, U., Freguglia, P. and Toti Rigatelli, L. *1992, Fonti per la storia della matematica*, Florence: Sansoni. [Covers arithmetic, algebra, geometry, mathematical analysis, probability and foundations.]

6.2 Calinger, R. (ed.) *1982, Classics of Mathematics*, Oak Park, IL: Moore.

6.3 Campbell, D. M. and Higgins, J. C. (eds) *1984, Mathematics: People, Problems, Results*, 3 vols, Belmont, CA: Wadsworth.

6.4 Midonick, H. (ed.) *1968, The Treasury of Mathematics: A Collection of Source Material in Mathematics*, 2nd edn, 2 vols, Harmondsworth, Middlesex: Penguin. [1st edn, 1965.]

6.5 Newman, J. R. (ed.) *1956, The World of Mathematics: A Small Library of the Literature of Mathematics from A'h-mosé the Scribe to Albert Einstein*, 4 vols, New York: Simon & Schuster.

6.6 *Ostwald's Klassiker der exakten Wissenschaften, 1889–1939*, Leipzig: Engelsmann. [Vast collection of German editions of original texts, with many for mathematics. Continued in both East and West German series after the Second World War.]

6.7 Smith, D. E. (ed.) *1929, A Source Book in Mathematics*, New York: McGraw-Hill. [Repr. in 2 vols, 1959, New York: Dover.]

6.8 Struik, D. J. (ed.) *1969, A Source Book in Mathematics 1200–1800*, Cambridge, MA: Harvard University Press. [Repr. 1986, Princeton, NJ: Princeton University Press.]

13.2

Chronology

ALBERT C. LEWIS

This table includes some of the more important publications and events in mathematics up to 1931. Very few developments outside mathematics have been listed. For compactness, modern names of some notions, and short titles of major works (in their original languages) have been used. Most entries carry a reference to an appropriate article by its number, and a few include more than one reference.

Where a personal name is given only once, the corresponding date is either that of the person's main work or an approximate mean date for the period of their highest mathematical activity. Births and deaths are omitted, as many are given in the biographical notes (§13.3). In the case of journals and institutions, the date of founding is given.

Chronologies can also be found in general histories of mathematics (listed in §13.1) such as those by D. E. Smith (6.7) and C. B. Boyer (4.5). In addition to these works, use has been made of Parkinson *1985* and Williams *1975*. Excluded are events of unclear dating (especially for ancient and medieval periods), and complicated developments not susceptible to short description.

BIBLIOGRAPHY

Parkinson, C. L. *1985, Breakthroughs: A Chronology of Great Achievements in Science and Mathematics 1200–1930*, London: Mansell.
Williams, N. *1975, Chronology of the Modern World, 1763–1965*, rev. edn, Harmondsworth, Middlesex: Penguin.

BC
50,000	Evidence of counting
4700	Probable introduction of Babylonian calendar (§1.1)
3000	Numerals in Egypt (§1.2)
2773	Probable introduction of Egyptian calendar
2500	Pyramids in Egypt (§1.2)
2400	Positional notation in Mesopotamia. Tablets of Ur (§1.1)

2000 Babylonian rules of mensuration in use (§1.1)
1850 Astronomical instruments in Egypt (§1.2)
 Moscow papyrus (§1.2)
1650 Rhind papyrus (§1.2)
1500 Topographical map in Italy (§8.15)
1400 Chinese decimal numeration in use (§1.9)
 650 Coinage used in China and Asia Minor
 585 Thales of Miletus on demonstrative geometry (§1.3)
 550 Pythagorean school (§1.3)
 500 Hindu tradition, Jainism, *Śulbasūtras* (§1.2)
 450 Zeno describes paradoxes of motion (§1.3)
 Hippodamus of Miletus makes land surveys (§8.15)
 400 Eudoxus develops theory of proportion (§1.3)
 Democritus studies laws of perspective vision (§12.6)
 380 Plato presents his philosophy of mathematics (§1.3)
 370 Eudoxus on proportion (§1.3)
 350 Conics discovered (§1.3)
 History of geometry by Eudemus (§1.3)
 300 Decimal numeration in use in India (§1.12)
 Euclid: *Elements* (§1.3)
 Heliocentric hypothesis by Aristarchus of Samos (§1.4)
 230 Eratosthenes on prime numbers (§1.3)
 225 Apollonius of Perga investigates conics (§1.3)
 Archimedes on measurement of areas and volumes (§1.3)
 Chinese knotted cords (§1.9)
 200 '*Nine Chapters*', compendium of Chinese mathematics (§1.9)
 140 Hipparchus develops beginnings of trigonometry (§1.3)

AD
 30s Vitruvius: *De architectura* (§2.8)
 50 Hero of Alexandria on geodesy (§1.5)
 150 Ptolemy's astronomical system (§1.5) and geography (§8.15)
 250 Indeterminate equations by Diophantus (§1.5)
 320 Pappus composes last large compendium of Greek mathematics (§1.5)
 400 Sun Tzu presents Chinese remainder problem (§1.9)
 Hypatia's commentaries on Greek mathematics (§1.5)
 625 Wang Xiaotong investigates cubic equation (§1.9)
 628 Brahmagupta's astronomical calculations (§1.12)
 650s Textbooks for training Chinese government officials (§1.9)
 766 Hindu works translated into Arabic; Hindu numerals (§1.6)
 800 Baneū Musā reconstruct Apollonius texts (§1.6)
 820 al-Khwārizmī codifies Arabic algebra (§1.6)
 870 Thābit ibn Qurra translates Greek works (§1.6)
 900 Abū Kāmil studies irrationals in solutions of quadratic equations (§1.6)

1543	Recorde's English arithmetic and abridgement of Euclid's *Elements* (§2.1)
	Copernicus: *De revolutionibus* (§2.7)
1545	Cardano on the cubic equation; negative and complex numbers in computations (§6.1)
1569	Mercator publishes map of the world (§8.15)
1572	Bombelli's algebra (§6.1) and continued fractions (§6.3)
	Commandinus produces Latin edition of Euclid's *Elements* (§2.1)
1580s	Tycho Brahe amasses and reduces accurate astronomical data (§2.7)
1590	Stevin introduces decimal fractions
1591	Viète makes systematic use of letters for unknowns (§6.9)
1593	Pitiscus publishes addition and subtraction trigonometry theorems (§4.2)
1599	Wright applies the Mercator chart to navigation (§8.18)
1604	Kepler's treatise on optics (§2.9)
1609	Elliptical planetary orbits discovered by Kepler (§2.7)
1610	Harriot makes innovatory analysis of shipbuilding (§8.17)
1611	Kepler: *Dioptrice*, geometrical analysis of lenses (§9.3)
1613	Witt's treatise on annuities (§10.3)
1614	Bürgi and Napier on logarithms (§2.5)
1619	Kepler: *Harmonices mundi*, includes tilings (§12.8)
1620	Slide rule (§5.11)
1621	Bachet's edition of Diophantus stimulates number theory (§6.10)
1623	Schickard constructs calculating machine (§5.11)
1624	Briggs: *Arithmetica logarithmica* (§2.5)
1631	Harriot on relationships between roots, factors, coefficients (§6.1)
1635	Mersenne arranges scientific conferences
	Physical basis of music (§2.10)
	Cavalieri's method of 'indivisibles' in the calculus (§3.1)
1636	Mersenne: *Harmonie universelle* (§7.10)
1637	Descartes's analytic geometry (§7.1)
	Fermat's Last Theorem (§6.10)
1639	Desargues on projective geometry (§7.6)
1650s	Pascal and Fermat correspond on probability (§10.1)
1654	Pascal's arithmetical triangle (§4.1)
1657	Huygens' treatise on games of chance (§10.2)
1658	Pascal on quadrature (§3.1)
1664	Newton begins work on fluxions (§3.2)
1666	Leibniz's symbolic logic (§5.1)
	The Académie des Sciences in Paris (§11.1)
	The Royal Society of London (§11.7)
1670s	Huygens on centripetal force and pendulum (§8.13)
	James Gregory presents the binomial theorem (§4.1)
1678	Hooke's law for elastic extension (§8.6)
1684	Leibniz's calculus published (§3.2)
1685	Tschirnhaus's transformation for quintic equation (§6.1)
1687	Newton: *Principia mathematica* (§8.1)

1690	James and John Bernoulli present early calculus methods (§3.2)
	The Hamburg Mathematical Society
	Huygens: *Traité de la lumière*, wave explanation of double refraction (§9.1)
1696	L'Hôpital's textbook on Leibnizian calculus (§3.2)
	Brachistochrone problem; calculus of variations initiated (§3.5)
1703	David Gregory publishes influential edition of Euclid's *Elements* (§2.1)
1704	*Ladies' Diary* journal (§11.12)
	Newton publishes calculus (§3.2) in an appendix to his treatise on optics (§9.1)
1705	James Bernoulli on flexure of beams (§8.6)
1711	Newton presents results by infinite series (§4.3)
1713	James Bernoulli: *Ars conjectandi*, on probability (§10.1)
1715	Calculus of finite differences by Taylor (§4.3)
1723	Riccati solves 'his' differential equation (§3.14)
1725	The Saint Petersburg Academy (§11.7)
1730	De Moivre's theorem on series (§6.2)
1733	Saccheri constructs a non-Euclidean geometry (§7.4)
1734	Berkeley: *Analyst*, criticizes the foundations of calculus (§3.2)
1735	Euler solves the Königsberg bridge problem (§7.13)
1736	Euler proves Fermat's little theorem in number theory
1737	Euler and Clairaut on exact differentials (§3.15)
	Euler introduces zeta function (§6.10)
1738	The Paris Academy sets prize for theory of tides (§8.16)
	Daniel Bernoulli: *Hydrodynamica* (§8.5)
1740s	French work on determining shape of the Earth; Lapland expedition (§8.14)
	Robins' treatise on gunnery; translated and extended by Euler (§8.11)
	Bouguer and Euler produce treatises on shipbuilding (§8.17)
1742	Maclaurin's treatise on calculus (§3.2), tidal theory (§8.16)
	Goldbach conjecture (§6.10)
1743	D'Alembert: *Traité de dynamique*, partial differential equations (§3.15)
1744	Euler states law of quadratic reciprocity (§6.10)
	Euler's treatise on curved lines (§3.5)
1747	D'Alembert and Euler on the vibrating string problem (§3.15)
1748	Euler's textbook on 'infinitesimal analysis' (§4.3)
1750s	*Encyclopédie* of Diderot and d'Alembert (completed 1772)
	Euler and d'Alembert on basic equations of hydrodynamics (§8.5)
	Euler studies planetary perturbations; he and others on lunar theory (§8.8)
1750	Cramer's treatise on algebraic geometry (§7.1)
	Euler proposes polyhedral formula (§7.10)
1753	Daniel Bernoulli proposes trigonometric series solution of wave equation (§3.11)
1755	Euler's differential calculus treatise (§3.2)
1759	The Turin Academy (§11.8)
1760s	Lagrange papers on propagation of sound (§9.8)
1760	Lambert's 'theory of errors' (§10.13)

1762	Lagrange introduces 'δ' into calculus of variations (§3.5)
1763	Bayes' paper on probability (§10.16)
1765	Mayer, following Euler, produces lunar tables for navigators (§8.18)
1768	Euler's treatise on integral calculus (§3.2)
1770s	Laplace papers on potential theory (§3.17)
	Lagrange on resolution of equations presages group theory (§6.4)
	Proof of four-squares theorem in number theory (§6.10)
	Coulomb's laws for torsion and stability of structures (§8.6–8.7)
1770	Hyperbolic trigonometry with Lambert (§4.2)
1772	Lambert surveys cartographic methods (§8.15)
1776	Lagrange on force as gradient of potential function (§8.8)
1778	The Amsterdam Mathematical Society (§11.4)
1780s	Coulomb on inverse square laws for electrostatics and magnetism (§9.10)
	Lagrange and Laplace study Jupiter and Saturn; inequality and stability of planetary system (§8.8)
1784	Legendre polynomials with Legendre and Laplace; exterior potentials (§3.17)
1788	Lagrange: *Méchanique analitique* (§8.1)
1790s	Méchain and Delambre determine of logitudinal line; metre defined from it (§8.14)
1790	De Prony: *Nouvelle architecture hydraulique* (§8.5)
1794	Founding of the Ecole Polytechnique in Paris (§11.1)
1795	Gauss proves law of quadratic reciprocity
	Lagrange's textbook *Fonctions analytiques*, 1st edn (§3.2)
	Monge: *Géométrie descriptive*, 1st edn (§7.5)
1796	Gauss shows 17-sided polygon constructible with straightedge and compass
1797	Lacroix's treatise on the calculus; also more elementary textbooks on mathematics (§11.2)
1798	Kramp's table for the error function (§10.13)
1799	Gauss proves the fundamental theorem of algebra (until 1850) (§6.1)
	Montucla: *Histoire des mathématiques* (completed by Lalande, 1802) (§12.13)
	Laplace: *Mécanique céleste* (completed 1825)
	Ruffini tries to resolve the quintic equation (§6.1)
1800s	Puissant's treatises on cartography and geodesy (§8.15)
	Lagrange and Poisson brackets solutions to equations of motion (§8.9)
	Legendre, Gauss and Laplace on least-squares criterion (§10.3–10.5)
1800	Arbogast: *Calcul des dérivations* (§4.7)
1801	Gauss: *Disquisitiones arithmeticae*, on number theory (§6.10)
1803	Lazare Carnot promotes projective geometry (§7.6)
	Poinsot introduces the couple into statics (§8.1)
1805	Legendre's treatise on number theory (§6.10)
1806	Argand's geometrical interpretation of $\sqrt{-1}$ (no reaction) (§6.2)
	Laplace mathematizes capillarity (§9.9)
1807	Fourier presents heat diffusion equation and Fourier series (§9.4)

1836	Liouville's *Journal* (§11.1); Sturm–Liouville theory (§3.14)
1837	Poisson's treatise on probability (§10.5)
1838	Cournot pioneers mathematization of supply and demand (§10.18)
1839	*Cambridge and Dublin Mathematical Journal* (§11.7)
1840s	Wilhelm Weber and Franz Ernst Neumann develop electromagnetism (§9.10)
1840	Airy studies servomechanisms (governors) (§8.4)
	Louis pioneers statistical methods in medicine (§10.12)
1841	Jacobi's memoir on functional determinants (Jacobians) (§6.6)
1843	Hamilton discovers quaternions (§6.2)
1844	Grassmann presents multilinear algebra (2nd edn 1862) (§6.2)
	Boole's calculus of operators (§4.7)
	Cayley on linear transformations (invariants) (§6.8)
1845	Weisbach's book on engineering mechanics (edns to 1891) (§8.5)
1845	Stokes derives equations for viscous fluids (§8.5)
1846	Chebyshev proves weak law of large numbers in statistics (§10.6)
	Discovery of Neptune from Le Verrier's calculations (§8.8)
1847	Boole on algebra of logic (§5.1)
	Helmholtz formulates law of conservation of energy (§8.1)
1850s	Froude studies stability of ships (§8.17)
1850	Clausius presents the second law of thermodynamics (§9.5)
	Wilhelmy introduces differential equations into chemistry (§9.16)
1851	Riemann and Cauchy–Riemann equations (§3.12)
1852	Calculus of forms, by Sylvester
1854	Cayley's definition of group (§6.4)
	Riemann surface for complex-variable analysis (§7.10)
	Stokes's theorem in potential theory (§3.17)
1858	Cayley's algebra of matrices (§6.7)
	Möbius strip (§7.10)
1859	Riemann on the hypergeometric series (§4.4)
	Chebyshev polynomials (§4.4)
1860s	Cayley's papers on quantics encourages invariant theory (§6.8)
1860	Riemann's hypothesis (§3.12) for complex variables
1863	Helmholtz's treatise on physiological acoustics (§9.8)
	Jevons remodels Boolean algebra (§5.1)
1864	Riemann–Roch theorem on rational complex-variable functions (§4.6)
1865	Weierstrass–Bolzano theorem from Weierstrass (§3.3)
	Plücker's line geometry (§7.7)
1866	Culmann's textbook on graphical statics (§8.2)
	Maxwell's distribution law for molecules of gas (§9.14)
1868	The London Mathematical Society (§11.7)
	Boncompagni starts *Bollettino*, on history of mathematics (ceased 1887) (§12.13)
	Beltrami's model of non-Euclidean geometry (§7.4)
	Abstracting journal *Jahrbuch über die Fortschritte der Mathematik* (§12.13)
	Gordan's theorem on binary forms (§6.8)

1869   *Ostwald's Klassiker*, series of German editions of scientific texts, begins (§12.13)

1870s   Boltzmann's *H* theorem in statistical mechanics (§9.14)
        Helmholtz and Gibbs on chemical thermodynamics (§9.16)
        Cayley and Sylvester on the mathematics of chemical notations (§9.16)

1870   Linear associative algebras by Benjamin Peirce (§6.2)
        Jordan's treatise on substitution groups (§8.4)
        Weierstrass's counter-example to Dirichlet's principle (§3.12)
        Bulletin des sciences mathématiques (§11.12)
        Charles Peirce's logic of relatives (§5.1)

1872   Klein's *Erlanger Programm* (§7.6)
        Dedekind cuts define irrational numbers (§3.3)
        The French Mathematical Society (§11.1)
        Sylow theorems on subgroups (§6.4)
        Boussinesq's treatise on water flow (§8.5)

1873   Maxwell: *Treatise on Electricity and Magnetism* (§9.10)
        Clifford's biquaternions (§6.2)
        Hermite proves 'e' is transcendental (§6.10)

1874   Georg Cantor distinguishes different infinities (§3.6)
        Walras models economics on Newtonian mechanics (§10.18)

1875   Reuleaux's textbook on kinematics (§8.3)

1877   Lord Rayleigh (Strutt): *Theory of Sound* (§9.8)
        Routh on servomechanisms (governors) (§8.4)

1880s   Poincaré studies topological manifolds (§7.10)
        Russian school in probability and statistics around Chebyshev (§10.6)
        Hill and Poincaré on the three-body problem

1880   Moritz Cantor: *Vorlesungen*, on history of mathematics (finished 1908) (§12.13)

1881   Gibbs: *Elements of Vector Analysis* (§6.2)

1882   Lindemann proves that π is transcendental (§6.10)
        Kronecker on algebraic fields (§6.4)
        Mittag-Leffler founds *Acta mathematica* (§11.5)

1883   Georg Cantor states the continuum hypothesis (§3.6)
        Mach's history of mechanics, 1st edn (§8.1)

1884   Frege: *Grundlagen der Arithmetik* (§5.2)
        Eneström starts *Bibliotheca mathematica* (ceased 1915) (§12.13)

1886   Weierstrass develops rigorous theory of functions (§3.3)

1887   Jordan's curve theorem (§7.10)

1888   Lie's treatise on transformations and differential equations (finished 1893) (§6.5)
        Hilbert's finite basis theorem for invariants (§6.8)

1889   Peano's postulates for the natural numbers (§5.2)
        Galton's *Natural Inheritance* stimulates biometry (§10.7)

1890s Basic papers on topological manifolds by Poincaré (§7.10)
       Heaviside publishes on telecommunications (§9.11); stimulates Laplace transform (§4.8)
       Schönflies and Fedorov apply group theory to crystallography (§9.17)
       E. H. Moore builds up mathematics at Chicago University (§11.10)

1890   Schröder's treatise on algebraic logic (finished 1905) (§5.1)
       Peano's space-filling curve (§7.11)

1891   The German Mathematical Society (§11.2)
       D'Ocagne: *Nomographie* (§4.12)

1892   New York (later American) Mathematical Society (§11.10)

1893   Bertini and Segre publish major papers on algebraic geometry (§7.9)

1895   Georg Cantor's transfinite numbers (§5.4)
       First edition of *Formulaire de mathématiques* by Peano and colleagues (last edition 1908)

1896   Volterra and Pincherle on functional analysis (§3.9)
       Frobenius's group representations (§6.7)

1896   Hadamard and De La Vallée Poussin prove the prime number theorem (§6.10)

1897   First International Congress of Mathematicians, Zurich
       Hensel introduces *p*-adic numbers (§6.10)
       Fechner: *Kollektivmasslehre*, probability and statistics in psychology (§10.9)

1898   *Encyklopädie der mathematischen Wissenschaften* begins to appear (finished 1935) (§12.13)

1899   Hilbert: *Grundlagen der Geometrie*, 1st edn (§7.4)
       Galton stresses correlation in statistics (§10.10)

1900s Harris charts of tidal waves (§8.16)
       Huntington and Veblen pioneer model theory (§5.8)
       Kutta, Prandtl and Zhukovsky develop aerodynamics (§8.12)
       Planck and Einstein analyse black-body radiation (§9.15)
       Markov chains (§10.6)
       Bowley pioneers sample surveys (§10.8)
       Study of Mendel's 1865 work stimulates mathematical genetics (§10.10)
       Klein takes initiatives in mathematical education in Germany (§11.2)
       Mathematical Circle of Palermo adopts international role (§11.8)

1900   Minkowski stresses convex sets (§6.11)
       Second Mathematical Congress in Paris; Hilbert's 23 problems

1901   Ricci and Levi-Città's treatise on tensor calculus (§3.4)
       Pearson founds *Biometrika* (§10.10)

1902   Lebesgue integration (§3.7)
       Bjerknes mathematizes atmospheric processes (§9.7)
       Tropfke: *Geschichte der Elementar-Mathematik*, 1st edn (§12.13)

1903   Fredholm on integral equations (§3.10)
       Russell: *Principles of mathematics*; publishes paradoxes of set theory (§5.2–5.3)

1904   Zermelo's axiom of choice in set theory (§5.4)

1905   Einstein publishes special relativity (§9.13)

1906	Heiberg rediscovers important Archimedes text (§1.3)
	Fréchet's doctoral thesis on functional analysis (§3.9)
1908	Minkowski presents four-dimensional geometry of relativity (§9.13)
	Zermelo axiomatizes set theory (§5.4)
1910s	Ramanujan records insights in number theory (§6.10)
	Brouwer develops intuitionistic foundations of mathematics (§5.6)
	Rey Pastor encourages mathematics in Spain and Argentina (§11.9)
19.10	Steinitz systematizes concept of algebraic field (§6.4)
	Whitehead and Russell: *Principia mathematica* (finished 1913) (§5.2)
	Plancherel theorem on Fourier transform (§3.11)
1911	Collaboration begins between Hardy and Littlewood
1914	Hausdorff: *Grundzüge der Mengenlehre* (§5.4)
1917	Einstein's theory of general relativity (§9.13)
1920s	Rapid expansion of topology at Princeton (§11.10)
	Establishment of Laplace transform (§4.8)
	Definitive work on meta-mathematics by Hilbert and others (§5.5)
	Leśniewski develops logical systems (§5.7)
	Brouwer, Urysohn and Menger on definitions of dimension (§7.11)
	Kron's general theory of electrical machines (§9.12)
	Volterra and Lotka on models of population dynamics (§9.18)
	Molina and Shewhart apply statistics to production engineering (§10.14)
1922	Richardson's numerical methods in meteorology (§9.7)
1924	Pólya and Niggli enumerate tilings (§12.8)
1925	Heisenberg invents matrix mechanics for quantum mechanics (§9.15)
	Fisher: *Statistical Methods for Research Workers* (§10.15)
1927	Weyl and Peter found modern harmonic analysis (§3.11)
1928	Von Neumann on the foundation of quantum mechanics (§9.15)
	Van der Pol's differential equation for heartbeat (§9.18)
1930	Wiener on general harmonic analysis (§3.12)
1931	Gödel's incompleteness theorem (§5.5)
	Neugebauer founds *Zentralblatt* review journal (§11.12)

13.3

Biographical notes

ALBERT C. LEWIS

Modern figures are included if they had made substantial contributions by the 1930s. Institutions are given only where there is a particularly strong association with the person. Only a few principal subject areas are listed for each person, and these are confined to mathematics. In the few instances, such as for Newton, where birth or death years are affected by the adoption of the Gregorian calendar, the Old Style date is given if it was in effect at the time. See §13.1 for general biographical sources.

B   = One or more major biographies or autobiographies.
W   = Complete works or substantial selections of mathematical works.
W + = More than one edition or collection of works.
C   = A substantial edition of correspondence, not necessarily complete, published in book form. In some cases this represents an edition of correspondence between only two mathematicians.

Abel, Niels Henrik (1802–29), Norwegian. Analysis, elliptic functions. general quintic equation. B W + C.

Abū Kāmil, Shujāᶜ ibn Aslam ibn Muḥammad (flourished *c.* 880–900), Arabic. Algebra. Diophantine equations. W.

Adams, John Couch (1819–92), English. Discoverer of Neptune. W.

Agnesi, Maria Gaëtana (1718–99), Italian. Algebra, analysis. *Istituzioni analitiche ad uso della gioventù italiana* (1748).

Airy, George Biddell (1801–92), English astronomer. Planetary theory.

al-Bīrūnī, Abū Rayhn Muḥammad ibn Aḥmad (973–1048), Islamic. Arithmetic, geometry, trigonometry, astronomy.

al-Karajī (al-Karkhī), Abū Bakr ibn Muḥammad ibn al Ḥusayn (*c.* 1000), Arabic. Algebra.

al-Kāshī, Ghiyāth al-Dīn Jamshīd Mas'ūd (d. *c.* 1429), Islamic. Arithmetic, numerical mathematics, calculation of π. *Miftāḥ al-ḥisāb* (1427). W.

al-Khayyāmī (al-Khayyām), Ghiyāth al-Dīn Abu l-Fatḥ ᶜUmar ibn Ibrāhīm (Omar Khayyam) (d. 1022), Islamic poet and mathematician. Cubic equations, parallel postulate. W.

al-Khwārizmī, Abū Jaᶜfar Muḥammad ibn Mūsā (c. 830), Arabic. Algebra, astronomy.

al-Kūhī (al-Qūhī, Abū Sahl Wayjan ibn Rustam (fl. c. 1000). Islamic. Conic sections, astronomy.

al-Samawᶜal, Abū Naṣr ibn Yaḥyā al-Maghribī (d. c. 1175). Arabic. Arithmetic, alegbra, astronomy.

al-Ṭūsī, Naṣīr al-Dīn (Nasir eddin) Abū Jaᶜfar Muḥammad ibn Muḥammad ibn al-Ḥasan (1201–74), Arabic. Trigonometry, parallel postulate.

al-Ṭūsī, Sharaf al-Dīn al-Muẓaffar ibn Muḥammad (c. 1145–1215), Arabic. Cubic equations. W.

Alberti, Leone Battista (1404–72), Italian. Laws of perspective.

Aleksandrov, Pavel Sergeevich (1896–1982), USSR. University of Moscow. Topology.

Alexander, James Waddell (1888–1971), American. Princeton. Homology theory.

Alhazen. See Ibn al-Haytham.

Ampère, André Marie (1775–1836), French. Electromagnetism, analysis. B C.

Antiphon (fl. c. 430 BC), Greek. Squaring the circle.

Apollonius of Perga (fl. 3rd century BC), Greek. Conic sections. W + .

Archibald, Raymond Claire (1875–1955), American. History.

Archimedes of Syracuse (?287–?212 BC), Greek. Integral calculus, creator of statics and hydrostatics. B W + .

Archytas of Tarentum (?428–?365 BC), Greek. Pythagorean number theory.

Argand, Jean Robert (1768–1822), Swiss. Complex numbers.

Aristaeus (fl. c. 330 BC), Greek. Conic sections.

Aristarchus of Samos (fl. c. 270 BC), Greek. Planetary theory.

Arnauld, Antoine (1612–94), French. Port Royal *Logic*. W.

Artin, Emil (1898–1962), Austrian, emigrated to USA. Ring algebra.

Āryabhaṭa I (b. 476), Indian. *Āryabhaṭīya* (499).

Āryabhaṭa II (fl. 10th century), Indian. *Mahāsiddhānta*.

Babbage, Charles (1792–1871), English. Functional equations, pioneer of machine computing. B W.

Baire, René-Louis (1874–1932), French. Theory of functions. W.

Ball, Walter William Rouse (1850–1925), English. Cambridge University. History.

Banach, Stefan (1892–1945), Polish. Functional analysis. W.

Banū Mūsā, the three sons of Mūsā ibn Shākir: Muḥammad, Aḥmad and al-Ḥasan (fl. c. 850), Arabic. Geometry, mechanics, astronomy.

Barrow, Isaac (1630–77), English. Calculus. W.

Bateman, Harry (1882–1946) English-born American. California Institute of Technology. Partial differential equations, general relativity, hydrodynamics, aerodynamics.

Bayes, Thomas (1702–61), English clergyman. Probability theory.

Bell, Eric Temple (1883–1960), American. Number theory, history.

Beltrami, Eugenio (1835–99), Italian. Differential and non-Euclidean geometry. W.

Bendixson, Ivar Otto (1861–1936), Swedish. University of Stockholm. Analysis.

Bernays, Paul (1888–1977), British-born Swiss. Mathematical logic, set theory.

Bernoulli, Daniel (1700–82), Dutch-born Swiss. Mechanics, differential equations, probability theory, B W(in progess) C(in progress).

Bernoulli, James I (1654–1705), Swiss. Calculus, mechanics, probability theory. W(in progress) C(in progress).

Bernoulli, John I (1667–1748), Swiss. Calculus, mechanics, teacher of Euler. W(in progress) C(in progress).

Bernstein, Sergei Natanovich (1880–1968), Ukrainian. Applications to biology. W.

Bertrand, Joseph Louis (1824–1900), French. Ecole Polytechnique. Analysis, probability theory.

Bessel, Friedrich Wilhelm (1784–1864), German astronomer. Bessel functions. B W C.

Betti, Enrico (1823–1892), Italian. Galois theory, topology. W.

Bézout, Etienne (1730–83). French. Textbooks.

Bianchi, Luigi (1856–1928), Italian. Differential geometry. W.

Binet, Jacques Philippe Marie (1786–1856), French. Mechanics. Perturbation theory, determinants, elasticity, calculus.

Biot, Jean-Baptiste (1774–1862), French. Astronomy, elasticity, electricity and magnetism, geometry, heat, optics.

Birkhoff, George David (1884–1944), American. Harvard University. Dynamics, ergodic theory. W.

Bliss, Gilbert Ames (1876–1961), American. University of Chicago. Calculus of variations.

Blumenthal, Ludwig Otto (1876–1944), German. Ecole Polytechnique, Aix-la-Chapelle. Editor of *Mathematische Annalen*.

Bôcher, Maxime (1867–1918), American. Harvard University. Linear algebra, analysis.

Bohr, Harald August (1887–1951), Danish. Quasiperiodic functions. W.

Bolyai János, (Johann) von (1802–60), Hungarian. Non-Euclidean geometry. B W C.

Bolzano, Bernard(us) Placidus Johann Nepomuk (1781–1848), Czech. Philosopher and mathematician. Foundations of real numbers, set theory. B W + C.

Bombelli, Rafael (1526 – after 1572), Italian. Algebra, complex numbers.

Boncompagni, Baldassarre (1821–94), Italian. History. Founded *Bulletino di bibliografia e di storia delle scienze matematiche e fisiche* in 1868.

Boole, George (1815–64), English, worked in Ireland. Differential operators, mathematical logic. B W C.

Borda, Jean Charles (1733–99), French. Engineering mechanics. B.

Borel, Emile Félix-Edouard-Justin (1871–1956), French. Theory of functions of real variables. B W.

Boscovich, Roger Joseph (Bošković, Rudjer Josip) (1711–87), Croatian. Mathematical physics. B W.

Bossut, Charles (1730–1814), French. Hydrodynamics and hydraulics, textbooks in engineering physics.

Boyer, Carl Benjamin (1906–76), American. History.

Bradwardine, Thomas (?1290–1349), English. Geometry. W.

Brahe, Tycho (1546–1601), Danish. Astronomy, tables. C.

Brahmagupta (598 – after 665), Indian. Geometry, Algebra. W.

Brianchon, Charles Julien (1783–1864), French. Ecole d'Artillerie de la Garde Royale. Geometry.

Briggs, Henry (1561–1631), English. Logarithms.

Brill, Alexander Wilhelm von (1842–1935), German. Algebraic geometry.

Brioschi, Francesco (1824–97), Italian. Algebra. W.

Brouncker, William (2nd Viscount Brouncker) (?1620–84), English. Series. C.

Brouwer, Luitzen Egbertus Jan (1881–1966), Dutch. Topology, intuitionism. W.

Burali-Forti, Cesare (1861–1931), Italian. Linear transformations, set theory.

Burkhardt, Heinrich (1861–1914), German. Analysis, history.

Burnside, William (1852–1927), English. Naval College, Greenwich. Group theory, probability.

Cajori, Florian (1859–1930), Swiss-born American. History.

Cantor, Georg Ferdinand Ludwig (1845–1918), German. Set theory, transfinite numbers. B W C.

Cantor, Moritz Benedikt (1829–1920), German. History.

Carathéodory, Constantin (1873–1950), German. Calculus of variations, theory of functions. W.

Cardano, Girolamo (Jerome Cardan) (1501–76), Italian. Astrologer, physician. Algebra, general solution of the cubic equation. B W.

Carnot, Lazare Nicolas Marguérite (1753–1823), French. Calculus, mechanics, algebra.

Cartan, Elie Joseph (1869–1951), French. Continuous groups. W.

Castelnuovo, Guido (1865–1952), Italian. University of Rome. Algebraic geometry. W.

Cauchy, Augustin Louis (1789–1857), French. Calculus, complex analysis, elasticity, Fourier analysis, heat, hydrodynamics, optics, potential theory, mechanics. B W.

Cavaillès, Jean (1903–49). French. History and philosophy of mathematics. B.

Cavalieri, Bonaventura (1598–1647), Italian. Method of indivisibles. C.

Cayley, Arthur (1821–95), English. Cambridge University. Theory of matrices, invariant theory, algebraic geometry. B(in progress) W C.

Cesàro, Ernesto (1859–1906), Italian. Differential geometry. W.

Chang Ch'iu-chien (c. 450), Chinese. Number theoretic problems.

Chang Heng (78–139), Chinese. Volume of sphere, astronomy.

Chao Chün-ch'ing (fl. AD 250), Chinese. Right-angled triangles.

Chasles, Michel (1793–1880), French. Algebraic geometry.

Chebyshev, Pafnuty Lvovich (1821–94), Russian. University of St Petersburg. Probability theory, number theory, theory of functions, calculating machines, approximation theory. B W.

Chisholm, Grace. See under Young, William.

Christoffel, Elwin Bruno (1829–1900), German. Differential geometry, invariant theory. W.

Church, Alonzo (1903–). American. Logic, recursion theory.

Clairaut, Alexis Claude (1713–65), French. Hydrodynamics, geodesy, differential equations. B.

Clausius, Rudolf Julius Emmanuel (1822–88), German (born in Pomerania). Theoretical physics, electrodynamics.

Clavius, Christoph (1537–1612), German. University of Rome. Geometry, algebra, trigonometry. W C(in progress)

Clebsch, Rudolf Friedrich Alfred (1833–72), German. Invariant theory. Founder, with F. Neumann, of *Mathematische Annalen* (1869). W

Clifford, William Kingdon (1845–79), English. Biquaternions, Clifford parallels, Clifford surfaces. Also a philosopher of science. W.

Condillac, Etienne Bounot, Abbé de (1714–60). French. Semiotics, algebra. W.

Condorcet, Marie Jean Antoine Caritat, Marquis de (1743–94), French. Probability, philosophy. W.

Coolidge, Julian Lowell (1873–1954), American. History, geometry.

Copernicus, Nicolaus (1473–1543), Polish. Astronomy. W.

Coriolis, Gustave Gaspard de (1792–1843), French. Energy mechanics, engineering mathematics.

Cossali, Pietro (1748–1815). Italian. History, algebra. B.

Cotes, Roger (1682–1716), English. Cambridge University. Calculus. W C.

Coulomb, Charles Augustin (1736–1806), French. Mechanics, electricity and magnetism.

Courant, Richard (1888–1972), German-born American. Göttingen and New York Universities. Mathematical physics. B.

Cournot, Antoine Augustin (1801–77), French. University administration. Probability, mathematical economics.

Cramer, Gabriel (1704–52), Swiss. Determinants, analysis.

Crelle, August Leopold (1780–1855), German. Founder of *Journal für die reine und angewandte Mathematik* (1826). Engineering.

Cremona, Antonio Luigi Gaudenzio Giuseppe (1830–1903), Italian. Geometry. W.

d'Alembert, Jean le Rond (1717–83), French. Mechanics, calculus. B W + C.

Darboux, Jean Gaston (1842–1917), French. Differential geometry. W.

Darwin, George Howard (1845–1912), English. Astronomy, geophysics. W.

De La Vallée-Poussin, Charles Jean (1866–1962), French. Analytic theory of numbers, analysis.

De Moivre, Abraham (1667–1754), French, lived in England. Trigonometry, probability theory.

De Morgan, Augustus (1806–71), British, born in India. University of London. Algebra, probability theory, history, logic. B C.

Dedekind, Julius Wilhelm Richard (1831–1916), German. Analysis, algebraic number theory. W C.

Dehn, Max (1878–1952), German-born American. Topology.

Delambre, Jean-Baptiste Joseph (1749–1822), French. Astronomy, tables, history of astronomy.

Delaunay, Charles Eugene (1816–72), French. Mechanics, astronomy.

Desargues, Girard (1593–1662), French engineer and mathematician. Projective geometry. W.

Descartes, René (1596–1650), French philosopher and mathematician. Analytical geometry. B W C.

Dickson, Leonard Eugene (1874–1954), American. University of Chicago. Number theory. W.

Dini, Ulisse (1845–1918), Italian. Theory of real functions. W.

Diophantus of Alexandria (fl. *c.* AD 250), Greek. Number theory. W + .

Dirichlet, Johann Peter Gustav Lejeune- (1805–59), German. University of Berlin. Number theory, analysis, dynamics. B(in progress) W C.

Dodgson, Charles Lutwidge (Lewis Carroll) (1832–98), English. Mathematical games and puzzles, logic. B.

Duhamel, Jean Marie Constant (1797–1872), French. Partial differential equations, theory of heat, acoustics.

Duhem, Pierre (1861–1916), French. Mechanics, physics, history, philosophy.

Dupin, Pierre Charles François (1784–1873), French. Naval engineering, geometry, ergonomics, education.

Eddington, Arthur Stanley (1882–1944), English. General theory of relativity.

Edgeworth, Francis Ysidro (1845–1926), English. Probability and statistics.

Einstein, Albert (1879–1955), German-born American. Relativity theory, quantum mechanics. B W(in progress) C(in progress).

Eisenstein, Ferdinand Gotthold Max (1823–52), German. Number theory.

Eneström, Gustav (1852–1923), Swedish. Founded *Bibliotheca mathematica*. History, editor of works of Euler.

Engel, Friedrich (1861–1941), German. Editor of works of Lie and H.G. Grassmann.

Enriques, Federigo (1871–1946), Italian. University of Rome. History, philosophy, geometry. W.

Eratosthenes of Cyrene (?276 BC–197 BC), Alexandrian. Geometry, prime numbers, geography.

Euclid (fl. *c.* 300 BC), Alexandrian. Geometry, optics. *Elements*. B W + .

Eudemus (*c.* 300 BC), Greek. History.

Eudoxus of Cnidus (?408 BC–355 BC), Greek. Theory of proportion, method of exhaustion, proposed heliocentric planetary spheres. W

Euler, Leonhard (1707–83), Swiss. Analytical mechanics, number theory, algebra, geometry, calculus, hydraulics, ship construction, astronomy, optics. B W(in progress) C(in progress).

Fermat, Pierre de (1601–65), French. Analytic geometry, probability, number theory. B W C.

Fibonacci (Leonardo of Pisa) (*c.* 1180 – *c.* 1250), Italian. Hindu-Arabic number system, alegbra, geometry. W.

Fisher, Ronald Aylmer (1890–1962), English. Statistics, biometry. B W.

Fontana, Niccolò. *See* Tartaglia.

Forsyth, Andrew Russell (1858–1942), Scottish. Analysis, textbooks.

Fourier, Jean-Baptiste Joseph (1768–1830), French. Differential equations, Fourier analysis, equations. B W.

Fraenkel, Adolf Abraham Halevi (1891–1965), German-born Israeli. Set theory. B.

Francesco, Piero della (1420–92), Italian. Perspective. B.

1702

Fréchet, René Maurice (1878–1973), French. Theory of real functions, functional analysis.

Fredholm, Erik Ivar (1866–1927), Swedish. Stockholm University. Mathematical physics, theory of integral equations. W.

Frege, Friedrich Ludwig Gottlob (1848–1925), German. Logic, foundations of mathematics, philosophy. W + C.

Fresnel, Augustin Jean (1788–1827), French. Optics. W.

Frobenius, Georg Ferdinand (1849–1917), German. University of Berlin. Group theory. W.

Fuchs, Lazarus Immanuel (1833–1902), German. University of Berlin. Differential equations, theory of functions. W.

Galileo Galilei (1564–1642), Italian. Classical mechanics and physics. B W.

Galois, Evariste (1811–32), French. Theory of algebraic equations, number theory. B W +.

Galton, Francis (1822–1911), English. Statistics, anthropometry. B.

Gauss, Carl Friedrich (1777–1855), German. Göttingen University. Number theory, differential geometry, astronomy, non-Euclidean geometry, algebra, complex numbers, geodesy, potential theory, magnetism. B W C.

Gergonne, Joseph Diez (1771–1859), French. Founded *Annales de mathématiques pures et appliquées* (1810). Philosophy.

Gerhardt, Carl Immanuel (1816–99), German. History, editor of the works of Leibniz.

Germain, Sophie (1776–1831), French. Number theory, elasticity. B C.

Gibbs, Josiah Willard (1839–1903), American. Statistical mechanics, vector analysis. B W.

Girard, Albert (1595–1632), French-born, lived mainly in The Netherlands. Algebra, trigonometry, arithmetic, editor of Stevin's works.

Gödel, Kurt (1906–78), Austrian-born American. Mathematical logic, foundations, W(in progress)

Goldbach, Christian (1690–1764), German, lived in Russia. Number theory. B.

Gordan, Paul Albert (1837–1912), Polish, lived in Germany. Invariant theory.

Gosset(t), William Sealey ('Student') (1876–1937), English. Guinness brewery in Dublin. Statistics.

Goursat, Edouard Jean-Baptiste (1858–1936), French. Analysis and its applications.

Grandi, Guido (1671–1742), Italian. Calculus. C.

Grassmann, Hermann Günther (1809–77), German. Linguist, Grassmann algebras, theory of colours. B W.

Green, George (1793–1841), English businessman. Potential theory. W B.

Gregory, David (1661–1708), Scottish. Series.

Gregory, James (1638–75), Scottish. Calculus.

Grosseteste, Robert (1168–1253), English. Geometry, optics, astronomy.

Hachette, Jean Nicolas Pierre (1769–1834), French. Descriptive geometry, machine theory.

Hadamard, Jacques (1865–1963), French. Theory of real and complex functions, psychology of mathematical invention.

Hahn, Hans (1879–1934), Austrian. Philosophy of mathematics, analysis.

Halsted, George Bruce (1853–1922), American. Geometry, history, translator and publisher.

Hamilton, William Rowan (1805–65), Irish. Quaternions, Hamilton's principle in mechanics. B W.

Hankel, Hermann (1839–73), German. Theory of functions, history.

Hardy, Godfrey Harold (1877–1947), English. Analysis, number theory. W.

Harriot (or Hariot), Thomas (1560–1621), English algebraist, astronomer and physicist. B.

Hausdorff, Felix (1868–1942), German mathematician and philosopher. Topology. W.

Heath, Thomas Little (1861–1940), English civil servant. Historian of Greek mathematics, editor of Euclid.

Heaviside, Oliver (1850–1925), English. Operational and vector analysis, electromagnetism. B W.

Heiberg, Johan Ludvig (1854–1928), Danish. History.

Heine, Heinrich Eduard (1821–81), German. Analysis.

Helmholtz, Hermann Ludwig Ferdinand von (1821–94), German. Physiology, acoustics, mathematical physics. B W C.

Herbrand, Jacques (1908–31), French. Mathematical logic. W + .

Hermann, Jacob (1678–1733), Swiss. Mechanics. C.

Hermite, Charles (1822–1901), French. Algebraic forms, analysis. B W C.

Hero (or Heron) of Alexandria (fl. *c*. AD 60), Alexandrian geometer and mechanician. W.

Herschel, John Frederick William (1792–1871), English. Analysis, physics, philosophy. B.

Hesse, Ludwig Otto (1811–74), German. Invariant theory. W.

Hilbert, David (1862–1943), German. Göttingen University. Algebraic number theory, calculus of variations, Dirichlet's principle, integral equations, foundations of mathematics, mathematical physics. B W C.

Hill, George William (1838–1914), American. Columbia University. Astronomy. W.

Hipparchus of Rhodes (fl. late 2nd century BC), Greek. Trigonometry.

Hofmann, Joseph Ehrenfried (1900–73), German. History. W.

Hooke, Robert (1635–1702), English. Optics, simple harmonic motion.

Hopf, Heinz (1894–1971), German. Topology. W.

Huygens, Christiaan (1629–95), Dutch. Dynamics, optics, astronomy, anticipation of the calculus. W.

Ibn al-Haytham, Abū ᶜAlī al-Ḥasan ibn al-Ḥasan (Alhazen) (?965 – *c*. 1041), Arabic mathematician, physicist and astronomer. Optics.

Ibrāhīm ibn Sinān ibn Thābit ibn Qurra (908–46). Arabic. Geometry, astronomy. W.

Jacobi, Carl Gustav Jacob (1804–51), German. Elliptic functions, determinants, dynamics. W + C.

Janiszewski, Zygmunt (1888–1920), Polish. A founder of *Fundamenta mathematicae* (1920). Topology. W.

Jevons, William Stanley (1835–82), English. Logic, mathematical economics. B W.

Jordan, Marie Ennemond Camille (1838–1922), French. Topology, analysis, group theory. W.

Jourdain, Philip Edward Bertrand (1879–1919), English. History, set theory.

Karpinski, Louis Charles (1878–1956), American. History.

Kelvin, Lord. *See* Thomson, William.

Kepler, Johannes (1571–1630), German. Physics, laws of planetary motion, optics, anticipation of the calculus. B W.

Keynes, John Maynard (1883–1946), English. Probability theory, mathematical economics. B

Killing, Wilhelm Karl Joseph (1847–1923), German. University of Münster. Lie algebras.

Kirchhoff, Gustav Robert (1824–87), German. Physics. Elasticity. W.

Klein, Christian Felix (1849–1925), German. Göttingen University. Mathematical physics, history, education, geometry, analysis. W C B(in progress).

Klingenstierna, Samuel (1698–1765), Swedish. Optics.

Kolmogorov, Andrei Nikolaevich (1903–87), Russian. Probability theory. Logic, analysis, topology, functional analysis.

Kovalevskaya, Sofya Vasilyevna (1850–91), Russian. Student of Weierstrass. From 1883 at the University of Stockholm. Analysis. W B C.

Kronecker, Leopold (1823–91), German. University of Berlin. Elliptic functions, ideal theory, arithmetic of quadratic forms. W C.

Kummer, Ernst Eduard (1810–93), German. University of Berlin. Number theory and geometry. W C.

La Hire, Philippe De (1640–1718), French. Conic sections treated projectively, astronomy, geodesy, physics.

Lacroix, Sylvestre François (1765–1843), French. Philosophy, textbooks.

Lagrange(-Tournier), Joseph Louis (1736–1813), Italian born, worked in Germany and France. Calculus of variations, analysis, algebra, number theory, mechanics, astronomy. B W C.

Laguerre, Edmond Nicolas (1834–86), French. Analysis, geometry. W.

Lamb, Horace (1849–1934), English. Applied mathematics, acoustics.

Lambert, Johann Heinrich (1728–77), German. Analysis, philosophy, non-Euclidean geometry, physics. W+ C.

Lamé, Gabriel (1795–1870), French. Elasticity, differential geometry, engineering mathematics, heat. B.

Landau, Edmund (1877–1938), German. Göttingen University. Number theory. W.

Laplace, Pierre Simon (1749–1827), French. Analysis, celestial mechanics, probability and statistics. W+ .

Le Verrier, Urbain Jean Joseph (1811–77), French. Paris Observatory. Celestial mechanics.

Lebesgue, Henri Léon (1875–1941), French. Theory of integration. W.

Lefschetz, Solomon (1884–1972), Russian-born American. Princeton University. Algebraic geometry and topology. W.

Legendre, Adrien Marie (1752–1833), French. Number theory, celestial mechanics, Euclidean geometry, elliptic integrals. C.

Leibniz, Gottfried Wilhelm (1646–1716), German. Philosopher. Invention of calculus; combinatorics, mathematical logic, computing machines. B W + (in progress) C.

Leonardo of Pisa. *See* Fibonacci.

Leśniewski, Stanisław (1886–1939), Russian-born Pole. Philosophy of mathematics, mathematical logic. W.

Levi-Città, Tullio (1873–1941), Italian. Absolute differential calculus. W.

l'Hôpital (l'Hospital), Guillaume François Antoine Marquis de (1661–1704), French. First calculus textbook. C.

Li Chunfeng (607–70), Chinese. Editor of 'Ten Mathematical Classics'.

Lie, Marius Sophus (1842–99), Norwegian. Continuous transformation groups. B W.

Lindemann, Carl Louis Ferdinand von (1852–1939), German. Proved transcendence of π, applied mathematics.

Liouville, Joseph (1809–82), French. Integral and differential equations, number theory, complex analysis. Founder of *Journal des mathématiques pures et appliqueés* (1836). B C.

Lipschitz, Rudolf Otto Sigismund (1832–1903), German. Quadratic differential forms. C.

Littlewood, John Edensor (1885–1977), English. Cambridge University. Analysis. W.

Lobachevsky, Nikolai Ivanovich (1792–1856), Russian. Non-Euclidean geometry. B W.

Loria, Gino (1862–1939), Italian. History. Founded *Bollettino* (1898).

Löwenheim, Leopold (1878–1940), German. Mathematical logic.

Łukasiewicz, Jan (1878–1956), Polish. Worked in Warsaw and Ireland. Mathematical logic. W.

Luzin, Nikolai Nikolaievich (1883–1950), Russian. Moscow University. Measure theory, foundations. W.

Lyapunov, Aleksandr Mikhailovich (1857–1918), Russian. Differential equations, potential theory, probability theory. B.

Maclaurin, Colin (1698–1746), Scottish. Geometry, calculus. C.

Markov, Andrei Andreevich (1856–1922), Russian. Probability theory. B W C.

Mathieu, Emile Léonard (1835–90), French. Mathematical physics.

Maxwell, James Clerk (1831–79), born in Scotland, lived also in England. Physics, electromagnetism, mechanics. B W.

Mengoli, Pietro (1626–82), Italian. Calculus. C.

Mercator, Gerardus (1512–94), born in Flanders, emigrated to Germany. Mathematical instruments, geography.

Mersenne, Marin (1588–1648), French. Mechanics, optics, music, acoustics. C.

Milne, Edward Arthur (1896–1950), English. Relativity.

Minkowski, Hermann (1864–1909), Lithuanian, taught at Zurich and Göttingen. Space-time geometry and physics, number theory. W C.

Mittag-Leffler, Magnus Gösta (1846–1927), Swedish. Founder of *Acta mathematica* (1882). Complex analysis, differential equations, history. Book and manuscript collector. B C.

Möbius, August Ferdinand (1790–1868), German. Astronomy, barycentric calculus, Möbius strip. W.

Monge, Gaspard (1746–1818), French. Ecole Polytechnique. Descriptive geometry, differential equations, mechanics, physics. B.

Montucla, Jean Etienne (1725–99), French. History.

Moore, Eliakim Hastings (1862–1932), American. University of Chicago. Analysis.

Moore, Robert Lee (1882–1974), American. University of Texas. Topology, education.

Morin, Artur Jules (1795–1880), French. Engineering, efficiency of machines.

Muir, Thomas (1844–1934), Scottish. Glasgow and Cape Town, South Africa. History of determinants.

Müller, Johann. *See* Regiomontanus.

Napier, John (1550–1617), Scottish. Logarithmic tables. B W.

Navier, Claude Louis Marie Henri (1785–1836), French. Engineering, elasticity, fluid mechanics, Fourier analysis.

Neugebauer, Otto (1899–1990), Austrian-born American. History, abstracting journals for mathematics.

Neumann, Carl Gottfried (1832–1925), German. University of Leipzig. Mathematical physics, potential theory, electrodynamics.

Neumann, Franz Ernst (1798–1895), German. Potential theory, electrodynamics. W.

Neumann, John von. *See* von Neumann, John.

Newcomb, Simon (1835–1909), born in Canada, lived in the USA. Celestial mechanics.

Newton, Isaac (1642–1727), English. Cambridge University. Inventor of calculus; foundations of mechanics and dynamics. Algebra, geometry, optics. *Philosophiae naturalis principia mathematica* (1687). B W + C.

Nicolaus of Cusa (1401–64), German. Philosophy. B W.

Noether, Amalie (Emmy) (1882–1935), German. Ring theory, non-commutative algebra. B W.

Noether, Max (1844–1921), German. Algebraic geometry, algebraic functions, history.

Ockham (Occam), William of (*c.* 1300–49), English. Logic, philosophy. B.

Ohm, Georg Simon (1789–1854), German. Physics. B W.

Omar Khayyam. *See* al-Khāyyamī.

Oresme, Nicole (*c.*1323–82), French. Anticipated coordinate geometry; infinite series.

Osgood, William Fogg (1864–1943), American. Harvard University. Theory of functions.

Ostrogradsky, Mikhail Vasilevich (1801–62), Russian. Partial differential equations, elasticity, algebra. B W + .

Oughtred, William (1575–1660), English. Trigonometry.

Ozanam, Jacques (1640–1717), French. Recreational mathematics, analysis.

Pacioli, Luca (445–1517), Italian. Arithmetic, algebra, geometry, trigonometry. B.

Painlevé, Paul (1863–1933), French. Differential equations, mechanics. W.

Pappus of Alexandria (fl. 320 AD), Greek. Geometer. W.

Pascal, Blaise (1623–62), French. Hydrostatics, calculating machines, probability, combinatorial analysis. B W + C.

Pasch, Moritz (1843–1930), German. Foundations of geometry.

Peacock, George (1791–1858), English. Cambridge University. Algebra.

Peano, Giuseppe (1858–1932), Italian. Turin University. Analysis, geometry, mathematical logic. B W + .

Pearson, Karl (1857–1936), English. University College, London. Statistics, founder and editor of *Biometrika* (1900). History. W.

Peirce, Benjamin (1809–80), American. Celestial mechanics, linear associative algebra, number theory, geodesy.

Peirce, Benjamin Osgood (1854–1914), American. Line and surface integration.

Peirce, Charles Sanders (1839–1914), American. Pendulum theory, logic, philosophy of mathematics. B W + .

Pell, John (1611–85), English. Algebra.

Péter, Róza (1905–77), Hungarian. Recursion.

Pfaff, Johann Friedrich (1765–1825), German. University of Halle. Analysis. C.

Picard, Charles Emile (1856–1941), French. Sorbonne. Analysis, analytic geometry. W.

Pincherle, Salvatore (1853–1936), Italian. Founder of Italian Union for Mathematics. Analysis, functional equations. W.

Plato (4th century BC), Greek. Philosophy of mathematics. B W + .

Playfair, John (1748–1819), Scottish. Geometry, physics, geology. W.

Plücker, Julius (1801–68), German. Analytical geometry, physics. W.

Poincaré, Jules Henri (1854–1912), French. Sorbonne. Philosophy, differential equations, automorphic functions, ergodicity, dynamics of the electron, celestial mechanics, topology, probability theory. B W C.

Poinsot, Louis (1777–1859), French. Statics and dynamics, analysis.

Poisson, Siméon-Denis (1781–1840), French. Elasticity, electrostatics, hydro-dynamics, mechanics, optics, potential theory, probability theory.

Pólya, George (1887–1985), Hungarian-born American. Function theory, probability, heuristics. W.

Poncelet, Jean Victor (1788–1867), French. Analytic and projective geometry, engineering mathematics, energy mechanics, hydraulics. B.

Poussin, Charles De La Vallée. *See* De La Vallée-Poussin, Charles.

Ptolemy (Claudius Ptolemaeus) (2nd century AD), Alexandrian astronomer and geographer. *Almagest*.

Puiseux, Victor Alexandre (1820–83), French. Celestial mechanics, infinitesimal geometry, complex analysis.

Puissant, Louis (1769–1843), French. Cartography, geodesy.

Pythagoras of Samos (*c*. 572 BC – *c*. 497 BC), Greek. Geometer and astronomer. Founder of a school of natural philosophy.

Quine, Willard Van Orman (1908–), American. Logic, philosophy of mathematics.
Qurra, Thābit ibn. *See* Thābit ibn Qurra.

Ramanujan, Srinivasa Aaiyangar (1887–1920), Indian. Self-taught number theorist.
B W +.
Ramsey, Frank Plumpton (1903–30), English, Logic, mathematical economics.
W +.
Rankine, William John Macquorn (1820–72), Scottish. Mathematical physics, engineering.
Rayleigh, Lord. *See* Strutt, John William.
Recorde, Robert (1510–58), Welsh. Arithmetic, translator of Euclid's *Elements*.
Regiomontanus (Johann Müller) (1436–76), German translator and publisher.
Trigonometry, astronomy. B W C.
Rheticus, Georg Joachim (1514–74), Austrian. Astronomy, trigonometric tables.
Riccardi, Pietro (1828–98), Italian. History.
Riccati, Jacopo Francesco (1676–1754), Italian. Differential equations.
Ricci-Curbastro, Gregorio (1853–1925), Italian. Absolute differential calculus (precursor of tensor analysis). W.
Richard of Wallingford (1291–1336), English. Horology. W.
Richardson, Lewis Fry (1881–1953), English. Meteorology, numerical methods. B W.
Riemann, Georg Friedrich Bernhard (1826–66), German. Göttingen University.
Complex functions, Riemannian geometry, foundations of analysis and geometry. W +.
Ries, Adam (*c.* 1492–1559), German. Algebra. B.
Riesz, Frigyes (Frédéric) (1880–1956), Hungarian. University of Budapest.
Functional analysis. Founder of *Acta scientiarum mathematicarum* (1922). W.
Riesz, Marcel (1886–1969), Hungarian, worked in Sweden. Analysis. W.
Robert of Chester (fl. 12th century), English. Translator of Arabic mathematics.
Roberval, Gilles Personne de (1602–75), French. Infinitesimal geometry, anticipated the calculus, algebra, mechanics. W C.
Robinson, Abraham (1918–74), Polish-born American. Non-standard analysis, aerodynamics. B(in progress) W.
Rolle, Michel (1652–1719), French. Diophantine analysis, algebra, geometry.
Ruffini, Paolo (1765–1822), Italian. Philosophy, equations, W C.
Runge, Carl David Tolmé (1856–1927), German. University of Göttingen.
Numerical analysis.
Russell, Bertrand Arthur William (1872–1970), Welsh. Mathematical logic, philosophy. B W(in progress) C.

Saccheri, Girolamo (1667–1733), Italian. Logic, anticipated non-Euclidean geometry.
Saint-Venant, Adhémar Jean Claude Barré de (1797–1886), French. Elasticity.
Sarton, George (1884–1956), American. Harvard University. History.
Schmidt, Erhard (1876–1959), German. Integral equations.
Schönflies, Arthur Moritz (1853–1928), German. University of Frankfurt. Set theory, crystallography.

Schröder, Friedrich Wilhelm Karl Ernst (1841–1902), German. Algebra, logic.

Schrödinger, Erwin (1887–1961), Austrian. General relativity, wave mechanics.

Schur, Friedrich Heinrich (1856–1932), Polish, worked in Germany. Geometry.

Schur, Issai (1875–1941), German. Algebra, group theory.

Schwarz, Karl Hermann Amandus (1843–1921), German. Conformal mappings, ordinary and partial differential equations, calculus of variations. W.

Severi, Francesco (1879–1961), Italian. University of Rome. Algebraic geometry, founder of Istituto Nazionale di Alta Matematica. W.

Sierpiński, Wacław (1882–1969), Polish. University of Warsaw. Topology, set theory, number theory, A founder of *Fundamenta mathematica* (1920). W.

Simpson, Thomas (1710–61), English. Interpolation, probability theory.

Skolem, Thoralf Albert (1887–1963), Norwegian. Diophantine equations, set theory, mathematical logic. W.

Sluse, René François Walter de (1622–85), Belgian. Cycloid, equations, calculus. C.

Smith, David Eugene (1860–1944), American. History.

Smith, Henry John Stephen (1826–83), Irish-born, lived in England. Number theory. W.

Sommerfeld, Arnold (1868–1951), German. Wave mechanics. W.

Sommerville, Duncan MacLean Young (1879–1934), British (born in India), New Zealand. Non-Euclidean geometry, history.

Staeckel, Paul Gustav (1862–1919), German. History, analytical mechanics, differential equations.

Steiner, Jacob (1796–1863), Swiss. Projective geometry. W C.

Steinhaus, Hugo (1887–1972), Polish. Functional analysis, heuristics. W.

Stevin, Simon (1548–1620), Dutch engineer. Decimal notation, hydraulics. W +.

Stieltjes, Thomas Jan (1856–1894), Dutch. Analysis, number theory. W C.

Stifel, Michael (*c*.1487–1567), German. University of Jena. Arithmetic, algebra, anticipated logarithms.

Stirling, James (1692–1770), Scottish. Cubic curves, method of differences, series. B C.

Stokes, George Gabriel (1819–1903), Irish-born, lived in England. Partial differential equations, elasticity, potential theory. W C.

Stolz, Otto (1842–1905), Austrian. University of Innsbruck. Algebraic geometry, real analysis, textbooks, history.

Strutt, John William (Lord Rayleigh) (1842–1919), English. Acoustics, physics. B W.

'Student'. *See* Gosset, William Sealey.

Study, Eduard (1862–1930), German. Hypercomplex numbers, invariants.

Sturm, Jacques Charles François (1803–55), Swiss. Infinitesimal geometry, differential equations, Roots of equations, caustics.

Sun Tzu (*c*. AD 400), Chinese. Chinese remainder problem.

Sylow, Ludwig Mejdell (1832–1918), Norwegian. University of Christiania. Group theory. W.

Sylvester, James Joseph (1814–97), English, worked in USA. Algebra, invariant theory. Founded *American Journal of Mathematics* (1883). B(in progress) W C.

Tait, Peter Guthrie (1831–1901), Scottish. Quaternions, theory of vortices, natural philosophy. W.

Takagi, Teiji (1875–1960), Japanese. University of Tokyo. Algebra. W.

Tannery, Jules (1848–1910), French. History and philosophy of mathematics.

Tannery, Paul (1843–1904), French. History of Greek mathematics, philosophy. W.

Tarski, Alfred (1902–83), Polish-born American. Mathematical logic, model theory. W.

Tartaglia, Niccolò (Fontana, Niccolò) (?1501–57), Italian. Solution of cubic equations, mechanics. W C.

Taylor, Brook (1685–1731), English. Calculus, series. C.

Thābit ibn Qurra, Abū'l-Ḥasan (c. 835–901), Arabic. Geometry, number theory, translator from Greek into Arabic, astronomy, medicine. W.

Thales of Miletus (fl. 585 BC), Greek. Traditionally the founder of Greek mathematics.

Theaetetus of Athens (4th century BC), Greek. Irrationals.

Thomson, Joseph John (1856–1940), English. Atomic structure, physics. B.

Thomson, William (Baron Kelvin of Largs) (1824–1907), Irish-born, lived in Scotland. Thermodynamics, dynamics, potential theory, telegraphy. B W C.

Tietze, Heinrich Franz Friedrich (1880–1964), Austria. Topology.

Tisserand, François Félix (1845–96), French. Astronomy.

Toeplitz, Otto (1881–1940), German. Infinite linear and quadratic forms, history, heuristics.

Torricelli, Evangelista (1608–47), Italian. Calculus, mechanics. W.

Tropfke, Johannes (1866–1939), German. Education, history.

Tschirnhaus(en), Ehrenfried Walter von (1651–1708), German. Solutions of equations. C.

Turing, Alan Mathison (1912–1954), English. Mathematical logic, computer science. B W.

Turnbull, Herbert Western (1885–1961), English. Algebra, history.

Urysohn, Pavel Samuilovich (1898–1924), Russian. Topology.

Vallée-Poussin, Charles De La. *See* De La Vallée-Poussin, Charles.

Vandermonde, Alexandre Théophile (1735–96), French. Equations.

Varignon, Pierre (1654–1722), French. Mechanics.

Veblen, Oswald (1880–1960), American. Institute for Advanced Study, Princeton. Topology, differential geometry, model theory.

Venn, John (1834–1923), English. Cambridge University. Logic, probability theory.

Veronese, Giuseppe (1854–1917), Italian. Geometry.

Viète, François (Vieta Franciscus) (1540–1603), French. Theory of equations. W.

Volterra, Vito (1860–1940), Italian. University of Rome. Functional analysis. B W.

von Mises, Richard (1883–1953), Austrian. Aerodynamics, statistics and probability theory. W.

von Neumann, John (1903–57), Hungarian-born American. Institute for Advanced Study, Princeton. Set theory, computer theory, game theory, quantum mechanics, functional analysis. B W.

von Staudt, Carl Georg Christian (1798–1867), German. Geometry.

Wallis, John (1616–1703), English. Oxford University. Analytic geometry, algebra. W C.

Wang Xiaotong (*c.* AD 600), Chinese. Cubic equation.

Waring, Edward (1736–98), English. Number theory, calculus.

Weber, Heinrich (1842–1913), German. Algebra and number theory. W.

Weber, Wilhelm Eduard (1804–91), German. Göttingen University. Electromagnetism, potential theory. W.

Wedderburn, Joseph Henry Maclagan (1882–1948), Scottish. Princeton University. Editor of *Annals of Mathematics*. Matrices, abstract algebra.

Weierstrass, Karl Wilhelm Theodor (1815–97), German. University of Berlin. Matrices, analysis, mechanics; influence through his lectures. B W C.

Wessel, Caspar (1745–1818), Norwegian. Geometrical representation of complex numbers.

Weyl, Hermann (1885–1955), Germany. Geometry, mathematical logic, philosophy of mathematics, Lie groups. W + .

Whewell, William (1794–1866), English. Cambridge University. Mechanics, tides, history and philosophy of science. B.

Whitehead, Alfred North (1861–1947), English. Mathematical logic, philosophy. B.

Whittaker, Edmund Taylor (1873–1956), English-born. University of Edinburgh. Analysis, celestial mechanics, history.

Wiener, Norbert (1894–1964), American. Massachusetts Institute of Technology. Cybernetics, analysis, stochastic processes. B W.

Wittgenstein, Ludwig Josef Johann (1889–1951), Austrian. Philosophy. B.

Wren, Christopher (1632–1723), English. Architect, geometry. B.

Wroński, Hoëné (*c.* 1776–1853), Polish. Equations. W.

Young, William Henry (1863–1942) and Grace Chisholm (1868–1944), English. Analysis. W(in progress).

Yule, George Udny (1871–1951), English. Statistics. W.

Zermelo, Ernst Friedrich Ferdinand (1871–1953), German. Axiomatic set theory.

Zeuthen, Hieronymous Georg (1839–1920), Danish. Enumerative geometry, history.

ACKNOWLEDGEMENT

For assistance with Arabic names, thanks are due to J. Hogendijk.

13.4

List of authors and board members

I. GRATTAN-GUINNESS

Normally the institutional affiliation (or that held upon retirement) is given; the corresponding address is summarized, in an English version. Articles (indicative titles only) are given in italics. Indications of *further* interests have been provided by the authors; the word 'history' is to be understood in a description, unless the topic is obviously new or the word 'modern' (without parentheses) restricts it solely to current work.

Aiton, Eric J. (deceased) Faculty of Education, Manchester Metropolitan University, Manchester, UK. *Tides.* Astronomy; optics; Kepler, Leibniz.

Andersen, Kirsti Department of History of Science, University of Århus, Århus, Denmark. *Seventeenth-century calculus*; *Descriptive geometry.* Perspective theory.

Archibald, Thomas Department of Mathematics, Acadia University, Wolfville, Nova Scotia, Canada. *Electricity and magnetism.* Analysis and its applications in the nineteenth and twentieth centuries.

Ascher, Marcia Department of Mathematics, Ithaca College, Ithaca, NY, USA. *Ethnomathematics.* Use of mathematics in archaeology and anthropology.

Ascher, Robert Department of Anthropology, Cornell University, Ithaca, NY, USA. *Ethnomathematics.* Anthropology and cinema.

Bagley, John A. Science Museum, London, UK. *Mathematics and flight.*

Bayart, Denis Centre of Management Research, Ecole Polytechnique, Paris, France. *Statistical control of manufacture.*

Binder, Christa Institute for Technical Mathematics, Technical University, Vienna, Austria. *Austria and Hungary.* Modern number theory.

Bottazzini, Umberto Department of Mathematics, University of Palermo, Palermo, Italy. *Complex-variable analysis; Solving high-degree equations; Italian States.* Mathematics in the eighteenth and nineteenth centuries.

Brentjes, Sonja Karl-Sudhoff Institute, University of Leipzig, Germany. *Linear optimization.* Islamic mathematics.

Brown, Laurie M. Department of Physics and Astronomy, Northwestern University, Evanston, IL, USA. *Quantum mechanics.* Nuclear and elementary particle physics.

Brush, Stephen G. Department of History and Institute for Physical Science and Technology, Institute for History and Philosophy of Science, University of Maryland, College Park, MD, USA. *Geophysics; Statistical mechanics.* Physical science since 1800.

Campbell-Kelly, Martin Department of Computer Science, University of Warwick, Coventry, UK. *Computing.* Babbage.

Cantor, Geoffrey Department of Philosophy, University of Leeds, Leeds, UK. *Velocity of light.* British science; Faraday.

Chabert, Jean-Luc Department of Mathematics, University of Picardie, Saint-Quentin, France. *Fractals.* Modern algebra.

Chapin, Seymour L. (deceased) Department of History, California State University, Los Angeles, CA, USA. *Geodesy.* Astronomy; navigation; expeditions; French science.

Chemla, Karine National Centre for Scientific Research, Paris, France. *Fractions.* Chinese mathematics; nineteenth-century Europe.

Closs, Michael P. Department of Mathematics, University of Ottawa, Ottawa, Canada. *Maya Mathematics.* Native American mathematics.

Cohen, H. Floris Department of History of Science, University of Twente, Enschede, The Netherlands. *Music.* The scientific revolution.

Cooke, Roger Department of Mathematics and Statistics, University of Vermont, Burlington, VT, USA. *Elliptic functions; Abelian integrals; Russia.* Russian analysis; Kovalevskaya.

Crépel, Pierre Laboratory of Numerical Analysis, Lyon University, Lyon, France. *Statistical control in manufacture.*

Crilly, Anthony J. School of Mathematics and Statistics, Middlesex University, London, UK. *Invariants.* British mathematics.

Cross, James J. Department of Mathematics, University of Melbourne, Melbourne, Australia. *Potential theory; Elasticity theory.* Dirichlet.

Crossley, John N. Department of Mathematics, Monash University, Melbourne, Australia. *Counting and number.* Chinese mathematics; mathematical logic.

Dauben, Joseph W. Graduate Center, City University of New York, New York, USA. *Point set topology; Dimensions.* G. Cantor; Chinese mathematics.

Dawid, A. Philip Department of Statistics, University College London, London, UK. *Foundations of probability.* Modern probability in expert systems.

Deakin, Michael A. B. Department of Mathematics, Monash University, Clayton, Victoria, Australia. *Laplace transform.* Modern biomathematics.

Deiman, J. C. University Museum, Utrecht, The Netherlands. *Optical instruments.* Astronomy; meteorology.

Detlefsen, Michael Department of Philosophy, University of Notre Dame, Notre Dame, IN, USA. *Constructivism.* Hilbert's philosophy; Brouwer.

Dmitriev, Igor S. Faculty of Chemistry, University of St Petersburg, St Petersburg, Russia. *Mathematics in chemistry.* Quantum chemistry.

Duffy, M. C. School of Engineering and Advanced Technology, University of Sunderland, Sunderland, UK. *Electrical machines.* Engineering; technology and economic growth; relativity theory.

Edney, Matthew H. Department of Geography, State University of New York, Binghamton, NY, USA. *Cartography.*

Edwards, Anthony W. F. Gonville and Caius College, Cambridge, UK. *Probability and statistics in genetics.*

Evans, Gillian R. Fitzwilliam College, Cambridge, UK. *Teaching mathematics in the Middle Ages and Renaissance.* Medieval intellectual history.

Evesham, H. A. Faculty of Applied Sciences, Luton College of Higher Education, Luton, UK. *Nomography.* Computing.

Farebrother, Richard William Department of Econometrics and Social Statistics, University of Manchester, Manchester, UK. *Linear statistical model.* (Modern) econometrics; statistics.

Fauvel, John Faculty of Mathematics, Open University, Milton Keynes, UK. *Women in mathematics*; *Mathematics and poetry.* History and education in mathematics.

Feigenbaum, Lenore Department of Mathematics, Tufts University, Medford, MA, USA. *Calculus in the seventeenth and eighteenth centuries.* Taylor.

Foulkes, Paul Freelance writer and translator, London, UK. *Pendulum.* History of philosophy.

Fowler, David H. Mathematics Institute, University of Warwick, Coventry, UK. *Continued fractions.* Greek mathematics.

Fraser, Craig G. Institute for the History of Science, University of Toronto, Toronto, Canada. *Calculus of variations*; *Classical mechanics.* Analysis and mechanics in the eighteenth and nineteenth centuries.

Frei, Günther Department of Mathematics, Laval University, Sainte-Foy, Canada. *Number theory.* Hilbert, Klein.

Fuller, A. Thomas Department of Engineering, University of Cambridge, Cambridge, UK. *Feedback control systems.* Maxwell.

Garber, Elizabeth Department of History, State University of New York, Stony Brook, NY, USA. *Meteorology.* Physics in the eighteenth and nineteenth centuries; Maxwell.

Garciadiego, Alejandro R. Department of Mathematics, National University, Mexico City, Mexico. *Paradoxes of set theory.* Foundations of mathematics; Russell.

Gigerenzer, Gerd Department of Psychology, University of Chicago, Chicago, IL, USA. *Probability and statistics in psychology.* Modern cognitive psychology; philosophy of psychology.

Gilain, Christian Department of Mathematics, University of Paris VI, Paris, France. *Ordinary differential equations.* Integral calculus.

Gillies, Donald A. Department of Philosophy, King's College London, London, UK. *Philosophies of probability.* Philosophy of logic and mathematics.

Goodman, Nicolas D. Department of Mathematics, State University of New York, Buffalo, NY, USA. *Modern philosophy of mathematics.* Mathematical logic.

Grattan-Guinness, Ivor School of Mathematics and Statistics, Middlesex University, London, UK. *Introduction*; *Measure theory*; *Overview of trigonometry*; *Special functions*; *Operator methods*; *Functional equations*; *Roots of equations*; *Matrix theory*; *Descriptive geometry*; *Ballistics*; *Shipbuilding*; *Heat diffusion*; *Acoustics*; *France*; *Scandinavia*; *Russia*; *British Isles*; *Numerology*; *History of the history of mathematics*. Logic; Russell, C.S. Peirce; philosophy of mathematics and science; history and education in mathematics.

Gray, Jeremy J. Department of Mathematics, Open University, Milton Keynes, UK. *Geometry in complex-function theory*; *Differential equations and groups*; *Geometry: algebraic and analytic*; *Curves*; *(Non-)Euclidean, projective, early modern algebraic geometry*; *Vector spaces*. History and education in mathematics.

Gross, Kenneth I. Department of Mathematics and Statistics, University of Vermont, Burlington, VT, USA. *Harmonic analysis*. Modern Lie groups; representation theory.

Grünbaum, Branko Department of Mathematics, University of Washington, Seattle, WA, USA. *Regular polyhedra*. Geometry; combinations.

Guicciardini, Niccolò Department of Philosophy, University of Bologna, Bologna, Italy. *Traditions in the calculus*. Quantum mechanics.

Hall, A. Rupert Department of History of Science, Imperial College of Science, Technology and Medicine, London, UK. *Ballistics*. The scientific revolution; Newton, Oldenburg.

Hayashi, Takao Science and Engineering Research Institute, Doshisha University, Kyoto, Japan. *Indian mathematics*.

Herz-Fischler, Roger Department of Mathematics, Carleton University, Ottawa, Canada. *Golden number and division in extreme and mean ratio*. Mathematics and culture.

Heyman, Jacques Department of Engineering, University of Cambridge, Cambridge, UK; cathedral consultant. *Structures*. Modern masonry construction and plastic theory.

Hogendijk, Jan P. Mathematics Institute, University of Utrecht, Utrecht, The Netherlands. *Pure Islamic mathematics*. Islamic astronomy; Greek mathematics.

Houser, Nathan Peirce Edition Project, Indiana University, Indianapolis, USA. *Algebraic logic*. C.S. Peirce.

Howse, Derek National Maritime Museum, London, UK. *Navigation*. Royal Observatory, Greenwich; Maskelyne.

Høyrup, Jens Roskilde University Centre, Roskilde, Denmark. *Babylonian mathematics*. Conceptual and cultural history of early mathematics.

Israel, Giorgio Department of Mathematics, University of Rome, 'La Sapienza', Rome, Italy. *Biomathematics*; *Mathematical economics*. Italian mathematics; mathematics and its applications in the nineteenth and twentieth centuries.

Jones, Alexander R. Institute for the History of Science, University of Toronto, Toronto, Canada. *Greek mathematics to AD 300*; *Greek applied mathematics*; *Later Greek and Byzantine mathematics*.

Jordan, Dominic W. Department of Mathematics, University of Keele, Keele, UK. *Telecommunication theory*. Electrotechnics in the nineteenth century; modern non-linear differential equations.

Joseph, George Gheverghese Department of Econometrics and Social Statistics, University of Manchester, Manchester, UK. *Tibetan mathematics.* Non-European mathematics.

Kaunzner, Wolfgang Regensburg, Germany. *Logarithms.*

Kilmister, Clive W. Department of Philosophy, Kings College, London, UK. *Astrophysics and cosmology; Relativity.* (Modern) quantum mechanics; Eddington.

Kim, Yong Woon Graduate School, Hangyang University, Seoul, South Korea. *Korean mathematics.* Modern topology.

King, David A. Institute for History of Science, Frankfurt University, Frankfurt am Main, Germany. *Mathematics applied to aspects of religious ritual in Islam.* Islamic astronomy and mathematics; medieval astronomical instruments.

Kipnis, Nahum Bakken Museum, Minneapolis, MN, USA. *Physical optics.* Physics, especially electricity.

Knobloch, Eberhard Institute of Philosophy, Technical University, Berlin, Germany. *Medieval technology; Determinants; Combinatorial probability.* Taccola; Leibniz's technical drawings.

Koetsier, Teun Department of Mathematics and Computer Science, Free University, Amsterdam, The Netherlands. *Kinematics.* Philosophy of mathematics.

l'Huillier, Hervé Total Oil Company, Paris, France. *Medieval practical geometry.* Medieval mathematics; Chuquet.

Langermann, Y. Tzvi National Library, Jerusalem, Israel. *Jewish mathematics.* Medieval Jewish science.

Laugwitz, Detlef Division of Mathematics, Technical University, Darmstadt, Germany. *Real-variable analysis.* Modern analysis and differential geometry.

Ledermann, Walter Department of Mathematics, University of Sussex, Brighton, UK. *Matrix Theory.* Modern algebra.

Lewin, Christopher G. Biggar, Lanarkshire, UK (practising actuary). *Actuarial mathematics.*

Lewis, Albert C. Bertrand Russell Project, McMaster University, Hamilton, Canada. *Complex numbers and vectors; Bibliography; Chronology; Biographical notes; Index.* American mathematics; J.G. and H. Grassmann, Russell.

Lloyd, E. Keith Faculty of Mathematical Studies, University of Southampton, Southampton, UK. *Combinatorics.* (Modern) graph theory.

Lützen, Jesper Department of Mathematics, University of Copenhagen, Copenhagen, Denmark. *Integral equations; Partial differential equations.* Distributions; Liouville.

Maanen, Jan A. van Department of Mathematics and Computer Science, University of Groningen, Groningen, The Netherlands. *The Netherlands.* Prehistory of the calculus; history and education in mathematics.

Mainzer, Klaus Faculty of Philosophy, University of Augsburg, Augsburg, Germany. *Symmetry.* Geometry; philosophy of mathematics.

Malkevitch, Joseph Department of Mathematics, York College, City University of New York, Jamaica, NY, USA. *Tilings.*

Martzloff, Jean-Claude National Centre for Scientific Research, Paris, France. *Chinese mathematics.* Chinese astronomy.

Matthews, J. Rosser Department of History, Duke University, Durham, NC, USA. *Probability and statistics in medicine*. History of medicine.

Mikhailov, Gleb K. Russian National Committee on Mechanics, Moscow, Russia. *Hydrodynamics and hydraulics*.

Moesgaard, Kristian P. Department of History of Science, University of Århus, Århus, Denmark. *Medieval and Renaissance astronomy*.

Molland, A. George Department of History of Science, King's College, Aberdeen, UK. *Medieval and Renaissance mechanics*; *Philosophical background to medieval mathematics*.

Moore, Gregory H. Department of Mathematics, McMaster University, Hamilton, Ontario, Canada. *Logic and set theory*. Hausdorff, Russell, Zermelo.

Murata, Tamotsu Department of Mathematics, St Andrews University, Osaka, Japan. *Japanese mathematics*. Set theory.

Neuenschwander, Erwin Department of Mathematics, University of Zurich, Zurich, Switzerland. *Journals*. Greek mathematics; nineteenth-century analysis; Riemann, Liouville.

Ortiz, Eduardo L. Department of Mathematics, Imperial College of Science, Technology and Medicine, London, UK. *Iberian and Ibero-American mathematics*.

Parshall, Karen Hunger Department of Mathematics, University of Virginia, Charlottesville, VA, USA. *USA and Canada*. Mathematical institutions; Sylvester.

Pensivy, Michel Centre for Science and Technology, University of Nantes, France. *Binomial theorem*. Divergent series.

Pingree, David Department of History, Brown University, Providence, RI, USA. *Non-Western traditions* (editorial advisor).

Porter, Theodore M. Department of History, UCLA, Los Angeles, CA, USA. *Probability and statistics in the social sciences, and in the professional community*; *English biometry*. Quantification in science and in public life; objectivity.

Purkert, Walter Leipzig, Germany. *Abstract algebras*. Mathematics and engineering; G. Cantor.

Pycior, Helena M. Department of History, University of Wisconsin, Milwaukee, WI, USA. *Philosophy of algebra; Mathematics and prose literature*. Nineteenth-century algebras; gender and science.

Reich, Karin Faculty of Mathematics, University of Stuttgart, Germany. *'Coss' (German algebra)*; *Differential geometry*.

Richards, Joan L. Department of History, Brown University, Providence, RI, USA. *Philosophy of geometry*. Non-Euclidean geometry; foundations of eighteenth-century mathematics.

Rider, Robin Office for History of Science, University of California at Berkeley, CA, USA. *Operational research*.

Rodríguez-Consuegra, Francisco A. Department of Science, Logic, History and Philosophy, University of Barcelona, Spain. *Mathematical logic*. History and philosophy of science; Russell.

Roero, Clara S. Department of Mathematics, University of Turin, Turin, Italy. *Egyptian mathematics*. Seventeenth-century calculus; probability theory.

Romanovskaya, Tatiana B. Institute of Philosophy, Academy of Sciences, Moscow, Russia. *Mathematics in chemistry.* Quantum mechanics.

Rowe, David E. Faculty of Mathematics, University of Mainz, Germany. *Line geometry*; *USA and Canada.* Geometry; Klein, Hilbert and Göttingen mathematics.

Rüger, Alexander Department of Philosophy, University of Alberta, Edmonton, Canada. *Capillarity.* Quantum mechanics.

Russ, Steve B. Department of Computer Science, University of Warwick, Coventry, UK. *Computing.* Bolzano.

Scholz, Erhard Bergische University, Wuppertal, Germany. *Topology*; *Graphical statics*; *Crystallography.* Algebra.

Schönbeck, Jürgen G. Pedagogical High School, Heidelberg, Germany. *Euclidean and Archimedean traditions.* Mathematics education.

Schreiber, Peter Section for Mathematics, Arndt University, Greifswald, Germany. *Numerical methods*; *Algorithmic mathematics*; *Art and architecture.* Mathematics and stamps; modern logic.

Schubring, Gert Institute for Mathematics Education, University of Bielefeld, Bielefeld, Germany. *Germany.* Interactions between conceptual and intellectual history (France and Germany).

Seneta, Eugene School of Mathematics and Statistics, University of Sydney, Sydney, Australia. *Russian probability.* Bienaymé; Markov chains.

Shafer, Glenn Graduate School of Management, Rutgers University, Newark, New Jersey, USA. *Early mathematical probability.* Probability in artificial intelligence.

Shapiro, Stewart Department of Philosophy, Ohio State University, Newark, OH, USA. *Metamathematics.* Modern logic, and relationship to reasoning.

Sheynin, Oscar B. Mathematical Institute, Cologne, Germany. *Theory of errors.* Probability and statistics.

Siegmund-Schulze, Reinhard Berlin, Germany. *Functional analysis.* Mathematics under the Nazis.

Simons, Peter Institute for Philosophy, University of Salzburg, Salzburg, Austria. *Polish logics.* Nineteenth- and twentieth-century logic and phenomenology.

Singmaster, David Department of Mathematics, South Bank University, London, UK. *Recreational mathematics*; *Monuments.* Computing; science in London.

Smith, A. Mark Department of History, University of Missouri, Columbia, MO, USA. *Optics to the seventeenth century.*

Smith, J.H.B. Royal Aerospace Establishment, Farnborough, UK. *Mathematics and flight.* Modern vortex theory.

Stewart, Ian Mathematics Institute, University of Warwick, Coventry, UK. *Lie groups.* Modern non-linear dynamics.

Swade, Doron D. Science Museum, London, UK. *Calculating machines.*

Swijtink, Zeno G. Department of History and Philosophy of Science, Indiana University, Bloomington, IN, USA. *Probability and statistics in agronomy, and in mechanics.* Probability and statistics; scientific instruments.

Thiele, Rüdiger Karl Sudhoff Institute, University of Leipzig, Germany. *Mathematical games.* Mathematical physics and calculus of variations; Euler.

Toti Rigatelli, Laura Department of Mathematics, Siena University, Siena, Italy. *Theory of equations*. Medieval algebra; Galois theory.

Van Egmond, Warren Department of Philosophy, Arizona State University, Tempe, AZ, USA. *Abbacus arithmetic*. Medieval and Renaissance algebra.

Wallis, Helena Map Library, British Library, London, UK. *Cartography*.

Weaver, George Department of Philosophy, Bryn Mawr College, Bryn Mawr, PA, USA. *Model theory*. Philosophy of mathematics.

Wilson, Curtis St John's College, Annapolis, MD, USA. *Dynamics of the Solar System; Three-body problem*. Medieval logic.

Wilson, Robin J. Faculty of Mathematics, Open University, Milton Keynes, UK. *Combinatorics*. Graph theory; mathematicians and stamps; Victorian mathematics.

Wussing, Hans Karl-Sudhoff Institute, University of Leipzig, Germany. *Abstract algebras; Stamps*. Biographies of mathematicians.

Yagi, Eri Faculty of Engineering, Toyo University, Kawagoe City, Japan. *Thermodynamics*. Japanese science; entropy.

Zaslavsky, Claudia New York City, USA (education consultant). *African mathematics*. Mathematics education.

13.5

Index

ALBERT C. LEWIS

Page numbers in italic type after a subject, as in 'numeration systems *150–57*', refer to an article as a whole, and the subentries under that subject are generally only for references outside the article. (Exceptions are made for longer articles, such as §8.5 on hydrodynamics and hydraulics.) Thus the reader should look first under the most specific term for a topic. Alphabetization is word by word, so that 'Le Roux' precedes 'Leather Roll' and 'al-Uqlīdisī' precedes 'Alan of Lille', for example. For personal names the reader should also consult the Biographical Notes, §13.3. Substantial descriptions of works are entered under the authors' names if known, otherwise they are entered under the title. Ordering under personal names is: general or career information in the main entry, subentries of related topics (alphabetically by principal word), followed by works. The index does not cover the bibliographies or reference and information (§13.1–13.4).

clinical trial 1374–5
clocks *1082–7*
  navigation 1134, 1137
  practical geometry 190
  stamps of 1653
closed forms 690
Codazzi, Delfino 335, 1499
Codiakoptics 1230
Cohen, Paul 1522
  continuum hypothesis 358, 681
  forcing method 642–3
Coimbra University 1505
Cole, Frank Nelson 1514
Colebrooke, H. T. 1667
Coleridge, Samuel Taylor 1647
Collège de France 1435–6, 1667
Collège Royale 1431
Collins, John: continued fractions 736
Colmar, Thomas de: calculator 697
Colossus computer 704
colours, theory of 1143
Colson, J.: binomial theorem 497
combinations: *see* permutations and
  combinations
combinatorial manifolds 935
combinatorial probability *1286–92*
combinatorics 725, *952–64*
  studied by Leibniz 311
  operators 546
Commandino, Federico 178–9, 1496
  edition of Archimedes 183
  statics 231
commerce
  abbacus arithmetic 201–2
  origins of Hindu–Arabic numerals
    in 195, 203
  logarithms 211
  see also business mathematics
communication techniques: for
  algorithms 688–9
commutative diagrams 757
commutativity of multiplication 802
compact groups: harmonic
  analysis 412–13
compass: proportional 695
completeness: categoricity 673
complex analysis *419–30*
  complex numbers 723
complex coordinates 904
complex curves 864, 922
complex functions 447–8
  elliptic functions 536

geometry in *432–8*
special functions 525
complex numbers *722–7*, 799
  absence from Arabic
    mathematics 73
  equation solving 719, 795
  vector spaces 947
complexity 692
Compton, Arthur: quantum
  mechanics 1256
computability 692
  metamathematics *644–54*
'Computational Prescriptions in Nine
  Chapters' (*Jiuzhang
  suanshu*) 94–6, 105, 163, 688
computational techniques: see
  nomography
computers
  in actuarial mathematics 1307
  algorithms 688
  digital *701–6*
  in meteorology 1193–4
  in operational research 841–2
  proof methods 685
  stamps of 1653, 1657
  von Neumann 1521
Comrie, L. J.: his table-making 704
Comte, Auguste 1168
  astrophysics 1063–4
  probability 1344
  rational mechanics 985
conchoid of Nicomedes 862
Condillac, Abbé de
  algebraic logic 613, 797
  chemistry 1261
conditional probability 1402–4
Condon, E. U.: games 1563
Condorcet, Marquis de
  integrability 445
  medicine 1372
  probabilistic rationalism 1343
  social mathematics 1415, 1417
  stamps of 1654
Conforto, F.
  Abelian varieties 543
  elliptic functions 538
Confucianism
  influence on Chinese mathematics 95
  see also neo-Confucianism
congresses
  World's Columbian Exposition,
    1893 1515

Gmunden, Johannes von 1457
stamps of 1652
gnomons
Chinese 94
Indian 122
Korean 112
see also shadows, calculations
involving
Gödel, Kurt 11, 1462, 1522
axiom of choice 358
Church's thesis 652
first-order logics 675
incompleteness theorems 626, 642,
648–9, 676–8
on the status of sets 681
'Über formal unentscheidbare Sätze
der *Principia mathematica* und
verwandter Systeme' 648, 676
Gödel number 648, 677
Gold, Thomas: expanding
universe 1067
Goldbach, Christian 1478–9
correspondence with Euler 739,
825
Goldbach conjecture 825
golden number 1580–3
Goldschmidt, H.: partial differential
equations 465
Golomb, S. W.: games 1561
Gomes Teixeira, Francisco 1508
Gompertz, Benjamin: actuarial
mathematics 1306
González Domínguez, Alberto 1509
Göpel, J. G.: theta functions 542
Gordan, Paul
algebraic geometry 921
chemistry 1265
invariant theory 790
optimization 829
projective geometry 904
Gosset(t), W. S. (pseudonym
'Student') 1361
biometrics 1339
quality control 1386
tests of significance 1313, 1368
Göttingen University 1443, 1453
Americans at 1519
applied mathematics at 585, 1078,
1452
Goudsmit, Samuel: quantum
mechanics 1256
Gould, B. A. 1382

Goursat, Edouard: Cauchy integral
theorem 421
government: *see* state, role of
Graaf, A. de 1467
grad 483, 485
Grad, Harold 1521
Graeffe, C. H.
numerical mathematics 588
roots of equations 565
Gram, J. P. 1475
least-squares method 1313
Grammateus, Henricas 196
Grandi, Guido 1497
Grant, Barnard: his difference
engine 698
Granville, Evelyn Lee 1529
graphical computation: *see*
nomography
graphical representation
cartography 1104–8
of complex numbers 724
meteorology 1191
navigation 1131–3
graphical statics 987–92
vector spaces 948
graphs and graph theory 958–61
chemistry 1265
Hebrew literature 141
Malekula 1551–3
Medieval and Renaissance 233
Grassmann, Hermann G. 1450
generalized complex
numbers 725–6
influenced by Lagrange 728
line geometry 911
influence on Peano 379
philosophy of algebra 803
philosophy of geometry 918
influence on Schröder 756
tensors 1229
vector spaces 948, 949
influence on A. N. Whitehead 600
Ausdehnungslehre 725–6, 755, 783,
803
Grassmann, Justus Günther:
crystallography 1270–71
Grassmann, Robert: influence on
Schröder 756
Grau, A. de 1467
Graunt, John
genetics 1358
actuarial mathematics 1304, 1416

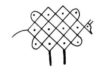